Lecture Notes in Computer Science 3036

Commenced Publication in 1973
Founding and Former Series Editors:
Gerhard Goos, Juris Hartmanis, and Jan van Leeuwen

Springer
Berlin
Heidelberg
New York
Hong Kong
London
Milan
Paris
Tokyo

Marian Bubak   Geert Dick van Albada
Peter M.A. Sloot   Jack J. Dongarra (Eds.)

Computational Science - ICCS 2004

4th International Conference
Kraków, Poland, June 6-9, 2004
Proceedings, Part I

 Springer

Volume Editors

Marian Bubak
AGH University of Science and Technology
Institute of Computer Science and Academic Computer Center CYFRONET
Mickiewicza 30, 30-059 Kraków, Poland
E-mail: bubak@uci.agh.edu.pl

Geert Dick van Albada
Peter M.A. Sloot
University of Amsterdam, Informatics Institute, Section Computational Science
Kruislaan 403, 1098 SJ Amsterdam, The Netherlands
E-mail: {dick,sloot}@science.uva.nl

Jack J. Dongarra
University of Tennessee, Computer Science Department
1122 Volunteer Blvd, Knoxville, TN 37996-3450, USA
E-mail: dongarra@cs.utk.edu

Library of Congress Control Number: Applied for

CR Subject Classification (1998): D, F, G, H, I, J, C.2-3

ISSN 0302-9743
ISBN 3-540-22114-X Springer-Verlag Berlin Heidelberg New York

Springer-Verlag is a part of Springer Science+Business Media

springeronline.com

© Springer-Verlag Berlin Heidelberg 2004
Printed in Germany

Typesetting: Camera-ready by author, data conversion by PTP-Berlin, Protago-TeX-Production GmbH
Printed on acid-free paper   SPIN: 11009306   06/3142   5 4 3 2 1 0

Preface

The International Conference on Computational Science (ICCS 2004) held in Kraków, Poland, June 6–9, 2004, was a follow-up to the highly successful ICCS 2003 held at two locations, in Melbourne, Australia and St. Petersburg, Russia; ICCS 2002 in Amsterdam, The Netherlands; and ICCS 2001 in San Francisco, USA.

As computational science is still evolving in its quest for subjects of investigation and efficient methods, ICCS 2004 was devised as a forum for scientists from mathematics and computer science, as the basic computing disciplines and application areas, interested in advanced computational methods for physics, chemistry, life sciences, engineering, arts and humanities, as well as computer system vendors and software developers. The main objective of this conference was to discuss problems and solutions in all areas, to identify new issues, to shape future directions of research, and to help users apply various advanced computational techniques. The event harvested recent developments in computational grids and next generation computing systems, tools, advanced numerical methods, data-driven systems, and novel application fields, such as complex systems, finance, econo-physics and population evolution.

Keynote lectures were delivered by David Abramson and Alexander V. Bogdanov, *From ICCS 2003 to ICCS 2004 – Personal Overview of Recent Advances in Computational Science*; Iain Duff, *Combining Direct and Iterative Methods for the Solution of Large Sparse Systems in Different Application Areas*; Chris Johnson, *Computational Multi-field Visualization*; John G. Michopoulos, *On the Pathology of High Performance Computing*; David De Roure, *Semantic Grid*; and Vaidy Sunderam, *True Grid: What Makes a Grid Special and Different?* In addition, three invited lectures were delivered by representatives of leading computer system vendors, namely: Frank Baetke from Hewlett Packard, Eng Lim Goh from SGI, and David Harper from the Intel Corporation.

Four tutorials extended the program of the conference: Paweł Płaszczak and Krzysztof Wilk, *Practical Introduction to Grid and Grid Services*; Grzegorz Młynarczyk, *Software Engineering Methods for Computational Science*; the *CrossGrid Tutorial* by the CYFRONET CG team; and the Intel tutorial.

We would like to thank all keynote, invited and tutorial speakers for their interesting and inspiring talks.

Aside of plenary lectures, the conference included 12 parallel oral sessions and 3 poster sessions. Ever since the first meeting in San Francisco, ICCS has attracted an increasing number of more researchers involved in the challenging field of computational science. For ICCS 2004, we received 489 contributions for the main track and 534 contributions for 41 originally-proposed workshops. Of these submissions, 117 were accepted for oral presentations and 117 for posters in the main track, while 328 papers were accepted for presentations at 30 workshops. This selection was possible thanks to the hard work of the Program

Committee members and 477 reviewers. The author index contains 1395 names, and almost 560 persons from 44 countries and all continents attended the conference: 337 participants from Europe, 129 from Asia, 62 from North America, 13 from South America, 11 from Australia, and 2 from Africa.

The ICCS 2004 proceedings consists of four volumes, the first two volumes, LNCS 3036 and 3037 contain the contributions presented in the main track, while volumes 3038 and 3039 contain the papers accepted for the workshops. Parts I and III are mostly related to pure computer science, while Parts II and IV are related to various computational research areas. For the first time, the ICCS proceedings are also available on CD. We would like to thank Springer-Verlag for their fruitful collaboration. During the conference the best papers from the main track and workshops as well as the best posters were nominated and presented on the ICCS 2004 Website. We hope that the ICCS 2004 proceedings will serve as a major intellectual resource for computational science researchers, pushing back the boundaries of this field. A number of papers will also be published as special issues of selected journals.

We owe thanks to all workshop organizers and members of the Program Committee for their diligent work, which ensured the very high quality of the event. We also wish to specifically acknowledge the collaboration of the following colleagues who organized their workshops for the third time: Nicoletta Del Buono (New Numerical Methods) Andres Iglesias (Computer Graphics), Dieter Kranzlmueller (Tools for Program Development and Analysis), Youngsong Mun (Modeling and Simulation in Supercomputing and Telecommunications).

We would like to express our gratitude to Prof. Ryszard Tadeusiewicz, Rector of the AGH University of Science and Technology, as well as to Prof. Marian Noga, Prof. Kazimierz Jeleń, Dr. Jan Kulka and Prof. Krzysztof Zieliński, for their personal involvement. We are indebted to all the members of the Local Organizing Committee for their enthusiastic work towards the success of ICCS 2004, and to numerous colleagues from ACC CYFRONET AGH and the Institute of Computer Science for their help in editing the proceedings and organizing the event. We very much appreciate the help of the Computer Science and Computational Physics students during the conference. We owe thanks to the ICCS 2004 sponsors: Hewlett-Packard, Intel, IBM, SGI and ATM, SUN Microsystems, Polish Airlines LOT, ACC CYFRONET AGH, the Institute of Computer Science AGH, the Polish Ministry for Scientific Research and Information Technology, and Springer-Verlag for their generous support.

We wholeheartedly invite you to once again visit the ICCS 2004 Website (http://www.cyfronet.krakow.pl/iccs2004/), to recall the atmosphere of those June days in Kraków.

June 2004

Marian Bubak, Scientific Chair 2004
on behalf of the co-editors:
G. Dick van Albada
Peter M.A. Sloot
Jack J. Dongarra

Organization

ICCS 2004 was organized by the Academic Computer Centre CYFRONET AGH University of Science and Technology (Kraków, Poland) in cooperation with the Institute of Computer Science AGH, the University of Amsterdam (The Netherlands) and the University of Tennessee (USA).

All the members of the Local Organizing Committee are the staff members of CYFRONET and/or ICS. The conference took place at the premises of the Faculty of Physics and Nuclear Techniques AGH and at the Institute of Computer Science AGH.

Conference Chairs

Scientific Chair – Marian Bubak (Institute of Computer Science and ACC CYFRONET AGH, Poland)
Workshop Chair – Dick van Albada (University of Amsterdam, The Netherlands)
Overall Chair – Peter M.A. Sloot (University of Amsterdam, The Netherlands)
Overall Co-chair – Jack Dongarra (University of Tennessee, USA)

Local Organizing Committee

Marian Noga
Marian Bubak
Zofia Mosurska
Maria Stawiarska
Milena Zając
Mietek Pilipczuk
Karol Frańczak
Aleksander Kusznir

Program Committee

Jemal Abawajy (Carleton University, Canada)
David Abramson (Monash University, Australia)
Dick van Albada (University of Amsterdam, The Netherlands)
Vassil Alexandrov (University of Reading, UK)
Srinivas Aluru (Iowa State University, USA)
David A. Bader (University of New Mexico, USA)

J.A. Rod Blais (University of Calgary, Canada)
Alexander Bogdanov (Institute for High Performance Computing and Information Systems, Russia)
Peter Brezany (University of Vienna, Austria)
Marian Bubak (Institute of Computer Science and CYFRONET AGH, Poland)
Rajkumar Buyya (University of Melbourne, Australia)
Bastien Chopard (University of Geneva, Switzerland)
Paul Coddington (University of Adelaide, Australia)
Toni Cortes (Universitat Politècnica de Catalunya, Spain)
Yiannis Cotronis (University of Athens, Greece)
Jose C. Cunha (New University of Lisbon, Portugal)
Brian D'Auriol (University of Texas at El Paso, USA)
Federic Desprez (INRIA, France)
Tom Dhaene (University of Antwerp, Belgium)
Hassan Diab (American University of Beirut, Lebanon)
Beniamino Di Martino (Second University of Naples, Italy)
Jack Dongarra (University of Tennessee, USA)
Robert A. Evarestov (SPbSU, Russia)
Marina Gavrilova (University of Calgary, Canada)
Michael Gerndt (Technical University of Munich, Germany)
Yuriy Gorbachev (Institute for High Performance Computing and Information Systems, Russia)
Andrzej Goscinski (Deakin University, Australia)
Ladislav Hluchy (Slovak Academy of Sciences, Slovakia)
Alfons Hoekstra (University of Amsterdam, The Netherlands)
Hai Jin (Huazhong University of Science and Technology, ROC)
Peter Kacsuk (MTA SZTAKI Research Institute, Hungary)
Jacek Kitowski (AGH University of Science and Technology, Poland)
Dieter Kranzlmüller (Johannes Kepler University Linz, Austria)
Domenico Laforenza (Italian National Research Council, Italy)
Antonio Lagana (Università di Perugia, Italy)
Francis Lau (University of Hong Kong, ROC)
Bogdan Lesyng (ICM Warszawa, Poland)
Thomas Ludwig (Ruprecht-Karls-Universität Heidelberg, Germany)
Emilio Luque (Universitat Autònoma de Barcelona, Spain)
Michael Mascagni (Florida State University, USA)
Edward Moreno (Euripides Foundation of Marilia, Brazil)
Jiri Nedoma (Institute of Computer Science AS CR, Czech Republic)
Genri Norman (Russian Academy of Sciences, Russia)
Stephan Olariu (Old Dominion University, USA)
Salvatore Orlando (University of Venice, Italy)
Marcin Paprzycki (Oklahoma State University, USA)
Ron Perrott (Queen's University of Belfast, UK)
Richard Ramaroson (ONERA, France)
Rosemary Renaut (Arizona State University, USA)

Organization

ICCS 2004 was organized by the Academic Computer Centre CYFRONET AGH University of Science and Technology (Kraków, Poland) in cooperation with the Institute of Computer Science AGH, the University of Amsterdam (The Netherlands) and the University of Tennessee (USA).

All the members of the Local Organizing Committee are the staff members of CYFRONET and/or ICS. The conference took place at the premises of the Faculty of Physics and Nuclear Techniques AGH and at the Institute of Computer Science AGH.

Conference Chairs

Scientific Chair – Marian Bubak (Institute of Computer Science and ACC CYFRONET AGH, Poland)
Workshop Chair – Dick van Albada (University of Amsterdam, The Netherlands)
Overall Chair – Peter M.A. Sloot (University of Amsterdam, The Netherlands)
Overall Co-chair – Jack Dongarra (University of Tennessee, USA)

Local Organizing Committee

Marian Noga
Marian Bubak
Zofia Mosurska
Maria Stawiarska
Milena Zając
Mietek Pilipczuk
Karol Frańczak
Aleksander Kusznir

Program Committee

Jemal Abawajy (Carleton University, Canada)
David Abramson (Monash University, Australia)
Dick van Albada (University of Amsterdam, The Netherlands)
Vassil Alexandrov (University of Reading, UK)
Srinivas Aluru (Iowa State University, USA)
David A. Bader (University of New Mexico, USA)

J.A. Rod Blais (University of Calgary, Canada)
Alexander Bogdanov (Institute for High Performance Computing and Information Systems, Russia)
Peter Brezany (University of Vienna, Austria)
Marian Bubak (Institute of Computer Science and CYFRONET AGH, Poland)
Rajkumar Buyya (University of Melbourne, Australia)
Bastien Chopard (University of Geneva, Switzerland)
Paul Coddington (University of Adelaide, Australia)
Toni Cortes (Universitat Politècnica de Catalunya, Spain)
Yiannis Cotronis (University of Athens, Greece)
Jose C. Cunha (New University of Lisbon, Portugal)
Brian D'Auriol (University of Texas at El Paso, USA)
Federic Desprez (INRIA, France)
Tom Dhaene (University of Antwerp, Belgium)
Hassan Diab (American University of Beirut, Lebanon)
Beniamino Di Martino (Second University of Naples, Italy)
Jack Dongarra (University of Tennessee, USA)
Robert A. Evarestov (SPbSU, Russia)
Marina Gavrilova (University of Calgary, Canada)
Michael Gerndt (Technical University of Munich, Germany)
Yuriy Gorbachev (Institute for High Performance Computing and Information Systems, Russia)
Andrzej Goscinski (Deakin University, Australia)
Ladislav Hluchy (Slovak Academy of Sciences, Slovakia)
Alfons Hoekstra (University of Amsterdam, The Netherlands)
Hai Jin (Huazhong University of Science and Technology, ROC)
Peter Kacsuk (MTA SZTAKI Research Institute, Hungary)
Jacek Kitowski (AGH University of Science and Technology, Poland)
Dieter Kranzlmüller (Johannes Kepler University Linz, Austria)
Domenico Laforenza (Italian National Research Council, Italy)
Antonio Lagana (Università di Perugia, Italy)
Francis Lau (University of Hong Kong, ROC)
Bogdan Lesyng (ICM Warszawa, Poland)
Thomas Ludwig (Ruprecht-Karls-Universität Heidelberg, Germany)
Emilio Luque (Universitat Autònoma de Barcelona, Spain)
Michael Mascagni (Florida State University, USA)
Edward Moreno (Euripides Foundation of Marilia, Brazil)
Jiri Nedoma (Institute of Computer Science AS CR, Czech Republic)
Genri Norman (Russian Academy of Sciences, Russia)
Stephan Olariu (Old Dominion University, USA)
Salvatore Orlando (University of Venice, Italy)
Marcin Paprzycki (Oklahoma State University, USA)
Ron Perrott (Queen's University of Belfast, UK)
Richard Ramaroson (ONERA, France)
Rosemary Renaut (Arizona State University, USA)

Organization

ICCS 2004 was organized by the Academic Computer Centre CYFRONET AGH University of Science and Technology (Kraków, Poland) in cooperation with the Institute of Computer Science AGH, the University of Amsterdam (The Netherlands) and the University of Tennessee (USA).

All the members of the Local Organizing Committee are the staff members of CYFRONET and/or ICS. The conference took place at the premises of the Faculty of Physics and Nuclear Techniques AGH and at the Institute of Computer Science AGH.

Conference Chairs

Scientific Chair – Marian Bubak (Institute of Computer Science and ACC CYFRONET AGH, Poland)
Workshop Chair – Dick van Albada (University of Amsterdam, The Netherlands)
Overall Chair – Peter M.A. Sloot (University of Amsterdam, The Netherlands)
Overall Co-chair – Jack Dongarra (University of Tennessee, USA)

Local Organizing Committee

Marian Noga
Marian Bubak
Zofia Mosurska
Maria Stawiarska
Milena Zając
Mietek Pilipczuk
Karol Frańczak
Aleksander Kusznir

Program Committee

Jemal Abawajy (Carleton University, Canada)
David Abramson (Monash University, Australia)
Dick van Albada (University of Amsterdam, The Netherlands)
Vassil Alexandrov (University of Reading, UK)
Srinivas Aluru (Iowa State University, USA)
David A. Bader (University of New Mexico, USA)

J.A. Rod Blais (University of Calgary, Canada)
Alexander Bogdanov (Institute for High Performance Computing and Information Systems, Russia)
Peter Brezany (University of Vienna, Austria)
Marian Bubak (Institute of Computer Science and CYFRONET AGH, Poland)
Rajkumar Buyya (University of Melbourne, Australia)
Bastien Chopard (University of Geneva, Switzerland)
Paul Coddington (University of Adelaide, Australia)
Toni Cortes (Universitat Politècnica de Catalunya, Spain)
Yiannis Cotronis (University of Athens, Greece)
Jose C. Cunha (New University of Lisbon, Portugal)
Brian D'Auriol (University of Texas at El Paso, USA)
Federic Desprez (INRIA, France)
Tom Dhaene (University of Antwerp, Belgium)
Hassan Diab (American University of Beirut, Lebanon)
Beniamino Di Martino (Second University of Naples, Italy)
Jack Dongarra (University of Tennessee, USA)
Robert A. Evarestov (SPbSU, Russia)
Marina Gavrilova (University of Calgary, Canada)
Michael Gerndt (Technical University of Munich, Germany)
Yuriy Gorbachev (Institute for High Performance Computing and Information Systems, Russia)
Andrzej Goscinski (Deakin University, Australia)
Ladislav Hluchy (Slovak Academy of Sciences, Slovakia)
Alfons Hoekstra (University of Amsterdam, The Netherlands)
Hai Jin (Huazhong University of Science and Technology, ROC)
Peter Kacsuk (MTA SZTAKI Research Institute, Hungary)
Jacek Kitowski (AGH University of Science and Technology, Poland)
Dieter Kranzlmüller (Johannes Kepler University Linz, Austria)
Domenico Laforenza (Italian National Research Council, Italy)
Antonio Lagana (Università di Perugia, Italy)
Francis Lau (University of Hong Kong, ROC)
Bogdan Lesyng (ICM Warszawa, Poland)
Thomas Ludwig (Ruprecht-Karls-Universität Heidelberg, Germany)
Emilio Luque (Universitat Autònoma de Barcelona, Spain)
Michael Mascagni (Florida State University, USA)
Edward Moreno (Euripides Foundation of Marilia, Brazil)
Jiri Nedoma (Institute of Computer Science AS CR, Czech Republic)
Genri Norman (Russian Academy of Sciences, Russia)
Stephan Olariu (Old Dominion University, USA)
Salvatore Orlando (University of Venice, Italy)
Marcin Paprzycki (Oklahoma State University, USA)
Ron Perrott (Queen's University of Belfast, UK)
Richard Ramaroson (ONERA, France)
Rosemary Renaut (Arizona State University, USA)

Alistair Rendell (Australian National University, Australia)
Paul Roe (Queensland University of Technology, Australia)
Hong Shen (Japan Advanced Institute of Science and Technology, Japan)
Dale Shires (U.S. Army Research Laboratory, USA)
Peter M.A. Sloot (University of Amsterdam, The Netherlands)
Gunther Stuer (University of Antwerp, Belgium)
Vaidy Sunderam (Emory University, USA)
Boleslaw Szymanski (Rensselaer Polytechnic Institute, USA)
Ryszard Tadeusiewicz (AGH University of Science and Technology, Poland)
Pavel Tvrdik (Czech Technical University, Czech Republic)
Putchong Uthayopas (Kasetsart University, Thailand)
Jesus Vigo-Aguiar (University of Salamanca, Spain)
Jens Volkert (University of Linz, Austria)
Koichi Wada (University of Tsukuba, Japan)
Jerzy Wasniewski (Technical University of Denmark, Denmark)
Greg Watson (Los Alamos National Laboratory, USA)
Jan Węglarz (Poznań University of Technology, Poland)
Roland Wismüller (LRR-TUM, Germany)
Roman Wyrzykowski (Technical University of Częstochowa, Poland)
Jinchao Xu (Pennsylvania State University, USA)
Yong Xue (Chinese Academy of Sciences, ROC)
Xiaodong Zhang (College of William and Mary, USA)
Alexander Zhmakin (Soft-Impact Ltd, Russia)
Krzysztof Zieliński (Institute of Computer Science and CYFRONET AGH,
Poland)
Zahari Zlatev (National Environmental Research Institute, Denmark)
Albert Zomaya (University of Sydney, Australia)
Elena Zudilova (University of Amsterdam, The Netherlands)

Reviewers

Abawajy, J.H.	Aluru, S.	Balogh, Z.
Abe, S.	Anglano, C.	Bang, Y.C.
Abramson, D.	Archibald, R.	Baraglia, R.
Adali, S.	Arenas, A.	Barron, J.
Adcock, M.	Astalos, J.	Baumgartner, F.
Adriaansen, T.	Ayani, R.	Becakaert, P.
Ahn, G.	Ayyub, S.	Belleman, R.G.
Ahn, S.J.	Babik, M.	Bentes, C.
Albada, G.D. van	Bader, D.A.	Bernardo Filho, O.
Albuquerque, P.	Bajaj, C.	Beyls, K.
Alda, W.	Baker, M.	Blais, J.A.R.
Alexandrov, V.	Baliś, B.	Boada, I.
Alt, M.	Balk, I.	Bode, A.

Bogdanov, A.
Bollapragada, R.
Boukhanovsky, A.
Brandes, T.
Brezany, P.
Britanak, V.
Bronsvoort, W.
Brunst, H.
Bubak, M.
Budinska, I.
Buono, N. Del
Buyya, R.
Cai, W.
Cai, Y.
Cannataro, M.
Carbonell, N.
Carle, G.
Caron, E.
Carothers, C.
Castiello, C.
Chan, P.
Chassin-de-
  Kergommeaux, J.
Chaudet, C.
Chaves, J.C.
Chen, L.
Chen, Z.
Cheng, B.
Cheng, X.
Cheung, B.W.L.
Chin, S.
Cho, H.
Choi, Y.S.
Choo, H.S.
Chopard, B.
Chuang, J.H.
Chung, R.
Chung, S.T.
Coddington, P.
Coeurjolly, D.
Congiusta, A.
Coppola, M.
Corral, A.
Cortes, T.
Cotronis, Y.

Cramer, H.S.M.
Cunha, J.C.
Danilowicz, C.
D'Auriol, B.
Degtyarev, A.
Denazis, S.
Derntl, M.
Desprez, F.
Devendeville, L.
Dew, R.
Dhaene, T.
Dhoedt, B.
D'Hollander, E.
Diab, H.
Dokken, T.
Dongarra, J.
Donnelly, D.
Donnelly, W.
Dorogovtsev, S.
Duda, J.
Dudek-Dyduch, E.
Dufourd, J.F.
Dumitriu, L.
Duplaga, M.
Dupuis, A.
Dzwinel, W.
Embrechts, M.J.
Emiris, I.
Emrich, S.J.
Enticott, C.
Evangelos, F.
Evarestov, R.A.
Fagni, T.
Faik, J.
Fang, W.J.
Farin, G.
Fernandez, M.
Filho, B.O.
Fisher-Gewirtzman, D.
Floros, E.
Fogel, J.
Foukia, N.
Frankovic, B.
Fuehrlinger, K.
Funika, W.

Gabriel, E.
Gagliardi, F.
Galis, A.
Galvez, A.
Gao, X.S.
Garstecki, L.
Gatial, E.
Gava, F.
Gavidia, D.P.
Gavras, A.
Gavrilova, M.
Gelb, A.
Gerasimov, V.
Gerndt, M.
Getov, V.
Geusebroek, J.M.
Giang, T.
Gilbert, M.
Glasner, C.
Gobbert, M.K.
Gonzalez-Vega, L.
Gorbachev, Y.E.
Goscinski, A.M.
Goscinski, W.
Gourhant, Y.
Gualandris, A.
Guo, H.
Ha, R.
Habala, O.
Habib, A.
Halada, L.
Hawick, K.
He, K.
Heinzlreiter, P.
Heyfitch, V.
Hisley, D.M.
Hluchy, L.
Ho, R.S.C.
Ho, T.
Hobbs, M.
Hoekstra, A.
Hoffmann, C.
Holena, M.
Hong, C.S.
Hong, I.

Peluso, R.
Peng, Y.
Perales, F.
Perrott, R.
Petit, F.
Petit, G.H.
Pfluger, P.
Philippe, L.
Platen, E.
Plemenos, D.
Pllana, S.
Polak, M.
Polak, N.
Politi, T.
Pooley, D.
Popov, E.V.
Puppin, D.
Qut, P.R.
Rachev, S.
Rajko, S.
Rak, M.
Ramaroson, R.
Ras, I.
Rathmayer, S.
Raz, D.
Recio, T.
Reichel, L.
Renaut, R.
Rendell, A.
Richta, K.
Robert, Y.
Rodgers, G.
Rodionov, A.S.
Roe, P.
Ronsse, M.
Ruder, K.S.
Ruede, U.
Rycerz, K.
Sanchez-Reyes, J.
Sarfraz, M.
Sbert, M.
Scarpa, M.
Schabanel, N.
Scharf, E.
Scharinger, J.

Schaubschlaeger, C.
Schmidt, A.
Scholz, S.B.
Schreiber, A.
Seal, S.K.
Seinstra, F.J.
Seron, F.
Serrat, J.
Shamonin, D.P.
Sheldon, F.
Shen, H.
Shende, S.
Shentu, Z.
Shi, Y.
Shin, H.Y.
Shires, D.
Shoshmina, I.
Shrikhande, N.
Silvestri, C.
Silvestri, F.
Simeoni, M.
Simo, B.
Simonov, N.
Siu, P.
Slizik, P.
Slominski, L.
Sloot, P.M.A.
Slota, R.
Smetek, M.
Smith, G.
Smolka, B.
Sneeuw, N.
Snoek, C.
Sobaniec, C.
Sobecki, J.
Sofroniou, M.
Sole, R.
Soofi, M.
Sosnov, A.
Sourin, A.
Spaletta, G.
Spiegl, E.
Stapor, K.
Stuer, G.
Suarez Rivero, J.P.

Sunderam, V.
Suzuki, H.
Szatzschneider, W.
Szczepanski, M.
Szirmay-Kalos, L.
Szymanski, B.
Tadeusiewicz, R.
Tadic, B.
Talia, D.
Tan, G.
Taylor, S.J.E.
Teixeira, J.C.
Telelis, O.A.
Teo, Y.M
Teresco, J.
Teyssiere, G.
Thalmann, D.
Theodoropoulos, G.
Theoharis, T.
Thurner, S.
Tirado-Ramos, A.
Tisserand, A.
Toda, K.
Tonellotto, N.
Torelli, L.
Torenvliet, L.
Tran, V.D.
Truong, H.L.
Tsang, K.
Tse, K.L.
Tvrdik, P.
Tzevelekas, L.
Uthayopas, P.
Valencia, P.
Vassilakis, C.
Vaughan, F.
Vazquez, P.P.
Venticinque, S.
Vigo-Aguiar, J.
Vivien, F.
Volkert, J.
Wada, K.
Walter, M.
Wasniewski, J.
Wasserbauer, A.

Watson, G.	Xiao, Y.	Zhang, J.W.
Wawrzyniak, D.	Xu, J.	Zhang, N.X.L.
Weglarz, J.	Xue, Y.	Zhang, X.
Weidendorfer, J.	Yahyapour, R.	Zhao, L.
Weispfenning, W.	Yan, N.	Zhmakin, A.I.
Wendelborn, A.L.	Yang, K.	Zhu, W.Z.
Weron, R.	Yener, B.	Zieliński, K.
Wismüller, R.	Yoo, S.M.	Zlatev, Z.
Wojciechowski, K.	Yu, J.H.	Zomaya, A.
Wolf, F.	Yu, Z.C.H.	Zudilova, E.V.
Worring, M.	Zara, J.	
Wyrzykowski, R.	Zatevakhin, M.A.	

Workshops Organizers

Programming Grids and Metasystems

V. Sunderam (Emory University, USA)
D. Kurzyniec (Emory University, USA)
V. Getov (University of Westminster, UK)
M. Malawski (Institute of Computer Science and CYFRONET AGH, Poland)

Active and Programmable Grids Architectures and Components

C. Anglano (Università del Piemonte Orientale, Italy)
F. Baumgartner (University of Bern, Switzerland)
G. Carle (Tubingen University, Germany)
X. Cheng (Institute of Computing Technology, Chinese Academy of Science, ROC)
K. Chen (Institut Galilée, Université Paris 13, France)
S. Denazis (Hitachi Europe, France)
B. Dhoedt (University of Gent, Belgium)
W. Donnelly (Waterford Institute of Technology, Ireland)
A. Galis (University College London, UK)
A. Gavras (Eurescom, Germany)
F. Gagliardi (CERN, Switzerland)
Y. Gourhant (France Telecom, France)
M. Gilbert (European Microsoft Innovation Center, Microsoft Corporation, Germany)
A. Juhola (VTT, Finland)
C. Klein (Siemens, Germany)
D. Larrabeiti (University Carlos III, Spain)
L. Lefevre (INRIA, France)
F. Leymann (IBM, Germany)
H. de Meer (University of Passau, Germany)
G. H. Petit (Alcatel, Belgium)

J. Serrat (Universitat Politècnica de Catalunya, Spain)
E. Scharf (QMUL, UK)
K. Skala (Ruder Boskoviç Institute, Croatia)
N. Shrikhande (European Microsoft Innovation Center, Microsoft
Corporation, Germany)
M. Solarski (FhG FOKUS, Germany)
D. Raz (Technion Institute of Technology, Israel)
K. Zieliński (AGH University of Science and Technology, Poland)
R. Yahyapour (University Dortmund, Germany)
K. Yang (University of Essex, UK)

Next Generation Computing

E.-N. John Huh (Seoul Women's University, Korea)

Practical Aspects of High-Level Parallel Programming (PAPP 2004)

F. Loulergue (Laboratory of Algorithms, Complexity and Logic,
University of Paris Val de Marne, France)

Parallel Input/Output Management Techniques (PIOMT 2004)

J. H. Abawajy (Carleton University, School of Computer Science, Canada)

OpenMP for Large Scale Applications

B. Chapman (University of Houston, USA)

Tools for Program Development and Analysis in Computational Science

D. Kranzlmüller (Johannes Kepler University Linz, Austria)
R. Wismüller (TU München, Germany)
A. Bode (Technische Universität München, Germany)
J. Volkert (Johannes Kepler University Linz, Austria)

Modern Technologies for Web-Based Adaptive Systems

N. Thanh Nguyen (Wrocław University of Technology, Poland)
J. Sobecki (Wrocław University of Technology, Poland)

Agent Day 2004 – Intelligent Agents in Computing Systems

E. Nawarecki (AGH University of Science and Technology, Poland)
K. Cetnarowicz (AGH University of Science and Technology, Poland)
G. Dobrowolski (AGH University of Science and Technology, Poland)
R. Schaefer (Jagiellonian University, Poland)
S. Ambroszkiewicz (Polish Academy of Sciences, Warsaw, Poland)
A. Koukam (Université de Belfort-Montbeliard, France)
V. Srovnal (VSB Technical University of Ostrava, Czech Republic)
C. Cotta (Universidad de Málaga, Spain)
S. Raczynski (Universidad Panamericana, Mexico)

Dynamic Data Driven Application Systems

F. Darema (NSF/CISE, USA)

HLA-Based Distributed Simulation on the Grid

S. J. Turner (Nanyang Technological University, Singapore)

Interactive Visualisation and Interaction Technologies

E. Zudilova (University of Amsterdam, The Netherlands)
T. Adriaansen (CSIRO, ICT Centre, Australia)

Computational Modeling of Transport on Networks

B. Tadic (Jozef Stefan Institute, Slovenia)
S. Thurner (Universität Wien, Austria)

Modeling and Simulation in Supercomputing and Telecommunications

Y. Mun (Soongsil University, Korea)

QoS Routing

H. Choo (Sungkyunkwan University, Korea)

Evolvable Hardware

N. Nedjah (State University of Rio de Janeiro, Brazil)
L. de Macedo Mourelle (State University of Rio de Janeiro, Brazil)

Advanced Methods of Digital Image Processing

B. Smolka (Silesian University of Technology, Laboratory of Multimedia Communication, Poland)

Computer Graphics and Geometric Modelling (CGGM 2004)

A. Iglesias Prieto (University of Cantabria, Spain)

Computer Algebra Systems and Applications (CASA 2004)

A. Iglesias Prieto (University of Cantabria, Spain)
A. Galvez (University of Cantabria, Spain)

New Numerical Methods for DEs: Applications to Linear Algebra, Control and Engineering

N. Del Buono (University of Bari, Italy)
L. Lopez (University of Bari, Italy)

Parallel Monte Carlo Algorithms for Diverse Applications in a Distributed Setting

V. N. Alexandrov (University of Reading, UK)
A. Karaivanova (Bulgarian Academy of Sciences, Bulgaria)
I. Dimov (Bulgarian Academy of Sciences, Bulgaria)

Modelling and Simulation of Multi-physics Multi-scale Systems

V. Krzhizhanovskaya (University of Amsterdam, The Netherlands)
B. Chopard (University of Geneva, CUI, Switzerland)
Y. Gorbachev (St. Petersburg State Polytechnical University, Russia)

Gene, Genome and Population Evolution

S. Cebrat (University of Wrocław, Poland)
D. Stauffer (Cologne University, Germany)
A. Maksymowicz (AGH University of Science and Technology, Poland)

Computational Methods in Finance and Insurance

A. Janicki (University of Wrocław, Poland)
J.J. Korczak (University Louis Pasteur, Strasbourg, France)

Computational Economics and Finance

X. Deng (City University of Hong Kong, Hong Kong)
S. Wang (Chinese Academy of Sciences, ROC)
Y. Shi (University of Nebraska at Omaha, USA)

GeoComputation

Y. Xue (Chinese Academy of Sciences, ROC)
C. Yarotsos (University of Athens, Greece)

Simulation and Modeling of 3D Integrated Circuits

I. Balk (R3Logic Inc., USA)

Computational Modeling and Simulation on Biomechanical Engineering

Y.H. Kim (Kyung Hee University, Korea)

Information Technologies Enhancing Health Care Delivery

M. Duplaga (Jagiellonian University Medical College, Poland)
D. Ingram (University College London, UK)
K. Zieliński (AGH University of Science and Technology, Poland)

Computing in Science and Engineering Academic Programs

D. Donnelly (Siena College, USA)

Sponsoring Institutions

Hewlett-Packard
Intel
SGI
ATM
SUN Microsystems
IBM
Polish Airlines LOT
ACC CYFRONET AGH
Institute of Computer Science AGH
Polish Ministry of Scientific Research and Information Technology
Springer-Verlag

Table of Contents – Part I

Track on Models and Algorithms

Track on Data Mining and Data Bases

Track on Networking

Poster Papers

Table of Contents – Part II

Track on Numerical Algorithms

Track on Finite Element Method

Track on Neural Networks

Track on Applications

Poster Papers

Table of Contents – Part III

Workshop on Programming Grids and Metasystems

Workshop on First International Workshop on Active and Programmable Grids Architectures and Components

Workshop on Next Generation Computing

Workshop on Practical Aspects of High-Level Parallel Programming (PAPP 2004)

Workshop on Parallel Input/Output Management Techniques (PIOMT04)

Workshop on OpenMP for Large Scale Applications

Workshop on Tools for Program Development and Analysis in Computational Science

Workshop on Modern Technologies for Web-Based Adaptive Systems

Workshop on Agent Day 2004 – Intelligent Agents in Computing Systems

Workshop on Dynamic Data Driven Applications Systems

Workshop on HLA-Based Distributed Simulation on the Grid

Workshop on Interactive Visualisation and Interaction Technologies

Workshop on Computational Modeling of Transport on Networks

Workshop on Modeling and Simulation in Supercomputing and Telecommunications

Workshop on QoS Routing

Workshop on Evolvable Hardware

Table of Contents – Part IV

Workshop on Advanced Methods of Digital Image Processing

Workshop on Computer Graphics and Geometric Modelling (CGGM 2004)

Workshop on Computer Algebra Systems and Applications (CASA 2004)

Workshop on New Numerical Methods for DEs: Applications to Linear Algebra, Control and Engineering

Workshop on Parallel Monte Carlo Algorithms for Diverse Applications in a Distributed Setting

Workshop on Modelling and Simulation of Multi-physics Multi-scale Systems

Workshop on New Numerical Methods for DEs: Applications to Linear Algebra, Control and Engineering

Workshop on Gene, Genome, and Population Evolution

Workshop on Computational Methods in Finance and Insurance

Workshop on Computational Economics and Finance

Workshop on GeoComputation

Workshop on Simulation and Modeling of 3D Integrated Circuits

Workshop on Computational Modeling and Simulation on Biomechanical Engineering

Workshop on Information Technologies Enhancing Health Care Delivery

Workshop on Computing in Science and Engineering Academic Programs

Optimization of Collective Reduction Operations

Rolf Rabenseifner

High-Performance Computing-Center (HLRS), University of Stuttgart
Allmandring 30, D-70550 Stuttgart, Germany
`rabenseifner@hlrs.de`,
`www.hlrs.de/people/rabenseifner/`

Abstract. A 5-year-profiling in production mode at the University
of Stuttgart has shown that more than 40% of the execution time of
Message Passing Interface (MPI) routines is spent in the collective communication routines MPI_Allreduce and MPI_Reduce. Although MPI
implementations are now available for about 10 years and all vendors
are committed to this Message Passing Interface standard, the vendors'
and publicly available reduction algorithms could be accelerated with
new algorithms by a factor between 3 (IBM, sum) and 100 (Cray
T3E, maxloc) for long vectors. This paper presents five algorithms
optimized for different choices of vector size and number of processes.
The focus is on bandwidth dominated protocols for power-of-two and
non-power-of-two number of processes, optimizing the load balance in
communication and computation.

Keywords: Message Passing, MPI, Collective Operations, Reduction.

1 Introduction and Related Work

MPI_Reduce combines the elements provided in the input vector (buffer) of each
process using an operation (e.g. sum, maximum), and returns the combined
values in the output buffer of a chosen process named root. MPI_Allreduce is
the same as MPI_Reduce, except that the result appears in the receive buffer
of all processes. MPI_Allreduce is one of the most important MPI routines and
most vendors are using algorithms that can be improved by a factor of more than
2 for long vectors. Most current implementations are optimized only for short
vectors. A 5-year-profiling [11] of most MPI based applications (in production
mode) of all users of the Cray T3E 900 at our university has shown, that 8.54 %
of the execution time is spent in MPI routines. 37.0 % of the MPI time is spent in
MPI_Allreduce and 3.7 % in MPI_Reduce. The 5-year-profiling has also shown,
that 25 % of all execution time was spent with a non-power-of-two number of
processes. Therefore, a second focus is the optimization for non-power-of-two
numbers of processes.

Early work on collective communication implements the reduction operation
as an inverse broadcast and do not try to optimize the protocols based on different buffer sizes [1]. Other work already handle allreduce as a combination of basic

M. Bubak et al. (Eds.): ICCS 2004, LNCS 3036, pp. 1–9, 2004.

routines, e.g., [2] already proposed the *combine-to-all* (allreduce) as a combination of *distributed combine* (reduce_scatter) and *collect* (allgather). Collective algorithms for wide-area cluster are developed in [5,7,8], further protocol tuning can be found in [3,4,9,12], and automatic tuning in [13]. The main focus of the work presented in this paper is to optimize the algorithms for different numbers of processes (non-power-of-two and power-of-two) **and** for different buffer sizes by using special reduce_scatter protocols without the performance penalties on normal rank-ordered scattering. The allgather protocol is chosen according the the characteristics of the reduce_scatter part to achieve an optimal bandwidth for any number of processes and buffer size.

2   Allreduce and Reduce Algorithms

2.1   Cost Model

To compare the algorithms, theoretical cost estimation and benchmark results are used. The cost estimation is based on the same flat model used by R. Thakur and B. Gropp in [12]. Each process has an input vector with n bytes, p is the number of MPI processes, γ the computation cost per vector byte executing one operation with two operands locally on any process. The total reduction effort is $(p-1)n\gamma$. The total computation time with optimal load balance on p processes is therefore $\frac{p-1}{p}n\gamma$, i.e., less than $n\gamma$, which is independent of the number of processes! The communication time is modeled as $\alpha + n\beta$, where α is the latency (or startup time) per message, and β is the transfer time per byte, and n the message size in bytes. It is assumed further that all processes can send and receive one message at the same time with this cost model. In reality, most networks are faster, if the processes communicate in parallel, but pairwise only in one direction (uni-directional between two processes), e.g., in the classical binary tree algorithms. Therefore $\alpha_{uni} + n\beta_{uni}$ is modeling the **uni**-directional communication, and $\alpha + n\beta$ is used with the **bi**-directional communication. The ratios are abbreviated with $f_\alpha = \alpha_{uni}/\alpha$ and $f_\beta = \beta_{uni}/\beta$. These factors are normally in the range 0.5 (simplex network) to 1.0 (full duplex network).

2.2   Principles

A classical implementation of MPI_Allreduce is the combination of MPI_Reduce (to a root process) followed by MPI_Bcast sending the result from root to all processes. This implies a bottle-neck on the root process. Also classical is the binary tree implementation of MPI_Reduce, which is a good algorithm for short vectors, but that causes a heavy load imbalance because in each step the number of active processes is halved. The optimized algorithms are based on a few principles:

Recursive vector halving: For long-vector reduction, the vector can be split into two parts and one half is reduced by the process itself and the other half is sent to a neighbor process for reduction. In the next step, again the buffers are halved, and so on.

Recursive vector doubling: To return the total result in the result vector, the split result vectors must be combined recursively. MPI_Allreduce can be implemented as a reduce-scatter (using recursive vector halving) followed by an allgather (using recursive vector doubling).

Recursive distance doubling: In step 1, each process transfers data at distance 1 (process P0 with P1, P2–P3, P4–P5, ...); in step 2, the distance is doubled, i.e., P0–P2 and P1–P3, P4–P6 and P5–P7; and so on until distance $\frac{p}{2}$.

Recursive distance halving: Same procedure, but starting with distance $p/2$, i.e., P0–P$\frac{p}{2}$, P1–P$(\frac{p}{2}+1)$, ..., and ending with distance 1, i.e., P0–P1, ...

Recursive vector and distance doubling and halving can be combined for different purposes, but always additional overhead causes load imbalance if the number of processes is not a power of two. Two principles can reduce the overhead in this case.

Binary blocks: The number of processes can be expressed as a sum of power-of-two values, i.e., all processes are located in subsets with power-of-two processes. Each subset is used to execute parts of the reduction protocol in a block. Overhead occurs in the combining of the blocks in some step of the protocol.

Ring algorithms: A reduce_scatter can be implemented by $p-1$ ring exchange steps with increasing strides. Each process computes all reduction operations for its own chunk of the result vector. In step i ($i=1\,..\,p\text{-}1$) each process sends the input vector chunk needed by $rank+i$ to that process and receives from $rank-i$ the data needed to reduce its own chunk. The allreduce can be completed by an allgather that is also implemented with ring exchange steps, but with constant stride 1. Each process sends its chunk of the result vector around the ring to the right ($rank+1$) until its left neighbor (($rank+p-1$) mod p) has received it after $p-1$ steps. The following sections describe the algorithms in detail.

2.3   Binary Tree

Reduce: The classical binary tree always exchanges full vectors, uses recursive distance doubling, but with incomplete protocol, because in each step, half of the processes finish their work. It takes $\lceil \lg p \rceil$ steps and the time taken by this algorithm is $T_{red,tree} = \lceil \lg p \rceil (\alpha_{uni} + n\beta_{uni} + n\gamma))$.

For short vectors, this algorithm is optimal (compared to the following algorithms) due to its smallest latency term $\lceil \lg p \rceil \alpha_{uni}$.

Allreduce: The reduce algorithm is followed by a binary tree based broadcast. The total execution time is $T_{all,tree} = \lceil \lg p \rceil (2\alpha_{uni} + 2n\beta_{uni} + n\gamma))$.

2.4   Recursive Doubling

Allreduce: This algorithm is an optimization especially for short vectors. In each step of the recursive distance doubling, both processes in a pair exchange the input vector (in step 1) or its intermediate result vector (in steps 2 ... $\lceil \lg p \rceil$) with its partner process and both processes are computing the same reduction redundantly. After $\lceil \lg p \rceil$ steps, the identical result vector is available in all processes. It needs $T_{all,r.d.} = \lceil \lg p \rceil (\alpha+n\beta+n\gamma))+$(if non-power-of-two $\alpha_{uni}+n\beta_{uni}$) This algorithm is in most cases optimal for short vectors.

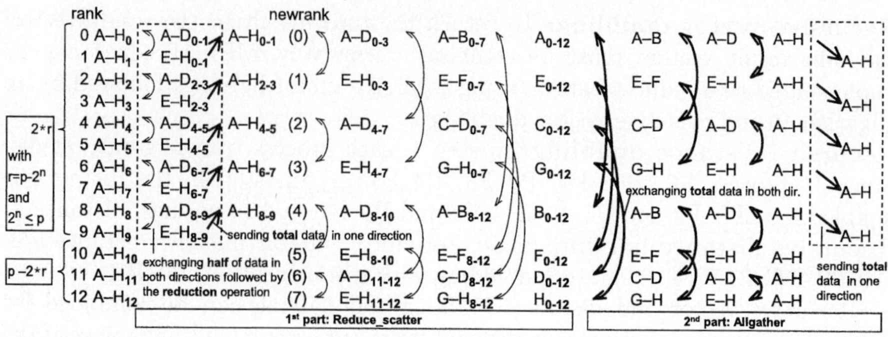

Fig. 1. Recursive Halving and Doubling. The figure shows the intermediate results after each buffer exchange (followed by a reduction operation in the 1st part). The dotted frames show the overhead caused by a non-power-of-two number of processes

2.5   Recursive Halving and Doubling

This algorithm is a combination of a reduce_scatter implemented with recursive vector halving and distance doubling[1] followed by a allgather implemented by a recursive vector doubling combined with recursive distance halving (for allreduce), or followed by gather implemented with a binary tree (for reduce).

In a first step, the number of processes p is reduced to a power-of-two value: $p' = 2^{\lfloor \lg p \rfloor}$. $r = p - p'$ is the number of processes that must be *removed* in this first step. The first $2r$ processes send pairwise from each even $rank$ to the odd $(rank + 1)$ the second half of the input vector and from each odd $rank$ to the even $(rank - 1)$ the first half of the input vector. All $2r$ processes compute the reduction on their half.

Fig. 1 shows the protocol with an example on 13 processes. The input vectors and all reduction results will be divided into p' parts (A, B,..., H) by this algorithm, and therefore it is denoted with A–H$_{rank}$. After the first reduction, process P0 has computed A–D$_{0-1}$, denoting the reduction result of the first half of the vector (A–D) from the processes 0–1. P1 has computed E–H$_{0-1}$, P2 A–D$_{2-3}$, ... The first step is finished by sending those results from each odd process $(1 \dots 2r - 1)$ to $rank - 1$ into the second part of the buffer.

Now, the first r even processes and the $p - 2r$ last processes are renumbered from 0 to $p' - 1$. This first step needs $(1 + f_\alpha)\alpha + \frac{1+f_{beta}}{2}n\beta + \frac{1}{2}n\gamma$ and is not necessary, if the number of processes p was already a power-of-two.

Now, we start with the first step of recursive vector halving and distance doubling, i.e., the even / odd ranked processes are sending the second / first half

---

[1] A distance doubling (starting with distance 1) is used in contrary to the reduce_scatter algorithm in [12] that must use a distance halving (i.e., starting with distance $\frac{\#\text{processes}}{2}$) to guarantee a rank-ordered scatter. In our algorithm, any order of the scattered data is allowed, and therefore, the longest vectors can be exchanged with the nearest neighbor, which is an additional advantage on systems with a hierarchical network structure.

of their buffer to $rank' + 1$ / $rank' - 1$. Then the reduction is computed between the local buffer and the received buffer. This step costs $\alpha + \frac{1}{2}(n\beta + n\gamma)$.

In the next $\lg p' - 1$ steps, the buffers are recursively halved and the distance doubled. Now, each of the p' processes has $\frac{1}{p'}$ of the total reduction result vector, i.e., the reduce_scatter has scattered the result vector to the p' processes. All recursive steps cost $\lg p'\alpha + (1 - \frac{1}{p'})(n\beta + n\gamma)$. The second part implements an allgather or gather to complete the allreduce or reduce operation.

Allreduce: Now, the contrary protocol is needed: Recursive vector doubling and distance halving, i.e., in the first step the process pairs exchange $\frac{1}{p'}$ of the buffer to achieve $\frac{2}{p'}$ of the result vector, and in the next step $\frac{2}{p'}$ is exchanged to get $\frac{4}{p'}$, and so on. A–B, A–D ... in Fig. 1 denote the already stored portion of the result vector. After each communication exchange step, the result buffer is doubled and after $\lg p'$ steps, the p' processes have received the total reduction result. This allgather part costs $\lg p'\alpha + (1 - \frac{1}{p'})(n\beta)$.

If the number of processes is non-number-of-two, then the total result vector must be sent to the r *removed* processes. This causes the additional overhead $\alpha + n\beta$. The total implementation needs

- $T_{all,h\&d,n=2^{exp}} = 2\lg p\alpha + 2n\beta + n\gamma - \frac{1}{p}(2n\beta + n\gamma)$
  $\simeq 2\lg p\alpha + 2n\beta + n\gamma$ if p is power-of-two,
- $T_{all,h\&d,n\neq2^{exp}} = (2\lg p' + 2 + f_\alpha)\alpha + (3 + \frac{1+f_{beta}}{2})n\beta + \frac{3}{2}n\gamma - \frac{1}{p'}(2n\beta + n\gamma)$
  $\simeq (3 + 2\lfloor\lg p\rfloor)\alpha + 4n\beta + \frac{3}{2}n\gamma$ if p is non-power-of-two (with $p' = 2^{\lfloor\lg p\rfloor}$).

This protocol is good for long vectors and power-of-two processes. For non-power-of-two processes, the transfer overhead is doubled and the computation overhead is enlarged by $\frac{3}{2}$. The *binary blocks* protocol (see below) can reduce this overhead in many cases.

Reduce: The same protocol is used, but the pairwise exchange with sendrecv is substituted by single message passing. In the first step, each process with the bit with the value $p'/2$ in its new rank identical to that bit in root rank must receive a result buffer segment and the other processes must send their segment. In the next step only the receiving processes continue and the bit is shifted 1 position right (i.e., $p'/4$). And so on. The time needed for this gather operation is $\lg p'\alpha_{uni} + (1 - \frac{1}{p'})n\beta_{uni}$.

In the case that the original root process is one of the *removed* processes, then the role of this process and its partner in the first step are exchanged after the first reduction in the reduce_scatter protocol. This causes no additional overhead. The total implementation needs

- $T_{red,h\&d,n=2^{exp}} = \lg p(1 + f_\alpha)\alpha + (1 + f_\beta)n\beta + n\gamma - \frac{1}{p}(n(\beta + \beta_{uni}) + n\gamma)$
  $\simeq 2\lg p\alpha + 2n\beta + n\gamma$ if p is power-of-two,
- $T_{red,h\&d,n\neq2^{exp}} = \lg p'(1 + f_\alpha)\alpha + (1 + f_\alpha)\alpha + (1 + \frac{1+f_{beta}}{2} + f_\beta)n\beta + \frac{3}{2}n\gamma - \frac{1}{p'}((1 + f_\beta)n\beta + n\gamma)$
  $\simeq (2 + 2\lfloor\lg p\rfloor)\alpha + 3n\beta + \frac{3}{2}n\gamma$ if p is non-power-of-two (with $p' = 2^{\lfloor\lg p\rfloor}$).

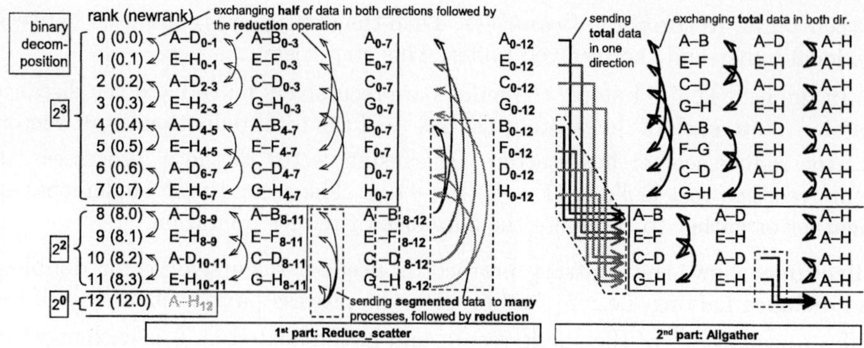

Fig. 2. Binary Blocks

2.6   Binary Blocks

Further optimization for non-power-of-two number of processes can be achieved with the algorithm shown in Fig. 2.

Here, the maximum difference between the ratio of the number of proccesses of two successive blocks, especially in the low range of exponents, determines the imbalance.

Allreduce: The 2nd part is an allgather implemented with buffer doubling and distance halving in each block as in the algorithm in the previous section. The input must be provided in the processes of the smaller blocks always with pairs of messages from processes of the next larger block.

Reduce: If the root is outside of the largest block, then the intermediate result segment of rank 0 is sent to root and root plays the role of rank 0. A binary tree is used to gather the result segments into the root process.

For power-of-two number of processes, the binary block algorithms are identical to the halving and doubling algorithm in the previous section.

2.7   Ring

While the algorithms in the last two sections are optimal for power-of-two process numbers and long vectors, for medium non-power-of-two number of processes and long vectors there exist another good algorithm. It uses the *pairwise exchange* algorithm for reduce_scatter and *ring* algorithm for allgather (for allreduce), as described in [12], and for reduce, all processes send their result segment directly to root. Both algorithms are good in bandwidth usage for non-power-of-two number of processes, but the latency scales with the number of processes. Therefore this algorithm can be used only for a small number of processes. Independent of whether p is power-of-two or not, the total implementation needs $T_{all,ring} = 2(p-1)\alpha + 2n\beta + n\gamma - \frac{1}{p}(2n\beta + n\gamma)$ for allreduce, and $T_{red,ring} = (p-1)(\alpha + \alpha_{uni}) + n(\beta + \beta_{uni}) + n\gamma - \frac{1}{p}(n(\beta + \beta_{uni}) + n\gamma)$ for reduce.

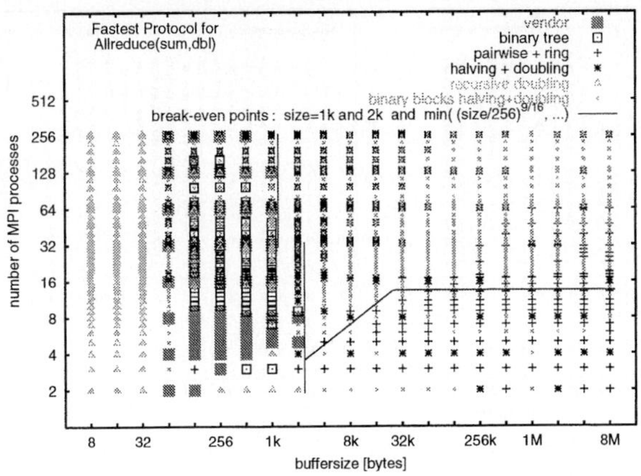

Fig. 3. The fastest protocol for Allreduce(double, sum) on a Cray T3E 900.

3   Choosing the Fastest Algorithm

Based on the number of processes and the vector (input buffer) length, the reduction routine must decide which algorithm should be used. Fig. 3 shows the fastest protocol on a Cray T3E 900 with 540 PEs. For buffer sizes less than or equal to 32 byte, *recursive doubling* is the best, for buffer sizes less than or equal to 1 KB, mainly *vendor's algorithm* (for power-of-two) and *binary tree* (for non-power-of-two) are the best but there is not a big difference to *recursive doubling*. For longer buffer sizes, the *ring* is good for some buffer sizes and some #processes less than 32 PEs. A detailed decision is done for each #processes value, e.g., for 15 processes, *ring* is used if length $\geq$ 64 KB. In general, on a Cray T3E 900, the binary block algorithm is faster if $\delta_{expo,max} < \lg(\frac{vector\ size}{1Byte})/2.0 - 2.5$ and *vector size* $\geq$ 16 KB and more than 32 processes are used. In a few cases, e.g., 33 PEs and less then 32 KB, *halving&doubling* is the fastest algorithm.

Fig. 4 shows that with the pure MPI programming model (i.e., 1 MPI process per CPU) on the IBM SP, the benefit is about 1.5x for buffer sizes 8–64 KB, and 2x – 5x for larger buffers. With the hybrid programming model (1 MPI process per SMP node), only for buffer sizes 4–128 KB and more than 4 nodes, the benefit is about 1.5x – 3x.

4   Conclusions and Future Work

Although principal work on optimizing collective routines is quite old [2], there is a lack of fast implementations for *allreduce* and *reduce* in MPI libraries for a wide range of number of processes and buffer sizes. Based on the author's algorithm from 1997 [10], an efficient algorithm for power-of-two and non-power-of-two number of processes is presented in this paper. Medium non-power-of-two number of processes could be additionally optimized with a special ring algorithm. The *halving&doubling* is already included into MPICH-2 and it is

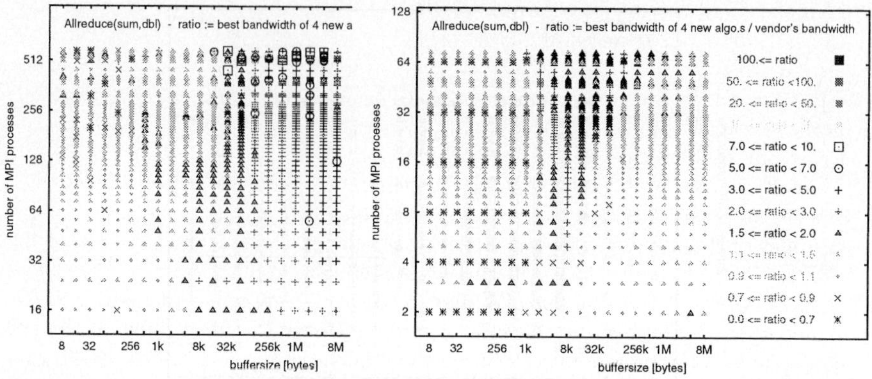

Fig. 4. Ratio of bandwidth of the fastest protocol (without *recursive doubling*) on a IBM SP at SDSC and 1 MPI process per CPU (left) and per SMP node (right)

planned to include the other bandwidth-optimized algorithms [10,12]. Future work will further optimize latency and bandwidth for any number of processes by combining the principles used in Sect. 2.3–2.7 into one algorithm and selecting on each recursion level instead of selecting one of those algorithms for all levels.

Acknowledgments. The author would like to acknowledge his colleagues and all the people that supported this project with suggestions and helpful discussions. He would especially like to thank Rajeev Thakur and Jesper Larsson Träff for the helpful discussion on optimized reduction algorithm and Gerhard Wellein, Thomas Ludwig, Ana Kovatcheva, Rajeev Thakur for their benchmarking support.

References

1. V. Bala, J. Bruck, R. Cypher, P. Elustondo, A. Ho, C.-T. Ho, S. Kipnis and M. Snir, *CCL: A portable and tunable collective communication library for scalable parallel computers*, in IEEE Transactions on Parallel and Distributed Systems, Vol. 6, No. 2, Feb. 1995, pp 154–164.
2. M. Barnett, S. Gupta, D. Payne, L. Shuler, R. van de Gejin, and J. Watts, *Interprocessor collective communication library (InterCom)*, in Proceedings of Supercomputing '94, Nov. 1994.
3. Edward K. Blum, Xin Wang, and Patrick Leung, *Architectures and message-passing algorithms for cluster computing: Design and performance*, in Parallel Computing 26 (2000) 313–332.
4. J. Bruck, C.-T. Ho, S. Kipnis, E. Upfal, and D. Weathersby, *Efficient algorithms for all-to-all communications in multiport message-passing systems*, in IEEE Transactions on Parallel and Distributed Systems, Vol. 8, No. 11, Nov. 1997, pp 1143–1156.
5. E. Gabriel, M. Resch, and R. Rühle, *Implementing MPI with optimized algorithms for metacomputing*, in Proceedings of the MPIDC'99, Atlanta, USA, March 1999, pp 31–41.

6. Message Passing Interface Forum. *MPI: A Message-Passing Interface Standard*, Rel. 1.1, June 1995, `www.mpi-forum.org`.
7. N. Karonis, B. de Supinski, I. Foster, W. Gropp, E. Lusk, and J. Bresnahan, *Exploiting hierarchy in parallel computer networks to optimize collective operation performance*, in Proceedings of the 14th International Parallel and Distributed Processing Symposium (IPDPS '00), 2000, pp 377–384.
8. Thilo Kielmann, Rutger F. H. Hofman, Henri E. Bal, Aske Plaat, Raoul A. F. Bhoedjang, *MPI's reduction operations in clustered wide area systems*, in Proceedings of the Message Passing Interface Developer's and User's Conference 1999 (MPIDC'99), Atlanta, USA, March 1999, pp 43–52.
9. Man D. Knies, F. Ray Barriuso, William J. Harrod, George B. Adams III, *SLICC: A low latency interface for collective communications*, in Proceedings of the 1994 conference on Supercomputing, Washington, D.C., Nov. 14–18, 1994, pp 89–96.
10. Rolf Rabenseifner, *A new optimized MPI reduce and allreduce algorithm*, Nov. 1997. `http://www.hlrs.de/mpi/myreduce.html`
11. Rolf Rabenseifner, *Automatic MPI counter profiling of all users: First results on a CRAY T3E 900-512*, Proceedings of the Message Passing Interface Developer's and User's Conference 1999 (MPIDC'99), Atlanta, USA, March 1999, pp 77–85. `http://www.hlrs.de/people/rabenseifner/publ/publications.html`
12. Rajeev Thakur and William D. Gropp, *Improving the performance of collective operations in MPICH*, in Recent Advances in Parallel Virtual Machine and Message Passing Interface, proceedings of the 10th European PVM/MPI Users' Group Meeting, LNCS 2840, J. Dongarra, D. Laforenza, S. Orlando (Eds.), 2003, 257–267.
13. Sathish S. Vadhiyar, Graham E. Fagg, and Jack Dongarra, *Automatically tuned collective communications*, in Proceedings of SC2000, Nov. 2000.

An extended version of this paper can be found on the author's home/publication page.

Predicting MPI Buffer Addresses*

Felix Freitag, Montse Farreras, Toni Cortes, and Jesus Labarta

Computer Architecture Department (DAC)
European Center for Parallelism of Barcelona (CEPBA)
Politechnic University of Catalonia (UPC)
{felix,mfarrera,toni,jesus}@ac.upc.es

Abstract. Communication latencies have been identified as one of the performance limiting factors of message passing applications in clusters of workstations/multiprocessors. On the receiver side, message-copying operations contribute to these communication latencies. Recently, prediction of MPI messages has been proposed as part of the design of a zero message-copying mechanism. Until now, prediction was only evaluated for the next message. Predicting only the next message, however, may not be enough for real implementations, since messages do not arrive in the same order as they are requested. In this paper, we explore long-term prediction of MPI messages for the design of a zero message-copying mechanism. To achieve long-term prediction we evaluate two prediction schemes, the first based on graphs, and the second based on periodicity detection. Our experiments indicate that with both prediction schemes the buffer addresses and message sizes of several future MPI messages (up to +10) can be predicted successfully.

1 Introduction

MPI (Message Passing Interface) is a specification for a standard library to address the message-passing model of parallel computation [11]. In this model, applications are divided into different tasks (or processes) that communicate by sending and receiving messages among them. A number of implementations of MPI are available like MPICH from Argonne National Laboratory [12], CHIMP from Edinburgh Parallel Computing Center (EPCC) [4], and LAM from Ohio Supercomputer Center [10].

Communication latencies have been identified as one of the performance limiting factors of message passing applications in clusters of workstations/multiprocessors [1]. On the receiver side, message-copying operations contribute to these communication latencies. In a standard implementation, there is at least one copy of the message from the buffer of the MPI implementation to the user space.

Zero message-copying on the receiver side of MPI has been indicated as a technique to reduce this communication latency [1]. One of the identified requirements to achieve zero message-copying is to predict the characteristics of the

---

* This work was supported in part by the Spanish Ministry of Education and Science under TIC2001-0995-C02-03, CEPBA and CIRI fellowship grants.

M. Bubak et al. (Eds.): ICCS 2004, LNCS 3036, pp. 10–17, 2004.

next MPI message. The prediction of only the next message, however, may not be enough to implement a zero message-copying mechanisms in real implementations, since messages do not arrive in the order in which they are requested. If the receiver could predict several future messages, then changes in the message arrival order could be handled, which enables the design of a zero message-copying mechanism.

The goals of this paper are the followings: We first describe the design of a zero message-copying mechanism, which requires several future messages to be predicted. Then we show that with the proposed prediction mechanisms this long-term message prediction can be achieved with high prediction rates.

The remainder or this paper is structured as follows: In section 2 we describe related work done on MPI message characterization and MPI message prediction. In section 3 we explain the design of a zero message-copying mechanism and the predictors we evaluate. Section 4 shows the achieved message prediction rates and compares both predictors. In section 5 we conclude the paper.

2   Related Work

Communication locality of MPI messages was studied by Kim and Lilja in [9]. The result of their work is important since if communication locality exits, then the future values of a stream will belong to the observed data set, and will allow prediction. In their experiments it was observed that the processes communicate only with a small number of partners (message-destination locality). Also, it was observed that the MPI applications usually have only 2-3 distinct message sizes (message size locality). The results showed the locality of MPI communication patterns in the spatial domain. The characteristics of the temporal patterns in MPI messages, however, were not reported. In our work, we examine the existence of temporal patterns in MPI communications, which is an important requirement to allow long-term prediction.

The prediction of MPI messages is proposed in [1]. In their work, predictors are based on heuristics and detect cycles in an observed data stream. The prediction heuristics predict the next value of the given data stream. It was shown that the predictors obtained very high hit rates for the studied benchmarks. The benchmarks were the NAS Bt, Cg, and Sp, and the PSTSWM benchmark. In [7], the prediction of MPI messages is explored with the goal to achieve a better scalability of MPI applications.

3   Our Approach

3.1   Requirement for a Zero Message-Copying Mechanism

Previous research [1] has proposed the idea of predicting the user buffer, where the data will be read, and to place it directly in its final location. This would avoid copying first in an MPI buffer and then copy it again to the user buffer. This idea, although interesting in its concept, it is not implementable. Predicting the destination buffer to use it by the MPI library is too dangerous because an error in the prediction may modify a memory location that has useful data for the application. It is also important to notice that previous work only predicted the location of the buffer for the next message, and this is not enough. Messages do not arrive in the same order they are requested and thus, the system has to be ready to receive messages out of order and still be able to implement the zero-copy mechanism with them. Otherwise, the applicability of the improvement will be very restricted.

Le us assume now that we are able to predict the buffers for the future messages (which we will show along this work), the challenge is how we can use this knowledge to avoid extra copies. As we have said, using the exact buffer is not a possibility because of miss predictions (among others), but the MPI library can place the data aligned in the same way as in the final destination buffer. If we hit in the prediction, then we can change the mapping of logical pages to physical pages to move the data to the user address space without a copy.

This proposal could avoid copies of full pages, but not the portions of the message that do not fill a full page. In the latter case, we will have to copy this information, but hopefully, long and costly messages will avoid the long copies. All messages longer than 2 pages, will at least avoid the copy of one page. Regarding miss predictions, they do not cause any problem because the message will be copied as if no prediction had been done. A miss prediction will mean that the improvement will not be possible but no miss function will appear.

The only requirement we have is a system call that allows us to switch the binding of logical pages to their physical pages (of course only among the pages of the process). Once proven its necessity, this implementation is simple and could be easily incorporated in new versions of the OS.

3.2 Graph-Based Predictor

Our first solution for long-term prediction is a graph-based predictor as described in [5]. This predictor is similar to the prediction heuristics used in [1]. The second prediction mechanism, which we evaluate for this task, is a periodicity-based predictor [6].

Graph-based predictors describe an observed sequence through a number of states with transitions between them. Each state represents an observation. A probability or counter values is associated with the transition between the states. The graphs are trained (and build) on the observed data. The number of states increases with the increase of different symbols in the observed sequence. The observations contribute to form the transition probability from one state to another. Cyclic behavior in a data sequence, for instance, can be easily represented with such graphs, as demonstrated in [1].

In order to evaluate long-term MPI message buffer address prediction, we implement the graph-based predictor following the description in [5]. Each state represents a sequence of three observations. The value of the transition between states is computed according to the observed sequence. Prediction can be achieved by selecting the most likely successor of a current state. We predict several future values by repeating this process on the predicted states.

3.3 Periodicity-Based Predictor

The approach of the periodicity-based predictor is different to the graph-based predictor, since prediction is based on the detection of iterative patterns in the temporal order of the sequence.

Previous results from [7] show that MPI messages contain repetitive sender and message size patterns. The knowledge of the periodic patterns allows predicting future values. We use the dynamic periodicity detector (DPD) from [6] to capture the periodicity of the data stream and modify it to enable the prediction of data streams.

The algorithm used by the periodicity detector is based on the distance metric given in equation (1).

$$d(m) = sign \sum_{i=0}^{N-1} |\, x(i) - x(i - m)\,| \tag{1}$$

In equation (1) N is the size of the data window, m is the delay (0<m<M), M<=N, x[i] is the current value of the data stream, and d(m) is the value computed to detect the periodicity. It can be seen that equation (1) compares the data sequence with the data sequence shifted m samples. Equation (1) computes the distance between two vectors of size N by summing the magnitudes of the L1-metric distance of N vector elements. The sign function is used to set the values d(m) to 1 if the distance is not zero. The value d(m) becomes zero if the data window contains an identical periodic pattern with periodicity m.

4 Evaluation

4.1 Benchmarks Used

In order to evaluate the long-term predictability of MPI buffer addresses we run several experiments with the NAS benchmark suite programmed with MPI. We use the NAS Bt, Cg, Ft, Is, Lu, and Sp benchmarks [2]. The MPI applications are executed on an IBM RS/6000 in a dedicated environment. We use the class A problem size of the NAS benchmarks. The MPI implementation we used is MPICH [12]. The applications are run with different numbers of processes, where the number of processes is from 4 to 32 processes.

In order to obtain the communication behavior of the applications, we instrument the MPICH implementation. To obtain the communication data we trace the MPI calls from the application code to the top level of the MPI library. We trace point-to-point and collective calls. Collective calls are represented in the trace as point-to-point calls from the different senders. The traces we extract correspond to the receiver side. We extract the buffer address, message size, and sender processes.

In Table 1 we summarize the characteristics we observed in these traces. Column 2 indicates the number of processes the application is executed with. Column 3 gives the number of messages received per process. Column 4 indicates the number of different buffer addresses in the pattern. Column 5 shows the number of different message sizes in the pattern. Column 6 indicates the size of the observed temporal pattern. The pattern size refers to the pattern formed by point-to-point and collective calls. We have obtained the size of the periodic patterns using the DPD of [6]. We can see that the results given in columns 4 and 5 confirm the data locality in MPI messages described in [9]. Column 6 indicates the existence and size of temporal patterns in MPI messages.

Table 1. Evaluated benchmarks and communication characteristics

Benchmark	# of processes	#of messages received	# of different buffer addresses in the pattern	#different message sizes in the pattern	detected pattern size
NAS BT	4	2425	7	4	18
	9	3630	7	3	12
	16	4835	7	3	24
	25	6034	7	5	30
NAS CG	4	1679	2	2	4
	8	2942	3	2	7
	16	2942	3	2	7
	32	4204	3	2	10
NAS FT	4	998	6	3	12
	8	70	8	1	8
	16	8992	6	7	24
	32	262	32	32	32
NAS IS	4	100	9	6	9
	8	188	17	9	17
	16	364	33	17	33
	32	716	65	34	65
NAS LU	4	31490	2	2	126
	8	31492	2	4	126
	16	31492	2	2	126
	32	47229	2	4	126
NAS SP	4	4823	6	3	12
	9	7226	6	6	18
	16	9000	6	7	24
	25	12000	6	12	30

4.2 Long-Term Buffer Address Predictability

The zero message-copying mechanism designed in section 3.1 requires the prediction of several messages. Therefore, we are interested in evaluating the long-term predictability of the buffer address together with the message size. The task is to predict the next (+1) and the tenth (+10) future buffer address and message size. The tenth message in the future (+10) is chosen as upper bound. The chosen mechanism for zero message-copying may require to advance less. In our experiments, the input stream of both predictors is a linear combination of the buffer address, messages size and sender process, such as used in [1] and [7].

In Figure 1 the prediction results are shown. In the graphics, we denote the prediction of the future values with +1, and +10. The letter "D" indicates the periodicity-based predictor and the letter "G" the graph-based predictor.

The prediction accuracy for messages larger than 8k is shown. As described in section 3.1, the zero-copying mechanism is effective for messages larger than 8k. It can be observed that the prediction accuracy is generally very high (many times > 90%) with both predictors and in the +10 scenario. We can see very similar performance of both predictors in the Bt, Cg, Ft, Is, and Sp benchmarks. An exception is the Bt executed with 25 processes. In this case the performance of the graph-based predictor decreases when performing the +10 prediction task.

A special behavior can be observed in the Lu benchmark for the graph-based predictor, when predicting messages larger than 8k on +10. Here, correct prediction is not achieved. The reason for failing in this prediction is discussed in detail in the next section.

4.3 Comparison of Predictors: Accuracy and Overhead

We observed very high prediction rates including for long-term prediction (+10) both with the graph and the periodicity-based predictor. In many benchmarks, the rates of both predictors are similar. These benchmarks include the Bt, Cg, Ft, Is, and SP, which all showed regular patterns of a rather small temporal size (see section 4.1). Furthermore, the sequences have many different elements, which is beneficial for the performance of the graph-based predictor.

Differently, the Lu benchmark showed a large pattern, of size 122 and 126, respectively, within which smaller (nested) pattern are repeated. In the Lu, a large message appears after observing a long sequence of small messages with identical values. In terms of prediction rates, the prediction of +10 in the Lu with the graph-based predictor goes down to zero, as it can be seen in Figure 1. The periodicity-based predictor, however, could predict such a message, even after having observed identical messages during a long time. The reason for this capability is that it captures the periodicity of 126 in the message stream.

We found that achieving the knowledge of such long periodicities with the periodicity-based predictor is computationally more expensive than using the graph. In our current implementation, the graph-based predictor is much faster than the periodicity-based predictor. In the periodicity-based predictor, the length of the history, which enables to compute the periodicity, strongly affects the execution time. In our study, we have used the default value of the periodicity-based predictor, which is a history of 256 samples (which allows to capture periodicities up to 256).

Although graphs are usually not used to predict several future values, we saw that predicting them by walking along the built graph provided accurate results for long-term prediction. We found that predictions of this type of sequences by statistical models such the graphs are computationally efficient combined with high prediction rates. On the other hand, the periodicity-based predictor showed its strength capturing large patterns such as observed in the Lu. This achievement, however, also involved a higher computational cost.

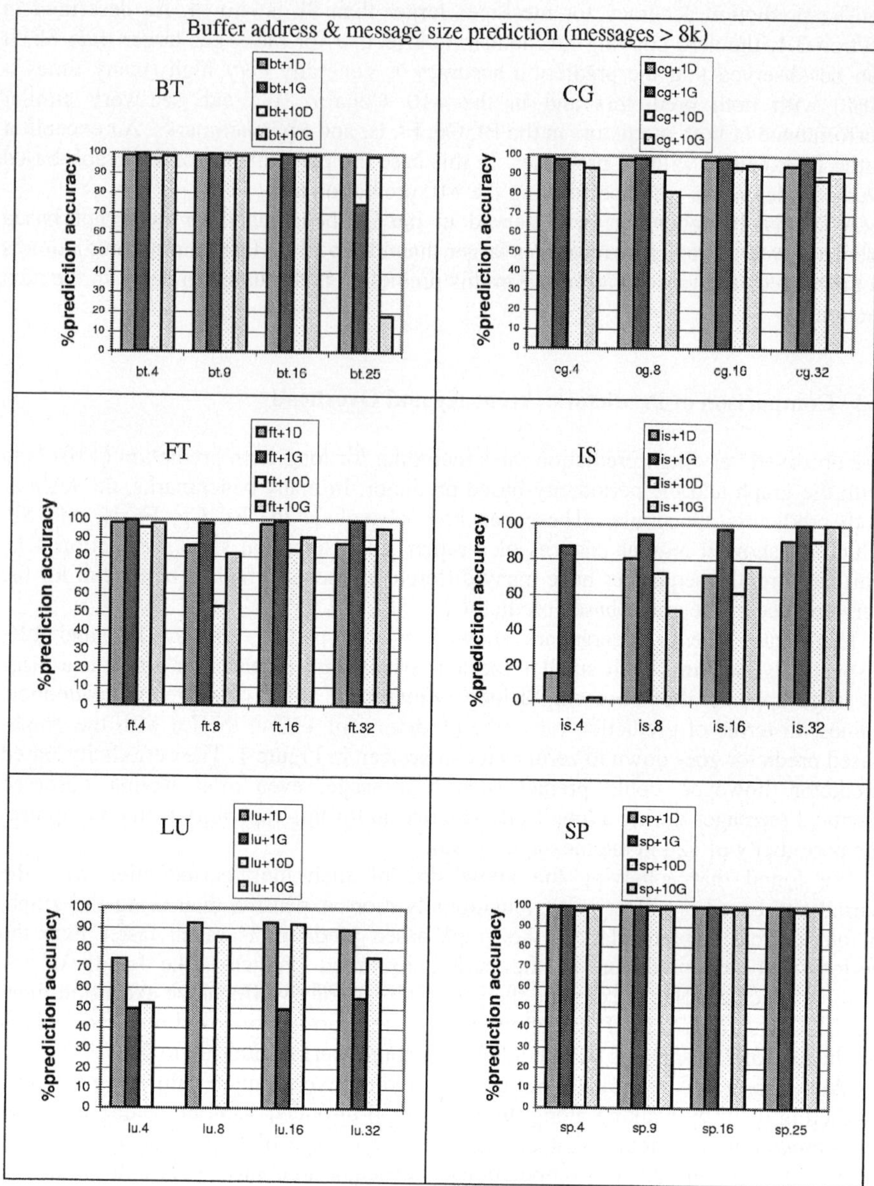

Fig. 1. Long-term buffer address and message size prediction

5   Conclusions

We indicated the need to predict several messages in order to implement a zero message-copying mechanism, which can cope with changes in the arrival order of messages. We described how this zero message-copying could be achieved. In traces of MPI communication data the existence of temporal patterns in the buffer addresses was observed. We evaluated two prediction schemes for prediction, one based on graphs, and the second based on periodicity detection. The predictors were used to predict the buffer addresses of future messages (+1, and +10). Our results indicate that the accuracy of long-term prediction is very high with both predictors. We identified an advantage of the periodicity-based predictor in capturing large patterns as in the Lu, but observed also a larger computational cost than in the graph-based predictor. If patterns are small and consist of different elements, the graph-based predictor showed to be computationally efficient combined with high prediction rates.

References

1. A. Afsahi, N. J. Dimopoulos. Efficient Communication Using Message Prediction for Cluster of Multiprocessors. *Concurrency and Computation: Practice and Experience* 2002; 14:859-883.
2. D. H. Bailey, E. Barszcz, L. Dagum, and H. D. Simon, NAS Parallel Benchmark Results, *Proceedings of the Scalable High-Performance Computing Conference*, 1994, pp. 111-120.
3. BlueGene home page: `http://www.rsearch.ibm.com/bluegene/`
4. CHIMP/MOI Project: `http://www.epcc.ed.ac.uk/epcc-projects/CHIMP`
5. K. M. Curewitz, P. Krishnan, and J. S. Vitter. ``Practical Prefetching via Data Compression," *Proceedings of the 1993 ACM SIGMOD International Conference on Management of Data (SIGMOD '93), Washington, DC*, May 1993, 257-266.
6. F. Freitag, J. Corbalan, J. Labarta. A dynamic periodicity detector: Application to Speedup Computation. In *Proceedings of International Parallel and Distributed Processing Symposium (IPDPS 2001)*, April 2001.
7. F. Freitag, J. Caubet, M. Farrara, T. Cortes, J. Labarta. Exploring the Predictability of MPI Messages. In *Proceedings of International Parallel and Distributed Processing Symposium (IPDPS 2003)*, April 2003.
8. W. Gropp, E. Lusk, N. Doss and A. Skjellum. A high-performance, portable implementation of the MPI message passing interface standard. In *Journal of Parallel Computing*, 22(6), pp. 789-828, September 1996.
9. J. Kim, D. J. Lilja. Characterization of Communication Patterns in Message-Passing Parallel Scientific Application Programs. In *Proceedings of the Workshop on Communication, Architecture, and Applications for Network-based Parallel Computing*, pp. 202-216, February 1998.
10. LAM/MPI home page: `http://www.lam-mpi.org/mpi`
11. MPI Forum. MPI: A message-passing interface standard. `http://www.mpi-forum.org`
12. MPICH home page. `http://www-unix.mcs.anl.gov/mpi/mpich`
13. Y. Sazeides, J. E. Smith. The Predictability of Data Values. In *International Symposium on Microarchitecture* (MICRO-30). 1997.

An Efficient Load-Sharing and Fault-Tolerance Algorithm in Internet-Based Clustering Systems

In-Bok Choi and Jae-Dong Lee

Division of Information and Computer Science, Dankook University,
San #8, Hannam-dong, Yongsan-gu, Seoul, 140-714, Korea
{pluto612, letsdoit}@dku.edu

Abstract. This paper proposes an efficient algorithm for load-sharing and fault-tolerance in Internet-based clustering systems. The algorithm creates a global scheduler based on the Weighted Factoring algorithm. And it applies an adaptive granularity strategy and the refined fixed granularity algorithm for better performance. It may also execute a partial job several times for fault-tolerance. For the simulation, the matrix multiplication using PVM is used in a Internet-based clustering system. Compared to other algorithms such as Send, GSS and Weighted Factoring, the proposed algorithm results in an improvement of performance by 55%, 63% and 20%, respectively. Also, this paper shows that it can process the fault-tolerance.

1 Introduction

Recently, most clustering systems are connected to high-speed network such as Myrinet, SCI, or Gigabit Ethernet for more performance elevation. However, these systems require additional expenses. Also, it is difficult to extend network[1].

With the rapid growth of the Internet, it is easy to build clustering system using computers that are connected to Internet without additional network. However, Internet consists of various networks and heterogeneous nodes. So, there are a lot of possibilities of imbalance and fault by cutting of network and breakdown of nodes. Therefore, in Internet-based clustering systems, a load-sharing algorithm must consider various conditions such as heterogeneity of nodes, characteristics of a network, imbalance of load, and so on.

This paper proposes an efficient algorithm called Efficient-WF algorithm for load-sharing and fault-tolerance in Internet-based clustering systems. The Efficient-WF algorithm creates a global scheduler based on the Weighted Factoring algorithm. Also, it applies an adaptive granularity strategy and the refined fixed granularity algorithm for better performance and executes a partial job several times for fault-tolerance.

The remains of this paper are organized as follows. Section 2 introduces related works about load-sharing algorithms. Section 3 describes the Efficient-WF algorithm.

M. Bubak et al. (Eds.): ICCS 2004, LNCS 3036, pp. 18–26, 2004.

Section 4 estimates the performance of the Efficient-WF algorithm. Finally, section 5 states some conclusions and plans for future work.

2   Related Works

Send and GSS algorithm show good performance for load-sharing in NOW(Network of Workstation) environment [3,5]. Weighted Factoring algorithm shows good performance in heterogeneous clustering system [2].

Send algorithm sends the first matrix and t columns of the second matrix together to the first slave, then the first matrix and the next t columns to the second slave, and so on. Each slave multiplies the columns that it receives by the first matrix it already has. When a slave finishes the multiplication, it sends the results back to the master. Master collects the results, writes it into appropriate place of the result matrix and then sends another t columns of the second matrix to the slave. This process continues until all columns of the second matrix have been sent [5].

In GSS(Guided Self-Scheduling) algorithm, a data size scheduled by a function of the remaining data. Given N data and P nodes, the ith data size is determined as follows [3,5]

$$G_i = \left\lceil\ (1-\tfrac{1}{P})^i \tfrac{N}{P}\ \right\rceil \tag{1}$$

Weighted factoring is identical to factoring except that the size of job in a batch is determined in proportion to the weight(W_j) of nodes. The size of the jth job size in the ith batch is determined as follows [2,3]

$$F_{ij} = \left\lceil\ (1-\tfrac{1}{x})^i \tfrac{N}{x}\ \frac{W_j}{\sum\limits_{k=1}^{k=p} W_k}\ \right\rceil = \left\lceil\ (\tfrac{1}{2})^{i+1} N \frac{W_j}{\sum\limits_{k=1}^{k=p} W_k}\ \right\rceil \tag{2}$$

3   Design of Efficient-WF Algorithm

This chapter describes the Efficient-WF algorithm for load-sharing and fault-tolerance in Internet-based clustering systems.

3.1   Design of Global Scheduler and Subroutines

A global scheduler is required to apply an adaptive granularity strategy to Weighted Factoring algorithm. Data structure for N data and P slave nodes is as follows.

```
struct  slave_node {
01:   int job[⌈ log₂N ⌉];       // assigned job
02:   int status[⌈ log₂N ⌉];   // 0:to do, 1:doing, 2:done.
03:   float weight;
```

```
04:  int remain;
05:  int doing;
} schedule[P];
```

Send function is usually used when master node assigns job to slave node. At this time, master node achieves following additional works to manage states of slave nodes.

Subroutine SEND(schedule[j].job[k], i)
- Input: schedule[j].job[k], received partial result; i, index of a slave node.

```
01:  send(job of schedule[j].job[k] to i^th node);
02:  schedule[j].status[k] = 1;
03:  schedule[j].remain -= size of schedule[j].job[k];
04:  schedule[j].doing += size of schedule[j].job[k];
```
End of SEND subroutine

Recv function is used when master node gets the partial result from a slave node. Master node also archives following works to manage the states of slave nodes.

Subroutine RECEIVE(sched[j].job[k], i)
- Input : schedule[j].job[k], received partial result; i, index of a slave node.

```
01:  recv(partial result schedule[j].job[k] from i^th node);
02:  schedule[j].status[k]=2;
03:  schedule[j].doing -= size of schedule[j].job[k];
```
End of RECEIVE subroutine

3.2 Load-Sharing Technique by an Adaptive Granularity Strategy

As Weighted Factoring algorithm shares loads by using weight that was evaluated earlier, it is difficult to cope with the change of slave nodes during work. Adaptive load-sharing policy that lowers priority order of slow slave node showed good performance in study of [4].

Therefore, it is desirable to reduce amount of jobs of some slow slave nodes by let some fast slave nodes that finished all their jobs execute the jobs of the slow slave nodes. The adaptive granularity strategy is described as follows.

Function AGS_LS(schedule[j].job[k], i)
- Input: schedule[j].job[k], received partial result; i, index of a slave node.
- Output: schedule[m].job[n], next partial job for i^th slave node.

```
01:  if(schedule[i].remain != 0) {
02:      m = i;
03:      n = k+1;
04:  }
05:  else if(Exist schedule[0...(P-1)].remain != 0) {
```

06:     m = index of the slowest slave node that have 'do do' job;
07:     n = the last 'to do' job index of m;
08: }
09: return (schedule[m].job[n]);
End of Function AGS_LS(schedule[j].job[k], i)

When master node receives a kth partial result of jth slave node from i^{th} slave node, if some jobs remain yet to the ith slave node, master node selects a job according to schedule (line 1-4). If all jobs of the ith slave node are finished (line 5) and 'to do' jobs remain yet in any other slave node (line 6), master node searches the slowest slave node and selects the last job that is not transmitted yet (line 7).

3.3 Load-Sharing Technique by the Refined Fixed Granularity Algorithm

The refined fixed granularity algorithm is to overlap between the time spent by slave nodes on computation and the time spent for network communication [1]. First, master node transmits two jobs in each slave node. Then master node transmits next job to a slave node that transmitted partial result. Therefore, a slave node can achieve next job without waiting for reception of next job from master node. The RFG_LS algorithm that uses the refined fixed granularity algorithm is as follows.

Algorithm RFG_LS
• Input: P, number of slave nodes; schedule[P], a scheduled array for P slave nodes.
• Output: result, merged partial result(e.g. array).
01: SEND(schedule[0...(P-1)].job[0], 0...(P-1));
02: SEND(schedule[0...(P-1)].job[1], 0...(P-1));
03: while(all partial results are not gathered) {
04:     RECEIVE (schedule[j].job[k], i);
05:     schedule[m].job[n] = AGS_LS(schedule[j].job[k], i);
06:     SEND(schedule[m].job[n], i);
07:     MERGE(partial result of schedule[j].job[k]);
08: }
End of RFG_LS

After master node send the first job in each slave node (line 1), send the second job while slave node achieve the first job without receiving partial result (line 2).

3.4 Fault-Tolerance Technique by Executing Jobs Several Times

Clustering systems such as NOW do not guarantee stability of nodes [3]. Algorithms proposed earlier do not consider fault of slave nodes, neither. The reason is that most clustering systems composed into stable regional network environment. However, transmission delay or connection cutting by problems of network can be in Internet.

This paper proposes fault-tolerance technique by executing jobs several times. The technique is that master node assigns 'doing' jobs of some slow slave nodes to fast slave nodes that finished all their jobs. This manner is as follows.

```
Function EJS_FT(schedule[j].job[k], i)
 • Input: schedule[j].job[k], received partial result; i, index of a slave node.
 • Output: schedule[m].job[n], next partial job for the i^th slave node.
01:  if(schedule[0...(P-1)].remain==0){
02:      m = index of the slowest slave node that have 'doing' job;
03:      n = the last 'doing' job index of m;
04:  }
05:  return(schedule[m].job[n])□
End of Function EJS_FT(schedule[j].job[k], i)
```

When master node receives a kth partial result of jth slave node from i^{th} slave node, if all jobs scheduled to i^{th} slave node are finished and any 'to do' job is not remain in any other slave node (line 1), master node searches the slowest slave node (line 2) and selects the last 'doing' job of the slave node (line 3).

3.5 Efficient-WF Algorithm

The Efficient-WF algorithm extends Weighted Factoring algorithm based on previous paragraph. This Efficient-WF algorithm is as follows.

```
Algorithm Efficient-WF
 • Input: N, size of job; P, number of slave nodes.
 • Output: result, merged partial results (e.g. array).
01:  CREATE(scheduler by Weighted Factoring algorithm);
02:  SEND(schedule[0...(P-1)].job[0], 0...(P-1));
03:  SEND(schedule[0...(P-1)].job[1], 0...(P-1));
04:  while(all partial results are not gathered){
05:     RECEIVE(schedule[j].job[k], i);
06:     if(Exist(schedule[0...(P-1)].remain != 0){
07:         schedule[m].job[n] = AGS_LS(schedule[j].job[k],i);
08:     }
09:     else{
10:         schedule[m].job[n] = EJS_FT(schedule[j].job[k],i);
11:     }
12:     SEND(schedule[m].job[n], i);
13:     MERGE(partial result of schedule[j].job[k]);
14:  }
End of Algorithm Efficient-WF
```

Master node creates a global scheduler for allocating of jobs (line 1). An adaptive granularity strategy is applied by using AGS_LS function (line 7). And, the Efficient-WF algorithm offers the fault-tolerance by using EJS_FT function (line 10).

3.6 Analysis of Efficient-WF Algorithm

The Efficient-WF algorithm overlaps between the time spent by slave nodes on computation and the time spent for network communication by transmitting previously next job in order to reduce total execution time. In this manner, the time gain is network communication time during slave nodes compute a previous job.

Let (t_s) is the transfer time between master node and slave node. Figure 1. shows an example of the time reduction by overlapping between computation time and network time when the execution times of master node and slave node are equal.

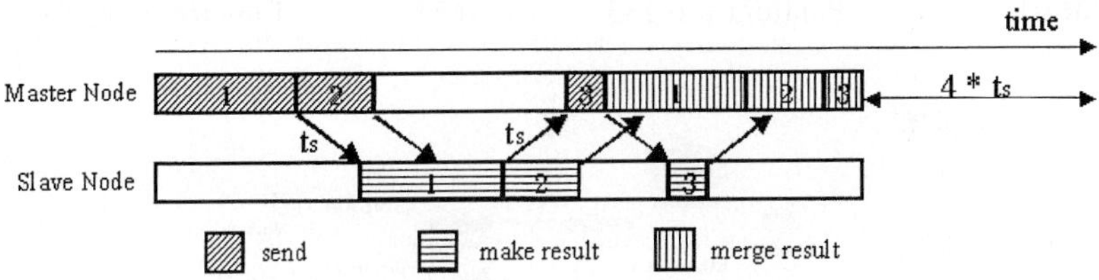

Fig. 1. Time reducing by overlapping between computation time and network time.

Execution time could be reduced by minimum when all network communication time overlap with computing time except the first transfer time from master node to slave node and the last transfer time from slave node to master node. $\lfloor \log_2 N \rfloor$ jobs are assigned to a slave node. Each job is transferred 2 times between master node and slave node. Also, the first transfer time and the last transfer time are excluded. Therefore, for the N data among P slave nodes, the maximum time we can reduce is as follows.

$$t_{profit} = P * ((\lfloor \log_2 N \rfloor * 2 t_s) - 2 t_s)$$
(3)

The Efficient-WF algorithm executes some last jobs several times for fault-tolerance. The slave nodes that have finished all their jobs are considered idle nodes. As these nodes could accomplish faster than the node that has been performing ahead, it could shorten whole execution time as well as process fault-tolerance.

4 Performance Evaluation of Efficient-WF Algorithm

This chapter evaluates the performance of Efficient-WF algorithm that has presented in chapter 3.

4.1   Environment for Performance Estimation

For the performance estimation, total eleven PCs were used. The clustering system composed by one master node and 10 slave nodes. The master node was participated in computing partial results.

Table 1. System configuration (Total 11nodes).

Node Name	CPU	Memory	OS
Master	Pentium3 450	320M	Linux(kernel2.4)
Node1,2	Pentium3 733	128M	Linux(kernel2.4)
Node3,6	Pentium3 450	128M	Linux(kernel2.4)
Node4,5	Pentium3 300	128M	Solaris8.0
Node7	Pentium-pro 133	64M	Linux(kernel2.0)
Node8-9	Pentium-pro 133	32M	Linux(kernel2.0)
Node10	Pentium-pro 133	16M	Linux(kernel2.0)

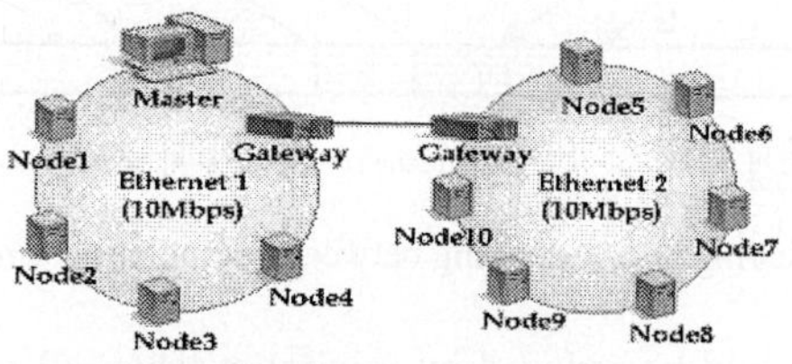

Fig. 2. Network configuration.

4.2   The Result of Performance Estimation

For the simulation, the matrix multiplication using PVM 3.4.4 library is performed on the environment of previous paragraph. All tests are performed 40 times respectively. In this paragraph, WF means the Weighted Factoring algorithm and E-WF means the Efficient-WF algorithm. The Mean value of performance evaluation is as follows.

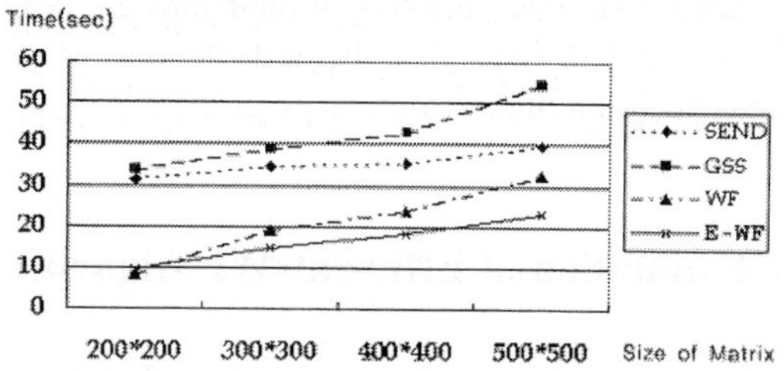

Fig. 3. Performance evaluation by size of matrix

Comparing to other algorithms such as Send, GSS and Weighted Factoring, Effective-WF algorithm results in an improvement of performance by 55%, 63%, 20%, respectively.

To measure fault-tolerance of the Efficient-WF algorithm, sixty-seconds delay is applied in an arbitrary slave node. Sixty seconds is longer than maximum execution time (54.59 seconds) of previous performance evaluation. Applying delay of 60 seconds means that fault happens in the slave node. The result of the test is as follows.

Table 2. Fault-Tolerance. (seconds)

Size of matrix	Fault-Tolerance	Normal state
200*200	9.74	8.75
300*300	15.38	14.49
400*400	18.93	18.38
500*500	23.26	22.81

If the Efficient-WF algorithm cannot offer Fault-Tolerance, the result of the experiment must take more than all 60 seconds. But, we can see that all results are almost equal in Table 4. Although performance of about 5% was fallen than normal state, the result of Table 4 explains that we can get the normal result despite fault happened in slave nodes. Therefore, we can see that the Efficient-WF algorithm can cope with fault of slave nodes.

5   Conclusions and Future Work

This paper has proposed the Efficient-WF algorithm for load-sharing and fault-tolerance in Internet-based clustering systems. The Efficient-WF algorithm uses adaptive granularity strategy and the refined fixed granularity algorithm for load-sharing, and it executes some jobs several times for fault-tolerance.

Comparing to other algorithms such as Send, GSS and Weighted Factoring, the Effective-WF algorithm resulted in an improvement of performance by 55%, 63%, 20%, respectively in experimental clustering system. Also, it offered stable execution time with Fault-Tolerance that could not offer at other algorithms.

The clustering environments in Internet-based clustering systems are more dynamic than current clustering environments. Therefore, adaptive load-sharing techniques in more various environments and compatible fault-tolerance techniques with existing tools will be studied in the future.

Acknowledgements. The present research was conducted by the research fund of Dankook University in 2004. This research was supported by University IT Research Center Project of Korea.

References

1. Bon-Geun Goo, "Refined fixed granularity algorithm on Networks of Workstations", KIPS, Vol.8, No.2, 2001.
2. S. F. Hummel, J. Schmidt, R. N. Uma, and J. Wein, "Load-Sharing in Heterogeneous Systems via Weighted Factoring", SPAA, 1997.
3. Yangsuk Kee and Soonhoi Ha, "A Robust Dynamic Load-Balancing Scheme for Data Parallel Application on Message Passing Architecture", PDPTA'98, pp. 974-980, Vol. II, 1998.
4. Jin-Sung Kim and Young-Chul Shim, "Space-Sharing Scheduling Schemes for NOW with Heterogeneous Computing Power", KISS, Vol.27, No.7, 2000.
5. A. Piotrowski and S. Dandamudi, "A Comparative Study of Load Sharing on Networks of Workstations", Proc. Int. Conf. Parallel and Distributed computing system, New Orleans, Oct. 1997.
6. G. Shao, "Adaptive Scheduling of Master/Worker Applications on Distributed Computational Resources", Ph.D. thesis, UCSD, June 2001.

Dynamic Parallel Job Scheduling in Multi-cluster Computing Systems

J.H. Abawajy

Deakin University
School of Information Technology
Geelong, Victoria, Australia

Abstract. Job scheduling is a complex problem, yet it is fundamental to sustaining and improving the performance of parallel processing systems. In this paper, we address an on-line parallel job scheduling problem in heterogeneous multi-cluster computing systems. We propose a new space-sharing scheduling policy and show that it performs substantially better than the conventional policies.

1 Introduction

In the last few years, the trends in parallel processing system design and deployment have been moving away from a single powerful supercomputers to cooperative networked distributed systems such as commodity-based *cluster computing* systems. Research in cluster computing has focused on tools that are useful for putting together a cost-effective off-the-shelf high-performance cluster computing systems as well as developing application programs and executing them remotely. Aggregation of many resources is not enough to guarantee good performance - careful scheduling must be employed to achieve the best performance possible [6]. Without the support from well-designed cluster job scheduling policy, resources are shared in an ad-hoc manner, limiting performance as well as the utilization of the resources. Hence, one of the most important problems that must be addressed in order to realize the advantages of cluster computing systems is that of *job scheduling* problem.

Job scheduling problem has been extensively studied on parallel computers (e.g., [8], [4], [11], [2]) and to a lesser extent on cluster computing systems (e.g., [1] and [6]). Existing job scheduling policies can be classified into *space-sharing* (e.g., [4], [10], [11]) and *time-sharing* (e.g., [8]). It is also possible to combine these two types of policies into a *hybrid policy* as in [6], [2]. In a time-sharing policy, processors are shared over time by executing different applications on the same processors during different time intervals, which is commonly known as time-slice or quantum. In the space sharing approach, processors are partitioned into disjoint sets and each application executes in isolation on one of these sets.

In this paper, we focus on space-sharing policy. In general, based on when processor partition is created, space-sharing policies can be classified as *fixed*, *static*, and *dynamic* approaches. In fixed policy, processors are partitioned into

M. Bubak et al. (Eds.): ICCS 2004, LNCS 3036, pp. 27–34, 2004.

a fixed size partitions at the time the system starts. The scheduler assigns one or more partitions to parallel jobs based on the size of the jobs. Most of the conventional cluster-based systems (e.g., LSF [1]) use fixed scheduling policy. The positive aspect of the fixed approach is its implementation simplicity. However, it does not adapt to changes both in the system load conditions and resource requirements of applications. Hence, it can lead to processor fragmentation problem [6], which in turn can lead to relatively low processor utilization and system throughput.

In static space sharing policies (e.g., [4], [11]), the partition size allocated to a job is determined at allocation time. Hence, it can adapt to the system load condition as such avoiding some problems associated with the fixed space-sharing approach. However, as in fixed approach, applications will hold the processors assigned to them until they terminate (i.e., for the lifetime of the application), which can also lead to processor fragmentation problem. The dynamic space-sharing approach eliminates most of the problems associated with fixed and static approaches. The basic idea behind the dynamic space-sharing policy is to make the jobs in the system share the processors equally as much as possible. This is achieved by varying the number of processors allocated to an application during its execution. This means processors may be reclaimed from a running job and distributed to newly arrived jobs, or additional processors may be added to an executing job when processors become available. Dynamic space-sharing policies are typically used in shared-memory systems since significant programming effort and execution overhead must be expended to change a job's processor allocation during execution. In a distributed memory environment especially, the costs of data repartitioning can overwhelm the scheduling benefits realized by malleable job support.

In this paper, we address the problem of scheduling parallel jobs in multiple cluster computing systems. Job scheduling is a challenging problem as the performance and applicability of the scheduling policy is highly sensitive to a number of factors such as machine architecture. In particular, cluster computing systems have several subtle but significant characteristics that influence and complicate scheduling decisions, which are not an issue in conventional parallel processing systems. Some of these factors include system heterogeneity, scale, interconnection technologies that typically exhibit high overhead and low bandwidth, and the availability of the system resources vary over time. This variation is both highly dynamic and unpredictable. Due to such inherent characteristics of cluster computing systems, scheduling strategies developed for the traditional distributed systems have to be extended significantly to support the dynamics of cluster computing systems. Dynamic scheduling is required since resources (i.e., machines and networks) may suffer dynamic load fluctuations or may be added or removed during the course of application execution.

The rest of the paper is organized as follows. In Section 2, the system model and the proposed scheduling policy are described. In Section 3, the performance evaluation of the proposed policy is described. It is shown that the proposed policy performs substantially better than a baseline policy. Conclusion and future directions are described in Section 4.

2   Dynamic Scheduling Policy

2.1   System Model

For the most part, job scheduling research in cluster computing environments has focused on a single cluster computing systems (e.g, [9]) under a common assumption that all processors in the system have equal processing capacity (i.e., homogeneous) (e.g., [10], [9]) and dedicated (e.g., [10]). In contrast, we focus on *shared*, *heterogeneous*, and *multi-programmed* cluster computing systems composed (see Figure 1) of *multiple clusters* as in [6], [3] spanning campus wide area.

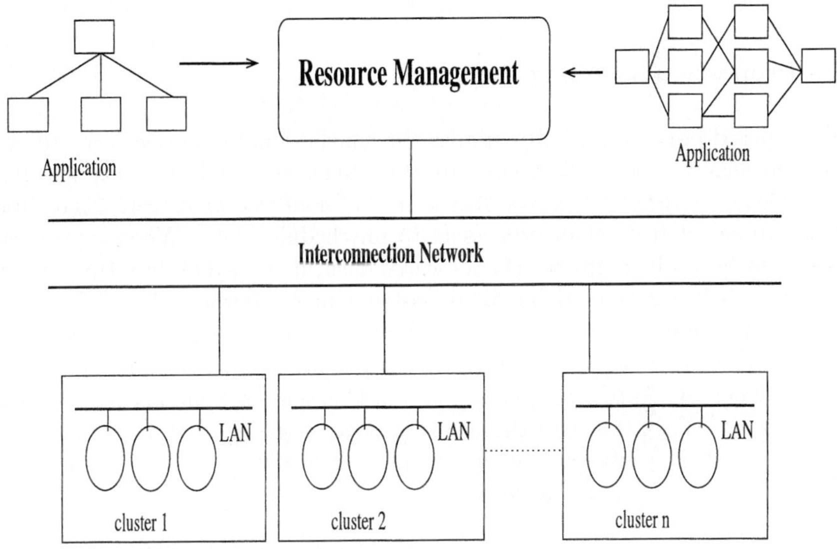

Fig. 1. Multicluster Computing System

We assume that a cluster is geographically compact and contains a set of workstations of similar architecture, connected by a fast Ethernet (100 Mbits/sec). Similarly, inter-cluster interconnection network is a fast Ethernet (100 Mbits/sec). We also assume that a network segment contains a single cluster only. Computers in different clusters do not share the communication bandwidth. Note that these assumptions can be easily relaxed, but allow us to ignore the mechanisms involved in getting jobs to run, and concentrate on policy issues for selecting which jobs as well as where and when they should run. In addition, we restrict our focus to two dimensions of heterogeneity: processor speed and cluster sizes as these two are expected to be the major components of system heterogenity in networked-systems [6].

Adaptive space-sharing policies for cluster computing systems should have at least two components: *processor allocation* (i.e., how many processors is allocated to a job) and *processor selection* (i.e., which processors should be allocated to a job). However, the classical space-sharing policies (e.g., [2], [10]), consider only the processor allocation aspect while they are oblivious to processor selection problem. An approach to partition a moldable parallel job on distributed systems is described in [11] but not address the processor selection problem. Also, they are characterized with *all-or-nothing* approach, which means that jobs will hold on to the processors allocated to them even if they do not need it. Existing adaptive policies do not actively correct such a resource imbalance state. As a result, some clusters may be overloaded while others may sit idle. This leads to poor resource utilization due to *processor fragmentation*. The next section discusses the proposed scheduling policy.

2.2   Proposed Scheduling Policy

The proposed policy essentially mimics the conventional *dynamic space-sharing* approach such that idling processors are allocated to other jobs that can fruitfully utilize them. To achieve this, we keep a *pool of available processors* from which the scheduler allots a set of processors to unscheduled jobs. When a processor is assigned to a job completes the allocated task, it immediately returns to the pool where it becomes available for re-assignment to another job.

The activation of the scheduler occurs in three instances: (1) when a job arrives; (2) when a job departs; and (3) when there are α processors, $0 \leq \alpha \leq \mathbf{P}$, in the *pool*, where α is a tunable parameter and $\mathbf{P}$ is the total number of processors in the system. When the algorithm is invoked, it assigns a set of processors equal to the lesser of the partition size and the job's maximum parallelism. The target partition size is determined as follows:

$$target = min(\text{maximum parallelism of the job}, \text{partition size}) \qquad (1)$$

where the maximum parallelism is given at the job arrival time and the *partition size* is computed as follows:

$$\text{partition size} = \left\lceil \frac{\mathbf{P}}{(\text{Queued Jobs}+1) + (0.5 \times \text{Executing Jobs})} \right\rceil \qquad (2)$$

The scheduler then goes into processor selection phase, which selects appropriate processors for the job. Processor selection phase is performed as follows:

1. Select processors within the same cluster if possible
2. Select processors taking cluster proximity into account

Next, the scheduler determines the exact number of tasks assigned to each processor, P_i, in the target partition as follows:

$$bunch = ff(P_i) \times \frac{\text{Job Maximum Parallelism}}{\textit{ff(System)}} \qquad (3)$$

where parameter ff is called a fitness factor for each processors P_i in the system that is defined as follows:

$$ff = \frac{\text{CPU speed} \times \text{Default MPL}}{\text{slowest processor speed in the system}} \qquad (4)$$

Finally, the tasks assigned to processor P_i are *folded* onto the target partition size allocated to the job. For example, if P_i is assigned 3 tasks, these tasks are folded into one large task assuming that jobs are malleable.

3   Performance Evaluation

We used discrete event simulation written in C programming language to study the performance of the proposed policy and compare it with the baseline policy [2]. For relative performance evaluation, we use the *mean response time (MRT)* and *average utilization*. The MRT is defined as the average job response time of completed jobs. The *average utilization* is defined as the job arrival rate times the mean service demand of the jobs divided by the number of processors in the system.

In this paper, we focus on moldable parallel workload. Moldable jobs are parallel jobs that are flexible in the number of processors at the time the job starts, but cannot be reconfigured during execution. The motivation for considering moldable parallel workload is that such workload constitutes a significant portion of the high-performance computing centers workloads and likely to increase in future. As in [2], we assume that jobs only request processors and we do not include in the model any other type of resources.

We used a system composed of 8 clusters, each cluster with 8 processors. In order to model the processors speed, we used the SPECfp2000 results for a family of Intel Pentium 3 and 4 processors with different clock speeds. We used $200\mu s$ for the context switch overhead, which is consistent with most of modern operating system for workstations. We set the space-sharing threshold ($\alpha = 2$), $MPL = 2$ and background job mean service demand=2.0. We model the communication overhead as follows:

$$T_{comm} = \text{Startup} + \frac{\text{Message size}}{\text{Bandwidth}} \qquad (5)$$

The communication network latency to be $50\mu sec$ with the transfer rate of 100Mbits/sec.

The abstract model for parallel programs consists of a set of input parameters along with a job structure. Each job is characterized by *arrival time, service demand time* in a dedicated environment, *maximum parallelism*, and the *size of the job* in Kbytes. Also, each can be decomposed into t tasks, $\mathbf{T}=\{T_1,...,T_t\}$ and each task T_i executes sequential code. The maximum parallelism of the jobs is uniformly distributed over the range of 1 to 64, while the service demands of the jobs are generated using hyper-exponential distribution with mean 14.06 [10], [2]. The default arrival CV is fixed at 1 (i.e., we assume Poisson arrivals) and the default service time CV is fixed at 3.5 as empirical observations at several supercomputer centers indicated this to be a reasonable value [10].

4   Results and Discussion

4.1   Relative Performance

We compared the relative performance of the proposed policy with the MPAP policy [2] in homogeneous environments. The result shows that at low system loads, there is no significant difference among the two policies as there are plenty of processors idle most of the time at this load level. However, as the load increases the dynamic policy performs better (by about 20% to 30%) than the baseline policy. This trend can be explained by the fact that in the MAP policy jobs tend to be allocated smaller partition-size as the system load increases and in the presence of local workload. As the number of processors allocated to a job decreases below its maximum parallelism, the service demand of the tasks also increases. This makes the jobs sensitive to the presence of the background load, which can increase the wait time of the jobs.

4.2   Sensitivity Analysis

We examined the impact of the background workload on the performance of the two policies. Note that the impact of the background load on the parallel job depends on the service demand of the parallel tasks; the longer a parallel task occupies the processor the more likely its execution is interrupted by the background workload. The result shows that the MAP policy is more sensitive than the dynamic policy. This is because the MAP policy suffers from a form of *processor fragmentation* induced by the presence of the background load and the way the partition-size is assigned to the jobs. Since partition size is computed based on the total number of processors in the system, the actual number of idle processors can be lower than the partition-size computed. This is because of the fact that some of the processors run background load at the time and the MAP policy does not take into account the background loads when computing the partition size. In this situation, the processors will remain idle as the scheduler will not assign them to jobs even if there are small jobs that can fruitfully utilize them.

Sensitivity of the two policies to the combined effects of background load and processor heterogeneity is also examined. At medium to high system loads, the MAP policy tends to be more sensitive than the dynamic policy. This is due to the fact that, in addition to processor heterogeneity, the jobs experience slowdown due to interference from the background jobs. Note that heterogeneity can lead to a load imbalance situation. In such situations, the completion time of the jobs increases. When the execution of jobs in an already imbalanced environment is interrupted due to background jobs, the finishing time of the jobs is further increased.

In summary, the relatively poor performance of the MAP policy is due to the space sharing nature of the policy. Note that in the extreme case, the MAP policy ends up allocating a single processor to each job. Therefore, all the optimizations proposed to alleviate the problems associated with MAP policy have no impact at this stage. In addition, the presence of the background load also exacerbates the situation by creating interference for currently running jobs. This interference is observed as an increase in the completion time of the parallel jobs due to resource contention. The MAP policy reaches a point at which performance dramatically falls off. When the load is sufficiently high with all the processors running at least 1 job, the interference increases rapidly and performance can be very poor. Also under the MAP policy a job may have to be executed in one of the slower processors, considerably degrading the execution time.

5   Conclusion and Future Direction

In this paper, we investigated parallel job scheduling problem on heterogeneous networks of workstations, where computing power varies among the workstations, and local and parallel jobs may interact with each other in execution. We proposed a scheduling policy based on a virtually rooted tree structure that employs a pull-push scheme for scheduling and load balancing parallel jobs over multiple clusters. The proposed policy allows multiple job streams to share the system concurrently in an environment where the actual load distribution is not completely predictable in advance and scheduling is done dynamically. Also, the policy integrates several approaches (i.e. task scheduling, load balancing, self-scheduling, and time-space sharing) into a simple framework for parallel job scheduling on a system composed of multiple clusters. We studied the performance of the proposed scheduling policy through simulation. The results indicate that the proposed scheduling policy significantly better than the other scheduling policies used in the study.

Acknowledgments. Financial help is provided by Deakin University. The help of Maliha Omar is also greatly appreciated.

References

1. Ming Q. X.: Effective Metacomputing using LSF MultiCluster, Proceedings of CCGrid (2001) 100-106.
2. Thyagaraj, T.K. and Dandamudi, S. P.: An Efficient Adaptive Scheduling Scheme for Distributed Memory Multicomputers, IEEE Transactions on Parallel and Distributed Systems, **12**, (2001) 758-768.

3. Abawajy, J. H. and Dandamudi, S. P.: A Unified Resource Scheduling Approach on Cluster Computing Systems, Proceedings of the PDCS'03, (2003) 43-48.
4. Rosti, E. Smirni, E. Serazzi, G. and Dowdy, L. W.: Analysis of Non-Work-Conserving Processor Partitioning Policies, Proceedings of JSSPP (1995) 165-181.
5. Abawajy, J. H. and Dandamudi, S. P.: Scheduling Parallel Jobs with CPU and I/O Resource Requirements in Cluster Computing Systems, Proceedings of the 11th IEEE/ACM MASCOTS'03 (2003) 336-351.
6. Abawajy, J. H. and Dandamudi, S. P.: Parallel Job Scheduling on Multi-Cluster Computing Systems, Proceedings of the IEEE Cluster (2003) 11-17.
7. Feitelson, D. G. and Rudolph, L.: Toward Convergence in Job Schedulers for Parallel Supercomputers, Proceedings of JSSPP (1996) 1-26.
8. Feitelson, D. G. and Jette, M. A.: Improved Utilization and Responsiveness with Gang Scheduling", Proceedings of JSSPP (1997) 238-261.
9. Ryu, K.D. and Hollingsworth, J.K.: Exploiting Fine-Grained Idle Periods in Networks of Workstations, IEEE Transactions on Parallel and Distributed System, **11**, (2000) 683-698.
10. Stergios, A. V. and Sevcik, K. C.: Parallel Application Scheduling on Networks of Workstations, Journal of Parallel and Distributed Computing, **43** (1997) 1159-66.
11. Zhengao, Z. and Dandamudi,S. P.: An Adaptive Space-Sharing Policy for Heterogeneous Parallel Systems, HPCN'01 (2001) 353-362.

Hunting for Bindings
in Distributed Object-Oriented Systems*

Magdalena Sławińska

Faculty of Electronics, Telecommunications and Informatics
Gdańsk University of Technology
Narutowicza 11/12, 80-952 Gdańsk, Poland
`magg@eti.pg.gda.pl`

Abstract. The paper examines the problem of finding a group of objects that are involved in a certain transitive relation. It is especially important when a group of related objects has to be identified, for example for monitoring. The article defines static and dynamic binding relations between objects in a distributed object-oriented system. It also presents an architecture for catching these relations since current operating systems do not support such mechanisms. In the paper the algorithm for finding bindings of a given object is described.

1 Introduction

In distributed object-oriented systems objects are scattered over the network. One of the main design goals of distributed systems is to provide different types of transparency for users, e.g., location or replication transparency [1][2]. However, sometimes such transparency can be very uncomfortable for testers and programmers, for example during debugging.

This paper presents the framework for finding objects that are involved in a certain transitive relation. For instance, consider the situation when a tester wants to examine method m of object o_1. However, method m invokes method $m1$ of object o_2. It means that object o_1 is in a *binding relation* with object o_2. The problem complicates when the so-called *native* and *foreign* objects are considered [3]. The former are objects with the full access to the source code by testers while the latter are those ones for which testers have only on-line access to objects' methods but no access to the source code.

Finding bindings among distributed objects is necessary when programmers want to isolate a group of *bound objects*, for example in order to limit the number of monitored objects only to relevant ones. It is also important in program replay or recovery [4]. The problem is similar to causality tracking, however it has a different flavor [5].

This paper is organized as follows: in Section 2 the model of distributed object-oriented system is presented. Section 3 introduces two important relations between objects: static and dynamic binding relations. In Section 4 the

---

* Funded in part by the State Committee for Scientific Research (KBN) under grant 4-T11C-004-22

M. Bubak et al. (Eds.): ICCS 2004, LNCS 3036, pp. 35–42, 2004.

architecture of the system for identifying bound objects is shown. Section 5 presents the algorithm for identifying bound objects and finally in Section 6 the paper is concluded.

2   The System Model

A distributed object-oriented system is a finite set of objects performing a given task. It is denoted by $S = \{o_1, o_2, ..., o_n\}$, where $n = 1, 2, 3, ...$. There is a finite set of hosts denoted by $H = \{h_1, h_2, ..., h_k\}$, where $k = 1, 2, 3, ...$. Objects are located in hosts. One object can be located in only one host but many objects may reside in one host. Objects and hosts are identifiable.

In contrast with classical procedural distributed systems like PVM (*Parallel Virtual Machine*) [6] or MPI (*Message Passing Interface*) [7] where processes communicate with each other by message passing, objects cooperate with other objects through method invocations. In fact, remote method invocation mechanisms only wrap the message-passing stuff. Messages are still sent and received over the network, however, it is done by special software entities automatically generated by compilers like *stubs* and *skeletons* [1][8]. They are responsible for object binding, (un)marshaling parameters and results, issuing method on a target object. In fact, considering the lower level of abstraction, a remote method invocation which returns a certain result consists of the following steps: (1) a client object sends request *req* for the method invocation of a server object, (2) the server object receives request *req*, (3) the server object carries out request *req*, (4) the server object sends reply *rep* to request *req*, (5) the client object receives reply *rep*. So invoking a remote method with a return value consists of four communication actions: two *sends* and two *receives*. Two kinds of messages are exchanged during a method invocation: *requests* (denoted by *req*) and *replies* (denoted by *rep*). Requests as well as replies are *messages* (denoted by *mesg*). A message consists of an identifier, a source object, a target object and contents. In case of requests the contents contains an identifier of a method to be invoked and parameters (if any). The contents of replies comprises a method identifier that has been invoked and a return value (if any). Figure 1 schematically shows a method call in terms of sending and receiving requests and replies. Please notice that it is also possible to model local invocations on a given object in the send/receive request/reply semantics (although it is inefficient). There are three objects in Figure 1: o_1, o_2 and o_3. Horizontal lines represent time and arrows denote messages. Symbols $s(\cdot)$ and $r(\cdot)$ stand for send and receive events, respectively. Subsequent events are numbered in the context of a given object (e_i). An identifier of a message consists of a source object identifier (unique in the system) and a number of an event in the context of the source object (unique in the object). Message identifiers contain a type of a message (reply or request).

For simplicity reasons it is assumed that all methods of an object are public and can be invoked remotely.

A sequence of events of a given object forms its history (h). For instance in Figure 1: $h_{o_1} = (e_1, e_2)$, $h_{o_2} = (e_1, e_2)$ and $h_{o_3} = (e_1, e_2, e_3, e_4)$. A history of an

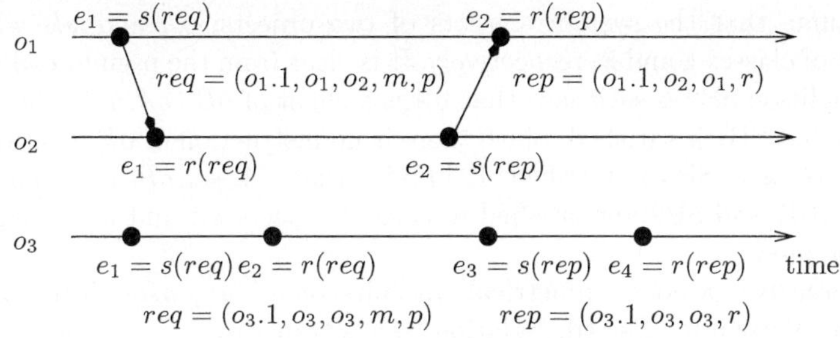

Fig. 1. Method invocations in the send/receive request/reply semantics in distributed object-oriented systems.

object is available for other objects. It may be implemented by interception of communication events. For example *interceptors* are defined in CORBA [8]. The Eternal system, which provides fault tolerance for distributed objects, intercepts method invocations in order to assure replica consistency [9].

3   Relations among Distributed Objects

Relations among objects concern the static or dynamic point of view.

3.1   Statically Bound Objects

In case of native objects, it is possible to deduce certain connections among objects by examining their source code.

Definition 1 (The static binding relation) *Object o_1 is in the "static binding" relation with object o_2 if the following conditions are satisfied:*

SB1. Object o_1 is a client of object o_2.
SB2. Condition SB1 can be deduced from the source code of object o_1.
SB3. Object o_1 does not create object o_2.

Notation $o_1 \leftrightsquigarrow_s o_2$ means that object o_1 is in the "static binding" relation with object o_2.

The simplest static binding relation is presented in Example 1.

Example 1

```
class A {                    class B {                class C {
  public void m1() {}          public void m1() {}      public void m1() {}
  public void m2() {           public void m2() {}      public void m2() {}
      oB.m1();                };                         public void m3() {}
  }                                                    };
  public void m3() {
      C c = new C();
      c.m1();
  }
};
```

Let's assume that the system consists of two objects: oA and oB which are instances of classes A and B, respectively. It is clear from the pseudo-code of class A (i.e. condition SB2 is satisfied) that oA is a client of oB ($oA.m2$ calls $oB.m1$). It means that SB1 is satisfied. Since there is no instruction of oB creation in the scope of oA, also SB3 is satisfied. It implies that $oA \leftrightsquigarrow_s oB$. Let's notice that although SB1 and SB2 are satisfied in case of objects oA and c, $oA \not\leftrightsquigarrow_s c$ since SB3 is violated.

Relation $\leftrightsquigarrow_s$ is not symmetrical. In spite of $oA \leftrightsquigarrow_s oB$, $oB \not\leftrightsquigarrow_s oA$ since condition SB1 is not satisfied. Relation $\leftrightsquigarrow_s$ is transitive.

Example 2

```
class A {                class B {                class C {
   public void m1() {}       public void m1() {}       public void m1() {}
   public void m2() {        public void m2() {        public void m2() {}
       objB.m1();                objC.m1();                public void m3() {}
   }                        }                        };
};                       };
```

Let's assume that there are three objects $objA$, $objB$ and $objC$ which are instances of classes A, B and C defined in Example 2. It is easy to notice that $objA \leftrightsquigarrow_s objB$ and $objB \leftrightsquigarrow_s objC$ since SB1, SB2 and SB3 are satisfied. It means that if $objA$ is to exist, also $objB$ should exist and if $objB$ is to exist also $objC$ should exist what implies that since $objA$ should exist, $objC$ should exist. The presented deduction is performed on the source code analysis (SB2 is satisfied). In fact, $objA$ is an indirect client of $objC$ (SB1 is satisfied). Also $objC$ is not created by $objA$ (SB3 is satisfied too).

The static binding relation is suitable for finding hypothetical relationships among objects. For instance, in Example 1 it is possible that method $oA.m2$ will be never invoked in real execution, i.e., oA will never be a client of oB. Relation $\leftrightsquigarrow_s$ only indicates that a given binding is probable but it does not guarantee that it really happens.

3.2   Dynamically Bound Objects

In case of foreign objects, the source code is unavailable. In order to identify bindings among foreign objects, or to identify bindings that happened in the past, it is necessary to analyze their histories.

Definition 2 (The dynamic binding relation) *Object o_1 is in the dynamic binding relation with object o_2 if the following conditions are satisfied:*

DB1. In the history of object o_1 exists event e which is a sending of a request of a method invocation of object o_2.
DB2. If o_1 is in the dynamic binding relation with object o_2 and o_2 is in the dynamic binding relation with object o_3 then o_1 is in the dynamic binding relation with object o_3.

Notation $o_1 \leftrightsquigarrow_d o_2$ means that object o_1 is in the dynamic binding relation with object o_2.

For example in Figure 1 $o_1 \leftrightsquigarrow_d o_2$ since $e_1 \in h_{o_1}$ and $e_1 = s(o_1.1, o_1, o_2, m, p)$. Let's notice that $o_3 \leftrightsquigarrow_d o_3$ since $e_1 \in h_{o_3}$ and $e_1 = s(o_3.1, o_3, o_3, m, p)$. In comparison to relation $\leftrightsquigarrow_s$, relation $\leftrightsquigarrow_d$ is transitive by definition (condition DB2). Please notice, that DB2 is not a property since relation $\leftrightsquigarrow_d$ may concern the same objects but different methods and transitivity will be not satisfied.

Relation $\leftrightsquigarrow_d$ extends $\leftrightsquigarrow_s$ since sometimes the source code of objects is unavailable (foreign objects) or even if it is available it is impossible to find out with which objects the interaction will be performed (e.g. a dynamic vector of object references).

3.3   Bound Objects

It is useful for further consideration to define more general binding relation.

Definition 3 (The binding relation) *Object o_1 is in the binding relation with o_2 if object o_1 is in the static binding relation with o_2 or object o_1 is in the dynamic binding relation with o_2, i.e., $(o_1 \leftrightsquigarrow o_2) \iff (o_1 \leftrightsquigarrow_s o_2) \vee (o_1 \leftrightsquigarrow_d o_2)$.*

Since relation $\leftrightsquigarrow_s$ is transitive and $\leftrightsquigarrow_d$ is transitive by definition then relation $\leftrightsquigarrow$ defined as a logical sum is transitive.

Binding relations can be represented by directed graphs where vertices depict objects in the system and directed edges denote binding relations (e.g. Figure 4).

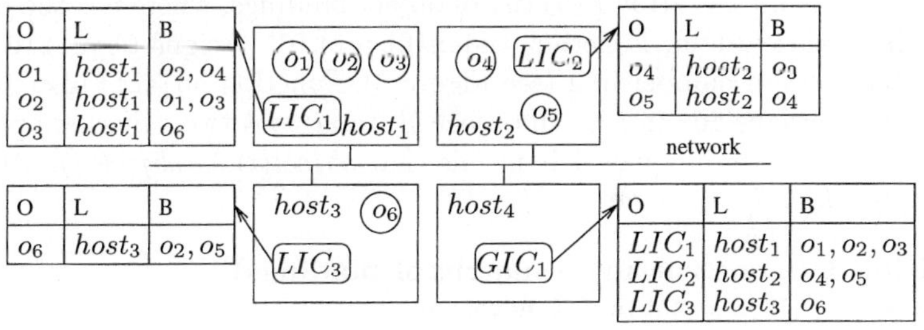

Fig. 2. The architecture for finding bound objects.

4   The Architecture for Finding Bound Objects

Figure 2 presents the architecture for finding objects which are involved in the binding relations defined in the previous section. There are three important entities in Figure 2: *Global Information Center (GIC)*, *Local Information Center (LIC)* and *Object-Location-Bindings (OLB)* tables. Both GIC and LIC objects keep the so-called OLB tables where they store necessary information. Column L has the same meaning for GIC-OLB or LIC-OLB tables. It shows a location of an object from column O. In case of a GIC-OLB table, column O contains LIC objects registered in a given GIC object whereas column O of a LIC-OLB table shows objects registered in a given LIC object. For example, in Figure 2 GIC_1 keeps information about LIC_1, LIC_2 and LIC_3, while LIC_3 stores information about object o_6. Column B has different meaning for GIC-OLB and LIC-OLB tables. A GIC-OLB table contains data about objects registered in a given LIC object. However, in case of LIC objects column B indicates the set of objects which a given object from column O is in relation $\leftrightarrow$ with. For instance, in Figure 2 from $GIC_1\text{-}OLB$ it is clear that two objects are registered in LIC_2 (o_4 and o_5). However, table $LIC_3\text{-}OLB$ indicates that $o_6 \leftrightarrow o_2$ and $o_6 \leftrightarrow o_5$.

GIC as well as LIC objects are responsible for keeping current information in their OLB tables. The presented architecture assumes that it exists a special layer (not depicted in the figure) that is responsible for recording histories of objects to the log. In order to keep a GIC-OLB table up-to-date different strategies can be used, e.g., *push* (changes are pushed to GIC objects), *pull* (GIC objects pull information from LIC ones) or *mixed* (pull or push when necessary) models [10].

The hierarchical structure of the architecture assures scalability [1]. In order to improve performance and availability GIC objects may be replicated [1][2][9].

In order to find out a list of objects bound with a given object a graph of bindings in the system can be constructed.

5   Constructing a Graph of Bindings

A graph of bindings in a system can be constructed with Algorithm 1. The result of Algorithm 1 is array `A[n][n]` of object bindings, where n is the number of objects in the system. Function `index(ObjectId)` assigns `ObjectId` in the system to a column number in `A` (see Figure 3). Function `object(idx)` (reverse to `index(·)`) assigns `idx` in `A` to an object id. In array `A` two values are possible: 0 and 1. If value A[i][j] equals 1 it means that $object(i) \leftrightarrow object(j)$, otherwise $object(i) \not\leftrightarrow object(j)$.

Algorithm 1 (Constructing a matrix of bindings)
```
1. Update the GIC-OLB table (if necessary).
2. Construct array A[n][n] and fill it with 0 values.
3. For each LIC in the GIC-OLB table do:
   (a) Get column O of the LIC-OLB table
   (b) For each object o in column O do:
           Get relevant values from column B of the LIC-OLB table
               For each value b from column B do:
                   A[index(o)][index(b)] = 1
```

The result of Algorithm 1 is the matrix of bindings in the system. For the system in Figure 2 the result of the algorithm is shown in Figure 3. In order to get a directed graph it is necessary to transform array `A` by, for example, Algorithm 2. The obtained graph is presented in Figure 4. In order to generate a list of bound objects Algorithm 3 may be applied.

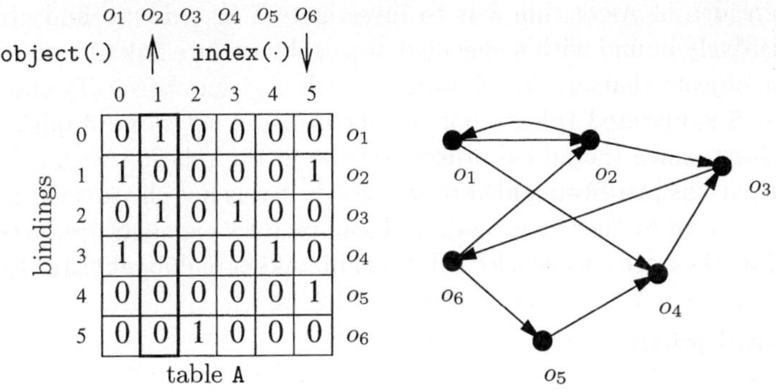

Fig. 3. The matrix of bindings in the system in Figure 2.

Fig. 4. The transformed matrix from Figure 3.

Algorithm 2 (Transforming A2G)

```
1. Put n vertices.
2. Label each vertex according to function object()
3. For each column in A do:
       For each value in column do:
          if(A[i][j] == 1) then {draw an arrow from object(i) to object(j)}
```

Algorithm 3 (Finding a list of bound objects)

```
1. A;                        // up-to-date table of bindings (Algorithm 1)
   Vector V;                 // the dynamic list of bound objects
   Set    S;                 // a result set of bound objects
   int I = index(O);         // for what object -- O we have to look for
   boolean isRemoved;        // indicates if an element was removed from V
2. do{
      isRemoved = false;     // nothing was removed from vec
      makeBoundObjectList(V, S, A, I); // see point 5
      if( V.size() > 0){
         I = V[0];           // take index of object to be checked
         V.remove(0);        // remove it from from vector V
         isRemoved = true;}  // indicate that the length of V was decreased
      }while ((V.size() != 0) || isRemoved );
3. check in A if O is in the relation with itself; update S if necessary
4. S contains a list of indexes which object O is bound with
5. Procedure makeBoundObjectList(V, S, A, I):
       for (i = 0; i < A[I].length; i++)
          if( A[I][i] == 1 ) then  // object(I) in relation with object(i)?
             if ( S.add(i) ) then  // check if elem. i in S and add if not
                V.add(i);          // we must check bindings of object(i)
                                   // so remember it in V
```

The result obtained from Algorithm 3 on table A from Figure 3 for finding bindings of o_1 is $S = \{1, 3, 2, 5, 4\}$ what implies (after $\text{object}(\cdot)$) $\{o_2, o_4, o_3, o_6, o_5\}$.

The main idea of Algorithm 3 is to investigate A in order to find objects that are transitively bound with a specified object. Procedure `makeBoundObjectList` looks for objects that are in relation $\looparrowright$ with a given object (I) and if it finds some, set S is updated (please notice that this is a *set* so no duplicates are allowed). Next, since the added object may be in the relation with other objects (and since it has just been added to the set, so it was not checked earlier) it must be added to V for further investigation. From vector V elements are systematically removed as they are checked for relations in `makeBoundObjectList()`.

6   Conclusions

The paper describes the algorithm for finding objects involved in the transitive binding relation with a given object. It is especially important if a tester wants to identify a group of objects. The article presents the framework architecture for maintaining information about bindings among distributed objects. The algorithm makes use of OLB tables, GIC and LIC services. It constructs the 2-dimensional table of bound objects. Having such a table it is possible to find out all objects that are bound with a given object. The situation complicates in the case of foreign objects since it is practically impossible to deduce relationships. It implies that special architectures for logging relevant information is necessary. The presented algorithms have been implemented in a prototype tool in order to verify the concepts in practice.

References

1. G. Coulouris, J. Dollimore, and T. Kindberg, *Distributed Systems. Concepts and Design.* Addison-Wesley Longman Limited, 1994.
2. A. S. Tanenbaum, *Distributed Operating Systems.* Prentice-Hall International, Inc., 1995.
3. M. Sławińska, "Testability of Distributed Objects," in *Proc. of the 5-th International Conference on Parallel Processing and Applied Mathematics*, Springer-Verlag, 2003. (to appear).
4. E. N. M. Elnozahy, L. Alvisi, Y.-M. Wang, and D. B. Johnson, "A survey of rollback-recovery protocols in message-passing systems," *ACM Computing Surveys*, vol. 34, no. 3, pp. 375–408, 2002.
5. L. Alvisi, K. Bhatia, and K. Marzullo, "Causality tracking in causal message-logging protocols," *Distributed Computing*, vol. 15, no. 1, pp. 1–15, 2002.
6. A. Geist, A. Beguelin, J. Dongarra, W. Jiang, R. Manchek, and V. Sunderam, *PVM:Parallel Virtual Machine: A Users' Guide and Tutorial for Networked Parallel Computing.* MIT Press, 1994.
7. Message Passing Interface Forum, ed., *MPI: A Message-Passing Interface Standard.* Message Passing Interface Forum, June 1995.
8. OMG, *Common Object Request Broker Architecture: Architecture and Specification, v3.0.* http://www.omg.org, December 2002.
9. P. Narasimhan, L. Moser, and P. M. Melliar-Smith, "State Synchronization and Recovery for Strongly Consistent Replicated CORBA Objects," in *Proc. of the IEEE Int. Conf. on Depend. Syst. and Net.*, IEEE Computer Society Press, 2001.
10. OMG, *Event Service Specification, v1.1.* http://www.omg.org, March 2001.

Design and Implementation of the Cooperative Cache for PVFS

In-Chul Hwang, Hojoong Kim, Hanjo Jung, Dong-Hwan Kim, Hojin Ghim,
Seung-Ryoul Maeng, and Jung-Wan Cho

Division of Computer Science, Dept. of Electrical Engineering & Computer Science, KAIST,
373-1 Kusung-dong Yusong-gu, Taejon, 305-701, Republic of Korea
{ichwang, hjkim, hanjo, dhkim, hojin, maeng, jwcho}
@calab.kaist.ac.kr

Abstract. Recently, there have been many efforts to get high performance in cluster computing with inexpensive PCs connected through high-speed networks. Some of them were to provide high bandwidth and parallelism in file service using a distributed file system. Other researches for distributed file systems include the cooperative cache that reduces servers' load and improves overall performance. The cooperative cache shares file caches among clients so that a client can request a file to another client, not to the server, through inter-client message passing. In various distributed file systems, PVFS (Parallel Virtual File System) provides high performance with parallel I/O in Linux widely used in cluster computing. However, PVFS doesn't support any file cache facility. This paper describes the design and implementation of the cooperative cache for PVFS (Coopc-PVFS). We show the efficiency of Coopc-PVFS in comparison to original PVFS. As a result, the response time of Coopc-PVFS is shorter than or similar to that of original PVFS.

1 Introduction

Recently, there have been many efforts to get high performance in cluster computing with inexpensive PCs connected through high-speed networks. It is necessary to connect PCs with efficient inter-connection network and to support applications efficiently in operating system to get high performance. Among these efforts, there have been many researches about distributed file systems which access disks much slower than any other component in cluster computing.

Among researches for distributed file systems, the cooperative cache [4,5,6] was proposed to reduce servers' load and to get high performance. Because the access time of another client's memory is faster than that of a server's disk, to get the block from another client's memory has faster response time than to get a block from a server's disk. In the cooperative cache, a client finds a block first from its own file system cache, and then the other clients' file system caches before finding the block from servers' disks.

M. Bubak et al. (Eds.): ICCS 2004, LNCS 3036, pp. 43–50, 2004.

In various distributed file systems, PVFS (Parallel Virtual File System) [1,2], which supports parallel I/O on Linux which is widely used in cluster computing, was developed in Clemson University. PVFS can get high bandwidth by stripping files over I/O servers. However, PVFS doesn't support any file system caching facility but only supports applications with transfer of data from/to I/O servers.

In this paper, we describe the design and implementation of the cooperative cache for PVFS (Coopc-PVFS). We also present various performance results with Coopc-PVFS and PVFS on the CAN cluster [8] at KAIST. We present the result of executing a simple matrix multiplication program. Then, we show the result for the BTIO benchmark programs [9].

The rest of this paper is organized as follows. In the next section, we discuss the related work of PVFS and cooperative cache. In section 3, we describe the design and implementation of Coopc-PVFS. In section 4, we present and discuss the performance results. In section 5, we summarize major contributions of this work and discuss future work.

2   Related Work

2.1   PVFS (Parallel Virtual File System)

PVFS [1,2], which supports parallel I/O on Linux widely used in cluster computing, was developed in Clemson university.

PVFS is composed of compute nodes, single metadata manager and I/O servers. The compute nodes are clients that use PVFS services. The metadata manager manages the metadata of PVFS files. The I/O servers store the actual data of PVFS files. In PVFS, a file data is stripped over I/O servers.

There are two schemes by which users can access files in PVFS. First, users can access files by recompiling their application codes with the PVFS user-level library – the PVFS library scheme. Another scheme is that users can access files through UNIX I/O system call using the PVFS kernel module – the PVFS kernel module scheme

There was a research for the file system caching effect of PVFS. Vilayannur et al. [3] designed and implemented a file system cache of a client in the PVFS library scheme. They showed that a file system cache in a client is efficient if many applications in the client share files among them. But their research was limited to a file system cache in a single node. Because many users share files in cluster environments, the cooperative cache is more appropriate than a file system cache in a client.

2.2   Cooperative Cache

The cooperative cache [4,5,6] was proposed to reduce servers' load and to get high performance. In the cooperative cache, if a file system cache in a client doesn't handle a request to a file, the client sends the request to the other client's cache that caches the file rather than to the server because the access time of another client's memory is faster than that of the server's disk. Servers' load can be reduced in the cooperative cache so that it is scalable as the number of clients increases. Because there is much more memory in the cooperative cache than in a single file system cache, the cooperative cache can handle more requests and improve overall system performance.

There have been many researches about the cooperative caching. Dahlin et al. [4] suggested the efficient cache management scheme called N-chance algorithm, Feeley et al. [5] suggested another efficient cache management scheme called modified N-chance algorithm in GMS (Global Memory Service). Sarkar et al. [6] suggested the hint-based cooperative caching to reduce the management overhead using hint. Thus, the hint-based cooperative cache is scalable and can be adopted in the large-scale system such as cluster computer. Because Sarkar's idea uses a file as a file system management unit, it can not be applicable to parallel file systems in which many users share large files concurrently.

3 Design and Implementation of Coopc-PVFS

3.1 Overview of Coopc-PVFS

In the PVFS kernel module scheme, an application reads a file from I/O servers through the PVFS kernel module without a file system caching facility. We added the cooperative caching to the PVFS kernel module scheme in computing nodes.

In figure 1, we present the workflow of Coopc-PVFS added in the PVFS kernel module scheme.

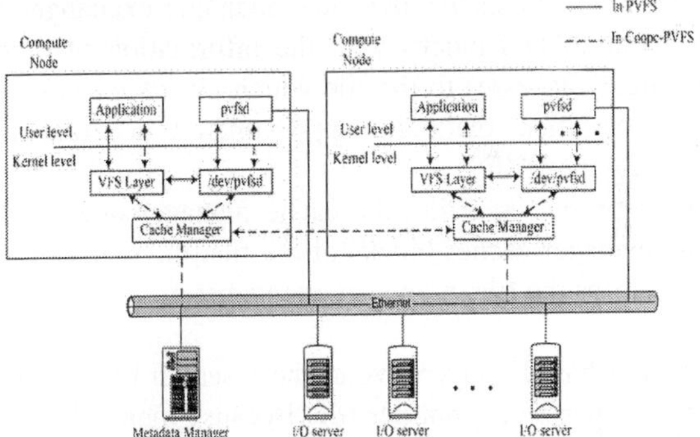

Fig. 1. Workflow of Coopc-PVFS

In the cooperative cache added to PVFS kernel module scheme, when an application reads a file, the cache manager in the client looks up whether the requested block is in its own cache. If the block is in the cache manager, the cache manager copies the block to the application. If the block is not found in its own cache manager, the cache manager looks up whether there is any client that caches the block. If the block is found in the other clients' cache managers, the cache manager gets the block from one of them, and then caches the block and copies the block to the application. If the block is not found in other clients' cache manager, the cache manager gets the block from I/O servers, and then caches the block and copies the block to the application.

3.2 Design of Coopc-PVFS

3.2.1 Information Management

Because of large overhead to maintain accurate information about cached blocks, we designed Coopc-PVFS as a hint-based cooperative cache. To maintain the hint – opened clients list, we added new function to the metadata manager to keep the clients list that contains information of clients that opened the file before. Whenever a client opens a file, the client gets both the metadata and the opened clients list of the file from the metadata manager.

To accurately look up a block whether other clients have it, the client must know the information about cached blocks in other clients. To maintain this information, we used the methods like below:

- In PVFS, when an application opens a file, the application gets metadata of the file from the metadata manager. Using this mechanism, the metadata manager manages the hint per file – the IPs of the clients which opened the file before, named opened clients list. When an application opens a file in Coopc-PVFS, the application gets not only the metadata of the file but also opened clients list.
- To maintain the information about cached blocks, when an application reads a block that is not in its own cache, the cache manager exchanges its own information (bitmap) about cached blocks with the information of other client's cache manager. After many accesses to the file which clients exchange the information with each other, the client maintains approximately accurate information about cached blocks in Coopc-PVFS.
- When an application closes a file, the cache manager doesn't do anything. Because the cache manager caches blocks of the file, the metadata manager remains the client which closes the file in opened clients list.

Unlike previous hint-based cooperative cache research [6], we managed information and cached blocks per block, not per file. Because many clients share large files among them in a parallel file system, it is more adaptable to manage information and to cache per block than per file in Coopc-PVFS.

3.2.2 Consistency

In PVFS, all the accesses to files go through the I/O servers. To preserve the consistency likewise in PVFS, the cache manager must invalidate blocks cached in other clients before writing the block to the I/O server in Coopc-PVFS. To do that, whenever an application writes a block, the cache manager sends the block invalidation propagation request to the metadata manager before sending the written block to the I/O server. When the metadata manager gets the block invalidation propagation request, it sends the block invalidation messages to the clients in opened clients list of the file then all of the clients that receive the block invalidation message invalidate the block. Therefore, all of cached blocks in other clients are invalidated before sending the written block to the I/O server and we can preserve the consistency in Coopc-PVFS the same as in PVFS.

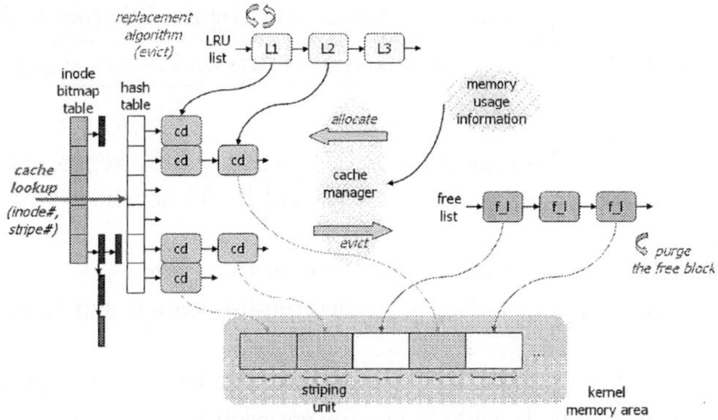

Fig. 2. Data structures used by a cache manager in Coopc-PVFS

3.3 Implementation of Coopc-PVFS

In figure 2, we present data structures used by the cache manager in Coopc-PVFS.

We implemented most of data structures using linked lists in Coopc-PVFS for dynamic addition and deletion of entries. For replacement of the cached block, we managed the LRU list. For unused blocks allocated from system memory, we managed the free list. Each cached block is managed by the size of stripping unit used in PVFS. In PVFS, the size of stripping unit is 64KB by default and it can be changed for various system configurations.

In the metadata manager, we manage opened clients lists of files using linked lists. Each opened clients list of a file has its own lock in order to access it concurrently.

For allocating cache blocks from system memory, we didn't use the page cache system in Linux for the cache manager. If we use the page cache system in Linux, we don't know which blocks are cached in the cache manager – If we use, cache blocks can be freed by the Linux memory manager so that we can't know about that event. Therefore, we must allocate cache blocks from kernel memory and do memory management for cache blocks. For doing memory management, we implemented the cache replacement manager which can do memory management according to the amount of free kernel memory in the system as a kernel thread.

4 Performance Evaluation

We used CAN cluster [8] in KAIST to evaluate the performance of Coopc-PVFS. The system configuration of CAN cluster is presented in table 1.

Table 1. System configuration

CPU	Pentium IV 1.8GHz
Memory	512MByte 266MHz DDR
Disk	IBM 60G 7200rpm
Network	3c996B-T(Gigabit Ethernet) 3c17701-ME(24port Gigabit Ethernet Switch)
OS , PVFS	Linux(Kernel version 2.4.18) , 1.5.3

The metadata manager was allocated in one node and the I/O server was allocated in another node. And one to four other clients were used to execute the test applications –a simple matrix multiplication program and BTIO benchmark programs. Each program operates like below:

– Matrix multiplication program: Applications in four clients read two input files of 1024*1024 matrix and calculate the matrix multiplication and write the result to the output file.
– BTIO benchmark programs: BTIO [9] is a parallel file system benchmark. BTIO contains four programs. In table 2, we present each four programs. We can evaluate the parallel I/O performance of Coopc-PVFS using four clients with smallest sized class s in BTIO.

Table 2. BTIO benchmark programs

Full (*mpi_io_full*)	MPI I/O with collective buffering
Simple (*mpi_io_simple*)	MPI I/O without collective buffering
Fortran(*fortran_io*)	Fortran 77 file operations used
Epi (*ep_io*)	Each process writes the data belonging to its part of the domain to a separate file

Table 3. Execution time of the matrix multiplication program

PVFS	I/O server doesn't caching data(*iod_cool*)	209.174 secs
	I/O server caches data(*iod_hot*)	128.283 secs
Coopc- PVFS	anyone doesn't cache data(*iod_cool*)	120.451 secs
	I/O server cache data(*iod_hot*)	120.005 secs
	Coopc-PVFS cache data(*coopc_hot*)	120.029 secs

4.1   Execution Time of Matrix Multiplication Program

The execution time of the matrix multiplication program is in table 3.

To analysis of total read/write time of this program, we present the time breakdown of average execution time in figure 3.

The matrix multiplication program is a read-dominant program so that total read time is much longer than total write time. In Coopc-PVFS, we can reduce the read time to approximately zero because the file is cached in Coopc-PVFS after once a file is read. When the I/O server doesn't cache the file in PVFS, the waiting time is much larger than any other case because the variation of read time is much larger than any other case. The write time in Coopc-PVFS is a little longer than that in PVFS because the write in Coopc-PVFS has slightly overhead than in PVFS.

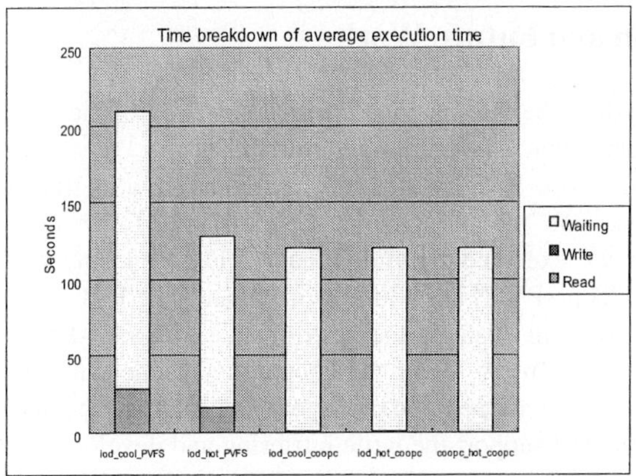

Fig. 3. Execution time breakdown of the matrix multiplication program

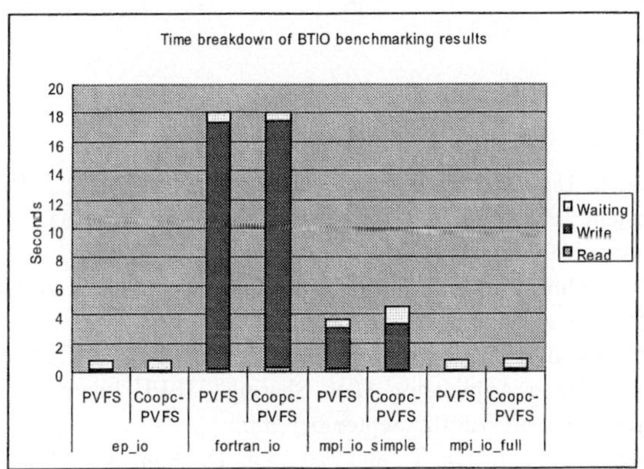

Fig. 4. Execution time breakdown of BTIO benchmark programs

4.2 Performance Evaluation using Benchmark Programs

In figure 4, we present the time breakdown of BTIO benchmarking results.

BTIO benchmark programs are write-dominant programs so that total write time is much longer than total read time in the results. Using MPI, we can get much shorter write time. Collective I/O reduces almost part of write time. In most cases, write time of Coopc-PVFS is longer than in PVFS because the write in Coopc-PVFS has more overhead than the write in PVFS and read time of Coopc-PVFS is shorter than in PVFS because cli-ents cache all files in Coopc-PVFS. Totally, the execution time in

Coopc-PVFS is a little longer than in PVFS. Therefore, we can know that the performance improvement of writing is needed in Coopc-PVFS.

5 Conclusion and Future Work

In this paper, we describe the design and implementation of the cooperative cache for efficient data sharing that is not supported in PVFS.

We evaluated Coopc-PVFS with many programs. In a matrix multiplication program, we can execute the program in Coopc-PVFS faster than in PVFS about 6%~50%. When we executed write-dominant BTIO benchmark programs, we can know that using Coopc-PVFS has a little worse than using PVFS.

In the future, we will evaluate the performance of Coopc-PVFS using many scientific applications in cluster. Using the cooperative cache can improve the performance of reading, but the cooperative cache can not improve the performance of writing. Therefore, we will support the write buffering and develop new write schemes to improve the write performance. And we will adopt the collective I/O request technique which the cache manager sends many requests at a time instead of sending a request at a time.

References

1. P. H. Carns, W. B. Ligon III, R. B. Ross, and R. Thakur, "PVFS: A Parallel File System For Linux Clusters", Proceedings of the 4th Annual Linux Showcase and Conference, Atlanta, GA, October 2000, pp. 317-327
2. R. B. Ross, "Providing Parallel I/O on Linux Clusters" ,Second Annual Linux Storage Management Workshop, Miami, FL, October 2000.
3. M.Vilayannur,M.Kandemir, A.Sivasubramaniam, "Kernel-Level Caching for Optimizing I/O by Exploiting Inter-Application Data Sharing", IEEE International Conference on Cluster Computing (CLUSTER'02),September 2002
4. Dahlin, M., Wang, R., Anderson, T., and Patterson, D. 1994. "Cooperative Caching: Using remote client memory to improve file system performance", In Proceedings of the First USENIX Symposium on Operating Systems Design and Implementation. USENIX Assoc., Berkeley, CA, 267-280
5. Feeley, M. J., Morgan, W. E., Pighin, F. H., Karlin, A. R., and Levy, H. M. 1995. "Implementing global memory management in a workstation cluster", In Proceedings of the 15th symposium on Operating System Principles(SOSP). ACM Press, New york, NY, 201-212
6. Prasenjit Sarkar , John Hartman, "Efficient cooperative caching using hints", Proceedings of the second USENIX symposium on Operating systems design and implementation, p.35-46, October 29-November 01, 1996, Seattle, Washington, United States
7. "Linux Kernel Threads in Device Drivers",
http://www.scs.ch/~frey/linux/kernelthreads.html
8. Can cluster, http://camars.kaist.ac.kr/~nrl
9. Parkson Wong, Rob F. Van der Wijngaart, NAS Parallel Benchmark I/O Version 2.4, NAS Technical Report NAS-03-002, NASA Ames Research Center, Moffett Field, CA 94035-1000

Towards OGSA Compatibility in Alternative Metacomputing Frameworks[*]

Gunther Stuer[1], Vaidy Sunderam[2], and Jan Broeckhove[1]

[1] Dept. of Math and Computer Science,
University of Antwerp, 2020 Antwerp, Belgium.
{gunther.stuer, jan.broeckhove}@ua.ac.be
[2] Dept. of Math and Computer Science,
Emory University, Atlanta, GA 30322, USA
vss@mathcs.emory.edu

Abstract. Lately, grid research has focused its attention on interoperability and standards, such as Grid Services, in order to facilitate resource virtualization, and to accommodate the intrinsic heterogeneity of resources in distributed environments. To ensure interoperability with other grid solutions, it is important that new and emerging metacomputing frameworks conform to these standards. In particular, the H2O system offers several benefits, including lightweight operation, user-configurability, and selectable security levels; with OGSA compliance, its applicability would be enhanced even further In this contribution, a framework is presented which will augment the H2O-system with the functionality to produce and publish WSDL and GSDL documents for arbitrary third party pluglets, and thereby enhance OGSA compatibility.

1 Introduction

The benefits of distributed computational systems are well established, and numerous software architectures and toolkits have evolved in recent years to support this mode of computing. Most of these systems however are rather limited in scope. Some, such as UD Patriot Grid [1] and SETI@Home [2] are targeted at specific research projects, while others, such as MPICH [3] and PVM [4] are usually confined to single administrative domains. More generalized metacomputing systems, or *grids*, have gained tremendous popularity in recent times because they enable secure, coordinated, resource sharing across multiple administrative domains, networks, and institutions. This model [5] has been realized in several software toolkits, such as Globus [6] and Legion [7].

However, many applications of smaller magnitude do not require explicit coordination and centralized services for authentication, registration, and resource brokering as is the case in traditional grid-systems. For these applications, a lightweight and stateless model, in which individuals and organizations share

[*] Research supported in part by U.S. DoE grant DE-FG02-02ER25537 and NSF grant ACI-0220183.

M. Bubak et al. (Eds.): ICCS 2004, LNCS 3036, pp. 51–58, 2004.

their superfluous resources on a peer-to-peer basis, is more suitable. Two examples of such lightweight peer-to-peer distributed computational systems are H2O [8,9,10] and JGrid [11].

H2O, the framework used in our research, is a novel component-based, service-oriented framework for distributed metacomputing. Adopting a provider-centric view of resource sharing, it emphasizes lightweight software infrastructures that maintain minimal state. Resource owners host a software backplane, known as the *kernel*, onto which owners, clients, or third-party resellers may load components or component-suites, known as *pluglets*, that deliver value added services without compromising owner security or control.

In the current phase of evolution, grid research has focused on interoperability and standards in order to facilitate resource virtualization and to accommodate the intrinsic heterogeneity of resources in distributed environments. To this end, OGSA [12] aims to define a new common and standard architecture for grid-based applications based on the concept of Grid Services, an extension of Web Services [13]. The formal and technical specification of these concepts can be found in the OGSI [14] specifications and a reference implementation is provided by the GTK3 [15].

Such standard frameworks, based on XML, are used to describe service specifications in a universally understood manner, thereby permitting clients to discover and utilize services across platforms and context domains. The functional description of a Web Service is written in the *Web Services Description Language* (WSDL) [16]. It can be published using various registration and discovery schemas, such as UDDI [17]. Grid Services are described using an extension of WSDL known as *Grid Services Description Language* (GSDL) [14].

Being compliant to these standards is not only important for heavyweight Grid-systems such as the Globus Toolkit, but also for the lightweight solutions referred to above. In H2O this is already partially the case. Through the use of RMIX [18] as the communications layer, pluglets can be exported using various remote bindings such as stub-less JRMP, IIOP and SOAP. When using the latter, the exported pluglets can be seen as Web Service instances that provide standardized SOAP-endpoints.

However, in H2O there is currently no provision for the creation and publication of WSDL and GSDL documents for pluglets that have already been deployed. At present, the WSDL/GSDL description file needs to be created manually and be published using third party tools. In this contribution, a framework is presented that will augment the H2O-system with the functionality to produce and publish WSDL and GSDL documents for arbitrary third party pluglets, thus enhancing OGSA compatibility and thereby grid interoperability.

2   Architecture

We present a framework that consists of three important components: two pluglets and one command-line tool.

The *SoapExportPluglet*, or *SEPluglet* for short, is the main component. It is responsible for the generation of WSDL / GSDL documents and SOAP / Grid-Service endpoints for a third party pluglet. In this contribution, we will denote such a pluglet as *pluglet-X*. The second pluglet, *PublisherPluglet* is responsible for publishing generated WSDL/GSDL documents to some registry. *ExportTool*, a command-line tool, serves two purposes. It is both a demonstration of how to use the *SEPluglet*'s API and a utility that allows H2O-users to export new or already uploaded pluglets as Web or Grid Services.

Figure 1 shows the architecture of the framework. A kernel can contain zero or more *SEPluglet* instances, which will all have the same name, but different unique identifiers. Each *SEPluglet* stores a *SoapExportInfo* object for every pluglet it exports. Each *SoapExportInfo* object contains the necessary information and business logic to create the endpoints and WSDL/GSDL documents for the pluglet it represents. Each *SEPluglet* can be monitored by zero or more *PublisherPluglets* and each *PublisherPluglet* can monitor zero or more *SEPluglets*. This design allows for substantial flexibility in the choice of registry that publishes the WSDL/GSDL description of the exported pluglets.

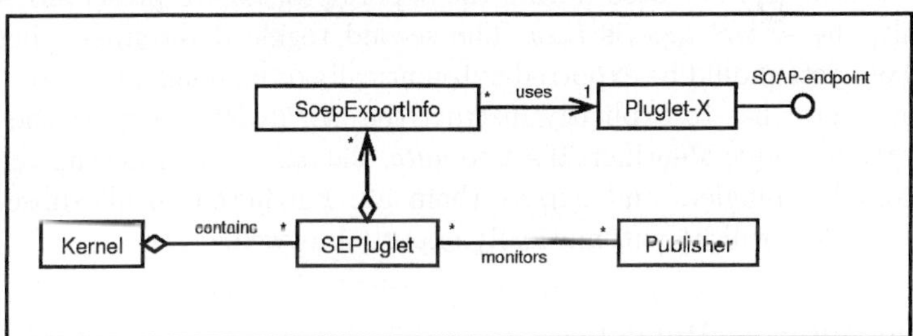

Fig. 1. Architecture of the framework.

3   The Components

3.1   The SoapExportPluglet

The *SEPluglet* is the most important component of our framework. It is responsible for the lookup, creation and destruction of WSDL/GSDL documents and SOAP/GridService endpoints. *SEPluglet* has the capability to deploy and export a pluglet in a single step, but it can also export pluglets that have already been deployed. To allow lookups, it contains a *SoapExportInfo* object for every pluglet it is currently exporting. This feature doubles as a cache: when multiple requests are made to export the same pluglet, the corresponding *SoapExportInfo* object is created only once. *SoapExportInfo* objects remain stored until either the corresponding pluglet is destroyed, or an actor explicitly requests that the export of a pluglet be revoked.

The steps required for uploading and exporting a pluglet are illustrated in the first part of figure 2. Of course, before any service can be exported, the *SEPluglet* and the target pluglet have to be deployed. After this, the *SEPluglet* can be instructed to export a given pluglet by specifying its unique identifier. If this pluglet is not already exported, its PlugletContext will be retrieved from the kernel and stored in a newly created *SoapExportInfo* object which will be returned.

Subsequently, the *SoapExportInfo* object can be instructed to create the endpoints and the WSDL / GSDL documents. Note that artifacts are created only once. Subsequent requests return the cached values. The creation of artifacts is a cascading activity. In order to create a GSDL document, one needs a WSDL document and a Grid Service-endpoint. In order to create a WSDL document, one needs a SOAP-endpoint and the remote Java interface of the pluglet to export. So, as a consequence of constructing the GSDL document, all other artifacts are built as well. This behavior is illustrated in the second part of figure 2.

There are two toggles which modify the behavior of the *SEPluglet*: the *export type* and the *export mode*. The fist determines when the artifacts of an exported pluglet will be created. If set to *lazy*, they will be constructed on first use. If set to *eager*, all artifacts are created during the export process. To preserve resources, by default, the *export type* is *lazy*. The second toggle determines whether all deployed pluglets should be exported automatically or manually. If set to *manual*, an external actor has to explicitly instruct the *SEPluglet* to export the pluglet with the given unique identifier. If set to *auto*, the *SEPluglet* will scan the kernel for all available pluglets and export them all. Furthermore, all subsequently deployed pluglets will be automatically exported as well.

3.2   The PublisherPluglet

Generating the WSDL/GSDL documents is only part of the process. It is also very important to publish them to all appropriate registries, such as UDDI and LDAP, from which they can be discovered by third parties. This allows clients, which are either compliant to the Web Services standard or the Grid Services standard, to discover and use the exported pluglets. Furthermore, when a pluglet is no longer available, the documents should be removed from the registry. These tasks are performed by the *PublisherPluglets*.

This component is designed in two layers, an abstract base class which takes care of all the bookkeeping and a concrete subclass which implements the business logic to actually publish or remove the WSDL/GSDL documents from a registry. Whether it is the WSDL, the GSDL or both that are published or removed depends on the concrete implementation. This design makes it very easy to add new types of publishers. One only needs to derive from the abstract base class and implement two abstract methods: `publish (SoapExportInfo)` and `unpublish (SoapExportInfo)`. For now, one derived class has been implemented which maintains a cache in memory of all published documents.

The *PublisherPluglets* can operate in three modes: *manual, semi automatic* and *automatic*. They can switch between any of these modes at runtime. In *man-*

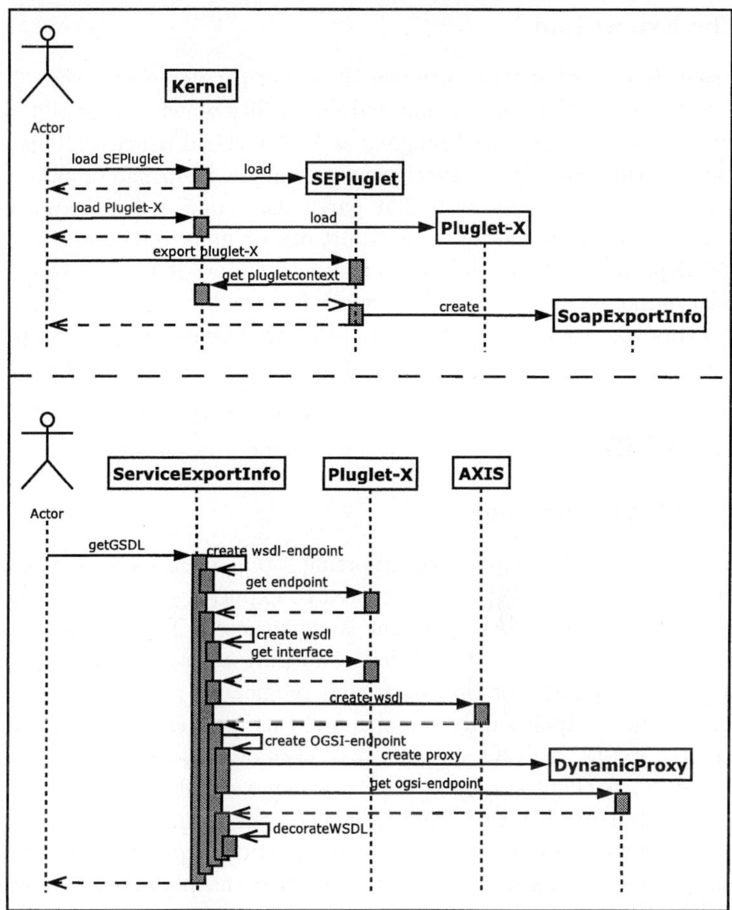

Fig. 2. Exporting and creating a GSDL

ual mode, an actor has to specify the unique identifier of the pluglet that needs its documents published or removed. The *PublisherPluglet* locates the exporting *SEPluglet* and retrieves the corresponding *SoapExportInfo* object. From this, the necessary information can be retrieved to publish or remove the WSDL/GSDL documents. In *semi automatic* mode, an actor can specify which *SEPluglets* to monitor for export events. If any of these *SEPluglets* (un)exports a pluglet, its documents automatically get (un)published as well. In *automatic* mode, the kernel is scanned for deployed *SEPluglets* and all are monitored for export events. Furthermore, all subsequently deployed *SEPluglets* will be automatically monitored as well.

3.3   The ExportTool

The (*ExportTool*) serves two purposes. It is both a demonstration of how to use the *SEPluglet*'s API and a command-line utility which allows H2O-users to export new or already deployed pluglets as Web or Grid Services. This tool can export an existing pluglet by specifying its unique identifier or name. In the second scenario, all pluglets with that particular name will be exported. Two similar operations are available to unexport one or more pluglets. If the pluglet is not yet deployed, it is possible to deploy and export it in one step. For this, the pluglet name, service class and classpath have to be specified. There are two version of this operation: one taking the list of arguments mentioned above, and one taking the name of a Java properties-file containing the required information.

4   The Artifacts

4.1   The SOAP-Endpoint

The RMIX subsystem is capable of exporting a pluglet using various communication protocols. When RMIX is instructed to export a given pluglet using the SOAP protocol, a new SOAP-endpoint is constructed. This endpoint is represented as a URL, pointing to the address and port where RMIX is listening for incoming SOAP-requests for that particular pluglet.

However, this endpoint only remains valid as long as the session in which it was created exists. In H2O, one has two types of sessions: client-sessions and pluglet-sessions. A client-session is created when a client logs into a kernel and the session is destroyed during logout, or when the connection to the kernel is broken. A pluglet-session is associated with a particular pluglet and it is created when the pluglet is loaded. It is destroyed when the pluglet is removed from the kernel. In the current H2O release (0.8.2), endpoints are associated with the originating session. When a pluglet is manually exported using the *ExportTool*, the originating session is a user-session and the endpoints will remain valid as long as this session is active, i.e., as long as the *ExportTool* is connected to the kernel.

To address this problem, an extension to the H2O-kernel is needed. It has to be possible to specify that an endpoint is associated with a session other than the originating one. This way, it can be associated to the *SEPluglet*'s session and the endpoint will remain valid as long as the *SEPluglet* is active. With this extension, the *ExportTool* could connect to a kernel, export a pluglet, create the endpoints and disconnect again without invalidating the endpoints. This feature will be incorporated into a future H2O release.

4.2   The WSDL-Document

The WSDL-document is created using the Apache AXIS tool[19]. Among other things, this tool has the functionality to convert Java interfaces in WSDL-documents. To do so, AXIS needs access to the class files of the remote interface defining the pluglet's available operations. In H2O this is not as straightforward because, for security reasons, each pluglet has its own classpath. This prevents the *SEPluglet*, and by extension, the AXIS-engine from reading the class files of the pluglet to be exported. This problem was solved by modifying the kernel's security policy file to allow the *SEPluglet* to construct new classloaders. For each pluglet that is exported, a new classloader is created which is the union of the *SEPluglet*'s classloader and the one from the pluglet to be exported. This new classloader is then passed to the AXIS-engine.

4.3   The Grid Service-Endpoint

An important step in making H2O OGSI-compliant is to ensure that all pluglets with a SOAP-compliant interface can be exported as a Grid Service. According to the OGSI-specifications [14], this means that they must be addressable using SOAP and must implement a number of pre-defined operations. However, since exporting a pluglet as a Grid Service is only one of several possible bindings, it is not desirable that pluglets implement these methods themselves, nor that they have to be derived from a superclass. For this reason, a mechanism has to be in place which will dynamically extend existing pluglets with the pre-defined OGSI-operations whenever they are exported as Grid Services.

A solution can be constructed using Java's dynamic proxies. They allow a new class to be constructed at runtime which will implement all the pluglet's original operations plus those defined by OGSI. This proxy object can subsequently be exported as a SOAP-endpoint and will serve as a method router, dispatching incoming method invocations to either the proxied pluglet or to some object implementing the methods defined by OGSI. There is one remaining problem though. The current version of RMIX-SOAP does not allow an object to export multiple remote interfaces. This is a necessary feature because the exported proxies implement two remote interfaces: one defining the operations on the proxied pluglet, and one defining the OGSI-operations. This feature will be incorporated in a future H2O release.

4.4   The GSDL-Document

GSDL is basically an extension of WSDL. Therefore, building a GSDL-document essentially involves building the WSDL-document and subsequently extending it with the GSDL-specific parts. Three additions have to be made to the WSDL-document. The namespace *girdservicesoapbinding* has to be defined, the file *ogsi_bindings.wsdl* has to be imported and an extra port, *GridServiceSOAP-BindingsPort*, has to be added. The algorithm that we use is based upon the *DecorateWSDL*-algorithm of GTK3. Since it uses files for input and output, which is not desirable in H2O-kernels, the GTK3-implementation could not be used as is. A slightly modified version has been implemented which uses strings for input and output.

5   Summary

In this contribution we have presented a framework to export the pluglets of the H2O computational system as Web/Grid Services, and thereby enhance OGSA compatibility. The core of the framework is the *SEPluglet*, which is responsible for generating the WSDL/GSDL artifacts. The second pluglet, *PublisherPluglet*, is responsible for the publication of the WSDL and GSDL documents of exported pluglets.

A kernel can contain multiple *SEPluglet* instances and each instance can be monitored by multiple *PublisherPluglets*. This design allows for great freedom as to the location at which the WSDL/GSDL descriptions of exported pluglets will be published.

References

1. UD. Project: The PatriotGrid. http://www.grid.org/projects/patriot.htm.
2. SETI@Home. http://setiathome.ssl.berkeley.edu.
3. W. Gropp, E. Lusk, N. Doss, and A. Skjellum: *A high-performance, portable implementation of the MPI message passing interface standard.* Parallel Computing, 22(6):789-828, Sept. 1996.
4. A. Geist, A. Beguelin, J. Dongarra, W. Jiang, R. Manchek, and V. Sunderam: *PVM: Parallel Virtual Machine: A User's Guide and Tutorial for Networked Parallel Computing.* MIT Press, Cambrdige, MA, USA, 1994.
5. I. Foster, C. Kesselman, and S. Tuecke: *The anatomy of the grid: Enabling scalable virtual organizations.* Int J. of Supercomputer Applications, 153), 2001.
6. I. Foster, and C. Kesselman: *A metacomputing infrastructure toolkit.* The Int. J. of Supercomputer Applications and High Performance Computing, 11(2):115-128, 1997.
7. A. Natrajan, M. A. Humphrey, and A. S. Grimshaw: *Grids: Harnessing geographically-seperated resources in a multi-organisationel context.* In 15th Annual Int. Symp. on High Performance Computing Systems and Applications, 2001.
8. V. Sunderam, D. Kurzyniec: *Lightweight Self-Organizing Frameworks for Metacomputing.* In 11th Int. Symp. on High Performance Distributed Computing, 2002.
9. The H2O project home page: http://www.mathcs.emory.edu/dcl//h2o/.
10. The H2O tutorial: http://www.mathcs.emory.edu/dcl//h2o/h2o-tutorial.pdf.
11. Z. Juhasz, A. Andics, S. Pota: *JM: A Jini Framework for Global Computing.* IEEE Int. Symp. on Cluster Computing and the Grid, 2002.
12. I. Foster, C. Kesselman, J. Nick, and S. Tuecke: *The physiology of the Grid: An Open Grid Services Architecture for distributed systems integration.* http://www.globus.org/research/papers/ogsa.pdf.
13. Web Services specifications: http://www.w3.org/2002/ws/.
14. Open Grid Service Infrastructure (OGSI): http://www.gridforum.org/ogsi-wg/drafts/draft-ggf-ogsi-gridservice-29_2003-04-05.pdf.
15. Globus Toolkit 3: http://www-unix.globus.org/toolkit/download.html.
16. Web Services Description Language specification: http://www.w3.org/TR/wsdl.
17. Universal Description, Discovery and Integration: http://www.uddi.org/.
18. D. Kurzyniec, T. Wrzosek, and V. Sunderam: *Heterogeneous Access to Service-based Distributed Computing: the RMIX Approach.* Int. Parallel and Distributed Processing Symp., 2003.
19. The Apache AXIS-tool: http://ws.apache.org/axis/.

DartGrid: Semantic-Based Database Grid

Zhaohui Wu , Huajun Chen, Changhuang, Guozhou Zheng, and Jiefeng Xu

College of Computer Science, Zhejiang University, Hangzhou, 310027, China
`{wzh,huajunsir,changhuang,zzzgz,xujf}@zju.edu.cn`

Abstract. In presence of web, one critical challenge is how to globally publish, seamlessly integrate and transparently locate geographically distributed database resources with such "open" settings. This paper proposes a semantic-based approach supporting the global sharing of database resources using grid as the platform. We have built a semantic query system, called DartGrid, with the following features: a) database providers are organized as an ontology-based virtual organization; by uniformly defined domain semantics, database could be semantically registered and seamlessly integrated together to provide database service, and b)we raise the level of interaction with the data base system to a domain-cognizant model in which query request are specified in the terminology and knowledge of the domain(s), which enable the users to publish, discovery ,query databases only at a semantic or knowledge level. We explore the essential and fundamental roles played by data semantics, and implement some innovative semantic functionalities such as semantic browse, semantic query and semantic registration. We also reports on application results from Traditional Chinese Medicine (TCM) that requires data-intensive collaboration.

1 Introduction

In the next evolution step of web, termed semantic web[1], vast amounts of information resources (databases, multimedia, programs) will be enriched with uniformed semantics for automatic discovery, seamless communication and dynamic integration. In presence of such semantics defined for database integration or sharing, one critical challenge is how to transparently translate a semantically enriched query into a distributed query plan, and then properly locate and access geographically distributed database resources with such "open" settings.

This paper proposes a semantic query system, called DartGrid, using grid as the platform. DartGrid is designed to support the building of large-scale ontology-based database Virtual Organization (DB-VO), in which databases are organized by uniformly defined semantics, namely, domain ontologies. In a DB-VO, databases are semantically registered to a web service called Semantic Registry Service(SeRS) and the user query the system only at a semantic and knowledge level. Usually, the user dynamically generates a visual conceptual query when browsing the ontolgies stemming from Ontology Service, meanwhile, a semantic query is generated and submitted to Semantic Query Service (SeQS). After inquiring of SeRS about the mapping

M. Bubak et al. (Eds.): ICCS 2004, LNCS 3036, pp. 59–66, 2004.

from shared ontologies to local database schemas, the semantic query is converted into a local database query languages (e.g., SQL , XQuery,etc.) . And then SeQS builds a distributed query plan to dispatch the query into proper database service .The result returned will be semantically wrapped again before they are presented for semantic browsing by the user.

Our work is essentially motivated and informed by requirements of communities of TCM researchers and professionals and our experience in building several TCM information systems [2]. Currently, in our deployed testbed, an ontology service, with about 10,000 records of TCM ontology instances contained, has been set up, and ten nodes with thirty TCM-related databases have been deployed. Reports from our partners, China Academy of TCM and its associated enterprises and institutes, show that our system significantly promotes the sharing and integration of their database resources and greatly facilitates their cooperation in their preferable web mode.

2 Ontology-Based Virtual Organization

As well-known, Virtual Organization enables disparate groups of organizations and/or individuals to share resource in a controlled fashion, so that members may collaborate to achieve a shared goal. We argue that ontology will play a significant role in constructing such Vos . Ontology defines the formal conceptuation model and standard terminology of the domain, which could significantly improve the sharing level within such VOs. In the following, we give a formal definition of the ontology-based VO as follows.

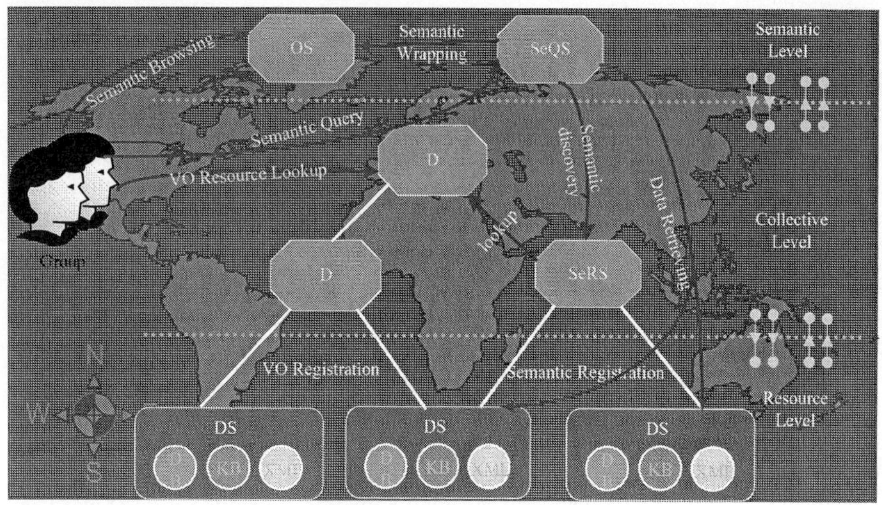

Fig. 1. A Formal Model of Ontology-based Virtual Organization

Definition 1: an Ontology-based Virtual Organization OntoVO is a four-tuple OntoVO = (O+, SeRS, DS, FS) and

- *O+ is a set of ontology services related to VO, "+"means every OntoVO must has at least one ontology;*
- *SeRS is Semantic Registration Service maintaining the mapping from the global semantics to local schemas, and is also a ontology-based index for classifying data objects*
- *DS is Data Service providing data objects*
- *FS is a set of optional functional services such as Semantic Query Service(SeQS) providing semantic query parsing and dispatching and address directory service (D in the figure) maintaining the physical address of all networked entities.*

__Definition 2:__ an Ontology Service OS is a two-tuple OS=(T_{os}, P_{os}), and
- *T_{os} is a set of terminologies which define the domain of this VO;*
- *P_{os} is service porttype which specifies a set of necessary knowledge-level operations on T_{os}.*

__Definition 3:__ an Data Service DS is a two-tuple DS= (M_{ds}, A_{ds}) where
- *M_{ds} is meta data about the data service ;*
- *A_{ds} is the data objects provided by data service;*

3 Implementation

DartGrid is a referential implementation of the OntoVO model. The principal technical characteristics of DartGrid are highlighted as below:

1. We develop it on Globus 3.0, the de facto standard platform to construct VO in Grid Computing research area.
2. RDF, the standard data model for web semantics defined by W3C, is adopted as the universal data model for defining protocols such as protocol for semantic registration.
3. Ontologies used in DartGrid comply with the syntax and semantic of OWL, the standard ontology description language proposed by W3C.

Firstly, we introduce the core components developed in DartGrid as follows.

3.1 Building Blocks of DartGrid

3.1.1 Semantic Browser

Current web browsers are designed for human to browse web documents, and they only know how to interpret the HTML tags and present it as a plain text document. We proposed and developed a general-purpose browser, called the Semantic Browser [4], as the uniform user interface that enables the user to manipulating data semantics in DartGrid. The semantic browser we developed has the following characteristics. Figure 2 is a snapshot of the semantic browser.

(1) Improved navigation. User could use semantic browser to visit an ontology service and visualize the ontologies maintained in it. Using of ontologies provides the improved navigation. The user gets easy access to relevant information by browsing through the modeled concepts and their relations. An example is the navigation from a medicine to its relevant diseases.

(2) Visual Semantic Query Generation. User could visually generate a conceptual level query by interacting with the semantic browser when browsing domain ontologies.

(3) Visual Semantic Registration. Semantic browser provides the data vendor with a tool for visually doing mapping from local data semantics to shared domain ontologies.

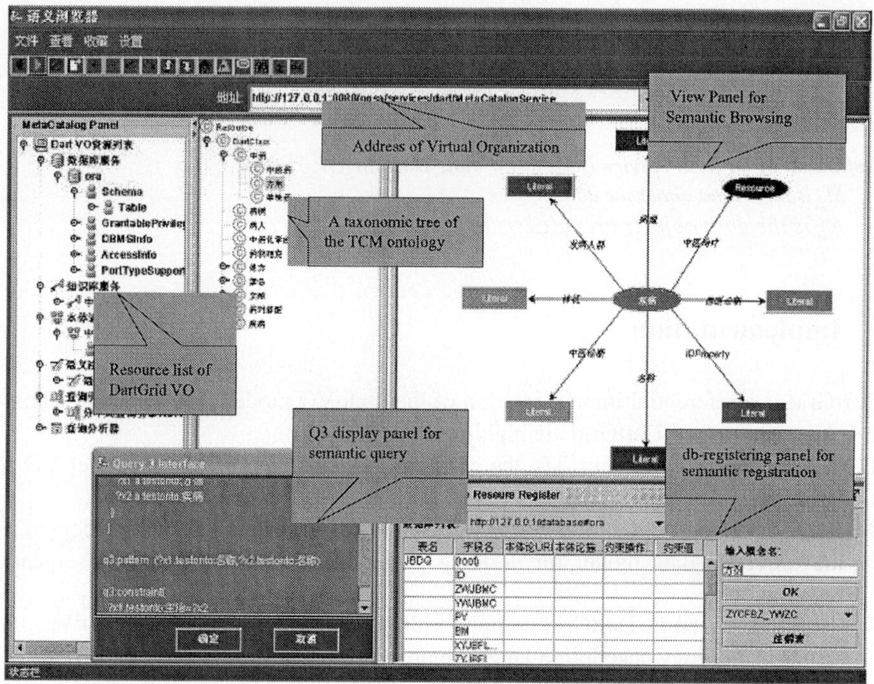

Fig. 2. A Snapshot of the Semantic Browser

3.1.2 Semantic Services

DartGrid has implemented several semantic-level services, they are:

(1) Data Semantics Service (DSS):

Data resource vendors publish the information about local data semantics by this service. For database resource, the local semantic information is right the schema information about the tables contained in that database. Others could inquire of this service about the local data semantics to fulfill some tasks such as doing semantic mapping or integration.

(2) Ontology Service (OS):

Ontologies define the standard vocabularies/ terminologies/concepts and models of the domain of a VO. Thereby, ontologies could be viewed as public-agreed global data semantics. In the new VO model we defined for DartGrid, a VO should have at least one shared ontology. If a data vendor wants to join in the VO, he/she should map his local data semantics to the ontology service, which guarantee the data sharing and integration.

(3) Semantic Registration Service (SeRS):

In current DartGrid prototype, SeRS is designed specially for database resource sharing use and distinguishes themselves by the following characteristics from other index solutions such as UDDI or Grid Index Service:

- It maintains the information about semantic mapping from local data schemas to global ontologies. Any data resource provider should register their local schema to SeRS and finish the mapping process.
- It maintains an ontology-based taxonomy and is responsible for classifying all data objects by this taxonomy. In this case, SeRS determines which data service could answer a specific query and where they locate.

3.2 Semantic Query

Informally, we specify a semantic query herein as:

(1) a query expressed at a knowledge level, namely, the query is specified by formally defined concepts and their relationships, for example, the concept of *Disease* and *Medicine*, and their relationship *curedBy*. A typical knowledge query could be expressed by LISP syntax :*(curedBy HeartDisease ?Medicine)*,Which means we want to query all medicines which could help treat heart diseases.

(2) a query whose terms are specified by uniform shared global semantics, namely, all terms used in a semantic query are defined by public-available, widely-agreed shared ontologies.

(3) a query whose result returned should also be semantically enriched.

Normally, a semantic query will be converted into some local query languages such as SQL or XQuery. Afterward, a distributed query plan will be generated and the query is dispatched to proper data service to retrieve data of interest. In DartGrid, three key components are developed to implement the semantic query, they are:

(1) Q3 language:

We devise a formal semantic query language called Query3(Q3). Q3 adopts RDF data model and N3(Notation3) syntax, and could also be used to query description-logic knowledge base.

Definition 4*: an Semantic Query is a triple SeQ=(Cxt, Pat, Cst) and*

- *Cxt is the context of the query and Cxt=NS U VB where NS is the namespace of the term used and VB is the variable biding and scoping;*
- *Pat is the concept pattern of the result returned;*
- *Cst is the constraint of the query and one constraint is a statement (S,P,O) in which S is subject, P is predicate and O is object, all of them could be bound with a variable*

Figure 3 illustrate an example of a typical Q3 query where :

- *q3:prefix* specifies the namespace *"http: //grid.zju.edu.cn/tcmonto#"* used in this query, and its corresponding QName *"tcm"*;

- *q3:variable* specifies the variables used in portions of *q3:pattern* and *q3:constraint*. For example, *?x1* stands for the concept tcm:方剂 (*tcm:CompoundMedicine*);
- *q3:pattern* specifies the concept pattern for the result. For example, the occurrence of the term ?x1.tcm:方剂名称 implies that the result should contain the name of the *tcm:CompoundMedicine*.
- *q3:constraint* specifies the query constraints that the result should satisfy.

(2) Visual Semantic Query Generator:

Although we have provided programming interface for user to write and issue a Q3 query, it is a non-trivial task to manually write a Q3query. We have developed a visual query generator, a core component of our semantic browser, to facilitate the query construction.

```
[q3:context
  [
    q3:prefix (tcm:http://grid.zju.edu.cn/tcmonto#)
    q3:variable {
      ?x1 a tcm:方剂
      ?x2 a tcm:疾病
      ?x3 a tcm:文献
    }
  ]

  q3:pattern (?x1.tcm:方剂名称, ?x1.tcm:用法用量,
              ?x2.tcm:疾病名称, ?x2.tcm:病理,
              ?x2.tcm:发病人群,?x3.tcm:文献名称)

  q3:constraint
  {
    ?x1.tcm:主治=?x2
    ?x2.tcm:各家论述=?x3
    ?x2.tcm:疾病名称 = "感冒"
  }
]
```

Fig. 3. A Q3 Example

Normally, user browses the ontologies graphically, select the concepts of interest, and specify the constraints dynamically, afterward, submit the query to a semantic query service.

(3) Semantic Query Service(SeQS):

In our implementation, SeQS plays two significant functionalities, they are:

- Receiving a Q3 query generated by Semantic Browser and then converting it into a distributed query plan in SQL-syntax;
- Wrapping the results returned from DB resources with semantics, which enables the users to browse the results semantically.

3.3   Semantic Registration

The following TCM scenario illustrates how to register a db-resource to SeRS. In this case, a TCM data provider wants to add his compound-medicine (方剂) database resources into the TCM-VO for sharing.

(1) Publish his databases as a Data Service;

(2) Visit the data service from semantic browser. This will retrieve the data schema of the databases and display it in the db-registering panel.

(3) At the same time, he opens the TCM ontology service and locates the concept 方剂 in it. Firstly, he maps the concept to the 方剂 table, and then maps the properties of the concept of 方剂 to the corresponding column name of the 方剂 table. This will construct a registration entry in XML format.

(4)Before the final submitting, the registration entry will be sent to the ontology service for semantic verification. This will verify that the concepts included in the entry are valid.

4 Related Work

In a wider technical context, the proposal presented in this paper is part of a collection of results on knowledge-based query processing in distributed information system such as SIMS[5], OBSERVER[6] , TAMBIS[7] etc.. However, the essential difference between such proposals and ours is that DartGrid enables semantic processing in such a world-wide open setting. Although the knowledge-based "global as view or model" is adopted in all of those proposals, there was no consideration with related to semantics.

In a technical context of semantic web, DartGrid is also significantly different with those semantic web query systems such as SESAME (RQL) [8], HP's Jena (RDQL) [9], or DARPA's DQL [10]. The other characteristic of an open system such as web is "dynamic"; all of the proposals aforementioned take no consideration with that issue. In our approach, the SeRS enables the data providers to join in or drop out from the VO dynamically. Particularly, SeRS maintain the current status of the providers, and if some providers are become unavailable at some time, SeRS will mark them as inactive. User is not aware of it at all and there is no need for him to care which data service could answer it or where they locate.

In a technical context of Data Grid efforts such as EU's DataGrid[11], GridPhyN project [12], GGF's DAIS working group and so on, the significant difference is the semantic-based approach adopted in DartGrid. We have not seen such approach adopted in those efforts.

5 Conclusion

DartGrid catches the aims of enabling to build ontology-based database virtual organizations in such open settings. The significance of semantic in DartGrid is reflected by the following notions:

(1) Semantics guarantees the scalability of the system. This is very important for a web-based open query system.

(2) Semantics enables that the user only need to interact with the system at a semantic level.

DartGrid has been successfully applied in data sharing for Traditional Chinese Medicine in China. In the future, more types of resources such as pictures, audios, etc. will be added into our prototype.

Acknowledgements. This work is supported in part by the project Data Grid for Traditional Chinese Medicine, subprogram of the Fundamental Technology and Research Program, China Ministry of Science and Technology, and the China 863 Research Program on Core Workflow Technologies supporting Components-library-based Coordinated Software Development under Contract 2001AA113142, and Chinese 211 core project: Network-based Intelligence and Graphics.

References

1. Tim Berners-Lee, James Hendler, Ora Lassila. The Semantic Web. Scientific American May 2001.
2. Huajun Chen, Zhaohui Wu, Chang Huang, Jiefeng Xu: TCM-Grid: Weaving a Medical Grid for Traditional Chinese Medicine. Lecture Notes in Computer Science, Volume 2659, Jan. 2003.
3. Ian Foster, Carl Kesselman, and Steven Tuecke. The Anatomy of the Grid: Enabling Scalable Virtual Organizations. Lecture Notes in Computer Science, 2001, Vol. 2150: 1-26.
4. Mao Yuxin, Wu Zhaohui, Chen Huajun: SkyEyes: A Semantic Browser For the KB-Grid. International Workshop on Grid and Cooperative Computing, 2003, Shanghai.
5. Y. Arens, C.A. Knoblock, and W-M. Shen. Query Reformulation for Dynamic Information Integration. J. Intelligent Information Systems,6(2/3):99–130, 1996.
6. E. Mena, A. Illarramendi, V. Kashyap, and A.P Sheth. OBSERVER: An approach for query processing in global information systems based on interoperation across pre-existing ontologies. Distributed and Parallel Databases, 8(2):223–271, 2000.
7. Nardi, and R. Rosati. Information integration: Conceptual modeling and reasoning N. W. Paton, R. Stevens, P. Baker, C. A. Goble, S. Bechhofer, and A. Brass. Query Processing in the TAMBIS Bioinformatics Source Integration System. In Proc. SSDBM, pages 138–147. IEEE Press,1999.
8. Gregory Karvounarakis Sofia Alexaki Michel Scholl: RQL: A Declarative Query Language for RDF*. WWW2002, May 7–11, 2002, Honolulu, Hawaii, USA.
9. Libby Miller, Andy Seaborne, Alberto Reggiori: Three Implementations of SquishQL, a Simple RDF Query Language. HP Technical Report :
    http://www .hpl.hp.com/techreports/2002/HPL-2002-10.htmll
10. Richard Fikes, Pat Hayes, Ian Horrocks : DQL – A Query Language for the Semantic Web WWW 2003, May 20-24, 2003, Budapest, Hungary.
11. Kunszt, P. (CERN, IT Division): European DataGrid project: Status and plans. Nuclear Instruments and Methods in Physics Research, Section A: Accelerators, Spectrometers, Detectors and Associated Equipment, v 502, n 2-3, Apr 21, 2003, p 376-381.
12. Ewa Deelman, Carl Kesselman et al. GriPhyN and LIGO, Building a Virtual Data Grid for Gravitational Wave Scientists. Proceedings of the 11 th IEEE International Symposium on High Performance Distributed Computing HPDC-11 2002 (HPDC'02).

A 3-tier Grid Architecture and Interactive Applications Framework for Community Grids

Oscar Ardaiz, Kana Sanjeevan, and Ramon Sanguesa

Polytecnic University of Catalunya
Campus Nord, Barcelona 08034 Spain
{oardaiz,sanji}@ac.upc.es, sanguesa@lsi.upc.es

Abstract. Grids originated within the scientific community where the benefits of utilizing an infrastructure that connects and shares resources that are geographically and organizationally dispersed was first realized. Community grids have very different characteristics and requirements to that of scientific grids – the most important being the heterogeneous resource offer and the application demands of the users. The users of this type of grid does not expect to know each and everyone of the resources available, nor does he expect to provide and install his own applications. Additionally, the applications that are usually run by community grid members have one special characteristic – they are very interactive in nature and require quick responses and a communication protocol between the user and the grid. Since the 2-tier architecture of current grids is unable to deal with these special requirements, we propose a 3-tier architecture and an interactive application framework for community grids. Our architecture and framework is able to effectively overcome the problems of application deployment, resource management and interactive execution.

1   Regional Community Grids

Many communities, besides the scientific community, can benefit from an infrastructure that connects and shares resources geographically and that are organizationally dispersed, such as the grid [6]. For example, a regional community can create a grid connecting its resources and share them for the execution of applications. These types of grids are different in many aspects compared to grids belonging to the scientific community. Differences are due to the heterogeneous resource offer, and different applications demanded by its members. Firstly, a regional community grid has very heterogeneous resources: some members might provide supercomputing resources while others provide small personal computers, some may provide their computing resources only at certain hours of the day, while others provide them any time provided their users can continue to run their local applications with no loss in the quality of service. Such heterogeneous resource offer requires an adaptive resource management that may employ 'agents' technology [8]. Secondly, regional community grid users have different application requirements to that of the scientific community. Applications are not domain or problem specific, on

M. Bubak et al. (Eds.): ICCS 2004, LNCS 3036, pp. 67–74, 2004.

the contrary many different groups use the same applications. Application providers will develop and provide such applications as a service for an economic incentive. Moreover, regional community grid members demand applications for business, education or local administration purposes. These applications have different requirements to that of scientific applications, the most important one being: interactivity. Business, educational and community applications produce results for rapid consumption by users, as compared to scientific applications whose results have to be carefully analyzed by scientists. Current grid architectures do not support application providers effectively and are not suited for interactive applications.

1.1   Application Provision and Access Requirements

Current grids have a 2-tier architecture. Such an architecture has three main problems. Firstly, each resource is configured so that each user has permission to access such resource, therefore it will not scale with the number of users. Secondly, each user must know every single resource he has access to, it is not scalable since every community member has to maintain a list of every grid resource. And thirdly, each user must provide his own applications; though one can reuse some one else's code, he must transfer, install and execute it. It is not realistic that every community member provides his own applications, because it is a task that requires complex technical skills. Such an architecture is not appropriate for a regional community grid. What is required is a grid architecture that permits resources to be configured to scale with the number of users; that permits users to access large number of resources in a transparent manner, and that permits application developers to provide applications and give access to end users in a straightforward way.

1.2   Interactive Applications Requirements

Computational grids are designed so that grid users minimize the time it takes them to obtain results from applications being executed in a grid. A distributed computational capacity-scheduling algorithm achieves such execution time reduction by selecting resources with large computational capacity. Data grids have an equal goal, but in addition have to deal with large data transfers among different computers. Transfer time must be taken into account, and included as a parameter of the scheduling mechanisms [2]. Interactive applications have other requirements. Users require an interactive application to provide results in a very short time, values that are in the order of hundreds of millisecond. Such a requirement demands interactive grid applications to be scheduled in resources that should be near to users, so as to minimize network latency. Also, scheduling resources in advance can reduce access time. Finally, an interactive protocol is required so that users can interact with grid applications.

2   Related Work

Grid portals aim at providing grid users with a simple way to access many resources. A grid portal provides a unifying interface for grid users to gain access to all of its resources from a centralized location, thereby users do not have to discover and maintain a list of every resource available. Several portals are being developed [10]. The GRASP project has proposed a grid business model for grids and application providers, though no implementation is [3]. An architecture for running interactive applications in grids has being proposed by Kumar [7], though its requirements have not been defined, and it has not been implemented.

3   Three-Tier Grid Architecture: Application Clients, Mediators, and Resources

A grid that accomplishes the three previously mentioned objectives of: scaling with number of resources, scaling with number of users and does not require each user to provide its applications must have a three-tier architecture, as show in figure 2. Such an architecture incorporates a mediator between the application clients and resources, acting as a user portal and application deployer. Such a mediator would facilitate the following: Firstly, the application user need not be aware of which resource are being used; they need not discover and monitor individual resources. Secondly, resources should not be aware of every application user - resources are not configured to give access to individual users but to the application provider.

3.1   Application Deployment

Application deployment encompasses every action required from the time an application is created till the users gain access to it [1]. It includes resource discovery, resource reservation, resource allocation, application code transfer, and application initialization. If the number of applications is low compared to the number of users it will be more efficient to deploy application once, since it will be used many times by different clients. Therefore initial set up costs due to resource discovery and allocation, code transfer, and application initialization will be shared. The mediator of a 3-tier grid has this role of interfacing with application providers to permit them to deploy new applications.

3.2   Users Portal

Users access an application through a portal. A mediator portal controls which users are allowed to use an application. It also maintains accounting information so that clients are charged for usage of applications and resources. Application providers and resource providers each receive a share of this income.

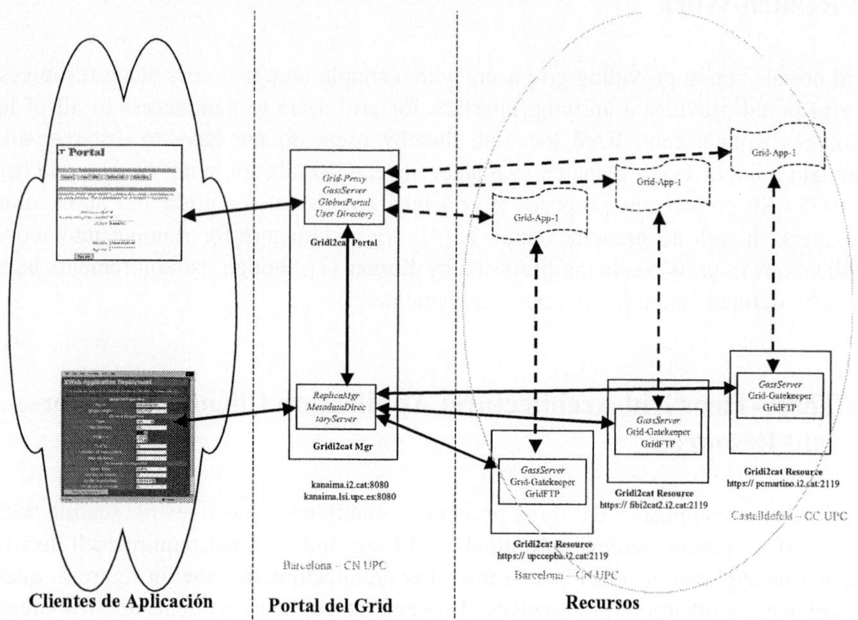

Fig. 1. 3-tier grid architecture: applications providers deploy applications once; multiple clients access the application through the portal many times

4   Interactive Applications on Grids

A grid infrastructure that provides interactive applications must meet a number of requirements: First, resources should be assigned physically close to the user so as to minimize network latency. Second, resource scheduling and application installation and start up time should be decreased as much as possible, it should preferably be made in advance. Also an interactive protocol is required so that users can interact with applications.

4.1   Proximity Based Scheduling and Advance Deployment

A grid that provides sinteractive applications must schedule resources which provide enough computational capacity, storage capacity and which are nearer to users. We have studied and simulated resource allocation mechanisms that take into account distance between resource and clients in a previous work. Our algorithms are also based in an economic model, which is being proposed as a basic model for resource allocation in grids. Results show that proximity based scheduling is feasible [4].

Advance deployment is an attractive option so that interactive applications do not suffer an initial delay due to application transfer and start up. However advance reservation and allocation mechanisms have an important drawback due to non-utilization of some allocations. But if the ratio between the number of application users and number of applications is high, then due to statistical multiplexing such cost will be decreased.

4.2 Interactive Protocols

Finally the last and most important requirement for a grid that supports interactive applications is an interactive protocol between grid nodes and application clients. The grid model assumes that applications employ minimal clients as possible, i.e. computational and data grids can operate with a simple asynchronous email client. Interactive protocols however require more complex clients, though there are different levels of complexity that come into play. The least demanding client-side interactive protocols are Virtual Network Computer VNC [9] and the streams protocol. VNC is designed to provide remote graphics visualization and remote control with the least demanding client. Streams provide a reliable transport mechanism for transfer of simple data.

5 Implementation

GridCat is the research prototype of the regional community grid of Catalonia. It consists of a testbed implementing the 3-tier grid architecture, and an interactive application framework that provides basic building blocks for gridifications of interactive applications. GridCat is a regional community grid, based on a local community, Catalonia. Its resources are being provided by all kinds of institutions of the Catalan society: universities, local companies, local administrations and civic associations.

5.1 GridCat Testbed

An intermediary tier, implemented as a mediator, facilitates a grid where application providers can provide applications and application users access such applications. The implementation of such a 3-tier testbed is based on the Globus Toolkit v2 implementation. It has been extended to facilitate application providers to deploy applications, and users to access applications transparently with a mediator module that functions as a user portal and application deployer.

The Globus Resource Manager GRAM has been installed at each resources node. Resources have being configured so that certain applications are permitted for execution. To provide a high level of security, X.509 certificates and GSI Globus Security Infrastructure are being used. Only grid execution request from authorized application providers are be permitted. Application providers delegate to the mediator such certificates at application deployment time.

Several applications have being "gridified": where the functionality of an application has been divided so that it can be executed in several grid nodes. For each application a certificate has been created. Resource nodes have been configured to accept deployment and execution requests from different applications identified by a Distinguished Name DN. For example `"O=Prous, CN=MoleculeSearcherApp"`.

Mediators implement a web based portal interface so that application users access and execute applications. Application users authenticate to portals with a username and password or with a proxy certificate provided by a proxy certificates server implementation of the Globus Portal Development Kit GPDK. Intermediates check which applications each user is allowed to use, and calls a resource manager to select which resources should be assigned to such a user.

Application providers request the mediators to deploy applications for usage by a estimated number of users. Mediators deploy applications in a number of resource nodes using GSI-FTP and GRAM Globus services. Intermediates maintain a table of nodes where an application has been deployed.

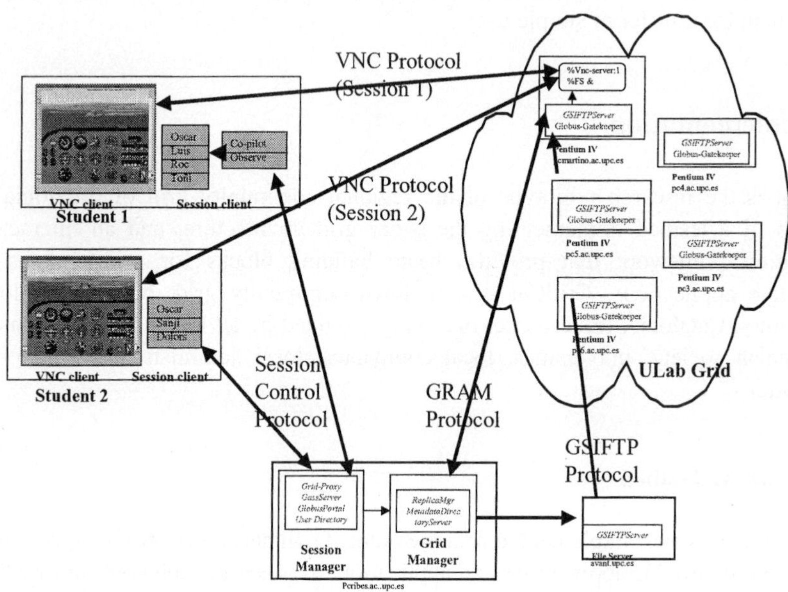

Fig. 2. Grid interactive applications framework: users access a flight simulator session that is being executed in grid nodes through VNC sessions

5.2   Grid Interactive Application Framework

We have selected two different interactive applications that are computationally intensive and that could take advantage of a grid. One of the applications is a flight simulator that has been "gridified". Flight simulators have high computational resource requirements because three dimensional graphics rendering is computationally intensive. The simulator used is FS [5], which is an open source flight simulator that renders graphics using the OpenGL library. The other application is a molecule finder and visualization application that has also been "gridified". Such applications search for molecules containing certain submolecules selected by users in a database, and present the results in a three dimensional view.

We have implemented a framework that provides all basic functionality for the "gridification" of future interactive applications. Such a framework consists of 3 modules: a VNC client and server, a stream client and server, and a connector that coordinates every interactive session of each application session.

A flight simulator session requires the grid mediator to bind a flight engine and a rendering engine to a VNC server, the VNC client is then connected to this VNC server. A molecules search session requires the grid mediator to bind the molecular databases and molecule search engines to a stream session, at the other end of the stream session there is a molecule rendering engine. Such an engine is bound to a VNC server.

6   Conclusions and Future Work

This work presents three contributions, first, the community grid, a distributed computational system built from very heterogeneous resources provided by a regional community, and with very specific application demand requirements. Second, the 3-tier grid architecture, that makes an efficient use of resources by employing a strategy of deploy once and run multiple times an application. And the third contribution is the framework for interactive applications running on grids.

We are currently taking measurements to evaluate the effectiveness of this approach, and also we plan to adapt this framework to Globus Toolkit version 3.

References

1.   O. Ardaiz, L. Navarro "Xweb a Framework for Application Network Deployment in a Programmable Internet Service Infrastructure", 12-th Euromicro Conference on Parallel, Distributed and Network based Processing. Febrery 2004, A coruña Spain.
2.   W. H. Bell, D. G. Cameron, R. Carvajal-Schiaffino, A. Paul Millar, K. Stockinger, and F. Zini. "Evaluation of an Economy-Based File Replication Strategy for a Data Grid" In Intl. Workshop on Agent based Cluster and Grid Computing at CCGrid 2003, Tokyo, Japan, May 2003.
3.   Theo Dimitrakos, Matteo Gaeta, Pierluigi Ritrovato,Bassem Serhan, Stefan Wesner, Konrad Wulf "Grid Based Application Service Provision", Euroweb 2002, Oxford, UK, Dec 2002

4.   T. Eymann, M. Reinicke, O. Ardaiz, P. Artigas, F. Freitag, L. Navarro, "Decentralized Resource Allocation in Application Layer Networks", In Intl. Workshop on Agent based Cluster and Grid Computing at IEEE 4CCGrid'2003, May, 12th-15th, 2003. Tokyo, Japan.

5.   "Flightgear Simulator", www.flightgear.org , 2003.

6.   I. Foster, C. Kesselman, S. Tuecke."The Anatomy of the Grid: Enabling Scalable Virtual Organizations" International J. Supercomputer Applications, 15(3), 2001.

7.   Raj Kumar, Vanish Talwar, Sujoy Basu: "A Resource Management Framework For Interactive Grids". Middleware Workshops 2003: 238-244.

8.   Luc Moreau. "Agents for the Grid: A Comparison with Web Services (Part 1: the transport layer)". Second IEEE/ACM International Symposium on Cluster Computing and the Grid (CCGRID 2002), pages 220-228, Berlin, Germany, May 2002. IEEE Computer Society.

9.   Richardson T., Staord-Fraser T., Wood K.R., Hopper A. "Virtual network computing". IEEE Internet Computing, 2(1):33-38, January-February 1998.

10.   M. Thomas, M. Dahan, K. Mueller, S. Mock, C. Mills. "Application Portals: Practice and Experience. Grid Computing environments": Special Issue of Concurrency and Computation: Practice and Experience. Winter 2001.

Incorporation of Middleware and Grid Technologies to Enhance Usability in Computational Chemistry Applications

Jerry P. Greenberg, Steve Mock, Mason Katz,Greg Bruno, Frederico Sacerdoti, Phil Papadopoulos, and Kim K. Baldridge

San Diego Supercomputer Center (SDSC)
University of California, San Diego (UCSD)
9500 Gilman Drive, Mail Code 0505, La Jolla, CA 92093-0505, USA
{jpg,mock,mjk,bruno,fds,phil,kimb}@sdsc.edu

Abstract. High performance computing, storage, visualization, and database infrastructures are increasing in complexity as research moves towards grid-based computing, often pushing breakthrough computational capabilities beyond the reach of scientists due to the time needed to harness the infrastructure. Hiding the underlying complexity of networked resources becomes essential if scientists are to utilize these resources in a time-effective manner. There are a myriad of solutions that have been proposed, ranging from underlying grid glue, to fully integrated problem solving environments. In this work, we discuss a workflow management system that is fully integrated with emerging grid standards but can be dynamically reconfigured. Through defined XML schema to describe both resources and application codes and interfaces, careful implementation with emerging grid standards and user-friendly interfaces, a "pluggable" event-driven model is created where grid-enabled services can be composed to form more elaborate pipelines of information processing, simulation, and visual analysis.

1   Introduction

High Performance Computing (HPC) has dramatically changed scientific research, and enabled advancements to be made at a rate far beyond that conceived a decade ago. HPC brings together the wealth of technology advancements not only in terms of hardware and raw computational speed, but also database technology, visualization infrastructure, algorithms, both efficient for the latest hardware as well as integrated for enhanced capability, networking and remote access through new portal web service technologies. The infrastructure complexity to make these logical components work efficiently together often overwhelms domain scientists. The ultimate goal is to extend and expand efforts in interface simplification in order to build easily reconfigurable web-based workflows that span a variety of technologies and enable complex science to be performed on the grid. Such endeavors uniquely integrate expertise in high-end integration and scientific domain experts to produce an infrastructure that meets the needs of the underlying science driver.

M. Bubak et al. (Eds.): ICCS 2004, LNCS 3036, pp. 75–82, 2004.

As scientific algorithms evolve and progress, the number and diversity of computational and grid resources options quickly multiplies. In terms of computational hardware, users may choose from a broad spectrum of possibilities from special purpose, loosely coupled platforms such as a network of PC's spread across an unknown local or remote area, to grids and cluster infrastructures, to highly coupled, highly parallel architectures. In reality, scientific users do not want to differentiate among the vast number of possibilities. Although advances in the capabilities of high-performance computers have made it possible for computational scientists and engineers to tackle increasingly challenging problems, at the same time, it has become considerably more difficult to build, manage, and integrate the software that can achieve the highest performance, use resulting data most efficiently, and/or make it production quality for the community at large. The rate-limiting step in pushing forward advanced research is often the user interface and underlying protocols required to connect to services such as hardware, database, visualization, and software.

2   Motivating Application

The most common way to process and analyze data obtained from computational software is via "cut and paste" from output files. Such a process is tedious and in the case that a whole class of analogous calculations is required, the effort may adversely affect the project. An alternative way is via processing data from sequential programs with command language scripts, but here as well, there are several obstacles. For example, a change in the output format of the software program may result in a failure of the script to process the output. Instead, it is more efficient to put structured data output facilities directly into the computational program (e.g., in this work, GAMESS [1]) as well as into any associated output analysis programs (e.g., in this work PLTORB). As well, similar incorporation into the output of other computational codes can be done, potentially opening up opportunities for hybrid integration. In this work, we give an example of such an integration using GAMESS and APBS [2], two molecular based software modeling tools. The result, though more laborious initially then parsing output files in the traditional way, is that dependency on data parsing is removed and the data put into a form that may be used for database storage, querying, and efficient transport over the grid in the form of JAVA objects.

We have begun this process by designing an XML schema specifically for our computational chemistry applications. Initially, the schema provided a template for storing only basic data, such as atomic coordinates, atom types, energies, molecular orbital coefficients and basic input options. We have now also added gradients, Hessian elements, and volumetric grids associated with properties of molecules.

The current schema may be viewed at http://www.sdsc.edu/~jpg/nmi/gamess.xsd. The production of XML data documents based on this schema was accomplished by putting into GAMESS calls to a library of C functions that produce a JAVA object. We do this by using the "Castor" SourceGenerator [3] to create JAVA source that maps XML elements to JAVA classes. By linking GAMESS to the XML/C library together with the JNI (Java Native Interface)[4] library which allows one to call JAVA methods from C, the data may be stored as a JAVA object and may be marshaled into an ascii XML file as well.

As an example, consider a call within GAMESS to send the coordinates and name of one atom to the GAMESS JAVA object:

```
call output_coord(ANAM(IAT),BNAM(IAT),ILEN,ZNUC,X,Y,Z)
```

The code which prints out an ascii XML document is as follows (see the schema):

```
fprintf("fp,"<ATOM_POSITION>\n");
fprintf(fp,"    <XCOORD>%s</XCOORD>\n",fixupdouble(*x,"%lf",string));
fprintf(fp,"    <YCOORD>%s</YCOORD>\n",fixupdouble(*y,"%lf",string));
fprintf(fp,"    <ZCOORD>%s</ZCOORD>\n",fixupdouble(*z,"%lf",string));
fprintf(fp,"  </ATOM_POSITION>\n");
```

The equivalent for writing to the JAVA object is:

```
callmethodn("g_add_atom ");
callmethodd("g_set_xcoord",x);
callmethodd("g_set_ycoord",y);
callmethodd("g_set_zcoord",z);
```

That is, the "g_add_atom" method instantiates a new "ATOM" and "ATOM_POSITION" object, and the subsequent calls add the data for this atom to the object.

The "g_add_atom" method is given below:

```
public static void g_add_atom() {
  atom = new ATOM();
  atom_position = new ATOM_POSITION();
  system_state.addATOM(atom);
  atom.setATOM_POSITION(atom_position);
}
```

Now we not only have a way to store our data in a rational fashion, but also a method for delivering it to other programs. Consider the GAMESS auxiliary program PLTORB3D, which uses the calculated wavefunction to create a 3D orthogonal grid of molecular orbital values. In the original implementation, separate files for the atomic basis set and input options, as well as molecular orbital coefficients had to be "cut out" of one of the GAMESS output files. Then, PLTORB3D produced a file which the visualization program QMView [5, 6] could read in order to display contours and isosurfaces. With modifications to PLTORB3D, an XML file is "unmarshalled", new data is added to the JAVA object and a new XML document is "marshaled" that contains all the original data plus the volume data. This file (or JAVA object) may be stored and retrieved in order to set up a new GAMESS run, or as part of a collection queried from a database.

3   Middleware Methodology

The work described above is rather specific to computational chemistry and in particular, to GAMESS related calculations. Below, we will describe our efforts to develop a general system for facilitating scientific workflows over the grid. For example, in the above discussion of possible GAMESS related workflows, we left out any details of submitting individual GAMESS jobs or PLTORB3D jobs to compute platforms or how jobs are submitted in succession. What is lacking is software that integrates the individual programs, submits jobs to particular platforms, monitors the whole process and stores results.

Previous efforts to facilitate the submission of jobs to remote platforms included the SDSC portals which are based on the NPACI Gridport [7] software.   The portals provides low level tools to GLOBUS[8] for secure access to remote platforms and to the SRB (Storage Resource Broker)[9] for storing data collections. Through the web sites built with Gridport, users were shielded from the complexity of the remote platforms used for computations and archival storage. However, building portals is nontrivial and involves the intricacies associated with writing html files and cgi scripts. Additionally, once built, portals are not easily reconfigurable.

To address these problems, we are creating a general Scientific Workflow as part of the National MiddleWare Initiative [10]. The Workflow project will facilitate the building of scientific workflows from smaller tasks using web and grid services. The project is designed to shield both users and application developers from the intricacies of the grid. Resulting infrastructure will provide "hooks" to connect to user interfaces thus exposing that interface to a variety of other user interfaces.

The workflows are defined by XML documents and the workflow is divided into layers that separate users and application developers from the underlying services. The whole process is managed by a "Workflow Engine" whereby the XML documents defining a job are instantiated as JAVA objects, resources are found, the jobs are executed, and their status is sent to the user (Figure 1).

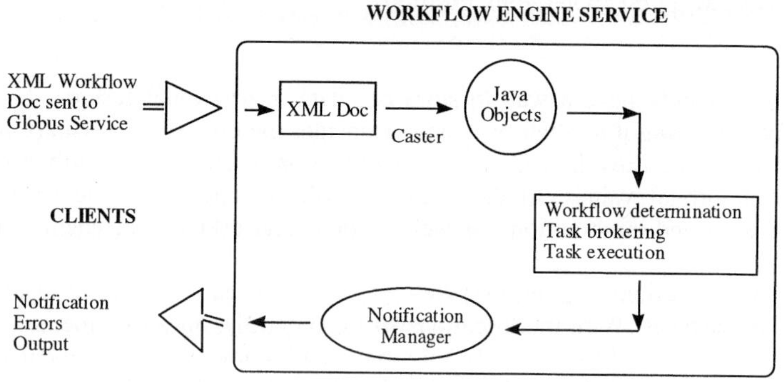

Fig. 1. Flowchart of a Scientific Workflow

The Workflow Engine service is not dependent on any particular application, data file format, or operating system. A user may substitute their own user interface as long they adhere to the protocols of the Workflow Engine.

4   Building Application Grids

4.1   Cluster Infrastructure and Maintenance

For decades, complex and expensive supercomputers were the only resources available for computational chemistry computations on complex molecular systems that involve large numbers of atoms, large basis sets, and/or higher order methods. Besides considerable investment in large-scale hardware itself, such platforms also require an inordinate amount of peripheral infrastructure, such as cooling infrastructure and system administration. In recent years, "commodity" clusters running Linux have become quite popular because they are often affordable by individual research groups, permit scalability to larger numbers of compute nodes, demand little in terms of special equipment or a surrounding facility, and in principle can be administered by the researchers themselves.

In practice however, cluster administration can be complex and time-consuming. There are n-fold copies of the operating system to update and maintain. If the software and the operating system is not maintained, the system may become unstable, endangering the completion of long calculations, security holes may not be patched, and software may not be updated.

The key to rapidly deploying cluster infrastructure is to automate the process of building and managing a cluster, which is itself a single grid end point. In our case the automation of building the grid-enabled clustered endpoint is achieved using the NPACI ROCKS cluster distribution. ROCKS is a cluster-aware Linux distribution that makes it possible to deploy a world class supercomputer in a matter of hours, something that has historically taken much longer.

The philosophy of ROCKS is to make the installation of the operating system the basic management tool. That is, it is easier, when automated, to reinstall all nodes to a known configuration then to determine which nodes are not synchronized. This is the opposite of the maintenance procedures on desktop systems where the operating system is rarely, if ever, reinstalled, or in configuration management tools such as CfEngine[11] that perform maintenance on exiting operating systems installations.

4.2   Grid Computing and Middleware

The ROCKS release also contains essential software elements of grid computing. From the end-user perspective, submitting a workflow to the grid involves interaction solely with the web portal interface, a user interface to the middlware described above, or a grid scheduler such as NIMROD[12]. That is, the underlying grid components are hidden from view for day-to-day application operations. However, this underlying system involves substantial complexity and is the focus of several grid efforts today. Figure 2 shows the high level architecture of an application grid. Once a resource is selected, GLOBUS is used to start the jobs on the end point. In addition to providing the connections from a web portal to the grid end points, the grid scheduler continuously monitors all the end points for status information such as CPU utilization, free disk space, and operation system version. This information is gathered us-

ing the GLOBUS Monitoring and Discovery Service (MDS) and allows the scheduler
to make reasonable scheduling decisions.

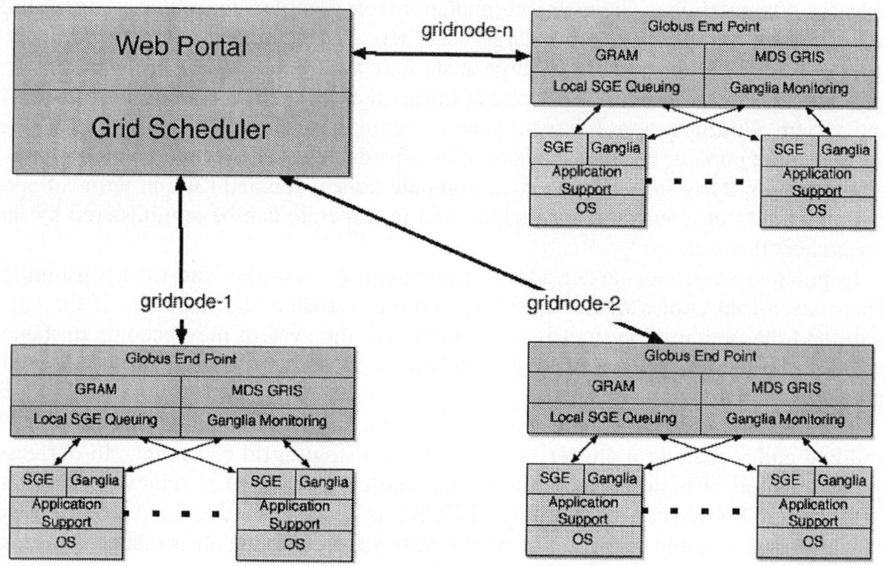

Fig. 2. Application Grid Architecture

Within the last year ROCKS has matured from a cluster tool to a grid tool, and
now includes the GNSF Middleware Initiative's (NMI) Globus and Certificate
Authority software. Because it takes only hours to build a grid-enabled cluster the
only remaining step to deploying a grid is the standard Globus certificate exchange
between all end points, and the grid scheduler – portal setup.

Figure 3 illustrates the basic system architecture of a ROCKS Linux cluster. The
grid scheduler is responsible for submitting jobs to the local Globus Resource Application Manager (GRAM), which next submits the job to the local cluster wide system[13] which contains a current snapshot of the "state" of the cluster. This Ganglia
information (memory availability, cluster size, cpu speed etc.) can then be used to
feed the local Grid Resource Information Service (GRIS) information about the state
and configuration of the cluster. This information is then given to the grid scheduler
(Figure 2) to aid in job scheduling decisions.

GAMESS is particularly well suited for running on Linux clusters and will be included in future releases of ROCKS. It does not require any type of parallel software,
is built with open source compilers, and scales reasonably well with conventional
network interconnects. The rapid building of a cluster, deployment of the operating
system and software via ROCKS, and the subsequent running of large scale calculations using GAMESS was demonstrated at Super Computing 2003 [14].

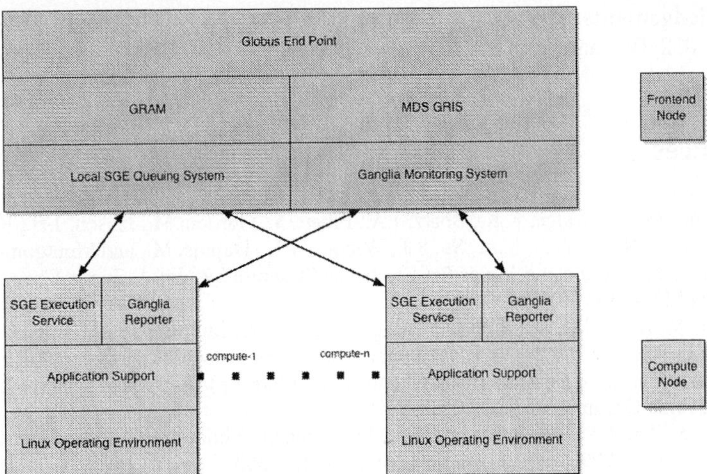

Fig. 3. Grid Endpoint Architecture

5   Conclusions

The successes of highly efficient, composite software, for molecular structure and dynamics prediction has driven the proliferation of computational tools and the development of first-generation computational chemistry grid-enabled infrastructure. The approach that we are taking, as illustrated in this work, is that of a layered architecture. The layers shield the developer from the complexity of the underlying system and provide convenient mechanisms of abstraction of functionality, interface, hardware, and software. A layered approach also helps with the logical design of a complex system by grouping together the related components while separating them into manageable parts with interfaces to join the layers together.

With such tools, researchers can begin to ask more complex questions in a variety of contexts over a range of scales, using seamless transparent computing access. As more and more realistic time simulations are enabled extending well into the nanosecond and even microsecond range at a faster turnaround time, and as problems that simply could not fit within the physical constraints of earlier generations of supercomputers become feasible the ability to integrate methodologies becomes more critical. Assembly and modeling can often utilize different computational approaches, and are best optimized for use on different computer architectures. Thus, considerations of porting and optimizing the problem-specific application for both the high-end supercomputers, and a commodity cluster of computers often factor in.

The described technology will help to tie computation with investigator intuition regardless of location, to facilitate scientific investigations by exploiting novel grid capabilities and teraflop hardware speeds, enabling direct user input and feedback. It is anticipated that such infrastructure will impact scientists that potentially need such tools for interdisciplinary research. This will in turn foster development of new modeling, data, computational technologies.

Acknowledgements. We acknowledge support from the NSF through DBI-0078296 and ANI-0223043 and from the NIH through NBCR-RR08605.

References

1. Schmidt, M., Baldridge, K.K., Boatz, J.A., Elbert, S., Gordon, M., Jenson, J.H., Koeski, S., Matsunaga, N., Nguyen, K.A., Su, S.J., Windus, T.L., Dupuis, M., and Montgomery, J.A.: The General Atomic and Molecular Electronic Structure System. J. Comp. Chem. 14 (1993) 1347-1363
2. Baker, N., Holst, M., and Wang, F.: Adaptive multilevel finite element solution of the Poisson-Boltzmann equation II. Refinement at solvent-accessible surfaces in biomolecular systems. Journal of Computational Chemistry 21 (2000) 1343 - 1352
3. The Exolab Group: Castor (2002)
4. Liang, S., The JavaTM Native Interface: Programmer's Guide and Specification. 1 ed. The JAVA series. 1999: Addison Wesley Longman, Inc. 303.
5. Baldridge, K.K. and Greenberg, J.P.: QMView: A Computational 3D Visualization Tool at the Interface Between Molecules and Man. J. Mol. Graphics 13 (1995) 63-666
6. Baldridge, K.K. and Greenberg, J.P.: QMView as a SupramolecularVisualization Tool In: J. Siegel (ed.): Supramolecular Chemistry Kluwer Academic Publishers, Dordrecht Norwell New York London. (1995) 169-177.
7. Thomas, M., Mock, S., Dahan, M., Mueller, K., Sutton, D., and Boisseau, J.R. The Gridport Toolkit: a System for Building Grid Portals. in 10th IEEE International Symp. on High Perf. Comp. San Francisco (2001).
8. Foster, I. and Kesselman C.: Globus: A Metacomputing Infrastructure Toolkit. Intl. J. Supercomputing Applications 11 (1997) 115-128
9. Rajasekar, A.K. and Wan, M. SRB and SRBRack- Components if a Virtual Data Grid Architecture. in Advanced Simulation Technologies Conference. San Diego CA (2002).
10. Papadopoulos, P., Baldridge, K., and Greenberg, J., Integrating Computational Science and Grid Workflow Management Systems To Create a General Scientific Web Service Environment. NSF award (2002)
11. Burgess, M.: Cfengine: a site configurable engine. USENIX Computing Systems 8 (1995)
12. Abramson, D., Lewis, A., and Peachy, T. Nimrod/O: A Tool for Automatic Design Optimization. in The 4th International Conference on Algorithms & Architectures for Parallel Processing. Hong Kong (2000).
13. Sacerdoti, F.D., Katz, M.J., Massie, M.L., and Culler, D.E. Wide Area Cluster Monitoring with Ganglia. in Proceedings of the IEEE Cluster 2003 Conference. Hong Kong (2003).
14. Gannis, M. and Lund, G.: SDSC/NPACI Rocks Team and Sun Create Supercomputer, Run Scientific Applications in Less than Two Hours. http://www.sdsc.edu/Press/03/112403_NPACIRocks.html (2003)

An Open Grid Service Environment for Large-Scale Computational Finance Modeling Systems

Clemens Wiesinger, David Giczi, and Ronald Hochreiter

Department of Statistics and Decision Support Systems, University of Vienna

Abstract. In this paper we present the basic concepts of our complex problem modeling and solving environment based on a state of the art component architecture. We propose a system where components exist as instances of meta-components carrying relevant semantic information about the application problem realm. The implementation of the system follows the Open Grid Service Environment (OGSE) Service Stack, also discussed in this paper. A motivating workflow example from the field of computational finance is given.

1   Introduction

In the last decade the structure of applications changed from large monolithic pieces of code with some internal structuring to workflow applications, see [10]. Recently, Grid and Web Service based applications emerged, which provide the basis for the adaptation of our AURORA Financial Management System (see [13] for a discussion about prior implementations). In general, this system is a complex problem modeling and solving tool for large-scale financial decision models. Many efforts have been undertaken to bridge gaps between computer science and computational management science (for operational research approaches see e.g. [12][4], for distributed computing approaches see [6]). However, most of the available solutions for tackling problems in this area focus either entirely on low-level specialized problem formulations or on special optimization problem solutions. There is practically no abstract layer that provides a common framework in which components are interchangeable due to clear interface definitions and service descriptions.

In this paper we outline the nature of large-scale financial problems in general in section 2 and give an example for an typical problem in this area. Furthermore, we use this example to show how to apply a Grid environment to enhance the performance by exploiting intra-component and workflow parallelisms. The development of the Open Grid Service Environment architecture motivated by general considerations of component-based architectures for problem solving environments is discussed in section 3.

M. Bubak et al. (Eds.): ICCS 2004, LNCS 3036, pp. 83–90, 2004.

2 Motivation for a Financial Problem Modeling and Solving Environment

In comparison to exact sciences like pure mathematics, decision science usually deals with incomplete information along with subjective models and even more subjective interpretation of solutions. Innovative models and solutions often exist but are spread throughout the scientific community. Therefore, the need for a common workflow platform for interchangeable components to reevaluate results and to extend large financial workflows in certain areas arises. Hence the proposed system should be capable of integrating and orchestrating different components for the realization of larger tasks, where maybe only a small part of the whole workflow is of interest to a specific researcher. In this manner components implemented by other people with different research focuses become useable and comparable.

It seems obvious, that such a flexible system needs higher level semantic descriptions of specific components. This is attained by defining meta-components and orchestrating these into a meta-workflow which gives a compound semantic description of what needs to be done and defines the steps to achieve this goal. Each concrete component is an implementation of a meta-component. The interchangeability of components derived from a meta-component arises from their common definition of their input and output structure.

Furthermore, it is important to emphasize that computational finance is an area where problems can be made arbitrarily complex in a computational sense. Like models from meteorology, chemistry, physics and material sciences models can be configured to consume all available computational power by making them finer and by that more realistic.

2.1 An Example from Computational Finance

As a prominent example for a computational heavy task, with possibly many subtasks, we present a specific multi-stage stochastic modeling and optimization problem, meaning that during the considered time period multiple consecutive decisions and possible recourse actions are modeled and subject to optimization. The goal of the optimization is to minimize the subjective risk of a financial portfolio, which is calculated from the return distribution, derived from historical development of the considered assets. The processing of this task can be broken into subtasks as described below. Each subtask exists as abstract description of the task to be performed and interchangeable actual implementations with the same input and output structure.

Below we present the meta-components that are involved in the workflow and also illustrate an example for an executable workflow formed from concrete implementations of the used meta-components. In contrast to the meta-workflow the concrete workflow is ready to be processed by workflow enactment mechanisms.

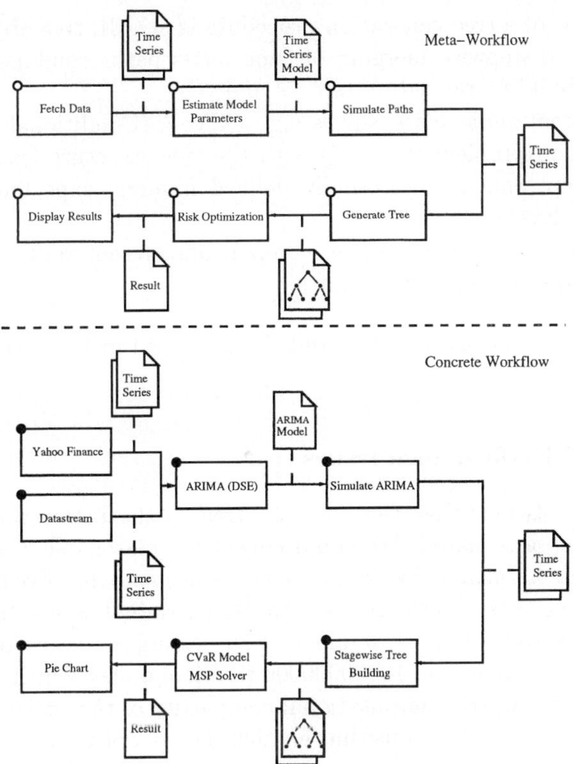

Fig. 1. OGSE workflow - multi-stage tree generation and portfolio risk minimization. Above the line the abstract meta-component workflow is described, while in the lower part the corresponding instanciation is shown. Every concrete implementation shown in the lower part has exactly one corresponding meta-description in the upper part. A rectangle with a circle in the upper left corner represents a meta-component, while a rectangle with a filled circle represents an implementation.

1. *Data fetching and converting components*: collect econometric (time series) data and convert to a suitable input format for consecutive steps. Differences between actual implementations arise from differences in data sources and data formats. All series are stored in the time series XML structure.
2. *Estimation components*: to capture inter-period dependencies in our data we use time series models to estimate the properties of our stochastic process. These components fit the data handed over from the previous step to a time series model and store the estimated parameters in a suitable format.
3. *Simulation components*: these components simulate a pre-defined number of trajectories according to the parameters of the model which are the output of the estimation step.
4. *Tree generation components*: in our example a tree is build out of the simulated trajectories. A tree can be viewed as a multi-dimensional filtration.

The output of a tree generation procedure is a XML tree object. We choose a method of stepwise merging of time series paths combined with a stage wise tree building procedure (see [5] and [7]).

5. *Optimization components*: solves e.g. a CVaR (Conditional Value at Risk) minimization problem (see [15]) with the tree structure from the tree generation step, and uses externally defined minimal expected return μ and confidence level α.

6. *Presentation components*: these components present results graphically in form of reports, tables, and charts.

The meta-component workflow and the described instanciation of the workflow are depicted in Fig. 1.

2.2 Parallel Performance Issues

It seems quite obvious that the finer the tree is, the better the true stochastic process is approximated. We could consider a reasonable realistic tree that models the time horizon as five stages where every node has five successors. The resulting tree consists of 3906 nodes with 3125 terminal nodes. If we double the amount of time steps and the number of successors in every node, we end up with 11111111111 nodes and 10000000000 terminal nodes, which accounts for an enormous increase in the computational complexity of the problem. However, it was shown in many publications that for this class of optimization problems parallel implementations can achieve a nearly linear speedup. Reported efficiency is usually larger than 90%, see [2] for a general overview of parallel optimization and [1] for multi-stage stochastic optimization, which is the common approach to solving large scale financial management problems. A first approach towards implementing multi-stage stochastic solvers on the Grid was successfully depicted in [11]. There is also scope for parallelization of other components besides the optimization itself, especially the tree generation methods, which are often computationally demanding, see e.g. [8] for a parallel clustering algorithm with super-linear growth, which can be used as the basis of many tree generation techniques.

Furthermore, it is possible to exploit not only intra-component parallelisms, but also concurrent work on different tasks of a workflow. The OGSE architecture is meant to provide the infrastructure for handling different types of parallelism in one common framework.

3   OGSE Architecture

3.1   OGSE Components

Figure 2 summarizes the ideas of [14] and [9] mapped to a component-based architecture which defines the ten main (software) building blocks of an open PSE.

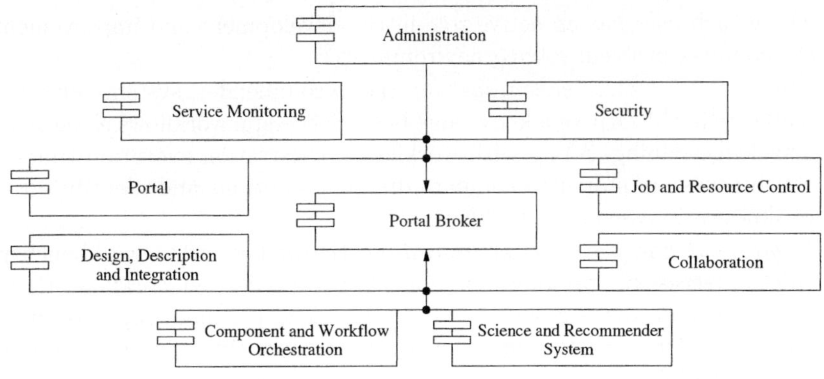

Fig. 2. OGSE Components for complex problem solving.

- *Portal:* portal contains user-centric presentations of the portal services. Groups of users have different roles such as scientific users, developers, and administrators with different views on the Grid-enabled problem solving environment.
- *Service monitoring:* the monitoring component provides facilities to compose monitoring operations, gather survey information about service activities and states, keep track of workflow execution, and include debugging and error recovery.
- *Administration:* these services enable members of privileged groups to perform maintenance and configuration of the problem solving infrastructure. Typical tasks are user management, service control, and service update.
- *Design, description and integration:* the system requires the usage and development of XML-based standards and specifications. Existing service standards, mainly lead-managed by the World Wide Web Consortium (W3C), Globus, and the Global Grid Forum (GGF), are exploited to provide a flexible (plug-in alike) service architecture for component integration (through semantic descriptions and rules).
- *Security:* the security component covers authentication, authorization, and confidentiality. Typical features are single sign-on, role-based user access, and data signing and encryption in the Grid environment.
- *Job and resource control:* every workflow consists of different jobs which have to be scheduled and are later submitted to several resources in the Grid system. According to the workflow orchestration, jobs have certain dependencies which have to be taken into consideration, when the workflow enactment is handled. Furthermore, resource monitoring and user role restrict the resource allocation and usage.
- *Collaboration:* is supported by user forums, Frequently Asked Questions (FAQs) and news bulletins, where relevant Grid and complex problem solving issues are discussed, common problems are listed and updates are announced. Collaboration services aim at creating a responsive scientific community,

where each user has an active role in the development and improvement of
the complex problem solving environment.

- *Science and recommender system:* the recommender system advises researchers in the form of a knowledge base. Successful workflows are stored for
  knowledge mining. Along with workflow orchestration execution times and
  benchmarks are stored to compare different solutions and identify possible
  weaknesses in assembled workflows.
- *Component and workflow orchestration:* the orchestration workbench provides all accessible workflow components and predefined workflows that are
  stored in repositories along the Grid sites. A visual modeling desktop supports the user in discovering of workflow components and assembling of a
  specific workflow.
- *Portal broker:* the portal broker is the missing link in the complex problem
  solving architecture which integrates and connects the above named architectural components. The broker handles all messages between services and
  provides events to the user portal.

The above enumeration lists most of the issues considered important in the
development of a PSE on the Grid. We understand that in a complete financial
management system all of these mentioned points are nearly equally important
and must be properly treated. As our research is mostly influenced by the financial application side we currently focus on the workflow orchestration, the
building of the portal, the portal broker, and the definition of appropriate XML-
based structures for service, problem, model, and data description. Other issues
will be treated superficially in the next phase and extended in a later stage of
the research project. If industrial partners start to use the system for consulting
purposes other issues especially security and accounting gain importance.

3.2   OGSE Service Stack

The component collection introduced in section 3.1 provides a complex compound of highly distributed architectural services over an open PSE. To establish
the distributed architecture, we propose an integrated service stack, the OGSE
Service Stack (see Fig. 3), based on well defined and commonly used standards.
The core building block of the OGSE Service Stack is the W3C Web Service
Stack [3] which covers XML specification work in the workflow orchestration
and enactment (workflow processing and monitoring services), service discovery,
service description, and messaging. Semantics, also elementary covered by a W3C
initiative, enhance the service core with PSE-specific descriptions in the area of
problems, models, and (I/O) data. This vocabulary provides the foundation for
an advanced arrangement of entities to an associative net, where constraints
and relations are modelled with an assertional language into an ontology. The
matching between a request for service through description data and the actual
semantic data is provided by the matchmaking service. The OGSE API allows
software developers to provide their own services.

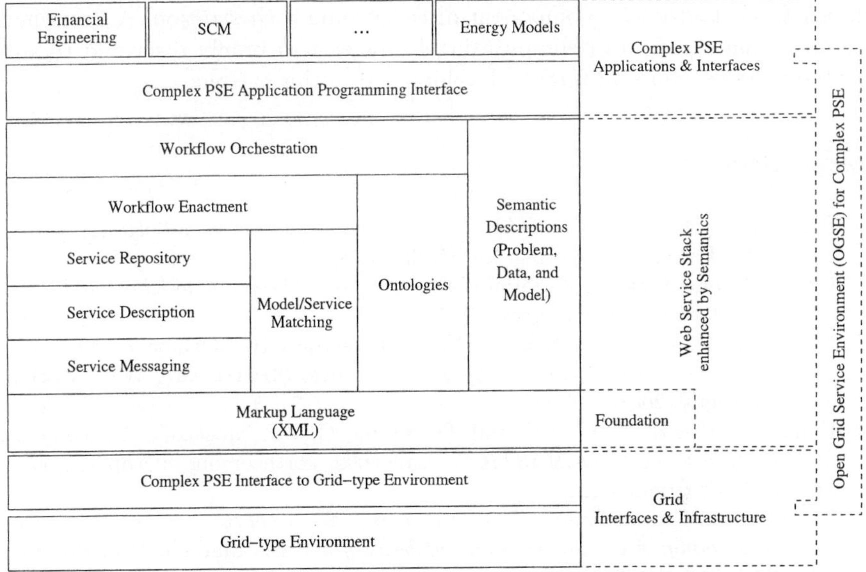

Fig. 3. OGSE Stack for complex problem solving applications.

The main goal of this stack is to establish an integration framework for a broad range of application areas with the main focus on an intelligent service discovery and workflow assembling in a computer processable way. The view on the service stack is application-centric from top down and Grid-centric from bottom up. The main ideas are the provision of

- *application-centric services*, that aim to cover the wrapping of existing legacy code, semantically describe facilities of the typical financial computation data, match the repository with the means of application data, and combine intelligent search features with the workflow orchestration.
- *grid-centric services*, that ensure compatibility requirements with computational grids (mainly covered by the Globus [16] initiative), workflow enactment on physical resources, and automatized discovery and allocation of computational resources.

4   Conclusion

In this paper we presented the basic concepts of our complex problem modeling and solving environment. This development grew out of the extension of the AURORA Financial Management System and has reached the level of being a unified framework for a broad range of application classes. The implementation of the core meta-workflow and the component architecture is based on the Open Grid Service Environment (OGSE). The meta-components within the system carry the semantic description which can be used for workflow composition and further (semi-)automatic component discovery and orchestration. A prominent example from the field of computational finance was briefly discussed to substantiate the relevance of further development on this system.

References

1. Blomvall, J. *A multistage stochastic programming algorithm suitable for parallel computing*. Parallel Computing. **29(4):** 431–445.
2. Censor, Y. and Zenios, S.A. *Parallel Optimization: Theory, Algorithms and Applications*. Oxford University Press. 1997.
3. Champion, M., Ferris, C., Orchard, D., Booth, D. Haas, H., McCabe, F., Newcomer E. *Web Services Architecture.* http://www.w3.org/TR/ws-arch/, W3C Working Draft. 8 August 2003.
4. Dempster, M.A.H, Scott, J.E. and Thompson, G.W.P. *Stochastic Modeling and Optimization using STOCHASTICS,* Draft 2002. Forthcoming in Applications of Stochastic Programming.,
5. Dupacova, J., Groewe-Kuska, N. and Roemisch, W. *Scenario reduction in stochastic programming: An approach using probability metrics*. Mathematical Programming **95A** (2003) 493–511
6. Ferris, M.C. and Munson, T.S. *Modeling Languages and Condor: Metacomputing for Optimization*. MetaNEOS Technical Report, 1998.
7. Gröwe-Kuska, N., Heitsch, H. and Römisch, W. *Scenario reduction and scenario tree construction for power management problems*. IEEE Bologna Power Tech Proceedings (A. Borghetti, C.A. Nucci, M. Paolone eds.), 2003
8. Iyer, L.S and Aronson, J. E. *A parallel branch-and-bound method for cluster analysis*. Annals of Operations Research **90** (1999) 65–86.
9. Laszewski, G., Foster, I., Gawor, J., Lane, P., Rehn, N. and Russell, M. *Designing Grid-based Problem Solving Environments and Portals*. Argonne National Laboratory, 2001.
10. Leymann, F. and Roller, D. *Workflow-based applications*. IBM Systems Journal **36**:1, 1997.
11. Linderoth, J. and Wright, S. *Decomposition Algorithms for Stochastic Programming on a Computational Grid*. Technical Report Computer Sciences Department, University of Wisconsin-Madison. 2002.
12. Kall, P. and Mayer, J. *SLP-IOR: An interactive model management system for stochastic linear programs*. Mathematical Programming **75** (1996) 221–240.
13. Pflug, G.Ch., Swietanowski, A., Dockner, E. and Moritsch, H. *The AURORA Financial Management System: Model and Parallel Implementation Design*. Annals of Operations Research **99** (2000) 189–206
14. Houstis, E. and Rice, J.R. *On the Future of Problem Solving Environments*. Department of Computer Sciences, Purdue University, 2000.
15. Rockafellar, R.T. and Uryasev, S.: *Optimization of Conditional Value-At-Risk*. The Journal of Risk **2(3)** (2000) 21-41
16. Tuecke, S., Czajkowski, K., Foster, I., Frey, J., Graham, S., Kesselman, C., Maguire, T., Sandholm, T., Vanderbilt, P., Snelling, D. *Open Grid Services Infrastructure (OGSI) Version 1.0*. Global Grid Forum Draft Recommendation, 6/27/2003.

The Migrating Desktop as a GUI Framework for the "Applications on Demand" Concept

Mirosław Kupczyk, Rafał Lichwała, Norbert Meyer, Bartosz Palak,
Marcin Płóciennik, Maciej Stroiński, and Paweł Wolniewicz

Poznań Supercomputing and Networking Center, Noskowskiego 10, Poznań, Poland
{miron,syriusz,meyer,bartek,marcinp,stroins,pawelw}@man.poznan.pl

Abstract. The Migrating Desktop is a ready to use GUI framework for making use of grid applications and putting into practice the "application on demand" concept. On-demand computing, contrary to the traditional approach of assigning resources to applications, refers to the concept of pooling system resources and dynamically allocating them to meet shifting demands. We introduce a ready-to-use framework for the integration of different computing systems into a comfortable working environment with support for "applications on demand". This work is done under the EU CrossGrid project IST-2001-32243.

1   Introduction

In this paper we describe the idea of using the Migrating Desktop as a ready-to-use GUI framework for making use of grid applications and also putting into practice a new generation concept of computing on demand. We focus on the architecture and functionality rather than technical details. This functionality refers to two different grid projects: the EU CrossGrid project IST-2001-32243 [1], and Progress (co-founded by Sun Microsystems and the Polish State Committee for Scientific Research) [2].

The "On Demand" concept is a term that covers several additional buzzwords like grid computing, utility computing and others. Utility computing is a combination of two approaches: according to the first one companies can call upon a third party to host and manage their IT infrastructure, and according to the second one, companies can pay for the resources they use. Grid computing is similar to utility computing but with a different approach. Grid computing is a form of virtualization that can handle computation-intensive tasks, using a large number of systems and combining them into one grid. Such grids can include widely distributed systems or systems within one data centers. Grid technology is enabling computing resources to be shared globally and easily managed, and the infrastructure becomes incredibly flexible. Nowadays the infrastructure is a pool of virtual resources that the user can call on as needed. Our main aim was to give the user a possibility of easy and flexible usage and gaining profits from new technology. We also introduce mechanisms that give an opportunity to look ahead to support future technologies.

M. Bubak et al. (Eds.): ICCS 2004, LNCS 3036, pp. 91–98, 2004.

We propose a transparent user work environment, independent of the system version and hardware. A flexible system structure enables defining and adaptation requirements of individual groups of users. We have to look ahead so that our users would not be bound to hardware configuration in order to have access to their working environment, applications, or resources. Our proposition is to give the user their own user-friendly environment with easy access to the applications and High Performing Computing (HPC) infrastructures on demand, independently of the location and hardware/software configuration. That is the reason why we have created the Migrating Desktop, which is a graphical user working environment (see Fig. 1.).

The Migrating Desktop is just a front end to the Roaming Access Server (RAS), which intermediates to communication and gives roaming access to different grid infrastructures. All user settings are stored in RAS therefore the Migrating Desktop started from another workstation looks the same. One of our aims is to have a common protocol for dealing with different infrastructures and to have one interface to the end users. The server implementing this interface is plug-in based, and plug-ins are responsible for communication with specific grid infrastructures. This architecture provides an important future direction with respect to the general acceptance of services and protocols. A grid infrastructure is being simultaneously developed in the framework of many academic and commercial projects. One of the biggest challenges for grid designers is managing the enormous complexity of Grid-based systems, making them interoperable with the existing systems and other emerging technologies and standards.

2   Architecture Overview

The architectures of the entire CrossGrid and Progress projects are presented in appropriate documents [3,4,5]. For the purpose of this paper we show only the basic components and interfaces between the corresponding modules.

The heart of the architecture is the Roaming Access Server that provides web services for accessing HPC resources.

The Roaming Access Server [3] offers a well-defined set of web-services that can be used as an interface for accessing HPC systems and services (based on various technologies) in a common, standardized way. All communication bases on web services technology.

The Roaming Access Server is a set of modules and plug-ins that provides interfaces to work with different grids. It consists of several independent parts responsible for application management, monitoring, user profile management, data management and authorization (see Fig. 2.). The provided functionality contains a wide range of different services that are common for many various grid projects. Currently the RAS provides plug-ins for interoperability with two separate grids - Polish Progress and EU CrossGrid; however, its infrastructure is open for expansion and attaching other HPC systems. The number of RAS servers depends on the amount of users and their requirements; it will be a fully replicated part of the system.

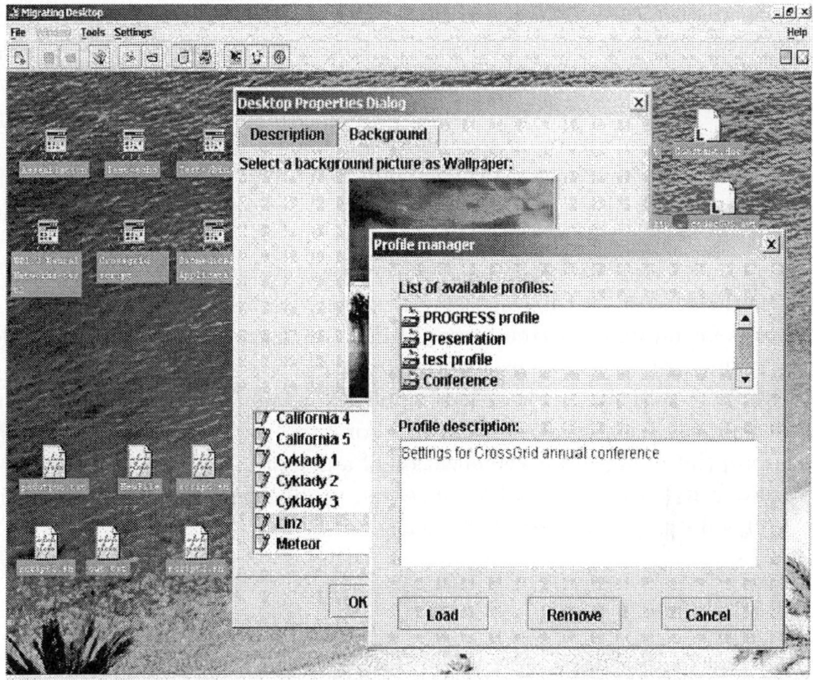

Fig. 1. Migrating Desktop main window and sample configuration dialogs

System architecture

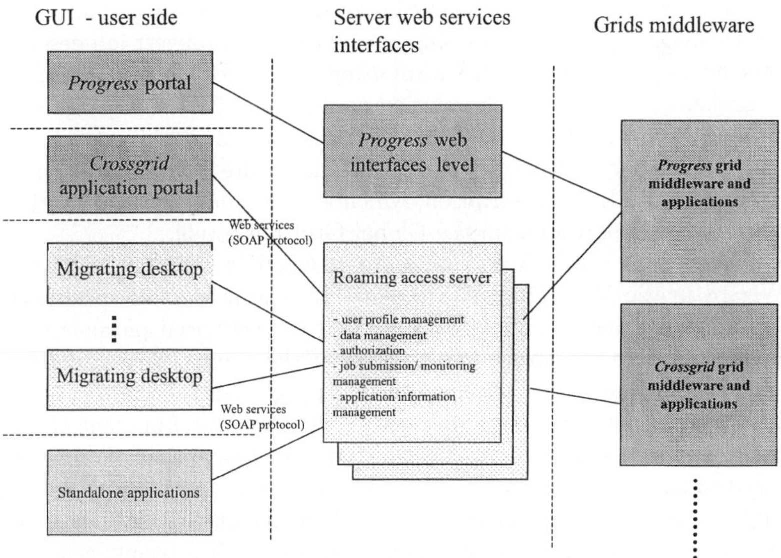

Fig. 2. RAS architecture overview

The Migrating Desktop is an advanced user-friendly environment that serves as a uniform grid working environment independent of specific grid infrastructure. Java based GUI is designed especially for mobile users and is independent of the platform (MS Windows, Linux, Solaris) and hardware. It is a complex environment that integrates many tools and allows working with many grids transparently and simultaneously. The main functionality concerns local and interactive grid and remote application support, local and grid file management, security assurance, authorization of access to resources and applications, and single sign-on technology based on X509 certificates.

A general concept of the Migrating Desktop was to provide the containers - frameworks for plug-ins written by application developers. Such approach allows increasing functionality in an easy way without the need of architecture changes. It is possible to add various tools and applications and support visualization of different formats using that mechanism. For example, the Migrating Desktop loads appropriate plug-in for visualization of a non-standard graphics format file on demand from the network. That makes our product independent of specialized tools designed only for a specific application.

3   "Applications on Demand" Services

Application Framework

The Migrating Desktop supports different kinds of accessing remote/HPC applications. We have focused mainly on supporting frameworks for grid applications, unifying different interfaces into one common solution. The Migrating Desktop calls appropriate web services on the RAS. The server job submission interface gives uniform access to different resource brokers. Clients can submit all jobs with this interface and then appropriate plug-ins are called to convert job description to the specific language. Currently we are using plug-in for CrossGrid/DataGrid resource broker to convert job description to Job Definition Language used in DataGrid and Progress plug-in to convert job description to Extended Resource Specification Language used in Progress. In the future Job submission interface will support XML job description, which is under development by the Job Submission Language group from the Global Grid Forum [6].

Grid applications available for users are grouped in a user-friendly way in a Job Wizard in the MD. This Wizard simplifies the process of specifying parameters and limits, suggesting user defaults or recently used parameters. The Wizard is responsible for proper preparation of the user's job and consists of several panels. One panel is application specific plug-in and the rest can be used to set job information, resource requirements, files and environment variables. Application plug-in asks user for application parameters and that parameters are than passed to the command line. Application plug-in is just a Java class inherited from the ApplicationPluginBase and must implement its abstract method. The most important methods are getArguments and setArguments called when the job is going to be submitted or when job is presented to the user. For application developers that do not like to prepare plug-in we created generic XML plug-in that interprets argument description given in XML to a graphical form. Example plug-ins are presented in Fig. 3.

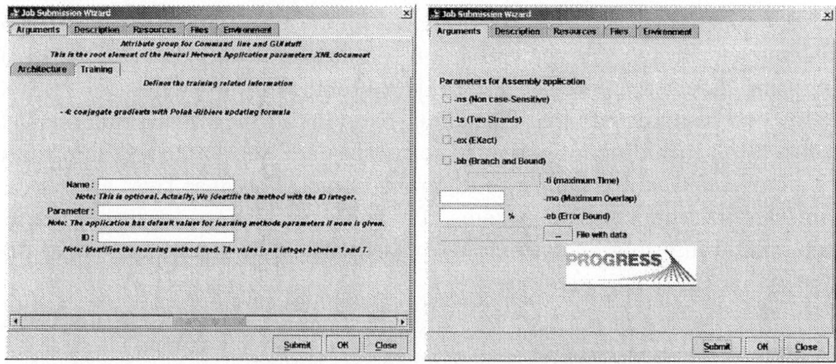

Fig. 3. Example Application plug-ins. Left - generated from XML, right - Java class

The Migrating Desktop also allows running remote Java tools. In contrast to grid applications that are submitted to the resource broker, tools are Java applets downloaded from different network localizations on demand.

Some applications require a graphical visualization of job results. In our approach visualization are just tools and can also base on the Tool Plug-in. Tool Plug-ins can visualize a single file or a single job. All available file and jobs visualization are registered in the Migrating Desktop database, and the appropriate Tool Plug-in is chosen depending on the file extension or type of the job.

We also provide a framework for interactive application as the VNC [7] plug-in. Interactive applications can be started and managed using the edg2.0 middleware. This middleware is developed in the DataGrid project and is still under enhancement in CrossGrid.

Using the VNC plug-in is also our proposed solution for other types of interactive remote applications from different providers. In our approach the user can have only a simple terminal, a freeware system with a web browser, and gain profits at an opportunity of using remote resources. An application instance run, for example, on a specialized application server could be transparently offered to the user on his/her demand. The user does not have to possess expensive products to benefit from the usage of its profits, he/she just uses time of that application making use of the next generation networks.

The features mentioned above give an easy way for including open and dynamic architectures of applications available on demand. Our framework supports a wide range of different remote applications types so it can fulfil growing demands of step-ahead technology users.

Application State Monitoring

Monitoring can be performed on different moments/levels of the application life. We assumed to present all monitoring types supported by application providers. At the moment we focus mainly on the grid jobs state monitoring framework.

Users can check the state of the launched application/submitted job, suspend or delete the selected job using the Job Monitoring tool in the MD. The Job Monitoring dialog presented in Fig. 4. is a useful tool for tracing the status of the previously submitted jobs. This dialog contains all the information about the submitted job including an extended job status and job log. Some information like the extended status can have different formats, because they are returned by different grid information systems. The RAS plug-ins read information from a specific grid information system and presents them to users in a text or XML format.

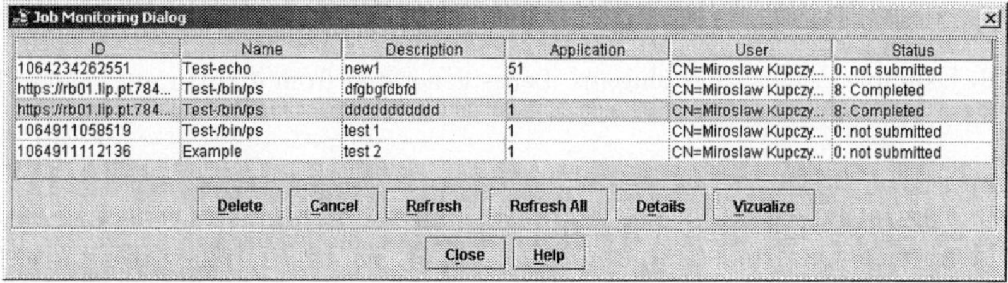

ID	Name	Description	Application	User	Status
1064234262551	Test-echo	new1	51	CN=Miroslaw Kupczy...	0: not submitted
https://rb01.lip.pt:784...	Test-/bin/ps	dfgbgfdbfd	1	CN=Miroslaw Kupczy...	8: Completed
https://rb01.lip.pt:784...	Test-/bin/ps	ddddddddddd	1	CN=Miroslaw Kupczy...	8: Completed
1064911058519	Test-/bin/ps	test 1	1	CN=Miroslaw Kupczy...	0: not submitted
1064911112136	Example	test 2	1	CN=Miroslaw Kupczy...	0: not submitted

Fig. 4. Job Monitoring Dialog

Data Management

Data management is a very important and complicated part in every system. After many analysis a common set of interfaces that allow operation on data and metadata was created. The designed framework allows extending MD/RAS infrastructure easily so that many management systems could be attached as a plug-in. The main file Management tool in the Migrating Desktop is called Grid-Commander (Fig. 5.). The Grid Commander is a two-panel application similar to the Commander family tools. A single panel can represent a local directory, gridftp or ftp directory or other protocol to the native storage. Each protocol is defined as a plug-in that makes file management in MD easy to extend to other protocols. From the technical points of view graphical elements operate on generic file systems that invoke appropriate functions within the supported plug-in.

Authentication Issues

We try to keep authentication as simple for the user as possible. Authentication to the Migrating Desktop is done with users X509 certificates that have to reside on a local disk, removable disk, or can be downloaded from MyProxy server.

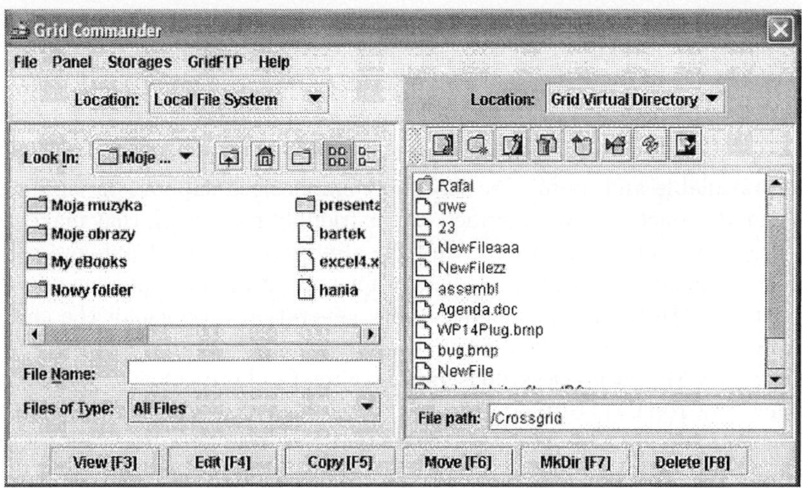

Fig. 5. Grid Commander Dialog

X509 proxy is passed to each RAS interface and is used to access grid resources.
For resources that require other methods of authentication (e.g. username and
password) the Migrating Desktop can ask about necessary information and then
store it in the user profile. Any subsequent authentication can be done automatically. In our current implementation we store username and password for
authentication to the Progress grid infrastructure which bases on the IPlanet
Portal Server.

User Profile Management

User profile management provides functionality that allows operations on user
profiles. User profiles contain all information that define the current user working
environment including information about the graphical configuration (e.g. desktop background, icon locations, and colors) and information needed to access
specific HPC infrastructures (e.g. username and password). The LDAP protocol
is currently used for saving and retrieving stored information.

Other Services

We are going to extend the functionality of the Migrating Desktop/Roaming
Access Server not only in the field of grid services. We will design a functionality
that will simplify the communication between users for data exchanging. In scope
of our research there are also ways of notifying the user via e-mails, SMS etc.
about important events connected with the status of submitted jobs.

To add support for a new grid infrastructure the MD/RAS plug-in is
necessary. Plug-ins provide systems specific information and are used by
job/application submission, monitoring and data management modules.

4   Summary

In this paper we presented an overview of the functionality of the Migrating Desktop and Roaming Access Server. It consists of several components that are using the Web Service technology. It gathers all facilities that make the grid resources available and useful. We especially emphasized support of "application on-demand" concepts. We described a user-friendly framework that makes work with grids easy and comfortable. We have also pointed out advantages and features of the general framework dedicated for presentation of different providers' applications. We have implemented a new generation product for the end user according to the growing demands and needs.

Currently Migrating Desktop is used for CrossGrid project as a graphical user interface for CrossGrid testbed. It was also used for Polish Progress project. After two years of development we can conclude that the idea of graphical environment for work with grid is accepted by users. "Application on demand" is our next goal we would like to achieve.

References

1. Development of Grid Environment for Interactive Applications. Annex 1, EUCross-Grid Project, IST-2001-32243,
   `http://www.eucrossgrid.org/CrossGridAnnex1_v31.pdf`
2. Polish Research on Grid Environment for SUN Servers, `http://progress.psnc.pl`
3. Bubak, M., Malawski, M., Zajac, K.: Current Status of the CrossGrid Architecture. Proceedings of the Cracow '02 Grid Workshop, Kraków, Poland, December 2002
4. Kosiedowski, M., Mazurek, C., Stroiński, M.: PROGRESS - Access Environment to Computational Services Performed by Cluster of Sun Systems. Proceedings of the Cracow '02 Grid Workshop, Kraków, Poland, December 2002
5. Kupczyk, M., Lichwała, R., Meyer, N., Palak, B., Płóciennik, M., Wolniewicz, P.: Mobile work environment for Grid Users. Proceedings of the Across Grids Conference, Santiago de Compostela, Spain, 2003
6. Job Submission Description Language Working Group,
   `http://www.epcc.ed.ac.uk/~ali/WORK/GGF/JSDL-WG/`
7. Richardson, T., Stafford-Fraser, Q., Wood, K.R. and Hopper, A.: Virtual Network Computing. IEEE Internet Computing, **2(1)**, (1998), 33–38

Interactive Visualization for the UNICORE Grid Environment

Piotr Bała[2,1], Krzysztof Benedyczak[2], Aleksander Nowiński[1],
Krzysztof S. Nowiński[1], and Jarosław Wypychowski[1]

[1] Warsaw University
Interdisciplinary Center for Mathematical
and Computational Modelling, `know@icm.edu.pl`
[2] Faculty of Mathematics and Computer Science
Nicolaus Copernicus University, `bala@mat.uni.torun.pl`

Abstract. A description of interactive visualization plugins for the UNICORE grid system is provided. The plugins allows for attaching to running UNICORE jobs, downloading current state of computation and visualization of the results. The plugin set contains a general purpose 2D visualization system Sapphire and an example plugin MolDyAna for 3D visualization of the progress of molecular dynamics jobs. The Sapphire subsystem contains user configurable tools for simple plots of different types and 2D data plotting tools, together with an extensive GUI. MolDyAna provides both standard methods of animation and graphing and several advanced postprocessing and visualization techniques

1   Introduction

Recent advances in the grid technology created number of tools which provide user's interface to distributed resources [1]. Computational grids allowed users to use variety of geographically distributed resources and to select and aggregation of them across multiple organizations for solving large scale computational and data intensive problems. In order to attract users, grid interfaces must be easy to install, simple to use and able to work with the different operating system and hardware platforms. The important feature required by the users is ability to perform scientific visualization.

Recent grid activity has been significantly oriented towards utilization of the computational resources for the data processing and number crunching rather than visualization. This approach was also influenced by the fact that most of computational intensive applications have no dedicated user interfaces. Some expensive commercial graphical tools are available but they are treated as separate applications run locally on the users workstation and can hardly be integrated with the grid environment. Using these applications the user can prepare input or perform advanced postprocessing. The results are stored in the files used as input for another job to be run on the grid. The grid middleware is used to transfer input and output files and to submit and control job.

M. Bubak et al. (Eds.): ICCS 2004, LNCS 3036, pp. 99–106, 2004.

In order to provide user with integrated environment attempts to integrate advanced visualization with the grid environment have been performed. Most of them are limited by the existing technology.

There are two most important barriers inhibiting creation Of effective grid visualization tools: lack of the visualization capabilities in mostly script based grid tools and the fact that the visualization software is build using traditional programming languages and libraries such as Tcl, Gtk or GL. In effect, existing visualization software is difficult to integrate with standard grid environment built in Java.

Different approach is presented by the UNICORE [2] middleware. Compared to other tools UNICORE provides access to the various types of resources, has advanced user interface, is flexible and allows for easy integration with external applications. Well-developed user interface, which can be easily extended by the user is another advantage. In particular, both generic Java visualization libraries can be used for visual analysis of job progress and specialized visual applications can be integrated with the help of plugin technology into the UNICORE platform.

2   Visualization Capabilities in the UNICORE Grid Environment

The details of the UNICORE can be found elsewhere [3] and we will summarize here its most important features and recent achievements and extensions.
UNICORE is uniform interface to the computer resources which allows users to prepare, submit and control application specific jobs and file transfers (see. Fig 2). Jobs to be run on the same system can be grouped in job-groups. The user specifies target system for job group as well as resource requirements for CPU time, number of CPUs amount of memory and disk space for job. Jobs can have complicated structure with various job subgroups which can be run at different systems. Subgroup jobs and file transfers can be ordered with user-defined dependencies. This includes conditional execution and loops with control which can be used as environment variables in the user's jobs. The user input is mapped to the target system specific commands and options by the UNICORE infrastructure.

The UNICORE client provides the plugin mechanism, which became very attractive and efficient method for integration of applications with grid middleware. Plugin is written in Java and is loaded to the UNICORE client during start, or on request. Once it is available in the client, in addition to the standard futures such as preparation of the script job, user gets access to the menu, which allows preparation application specific jobs.

Dedicated plugins can be used for a database access or postprocessing. Since plugins are written in Java, they can be easily integrated with external applications or Java applets. Number of plugins have been developed to provide users with interface to the most popular applications. The visualization has been however limited to the traditional textual GUI.

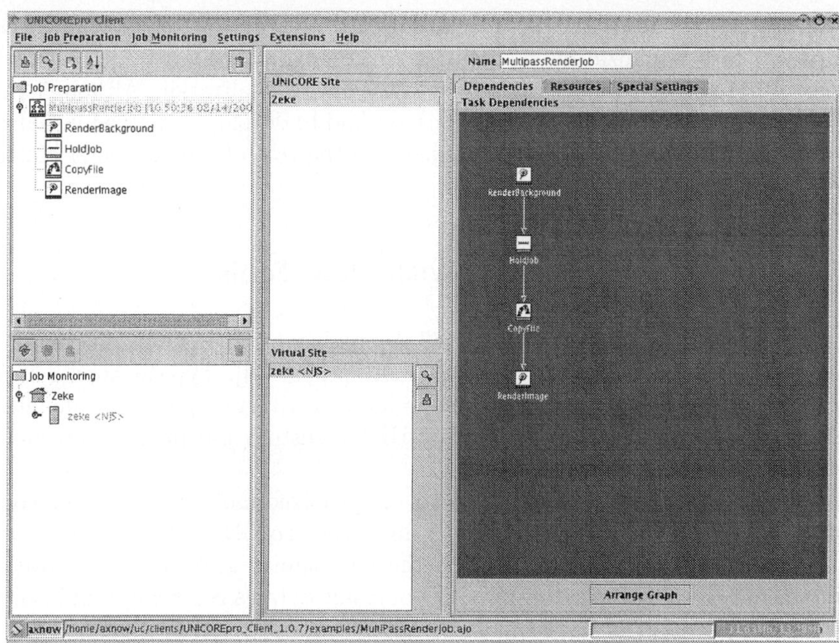

Fig. 1. UNICORE Client

During last two years in ICM a couple of plugins for UNICORE client has been developed providing advanced visualization capabilities. Those plugins show that there is large potential area of integration visualization services into the grid technology. Plugins typically provide support for building custom application tasks, but so called 'outcome panel' provides ability to include GUI for data analysis and visualization. Effectively it gives to user opportunity to run custom job, acquire results and visualize it in one application. We have found the opportunity of early result examination very important and comfortable to the user.

The advanced visualization capabilities were added to the UNICORE Client in the different ways. The first approach was to extend existing plugins and build in some visualization routines dedicated to the particular type of application. This method is straightforward, but in log scale is time consuming because requires redesign of the code for each changes in the application. This method was used by us to provide graphic capabilities to Gaussian plugin.

The another approach is to develop dedicated visualization plugin which can be used along with other plugins to construct jobs with complicated workflow. Such plugin can be used in various application areas and is not limited to the particular application. One should note that application specific data formats can be easily translated to the visualization suitable one with a simple command or script task which can be included in the UNICORE workflow.

In both cases the UNICORE Client must be integrated with the application, which provides visualization capabilities. Since plugins are written in Java, the easiest way to create uniform system is to use Java visualization. Although Java provides reasonable graphical extensions, we had to develop dedicated visualization system, which fulfills user requirements in the area of the molecular biology and quantum chemistry.

3   Job Progress Watching with New Tools

And how does it work in a real world? A number of UNICORE plugins have been developed in ICM during EUROGRID project [4]. Typical approach for plugin programming is represented by a Gaussian Plugin - plugin for quantum chemistry code. This plugin contains GUI for custom job preparation and a visualization panel.

Typically, a Gaussian output consists of optimized molecule geometry, some simple numerical data like energy values and a series of 3D fields containing data on electron density, potential, etc. on the rectangular grid. The visualization component allows the user to view slices of result data as color maps and isoline maps with atoms of analyzed molecule giving an opportunity to examine the results of computation in progress.

The access to data produced by a current (running) job is provided by so-called "Filter plugin". It is able to find the job directory on the target system, and then filter job output and send it back to the user allowing thus to analyze computation results in the middle of processing. This opportunity is very helpful, especially if it is known that program may loose right path of computation without general failure as it is often the case of molecular dynamics computations. Knowing intermediate results allow user to make decision about aborting or continuing computation — in general lots of impatient people find this a very nice feature. Filter plugin analyzes job output files using regular expression mechanism and retrieves interesting part of computation results. Then visualization panel can read the data and present it to the scientist in graphical form providing immediate feedback, and allowing in-time result control. At this point UNICORE does not support direct connection to the job required by advanced visualization systems (like VisIt [5]), and this kind of job connection can be done only in very limited way, but we do find it potentially large area of future development.

The final component, that is the visualization itself, can be provided by a general visualization plugin, allowing user to choose among several generic visualization techniques. The so called Vis plugin, and its descendant — Sapphire plugin provide this functionality. Any file in recognized format may be downloaded from execution space while job has been finished, and then presented in custom way in client's window. The sapphire plugin is a sample application of visualization components that can be used in a standalone visualization application. The result is fully featured 2D data viewer, allowing visualize and analyze data quickly in single user environment.

In the case of grid molecular dynamics simulations a specialized plugin using Java3D technology can be used for extensive visual analysis of the MD results. The MolDyAna plugin provides standard animation and graphing functions as well as advanced data filtering and visualization methods revealing deeper features of a molecular dynamics trajectory.

4   Generic Tools for 1D and 2D Data Visualization – The Sapphire Library

Custom components for data visualization are grouped into a special library called "Sapphire". It provides easily extensible and quite powerful toolset for building 2D data presentation together with an extended GUI for graph customization, allowing easy modifications for all properties of plots.

Sapphire provides all important data visualization methods in 2D environment. It includes line, bar and area plots, mesh plots, colormaps, isoline maps, vector fields and glyph plots. Some specialized features like barb plots for meteorological applications are also included. All those types of visualization can be mixed in any way on the single graph, using more then one axis for each coordinate. Java2D PostScript and off-screen image rendering capabilities ensure very high publication quality of hardcopies and image exports.

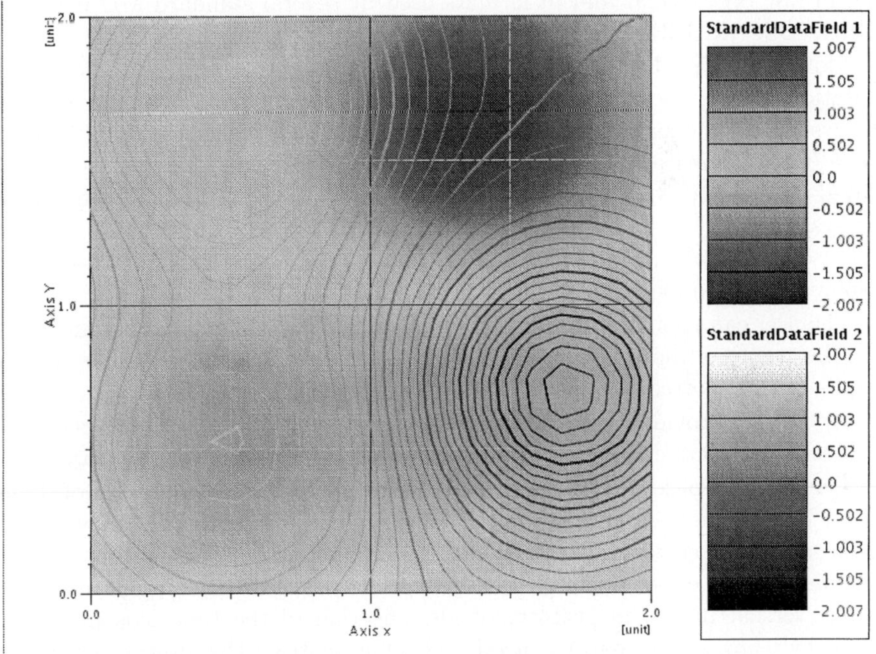

Fig. 2. Example of 2D Sapphire plot of electrostatic field in vicinity of a molecule

The Sapphire plot system is organized in an object tree that can be browsed and modified by adding new graph components (datasets, plots, axes etc.) from a library with the help of a GUI component built on top of standard Java tree GUI component.

Each graph component has its own set of properties controllable by 'property sheets', that are feature of java basically intended to be used in IDE packages. Property sheet for object shows object properties values and allows editing them in customizable way. For example, property editor for line renderer (object responsible for rendering data as a poly line) allows user to modify both line properties (color, thickness, style etc.) and rendering method (standard polyline or staircase). What is important in this method, all newly added components do not require to provide custom GUI. For a programmer wishing to add own components into library it is enough to follow java code guidelines and mark through special method, which object properties shall be available for user.

5   Molecular Dynamics Analyzer (MolDyAna) and Its Use in Examination of MD Trajectories

MolDyAna is a toolkit for interactive graphical analysis of the molecular dynamics trajectories build as a Java standalone application or as UNICORE plugin with an extensive use of Java3D technology. It can accept ASCII trajectory files in raw .xyz format and in formats used by several standard MD packages (GROMOS, AMBER etc.).

MolDyAna can display molecule geometry using standard methods of presentations (lines, sticks, ball-and-stick and space fill modes). Different parts of the system can be assigned to up to 4 classes with each class shown in its own mode (e.g. reacting atoms as large spheres in space fill mode, active center displayed as ball-and stick with its neighborhood shown as sticks and the rest of the system with lines only presentation. Tube/ribbon plot of protein backbone will also be available. MolDyAna allows the user to browse trajectory frames or display animation of the trajectory.

The user can pick interatomic distances, angles and dihedral angles and analyze the graphs of these measurements versus time. An arbitrary number of such graphs can be plotted.

MolDyAna provides in addition several non-standard interactive visual analysis techniques that can be used to obtain better insight into the MD trajectory.

The user can pick an atom or a set of atoms and redisplay the paths of these atoms in the whole simulated time. Moments of particularly fast movements of the atoms or correlated movements in various parts of the system can be pinpointed much easier this way than by the analysis of distance/angle/dihedral graphs versus time. The problem of identification of the time axis has been solved by using a colormap to encode the time, plotting the atom positions at the beginning of simulation in blue and corresponding position at the end of simulation in red with interpolated color values in between.

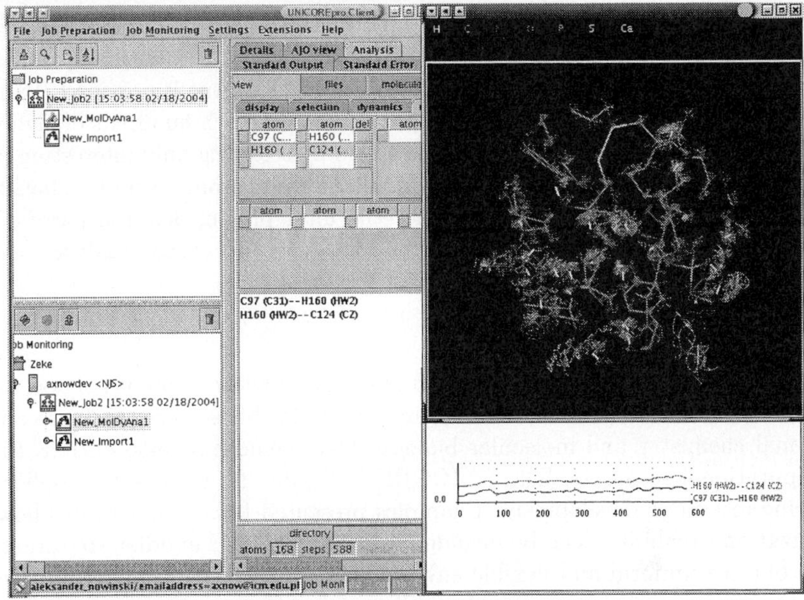

Fig. 3. MolDyAna plugin example

In addition to standard plots of distances or angles versus time MolDyAna offers an abstract 3D plot of three user chosen measurements again color coded with the simulated time. The complicated 3D trajectory can often reveal e.g. correlations between various geometric features and changes of such correlations in time. For example, a correlation emerging at some moment may indicate creation of some sort of a bond between observed subsystems.

Fast vibrations of relatively rigid subsystems or movement of low inertia fragments often visually dominate MD trajectories . To hide such fast modes and to present some generalized picture of a MD trajectory, MolDyAna includes a possibility of Sawitzky-Golay filtering [6], [7].

In order to obtain for a function $f(t)$ its smoothed approximation value $g(t_0)$ one computes best approximation of $f(t)$ on the interval $< t_0 - \Delta t, t_0 + \Delta t >$ by a quadratic polynomial $p(t) = a_{t_0}(t - t_0)^2 + b_{t_0}(t - t_0) + c_{t_0}$ and takes $g(t_0) = p(t0) = c_{t_0}$. (MolDyAna computes a Gaussian weighted approximation here). In addition to good quality of trajectory smoothing without excessive deformation of molecule geometry one obtains physically meaningful values of b_{t_0} (average velocity component) and a_{t_0} (average acceleration component) that can be used for further visual analysis.

The fast vibrational models are neglected by the smoothing procedure but are still present in the picture: the statistics of difference $f(t)_g(t)$ on $< t_0 - \Delta t, t_0 + \Delta t >$ representing the neglected vibrations is visualized as principal moments ellipsoid.

The MolDyAna panel can be used both in a stand alone application and (with a suitable wrapper) in an UNICORE plugin. In the later case, some two-way communication between a grid MD job and MolDyAna is provided. To avoid excessive transfer of MD trajectories that can easily reach hundreds of gigabytes the user can start a MolDyAna plugin session by selecting only interesting parts of the trajectory (e.g. given time interval, reaction site atoms or protein backbone only) and generate a suitable script for the filter plugin. Selected parts of the trajectory file will be then transferred to MolDyAna for visual analysis.

6   Conclusions

The UNICORE software was used as framework providing uniform access to the number of resources and applications important for life sciences such as computational chemistry and molecular biology. This includes seamless access to the computational resources. The UNICORE is flexible framework for application specific interfaces development. Examples presented here demonstrates how visualization capabilities can be included in the UNICORE middleware. In results user obtains uniform and flexible environment for scientific simulations. Developed extensions allows to use grid middleware in the scientific discovery process which require significant visualization capabilities both during pre- and post-processing. Modular architecture based on the plugin concept extended now with visualization plugins opens number of possible applications in the different areas.

One should note, that UNICORE middleware can be used also together with other grid middleware, especially it can be used as job preparation, submission and control tool for Globus.

Acknowledgements. This work is supported by European Commission under IST grants EUROGRID (no. 20247) and GRIP (no. 32257).

References

1. C. Kesselman I. Foster, editor. *The Grid: Blueprint for a Future Computing Infrastructure.* Morgan Kaufman Publishers, USA, 1999.
2. UNICORE. Unicore Forum. http://www.unicore.org.
3. P. Bała, B. Lesyng and D. Erwin EUROGRID - European Computational Grid Testbed. *J. Parallel and Distrib. Comp.* 63(5):590–596, 2003
4. J. Pytliński, L. Skorwider, P. Bała, M. Nazaruk, V. Alessandrini, D. Girou, G. Grasseau, D. Erwin, D. Mallmann, J. MacLaren, J. Brooke. BioGRID - An European grid for molecular biology. *Proceedings 11th IEEE International Symposium on Distributed Computig,* IEEE Comput. Soc. 2002, p. 412
5. Eickermann, Thomas; Frings, Wolfgang VISIT - a Visualization Interface Toolkit Version 1.0 Interner Bericht FZJ-ZAM-IB-2000-16, Dezember 2000
6. Savitzky A., and Golay, M.J.E. 1964, Analytical Chemistry, vol. 36, pp. 1627–1639.
7. Hamming, R.W. 1983, Digital Filters, 2nd ed. (Englewood Cliffs, NJ: Prentice-Hall).

Efficiency of the GSI Secured Network Transmission[*]

Bartosz Baliś[1,2], Marian Bubak[1,2], Wojciech Rząsa[3], and Tomasz Szepieniec[2]

[1] Institute of Computer Science, AGH, al. Mickiewicza 30, 30-059 Kraków, Poland
{balis,bubak}@uci.agh.edu.pl
[2] Academic Computer Center – CYFRONET, Nawojki 11, 30-950 Kraków, Poland
t.szepieniec@cyf-kr.edu.pl,
[3] Rzeszów University of Technology, Wincentego Pola 2, 35-959 Rzeszów, Poland
wrzasa@prz-rzeszow.pl
phone: (+48 12) 617 39 64, *fax:* (+48 12) 633 80 54

Abstract. The security of information transmitted over the computer networks is a frequently raised problem. The more information is transmitted and the more important it is, the more important security issues become. Network protocols used in the Internet and local networks were not designed to meet high security demands of current applications. Though cryptography algorithms enabled security of data transfer over insecure network protocols, they introduce additional overhead. Cryptography became important part of the security policy of Grid systems. In this paper, we perform experiments to estimate the overhead introduced by this solution. The obtained results are of great importance for the OCM-G – Grid enabled monitoring system.

Keywords: Grid, security, GSI, application monitoring, OCM-G

1 Introduction

The Grid concept proposed by Foster, Kesselman and Tuecke [8] was designed to enable coordinated resource sharing between distributed entities. In a grid system, no existing relationships or trust between the entities are assumed. There are also no single control points, though the sharing should be strictly controlled. Available resources (computing capability, disk capacity, etc.) are shared by means of the existing network infrastructure provided by the Internet, thereby a global distributed heterogenous system is created and may be used by different people for various purposes. In addition, the shared resources can be more efficiently exploited.

Grid applications are complex, therefore tools facilitating their development process are required [4]. The OCM-G application monitoring system is designed as an agent between such tools and application processes running on nodes belonging to distributed Grid sites [1]. The OCM-G is designed as a distributed

---

[*] This work was partly funded by the European Commission, project IST-2001-32243, CrossGrid [6]

M. Bubak et al. (Eds.): ICCS 2004, LNCS 3036, pp. 107–115, 2004.
© Springer-Verlag Berlin Heidelberg 2004

and decentralized system, thus scalability required in the Grid environment is achieved. Monitoring system consists of two parts – permanent, handling multiple applications of numerous users and transient, belonging to the monitored application owner. The OCM-G works in on-line monitoring mode to address the requirements concerning delivery time between the user and the application processes.

Security issues are essential for the OCM-G as it supports multiple users and may control applications processes. It is obvious that the monitoring system should not lower security of the site and, at the same time, the solutions introduced to achieve the security should not degrade the scalability of the monitoring system [2].

Application of security aspects to network communication results in noticeable overhead. However as we mentioned above security cannot be omitted while considering the system as OCM-G despite of the strict demands concerning transmission time in on-line monitoring system. Therefore we decided to perform tests in order to estimate unwanted effects connected with security policy. However the experiments were designed to fit OCM-G realities, we believe the results can be useful for developers of the other Grid systems. Before detailed description of our tests we briefly introduce GSI that is security solution designed for the Grid systems.

2   GSI – Security Solution for the Grid Environment

The Grid concept assumes the use of existing network infrastructure to perform communication between resources and users. However, this implies some problems, security being one of the most important ones.

The security requirements concerning network data transmission can be divided into the following aspects.

- Authentication: the peers of the connection should be identified upon connection establishment.
- Authenticity and integrity of transmitted information should be ensured in order to avoid accidental or deliberated alteration.
- Confidentiality of transmitted data prevents eavesdropping.

Vulnerabilities of protocols used in the Internet or local networks are well known. Primary threats related to network transmission are briefly described below.

Sniffing or *eavesdropping* is possible in some low level communication protocols. It is a significant threat to confidentiality of the transmission. *Spoofing* is possible for each protocol commonly used in the Internet. Depending on the protocol it allows to impersonate host or makes attacker capable of deceiving authentication methods based on the source address of the packet or even allows a third host to become an agent between two other hosts, and fully control connections. Different varieties of spoofing threaten all aspects of secure data transmission. *Session take over* (or session hijacking) allows an attacker to intercept an already established TCP/IP session. Since the authentication is usually performed only on initialization of a connection this is significant threat to authenticity of transmitted data [3,5].

One may conclude that security problems cannot be solved by the means provided by network protocols. Therefore, cryptography algorithms [11] are applied in order to ensure the desired level of security. Asymmetric cryptography is used while secure connection establishment in order to perform reliable authentication and exchange symmetric keys between the peers. Thereafter efficient symmetric algorithms are used to encrypt transmitted data. Authenticity and integrity are ensured by the use of digital signature that is computed for each fragment of transmitted data. TLS is widely used example of protocol that can be used to securely transfer data over insecure networks [7].

Security of the Grid applications is also achieved by the use of cryptography. However the issues that we encounter in Grid environments are significantly more complex and cannot be solved by any existing protocol. Therefore, *Global Grid Forum Grid* Security Working Group [10] was formed in order to define Grid security requirements and find appropriate solutions. Since there are various existing concepts and well tested security solutions addressing different security issues, the GSI WG decided to develop the security solution for the Grid systems on the basis of existing ones. Thus, cryptography is used to ensure security in the Grid environment.

3   Tests

Applying cryptography algorithms to data transmission satisfies security requirements of the distributed systems. However, though these algorithms are becoming more and more efficient, they still introduce a significant overhead. In this section, we present results of tests performed in order to estimate this overhead. We describe three tests to evaluate resource consumption and other undesirable effects of cryptographic algorithms in different stages of the connection.

The security requirements of different Grid applications may vary. Therefore we performed the tests for different security levels applied: **CLEAR** – no security mechanisms were enabled, **AUTH** – authentication and authorization were performed upon connection establishment, **PROTECT** – authenticity and integrity of the transmitted data via digital signatures were ensured, **ENCRYPT** – transmitted data were encrypted. Note that the subsequent security level include all aspects of the previous ones.

We found it convenient to use two different execution environments for two kinds of tests. Experiments concerning the transmission overhead were carried out on slower machines, therefore it was easier to observe the CPU time usage. DoS vulnerability tests required possibly large cluster of machines, while high computing capability did not present a problem.

All test programs were implemented using `Globus_IO` – a communication library being part of the *Globus Toolkit 2* [9]. `Globus_IO` implements the GSI security solutions and is designed specifically for the Grid systems.

3.1   Transmission Overhead Test

The first experiment we performed in order to estimate the transmission overhead related to secure data transfer. The aim of this test is to evaluate the overhead resulting from the security mechanisms applied, depending on the security level ensured while data transmission.

Implementation and Testbed. In order to carry out the test we have implemented two programs `sender` and `responder`. The `sender` was designed to initialize network connection, transmit data over the network and receive them back. The `responder` was to receive data and transmit them back to the `sender`.

The test was performed on a PC cluster. The sender process was executed on an Intel Celeron 300 MHz machine, the responder on an Intel PIII 600 MHz one. The hosts were connected with 100Mb switched LAN.

Experiment and Results. The measurement consists in transmitting data through the network between the `sender` and the `responder`. The `sender` measured CPU time of the two-way transmission. The test was carried out for different quantities of 100-bytes packets with different security levels applied. The results are presented on Fig. 1.

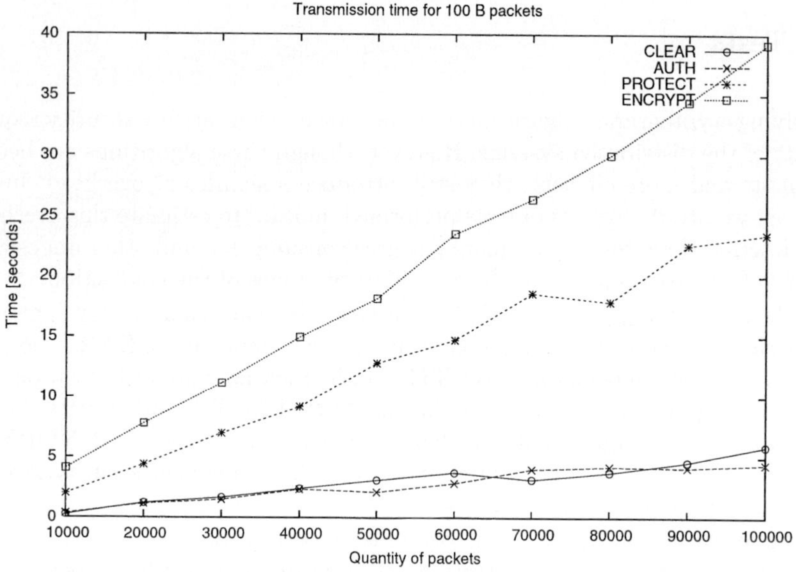

Fig. 1. Results of the security overhead test

We may notice a linear relationship between CPU time and quantity of packets for all security levels, however, for higher security levels the CPU time increases faster. In order to estimate the overhead resulting from the proposed solution, we present an average transmission time for discussed security levels (see Fig. 1).

3.2   Transmission Time and Packet Size

The previous test was performed for a fixed size of the packet. Now we will study the dependency between transmission time and the size of transmitted packet. We should expect that it is more efficient to transmit data in large packets.

For this test we used the **sender** and the **responder** processes configured as in the previous one. We performed two-way transmission of 10,000 packets of different size, and measured CPU time consumption. In order to compare obtained results, we computed average CPU time required to transmit 100B in each size of the packet. The results of the experiment are presented on Fig. 2.

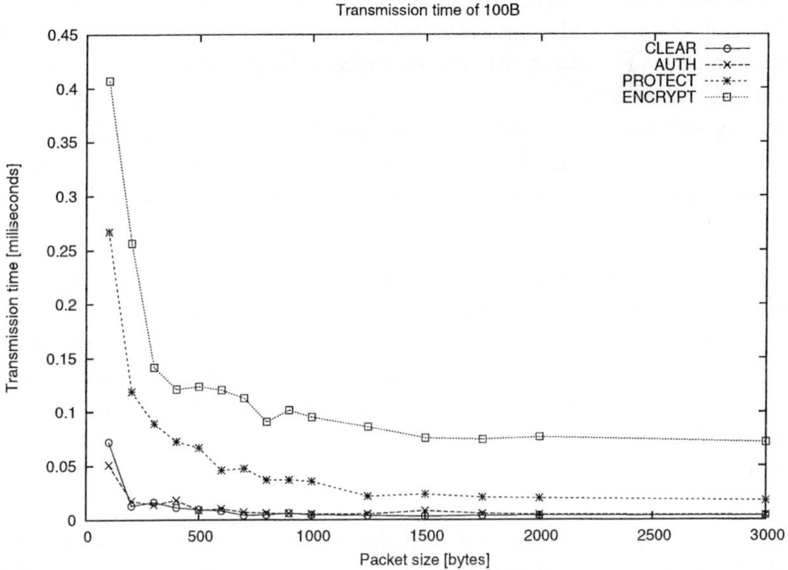

Fig. 2. Size of the packets and CPU time consumption for different security levels

3.3   Suspectibility to Denial of Service Attack

It is widely known that the establishment of a secure connection is resource consuming. After connection establishment transmitted data is encrypted with symmetric cryptography algorithms, however, before parameters of the connection are negotiated numerous asymmetric-cryptography operations should be performed. The asymmetric algorithms use longer keys and more complex mathematical operations are performed therefore they are less efficient than symmetric ones and consume significantly more resources.

From above considerations we may conclude that servers may not be able to simultaneously establish numerous secure connections. Moreover the port listening for secure connections may be easily used as a target for *Denial of Service* attack.

The aim of the test described below is to estimate vulnerability of the secure connections server as regards DoS attack. We try to find the practical limit of secure connections that can be handled by the server, verify the results of server overload and estimate number of connections that can be safely handled.

Implementation and Experiment Environment. In order to perform the experiment we implemented two short programs: `client` and `server`. The `server` was listening for connections, and after a connection arrived, it logged the time of the arrival. Then it tried to accept the connection and logged the time of establishment or failure. After the `client` started it waited until specified clock time and tried to establish connection with the specified host on specified port. The start time and connection establishment or failure time were logged. In this scenario, the `client` was the intended attacker while the server was the "victim" of this attack.

The experiment was performed on another PC cluster with machines equipped with two 2.4 GHz Intel Xeon processors and 1GB RAM each. Eighteen machines connected with 100 Mb switched LAN were involved in the test.

Description of Experiment and Results. The experiment consists in simulated Denial of Service attack that was performed by the `clients` to the `server`. The connections to the `server` running on one of the machines were requested from the other seventeen hosts. On the hosts we spawned numerous `client` processes. Each process tried to establish connection with the `server` located on specified node. The clocks of particular nodes were synchronized with the NTP protocol [12], thus we were able to achieve the required accuracy while triggering the `clients`.

During the experiment we increased the number of `client` processes on each node to reach maximum number of connections incoming in one second that can be handled by the `server`. The test was performed for the connections with different security levels applied. We present maximum number of connections that were requested in one second and properly established (see: Tab. 1). We have observed that in case of the connection failure `client` returned *Connection timed out* error.

We did not provide results of a DoS attack for *clear text* transmissions since a successful DoS attack using raw TCP protocol may strongly affect server's availability and we did not perform such a test. We only present results of a test which shows that *clear text* transmissions are significantly more efficient that secured (see: Tab. 2).

Table 1. Results of the simulated DoS attack

| Security level | Connections | | | | Failed |
| | Requested | | Established | | |
	Overall	In 1 second	Overall	In 1 second	
AUTH	901	899	881	30	20
PROTECT	901	894	871	30	30
ENCRYPT	901	896	897	30	4

Table 2. TCP connections established in one second

| Security level | Connections | | | | Failed |
| | Requested | | Established | | |
	Overall	In 1 second	Overall	In 1 second	
CLEAR	1700	1692	1700	1691	0

4 Conclusion

The first presented experiment has shown that the average CPU time required to perform data transmission with integrity checking is over four times greater, than the time required for the *clear text* transmission. Transferring data via encrypted connections requires even more resources: the average CPU time for this connection is over seven times greater than for insecure ones. However, we should realize that the CPU time required for the most resource consuming connection is in the order of 1/10 milliseconds even though we did not perform the experiment on an extremely fast hardware. Thus, even the connection that requires the most computing resources should not cause significant overhead, neither in the CPU consumption nor in the delay of message delivery.

However, while designing the security policy of a system we should consider which security level is really required for data transfer. From the results of the first test we can see that the differences in CPU time consumption between particular levels of security is significant. Therefore, it is desirable to restrict the security aspects only to those which are really required by the system.

Considering the results of the second test we may conclude that the size of the transmitted packet is significant for the transmission efficiency. The average CPU time required to transmit 100 bytes significantly decreases with packet size increasing to the size of 1.5 kilobytes. Thus, it appears to be more efficient to transmit small amount of large packets than huge amount of small packets. This is also true for the raw TCP connections, however, this factor seems to be more significant for the secured ones.

The establishment of a secure connection is an exceptionally resource consuming process. Therefore, secure connections servers can handle fewer connections than those which use raw TCP sockets. However, we can conclude from the results of the third test that the amount of connections which can be handled by such a server in one second is large enough to respond requirements of the OCM-G. Also, we have found important the fact that in our test, the client that could not establish a connection with an overloaded server received the *connection timed out* network error. Thus it was possible to conclude the reason for which the connection could not be established.

5   Summary and Future Work

In this paper, we have shown the results of three experiments which tested the efficiency of different aspects of connection secured with cryptographical algorithms. We have evaluated the overhead resulting from secure data transfer as well as connection establishment.

It can be seen from the presented tests that applying security mechanisms to secure network connections results in significant overhead. The increase of CPU time required to transmit protected information in comparison to clear data transfer, as well as increased vulnerability of secured connections to the DoS attack cannot be passed over. Security offered by asymmetric cryptography algorithms results in significant resource consumption.

For the reasons described above, security mechanisms applied to data transmission in a system should always match the real security requirements. It seems to be useful to introduce optional lower security level for the less vulnerable network connections whenever it is permitted by the system security policy, especially when we desire a highly efficient data transmission.

On the basis of the results we have presented we can conclude, the overhead resulting from the security aspects applied to network connections are acceptable for the OCM-G as well as increased vulnerability to the DoS attack. However to make our results widely useful it seems to be necessary to work out an analytic model, that would make possible to estimate the overhead for the other, similar cases.

References

1. Bališ, B., Bubak, M., Funika, W., Szepieniec, T., and Wismuüller, R.: An Infrastructure for Grid Application Monitoring In: D. Kranzlmueller, P. Kacsuk, J. Dongarra, J. Volker (Eds.) Recent Advances in Parallel Virtual Machine and Message Passing Interface, Proc. 9th European PVM/MPI Users' Group Meeting, Linz, Austria, 2002, LNCS 2474, pp 41-49
2. Bališ B., Bubak M., Rząsa W., Szepieniec T., Wismüller R.: Security in the OCM-G Grid Application Monitoring System. In proc. of 5th International Conference on Parallel Processing and Applied Mathematics, September 2003.
3. Bellovin, S.: Security Problems in the TCP/IP Protocol Suite. Computer Communication Review **19** (2) 1989 pp 32-48
   http://www.research.att.com/~smb/papers/ipext.ps
4. Bubak, M., Funika, W., and Wismüller, R.: The CrossGrid Performance Analysis Tool for Interactive Grid Applications. In: D. Kranzlmueller, P. Kacsuk, J. Dongarra, J. Volker (Eds.) Recent Advances in Parallel Virtual Machine and Message Passing Interface, Proc. 9th European PVM/MPI Users' Group Meeting, Linz, Austria, 2002, LNCS 2474, pp 50-60

5. Bellovin S.: Defending Against Sequence Number Attacks. RFC 1948.
6. The CrossGrid Project. http://www.eu-crossgrid.org
7. Dierks, T., Allen, C.: The TLS Protocol Version 1.0. RFC 2246.
8. Foster, I., Kesselman, C., Tuecke, S.: The Anatomy of the Grid. International Journal of Supercomputer Applications **15** (3) 2001
9. Globus Alliance homepage: http://www.globus.org
10. GGF GSI Working Group: http://www.ggf.org/security/gsi/index.htm
11. Menezes, A., van Oorschot, P., Vanstone, S.: Handbook of Applied Cryptography. CRC Press, 1996
http://www.cacr.math.uwaterloo.ca/hac/
12. Network Time Protocol project homepage http://www.ntp.org/

An Idle Compute Cycle Prediction Service for Computational Grids

Suntae Hwang[1], Eun-Jin Im[1], Karpjoo Jeong[2], and Hyoungwoo Park[3]

[1] School of Computer Science, Kookmin University, Seoul, Korea
{sthwang, ejim}@kookmin.ac.kr
[2] College of Information and Communication, Konkuk University, Seoul, Korea
jeongk@konkuk.ac.kr
[3] Supercomputing Center, KISTI, Daejon, Korea

Abstract. The utilization of idle compute cycles has been known as most promising and cost-effective way to build a large scale high performance computing system, but not widely used because of the lack of effective idleness prediction techniques. In this paper, we argue PCs at university computer labs have a great potential for the utilization of idle CPU cycles, and propose two techniques for predicting idle cycles of those PCs: heuristic and statistical. Based on these techniques, we present the design and implementation of an idle compute cycle prediction service for computational grids. Our experimental results show that the utilization of idle compute cycles is a viable approach to cost-effective large scale computational grids.

1   Introduction

Scientific applications such as Molecular Simulation, High Energy Physics and Genome Informatics have challenging requirements for computation which can not be satisfied by the conventional supercomputing technology in the near future. Recent technical advances in grid computing provide us with an opportunity to tackle those problems by aggregating a large number of computers at organizations around world [3,6,7]. Finding idle resources is crucial for such approach; otherwise, organizations have to invest new computing platforms for grid computing. However, to predict the idleness of computing resources is generally very difficult because the schedulers do not have full knowledge about future tasks or take full control over them.

In this paper, we argue that university computer labs have a great potential for grid computing platforms for the following reasons. First, university computer labs have hundreds or thousands of PCs(Personal Computers) whose aggregate computing power can equal that of supercomputers. Second, those PCs are idle most of the time (e.g., at night). Moreover, they are almost completely idle during vacations. Third, PCs that do not have owners or valuable data are much less sensitive to the privacy issue that other computing resources. Finally, students' PC usage patterns are more regular and predictable than usage patterns of other computing resources because class schedules and school administration

M. Bubak et al. (Eds.): ICCS 2004, LNCS 3036, pp. 116–123, 2004.

schedules which are often repeatable and predictable determine students' activities in schools to a large extent[5].

In this paper, we propose two techniques for predicting idle compute cycles of PCs at university computer labs: *heuristic* and *statistical*. The heuristic technique uses the usage patterns of the last week for predicting those of the current week, but the statistical technique uses the *Hidden Markov Model* (HMM) [1] to predict future idle cycles. We compare these two techniques by real monitoring data collected for six months. Based on these two techniques, we present the design and implementation of an idle cycle prediction service for computational grids.

2   Idle Compute Cycle Prediction Models for University Computer Labs

Before we go on to data collection and analysis, first we clarify assumptions on our university PC labs.

- We do not consider the remote login or job queue length in both monitoring systems, because Microsoft Windows 2000 Professionals are running on the PCs. This is strict, but in this way, we can be sure that the user is not disturbed by other users harvesting his idle cycles. That means Grid jobs are submitted to those PCs by special scheduler, while other users can not submit remote jobs to those PCs.
- The PC rooms are classified into two types. One type of rooms, which we call *lecture rooms*, are assigned to school classes. The other type of rooms, which we call *open rooms*, are always open to students during office hours. Computers in *open rooms* are freely accessible to the users with authorized cards, while computers in *lecture rooms* are exclusively used by the enrolled students during lecture time but also freely accessible to the users between lecture times. Open hours are from 9AM to 9PM for both type of rooms.
- In this analysis, we treat one hour as a basic unit of idle-ness; that is, although a PC is used for a short period of time, it is assumed to be busy for the entire hour.

2.1   Data Collection

These days, most of PC rooms in universities adopt card system, where individual PC in the facility has magnetic card reader and the user gets authorization with their card to get access to a computer. The card system keeps track of data about when a PC is occupied by a user.

In our institution, we have collected card system log for the open rooms, but the card system log was not available for the lecture rooms, because the card system is off when the room is used for the lecture. For that reason, we ran a simple program on PCs in our lecture room; it infinitely repeats collecting load information in one minute's interval. We collected approximately 9.2MB of data on 54 PCs from April 2002 to September 2002. The raw monitoring data is converted into a sequence of idle, busy, and unknown periods. In this latter monitoring system for lecture rooms, a computer is defined to be idle if the keyboard or the mouse has not been touched for the last minute; otherwise, it is busy. In the card system for the open rooms, the definition of idle state is slightly different, where it means the previous user logged out, and none has logged in yet.

Because our primary concern was long running jobs, the data is again converted at hourly interval, where the state of a computer is marked *busy* if it was marked busy at least one minute in the hour, and marked *idle* otherwise. Since unknown state is the case when the PC is turned off, we marked unknown state idle because all PCs can be actually used if the administration policy is changed to turn on all PCs for 24 hours.

2.2   A Weekly Usage Pattern-Based Model and a HMM-Based Model

First, we designed a heuristic model based on our intuition that students' PC usage patterns are largely affected by class and school administration schedules usually repeated in a weekly basis. In this model, we simply assume that idle-ness patterns are repeated in a weekly basis. That is, we predict the idle-ness patterns of the current week based on those of the previous two weeks. For example, we predict a PC will be idle from 1:00pm to 2:00pm this Tuesday, only if the PC was idle during the same period of time last Tuesday and the previous Tuesday. For this reason, we call this model a weekly usage pattern-based or schedule-based model.

We also modeled the usage pattern of PCs using a *Hidden Markov Model* (HMM) In this model, we build a HMM for each PC in computer rooms, using two week's history of the computer usage. The number of states in the HMM was arbitrarily chosen to be 10, since, after testing different number of states, we concluded that the accuracy of the prediction was not much different in HMM with larger number of states. Then in the prediction step, we generate a most probable sequence of length equal to one week's usage sequence, using a Viterbi algorithm [8]. The generated sequence is used as a predicted usage pattern for the next week. The sequence length was in the unit of weeks, based on the observation in the previous subsection. With two week's usage pattern, we expect the HMM learns the hidden states of weekly pattern.

2.3   Comparison of Two Prediction Models

We compared the effectiveness of the two prediction models for various circumstances. We calculated *hit ratios* (ratios of correct predictions) of both prediction models for each PC on real monitoring data.

In this experiment, we assumed that the job length is two hours long because we are interested in long-running jobs. Each prediction is either hit (correct) or miss (wrong). We call a miss failure. Failures are again classified into two categories: *critical* and *non-critical*. The critical failure is the case where the model predicts the busy state of PC in the real world to be idle. The non-critical failure is vice versa. The former is more serious than the latter because that prediction causes a busy computer to be interrupted and scheduling work to be wasted.

The figures 1, 2 and 3 compare the hit ratios and critical failure ratios of the two prediction models. The prediction was again tested in three modes. In the first mode shown in figure 1, the prediction was performed using the whole data. In the second and third modes shown in figures 2 and 3, the prediction was performed respectively using the data for the daytime (9AM-9PM) and for the nighttime (9PM-9AM). For the schedule-based heuristic model, it only means that the times of the day of collected data are different, but for the statistical model based on HMM, it also means that the constructed HMM in each of three modes are different, because the training data were different. That is, the day model was built only with history of the daytime in the previous two weeks.

The left graphs in the figures shows the hit ratio, and the right graphs are for the critical failures. The data are again collected for different period, first two groups of bars for the whole 21 weeks excluding the first two weeks, the next two groups for the first 7 weeks (weeks 2-9) which belongs to the spring semester, and the final two groups for the remaining weeks in summer vacation. The bars in the graphs are paired for the two different prediction models, and the pairs of two different type of rooms, open room and lecture room, are again shown next to each other. For each pair of bars, the usage rate of the corresponding period is shown on the top of bars, so that we can relate the effectiveness of the models to the usage rate of computers.

Overall, the hit ratios are comparable in the two prediction models, but as the usage rate of the computer increases, the accuracy of the HMM-based model tends to be worse than that of the heuristic model. More importantly, we note that the rate of critical failures of the heuristic model is consistently lower than that of the HMM-based model.

3   Idle Compute Cycle Prediction Service

A computation grid consists of various services. To utilize idle compute cycles in university computing room for computational grid, we need local resource management system that can continuously update PC's load state in computing room, and monitor and control allocated jobs.

Applications designed to execute on computational grids frequently require the simultaneous co-allocation of multiple resources in order to meet performance requirements. Since this paper focuses on utilizing of idle compute cycles, we mention to computational resources only although the definition of resources includes all devices that an application might require, including networks, memory, storage, rendering hardware, and display devices. Because it is not reasonable to assume a centralized global scheduler in Grid environment, scheduling conflicts may result in over all degraded performance. One approach to enhance the local resource management system is to incorporate advance reservation capabilities

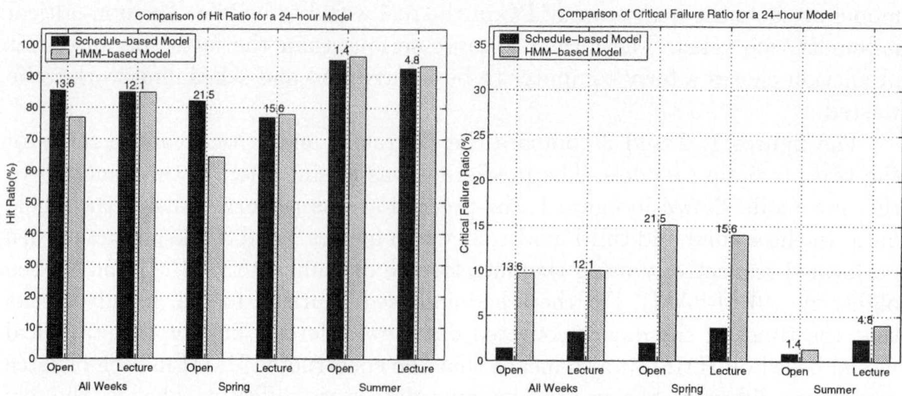

Fig. 1. Prediction Result of two Models in a 24-hour mode. The numbers on the top of bars are the usage rates of the corresponding period.

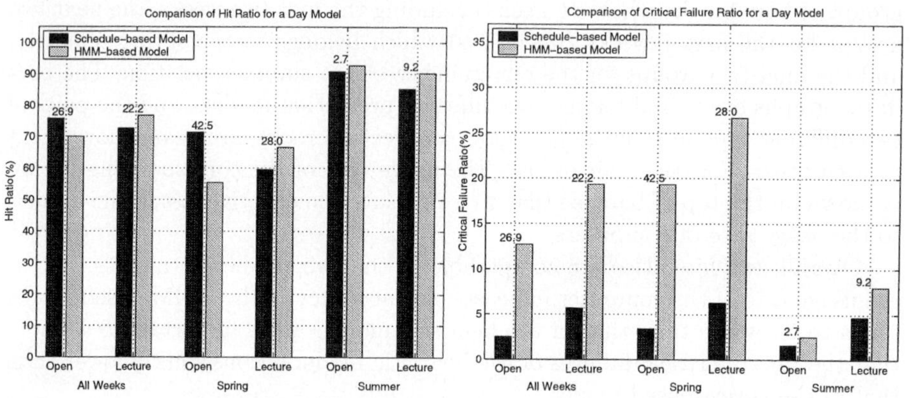

Fig. 2. Prediction Result of two Models in a day mode.

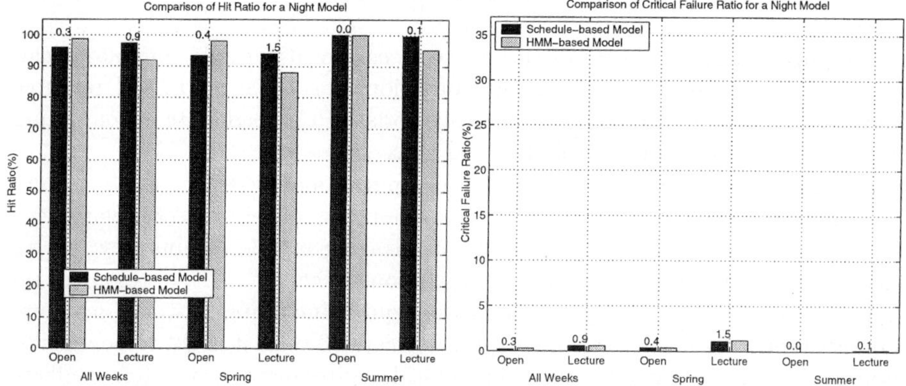

Fig. 3. Prediction Result of two Models in a night mode.

into a local resource manager. Then, a co-allocator can obtain guarantees that a resource will deliver a required level of service when required[2,4].

Therefore forecasting of the load of computational resource is very important for resource reservation and co-allocation. Therefore, local resource management system needs to provide continuously prediction service. Grid system on OGSA foundation is formed in various service form. Among these, scheduler or co-allocator will also exist as OGSA service form. To properly operate such services, local resource state should be predicted by prediction service. And based on this, resource will be either allocated or reserved.

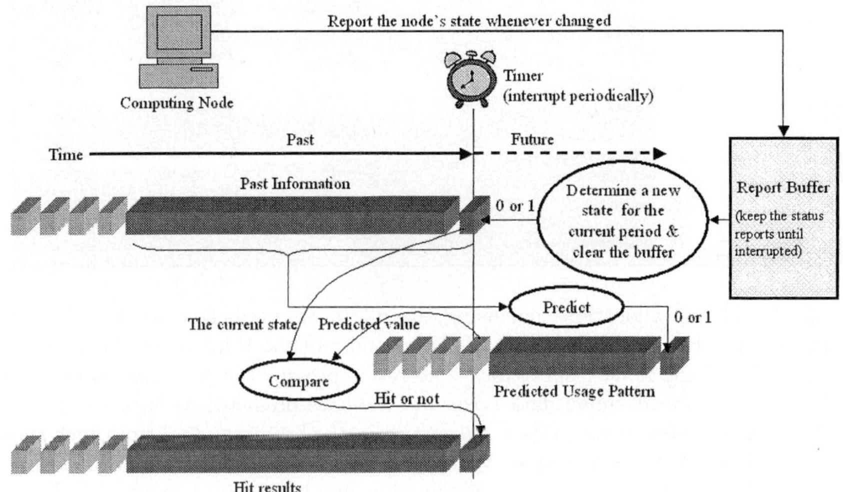

Fig. 4. Prediction service implementation. Whenever timer periodically interrupts, circled actions will be processed. Here, we have used schedule-based prediction model which was described in Section 2.

4   Implementation

Based on prediction model, which was described in Section 2, we have implemented and added prediction service to our local resource management system. States of nodes that are reported from each PC in compute room, are recorded in resource management system's buffer. And current period's state is determined whenever periodical (e.g. one hour) interruption occurs based on the state reports which were recorded in buffer. In other words, it is *idle* only reported states in buffer are all idle, otherwise it is regarded as *busy* if there is even one busy state was reported during the interval. As shown in Figure 4, if timer interrupts, prediction server determines the new state of the node, predicts a state of a week ahead according to the updated past information, and recalculates the usage ratio and hit ratio of the given duration(for example two weeks in our case).

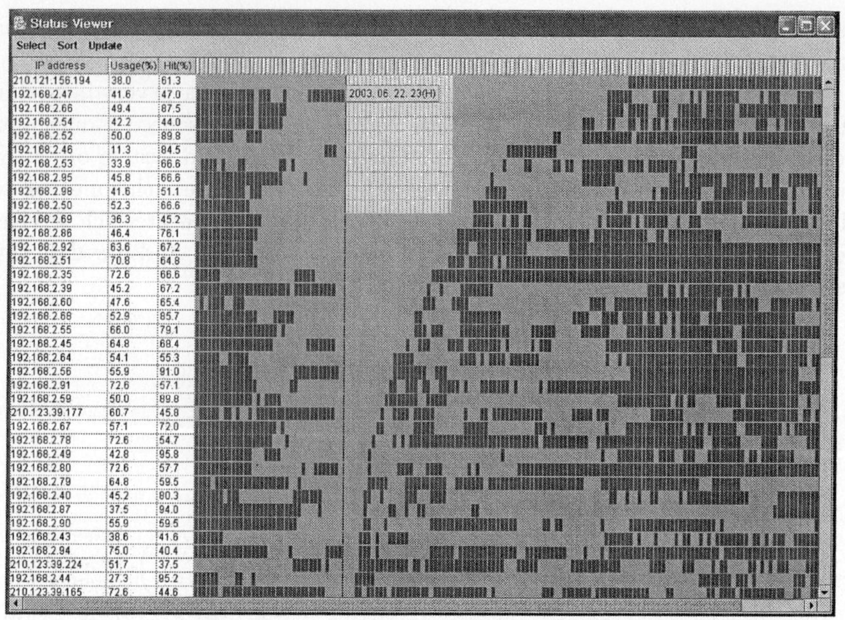

Fig. 5. A resource allocation tool using the idle cycle prediction service. The first column represents host IP address and the second and the third represent usage rate and hit ratio of the node respectively. The rest represent the predicted state of the node for a week ahead. Small dark boxes are the ones predicted as *busy* and the rest are the ones predicted as *idle*. If you put mouse on the head row, tool tip, which shows applicable date and time, will show. When you click the mouse button on the tool tip, data will be sorted in order which is predicted as will be idle for longest period based on that moment(indicated by the vertical line). And little more lightly marked block is selected part for reservation.

Figure 4 shows a data structure for a single node. The prediction server returns predict patterns, host IP addresses, usage ratios and hit ratios as many as number of nodes. Figure 5 shows a snapshot of a tool which can be used together with job preparation tool.

5   Conclusions and Future Work

The lack of effective techniques for predicting long idle cycles is one of major obstacles to building a large scale cost-effective high performance computational grid. In this paper, we proposed two techniques to predict long idle cycles for PCs at university computer labs. One technique is designed based on our intuition that students' PC usage patterns are largely affected by class and school administration schedules usually repeated in a weekly basis. The other technique is based on a statistical model called Hidden Markov Model (HMM).

We examined the effectiveness of these two techniques based on real monitoring data for PCs at our university computer labs. For these experiments, PCs were monitored for about six months. Overall, both of these two techniques consistently showed high correct prediction rates for various circumstances. But our intuition-based heuristic technique outperforms the HMM-based technique in most cases. For the crucial case where busy states are predicted to be idle, the effectiveness of the heuristic technique is significantly better than that of the HMM-based technique. That is, our intuition about idleness patterns of PCs is proven to be better.

Based on these techniques, we presented the design and prototype implementation of a idle cycle prediction service for computational grids. Currently, we have been integrating this service into Globus toolkit 2.0. In addition, we plan to reimplement this as a service for Globus toolkit 3.0 and to develop a grid meta scheduling technique based on this service.

Acknowledgments. This work was partially supported by MIC (Ministry of Information and Communication) through National Grid Infrastructure Implementation Project of KISTI (Korea Institute of Science and Technology Information). It was also in part supported by University IT Research Center Project and by Grant No. R04-2002-000-20066-0 from Korea Science and Engineering Foundation.

References

1. Jeff Bilmes. What HMMs Can Do. UWEE Technical Report UWEETR-2002-0003, University of Washington, January 2002.
2. Karl Czajkowski, Ian Foster, and Carl Kesselman. Resource co-allocation in computational grids. In *The Eighth IEEE International Symposium on High Performance Distributed Computing*, August 1999.
3. I. Foster and C. Kesselman. Globus: A Toolkit-based Grid Architecture. In I. Foster and C. Kesselman, editors, *The Grid: Blueprint for a New Computing Infrastructure*, pages 259–278. Morgan Kaufmann, 1999.
4. I. Foster, C. Kesselman, C. Lee, R. Lindell, and K. Nahrstedt. A distributed resource management architecture that supports advance reservations and co-allocation. In *International Workshop on Quality of Service*, 1999.
5. Suntae Hwang, Karpjoo Jeong, Eun-Jin Im, Chongwoo Woo, Kwang-Soo Hahn, Moonhae Kim, and Sangsan Lee. An Analysis of Idle CPU Cycles at University Computer Labs. *Lecture Notes in Computer Science*, 2667(1):733–741, 2003.
6. C. Liu, L. Yang, I. Foster, and D. Angulo. Design and evaluation of a resource selection framework for Grid applications. In *Proc. of 11th IEEE Symposium on High Performance Distributed Computing*, July 2002.
7. V. Subramani, R. Kettimuthu, S. Srinivasan, and P. Sadayappan. Distributed job scheduling on computational grids using multiple simultaneous requests. In *Proc. of 11th IEEE Symp. On High Performance Distributed Computing*, july 2002.
8. A. J. Viterbi. Error bounds for convolutional codes and an asymptotically optimal decoding algorithm. *IEEE Trans. Informat. Theory*, IT-13:260–269, April 1967.

Infrastructure for Grid-Based Virtual Organizations

L. Hluchy, O. Habala, V.D. Tran, B. Simo, J. Astalos, and M. Dobrucky

Institute of Informatics, Slovak Academy of Sciences
Dubravska cesta 9, 84507 Bratislava, Slovakia
hluchy.ui@savba.sk

Abstract. This paper presents architecture of a collaborative computation environment based on a Grid infrastructure, used as a support for large scientific virtual organizations. The environment consists primarily of a collaboration-supporting user interface, workflow system capable of submission of jobs to the Grid and a Grid-based data management suite. A prototype of such an environment is deployed and tested for a flood forecasting system. The system consists of workflow system for executing simulation cascade of meteorological, hydrological and hydraulic models, data management system for storing and accessing different computed and measured data, and a set of web portals.

1 Introduction

In recent years a number of scientific projects with international (even global) participation emerged as an answer to increasingly complicated problems of modern science – a well-organized business, with dense network of cooperation between people, organizations and countries. Such cooperation also requires an effective toolset for communication, experiment management and results sharing. The natural way to produce such a toolset is to develop a network-enabled software suite. Such software suites exist – although mainly incomplete and not mature – and in recent years are becoming more oriented toward the paradigm of virtual world-wide resource sharing – Grid computing.

2 The Architecture of a Grid Infrastructure for Virtual Organization

The basic perpetual cycle of work in a scientific virtual organization is simple – data is processed, another data is created. Of course, new data also enters the virtual organization and produced data is used, viewed, analyzed and interpreted to obtain results as the main purpose of the actual existence of the organization.

So we begin the analysis of the software infrastructure with the word data, and this tells us that the support of work with this data is one important part of the infrastructure – the data management. Another part of the infrastructure is in the core of the

M. Bubak et al. (Eds.): ICCS 2004, LNCS 3036, pp. 124–131, 2004.

cycle – the processing facility. This is the oldest part of any software and the very first PSEs were just a layer of control above such a computational core. Current efforts for computation management widely employ workflows (sequences of simpler tasks) and workflow management. This becomes important with the steadily increasing complexity of scientific computation, making a single integrated computational module too robust and inflexible.

To actually enable two or more scientists to work together towards achieving a common goal, a collaboration and communication suite integrated in the infrastructure is necessary. The word „integrated" means that this part is connected together with other controls available to the user.

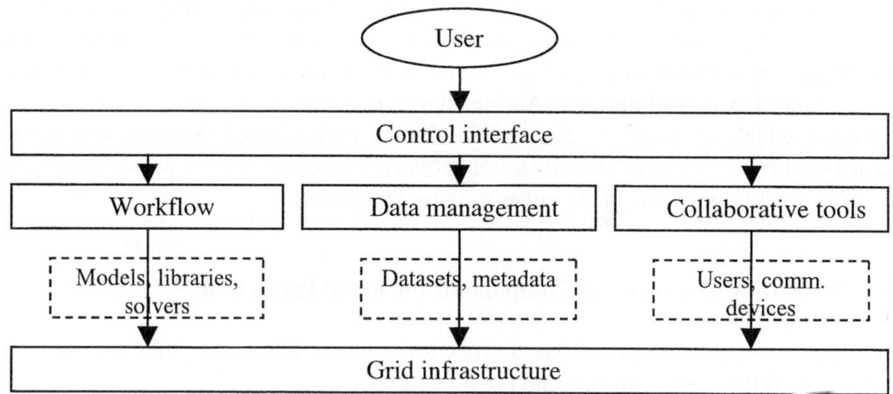

Fig. 1. Architecture of a Grid infrastructure for scientific VO

So we may say that the Fig. 1 is a basic and general architecture of a modern Grid-based collaboration environment. Many concepts inherent to the environment are abstracted and the data flow is also much more complicated, with parts of the workflow requiring access to data and data management tools, users accessing the stored datasets and all parts actively communicating with users.

3   Workflow Management

As the grid infrastructure matures it is being used by scientists for more and more complex computations. Each such computation can include executions of several applications and transfers of required data. The complexity of the process is becoming too high to be handled manually. Therefore the employment of workflow concept seems quite natural. Similar to the definition of a workflow in a business process management [2], a grid workflow is an automation of a grid process, during which documents, information or data are passed from one grid service to another for action, according to set of procedural rules.

3.1 Existing Systems

Most of grid workflow systems being developed focus on the web services, although there are older systems that do not use the web services paradigm.

One example of system not using web services can be the Condor DAGMan [3] – a meta-scheduler for the Condor workload management system. It uses a directed acyclic graph (DAG) to represent a set of programs where the input, output, or execution of one or more programs is dependent on one or more other programs. Pegasus [4] is a system for transforming abstract workflow descriptions into concrete workflows, which are then executed using the Condor's DAGMan.

Web services oriented workflow systems are mostly in early stage of development as can be seen in the Scientific Workflow Survey [5] web page. There are two main specifications of workflow languages: Web Services Flow Language (WSFL) [14] targeting web services and the Grid Services Flow Language (GSFL) [15], which builds upon Open Grid Services Architecture (OGSA) [16]. OGSA is based on grid services, which are web services with additional grid-oriented features, and allows distributed resources to be shared over a network. Currently, we are not aware of any existing workflow system using either of these languages.

3.2 Workflow Management Components for Flood Prediction

We need an interactive portal-based workflow system that enables the user to construct a workflow or to choose from predefined ones.

As for the interactivity, it means the possibility to view the results of each task (activity) instantly after it has finished without waiting for the whole workflow to finish and ability to clone existing (possibly running) workflow and submit it with modified parameters. The modification may cover one or more tasks.
Important feature is the ability to replace selected step or steps in the workflow with user selected or defined "output" in order to let the user perform various parameter studies. Such replacement must be possible both during workflow definition and during workflow execution.

4 Collaborative Tools

The need of cooperation between scientists and users from many organizations in Grid projects requires sophisticated tools for collaborations in portals. The scientists need to access and share data, analyze them, and discuss with other scientists via the collaborative tools. Therefore, collaborative tools are one of the key elements of virtual organizations.

Collaborative tools may be mailing lists, instant messaging, file-sharing tools, discussion groups, etc. However, one single tool cannot provide all features necessary for the collaborations. Therefore, there are several projects that aim to provide an integrated and extensible collaborative environment via portals. One of such projects, used also by us in the FloodVO portal design is CHEF [17].

CHEF (CompreHensive collaborativE Framework) is a collaborative environment based on Jespeed portal engine [6]. The collaborative tools (teamlets) are written as portlets in Jetspeed that are extended to special features for multi-user group work nature of collaborative tools:

- Resource access security: The users can only view and modify what they have permissions to.
- Automatic updating of displays: as a user makes changes that effects the display other users are viewing, their display is automatically updated.
- Multi-user safe: if several users are using the same tool at the same time, they work together to avoid conflicts.
- Presence: every user can see who else is using the same tool in the same area at the same time.
- Notification: every user can request to be notified of changes made through the tool by other users.

For accessing to the collaborative tools in CHEF, users need a standard web browser and access to the portal.

5   Data Management for Scientific Virtual Organizations

The increasing needs for volume and accessibility of data in scientific computations in the last decade leads also to increased demands on better data management tools. The main responsibilities of such software are:

- To track available datasets in the virtual organization.
- To store and maintain these datasets in a coherent fashion.
- To publish their properties and enable their discovery.
- To enable their download and usage.

The data stored in a virtual organization's data storage facilities has two main parts - the actual datasets and their metadata (meaning their description by another layer of data). Thus, also the data management efforts are divided into two main streams – replica management and metadata storage/lookup.

5.1 Replica Management

The actual storage and maintenance of a coherent dataset collection is performed by replica management software. It keeps track of the datasets, potentially stored at multiple places duplicitly (replicated). The creation of replicas of a single dataset may be well used for better security and protection against an unwanted loss of the dataset because of a sudden storage device failure, as well as for better access to the file by making it more local to the place that requires it. The software developed in work package 2 of the EU IST DataGrid project [10][18] covers the registration, lookup, transfer and replication tasks of a mature replica management suite, with sufficiently distributed control. Its last implementation is based on the modern paradigm of web services [19] and OGSA [16] architecture. Anyway, it is lacking a modern and scalable metadata repository.

5.2 Metadata Management

The problem of managing and searching the descriptive information of the dataset collection of a virtual organization (especially for large international scientific Grid-based virtual organizations) is in its nature very similar to the problems of recent peer-to-peer computing efforts. Potentially, the space of storage nodes in such an organization is very large and the especially the distributed lookup is a non-trivial problem. Various solutions have been proposed and evaluated [20], but the more efficient of them pose severe restrictions on the stored metadata. But several peer-to-peer infrastructure problems, connected with the high instability of the whole network may be disregarded in Grid computing, and in such a controlled environment a decentralized and efficient metadata registry may be deployed. Also, considering the better and more reliable network infrastructure available in grids, a certain level of centralization may be tolerable, without the fear of creating a single point of failure of a bottleneck in the metadata lookup middleware.

6   Test Case – Virtual Organization for Flood Prediction in the CROSSGRID Project[1]

The efforts presented above are applied and tested in an international IST project CROSSGRID [13]. Collaborative tools, data management as well as workflow control have been employed and used in a way needed for flood prediction in the Grid-based Virtual Organization for Flood Prediction (FloodVO).

6.1   Workflow in FloodVO

A workflow system that we designed for our flood prediction system enables the user to define whole cascade execution in advance as a workflow and run it with the possibility to inspect every step.

The whole flood simulation uses three main steps – meteorology, hydrology and hydraulics - to produce the final result – the prediction of the parts of the target area that are going to be flooded. When the expert wants to use already computed results or does not need to compute the last step of the cascade, just parts of the cascade are required. The run of a single simulation model represents the simplest case.

---

[1] This work is supported by EU 5FP CROSSGRID IST-2001-32243 RTD project and the Slovak Scientific Grant Agency within Research Project No. 2/7186/20

We have decided to constrain the workflow selection to several predefined workflows in the first version. Workflow is defined for each target area based on the computation dependencies for that particular area. The changing part of the workflow is mainly hydrology because the run-off in the target catchment is computed from several subcatchments.

An expert who wants to perform a simulation chooses a target area and time for which to make the prediction, then the workflow template from the list of templates available for the area of interest and a model to be used in each step. The possibility to select more models for the same step or even to enter user defined values instead of running a particular simulation step makes it possible to have several parallel instances of a workflow giving several results for the same time and area.

6.2   Data Management in FloodVO

The general schema of possible data sources for FloodVO operation was described in previous work [9]. The most important data in FloodVO storage are the boundary condition for the operation of our meteorological prediction model ALADIN. The second type of data implemented are radar images of current weather conditions, and the third type are the ground-based water level, precipitation and temperature measurements provided by SHMI's (Slovak Hydrometeorological Institute) network of measurement stations.

Data management in the prototype of FloodVO was implemented using the DataGrid software. The metadata database was implemented using the MySQL [11] RDBMS and an OGSA frontend. A service and a client application have been implemented. The client enables to add, modify, locate and delete metadata for given file in the FloodVO. The data management will be incorporated into the workflow and portal system in the near future. Users will be able to locate data and construct workflows based on metadata descriptions.

6.3   User Interfaces for Collaboration

There are three different user interfaces for the flood application. We have developed GridPort [22] based application portal, we are developing flood application specific portlets for the Jetspeed portal framework based application portal and we are being integrated with Java based client called Migrating Desktop.

Application Portal Based on the Jetspeed Portal Framework
The Jetspeed [6] portal framework has been chosen in the CrossGrid project as a modern powerful platform for creating grid application portal for the applications in the project (Fig. 2). This framework is also being used by other Grid projects such as Alliance portal [7] and the new version of the GridPort toolkit – GridPort 3.0 [8].

Jetspeed provides framework for building information portals (pluggable portlets mechanism, user interface management, security model based on permissions, groups and roles, persistence of information etc.) but does not provide any support for grid services and applications. Common Grid portlets that can be used in Jetspeed are being developed in CROSSGRID and other projects. Portlet for submission of specific simulation models of flood application has been developed and now we are focusing on automatization of a computation of the flood simulation cascade by employing workflows. We are also investigating the possibility of using groupware portlets from the CHEF project.

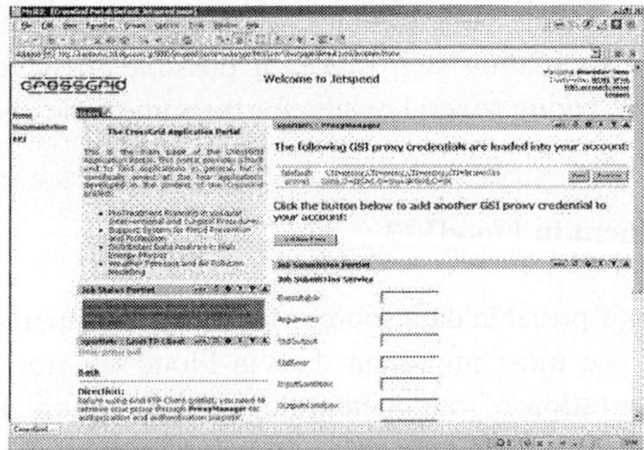

Fig. 2. Screenshot of the Jetspeed based application portal

Migrating Desktop
Migrating Desktop is a Java client being developed in the CrossGrid project. The idea was to create user interface with greater interactivity than could be possible to achieve by using web technology. Current version provides access to basic Grid services. Support for specific application features is addressed by application and tool plugin interfaces that enable to plug in code handling application specific parameter definition and visualization. We have implemented both plugins for the flood application.

7   Future Work

Many simulation models, in some cases, are not very reliable and are also dependent on many other factors (physical phenomena), which are not included in evaluation process of the models. Knowledge based treatment of historical data could provide enhanced functionality for the simulation models which strictly relies on the recent data sets. It also allows constructing several predicates of states according to knowledge evaluation with simulation run. More work will be also done in the area of workflow control and data management, where a more distributed and robust approach will be necessary.

References

1. Gallopoulos, S., Houstis, E., Rice, J.: Computer as Thinker/Doer: Problem Solving Environments for Computational Science. IEEE Computational Science and Engineering Magazine, 1994, Vol. 2, pp. 11-23.
2. Workflow Management Coalition Terminology & Glossary. http://www.wfmc.org/standards/docs/TC-1011_term_glossary_v3.pdf (visited December, 2003)
3. Condor DAGman. http://www.cs.wisc.edu/condor/dagman (visited December, 2003)
4. Pegasus. http://www.isi.edu/~deelman/pegasus.htm (visited November, 2003)
5. Scientific Workflows Survey. http://www.extreme.indiana.edu/swf-survey/ (visited December, 2003)
6. Jetspeed. http://jakarta.apache.org/jetspeed/site/index.html (visited October, 2003)
7. Alliance portal. http://www.extreme.indiana.edu/alliance/ (visited October, 2003)
8. Grid Port 3.0 Plans presentation. http://www.nesc.ac.uk/talks/261/Tuesday/GP3%20HotPage%20Combined%20Edinburgh%20Presentation.ppt (visited December, 2003)
9. Hluchý L., Habala O., Simo B., Astalos J., Tran V.D., Dobrucký M.: Problem Solving Environment for Flood Forecasting. Proc. of The 7th World Multiconference on Systemics, Cybernetics and Informatics (SCI 2003), July 2003, Orlando, Florida, USA, pp. 350-355.
10. Hoschek, W., et. al.: Data Management in the European DataGrid Project. The 2001 Globus Retreat, San Francisco, August 9-10 2001.
11. Widenius, M., Axmark, D.: MySQL Reference Manual. O'Reilly and Associates, June 2002, 814 pages.
12. Bell, W., et. al.: Project Spitfire - Towards Grid Web Service Databases. Technical report, Global Grid Forum Informational Document, GGF5, Edinburgh, Scotland, July 2002
13. Development of Grid Environment for Interactive Applications. IST-2001-32243. http://www.eu-crossgrid.org/ (visited December, 2003)
14. Web Services Flow Language. www.ibm.com/software/solutions/webservices/pdf/WSFL.pdf (visited October, 2003)
15. The Grid Services Flow Language (GSFL).
16. http://www-unix.globus.org/cog/projects/workflow/ (visited October, 2003)
17. Foster, I., Kesselman, C., Nick, J. M., Tuecke, S.: The Physiology of the Grid; An Open Grid Services Architecture for Distributed Systems Integration. http://www.globus.org/ogsa/ (visited December, 2003)
18. CHEF Information site. http://www.chefproject.org/ (visited December, 2003)
19. Peter Kunszt, Erwin Laure, Heinz Stockinger, and Kurt Stockinger. Advanced Replica Management with Reptor . In 5th International Conference on Parallel Processing and Applied Mathematics, Czestochowa, Poland, September 7-10, 2003. Springer Verlag.
20. W3C Web Services Activity web site. http://www.w3.org/2002/ws/ (visited December, 2003)
21. Joseph, S., Hoshiai, T.: Decentralized Meta-data Strategies: Effective Peer-to-Peer Search. IEICE Trans. Commun., Vol. E86-B, No. 6. June 2003.
22. W3C Resource Description Framework web site. http://www.w3.org/RDF/ (visited December, 2003)
23. Thomas, M., Mock, S., Boisseau, J., Dahan, M., Mueller, K., Sutton, D.: The GridPort Toolkit Architecture for Building Grid Portals. Proceedings of the 10th IEEE Intl. Symp. on High Perf. Dist. Comp. August 2001.

Air Pollution Modeling in the CrossGrid Project[*]

J. Carlos Mouriño, María J. Martín, Patricia González, and Ramón Doallo

Computer Architecture Group. Department of Electronics and Systems
University of A Coruña, Spain
`jmourino@mail2.udc.es`

Abstract. The CrossGrid project develops, implements and exploits new Grid components for interactive compute and data intensive applications like simulation and visualization for surgical procedures, flooding crisis team decision support systems, distributed data analysis in high-energy physics, and air pollution combined with weather forecasting. We present in this paper the integration of the air pollution application with other components of the project and show the benefits of using a Grid platform for running this application.

1 Introduction

Grid technologies have been under development since the late 1990s, extending cluster and distributed computing concepts to allow organizations to share and combine resources. The purpose of this work is to show the usefulness of the Grid in air pollution modelling. STEM-II, an Eulerian air quality model, is used to simulate the environment of As Pontes Power Plant in A Coruña (Spain). The STEM-II program, forced by meteorological data generated by an atmospheric model, simulates gaseous and aqueous concentrations fields of different pollutant species, reaction rates, amount of deposited species and ionic concentrations. The code is computationally intensive, thus, Grid computing should be potentially applied to achieve a reasonable response time.

The CrossGrid Project [1] offers us a Grid environment oriented towards compute and data–intensive applications that need interaction with an external user. Our model needs the interaction of an expert in order to make decisions about modifications in the industrial process. For this reason, besides the parallelization of the model, a Graphical User Interface (GUI) was developed with the possibility of testing the results and modifying parameters in real time.

The structure of this paper is as follows. Section 2 presents a brief overview of the CrossGrid Project. In Section 3, we briefly describe the air pollution model STEM–II. The GUI designed to support the interactivity needed for the application and its functionality is also described. In Section 4 the interactions with other CrossGrid components are shown. Finally, in Section 5, we briefly comment the future work.

---

[*] This work is supported by the CrossGrid European Project (Ref: IST-2001-32243)

M. Bubak et al. (Eds.): ICCS 2004, LNCS 3036, pp. 132–139, 2004.
© Springer-Verlag Berlin Heidelberg 2004

2   The CrossGrid Project

The project is developed with the purpose of supporting applications that require high computational resources and a great amount of data, with the characteristic that real time interaction with users is needed. Examples of these applications are: interactive simulation and visualization for surgical procedures [2], flooding crisis team decision support systems [3], distributed data analysis in High-Energy Physics (HEP) [4], and air pollution [5] combined with weather forecasting [6].

For the efficient development of this kind of applications into the Grid, new tools for verification of parallel source code, performance prediction, performance evaluation and monitoring are needed and are also developed into the project. All these applications and tools require new components into the Grid for application-performance monitoring, efficient distributed data access, and specific resource management. The final users should run the applications on the Grid in an easy and transparent way, without the need of knowledge on technical details or on how the Grid is implemented and how it works. Thus, the project develops user-friendly portals and personalized environments to access the applications and developed tools.

The project also offers a testbed architecture to test, validate and execute tools, applications and the Grid services developed. Testbed sites are distributed among ten different institutions in nine European countries. All this individual elements, together with others achievements from other Grid projects such as DataGrid and EuroGrid, are integrated in a new Grid environment.

3   The STEM-II Air Quality Model

Coal-fired electrical power plants constitute one of the most significant sources of air pollution, and for that reason, its study is one of the keys in pollution control. The aim of our work is to control the emissions produced by the Endesa Power Plant sited at As Pontes (A Coruña, Spain). This power plant generates 1400 MW of electrical energy by means of coal combustion. A coal mixture made of a local lignite, with a high content in sulphur, and other foreign coals, of greater calorific power and free of sulphur, are consumed. One of the objectives of the technicians of the power station is to know the optimal mixture so that the maximum yield is obtained fulfilling, at the same time, the norm on emissions and quality of the air.

In order to carry out the complete characterization on a regional scale of the dispersion of atmospheric pollution, it is necessary to simulate different pollutant emissions, transport by advection, convection and turbulence, with dynamic meteorological conditions; chemical transformation; and pollutant removal by deposition. STEM-II (Sulphur Transport Eulerian Model 2) is an Eulerian air quality model which simulates transport, chemical transformations, emissions and deposition processes in an integrated framework [7]. The model is computationally intensive because the governing equations are nonlinear, highly coupled and stiff. The speedup to be achieved in the simulation process is really important in order to save time when making decisions about modifications in the

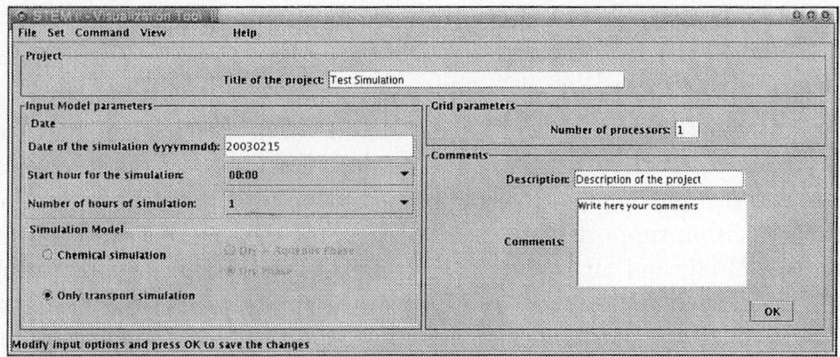

Fig. 1. Input parameters window

industrial process to fulfill the European regulations referring to the emission of pollutants. As with other computationally intensive problems, the ability to fully utilize these models remains severally limited by today's computer technology. Grid computing will be applied to achieve a reasonable response time.

3.1   Supporting STEM–II Interactivity in a Grid Platform

With the aim of helping the experts to make decisions, we have developed a user-friendly interactive visual tool which allows to set the input/output parameters, and watch the concentrations of pollutants graphically. This interface is interactive, that is, the technicians can vary parameters in real time and follow graphically the impact of these changes.

One of the features we had in mind to develop the graphical user interface (GUI) was the portability of the final tool. The Java Foundation Classes (JFC) are a set of Java class libraries provided as part of J2SE to support building graphics user interface and graphics functionality for Java technology-based client applications. Apart from these features, JFC is also used by other developers in the CrossGrid project. So the use of Java technology to develop the Air Pollution Visualization Tool will simplify its future integration with other tools of the project.

Another important objective to be considered was the abstraction of the graphical interface. This objective aims to present the Grid execution as transparent as possible for the user. To achieve it the Visualization Tool consists of three main components:

- User Interface, in charge of building the user environment: windows, panels, buttons, messages, etc.
- Data processing module, in charge of the image processing from the original data files. A module to process and convert the data in these matrices into image files is also provided in the Visualization Tool.
- Interface with the Grid platform, includes the construction of the RSL files and the job submission to the Grid platform via Globus. This is the most complex part of the tool. It involves relations both with the User Interface and with the data processing module.

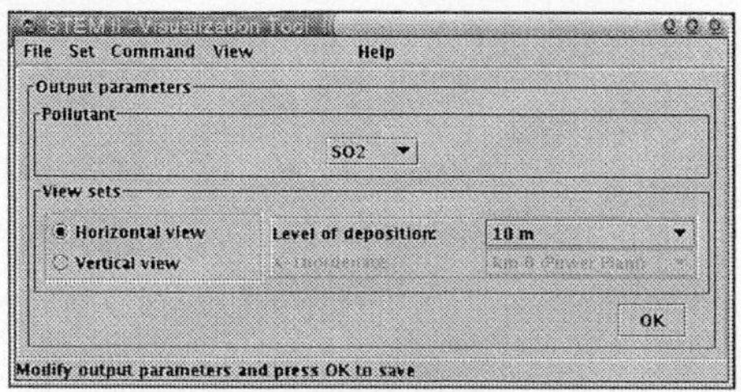

Fig. 2. Output parameters window

An initial window of the tool including the logo of the project and a menu bar with several options (`File`, `Set`, `Command`, `View` and `Help`) is first presented to the user. In the `File` item the user can choose the typical options: start a new project, load a saved project, store the actual project or finish the application.

The input and output parameters of the model can be set in the `Set` item (see Figure 1). This parameters are the date, the start hour and the number of hours of the simulation. Besides, some flags for the execution of an only transport simulation, or a simulation with chemical reactions in dry, or dry plus aqueous phases can be set.

In the `View` item we can see the specific graphical representation chosen in the output parameters window (see Figure 2). In this window the pollutant to visualize (SO2, O3, SO4-) can be set, as well as the type of view (horizontal or vertical), and the level or coordinate of representation in function of the view.

From the `Command` item the program can be submitted to the Grid. A graphical window with the selected output will be generated. Figure 3 shows an example with an horizontal view. The `type` box shows the pollutant that is being represented while the `level` box presents the altitude of the representation. The concentrations are represented with a background map for a better understanding of the results. The location of the power plant is also indicated. Moreover, a scale with a color code is shown on the right part of the image. The image is refreshed on real time when new concentration data are available. We can change interactively any output parameter, and the new representation will be shown.

4   The Air Pollution Application into the CrossGrid Framework

The STEM–II model requires as input data meteorological data such as temperature, wind fields, precipitation, etc. This meteorological data are provided by the COAMPS [6] meteorological prediction model maintained in the Interdisciplinary Centre for Mathematical and Computational Modelling (ICM) at Warsaw. This application generates data in the format and resolution needed. The CrossGrid project provides mechanisms to share these data. The file management policy is based on replication. Replica catalogues are used for cataloging and locating replicated files in the distributed CrossGrid environment.

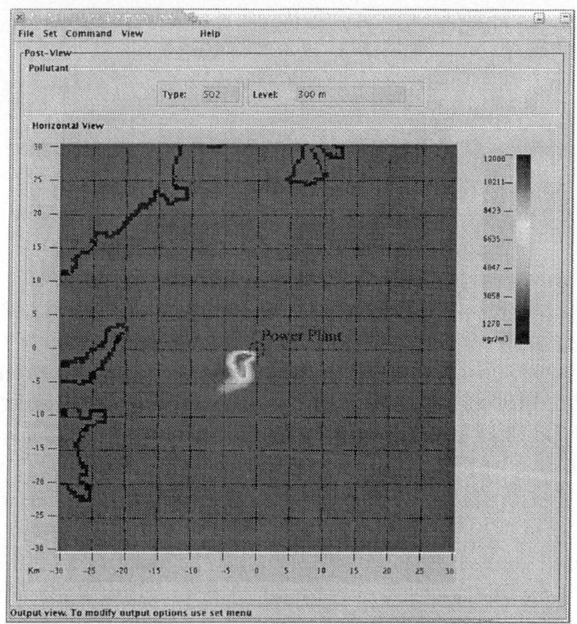

Fig. 3. Horizontal view of SO2 concentration at 300 m

Previously to the Grid-enable version of the code we have developed a MPI parallel version of the program. Details about this parallelization and experimental results in different platforms can be obtained from previous works [5]. Additionally, an experimental Grid built in our laboratories has also been used for testing the parallel code [8].

The CrossGrid project offers the possibility of testing the MPI implementation using MARMOT [9], an MPI verification tool developed at the High Performance Computing Center Stuttgart (HLRS). The objective of MARMOT is to verify the correctness of parallel, distributed Grid applications using the MPI paradigm. Thanks to this tool some "holes" were detected in our communications, which introduced unnecessary overhead in the parallel program. We packed non–consecutive data in each communication using complex datatypes. These datatypes have useful data and "holes" that must be skipped in the communications. MARMOT warned us that these "holes" were communicated as empty data. Now, that "holes" have been eliminated, changing the datatypes and increasing the number of communications, and the performance of the communications has been improved.

Besides, there are another tools in the project that propose a set of performance metrics to describe concisely the performance capacity of Grid applications. One of them is GridBench [10], developed in the High Performance Computing Laboratory (HPCL) of Dept. of Computer Science at University of Cyprus. It develops and implements benchmarks that are representative of typical Grid workloads. Such benchmarks are used for estimating the values of performance metrics for different Grid configurations, identifying important factors that affect end-to-end application performance, and providing application developers with initial estimates of expected application performance. Another task in the project develops an on-line tool, PPC [11], that allow application developers to measure, evaluate and visualize the performance of Grid applications with respect to data transfer, synchronization and I/O delay as well as CPU, network and storage use.

The MPI parallel code was transported to the Grid environment fairly easily by introducing appropriate components from the Globus bag of services [12]. Information about available Grid resources is essential for job scheduling. The Globus toolkit uses the MDS (Monitoring and Discovering Service) to publish static and dynamic information about existing resources. This implementation is based on the LDAP protocol. The Workload Management System (WMS) manages the Grid resources guarantying that jobs are executed in the "best" resources. The WMS consist of: the Resource Broker (RB) [13], that is the responsible of matching job requests, written in JDL, with the available resources; the Job Submission Service (JSS), that performs the job submission, using the Globus GRAM service, to the remote Computing Element found by the RB; the User Interface (UI), which is the component between the user and the RB and allows the user to obtain information about the jobs, transfer data files, find resources, cancel jobs, etc; and the Logging and Bookkeeping (LB), that keeps information about the executions.

Some results have been obtained submitting directly the program to specific nodes of the CrossGrid testbed (See next Section 4.1). Currently, we are working in the integration of our application into the portal and the migration desktop. Both elements are developed with the support of the roaming access subtask, that allows the users to access the Grid from different locations and different platforms. This is mainly developed by the Poznan Center of Supercomputing (PSNC) in Poland, and Datamat, Italy.

The portal is developed, by Algosystems, with the Jetspeed [14] technology, using Java and XML. A portal makes network resources available to end-users and the user can access the portal via a web browser. Each application must create its own portlet with the fields for the parameters needed to submit the application, in an XML specification. The migration desktop, however, is a personalized graphical desktop developed using Java technologies by the PSNC. The applications must provide specific plugins in order to be submitted to the Grid. From these plugins the parameters of the application and other general parameters as the number of processors, the data files, etc can be specified. The application developers can also provide output plugins to visualize the results in a graphic and interactive manner.

4.1   Preliminary Results in the CrossGrid Testbed

We have obtained some experimental results executing the air pollution application in the CrossGrid Testbed. Table 1 shows the results of executing a simulation of one hour among nodes that belong to different sites and nodes that belong to the same site of the testbed. Sites used to build Table 2(a) are placed at Univ. Valencia (Spain), Univ. Varsaw, Institute of Nuclear Physics (Poland) and Algosystems (Greece) while executions for Table 2(b) were carried out on nodes of the same site at Cyfronet (Poland). In this last case the obtained results are much better due to the lower communication cost. This should be taken into account in the design of the scheduler tool of the CrossGrid project.

Table 1. Times and speedup of 1 hour of real time (60 iterations)

proces.	tot. time	com. time	speedup
1	2133	20	-
2	5107	485	0.42
3	20212	13126	0.11
4	37874	24359	0.06

(a) executed in different sites

proces.	tot. time	com. time	speedup
1	4952	11	-
2	3316	73	1.49
3	2480	98	2.00
4	1865	112	2.65
5	2535	122	1.95
6	2466	125	2.01
7	2198	132	2.25
8	1304	136	3.80

(b) on nodes of the same site

It can be observed that the speedup does not grow linearly. This is because different iterations of the simulated mesh have different computational weight. We have observed experimentally that the most time-consuming iterations are due to presence of rain in that part of the simulated space. We are working on finding some different data distributions to increase the performance. Besides, the project offers tools for application monitoring, helping developers to detect bottlenecks and optimize the code. We plane to use some of these tools to improve the parallel performance.

5   Future Work

At this moment the parallel Resource Broker is in a testing stage. For this reason we are working with two versions of the program, one sequential for the automatic submission (through the RB using JDL) and a parallel version that currently is used for manual submission (via a RSL script and Globus). When the resource broker will be fully available we will submit our parallel version through the RB, and we will analyze its performance.

Currently, we are working on the integration of the air pollution application and its GUI into the portal and migration desktop, emphasizing in the interactivity feature. We also intend to improve the performance of the parallel STEM-II model using the tools that the CrossGrid project offers for performance prediction and monitoring. Moreover, we are working on a better data distribution to achieve large efficiency in the testbed.

References

1. Crossgrid home page. http://www.crossgrid.org/.
2. R.G. Belleman and P.M.A. Sloot. The Design of Dynamic Exploration Environments for Computational Steering Simulations. In *Proceedings of the SGI Users' Conference 2000*, volume 1, pages 57–74. Academic Computer Centre CYFRONET AGH, Krakov, Poland, October 2000.
3. L. Hluchy, O. Habala, B. Simo, J. Astalos, V.D. Tran, and M. Dobrucky. Problem Solving Environment for Flood Forecasting. In *1st International NAISO Symposium on Information Technologies in Environmental Engineering (ITEE'2003)*, Gdansk, Poland, June 2003.
4. S. Bethke, M. Calvetti, H.F. Hoffman, D. Jacobs, M. Kasemann, and D. Linglin. Report of the Steering Group of the LHC Computing Review. In *CERN/LHCC/2001-004, CERN/RRB-D 2001-3*, February 2001.
5. María J. Martín, David E. Singh, J. Carlos Mouriño, Francisco F. Rivera, Ramón Doallo, and Javier D. Bruguera. High Performance Air Pollution Modeling for a Power Plant Environment. *Parallel Computing*, 29 (11–12):1763–1790, 2003.
6. M. Niezgodka and B. Jakubiak. Numerical Weather Prediction System: Scientific and Operational Aspects. In *Proceedings of the III Symposium on Military meteorology*, pages 191–197, 1998.
7. G.R. Carmichael, L.K. Peters, and R.D. Saylor. The STEM-II Regional Scale Acid Deposition and Photochemical Oxidant Model - I. An Overview of Model Development and Applications. *Atmospheric Environment*, 25A(10):2077–2090, 1991.
8. José C. Mouriño, María J. Martín, Patricia González, Marcos Boullón, José C. Cabaleiro, Tomás F. Pena, Francisco F. Rivera, and Ramón Doallo. A Grid-enable Air Quality Simulation. In *First European Across Grids Conference*, 2003.
9. Bettina Krammer, Katrin Bidmon, Matthias S. Müller, and Michael M. Resch. MARMOT: An MPI Analysis and Checking Tool. In *Parallel Computing 2003*, Dresden, Germany, September 2003.
10. G. Tsoloupas and M. Dikaiakos. GridBench: A Tool for Benchmarking Grids. In *4th International Workshop on Grid Computing (Grid2003)*, Phoenix, Arizona, 2003. Accepted.
11. V. Blanco, P. González, J.C. Cabaleiro, D. Heras, T.F. Pena, J.J. Pombo, and F.F. Rivera. Visualizing the Performance Prediction of Parallel Iterative Solvers. *Future Generation of Computer Systems*, 19:721–733, 2003.
12. I. Foster and C. Kesselman. GLOBUS: a Metacomputing Infrastructure Toolkit. *International Journal Supercomputing Applications*, pages 115–128, 1997.
13. Elisa Heymann, Miquel A. Senar, Emilio Luque, and Miron Livny. Adaptive Scheduling for Master-Worker Applications on the Computational Grid. In Lecture Notes in Computer Science, editor, *Proceedings of the Grid Computing - GRID 2000: First IEEE/ACM International Workshop*, volume 1971, page 214, Bangalore, India, December 2000.
14. D. Engh, S. Smallen, J. Gieraltowski, L. Fang, R. Gardner, D. Gannon, and R. Bramley. GRAPPA: Grid Access Portal for Physics Applications. June 26 2003. Talk from the 2003 Computing in High Energy and Nuclear Physics (CHEP03), La Jolla, Ca, USA, March 2003.

The Genetic Algorithms Population Pluglet for the H2O Metacomputing System*

Tomasz Ampuła[1], Dawid Kurzyniec[1], Vaidy Sunderam[1], and Henryk Witek[2]

[1] Dept. of Math and Computer Science
Emory University, Atlanta, GA 30322, USA
{tamp,dawidk,vss}@mathcs.emory.edu
http://www.mathcs.emory.edu/dcl/
[2] Cherry L. Emerson Center of Scientific Computation and Dept. of Chemistry
Emory University, Atlanta, GA 30322, USA
hwitek@emory.edu

Abstract. This paper describes *GAPP* – a framework for the execution of distributed genetic algorithms (GAs) using the H2O metacomputing environment. GAs may be a viable solution technique to intractable problems; *GAPP* offers a distributed GA framework that can lead to rapid and efficient parallel execution of GAs from a variety of domains, with very little effort on behalf of the application scientist. It is premised upon the common phases embodied in GA lifecycles and contains modular implementations to handle each of them, whereas end applications simply provide domain-specific functions and parameters. *GAPP* is built for H2O, a component-oriented metacomputing system that enables cooperative resource sharing and flexible, reconfigurable concurrent computing on heterogeneous platforms. Experiences with the use of *GAPP* on H2O are described and preliminary results are very encouraging.

1 Introduction

Genetic algorithms (GAs) are known to be an effective approach for finding approximate solutions to complex (including NP-hard) problems. Often, when the theoretical nature of the problem is not well known, or no analytic algorithm is readily available, the use of nature-inspired techniques such as evolutionary algorithms or simulated annealing may be the only way to obtain near-optimal results in reasonable time [6].

Genetic algorithms are inspired by natural evolution, where stronger individuals in a population dominate, and eventually eliminate the weaker ones. The fitness of each individual depends on its unique set of chromosomes, i.e. a *genotype*. In the computational model, individuals represent candidate solutions to a given optimization problem. An individual genotype is represented by a data structure (a bit-vector in a canonical case) which is evaluated using a given fitness function. The evolution is simulated as an iterative process yielding a sequence of generations which are constructed by performing certain operations on individuals of the previous generation. The canonical set of operations include rating, selection, crossover, and mutations.

* Research supported in part by U.S. DoE grant DE-FG02-02ER25537 and NSF grant ACI-0220183.

M. Bubak et al. (Eds.): ICCS 2004, LNCS 3036, pp. 140–147, 2004.

For non-trivial problems, when individuals are complex or the computation of a fitness function is time-consuming, it is desirable to explore the potential for parallelism. There are four major models of parallel GA's [8,1]:

- *Single Population Farmer-Worker Model*: The farmer stores the whole population. Individuals are split into groups which are then sent to worker machines. Workers evaluate the fitness function and send results back to the farmer.
- *Fine Grained Single Population*: The single, large population is distributed among machines; however, the population has a spatial structure which limits interactions between individuals so that they can mate only with their neighbors.
- *Coarse Grained Parallel GA*: Many populations on many machines evolve independently, but individuals are allowed to migrate between populations (*The Island Model*)
- *Hybrid Parallel GA*: Various combinations of the above models.

In addition to the potential for speedup, the question of *qualitative* differences between distributed GAs and serial GAs has been studied in the past. A pioneering work involves Grosso's experiments of dividing a population into five demes [5], in which the rate of improvement was found to be faster in smaller demes than in one big population, and the rate depended on the model of migration between demes. Tanese proposed a 4D hypercube topology of subpopulations and reported that results as good as in a serial population model were found, with the added advantage of non-linear speedup [11]. An important factor in distributed GAs that may affect performance is the topology of the demes [2,4]. Tanese also found that migrating too many or too few individuals between subpopulations degrades performance. However good results were found faster that in serial GA even with no migration at all [12]. Other experiments showing differences between migration policies were also performed by Cantú-Paz [3]. It has also been shown that a parallel or distributed GA may outperform a serial GA in cases where partial solutions can be combined to form a better solution [9].

2   Design of the *GAPP* Framework

GAPP is a framework for coarse-grained distributed GA [8,1] computations and simulations on collections of heterogeneous machines. The framework has several objectives. First, it aims to facilitate the rapid development and deployment of distributed GA applications, and to harness the resources of multiple computer systems, regardless of their operating system or CPU speed. In addition, it is designed to enable seamless *resource sharing* and support ad-hoc collaborations. In this model, the end-user (a domain scientist) executes her GA application transparently on a dynamic, possibly large, and geographically distributed collection of machines to which access is supplied by independent contributors, without the need to have explicit login accounts established. Last but not least, the intention of the *GAPP* framework is to clearly separate aspects of GA application development, distributed data exchange, resource administration, and the actual end-user interface, so that the framework can be programmed easily and used by non-computer scientists. In order to support this model, four classes of actors in *GAPP* have been defined (while actual users may assume single or multiple roles), as illustrated in the Figure 1:

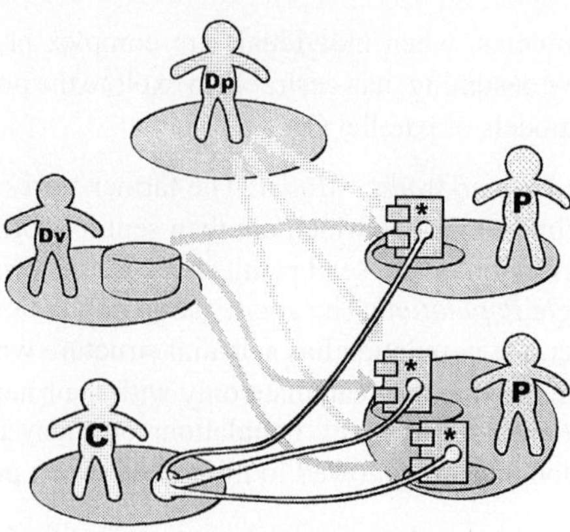

Fig. 1. Actors in *GAPP*: Providers (P), Deployer (Dp), Developer (Dv), Client (C), *GAPP* container(*) inside H2O kernel

- *Providers* (labelled **P** in the figure) supply computational power, by furnishing computer resources that host an environment suitable for running *GAPP* (although not necessarily exclusively). Providers define access privileges and grant security permissions to *deployers* (**Dp**) by specifying coarse- or fine-grained access policies (but need not grant login or other liberal access to their resources);
- *Deployers*: this group consists of users who dynamically install *GAPP* framework on the hosting environment supplied by providers, thus placing a layer over raw resources and enabling their use for GA computations;
- *Developers* (**Dv**)write the actual GA application code, which defines the fitness function and the population behavior with respect to standard GA operations. Importantly, the code is not concerned with aspects of population distribution, which are determined at run time. Once written, the code is published in a software repository (e.g. a Web server) to be available to *clients* (**C**);
- *Clients* are the end-users of *GAPP*. They harness resources supplied by providers, deployers and developers in order to solve their specific problems. Operationally, clients launch application codes (written by developers) within the *GAPP* framework (installed by deployers) on a collection of distributed resources (supplied by providers). These application codes operate on data specified by the client to solve a particular problem.

For example, in a GA approach to determine the shortest spanning tree in a graph, a developer implements an object that represents a spanning tree. The object must define methods required by *GAPP* to perform GA operations, such as crossover of two spanning trees. The deployer installs the *GAPP* container, and the client initializes it by providing the location of the application code in a software repository, and the data describing the graph in which the shortest spanning tree is to be found (e.g. a list of vertices). Note that end-clients do not have to have expertise in distributed GA programming. Having initialized sub-populations within *GAPP* containers, the client organizes them into a desired topology and initiates the evolution process. The populations evolve independently; however, after each lifecycle, individuals migrate between populations. Every source population designates a specified number of its best-fit individuals and sends them to neighboring populations. Every destination population chooses, according to a specified policy, whether (or not) to accept incoming individuals and adds them to its list.

3   The H2O Hosting Environment

H2O is a lightweight, component-oriented framework for distributed computing [10]. It is based on a container-component model that readily hosts components written or wrapped in Java, and provides all of the necessary scaffolding and infrastructure for resource sharing, with high levels of security and control. Using this feature, users may avail of resources across multiple administrative domains without the need for login access or for cumbersome pre-arranged batch execution of distributed codes. H2O defines both an architecture and a methodology for the construction of application-specific or application-category-specific pluglets; *GAPP* follows this methodology that leverages the resource sharing framework support of H2O to facilitate distributed GA algorithm execution.

The distinction among actors in *GAPP*, presented in the previous Section, is in fact an instance of a more general model defined by H2O. In that model, *providers* supply computational resources equipped with *H2O kernels* (containers) and make them remotely available by defining appropriate access policies. *Developers* create code of the H2O-compatible components (so called *pluglets*) and distribute it or place it in software repositories. Subsequently, *deployers* instantiate these components within specific kernels. Finally, *clients* connect to those deployed pluglets and take advantage of services provided by them. One of the crucial, distinguishing features of the H2O container model is the separation of provider and deployer roles. This separation enables providers to share raw resources and to allow third-parties to provide added value (i.e. configure the raw resources) remotely by dynamically deploying or hot-swapping appropriate pluglets.

Due to the fact that the H2O design does not introduce global state (i.e. relationships between actors are defined as pairwise rather than group-oriented), the natural fault tolerance of the distributed island model can be leveraged. In an event of kernel crash, resource revocation, or a network partition, GA applications may continue to run (perhaps with small, but acceptable, decrease in efficiency) despite the loss of some subpopulations or connectivity between subpopulations.

4   Implementation of *GAPP*

The *GAPP* container has been implemented as a *population pluglet* – a generic GA-enabling component which may be plugged by deployers into H2O kernels. Once deployed, a population pluglet is responsible for instantiating, initializing and controlling populations, upon request from clients. At first, the client initializes his population by specifying its behavior (via a pointer to a code in a software repository) and initial data, such as the population size, simulation starting point, etc. Then, the user may initiate the evolution process. Once the evolution has started, the client can stay connected and observe its progress, or may detach from the H2O kernel, and then reconnect and gather results at some later time. It is also possible to suspend and resume the running simulation, which may be useful for suppressing resource usage during peak hours.

To take advantage of the concurrent processing and distributed island model, it is necessary to interconnect population pluglets and to let individuals migrate between them. Every pluglet is connected to a generic *bridge* object, which is solely responsible for the distributed communication. That way, data exchange is abstracted away from the main course of the simulation, and does not have to be reimplemented within each GA code. Bridges between independent population pluglets (that usually run on distributed resources) may form one-one, one-many, or many-many, uni- or bidirectional connections. Therefore, any desired interconnect topology can be formed (see e.g. Figure 2).

Fig. 2. An example of population pluglets connected into a ring.

The initial topology is specified by the client at startup time; several predefined canonical topologies include star, n-dimensional torus, and hypercube. Furthermore, the topology can be modified at run time: new islands can be added, existing ones removed, or the entire topology can be reconfigured e.g. in order to respond to varying network conditions or CPU loads, without disrupting the ongoing computations. During each lifecycle of the population, a fitness function is evaluated for every member of the population. Then, according to calculated fitness values, a ranking of individuals is created. Subsequently, the bridge is asked to copy a fixed number of the best-fit individuals and to distribute them among the bridges it is connected to. To increase efficiency and to avoid latency-related delays, communication is asynchronous so that the bridge does not wait for (or care about) delivery confirmation from other pluglets. Since migrations are not synchronized, faster machines are not constrained by slower ones, which gives users the freedom to configure topologies without regard to relative machine speeds.

5   Examples and Tests

GAPP has been applied in practice to solve several textbook-problems, as well as state-of-the-art research GA problems, and our results so far have been extremely positive. Brief descriptions are given below.

Traveling Salesman Problem: We applied the *GAPP* to solve TSP on a 380-node graph for which the optimal tour length is known to be 1621 units, according to an integer-rounded Euclidean distance norm [7]. The first example, illustrated in Figure 3, compares the performance of a single population of 100 individuals and a ring of 5 populations of 100 individuals each. As shown, the single population found a tour length of 1630, while in the same time the ring of 5 populations reached 1622, only a single unit worse than the optimal value.

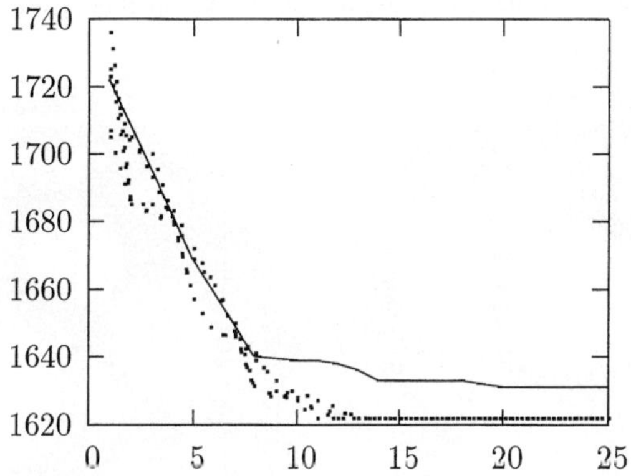

Fig. 3. Single population (solid line) versus five interconnected populations (dotted lines)

The second test compared the performance of a single large population (500 individuals) to five populations of 100 individuals each, again connected in a ring. Figure 4 shows that although both approaches yielded the same near-optimal value of 1622, it was reached by one of the ring members about 15 iterations sooner than by the single large population.

Buckminster fullerene (C_{60}) is a molecule consisting of 60 carbon atoms forming a truncated icosahedron. Among all possible systems consisting of 60 atoms of carbon distributed on a sphere, C_{60} is well known to be a global minimizer of a potential energy. The optimization of potential energy of chemical structures is a difficult problem due to the fact that the number of local minima grows exponentially with the number of atoms, and because the function itself is expensive to calculate. We have applied *GAPP* to this problem, using a surrogate form of the actual potential [13] as a fitness function and starting from a set of randomly generated structures. Although the algorithm was not able to reproduce C_{60} exactly, it obtained structures fairly close to it, as suggested by Figure 5. We are currently investigating the feasibility of using hybrid approaches

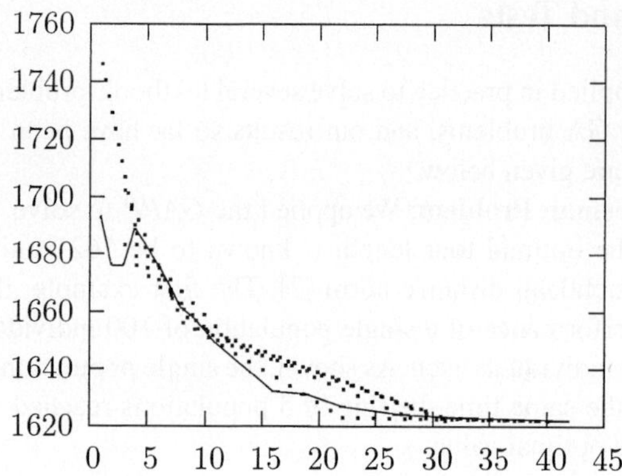

Fig. 4. Single large population (solid line) versus five small populations (dotted lines)

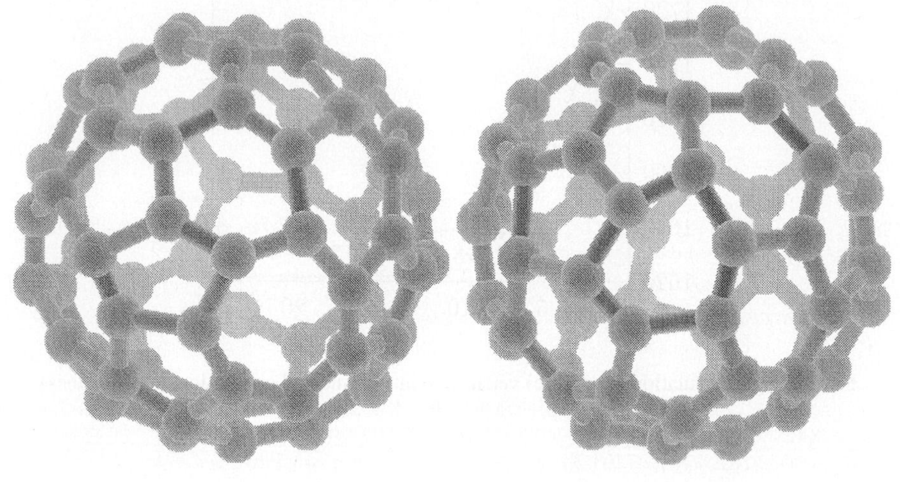

Fig. 5. Two examples of carbon nanoclusters generated by *GAPP*

(combining GAs with numerical optimization) to find exact global minima of potential energy.

The examples studied in this section indicate the potential of GA methods to find near-optimal solutions to NP-hard optimization problems commonly encountered in applied sciences. Also, they show that the distributed approach combined with the island model can improve performance and produce results much better than the local case. The natural scalability and fault tolerance of the island model are features which make it particularly appealing as a distributed computing application.

6 Conclusions

The *GAPP* framework offers an elegant solution to the parallel and distributed execution of GAs. Much of the details of distribution, managing populations in islands, and iterative evolution process is handled by the framework, thus necessitating only minimal, problem-specific effort on behalf of the domain scientist. Second, by leveraging the H2O system architecture, the role of developers and clients is decoupled, thus enabling the potential for increased exchange and cooperation among different groups of researchers who may leverage each other's efforts. Further, for applications dictated by the need for large computational resources, fault tolerance, steering, etc., H2O resource sharing across multiple administrative domains can be leveraged to significant advantage. These facilities and models for the execution of distributed GAs have been tested on several problems, and initial results are very encouraging. Hybrid techniques to combine GAs and traditional numerical methods, adjustments to improve the results in C_{60} energy minimization, a better understanding of result-quality in the island model, and characterization of performance gains through parallel execution comprise the focus of current and future work on this project.

References

1. T. C. Belding. The distributed genetic algorithms revisited.
2. E. Cantú-Paz. Topologies, migration rates, and multi-population parallel genetic algorithms.
3. E. Cantú-Paz. Migration policies, selection pressure, and parallel evolutionary algorithms. *IlliGAL Report*, (99015), June 1999.
4. R. Gaioni and R. Davoli. Communication topologies for parallel genetic algorithms: A comparative study on cray t3d.
5. P. Grosso. Computer simulations of genetic adaptation: Parallel subcomponent interaction in a multilocus model, 1985.
6. W. Hart. A theoretical comparision of evolutionary algorithms and simulated annealing.
7. A. Rohe. http://www.math.princeton.edu/tsp/vlsi.
8. O. T. Sehitoglu. Gene reordering and concurrency in genetic algorithms.
9. T. Starkweather, D. Whitley, and K. Mathias. Optimization using distributed genetic algorithms. In H. Schwefel and R. Maenner, editors, *Parallel Problem Solving from Nature*, Berlin, Germany, 1991. Springer Verlag.
10. V. Sunderam and D. Kurzyniec. Lightweight self-organizing frameworks for metacomputing. In *The 11th International Symposium on High Performance Distributed Computing*, Edinburgh, Scotland, July 2002.
11. R. Tanese. Parallel genetic algorithms for a hypercube. In *Proceedings of the Second Conference on Genetic Algorithms.*, 1987.
12. R. Tanese. Distributed genetic algorithms for function optimalization, 1989.
13. Y. Yamaguchi and S. Maruyama. A mollecular dynamics simulation of the fullerene formation process. *Chemical Physics Letters*, 286:336–342, April 1998.

Applying Grid Computing to the Parameter Sweep of a Group Difference Pseudopotential

Wibke Sudholt[1]*, Kim K. Baldridge[1], David Abramson[2], Colin Enticott[2], and Slavisa Garic[2]

[1] Department of Chemistry & Biochemistry and San Diego Supercomputer Center (SDSC), University of California, San Diego (UCSD), 9500 Gilman Dr., La Jolla, CA 92093-0505, USA
{wibke, kimb}@sdsc.edu
[2] Center for Enterprise Distributed Systems (DSTC) and School of Computer Science and Software Engineering, Monash University, Clayton, Victoria, 3800 Australia
{davida, Colin.Enticott}@csse.monash.edu.au,
Slavisa.Garic@infotech.monash.edu.au

Abstract. Theoretical modeling of chemical and biological processes is a key to understand nature and to predict experiments. Unfortunately, this is very data and computation extensive. However, the worldwide computing grid can now provide the necessary resources. Here, we present a coupling of the GAMESS quantum chemical code to the Nimrod/G grid distribution tool, which is applied to the parameter scan of a group difference pseudopotential (GDP). This represents the initial step in parameterization of a capping atom for hybrid quantum mechanics-molecular mechanics (QM/MM) calculations. The results give hints to the physical forces of functional group distinctions and starting points for later parameter optimizations. The demonstrated technology significantly extends the manageability of accurate, but costly quantum chemical calculations and is valuable for many applications involving thousands of independent runs.

1 Introduction

Efforts in cyberinfrastructure, which offer new research avenues through high-performance grid and information technologies, enable a better coupling of the science and engineering communities. With grid computing, we see a paradigm shift away from large-scale hardware and compute-intensive use to that of end-to-end performance, coordinating software and interfaces, as well as on data, and remote access. This requires multidisciplinary expertise and a deeper level of collaboration. Grid technology promises novel modes of coupling scientific models and unique strategies of sharing data, which help bridging the gaps in our knowledge of natural complexity. Understanding structure/function relationships and molecular processes in biological systems can leverage computer-based information tools, and the level of content generated by high-throughput technologies pushes research developments to new heights.

---

* Now: Institute of Organic Chemistry, University of Zurich, Winterthurerstr. 190, CH-8057 Zurich, Switzerland, {wibke, kimb}@oci.unizh.ch

M. Bubak et al. (Eds.): ICCS 2004, LNCS 3036, pp. 148–155, 2004.

Grid technology has already dramatically changed biomedical research, enabling a high rate of knowledge and application advancements. Theoretical studies mimic biological and chemical processes on a level of complexity such that their fundamental details can be extracted, which cannot be obtained otherwise. Experimental and clinical information then refines these empirical models, which in turn enhances their predictive power. Major components of this iterative procedure can be facilitated by grid resources, which enable the transparent interoperability of research advancements not solely in hardware and computational speed, but also in database/storage technologies, visualization/environment infrastructure, and computational algorithms, all under high-speed networks and remote access (portals). The ultimate goal of exploiting grid technologies for the life sciences is to facilitate knowledge acquisition by harnessing computational tools to create new algorithmic strategies, ease complex multi-step procedures, and organize, manage, and mine data, all in a seamless manner.

Computational modeling in the life sciences is still very challenging and much of the success has been despite the difficulties in integrating all the technologies. For example, many simulations still use a simplified physics/chemistry, or restrict the spatio-temporal dimension or resolution of the model systems. Grid technologies offer to create new paradigms for computing, enabling access to resources which could span the biological scale. In this work, we illustrate a conceptual approach for computational investigations that involve many steps of processing, bookkeeping, and repetitive computation over several variant parameters. Such investigations might be for the generation of algorithm pieces, or for the substantiation of a chemical hypothesis. The approach, which invokes new robust grid technologies, is illustrated for a particular case in the former category here, but a more general implementation is possible.

2 Motivation

2.1 Science Methodology

Biomedical research at every scale from molecules to organisms often involves numerical experimentation and hypothesis testing. When a parameter space is searched for optimal solutions, the computational requirements are amplified by several orders of magnitude. The simulation of extended molecular systems such as solutions, materials, and biomolecules is challenging by itself: The large number of atoms, dynamical sampling of the conformational space, and fine spatial and temporal resolution all require sophisticated techniques to correctly model physical and chemical behavior.

Hybrid quantum mechanics-molecular mechanics (QM/MM) methods, however, just describe a small, "active" region by accurate techniques based on the Schroedinger equation, while the surrounding larger, "inactive" region is treated with more approximate classical force fields. Unfortunately, these two physical concepts are so different that they cannot be easily coupled. In particular, when chemical bonds are cut between both parts, dangling bonds in the MM region can simply be eliminated, but the outermost atoms of the QM region would become radicals and behave completely different than required. One way to saturate these atoms includes the first atom of the MM part as a capping atom in the QM computation, parameterized such that it reflects the properties of the cut bond. This method does not require extended changes in the source code and does not lead to problems with artificial link atoms.

Zhang et al. recently developed such a "pseudobond" approach by adding an effective core potential (ECP) to a fluorine atom to model the methyl group in a carbon-carbon single bond [1]. Their parameterizations were done for the ethane molecule

$$
\begin{array}{ccc}
\text{H}_2\text{C}-\text{CH}_3 & \longrightarrow & \text{H}_2\text{C}-\text{C}_{ps}
\end{array}
\tag{1}
$$

(C_{ps} = F with pseudopotential), then tested on ethane derivatives, and later applied to enzymatic reactions. Unfortunately, we found serious instabilities in this pseudoatom ECP. Furthermore, it appears to be rather difficult to parameterize due to the diversity of the target properties and the multi-dimensionality of function and parameter space.

Therefore, we are developing a new effective pseudoatom potential [2], which only deals with the discrepancies between the isoelectronic CH_3 and F groups without exchanging the core. It is thus named "group difference potential" (GDP). This also provides direct information about substituent effects of functional groups, of interest from synthetic chemistry to drug development, for an important example case. In addition, the potential may be gradually switched on or off later, facilitating QM free energy difference determinations. A superposition of two Gaussian functions

$$
U_{\text{eff}}(r) = A_1 \exp\left(-B_1 r^2\right) + A_2 \exp\left(-B_2 r^2\right)
\tag{2}
$$

turned out to be the most appropriate functional, a format already implemented in many quantum chemical program codes. This form corrects for differences in electron interaction and basis set through negative (attractive) and positive (repulsive) values for the amplitude coefficients A_1 and A_2. The positive exponential prefactors B_1 and B_2 specify the radial extent of each term around the fluorine atom; the smaller their value, the less compact the corresponding function, and vice versa.

Again, the prototype molecule ethane is examined here, but we plan to extend this concept to more complicated systems later. To analyze the results for each parameter set, a cost function is required that reduces the differences between the properties of ethane and pseudoethane to a single number. Bond lengths, bond angle, dissociation energy, Mulliken overlap populations and atomic charges were selected as independent properties. We applied the B3LYP/6-31G(d) level of theory with both spherical and cartesian basis set formats. This results in four calculations for every A_1, A_2, B_1, B_2 set, two on each pseudoethane and the corresponding pseudomethyl radical. All computations were done with the GAMESS program package [3]. The 32 properties identified (hydrogen values appear three times) have diverse units, accuracy, and importance. As such, the differences between actual and target properties x_i and X_i must be weighted appropriately by applying the normalized least squares expression

$$
f(A_1, A_2, B_1, B_2) = \frac{1}{\sum_{i=1}^{32} w_i} \sum_{i=1}^{32} w_i \left(\frac{x_i - X_i}{u_i}\right)^2 .
\tag{3}
$$

The "weighting" factors w_i correct for the number of occurrences of each feature. The "unifying" factors u_i reflect their individual accuracy and are chosen from chemical intuition. Equation (3) is evaluated after completion of each tuple of GAMESS jobs, so that the w_i and u_i values can be easily adjusted for later parameter optimizations.

To identify interesting low-cost regions and avoid trapping in local minima, we first scanned a portion of the parameter space in its entirety. This task consists of huge numbers of short, uncoupled QM calculations und hence is a perfect computing grid application. In the initial experiment on the 4th Pacific Rim Applications and Grid Middleware Assembly (PRAGMA) workshop, the four variables were varied between -10 and 10 a.u. in steps of 1 a.u. for A_1 and A_2, and between 0 and 10 a.u. in steps of 2 a.u. for B_1 and B_2. Ignoring function symmetry for now, this leads to 15,876×4 jobs. On the Supercomputing (SC) conference 2003, we performed even larger parameter sweeps with 53,361×4 and 60,016×4 individual calculations.

2.2 Grid Methodology

The problem formulation above dictates a A_1, A_2, B_1, B_2 parameter set that minimizes the cost function value (3). This implies running GAMESS repetitively over a cross product of all values under consideration, resulting in tens of thousands of independent jobs. To perform this by hand on a single machine would be lengthy, and manually on a distributed computational grid, nearly unfeasible and error prone. Therefore, we invoked the Nimrod/G tool [2,4], which has been specifically designed to perform parameter sweeps using resources distributed across a computational grid [5]. Nimrod/G manages the experiment by finding suitable machines, sending input files to them, running a computation, and shipping the output files back to a central computer. The software also handles common events such as network and node failures.

Nimrod/G targets wide area networks as characterized by the global grid. At the system core is a database that stores all of the experiment details. Jobs are scheduled by considering constraints such as a soft real time deadline and the costs of various resources, and notionally allocating jobs to machines. They are actually executed by "agents", which themselves run on the various resources and request jobs. This architecture hides the latency for scheduling and invoking a remote computation. Nimrod/G is built on a variety of middleware layers, such as Condor, Legion, and Globus. Thus it only needs to interact with the uniform scheduling interface and security layer provided to the testbed resources regardless of their architecture, operating system, or configuration by, for example, the most widely deployed toolkit Globus.

```
parameter A1 float range from -10.0 to 10.0 step 1.0;
parameter A2 float range from -10.0 to 10.0 step 1.0;
parameter B1 float range from 0.0 to 10.0 step 2.0;
parameter B2 float range from 0.0 to 10.0 step 2.0;

task main
 copy cart_eth.inp.sub node:.
 copy pragma4 node:.
 copy pragma4.dat node:.
 node:substitute cart_eth.inp.sub cart_eth.A1=$A1.A2=$A2.B1=$B1.B2=$B2.inp
 node:execute $HOME/bin/rungms cart_eth.A1=$A1.A2=$A2.B1=$B1.B2=$B2 > cart_eth.out
 node:execute ./pragma4 > pragma4.out
 copy node:cart_eth.out results/cart_eth.out.A1=$A1.A2=$A2.B1=$B1.B2=$B2
 copy node:pragma4.out results/pragma4.out.A1=$A1.A2=$A2.B1=$B1.B2=$B2
endtask
```

Fig. 1. Shortened plan file used for the PRAGMA4 pseudopotential scan (one instead of four GAMESS jobs)

To create a computational experiment, the user builds a "plan" file like the one in Fig. 1. Plan files are fairly small and declarative in nature; thus new parameter sweeps can be set up very quickly. In the first part, they contain a definition of the parameters and their ranges. Parameters may be integers, floating point numbers or text. The second part is the "task" block. This set of commands is executed by the "agent" component for each parameter set; it includes the call of GAMESS here for example.

From the implementations of the Nimrod/G user interface we used the portal for this work. This web site enables the user to specify the available resources, as well as to setup and control an experiment through a conventional browser, without porting Nimrod/G to the client machine. Since GAMESS is available on various platforms, we were able to we built testbeds containing conventional workstations, clusters as well as vector supercomputers. These spanned a range of countries, organizations, administrative domains, queue managers, operating systems and architectures (for details, see Ref. [2]). In practice, some of the machines did not perform any computations, either due to their work load or software configuration problems. Fortunately, the dynamic nature of the grid allows deferring the decision about resource usage until execution time. Once all jobs have completed and the output files are returned to the user, the results need to be collapsed into an interpretable form. Here we used the scientific visualization package OpenDX. To explore the entire surface, we produced a sequence of visualizations, each showing isosurfaces of cost function value across three of the parameters, and a different frame for each value of the fourth parameter.

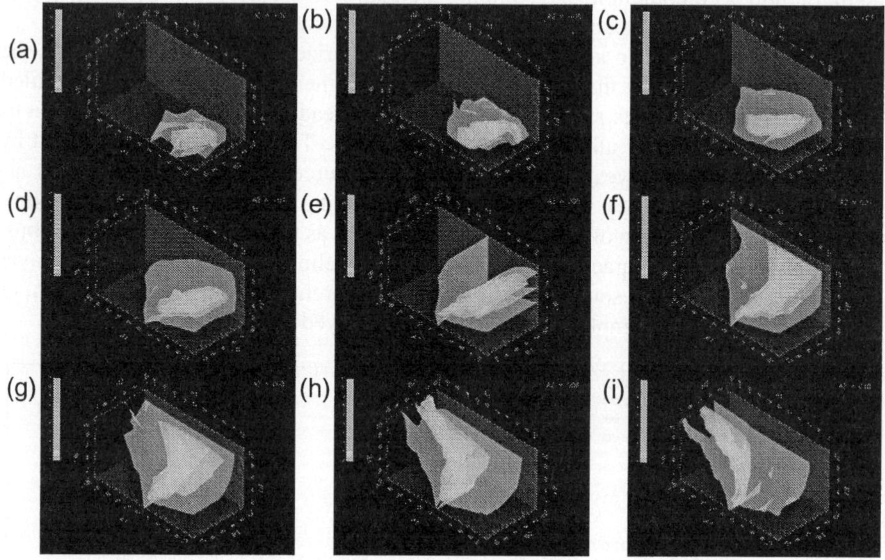

Fig. 2. Selected pictures of the GDP parameter space scanned in the PRAGMA4 experiment. A_1, B_1, and B_2 are displayed on the axes; $A_2 = \pm10, \pm6, \pm3, \pm1, 0$ evolves with the snapshots. Successively better isosurfaces of the cost function f are drawn (*red* = "zero" model)

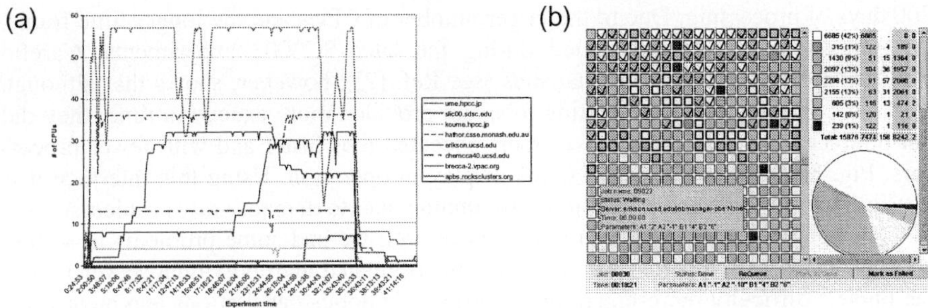

Fig. 3. (a) Distribution of CPU usage over the grid resources in the PRAGMA4 experiment. (b) Nimrod/G portal status display

3 Results and Discussion

3.1 Science Results

Images from the initial parameter scan at PRAGMA4 are displayed in Fig. 2. They demonstrate the complexity of the cost function hypersurface. To find starting points for subsequent optimizations, regions with values lower than the "zero" model (no or canceling GDP) are of special interest. Several such "local" minima are scattered over the parameter space with no apparent pattern, although further analysis suggests partial linear dependence. The most significant minimum here shows up in the middle of the scan around Fig. 2(g). However, all cost function values are still too high.

Therefore, in the SC2003 experiments we also performed sweeps with logarithmic point distribution and higher density in the most interesting region. The deepest minimum appears when a medium-size repulsive and a diffuse attractive Gaussian function are combined to build a maximum at the fluorine core and a shallow depression closer to the bound carbon atom. We tentatively attribute this to the larger size and smaller electron attraction of a methyl compared to the fluoro group. With these data collections in hand it also turned out that the significance of the currently best "global" minimum can be remarkably improved by reducing the bond angle unit factor, a fact that cannot be easily deducted from chemical reasons. This thus reveals the most promising parameter region and weighting for later GDP optimizations.

3.2 Grid Results

Fig. 3(a) visualizes the resource utilization during the PRAGMA4 experiment. Each curve represents a different machine, and shows the number of jobs running at any instant. The graph conveys the ability of the grid to dynamically adjust which resource provides a particular service to the available capacity. Nimrod/G leverages this by incorporating scheduling heuristics that allow moving load to meet soft deadlines. Most importantly, we were not able to accumulate the number of processors required to complete this work within 42 hours at any one of the sites. Overall, we executed over

200 days of processing. Due to the larger number of CPUs and the longer time frame, this amount was even multiplied during the later SC2003 experiments. Careful evaluation of the resource job statistics (see Ref. [2]), however, shows that although some machines had more executing jobs and provided more execution time, they did not produce more results, because others executed jobs faster and with fewer processors. Fig. 3(b) shows the portal status display in operation. Using this interface it is possible to see where individual jobs are running, and to diagnose any problems.

Although all experiments were very successful, we had some problems in setting up these large grid testbeds. Apart from network related and individual server issues, the biggest difficulty represented miss-configured Globus installations and bugs within Globus itself. We developed workarounds for these issues, described in Ref. [2].

4 Conclusions

The shift in science towards information-driven research enables computational studies coupled more tightly to experiment. The rapid growth of grid technologies facilitates the combination of software, data and analysis tools, and the development of grid-enabled chemistry and biology codes for complex problem solving. Linking together sophisticated methodologies as exemplified in this work facilitates new integration pathways to discovery, which can be automated and repetitively performed with variant input datasets. Additionally, end-to-end audit of the process is an implicit deliverable, i.e., the scientist has a record of every action performed on the data.

In collaboration with several international groups, the highlighted project illustrates access to global resources and application technologies via web interfaces. The goal is to develop computational capabilities which integrate our knowledge in (bio)chemistry, molecular modeling, experimental characterization, visualization and grid computing. Here, the GAMESS quantum chemistry software and the Nimrod/G grid distribution tool were coupled. Our purpose was the parameterization of a GDP pseudopotential, which describes the differences of a fluoro compared to a methyl group in the pseudoethane molecule for ultimate use as capping atom potential in QM/MM calculations. We developed a simple GDP formulation with four parameters and constructed a flexible cost function to measure their goodness. It was scanned on an array of points within a defined parameter space region. The resulting cost function hypersurface was further refined by parameter sweeps with different point distributions.

Subsequently, we plan to use the most suitable parameter combinations to start minimization runs. The Nimrod/O tool that performs automatic optimization [4], will be incorporated to search the parameter space and to find the final GDP. We already began to vary variables in the least squares procedure to generate a more funnel-like hypersurface. This will allow minimizations commencing from any remote place to travel towards the global minimum more directly. Overall, this procedure considerably reduces the development time of GDPs for further molecules and groups.

A second purpose of this study was to show how the middleware tool, Nimrod/G, significantly enhances scientific options. The up to $60,016\times4$ uncoupled QM calculations are systematically generated for a multidimensional grid of points, optimally distributed over several computing clusters, within a few days. The completion of such numbers of runs would not have been possible in a reasonable timeframe without such technology. The results allow a better conceptualization of the parameter optimizations, thereby providing more insight into the physics. Late failure of parameterizations can be improved and even the optimization procedure itself can be streamlined.

The described technology has been previously applied in other sciences, but is relatively new to quantum chemistry. One can imagine wider application, such as analysis of reactions, generation of algorithms, or cross-correlation of data. Related examples include parameterization of basis sets, force fields, and similar entities in computational chemistry. A classical case is the examination of potential energy surfaces. Similar approaches can also be used to scan large compound databases in high-throughput virtual screening. By integration into grid middleware architecture QM applications previously not feasible become doable. Furthermore, such infrastructure will help to tie computation with investigator intuition regardless of location, to facilitate scientific investigations by exploiting novel grid capabilities and teraflop hardware speeds, enabling direct user input and feedback. Such infrastructure will impact scientists that need such tools for interdisciplinary research. This will in turn foster development of new modeling, data, and computational science technologies.

Acknowledgements. We thank PRAGMA and all involved organizations, in particular the SDSC ROCKS group. We are grateful to J.P. Greenberg and K. Thompson, SDSC, for installing GAMESS and Globus. W.S. acknowledges support by J.A. McCammon, UCSD, and by a postdoc fellowship of the German Academic Exchange Service (DAAD). A part of the machines was sponsored by the National Biomedical Computational Resource (NBCR) and the W.M. Keck Foundation. K.B. acknowledges support from the NSF through DBI-0078296 and ANI-0223043 and from the NIH through NBCR-RR08605. We also thank D. Kurniawan from Monash University for the visualizations of the GAMESS output. The Nimrod project is supported by DSTC and GrangeNet, both of which are funded in part by the Australian Government. D.A. received financial support from the Australian Partnership for Advanced Computing (APAC) whilst on leave at UCSD.

References

1. Zhang, Y., Lee, T.-S., Yang, W.: A Pseudobond Approach to Combining Quantum Mechanical and Molecular Mechanical Methods. J. Chem. Phys. **110** (1999) 46-54
2. Sudholt, W., Baldridge, K.K., Abramson, D., Enticott, C., Garic, S.: Parameter Scan of an Effective Group Difference Pseudopotential Using Grid Computing. New Generation Computing **22** (2004) 125-136
3. Schmidt, M.W., Baldridge, K.K., Boatz, J.A., Elbert, S.T., Gordon, M.S., Jensen, J.H., Koseki, S., Matsunaga, N., Nguyen, K.A., Su, S.J., Windus, T.L., Dupuis, M., Montgomery, J.A.: General Atomic and Molecular Electronic-Structure System. J. Comput. Chem. **14** (1993) 1347-1363; http://www.msg.ameslab.gov/GAMESS/GAMESS.html
4. Abramson, D., Sosic, R., Giddy, J., Hall, B.: Nimrod: A Tool for Performing Parametised Simulations Using Distributed Workstations. The 4th IEEE Symposium on High Performance Distributed Computing, Virginia (August 1995); Abramson, D., Giddy, J., Kotler, L.: High Performance Parametric Modeling with Nimrod/G: Killer Application for the Global Grid? International Parallel and Distributed Processing Symposium (IPDPS), Cancun, Mexico (May 2000) 520- 528; http://www.csse.monash.edu.au/~davida/nimrod/
5. Foster, I., Kesselman, C. (eds.): The Grid: Blueprint for a New Computing Infrastructure. Morgan Kaufmann Publishers, USA (1999)

A Grid Enabled Parallel Hybrid Genetic Algorithm for SPN

Giuseppe Lo Presti[1,2], Giuseppe Lo Re[2], Pietro Storniolo[2], and Alfonso Urso[2]

[1] Dinfo – Università di Palermo.
[2] ICAR - Istituto di Calcolo e Reti ad Alte Prestazioni
C.N.R. – Consiglio Nazionale delle Ricerche, Palermo, Italy
{lopresti, lore, storniolo, urso}@icar.cnr.it

Abstract. This paper presents a combination of a parallel Genetic Algorithm (GA) and a local search methodology for the Steiner Problem in Networks (SPN). Several previous papers have proposed the adoption of GAs and others metaheuristics to solve the SPN demonstrating the validity of their approaches. This work differs from them for two main reasons: the dimension and the features of the networks adopted in the experiments and the aim from which it has been originated. The reason that aimed this work was namely to assess deterministic and computationally inexpensive algorithms which can be used in practical engineering applications, such as the multicast transmission in the Internet. The large dimensions of our sample networks require the adoption of an efficient grid based parallel implementation of the Steiner GAs. Furthermore, a local search technique, which complements the global search capability of the GA, is implemented by means of a heuristic method. Finally, a further mutation operator is added to the GA replacing the original genome with the solution achieved by the heuristic, providing thus a mechanism like the genetically modified organisms in nature. Although the results achieved cannot be applied directly to the problem we investigate, they can be used to validate other methodologies that can find better applications in the telecommunication field.

Keywords: Steiner Problem, Parallel Genetic Algorithm, Grid Computing.

1   Introduction

The Steiner Problem in Networks (SPN) [5] is a classic combinatorial optimization problem which, in its general case decision version, has been demonstrated [2] NP-complete. Its applications cover many scientific fields such, for instance, the VLSI and pipeline design, the Internet multicast routing, the telephone network design, etc. Many efforts have been produced in the last years to design polynomial-time algorithms to determine sub-optimal solutions: several heuristics have been developed capable of providing approximate solutions [3], [4]. Mathematical proofs constrain the solutions determined by these heuristics to

M. Bubak et al. (Eds.): ICCS 2004, LNCS 3036, pp. 156–163, 2004.

the optimal solution, binding them by some multiplicative factors. This property allows their adoption for many applications. Among the practical applications of the SPN there is the construction of a minimal distribution tree to connect a set of Internet routers involved in a multicast transmission. The extremely dynamic nature of this application imposes the development of efficient heuristics capable of determining, in a very short time, sub-optimal solutions that however may represent good approximations. In order to validate the effectiveness of a given algorithm it is useful to compare the approximations obtained with the exact solutions. However, the NP-complete nature of the problem, at least to the current knowledge, does not allow to perform complete algorithms for graphs whose dimensions are comparable with the current size of the Internet. Among the most efficient approximating algorithms, recently some metaheuristics such as Genetic Algorithms [9], tabu-search [10], and Simulated Annealing [7] have been proposed. Although these approaches can be considered the best approximating methodologies, they suffer the disadvantage of their non-deterministic behavior that does not allow their adoption in fields requiring a distributed coordination among several independent entities. However, the good performances produced by these evolutionary methods suggests the idea to exploit their results as an assessment term. The best suitable approach for exploiting the coarse grain parallelism available in our laboratory is a parallel implementation of genetic algorithm. This technique results extremely scalable and the software implementation, we carried out, allows us to extend its execution to very large grid computing systems, which currently are becoming available on the Internet. The relevant computing power available allowed us to solve very large instances of the problem, and in most of the cases to determine the best solutions ever obtained. The experiments have been carried out on several different sets of graphs, characterized by different topological features, with the aim to effectively evaluate and compare the performances over a wide range of samples. Furthermore, to demonstrate the general validity of the methodology we tested our implementation over a classical public library set of experiments, SteinLib [13], which represents a commonly accepted assessment term for the Steiner problem. The remainder of the paper is organized as follows. The Steiner Problem in Network is formulated in section II. Section III contains the description of the parallel hybrid genetic algorithm, and section IV describes the experimental results. Finally, section V concludes this work and discusses future directions.

2   The Steiner Tree Problem in Networks

Formally, the Steiner Tree Problem in Networks can be formulated as follows. Let $G = (V, E)$ be an undirected graph, $w : E \to R^+$ a function that assigns a positive weight to each edge, and $Z \subseteq V$ be a set of multicast or terminal nodes. Determine a connected subgraph $G_S = (V_S, E_S)$ of G such that:

- $Z \subseteq V_S$;
- the total weight $w(G_S) = \sum_{e \in E_S} w(e)$ is minimal.

The $V_S - Z$ set is called the Steiner nodes set and is denoted by S. Since the weight function assumes positive values, the resulting subgraph is called the Steiner minimum tree T, which spans each node in V_S. Throughout this paper, let $n = |V|, m = |E|, p = |Z|$. Many heuristics proposed in the past years are capable of identifying sub-optimal solutions with polynomial time complexities: among these, the Distance Network Heuristic (DNH) [5], the Shortest Path Heuristic (SPH) [3], the K-Shortest Path Heuristic (K-SPH) [1], the Average Distance Heuristic (ADH) [4], and the Stirring heuristic [11], which were used in the experimental tests. In particular, K-SPH builds a forest of subtrees joining together the closest nodes or subtrees until a single solution tree has been obtained. ADH is a generalization of K-SPH. It repeatedly connects nodes or subtrees through the most central node, which is determined by a heuristic function, and terminates when a single tree remains, spanning all the Z-nodes. The ADH algorithm is the most effective among these heuristics, though the better performances involve a higher computational cost, $O(n^3)$ versus the upper bound of $O(pn^2)$ of all other heuristics. The Stirring heuristic is a local search optimization method, constrained to assume a deterministic behavior, which uses a solution found from the above heuristics to determine better solutions.

3   Parallel Genetic Algorithm

A Genetic Algorithm (GA) provides a universal optimization technique that imitates processes of genetic adaptation that occur in natural evolution. By using this analogy, GA is able to evolve towards a solution for real-world optimization problems. The main advantage of the GA is its capability of achieving global optimization solution even for nonlinear, high-dimensional, multimodal and discontinuous problems [6].

Genetic Algorithms are naturally suited to be implemented on a parallel architecture. A survey on parallel GAs can be found in [8]. Several approaches to parallel implementations of GAs have been proposed ([15]). Among these, for the solution of the SPN we consider the two basic ones: the simple *global* model and the *coarse grained* model. In a previous work [17], the first approach has been used; in this implementation a master process is responsible of the main execution of the genetic algorithm and exploits the availability of different processors by allocating a slave process on each of them. Each slave is required to execute the evaluation function for some individual of the current population on the basis of its availability. In this paper the coarse-grained model will be studied and compared with the previous one. The coarse-grained model divides the population into smaller subpopulations, termed *demes*, constituting a given number of islands. A standard GA is executed on each island and is responsible for initializing, evaluating and evolving its own individuals. Furthermore, the standard GA is enforced by a migration operator, which periodically involves the transfer of individuals among the different subpopulations.

To perform the analysis of the solution space, a GA needs the representation of the problem solutions as basic individuals of its population, which are called genomes. During the execution of the algorithm new individuals will be generated by means of the mutation and crossover operators. To encode the feasible solutions of the SPN as binary genomes, we adopted the following representation: for each instance of the problem we define the genome as a binary array whose length corresponds to the dimension of the set $V - Z$, i.e. the set of all the nodes which are potential candidates for belonging to a given solution. The value of the i_{th} bit represents if the correspondent node in the set $V - Z$ should be considered as complementary node to generate a tree which connects the multicast Z nodes. To follow the genome indication of including the correspondent nodes in a solution tree we map each genetic individual in a new instance of the problem where the original Z nodes are extended with the nodes coded by the genes. This new instance of the problem is solved using the K-SPH or ADH heuristics, and the solution is pruned with regard to the original multicast set. K-SPH is a $O(pn^2)$ algorithm which is capable of isolating good solutions, although it uses only nodes which are along the shortest paths between the multicast nodes. ADH is a $O(n^3)$ algorithm which is capable of determining better solutions because it considers all the nodes in the network. To obtain a trade-off between execution time and competitiveness we adopted alternatively both heuristics in order to exploit their different features. The fitness value is straightforwardly calculated as the inverse of the tree cost, thus to restrict the range of the fitness function to the interval $(0, 1]$. The adoption of the heuristic methods on individuals provided by the GA represents a local search technique which complements the global search capability naturally owned by the GA. This way a hybrid optimization algorithm is obtained. Furthermore, considering that the evaluation process described above determines the minimal set of genes which forms the current solution, we introduce a further mutation mechanism which replaces the original genome with the solution achieved by ADH. This process could be viewed as the implementation in the GA of the procedure that leads to a Genetically Modified Organism in nature. This technique introduces the advantage of faster convergence towards the optimal solution.

3.1 Grid Based Implementation

The parallel implementation has been carried out on a cluster, simulating a grid environment of distributed machines in order to exploit further computing resources. The experimental cluster is composed by forty workstations managed by the Globus 3.0 toolkit [19]. The communication activities are carried out using MPIGH-G2 [16], the grid-enabled implementation of the Message Passing Interface (MPI). All nodes are equipped with an Intel Pentium 4 1.7 GHz CPU, 256 Mbytes of RAM, four 100Mbps Ethernet cards and they are managed by the version 7.2 of the Red Hat Linux distribution. A redundant degree of connectivity is achieved by means of eight 100Mbps Ethernet switches. The software system exploits the facilities provided by the GAlib, a C++ Library of Genetic Algorithm Components [18]. The master-slave paradigm has been adopted to implement the parallel version, embedding the MPICH-G2 primitives and GAlib object-oriented classes.

4   Experimental Results

In this section we discuss the experimental results obtained on three different
test sets of sample graphs, taken respectively from the public SteinLib library
[13], the BRITE [14] topology generator, and the Mercator project [12]. On this
experimental testbed, we execute the two models of the parallel Genetic Algorithm, the classical heuristics SPH, DNH, K-SPH, and ADH, and the stirring
heuristic. The GA parameters for the two different parallel implementations are
shown in Table 1.

Table 1. GAs parameters

Global	Coarse Grained
number of generations = 25	number of generations = 4
population size = 120	population size = 50
crossover probability $PC = 0.7$	crossover probability $PC = 0.7$
mutation probability $PM = 0.001$	mutation probability $PM = 0.001$
	number of demes = 5

We maintain these values constant for all the executions in order to compare
all problems on a homogeneous basis. Furthermore, the local search technique
and the additional mutation mechanism described in the above section are implemented. Figure 1 shows the fitness function values distributions for the *global*
and *coarse grained* parallel implementations. The charts plot the score obtained
by the first 1250 individuals of the *global* model and by all the individuals of the
coarse grained model; it is possible to note a faster convergence of the *coarse
grained* model. Moreover, the speedup achieved by the *coarse grained* model is
significantly better than that of the *global* model, because the first one has few
synchronization points due to its nature, and thus it can be run on a wide-area
grid environment without affecting its performances.

The better performances obtained by the *coarse grained* model, together with
its better speedup, motivate the adoption of it in all the following experiments.
The first test set is a subset of the SteinLib library, a public collection of Steiner
tree problems in graphs with different characteristics, taken from VLSI applications, genetic contexts, computer networks applications, etc. More specifically
we adopt the subset constituted by *Beasley*'s series C, D, E, formerly known as
the OR-library, which are random-weights graphs with sizes ranging from 500 to
2,000 nodes. The connection degree is relatively high, ranging from 0.1% up to
10%. The networks in this sample do not present any similarity with the Internet
like topologies [17]. However, we adopted it as test for our parallel implementation of GA, because it represents a commonly accepted assessment term since
the optimal solutions are known. Figure 2 shows the cumulative cost competitiveness of parallel GA and the classical heuristics over the above graphs. The
competitiveness is determined as the ratio between the costs of trees produced
by heuristics and the optimal ones. From the comparison of the solutions obtained by the GA with the optimal values, it can be observed that in about 80%

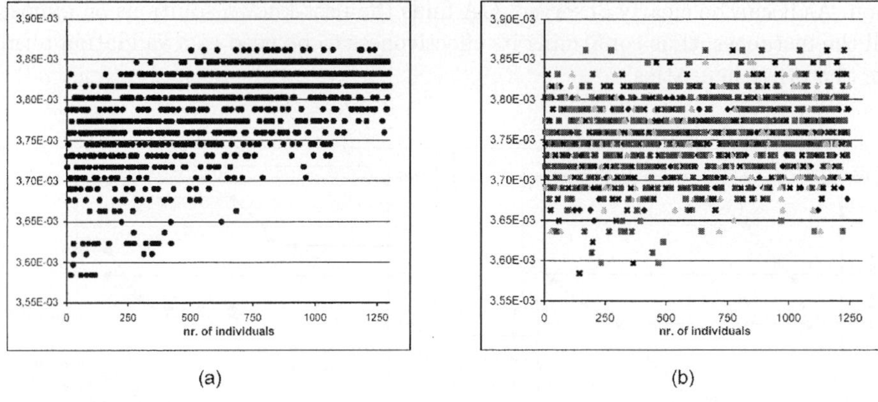

(a)   (b)

Fig. 1. Fitness function values distribution for the global (a) and coarse grained (b) parallel implementations

of the cases both the GAs are able to determine the optimal solution, and for the 90% of the instances the obtained solution is at most 1% larger than the optimal value.

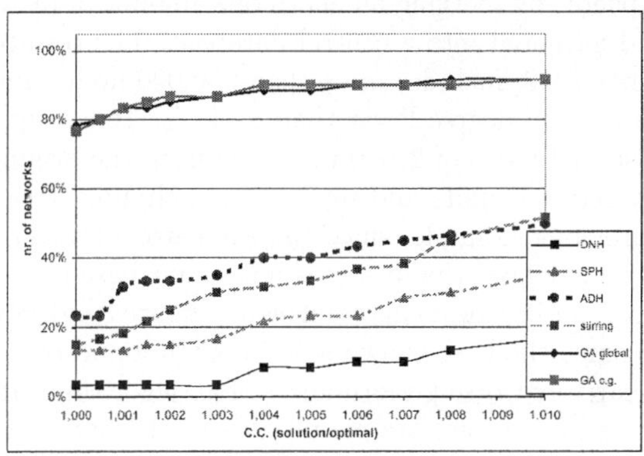

Fig. 2. Cumulative Cost Competitiveness on C, D, E SteinLib nets.

The following set of experiments is devoted to investigate the graphs with topological features similar to the Internet graphs. BRITE (Boston university Representative Internet Topology gEnerator) was developed to investigate the growth of large computer networks, and to compare several topology generation models. In our experiments, we tested several networks ($\sim$ 400) with homogeneous topological characteristics and sizes ranging from 1,000 to 2,000 nodes. Figure 3 (a) shows the cumulative cost competitiveness curves for a test set composed of fifty networks, each of them with 2,000 nodes. In this and in the following experiments, the competitiveness is determined as the ratio between the costs of trees produced by heuristics and the best-known sub-optimal solution. As it can be clearly observed, GA finds the best-known solutions on almost all the instances, thus confirming its effectiveness to be used as a validation term for the other heuristics.

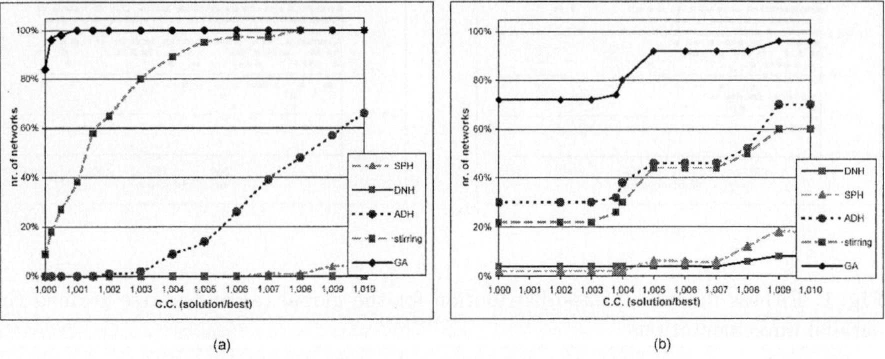

(a)   (b)

Fig. 3. Cumulative Cost Competitiveness on (a) Brite nets and (b) Mercator subnets.

In the last experiment, the test set is created starting from the real Internet data description produced by the Mercator project. This project has produced a real Internet snapshot, by merging an enormous amount of measurements taken over the time and gathered into a central database. The resulting network, obtained in November 1999, includes more than 280,000 nodes and nearly 450,000 edges, with a connection degree lower than $0,001\%$. To set up our experiment, we extracted 50 subnetworks of $2,000$ node size from the original map, starting from a randomly selected node and repeatedly including its neighbors. Differently from the previous example, since the Mercator data do not provide any cost associated to the edges, the metric is hop count based. The analysis of the cumulative cost competitiveness curves, shown in figure 3 (b), reveals the parallel GA effectiveness since the best-known solutions are found on about 70% of the instances. The relatively worse performances of all algorithms and the particular shape of the curves in this case are mainly due to the hop count metric, which leads to a higher quantization of the input data.

5   Conclusions

In this work we proposed the adoption of a parallel implementation of genetic algorithm and local search methodologies to obtain near-optimal solution to the Steiner Problem in Networks for large graphs with topological features similar to the Internet ones. The results have shown that our implementations achieved high competitiveness in all the experimented test sets, differentiated for topological characteristics. In most of the well known examples of the SteinLib library we found the optimal solutions. On the sample networks generated by the Brite tool or extracted from the Mercator graph, which simulate the Internet structure with the best accuracy, we almost always obtained the best calculated sub-optimal solutions, thus achieving a useful result for the comparison of the competitiveness of the polynomial and deterministic heuristics. As regards the future directions, we are currently developing more sophisticated parallel models, with the aim of further improving the GA performances and optimizing the total execution times.

References

1. J. Kruskal, On the Shortest Spanning Subtree of a Graph and the Traveling Salesman Problem, Proc. Amer. Math. Soc., vol. 7, pp. 48 - 50, 1956.
2. R. M. Karp, Reducibility among Combinatorial Problems, in R. E. Miller, J. W. Thatcher, Complexity of Computer Computations, Plenum Press, New York, pp.85-103, 1972.
3. H. Takahashi, A Matsuyama, An approximate solution for the Steiner problem in graphs, Math. Japan, 1980, pp. 573-577.
4. V. J. Rayward-Smith, The computation of nearly minimal Steiner trees in graphs, Int. Math. Ed. Sci. Tech. 14, 1983, pp. 15-23.
5. P. Winter, Steiner problem in networks: a survey, Networks, 17, 1987, pp. 129-167.
6. D. E. Goldberg, Genetic algorithm in Search, Optimization, and Machine Learning, Reading Ma: Addison Wesley 1989.
7. K. A. Dowsland, Hill-climbing, Simulated Annealing and the Steiner Problem in Graphs, Engineering Optimisation, 17, 1991, pp. 91-107.
8. E. Cantu-Paz, A summary of research on parallel genetic algorithms, Illinois GALab, Univ. Illinois Urbana-Champaign, Urbana, IL, Tech. Rep. 950076, July 1995.
9. H. Esbensen Computing Near-Optimal Solutions to the Steiner Problem in a Graph Using a Genetic Algorithm, Networks: An International Journal 26, 1995.
10. M. Gendreau, J. F. Larochelle, B. Sanso A Tabu Search Heuristic for the Steiner Tree Problem Networks 34: 162?172, 1999 1999 John Wiley & Sons, Inc.
11. G. Di Fatta, G. Lo Re, Efficient tree construction for the multicast problem, Special issue of the Journal of the Brazilian Telecommunications Society, 1999.
12. R. Govindan, H. Tangmunarunkit, Heuristics for Internet Map Discovery, Proc IEEE Infocom 2000, Tel Aviv, Israel, www.isi.edu/scan/mercator/mercator.html.
13. S. Voss, A. Martin, T. Koch, SteinLib Testdata Library, February 2001, elib.zib.de/steinlib/steinlib.php.
14. A. Medina, A. Lakhina, I. Matta, J. Byers, BRITE Topology Generator, April 2001 cs-pub.bu.edu/brite.
15. G. Folino, C. Pizzuti, G. Spezzano, Parallel Hybrid Method for SAT That Couples Genetic Algorithms and Local Search, IEEE Transactions on Evolutionary Computation, Vol. 5, No. 4, pp. 323-334, August 2001.
16. N. Karonis, B. Toonen, I. Foster, MPICH-G2: A Grid-Enabled Implementation of the Message Passing Interface, Journal of Parallel and Distributed Computing (JPDC), Vol. 63, No. 5, pp. 551-563, May 2003.
17. G. Di Fatta, G. Lo Presti, G. Lo Re, A Parallel Genetic Algorithm for the Steiner Problem in Networks, Proc. of the 15th IASTED Int. Conference on Parallel and Distributed Computing and Systems, Marina del Rey (CA), USA, November 2003.
18. GAlib: a C++ Library of Genetic Algorithm Components, http://lancet.mit.edu/ga/.
19. T. Sandholm, J. Gawor, Globus Toolkit 3 Core – A Grid Service Container Framework, http://www-unix.globus.org/toolkit/3.0/ogsa/docs/gt3core.pdf.

An Atmospheric Sciences Workflow and Its Implementation with Web Services

David Abramson [1], Jagan Kommineni [1], John L. McGregor [2], and Jack Katzfey [2]

[1] School of Computer Science and Software Eng., Monash University, 900 Dandenong Rd, Caulfield East, 3145, Australia
[2] Division of Atmospheric Science, CSIRO, PMB 1, Aspendale,Vic, 3195, Australia

Abstract. Computational and data Grids couple geographically distributed resources such as high performance computers, workstations, clusters, and scientific instruments. Grid Workflows consist of a number of components, including: computational models, distributed files, scientific instruments and special hardware platforms. In this paper, we describe an interesting grid workflow in atmospheric sciences and show how it can be implemented using Web Services. An interesting attribute of our implementation technique is that the application codes can be adapted to work on the Grid without source modification.

1   Introduction

Computational and data Grids couple geographically distributed resources such as high performance computers, workstations, clusters, and scientific instruments. Accordingly, they have been proposed as the next generation computing platform for solving large-scale problems in science, engineering, and commerce [4][5]. Unlike traditional high performance computing systems, such Grids provide more than just computing power, because they address issues of wide area networking, wide area scheduling and resource discovery in ways that allow many resources to be assembled on demand to solve large problems.

Grid applications have the potential to allow real time processing of data streams from scientific instruments such as particle accelerators and telescopes in ways which are much more flexible and powerful than are currently available. Of particular interest are applications, called "Grid Workflows", that consist of a number of components, including: computational models, distributed files, scientific instruments and special hardware platforms (such as visualisation systems) [3]. Importantly, such workflows are interconnected in a flexible and dynamic way to give the appearance of a single application that has access to a wide range of data, running on a single platform. Grid workflows have been specified for a number of different scientific domains including physics [6] and gravitational wave physics [2].

In this paper we describe a Grid workflow for solving problems in atmospheric sciences. The workflow supports the coupling of a number of pre-existing legacy computational models across distributed computers. An important aspect of the work

M. Bubak et al. (Eds.): ICCS 2004, LNCS 3036, pp. 164–173, 2004.

is that we do not require source modification of the codes. In fact, we don't even require access to the source code. In order to implement the workflow we overload the normal file IO operations to allow them to work in the Grid. We also leverage existing Grid middleware layers like Globus [4] [5] to provide access to control of the underlying resources.

In Section 2 we describe an atmospheric science workflow, with some detail of the functions of the various components. Section 3 discusses the implementation techniques, and Section 4 provides some experimental results.

2   An Atmospheric Sciences Workflow

Global climate models: A global climate model is a computer model representing the atmosphere, oceans, land and sea-ice. By solving mathematical equations based upon the laws of physics, a GCM simulates the behaviour of the climate system. The model divides the planet into a number of vertical layers representing levels in the atmosphere and depths in the oceans, and divides the surface of the planet into a grid of horizontal boxes separated by lines which may be similar to latitudes and longitudes. In this way, the planet is covered by a three-dimensional grid of boxes (Figure 1).

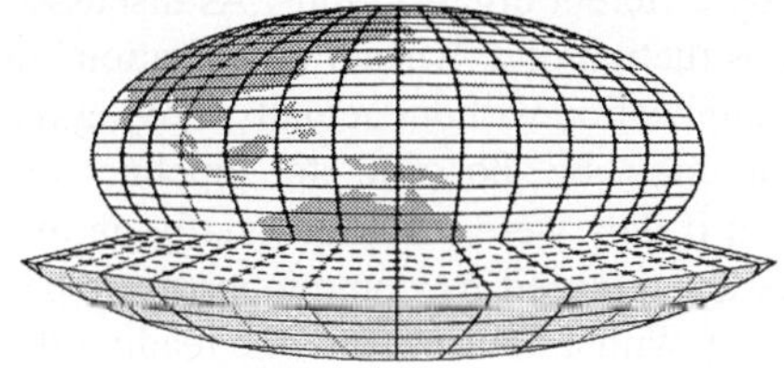

Fig. 1. Representation of the Earth's surface and atmosphere in a typical global climate model.

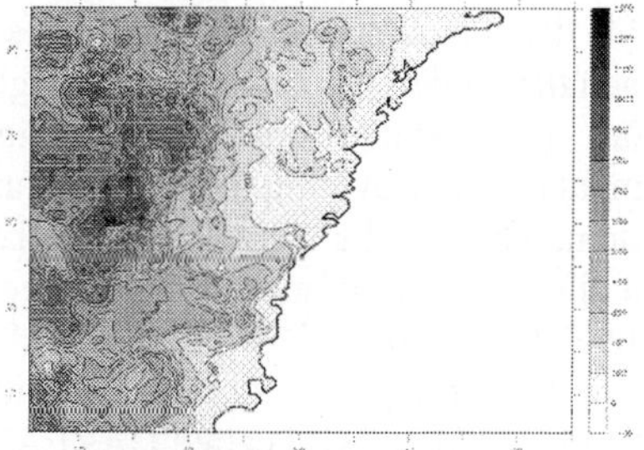

Fig. 2. DARLAM model domain for simulations using a 3 km grid over Sydney.

Global climate models capture large scale features like the deserts and tropics very well, but have difficulty capturing smaller features like cyclones and thunderstorms because they occur at scales much smaller than the grid boxes.

Regional climate models: To improve regional detail in climate models, it is desirable to reduce the spacing between grid points. However, due to the complexity of global climate modeling, computational requirements become prohibitive if the horizontal grid resolution is less than a few hundred kilometers. At this resolution, vitally important small-scale phenomena, like tropical cyclones and cold fronts, are poorly captured. This affects simulated patterns of temperature and rainfall, and hence the ability to realistically simulate observed regional climate features in GCMs.

A computationally feasible alternative to a coarse resolution global climate model is to use a finer resolution model over a small part of the globe. A regional climate model (RCM), with a horizontal resolution of about 100 km or less, is able to simulate regional weather patterns better than most GCMs [7][8][11]. Part of the reason for the improved climate simulation relative to GCMs is the fact that coastlines and mountains are represented in more detail in RCMs. Since topographic features strongly influence regional temperature and rainfall, more detailed features are likely to give a better climate simulation.

A regional climate model requires meteorological information at its lateral boundaries in order to simulate weather within its boundaries. For climate change studies, an RCM is typically driven at its boundaries by information from a coarser-scale GCM. This is commonly called nesting an RCM inside a GCM. One-way nesting allows information to flow from the GCM to the RCM each simulated day, but the weather simulated by the RCM does not affect the GCM interactively. This means that the RCM can be run after the GCM experiment has been completed.

A Grid Workflow: Traditionally, the GCM and the RCM have been executed on the same computer system, and data is passed between them using conventional files. However, the computational Grid discussed in Section 1 provides an ideal framework for executing these models on different machines that are physically distributed. There are many reasons why one might wish to do this. First, both models may have different computational requirements and be suited to different types of platforms. For example, one may execute well on a vector supercomputer and the other may be efficient on a parallel processor. Second, both models may not have been ported to the same hardware. Thus, it may be time consuming and expensive to couple them on one machine. Third, the models may be "owned" by different organizations. As discussed in [4] the computational grid facilitates the construction of a "virtual organization" in which the models are linked into a single grid application without actually moving the codes to a single organization. Finally, it may be possible to pipeline the computations, providing a quicker solution than if they were run sequentially on one system. This can be achieved if the data files are replaced by communication pipes that allow one program to write data concurrently with a downstream one reading the same data. Figure 3 shows such a grid workflow based around the models discussed above. In this paper we discuss a particular system involving three models – a GCM called C-CAM, a RCM called DARLAM and a data filter called cc2lam.

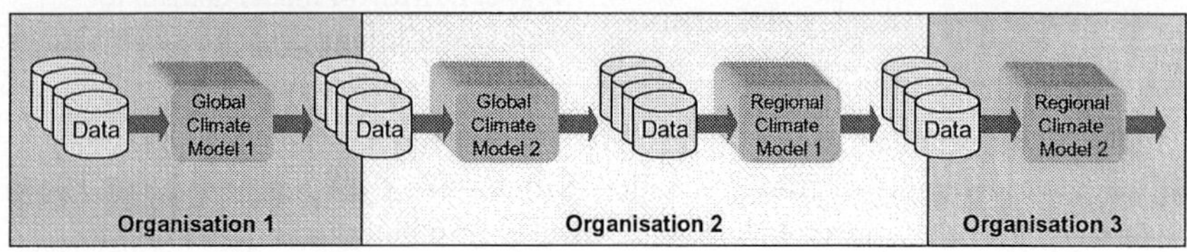

Fig. 3. An Atmospheric Sciences Grid Workflow

3   An Implementation with Web Services

Clearly it is possible to implement the system described in the previous section in a number of ways. For example, the programs could be run unmodified, and some system could be responsible for copying files from one machine to another. This is effectively the practice that atmospheric scientists have employed manually for some time now. Or, the files could even be shared by a single distributed file system like NFS or AFS, removing the need to explicitly copy them from one local file system to another. The major disadvantage of this general approach is that it does not take advantage of any potential overlap in the computations. Another option is to modify the programs so that they do not read directly from files, but instead they use a message passing library like PVM or MPI to send data from one model to another. This allows the computations to be pipelined, however the programs would need to be modified at the source level. Further, once modified that would no longer work as stand alone codes, limiting the flexibility of the individual components. A third alternative is to modify the file system library so it performs message passing rather than writing to and reading from local files. This achieves the advantages of both of the previous alternatives, requires no source modification (only relinking the application) and does not permanently fix the way the programs operate. For example, by linking the normal file system primitives, the programs behaves as normal, reading and writing local files. However, when the message passing library is linked, the program sends messages for writes, and receives messages for read. In a previous paper we proposed such a mechanism, called NetFiles, for implementing parallel master-slave programs [1]. NetFiles were implemented within a single cluster or parallel machine and could not cross administrative domains. Here we have broadened the approach to support interprocess communication across the computational Grid. Accordingly we have called this mechanism GridFiles. Figure 4 shows how two legacy computations can be coupled using GridFiles.

In the GridFiles approach, the conventional system calls (like open, read, write etc.,) are replaced (transparently) by a call to a module called a "File Multiplexer". The File Multiplexer is responsible for passing the file system operations onto an appropriate service, and has the flexibility to change the mappings dynamically. Thus, the File Multiplexer can redirect IO requests to local files, local processes, remote files or remote processes. This means that a program can perform a READ operation, and this might read a local file, or a remote one, or even be connected directly to a WRITE operation on a remote machine. The latter mode is the equivalent of connecting the two programs by sockets, and allows complete overlap of IO operations, and is called the buffer service. This structure is depicted in Figure 4. The File Multiplexer is composed of three clients. The Local File Client performs local file operations. The Grid File Client communicates with the Grid Buffer Service, and the GNS Client communicates with the GriddLeS Name Service. The Grid Buffer and Name Services are both implemented using Web Service technologies such a SOAP and XML [9][10].

The GridBuffer service acts as a sink for WRITE operations and a source for READs. In order to support random read and write operations, data is stored in a hash table rather than a sequential buffer. Thus, if a read is issued for a block that has not been written yet, the read waits for the data to arrive. After a block has been read from the hash table it is written to a cache file and then deleted from the hash table. The cache file is provided for two main reasons. First, it allows a block to be reread even after it has been deleted from the hash table. This occurs when the reader seeks back to a previous block. Second, it provides a mechanism for implementing broadcast operations to more than one process. When this happens, the first reader obtains the data from the hash table, but subsequent readers retrieve the data from the cache file. It is possible to have an arbitrary number of readers using the approach without the need to inform the Grid Buffer Service before hand.

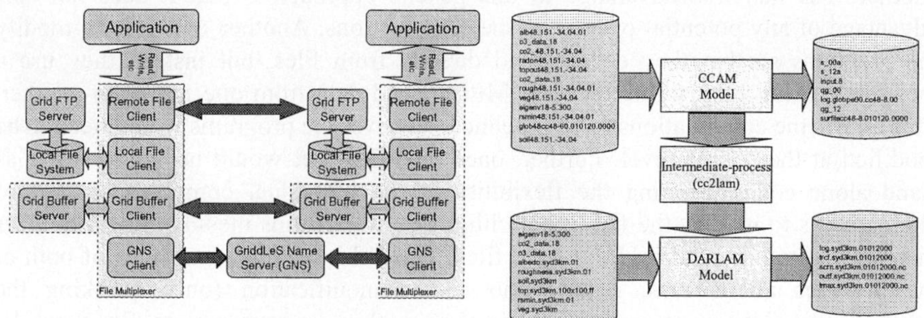

Fig. 4. Using GridFiles to link apps   **Fig. 5.** File dataflow

The GriddLes Name Service (GNS) is responsible for configuring the grid application. Each entry in the GNS indicates what should happen when a particular file is opened on a particular resource. At the time of the deployment the GNS reads the data from the configuration file and keeps it in a hash table. These keys and values correspond to the global naming schema. The reader and writer applications can use different keys which are mapped to the same entry in the GNS, indicating both are linked to either the same file or buffer. Buffers are distinguished from files by the word "buffer" in the value. If an entry in the GNS represents a buffer, then additional configuration information is contained. The GNS can be updated at any time, a this reconfigures the Grid application dynamically.

All the applications and web services can run on the same system or each one can be distributed to the different Grid nodes which are located geographically at different locations.

4   Experimental Results

Figure 5 shows a particular configuration of the three atmospheric models discussed in section 2, namely C-CAM, cc2lam and DARLAM. Importantly, most of the computation is performed by C-CAM and DARLAM, and cc2lam provides simple data manipulation and filtering between the two codes. In this section we describe an experiment in which C-CAM, cc2lam and DARLAM are coupled by Grid Buffers, that is, the output of C-CAM is streamed into DARLAM (via cc2lam). Because we used a file multiplexer as discussed in section 3, *it was possible to do this without source modification.* This is important because C-CAM and DARLAM are legacy codes written in Fortran, and we did not wish to make significant modifications to their structure. This scheme is very flexible. All models and services can run on a single machine, or on different nodes of a cluster or on different machines in a computational grid. Further, these configuration changes can occur without any modification to the atmospheric models.

When the first writer application (C-CAM) writes block of data (typically of 4096 bytes) using the write statement, then the data need to be transferred over to the GridBuffer service by using the client component of GridBuffer service which is in FMP underneath the application layer without the user interaction. Once the data arrives, the reader application (in this context cc2lam) uses the GriddBuffer client and obtain the block of data and writes into the other GridBuffer in a similar way, so that the other reader application (in this context the DARLAM Model) reads the block of data. In some instances, DARLAM rereads some of the input data. Because the data has already been deleted from the hash table in the Grid buffer Service, it is read form the cache file instead. This occurs transparently to the DARLAM model.

In this experiment we utilised a number of machines shown in Table 1, in three countries (AU, US and UK).

Table 1. Machine List

Name	Address & Details	Country	Name	Address & Details	Country
dione	dione.csse.monash.edu.au Pentium 4, 1500 MHz, 256 MB, Redhat Linux 7.3	AU	brecca	brecca-2.vpac.org Intel Xeon, 2.8 GHz, 2048MB, Redhat Linux 7.3	AU
freak	freak.ucsd.edu Athlon Processor, 700 MHz, 256 MB, i386, Debian	US	bouscat	bouscat.cs.cf.ac.uk Pentium 3, 1 GHz, 1544 MB, Red Hat Linux 7.2	UK
vpac27	vpac27.vpac.org Pentium 3, 997 MHz, 256 MB, Red Hat Linux 7.3	AU	dragon	dragon.vpac.org Intel(R) Pentium(R) 4 CPU 1.70GHz, 1006MB Red Hat Linux 8.0 3.2-7	AU

In this case study, three different experiments were performed.

Case 1: All models are executed concurrently on the same machine. There are two variations on this experiment. First, the programs read and write conventional local files. The results of this experiment are shown in the "Files" column in Table 2. Second, the programs use Grid Buffers instead of files. These times are shown in the Buffers column in Table 2. All times are cumulative, and thus the DARLAM time also indicates the total time taken. In this experiment the code was run for 480 and 960 time steps.

The results shown in Table 2 highlight that using buffers is always faster than using files when the codes are run on the same system. This is interesting because the models are multiprocessing the single CPU on these machines, and thus it suggests that buffers are allowing some overlap of computation and communication. Even more interesting is that most of the runs performed with buffers were actually faster than running the codes sequentially on the same platform, again because of the overlap in IO and computation. The exceptions to this are those run on dione and vpac27, which are presumably because of the relative speed of the computation and the IO on these two machines.

Case 2: C-CAM and DARLAM are executed on different computers at different locations, whilst cc2lam is run on the same machine as C-CAM. Because there are a large number of potential pairings of machines, we have selected a few interesting

Table 2. Cumulative concurrent runs on the same system (time in hr:min:sec)

Computer	Model	480 time steps		960 time steps	
		Files	Buffers	Files	Buffers
dione	C-CAM	00:41:18	00:44:10	01:23:57	01:29:59
	cc2lam	00:41:56	00:44:15	01:25:13	01:30:09
	DARLAM	01:08:17	00:49:12	02:19:13	01:35:00
brecca	C-CAM	00:18:13	00:20:05	00:36:29	00:41:21
	cc2lam	00:18:25	00:20:12	00:36:50	00:41:24
	DARLAM	00:27:58	00:22:57	00:56:35	00:44:11
freak	C-CAM	00:34:35	00:35:21	01:22:28	01:28:15
	cc2lam	00:35:26	00:35:33	01:23:15	01:30:01
	DARLAM	00:52:39	00:40:30	02:14:22	01:36:15
bouscat	C-CAM	01:10:22	01:17:51	02:44:03	02:42:42
	cc2lam	01:10:39	01:18:10	02:44:39	02:43:00
	DARLAM	01:55:27	01:29:59	04:14:54	02:54:55
dragon	C-CAM	00:41:25	00:41:28	01:23:01	01:27:57
	cc2lam	00:41:46	00:41:41	01:23:43	01:28:08
	DARLAM	01:06:17	00:46:24	02:13:12	01:32:57

ones and present the timing results in Table 3. Again all times are cumulative, and thus the DARLAM time also indicates the total time taken. In the case where local files are written we have also included the time taken to copy the files in the cumulative totals. The runs were done for both 480 and 960 time steps.

Table 3. Cumulative concurrent runs (Time in hr:min:sec)

Model	Mach	Close Systems 480 Time steps Files	Buffers	960 Time steps Files	Buffers	Mach	Distant Systems 480 Time steps Files	Buffer	960 Time steps Files	Buffer
C-CAM	dione	00:28:21	00:34:20	00:55:18	01:08:14	brecca	00:16:34	00:18:39	00:32:32	00:38:14
cc2lam	dione	00:28:29	00:34:32	00:55:40	01:08:25	brecca	00:16:42	01:00:48	00:32:49	01:48:42
File Copy		00:29:19		00:56:52			00:24:12		00:40:25	
DARLAM	vpac27	01:00:29	00:48:47	02:00:59	01:26:04	bouscat	00:56:04	01:05:03	01:46:19	01:52:58
C-CAM	brecca	00:16:34	00:18:37	00:32:32	00:38:55		01:07:29	01:18:18	02:29:58	02:22:44
cc2lam	brecca	00:16:42	00:18:44	00:32:51	00:39:01	bouscat	01:07:41	01:30:34	02:30:23	02:34:12
File Copy		00:16:57		00:33:04			01:17:11		02:37:59	
DARLAM	vpac27	00:47:57	00:40:43	01:37:11	01:14:35	bouscat	01:24:57	01:31:55	02:55:02	02:35:20
C-CAM	brecca	00:16:34	00:18:05	00:32:32	00:38:55	brecca	00:16:34	00:18:20	00:32:32	00:37:10
cc2lam	brecca	00:16:42	00:18:12	00:32:48	00:39:01	brecca	00:16:42	00:33:49	00:32:49	01:05:40
File Copy		00:17:32		00:33:59			00:20:17		00:37:14	
DARLAM	dione	00:30:48	00:25:10	00:59:53	00:44:27	freak	00:33:55	00:41:45	01:07:01	01:07:19
C-CAM	brecca	00:16:34	00:18:37	00:32:32	00:39:03	freak	00:30:31	00:36:57	01:16:56	01:14:03
cc2lam	brecca	00:16:42	00:18:40	00:32:48	00:39:18	freak	00:31:01	00:44:08	01:17:23	01:21:07
File Copy		00:17:32		00:33:04			00:34:36		01:21:48	
DARLAM	dragon	00:29:44	00:23:10	01:00:09	00:43:53	brecca	00:42:22	00:45:17	01:38:51	01:22:16

The results show that buffers are always faster for systems which have good network connections than when files are used. Interestingly, the results for 960 time steps are less than double those for 480. This is because the startup overheads are masked in the longer runs, and thus the parallel efficiency is higher.

The results for distant machines tell a different story. Because these machines have poorer networks between them, it is not always faster to use buffers than to copy the files. For example for the shorter runs of 480 time steps, it is always better to use file copies. On the other hand, buffers are more efficient for some of the longer 960 time step runs. The results shown here highlight the importance of being able to reconfigure the application dynamically because it is not always possible to know which configuration will be more efficient.

Case 3: In this experiment C-CAM and cc2lam are executed on brecca and DARLAM model is run on dragon. We run models for 480 time but data is exchanged at different intervals from 60 to 15 time steps. The results are shown in Table 4.

Table 4. Varying the output frequency (Time specified in hr:min:sec format)

Print Interval in time steps	Amount of Output Data produced by different models in MB			Total Time with Files	Total time with Buffers
	CCAM	cc2lam	DARLAM		
60	72.663	42.485	43.000	00:29:38	00:23:23
30	137.253	80.249	83.599	00:35:50	00:26:30
15	264.428	155.771	163.760	00:38:54	00:26:21

As expected, as the write interval reduces, the magnitude of data produced by each model increases. Even though there is an increase in computation time as the write interval reduces, the total computation time with the buffers is less than with files because more overlap is possible.

Table 5 analyses the degree of overlap in the computations, and therefore the efficiency, for a few different machine configurations. Here we calculate the best theoretical time that could have been achieved assuming no startup costs and perfect networking, and compare this to the actual time. The results indicate that the efficiency can be quite high for low latency, high bandwidth networks, especially for the longer runs.

Table 5. Overlap comparisons (Time specified in hr:min:sec format)

Machines	Time Steps	C-CAM with Files	DARLAM with Files	Total time with Buffers	Best Theoretical	Efficiency
Brecca-2 & vpac27	480	0:16:34	0:31:00	0:40:43	0:31:00	76%
	960	0:32:32	1:04:07	1:14:35	1:04:07	86%
dione & vpac27	480	0:28:21	0:31:00	0:48:47	0:31:00	64%
	960	0:55:18	1:04:07	1:26:04	1:04:07	74%
Brecca-2 & vpac27	480	0:16:34	0:13:16	0:24:58	0:16:34	66%
	960	0:32:32	0:25:54	0:44:27	0:32:32	73%

5   Conclusions

In this paper we have discussed the implementation of a grid workflow that couples two legacy atmospheric science applications, namely a global climate model and a regional weather model. *One of the more significant achievements of the experiment is that we managed to do this without any changes in the source code of the two models.* This is no mean feat since they are legacy codes written in Fortran and were designed without any knowledge of the underlying grid infrastructure. Inter-process communication is provided by a software device called a File Multiplexer, and we have chosen to implement this with web services.

The performance results indicate that there are cases when it is advantageous to couple the models tightly using pipes, and other cases where it is more efficient to write files locally, copy them to the receiving node and read them locally. An important feature of our implementation is that the decision about whether to copy or use buffers can be delayed until *the time that the application is configured* and does not need to be integrated into the source code of the models. We are not aware of other systems with equivalent flexibility.

The case study discussed here is actually a small fragment of a much larger system currently being constructed. Rather than just couple 2 models, we plan to couple a number of atmospheric science models including air pollution codes. We also plan to retrieve data directly from scientific instruments like temperature and pressure sensors. Such an application would form an interesting grid application because it would involve integration of a number of separate computational models, running on different computer systems (possibly owned by different organizations) and taking data from real time scientific instruments.

Acknowledgements. The Australian Research Council and Hewlett Packard support this work under an ARC linkage grant. Kommineni was supported by an Australian Postgraduate Research Award.

References

1.   Chan, P. and Abramson, D. "NetFiles: A Novel Approach to Parallel Programming of Master/Worker Applications", HPC Asia 2001, 24-28 September 2001 • Royal Pines Resort Gold Coast, Queensland, Australia.
2.   Deelman, E., Blackburn, K. et al., "GriPhyN and LIGO, Building a Virtual Data Grid for Gravitational Wave Scientists," presented at 11th Intl Symposium on High Performance Distributed Computing, 2002.
3.   Deelman, E., Blythe, J., Gil, Y., Kesselman, C., Mehta, G., Vahi, K., Lazzarini, A., Arbree, A., Cavanaugh, R. and Koranda, S. "Mapping Abstract Complex Workflows onto Grid Environments", Journal of Grid Computing, Vol. 1, No. 1, pp 9--23, 2003.
4.   Foster, I. and Kesselman, C., Globus: A Metacomputing Infrastructure Toolkit, International Journal of Supercomputer Applications, 11(2): 115-128, 1997.
5.   Foster, I., and Kesselman, C. (editors), The Grid: Blueprint for a New Computing Infrastructure, Morgan Kaufmann Publishers, USA, 1999.

6. GriPhyN 2003, `www.griphyn.org`

7. McGregor, J. L., Nguyen, K. C., and Katzfey, J. J. Regional climate simulations using a stretched-grid global model. In: Research activities in atmospheric and oceanic modelling. H. Ritchie (ed.). (CAS/JSC Working Group on Numerical Experimentation Report; 32; WMO/TD - no. 1105) [Geneva]: WMO. p. 3.15-3.16, 2002.

8. McGregor, J.L., Walsh, K.J. and Katzfey, J.J. Nested modelling for regional climate studies. In: A.J. Jakeman, M.B. Beck and M.J. McAleer (eds.), Modelling Change in Environmental Systems, J. Wiley and Sons, 367–386, 1993.

9. Robert van Engelen, "The gSOAP toolkit 2.0.", Technical report, Florida State University, `http://www.cs.fsu.edu/~engelen/soap.html`, 2001.

10. Sun Microsystems, "Web Services Made Easier: The Java APIs & Architectures for XML", 2002, `http://java.sun.com/webservices/white/` & `http://java.sun.com/webservices/webservicespack.html`

11. Walsh, K.J. and McGregor, J.L. January and July climate simulations over the Australian region using a limited-area model. J. Climate, 8 (10), 2387–2403, 1995.

Twins: 2-hop Structured Overlay with High Scalability*

Jinfeng Hu, Haitao Dong, Weimin Zheng, Dongsheng Wang, and Ming Li

Computer Science and Technology Department, Tsinghua University, Beijing, China
{hujinfeng00, dht02, lim01}@mails.tsinghua.edu.cn,
{zwm-dcs, wds}@tsinghua.edu.cn

Abstract. How to build an efficient P2P overlay network on a large-scale system is still in the air. Pastry-like p2p overlays have low maintenance costs because of their log(N)-sized routing tables. However their lookup efficiency is quite low. One-hop overlays, although having high routing efficiency, can not scale to large systems because of its high maintenance cost. In this paper, we present a novel structured overlay network, Twins. Routing in Twins can be accomplished in 2 hops in very high probability. With a report-based multicast maintenance algorithm, the overlay network consumes very low maintenance cost in presence of large-scale and highly dynamic network environments. The experimental results indicate that, when the system running over a network of 5,000,000 peers, each peer consumes only 6 messages per second for maintenance, and the routing latency is only 2 hops in a very high probability of 0.99.

1 Introduction

With the introduction of Napster in 1999, the peer-to-peer system became "the fastest growing Internet application ever". Currently, the number of concurrent users of Kazaa has rapidly increased to about 3,000,000[3], and seemingly this trend has no sign to cease in the near future. The increasing system scale has become one of the main challenges for the design of structured P2P overlay network.

Structured overlays can be classified into two categories: Pastry-like overlays and one-hop overlays. The size of Pastry-like overlays' routing table is O(logN) (such as Pastry[6], Tapestry[9], and Chord[8]). With a relatively small-scale routing table, Pastry-like protocols need little amount of maintenance cost to deal with the system membership changes, which means that these protocols can achieve high scalability. But the price of this approach is that on average lookup operation requires O(logN) steps to converge. For example, a typical 16-based(b=4) Pastry network needs about $\log_{16}3{,}000{,}000 \approx 5.38$ hops for message forwarding, which can not be tolerated by many particular applications. The overlay route latency is becoming another challenge for P2P protocol.

The other approach of structured overlays is one-hop overlay [1]. Every overlay node keeps complete membership information of all the other nodes in the system. Its

---

* This work is granted by The National High Technology Research and Development Program of China (G2001AA111010), Chinese National Basic Research Priority Program (G1999032702), and a joint research grant from National Science Foundation of China (project No.60131160743) and Hong Kong Research Grant Council.

M. Bubak et al. (Eds.): ICCS 2004, LNCS 3036, pp. 174–183, 2004.

lookup operation accomplishes in one hop. But the one-hop overlay's routing table is too large and consumes too much bandwidth for updating. Nowadays, the average lifetime of peers is about one hour [7], that is to say, given a P2P overlay network of 3,000,000 nodes, every node must receive information of 3,000,000*2 =6,000,000 member-changing events per hour. Typically, the data structure of a membership-changing event is about 200 bits long, include corresponding node's nodeId, IP address and port. Thus the bandwidth cost is no less than 33.3 kbps, which is too heavy a burden to most modem-linked peers. So the scalability of existing one-hop overlay is very poor.

It is the two challenges mentioned above that motivates us to design a new scalable structured overlay, Twins, which simultaneously obtains high routing efficiency (2-hop routing in 99% probability) and low bandwidth overhead (6 messages per second in a 5,000,000-node network).

Twins' routing table consists of two parts, one containing all nodes sharing a h-bits prefix and the other containing all nodes having a b-bits common suffix(h and b are systematic parameters), with a simple routing algorithm it can route messages to their destinations in just 2 hops, in a very high probability of 0.99. Twins adopts a report-based routing table maintenance algorithm. When a membership change event occurs, the overlay will multicast this event in a report-based mechanism, which helps the overlay consumes low bandwidth to deal with node's join and crash. Our experimental result shows that when running over a P2P network of 5,000,000 peers, each Twins node consumes only 6 messages per second for maintenance. This cost, as well as routing table size, varies as a $O\left(\sqrt{N}\right)$ function to the overlay scale N, so Twins can also run well in an even larger environment. Moreover, Twins introduces probabilistic routing into structured overlay. How many hops a message passes before reaching its destination is not strictly determined. This makes room for scalability design: we can raise the expectation of hops to keep a low overhead by adjusting system arguments.

The rest of this paper is organized as follows. Section 2 presents the design of Twins protocol. In section 3 we give a formalized analysis of the routing performance and the maintenance cost. Experimental results are presented in section 4. Final conclusion is given in section 5.

2 The Twins Protocol

This section describes the Twins protocol. The Twins protocol specifies how to locate keys, how to construct the routing table, and when new node join or existing nodes fail how to maintain the structure of the system.

Like Pastry and Chord, Twins assigns every node and key an identifier using SHA-1, typically 128-bit long. They are ordered in an identifier ring modulo 2^{128}, and we assume that all these identifiers are uniformly distributed in the identifier space. For a node, say node M, we call the first p bits of M's node ID the node's prefix, and the last s bits the node's suffix, we assume that p(s)<128, and the scale of the overlay network is N(N $\ll 2^{128}$).

2.1 Routing Table

Each Twins node has a routing table consisting of two parts. The first one contains all the nodes having the same prefix as M, called *prefix set*. While the second one contains all the nodes having the same suffix as M, called *suffix set*. In all, a Twins node's routing table is the union of its prefix set and suffix set.

Obviously if two nodes have the same prefix, their prefix set must also be the same. In this way, the set of all the peers can be partitioned into 2^h groups according to their different prefix. We call these groups *prefix groups*. Nodes within the same prefix group are fully interconnected with each other. Since node IDs are distributed evenly in the ID space, averagely each prefix-group includes about $N/2^h$ nodes. The *suffix group* is defined similarly, i.e., all the nodes are partitioned into 2^b different suffix groups which do not intersect with each other, and the expected value of the size of every suffix-group is $N/2^b$.

Given a 128 bits id N, we define prefix-group$_h$(N)={M | M is a 128 bit identifier, the first h bits of M is the same as N's first h bits}, suffix-group$_b$(N)={M | M is a 128 bit identifier, the last b bits of M is the same as N's last b bits}, h and b are system's parameters.

Figure 1 shows the routing table of a hypothetical node with a 12-bit node ID 011100101110.

2.2 Routing

Every message has a destination ID that is also 128-bit long. The first h bits of the message's destination ID is also called the message's h-bit prefix, and the unique prefix-group corresponding to this prefix is called its prefix-group. Similarly, we define the message's suffix-group as the group consisting of all the nodes whose ID's b-bit suffix is the same as the message's b-bit suffix.

Slightly different with Pastry, the destination node of a message is defined as following:

The destination node of a message M is the node in the M's prefix-group whose ID is the numerically closest to the destination ID of M.

Notice that all the nodes in same prefix-group are fully interconnected with each other, and the basic task of message routing is forwarding messages to the corresponding prefix-group. After that the message will directly reach its destination node in one hop.

When routing a message with destination ID M, node N first checks whether M and N have the same *prefix* (that is to say if M $\in$ prefix-group$_h$(N)). If so, N can directly forwards it to the destination node, which must be in N's prefix set; otherwise N inspects its suffix set, tries to seek out a node E who has M's prefix(the node E, such that M $\in$ prefix-group$_h$(E) and E $\in$ suffix-group$_b$(N)) and forwards the message to E. For example, when the node shown in Fig.1, named N, will route a message M. If M's destination ID is 011100000000, according to our routing algorithm, M $\in$ prefix-grouph(N), then N forward M to 011100100101, which is numerically closest to M in N's prefix set. And if M' destination ID is 001000000000, node N will choose node 001011001110 and forward M to it, because this node has the same 4-bit prefix as M.

<table>
<tr><td colspan="2" align="center">Node ID 011100101110</td></tr>
<tr><td colspan="2">Prefix set</td></tr>
<tr><td align="center">011101110001</td><td align="center">011110110011</td></tr>
<tr><td align="center">011100100101</td><td align="center">011111000110</td></tr>
<tr><td align="center">011111001001</td><td align="center">011100101100</td></tr>
<tr><td align="center">011101101110</td><td align="center">011111100111</td></tr>
<tr><td colspan="2" align="center">. . . . .</td></tr>
<tr><td colspan="2">Suffix set</td></tr>
<tr><td align="center">000101001110</td><td align="center">001011001110</td></tr>
<tr><td align="center">010011101110</td><td align="center">011101101110</td></tr>
<tr><td align="center">100100101110</td><td align="center">110001101110</td></tr>
<tr><td colspan="2" align="center">. . . . .</td></tr>
</table>

Fig. 1. State of Twins's Routing Table. Node's ID is 12 bit long, and we assume h=4, b=4. Node M's ID is 011100101110. M's Routing Table consists of Prefix set, which contains all the nodes whose ID has the same 4-bit prefix as M, and Suffix set, which contains all the nodes whose ID has the same 4-bit suffix as M. For all the nodes, their prefix and the suffix are shown in boldface. Because of space, we can not list M's Routing Table completely. And it is easily to discover that the prefix set may intersect with suffix set, in this case they have the common node ID **011101101110**.

If there is no node in N's suffix set which has the same prefix as M, N forwards the message to a node randomly chosen from its prefix set whose suffix is different from M.

2.3 Maintenance

Like one-hop overlay, Twins introduces a report-based multicast mechanism to maintain routing table. Note that prefix-groups are independent to one another, and so do suffix-groups. It allows us to simply consider how to maintain nodes' prefix set/suffix set within a prefix/suffix-group.

```
route(M)
   if  M ∈ prefix-group_h(N)
       then  Forward M to its destination node;
             Return;
   else  for every node E in N's suffix set do
       if  M ∈ prefix-group_h(E)
             then      Forward M to E;
                       Return;
   Random select a node E from N's prefix set;
   Forward M to E;
```

Fig. 2. Pseudo code of Twins' routing algorithm

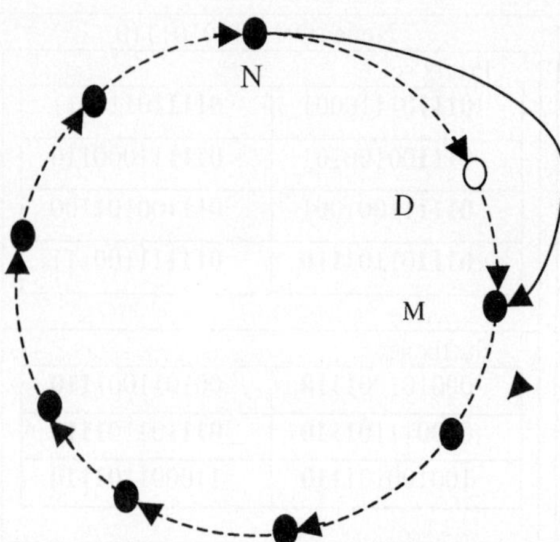

Fig. 3. This is a hypothetical prefix-group. All the nodes in this group are organized as a ring, in which every node must send periodical heartbeat message. We can see if there is a node, say M, which fail to receive heartbeat message from its frontal node, say D, for several heartbeat period, M will take node D for death, initiate the event of D's death and then multicast this event message in the prefix-group. The multicast algorithm is shown in fig.4. When node N receives this event message, it resets its next node as M and begins to send heartbeat message to M.

When a new node X joins Twins system, firstly it should contact an existing Twins node B, which is named as X's bootstrap node. On receiving X's join request, B initiates a lookup of X's ID to find two nodes, say P and S, in X's prefix group and suffix group respectively, and then X requests prefix set from P, and suffix set from S. After the receiving of X's prefix set and suffix set, X can establish its routing table to complete its joining process. And node P and node S multicast the event of X's joining to all the nodes in X's prefix-group and suffix-group respectively.

All the nodes in a group are ordered in an identifier ring. Every node sends periodical heartbeats to the first node next to it in the identifier ring. If a node D has not send heartbeats for several periods, then D is considered as dead, and the node next to D in the ring will multicast the event of D's death to all the nodes in this group. The message used to inform other peers that there occurs a member-changing event is called an event message. Fig.3 shows the maintenance mechanism of a hypothetical prefix/suffix-group.

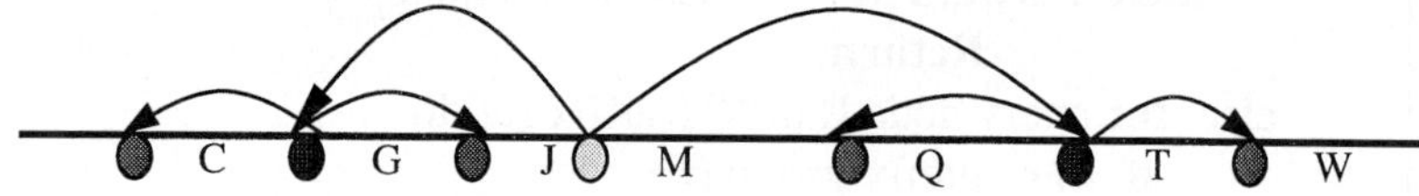

Fig. 4. An instance of tree-based multicast

Notice that all the nodes within a group are fully interconnected, there are various algorithms to carry out the multicast process. Here we adopt a simple tree-based multicast, illustrated in Fig. 4. When a node M initiates a multicast process , it sends the event to a node before it (say G) and another behind it (say T). Then alike, G and T each sends the event to two nodes, one ahead and the other behind. This procedure continues. At each step, every node should ensure that once it sends the event to another node, there is no other node between them who has already received this event.

If more efficient multicast is desired, the multicast tree can be modified to be 2^b-based. And if more reliable multicast is desired, response-redirect mechanism can be deployed, which will double the maintenance cost.

In this manner, except for the changing member's node ID, IP address and port, an event message (e.g. message from T to Q in Fig.4) should additionally include node ID of the first node before the receiver who has already received the message (that is M), as well as node ID of the first such node behind it (that is T). Plus UDP head (64 bits) and IP head (160 bits), an event message will not exceed 500 bits.

Assuming that the average lifetime of nodes is 1 hour, which is the statistic result from [7], all the items in the routing table have to be refreshed in a period of 1 hour. It means that for a Twins system, which consists of 5,000,000 nodes, and h and b are both set as 10, on average every prefix/suffix-group contains $5,000,000/2^{10}=4883$ nodes. Then on average, every node receives (4883+4883)*2=19532 event messages per hour, 5.43 messages per second in other words. Plus heartbeats and their responses, the total message count per second will not exceed 6, i.e., the bandwidth cost is lower than 6message/second*500bit=3kbps.

3 Performance Evaluation

In this section, we give a formalized evaluation of Twins protocol, and describe how to determine the length of prefix and suffix in a given system environment, and then estimate routing table size and maintenance cost.

Assuming that in a Twins overlay network of N nodes, every node has a prefix with h bits and a suffix with b bits. Then there are 2^h prefix-groups and 2^b suffix-groups totally. On average, each prefix-group contains $H=N/2^h$ nodes and each suffix-group contains $B=N/2^b$ nodes.

3.1 Routing Performance

For a certain node M, it has a suffix set with 2^b routing table entries. M wishes that these entries would distribute over all the prefix-groups. But unfortunately it is not the reality: there must be some prefix-groups which there are no routing table entries in M's suffix set belonging to. We note the number of such prefix-group for the given node M as S. The expected value of S can be defined as follows:

$$E(S) = \sum_{k=0}^{H-1} k \times \frac{\dbinom{2^h}{k}\dbinom{B-1}{2^h-k-1}}{\dbinom{2^h+B-1}{2^h-1}}$$

For a given message destination ID D, We call the probability of there are no less than one routing table entry in M's suffix set which has the same prefix as D as *hit ratio*. Obviously, the hit ratio can be defined as $hr = 1 - \dfrac{E(S)}{H}$. Consequently, hr is related to the ratio of size of suffix set (B) to the number of prefix-groups(2^h), we define $R = S / 2^h$, and then we can see that when 2^h is larger than 50, choosing R as 4.7 will ensure a hit ratio more than 0.99. So in common cases we demand that $R > R_0 = 4.7$. That is to say, $N/(2^b \cdot 2^h) > R_0$, namely

$$b + h \leq \log_2(N/R) \tag{1}$$

Next we estimate maintenance cost. Heartbeat cost is a fixed small value, so for simplicity we can ignore it. The substantial cost is for maintaining prefix set and suffix set, with size of $H+B$. Assuming that nodes' average lifetime is L seconds, each node triggers two events in a period of L seconds on average. So every node receives $2 \cdot (H+B)$ events during every L seconds. If redundancy of the multicast algorithm we adopt is f, then the number of messages a node receives per second is

$$m = (H + B) \times 2 \times f \Big/ L = \left(\frac{N}{2^h} + \frac{N}{2^b}\right) \times 2f \Big/ L$$

When $h = b$, m reaches its minimum value:

$$m_0 = \frac{2N}{2^{\frac{1}{2}(h+b)}} \times 2f = \frac{4N \cdot f}{2^b \cdot L} \tag{2}$$

$$b = h = \left\lfloor \frac{\log_2(N/R)}{2} \right\rfloor$$

Considering both (1) and (2), we should set ⟨ ⟩ to ensure a 2-hop routing hit ratio larger than 0.99 using a minimal maintenance cost. Then we can get the following results:

a) Routing table size R_{size} satisfies

$$2\sqrt{N \cdot R} \leq Rsize < 4\sqrt{N \cdot R} \tag{3}$$

b) Maintenance cost m_0 satisfies

$$\frac{\sqrt{N \cdot R} \times 4f}{L} \leq m_0 < \frac{\sqrt{N \cdot R} \times 8f}{L} \tag{4}$$

Figure 5, 6 and 7 show the variation of b, R_{size} and m_0 as functions of N using $f=1$ and $L=3600$.

3.2 Scalability

Inequation (3) and (4) show that a Twins node with a routing table with $O(\sqrt{N})$ entries consumes $O(\sqrt{N})$ bandwidth to maintain it. This means a good scalability property of Twins overlay network. In addition, when the maintenance cost is not acceptable by peers, Twins also can trade hops for bandwidth consumption like other

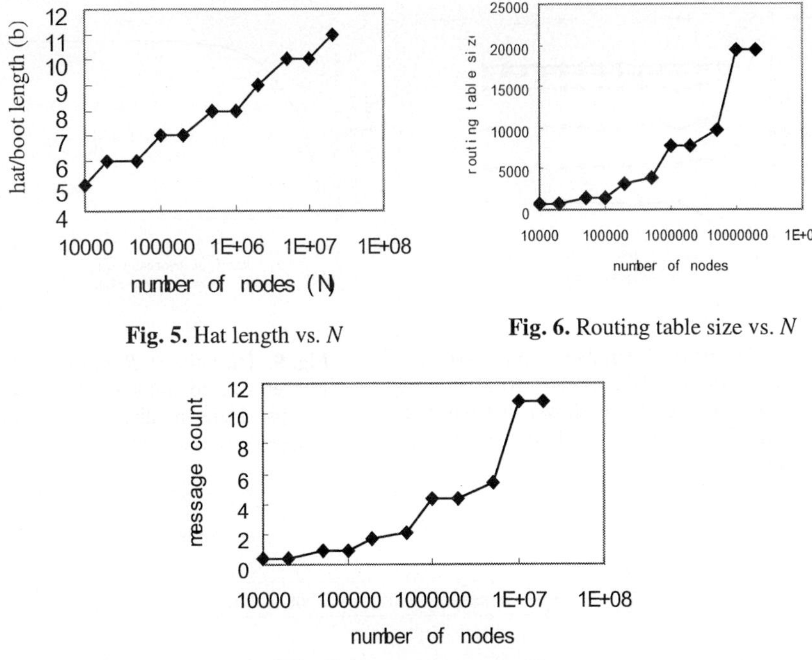

Fig. 5. Hat length vs. N

Fig. 6. Routing table size vs. N

Fig. 7. Message count vs. N

overlay protocols. To illustrate it, we put Twins into a stricter environment where N=10,000,000, f=2 and L=2,400. Using inequation (4) we know that if keeping 2-hop routing, a node should receive at least 22 messages per second. Twins can reduce this cost by decreasing R, which will raise b and h, decrease B, and then reduce hr. Expected value of hop counts can be calculated as:

$$E(hops) = \sum_{i=2}^{\infty} i \cdot (1-hr)^{i-2} hr = 1 + \frac{1}{hr}$$

Even when hr drops to 0.25, average hop counts only rise to 5.

4 Simulation and Experimental Result

In this section, we present experimental results obtained with a simulator for Twins protocol. The simulator was implemented on ONSP, an overlay networks simulation platform, which can parallel simulating the function of most off-the-shelf peer-to-peer protocols. By implementing the event logic according to the protocol's definition, the user can easily simulate various protocols. The detailed description of ONSP exceeds the scope of this paper, we will present it in another paper.

Our experiments were performed on a 32 processors cluster (Pentium IV CPU and 2G memory), running Linux Redhat 7.0.

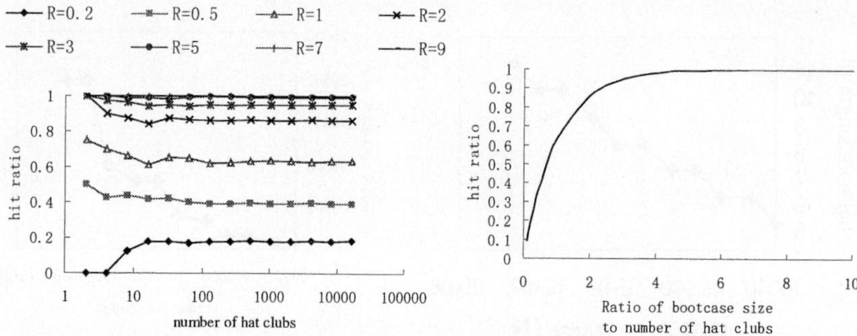

Fig. 8. Hit ratio vs. number of prefix-groups. R is the ratio of suffix set size (B) to number of prefix-groups (2^h). It shows that when 2^h exceeds 50, hit ratio is almost a constant number.

Fig. 9. Hit ratio vs. R, ratio of suffix set size to number of prefix-groups. Here number of prefix-groups is fixed at 1024.

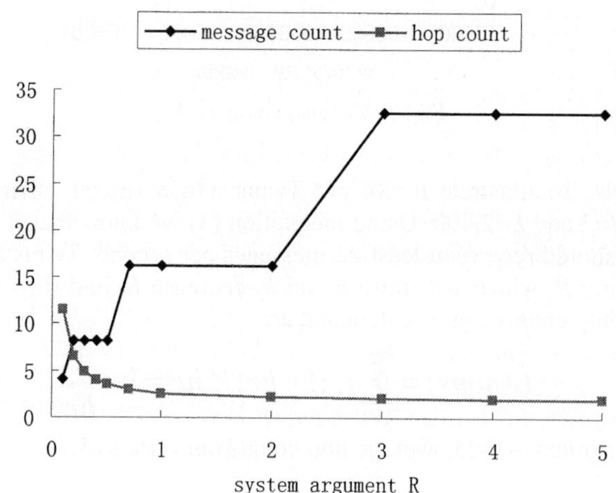

Fig. 10. Message cost and hop count in relation to the argument R, where N=10,000,000, f=2 and L=2,400

To simulate the real network topology and latency, we implemented a network topology generator based on GT-ITM[10]. The generator produces a transit-stub network topology model. We also obtained a trace of node joins and crashes from a measurement study of Gnutella[7]. The average living time of a node over the trace was about 2.3 hours. All the result is showed in the figure8, 9, and 10.

5 Conclusion

In fact, all the structured overlay networks are the compromise of size of routing table(degree of overlay node) and routing efficiency(diameter of overlay graph). We believe that the diameter of overlay graph is the most important object of peer-to-peer system design, because it impacts on the routing performance of overlay networks while degree of overlay node only has effect on maintenance cost.

The paper considers the trade-off between the size of routing table and the routing efficiency. We present a new structured overlay network that can route message in 2 hops in very high probability and scale to large membership changes. The main feature of Twins is the design of its routing table, the two parts of routing table insure each node keep a small-world manner routing state, which help our system to get high routing efficiency.

Our future works will be focused on exploring great heterogeneity of nodes in real P2P systems, to make Twins protocol more efficiency and scalable.

References

1. Anjali Gupta, Barbara Liskov, Rodrigo Rodrigues. One Hop Lookups for Peer-to-Peer Overlays. HOTOS IX. May 2003.
2. Indranil Gupta, Ken Birman, Prakash Linga, Al Demers, Robbert van Renesse. Kelips: building an efficient and stable P2P DHT through increased memory and background overhead. IPTPS '03. February 2003.
3. Kazaa. http://www.kazaa.com. November 2003
4. Petar Maymounkov and David Mazieres. Kademlia: A Peer-to-peer Information System Based on the XOR Metric. IPTPS '02. March 2002.
5. S. Ratnasamy, P. Francis, M. Handley, R Karp, and S. Shenker. A Scalable Content-Addressable Network. SIGCOMM 2001. August 2001.
6. A. Rowstron and P. Druschel. Pastry: Scalable, distributed object location and routing for large-scale peer-to-peer systems. Middleware 2001. November 2001.
7. Saroiu, S., Gummadi, P. K., and Gribble, S. D. A Measurement Study of Peer-to-Peer File Sharing Systems. MMCN '02. January 2002.
8. I. Stoica, R. Morris, D. Karger, M. F. Kaashoek, and H. Balakrishnan. Chord: A scalable peer-to-peer lookup service for internet applications. SIGCOMM 2001. August 2001.

Dispatching Mechanism of an Agent-Based Distributed Event System

Ozgur Koray Sahingoz[1] and Nadia Erdogan[2]

[1] Air Force Academy, Computer Engineering Department, Yesilyurt, Istanbul, Turkey,
`sahingoz@hho.edu.tr`
[2] Istanbul Technical University, Electrical-Electronics Faculty,
Computer Engineering Department, Ayazaga, 34469, Istanbul, Turkey
`erdogan@cs.itu.edu.tr`

Abstract. In recent years, event-based communication paradigm has been extensively studied and it is considered a promising approach to develop the communication infrastructure of distributed systems. In most of previously developed event systems, events are defined as simple messages, such as records tuples, or simple objects. Our work introduces a new approach which allows events to be represented as mobile intelligent agents that are called **agvents** (**ag**ent e**vent**) in the context of an agent-based distributed event system, the Agvent System. In this paper, we present the dispatching mechanism implemented by the Agvent System, where agvents are responsible for determining their own paths through the network, selecting target nodes and moving themselves to these targets.

1 Introduction

Event-based communication model represents an emerging paradigm for middleware that asynchronously interconnects the components that comprise an application in a potentially distributed and heterogeneous environment [1], and has recently become widely used in application areas such as large-scale Internet services. Event-based communication model supports either one-to-many or many-to-many communication pattern that allows one or more application components to react to a change in the state of another application component. Event-based communication generally implements what is commonly known as the publish/subscribe protocol.

The publish/subscribe protocol is very well suited for connecting loosely coupled large-scale applications in the Internet. In this model, receivers of messages express their interest by subscribing to a class of events, and they are asynchronously notified if a sender publishes an event which matches their subscription. In this way, the model allows a flexible n-to-m communication among the communicating parties.

As pointed out in [2], two open problems of distributed publish/subscribe systems (especially content-based systems) are security issues of the system and routing of events. Security is an important issue, because it is possible for a subscriber to be interested in an event (have a subscription for a particular event) but not be authorized to read that event (because of restrictions from a publisher). Although defining security architecture is important, it is outside the scope of this paper.

M. Bubak et al. (Eds.): ICCS 2004, LNCS 3036, pp. 184–191, 2004.

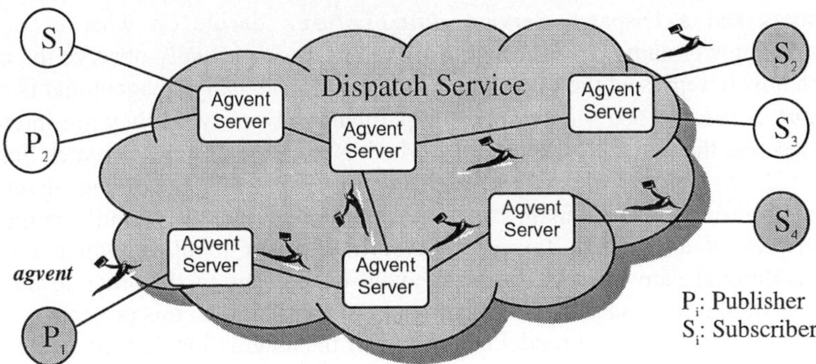

Fig. 1. Framework of Agvent System

For the scalability of distributed event systems, routing strategy is an important issue also. Generally a routing table is used in an event server to select the outgoing servers and clients, to which incoming messages will be sent. Different strategies are used for composing routing tables such as flooding, simple routing and content based routing [3]. In all these strategies, an event is defined as a simple message and it is dispatched to selected targets by a manager agent in the event server. In this paper, we propose a new approach for routing incoming events to targets. An important feature of our model is that, it allows events to be represented as mobile and intelligent agents that are called **agvents.** An agvent is an autonomous entity that acts on its own to reach its goal. A significant capability of an agvent is its ability to discover target nodes, namely subscribers, which need to be notified of the occurrence of an event, and to route itself to them. This is accomplished by enabling the published agvent itself search the knowledge base of an Agvent Server, select the registered subscribers, clone itself and send each agent clone to a subscriber on the selected list.

In this paper, we describe the dispatching mechanism of Agvent System, which provides high flexibility, scalability, and resilience to adverse network conditions. The goal of the agvent model is to keep the support required from agvent servers in the network to a minimum, placing intelligence in agvents rather than in individual nodes. This approach provides flexibility and obviates the need for the potentially impossible task of updating all nodes in a network for the implementation of a new application or protocol.

In the following, first Agvent System is described in Section 2. Next, the dispatching mechanism of Agvent System is explained in detail in Section 3, and finally, we present our conclusion and plans for future work in Section 4.

2 Agvent System

The framework depicted in Figure 1 represents the infrastructure of Agvent System [4] which is an agent based distributed event system that aims to meet the communication requirements of distributed and cooperative applications through the event-based design style. Agvent System combines two developing technologies, mobile agents and the publish/subscribe communication paradigm, in order to benefit the advantages of both. Its framework consists of three main components: Publishers, Subscribers and a Dispatch Service. *Publishers* decide on what events are observable, how to name or describe those events, how to actually observe the event, and then how to represent the event as a discrete entity, actually an agent that is called an *agvent*. *Subscribers* determine the particular agvent types they are interested in and describe them in the form of a rule which is processed by the Dispatch Service. *Dispatch Service,* which is responsible for dispatching incoming agvents to registered subscribers, is the main component of the Agvent System and is comprised of a network of distributed *Agvent Servers.* In a server topology which contains cycles, additional care must be taken to avoid cyclic routing. In order to simplify implementation issues, only *acyclic* topologies are considered in this project.

The entire Agvent system model is built under the assumption that the node/server architecture must be kept as simple and flexible as possible. Routing for instance, is agvent specific and not node/server specific. In this way, different agvents can execute different selection algorithms simultaneously on the same server. The Agvent Server, on the other hand, is responsible for storing subscription information in its knowledge base and provides an *Agvent Operation Platform* where agvents can migrate to and pursue their execution. The inner structure of an Agvent Server is depicted in Figure 2. Every Agvent Server processes incoming subscription and advertisement requests according to a protocol, which includes their propagation to adjacent/neighbor Agvent Servers. An Agvent Server communicates with publishers, subscribers and other Agvent Servers through the set of methods shown in Figure 2. Agvents move themselves to adjacent Agvent Servers and/or subscribers similarly to reach target nodes.

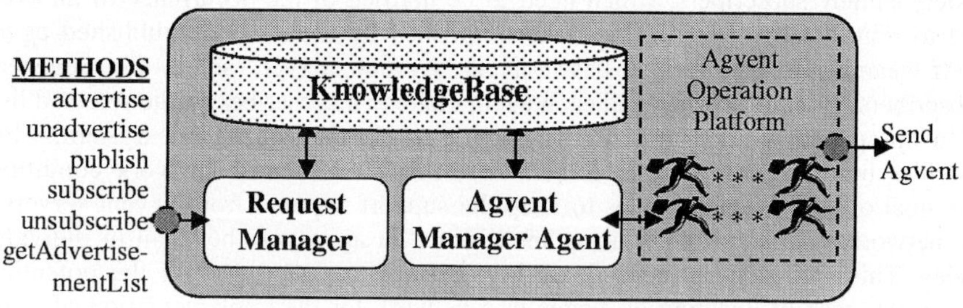

Fig. 2. Inner Structure of an Agvent Server

Agvent System differs from other distributed event systems with its distinct characteristics that are described below.

– **Autonomous Events:** Events are not viewed as simple messages. On the contrary, they are represented as mobile agents that have their own goals, beliefs and behaviors that they acquire at creation. This approach adds to the flexibility of the system and reduces the load and complexity of agvent servers as well.

– **Self-Routing Agvents:** Agvents are self-routing, that is, they are responsible for determining their own paths through the network, utilizing a minimal set of facilities provided by Agvent Servers. Agvent Servers in the network support incoming agvents by providing a simple, architecturally independent environment for the receipt and execution of agvents.

- **Agvent Based Subscription:** Subscribers register on agvent types. For example, a subscriber can register on an agvent, which is an instance of "BookAgvent" class, specifying certain constraints based on its advertised attributes and behaviors.
- **Information Hiding:** The published agvent itself searches the knowledge base of the agvent server *by talking with the* Agvent Manager Agent, selects the registered subscribers, clones itself and sends each agent clone to a subscriber on the selected list. Therefore, an agvent server has no access to the content of the published event data, which simplifies its role and consequently facilitates the server development process. Information hiding also meets requirements of certain applications where confidentiality of event data is essential.
- **User/Application defined agvent types:** A publisher creates its own agvent type and declares its properties and behavior through an advertisement message sent to the Dispatch Service. Once an agvent type is announced on the Dispatch Service, subscribers can register on agvents of that type.

2.1 Agvents

An agvent is a collection of code and data that migrates through the network, routing itself at each node on its path towards its target node. Agvents are created by publishers and sent out to subscribers over the Dispatch Service which is comprised of a network of distributed nodes. Each node hosts an Agvent Server which provides an operation platform for agvent execution, where incoming agvents execute their code in order to achieve their prescribed objectives. An agvent carries its routing procedure and routes itself at each node on the path toward a node of interest. To perform routing, an agvent communicates with Agvent Manager Agent (AMA) (Figure 2), using FIPA Agent Communication Language [5] in order to obtain the required information from the knowledge base of the Agvent Server.

After receiving the required information from AMA, an agvent may clone itself to create new agvents to be sent out to other Agvent Servers or subscribers on the network. Thus, an agvent can eventually generate multiple agvents, even though it starts out as a single one.

```
class BookAgvent extends Agvent
        {       private String   Author;
                private String   Name;
                private float[]   Dimensions ;
                private String   Publisher;
                private Date     Publish ;
                private int      ISBN;
                private float    ListPrice;
                private boolean ReferenceContains(String AuthorName) {...}
                private boolean TOCContains(String topic)  {...}
                privatefloat WholesalePrice(int amount,String destination) {...}
                .........
                .........}
```

Fig. 3. Class definition of BookAgvent

The following is a sample scenario, an event which involves the introduction of a new product. A publishing house wishes to announce a new book to interested subscribers, according to their subscription criteria. It first creates the class for an agvent, namely *BookAgvent* (see Figure 3), which contains only the necessary information for subscribers, however restricting access to detailed information of the book from both servers and subscribers of the system. Next, the publishing house creates an agvent instance of *BookAgvent*, embeds the necessary information of this new book into this agvent, and publishes this agvent to the system.

2.2 Life Cycle of an Agvent

An agvent has a four-stage execution life cycle on different nodes of the Agvent System. Flow of transition between stages, depicted in Figure 4, is as the following: (1) being created on a publisher, (2) migrate to an agvent server/subscriber of interest, (3) execution upon a target subscriber and (4) disposal upon completion of mission.

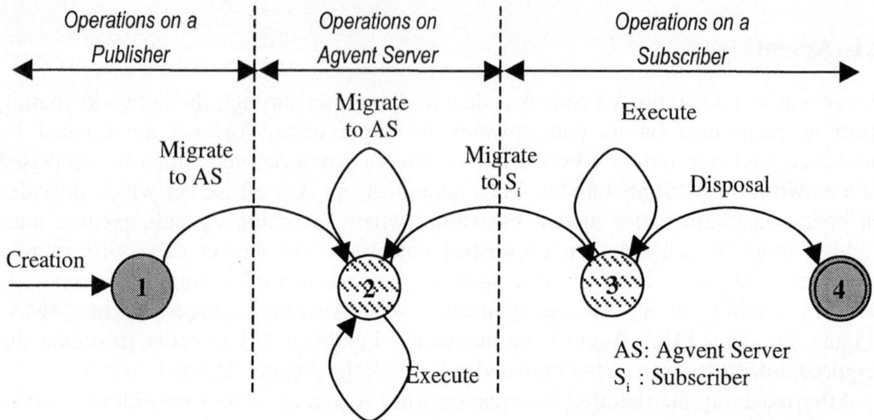

Fig. 4. Life Cycle of an Agvent

Publishers which decide on what events are observable and how to describe them as agvents. The life cycle of an agvent starts with its being created by a publisher on the observation of an event. Next, it routes itself to an Agvent Server where it identifies its target set of subscribers and a path which will take it to its destination nodes. At this point, new agvents may be spawned to be sent to different target nodes. After arrival on a target, actually a subscriber node, the agvent carries out a prespecified set of actions to reach its goals, and upon completion, disposes itself.

3 Dispatching Mechanism of Agvent System

The Agvent System implements three different dispatching mechanisms for different entities of the system, namely advertisements, subscriptions and agvents. To carry out the dispatching process correctly, each Agvent Server maintains a knowledge base to keep information to be used to route incoming request messages to local clients or neighboring Agvent Servers, the decision being based on the content of the message.

3.1 Dispatching of Advertisements

Every publisher knows of an Agvent Server to which it issues an advertisement request to advertise intent to publish a particular kind of agvent and make it visible to all subscribers of the system. An advertisement describes the properties of the relevant agvent, containing not only the attributes (with necessary descriptions) but also its behaviors that are to be exposed to subscribers. By getting this information, a subscriber can register on an agvent with constraints on these properties, as predicates asserted on the attributes or on the return values of behaviors on a list of input parameters prespecified by the subscriber.

Returning to the previous scenario, after developing the "BookAgvent" class, the publishing house issues an advertisement message for this agvent and sends it to the Dispatch Service. This message is structured as an instance of *"Advertisement"* class, which contains the necessary data structures to store information on filterable attributes and behaviors of the agvent, as depicted in Figure 5. The Dispatch Service distributes this message to all its constituent Agvent Servers through flooding and accepts subscriptions on this advertisement.

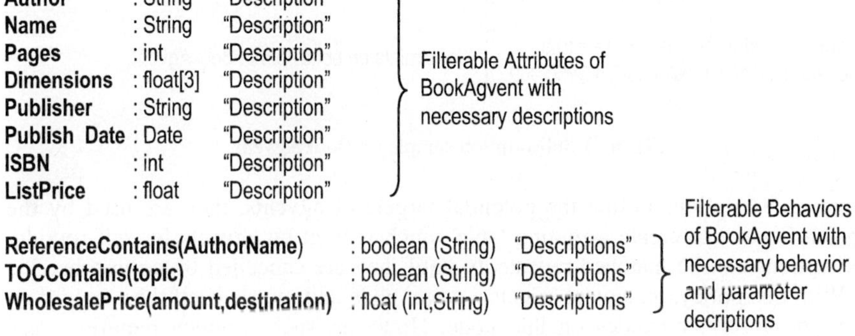

Fig. 5. An Advertisement of a BookAgvent

When an Agvent Server receives an advertisement, it stores the content of the message in an *advertisement table* along with the addresses of both the Agvent Server from which it received the message and the publisher node, that is, the owner of the advertisement. Next, it forwards the message to adjacent Agvent Servers. Thus, with this approach, the advertisement message is propagated to all Agvent Servers on the Dispatch Service and a trail of backward pointers from Agvent Servers to the publisher node is created. Consequently, it becomes possible to reach the publisher of the advertisement from any Agvent Server when successive locations on this trail are visited. Subscriptions are also dispatched over these routes created during the advertisement process. When a subscriber asks for the whole advertisement list or enquires a specific type of advertisement, the Agvent Server replies with the requested information and waits for subscriptions. Advertisements remain in effect until they are cancelled by a call to unadvertise.

3.2 Dispatching of Subscriptions

Subscriptions express the interests of applications/subscribers. With a subscription, an application can instruct the Dispatch Service its request to receive a certain agvent type through a filtering function. Setting a filter with a subscription means defining a predicate that is stored in the Agvent Server and that will be evaluated by every incoming agvent of that type. It is of fundamental importance to define the domain of these operations. In other words, it is crucial to determine

- which attributes of an agvent are filterable for the evaluation of subscriptions
- what kinds of primitive predicates and connectors are available (such as ">, >=,<,<=,!,!=,=…etc").

Agvent system uses the *Subscription* class which contains the necessary data structures to describe constraints on both attributes and behaviors of an agvent. A sample subscription for the previously advertised BookAgvent may contain the constraints depicted in Figure 6.

```
Author          =" Valentina Plekhanova"  ⎫
Pages           > 200                      ⎬  Constraints on attributes of BookAgvent
Publish  Date   > "January 1, 200"2        ⎭


ReferenceContains("Alonso, E.") == true       ⎫
WholesalePrice(1000, "Istanbul") < $150.000   ⎬  Constraints on behaviors of BookAgvent
                                               ⎭
```

Fig. 6. A Subscription sample on BookAgvent

Since subscriptions define the potential targets of agvents, they are used by the Dispatch Service to create a routing table which is used by agvents for self-routing. Subscriptions can be matched repeatedly until they are cancelled by an unsubscribe call. With this policy, an agvent selects a target node and moves itself there only if an interested subscriber resides on that node. However, such a policy requires every subscription to be propagated to every Agvent Server in the system.

When a subscription message reaches an Agvent Server, either from a subscriber client or from another Agvent Server, the server adds the new subscription information to its routing table in the knowledge base and propagates that subscription message to adjacent servers from which it has received the advertisement of the particular type of agvent the subscription is on. Every subscription is stored and forwarded from the originating Agvent Server to all other Agvent Servers in the network.

3.3 Dispatching of Agvents

An agvent is a collection of code and data that migrates through the network, routes itself at each node on the path, and executes on nodes of interest. An agvent determines its own path through the network, utilizing the minimal set of facilities provided by nodes. A key challenge in this model is the ability to discover target nodes, and to route itself to them. An agvent specifies target nodes with matching subscribers after applying filtering specifications. The next step requires the migration of clones of the agvent to each target. This execution cycle is completed in the following steps.

- First, an agvent has to be admitted at the destination node. A launcher task at an agvent server is to continuously receive agvents arriving from other servers or publishers.
- Second, upon acceptance, this agvent is activated on Agvent Operation Platform and scheduled for execution as a thread. During its execution the agvent may yield the processor and wait for data from Agvent Manager Agent and the routing behavior of an agvent chooses target nodes(agvent servers or subscribers). The execution performed at each step may differ based on peculiar properties of that node.
- Third, when the agvent completes its execution on the current node, it clones itself for transport to each of these target nodes. Agvent Manager Agent sends these clones to the destinations as dictated by agvent's behaviors.

4   Conclusion

This paper presents a new model for agent based distributed events systems, the Agvent System, which combines the advantages of publish/subscribe communication and mobile agents into a flexible and extensible distributed execution environment. The major novelty of the model is that an event is represented as a mobile intelligent agent, an **agvent**, which is treated as a first class citizen of the system and given autonomy and mobility features to select and travel between system components. Agvents have their own identity and behavior which permit them to actively navigate through the underlying dispatching system, and carry out various tasks at the nodes they visit. Agvents operate independently of the sending application, therefore removing any dependencies with the application. The 'intelligence' is in the agvents themselves rather than in the network. We think the new model will serve as an effective choice for several information-oriented applications, such as e-commerce or information retrieval, for its benefits stated above. Currently a prototype of system is being implemented in Java.

References

1. Bacon, J., Moody K., Bates J., Hayton R., Ma C., Mcneil A., Seidel O., and Spiteri M. "Generic support for distributed ap-plications." IEEE Computer 33, 3 (2000), 68–76.
2. Carzaniga, A. and Wolf, A. "Content-based Networking: A New Communication Infrastructure." NSF Workshop on Infrastructure for Mobile -Wireless Systems. Oct., 2001.
3. Mühl, G. Large-scale content-based publish/subscribe systems. PhD thesis, Darmstadt University of Technology, September 2002.
4. Sahingoz, O. K. and Erdogan, N. "AGVENT: Agent Based Distributed Event System", accepted for presentation in 30th Conference on Current Trends in Theory and Practice of Computer Science, (SOFSEM-2004), Czech Republic, 2004
5. FIPA Agent Communication Language Specifications. 2000. HTML, http://www.fipa.org/repository/aclspecs.html.

An Adaptive Communication Mechanism for Highly Mobile Agents

JinHo Ahn

Dept. of Computer Science, Kyonggi University
San 94-6 Yiuidong, Paldalgu, Suwonsi Kyonggido 442-760, Republic of Korea
`jhahn@kyonggi.ac.kr`

Abstract. Agent mobility causes reliable inter-agent communications to be more difficult to achieve in distributed agent based systems. To solve this issue, three representative agent tracking and message delivery mechanisms, broadcast-based, home-based and forwarding pointer-based, were previously proposed. However, due to their respective drawbacks, none of them is suitable for efficient delivery of messages to highly mobile agents, which move frequently between service nodes. This paper introduces an adaptive forwarding pointer-based agent tracking and message delivery mechanism to alleviate their disadvantages. The proposed mechanism allows each mobile agent to autonomously leave tails of forwarding pointers on some few of its visiting nodes depending on its preferences. Thus, it is more efficient in terms of message forwarding and location management than the previous forwarding pointer-based one. Simultaneously, it considerably reduces the dependency on the home node in agent location updating and message delivery compared with the home-based mechanism.

1 Introduction

Mobile agent is an autonomously running program, including both code and state, that travels from one node to another over a network carrying out a task on user's behalf[4]. Due to its beneficial characteristics, i.e., dynamicity, asynchronicity and autonomy, it has been primarily used as an enabling programming paradigm for developing distributed computing infrastructures in various application fields such as e-commerce, telecommunication, ubiquitous computing, active networks and the like[2,4]. However, as the size of these fields is rapidly increasing, several research issues related to the mobile agent technology such as communication, security, dynamic adaptation, etc., should be reconsidered to be suitable for their scale. Among them, it is most important to enhance the performance of the agent communication in Internet-scale infrastructures. For this purpose, some effective and efficient inter-agent communication mechanism is required in distributed agent-based systems. Agent mobility may lead to the loss of messages being destined to an agent on its migration. Thus, it causes reliable inter-agent communications to be not easy to achieve in the distributed agent based systems. Especially, guaranteeing the delivery of messages

M. Bubak et al. (Eds.): ICCS 2004, LNCS 3036, pp. 192–199, 2004.

to highly mobile agents, which move frequently among service nodes, is a more challenging problem, which this paper attempts to address. To consider the agent mobility, three representative agent tracking and message delivery mechanisms, broadcast-based, home-based and forwarding pointer-based, were previously proposed. The broadcast-based mechanism[7] guarantees transparent and reliable inter-agent communication and can also provide multicast communication to a set of agents. But, to locate the message destination, the mechanism has to contact every visiting node in the network. Thus, its large traffic overhead makes broadcasts impractical in large-scale distributed agent systems.

The home-based mechanism[5] is borrowed from the idea of Mobile IP[8]. It requires that each mobile node has a home node, and forces the mobile node to register its current temporary address, called care-of-address, with its home node whenever it moves. Thus, when some messages are sent to a mobile node currently located at a foreign network, the messages are first directed to its home node, which forwards them to the mobile one. This mechanism is simple to implement and results in little mobile node locating overhead. However, it is unsuitable for highly mobile agents in distributed agent based systems because every agent location updating and message delivery are all performed around the home agent, which introduces centralization. Moreover, the Mobile IP generally assumes each mobile node's home node is a static one whereas distributed agent based systems don't have this assumption, i.e., the home node may be disconnected from the network. Thus, this mechanism cannot address the disconnection problem.

In the forwarding pointer-based mechanism[3,6], each node on a mobile agent's movement path keeps a forwarding pointer to the next node on the path. Thus, if a message is delivered to an agent not being at the home node, the message must traverse a list of forwarding nodes. Thus, this mechanism can avoid performance bottlenecks of the global infrastructure, and therefore improve its scalability, particularly in large-scale distributed agent-based systems, compared with the home based one. Additionally, even if a home node is disconnected from the rest of the network, the forwarding pointer based mechanism allows agents registering with the node to communicate with other agents. However, as highly mobile agents leads to the length of their chains of pointers being rapidly increasing, its message forwarding overhead may be significantly larger. Furthermore, the number of forwarding pointers each service node needs to keep on its storage may exponentially increase if a large number of mobile agents are running in the systems. In a previous work[6], a type of update message called *inform* message was introduced to include an agent's current location for shortening the length of trails of forwarding pointers. In this case, a node that receives the message is allowed to update its table if the received information is more recent than the one it had. However, it introduces no concrete and efficient solutions for this purpose, for example, when update messages should be sent, and which node they should be sent to.

Therefore, we observe these respective drawbacks of the three previous mechanisms may be critical obstacles to efficient communications between highly mobile agents in large-scale distributed agent systems. This paper introduces an adaptive forwarding pointer-based agent tracking and message delivery mechanism to avoid their disadvantages. The proposed mechanism allows each mobile agent to autonomously leave trails of forwarding pointers only on some few of its visiting nodes depending on its preferences such as location updating and message delivery costs, security, network latency and topology, communication patterns, etc.. Thus, it is more efficient in terms of message delivery and location management than the previous forwarding pointer-based one. Additionally, it considerably decentralizes the role of the home node in agent location updating and message delivery. This feature alleviates the two problems of the home-based mechanism.

Due to space limitation, our system model, formal descriptions and correctness proof of the proposed mechanism, and related work are all omitted. The interested reader can find them in [1].

2   The Adaptive Communication Mechanism

As mentioned in section 1, the proposed adaptive communication mechanism is designed to have the following features unlike previous ones.
• Require small size of storage for location management per service node.
• Result in low message forwarding overhead.
• Reduce considerably the dependency on home node in agent location updating and message delivery.

First of all, let us define two important terms, *forwarder* and *locator*. Forwarder of an agent means a service node keeping a forwarding pointer of the agent on its storage. Thus, depending on the behavior of agent communication mechanisms, there may exist various number of forwarders of each agent in the system. Locator of an agent is the forwarder managing the identifier of the node where the agent is currently located. In this paper, it is assumed that there is only one locator in the system. To satisfy all the three requirements, our adaptive mechanism forces only some among all visiting nodes of each agent to be forwarders. This behavior can considerably reduce both the amount of agent location information each node needs to maintain and the delivery time of each message because the length of its forwarding path may be much more shortened. Also, since there exist multiple forwarders, not only one, in this mechanism, the home node centralization can be avoided. But, as a part of the vising nodes are forwarders in this mechanism, a new method is required to consistently manage location of each agent and enable each sent message to be delivered to the agent despite its migrations unlike in the previous forwarding pointer-based one. In the following, two components of the proposed mechanism, agent location management and message delivery algorithms, are explained in detail respectively.

Agent Location Management.   For the agent location management algorithm, every node N_i should maintain the following data structures.

- $RunningAgents_i$: A table for saving location information of every agent currently running on N_i. Its element is a tuple ($agent\_id$, $locmngr\_n$, $agent\_t$). $locmngr\_n$ is the identifier of agent $agent\_id$'s locator. $agent\_t$ is the timestamp associated with agent $agent\_id$ when the agent is located at N_i. Its value is incremented by one every time the corresponding agent migrates. Thus, when agent $agent\_id$ migrates to N_i, N_i should inform $locmngr\_n$ of both its identifier and $agent\_t$ of the agent so that $locmngr\_n$ can locate the agent.
- $AgentLocs_i$: A table for saving location information of every mobile agent which is not currently running on N_i, but of which N_i is a forwarder. Its element is a tuple ($agent\_id$, $destination\_n$, $agent\_t$, $ismanaging\_f$, $ismigrating\_f$). $destination\_n$ is the identifier of the node where N_i knows agent $agent\_id$ is currently located and running. $agent\_t$ is the timestamp associated with the agent when the agent is located at node $N_{destination\_n}$. It is used for avoiding updating recent location information by older information[6]. $ismanaging\_f$ is a bit flag indicating whether N_i is agent $agent\_id$'s locator or not. In the first case, its value is $true$ and otherwise, $false$. $ismigrating\_f$ is a bit flag designating if the agent is currently migrating to another node(=$true$) or not(=$false$).

The algorithm for managing each agent's location on its migration is informally described using figure 1. This figure shows message interactions between nodes occurring in a's location updating and location information maintained by each node while migrating from its home node to N_1 through N_5. In figure 1(a), a is created on N_{home} and then an element for a (id_a, $home$, 0) is saved into $RunningAgents_{home}$ in the first step. If a attempts to move to N_1 after having performed its partial task, in the second step, it inserts into $AgentLocs_{home}$ a's element (id_a, 1, 1, $true$, $true$) indicating N_{home} is a's locator and a is currently moving to N_1. Then, N_{home} dispatchs the agent with the identifier of the node and a's timestamp to N_1. When receiving these, N_1 increments the timestamp by one. In this case, as a wants N_1 to be its locator, it inserts a's location information (id_a, 1, 1) into $RunningAgents_1$ in the third step. At the same time, N_1 sends N_{home} a message $changelmngr$ including a's timestamp in order to inform N_{home} that N_1 is a's locator from now. On receiving the message, N_{home} updates a's location information on $AgentLocs_{home}$ using the message and sets two fields of a's element, $ismanaging\_f$ and $ismigrating\_f$, all to $false$. If the messages destined to a have been buffered in N_{home}'s message queue due to the migration, they are transmitted to N_1. When a attempts to migrate to N_2 after a has performed a part of its task in the forth step, N_1 puts a's element (id_a, 2, 2, $true$, $true$) into $AgentLocs_1$ and then dispatchs agent a to N_2. In this case, N_2 increments a's timestamp by one and then inserts a's element (id_a, 1, 2) into $RunningAgents_2$ like in the fifth step because a wants N_2 to be just a visiting node. Also, the node registers a's current location with its locator N_1 by sending a message $update$ including the timestamp to N_1. If there are any messages sent to the agent in the queue of N_1, they are forwarded to N_2.

Figure 1(b) illustrates that agent a moves from N_2 to N_3. In this example, N_2 first sends N_1 a message $m\_initiated$ indicating that a's migration process begins from now. When receiving the message, N_1 sets one field of a's element $ismigrating\_f$ to $true$ in the second step and then send a message m_reply to N_2. Suppose the migration is started without the execution of this invalidation procedure. If N_1 receives any message destined to a in this case, it forwards the message to N_2 because it doesn't know whether the migration continues to be executed. But, neither a may be currently running on N_2 nor N_2 keep a's location information on $AgentLocs_2$ because N_2 isn't a's forwarder. In this case, the message cannot be delivered to a. Therefore, N_2 has to push a to N_3 after having received the message m_reply from a's locator, and then remove a's element from $RunningAgents_2$. Afterwards, a's visiting node N_3 increments a's timestamp and saves a's element $(id_a, 1, 3)$ into $RunningAgents_3$ in the third step, and then sends a message $update$ to N_1. On the receipt of the message, N_1 updates a's element in $AgentLocs_1$ to $(id_a, 3, 3, true, false)$ using the message.

Figure 1(c) shows an example that N_5 becomes a's locator when agent a moves from N_4 to N_5. In this case, after N_5 creates a's location information $(id_a, 5, 5)$ and inserts it into $RunningAgents_5$ in the third step, the node sends N_1 a message $changelmngr$ for notifying the previous locator N_1 that N_5 is a's locator from now. Also, a attempts to register its current location with N_{home} in order to reduce the message delivery time incurred when another agent initially sends a message to a via N_{home}. For this purpose, N_5 sends a message $update$ to N_{home}. If a recognizes this consideration helps no performance improvement, it doesn't perform the home node update procedure. After that, N_{home} and N_1 update a's location information respectively like in the third step.

Message Delivery.   For the message delivery algorithm, every node N_i should contain an agent location cache, $ALocsCache_i$, as follows.

• $ALocsCache_i$: A cache for temporarily storing location information of each mobile agent which agents running on N_i communicate with. Its element is a tuple $(agent\_id, forward\_n, agent\_t)$. $forward\_n$ is the identifier of the node where N_i knows agent $agent\_id$ is currently located and running. Thus, when attempting to deliver messages to agent $agent\_id$, each agent on N_i forwards them to $forward\_n$ regardless of whether this address is outdated. $agent\_t$ is the timestamp assigned to agent $agent\_id$ when the agent was located at node $N_{forward\_n}$.

We intend to use an example in figure 2 to clarify the algorithm to enable every sent message to be reliably delivered to its target agent despite agent migrations. This example illustrates agent b sends three messages, $msg1$, $msg2$ and $msg3$ to agent a in this order while a is migrating from its home node to N_1 through N_6 according to its itinerary. In figure 2(a), after a has moved from N_{home} to N_2, b at N_{sender} will deliver the first message $msg1$ to a. In this case, N_{sender} has no location information for a in its location cache $ALocsCache_{sender}$. Thus, N_{sender} creates and saves a's element $(id_a, home, 0)$ into $ALocsCache_{sender}$. Then, it sends the message $msg1$ to N_{home}. Receiving the message, N_{home} retrieves a's element from $AgentLocs_{home}$. In this case, as the value of the bit flag $ismanaging$ in the element is $false$, N_{home} isn't a's

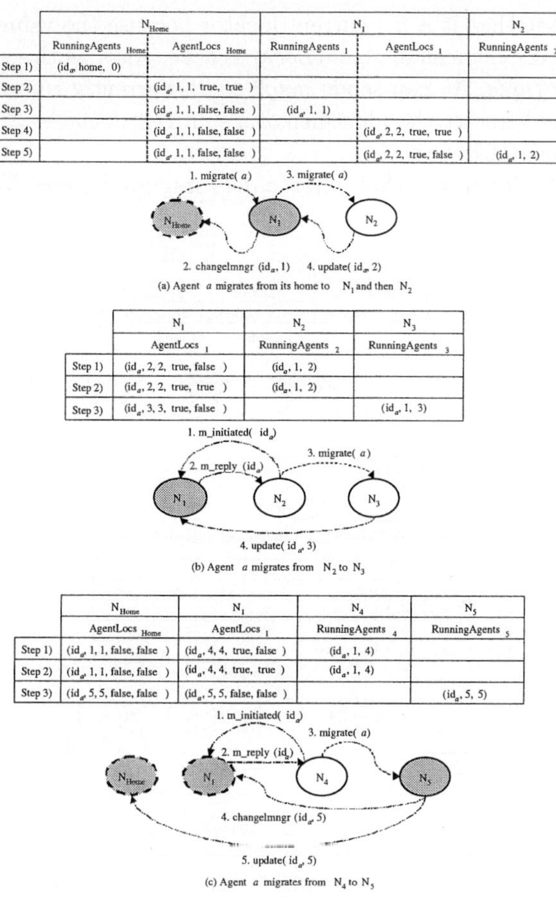

	N_{Home}		N_1		N_2
	RunningAgents $_{Home}$	AgentLocs $_{Home}$	RunningAgents $_1$	AgentLocs $_1$	RunningAgents $_2$
Step 1)	$(id_a,$ home, 0)				
Step 2)		$(id_a, 1, 1,$ true, true $)$			
Step 3)		$(id_a, 1, 1,$ false, false $)$	$(id_a, 1, 1)$		
Step 4)		$(id_a, 1, 1,$ false, false $)$		$(id_a, 2, 2,$ true, true $)$	
Step 5)		$(id_a, 1, 1,$ false, false $)$		$(id_a, 2, 2,$ true, false $)$	$(id_a, 1, 2)$

(a) Agent a migrates from its home to N_1 and then N_2

	N_1	N_2	N_3
	AgentLocs $_1$	RunningAgents $_2$	RunningAgents $_3$
Step 1)	$(id_a, 2, 2,$ true, false $)$	$(id_a, 1, 2)$	
Step 2)	$(id_a, 2, 2,$ true, true $)$	$(id_a, 1, 2)$	
Step 3)	$(id_a, 3, 3,$ true, false $)$		$(id_a, 1, 3)$

(b) Agent a migrates from N_2 to N_3

	N_{Home}	N_1	N_4	N_5
	AgentLocs $_{Home}$	AgentLocs $_1$	RunningAgents $_4$	RunningAgents $_5$
Step 1)	$(id_a, 1, 1,$ false, false $)$	$(id_a, 4, 4,$ true, false $)$	$(id_a, 1, 4)$	
Step 2)	$(id_a, 1, 1,$ false, false $)$	$(id_a, 4, 4,$ true, true $)$	$(id_a, 1, 4)$	
Step 3)	$(id_a, 5, 5,$ false, false $)$	$(id_a, 5, 5,$ false, false $)$		$(id_a, 5, 5)$

(c) Agent a migrates from N_4 to N_5

Fig. 1. An example of agent a's location updating on its migration according to its itinerary

locator. Thus, it consults the element and forwards the message $msg1$ to the next forwarder N_1. On the receipt of the message, N_1 obtains a's element from $AgentLocs_1$ and then checks the flag *ismanaging* in the element. In this case, N_1 is a's locator because the value of the flag is *true*. Also, as the value of the second flag *ismigrating* is *false*, it forwards the message to a's currently running node N_2 by consulting the element. At the same time, as N_{sender} has the outdated identifier of a's locator, N_1 sends N_{sender} a message *updateCache* containing the identifier of a's current locator$(=N_1)$ and timestamp$(=2)$ in figure 2(a). When receiving the message, N_{sender} updates a's element in $ALocsCache_{sender}$ using the message in the second step.

Figure 2(b) shows that a migrates from N_2 to N_4 and then b at N_{sender} sends the second message $msg2$ to a. In this case, N_{sender} finds a's element from $ALocsCache_{sender}$ and then forwards $msg2$ to N_1. On the receipt of

$msg2$, N_1 can see that it is a's current locator because the value of the bit flag *ismanaging* of a's element in $AgentLocs_1$ is *true*. Also, as the value of the flag *ismigrating* is *false*, N_1 can sends $msg2$ to a's currently running node($=N_4$). At this time, as b knows the identifier of a's current locator, N_1 doesn't sends any message *updateCache* to b.

Figure 2(c) illustrates that after a has migrated from N_4 via the next

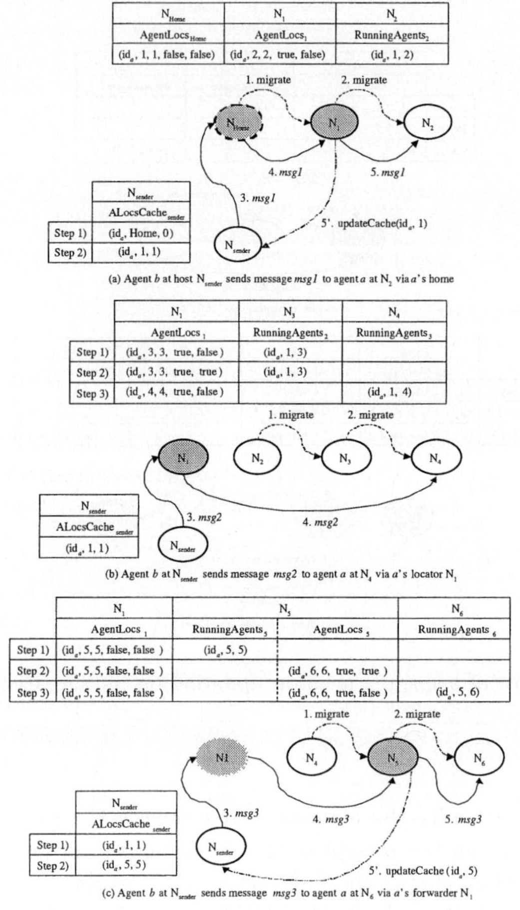

N_{Home}	N_1	N_2
$AgentLocs_{Home}$	$AgentLocs_1$	$RunningAgents_2$
(id_a, 1, 1, false, false)	(id_a, 2, 2, true, false)	(id_a, 1, 2)

	N_{sender}
	$ALocsCache_{sender}$
Step 1)	(id_a, Home, 0)
Step 2)	(id_a, 1, 1)

(a) Agent b at host N_{sender} sends message $msg1$ to agent a at N_2 via a's home

	N_1	N_3	N_4
	$AgentLocs_1$	$RunningAgents_2$	$RunningAgents_3$
Step 1)	(id_a, 3, 3, true, false)	(id_a, 1, 3)	
Step 2)	(id_a, 3, 3, true, true)	(id_a, 1, 3)	
Step 3)	(id_a, 4, 4, true, false)		(id_a, 1, 4)

	N_{sender}
	$ALocsCache_{sender}$
	(id_a, 1, 1)

(b) Agent b at N_{sender} sends message $msg2$ to agent a at N_4 via a's locator N_1

	N_1	N_5		N_6
	$AgentLocs_1$	$RunningAgents_5$	$AgentLocs_5$	$RunningAgents_6$
Step 1)	(id_a, 5, 5, false, false)	(id_a, 5, 5)		
Step 2)	(id_a, 5, 5, false, false)		(id_a, 6, 6, true, true)	
Step 3)	(id_a, 5, 5, false, false)		(id_a, 6, 6, true, false)	(id_a, 5, 6)

	N_{sender}
	$ALocsCache_{sender}$
Step 1)	(id_a, 1, 1)
Step 2)	(id_a, 5, 5)

(c) Agent b at N_{sender} sends message $msg3$ to agent a at N_6 via a's forwarder N_1

Fig. 2. An example of agent b sending three messages $msg1$, $msg2$ and $msg3$ to agent a in order while a migrating to several nodes according to its itinerary

locator N_5 to N_6, b transmits the third message $msg3$ to a. In this figure, when the migration has been completed, three nodes N_1, N_5, N_6 have each a's location information like in the step 3. First, N_{sender} sends $msg3$ to N_1 by consulting a's element in $ALocsCache_{sender}$. In this case, as the value of the flag *ismanaging* of a's element in $AgentLocs_1$ is *false*, N_1 forwards $msg3$ to the next forwarder N_5. When N_5 receives the message, it transmits the message to N_6 because it recognizes that it is a's locator and a isn't currently migrating. Concurrently, N_5 informs N_{sender} that a's current locator is N_5 by sending a message *updateCache* to N_{sender}.

3   Conclusion

In this paper, an adaptive agent communication mechanism was introduced to considerably reduce both the amount of agent location information maintained by each service node and the delivery time of each message while reducing the dependency on the home node. To achieve this goal, this mechanism forces only a small part of all visiting nodes of each agent to becomes forwarders. In this mechanism, each agent should register its location with its current locator on every migration until it arrives at the next locator of the agent. This method may result in slightly higher location update cost per agent migration compared with that of the previous forwarding pointer-based mechanism. However, if each agent determines some among its visiting nodes as forwarders by properly considering several performance factors, the gap between the two costs may be almost negligible. Moreover, the mechanism allows the identifier of each agent's locator to be kept on the agent location cache of a node. This behavior highly reduces the cache update frequency of the node compared with the previous mechanism.

Future work is focused on selecting appropriate forwarders of each agent according to the changes which the agent senses in its environment related to location updating and message delivery costs, security policies, network latency and topology, communication patterns.

References

1. J. Ahn. Adaptive Communication Mechanisms for Mobile Agents. *Technical Report KGU-CS-03-050*, Kyonggi University, 2003.
2. P. Bellavista, A. Corradi and C. Stefanelli. The Ubiquitous Provisioning of Internet Services to Portable Devices. *IEEE Pervasive Computing*, Vol. 1, No. 3, pp. 81-87, 2002.
3. J. Desbiens, M. Lavoie and F. Renaud. Communication and tracking infrastructure of a mobile agent system. *In Proc. of the 31st Hawaii International Conference on System Sciences*, Vol 7., pp. 54-63, 1998.
4. A. Fuggetta, G.P.Picco and G. Vigna. Understanding Code Mobility. *IEEE Transactions on Software Engineering*, Vol. 24, No. 5, pp. 342-361, 1998.
5. D. Lange and M. Oshima. *Programming and Deploying Mobile Agents with Aglets*. Addison-Wesley, 1998.
6. L. Moreau. Distributed Directory Service and Message Router for Mobile Agents. *Science of Computer Programming*, Vol. 39, No. 2-3, pp. 249-272, 2001.
7. A. L. Murphy and G. P. Picco. Reliable Communication for Highly Mobile Agents. *Journal of Autonomous Agents and Multi-Agent Systems*, Vol. 5, No. 1, pp. 81-100, 2002.
8. C. Perkins, IP Mobility Support. *RFC 2002*, October 1996.

Knapsack Model and Algorithm for HW/SW Partitioning Problem

Abhijit Ray, Wu Jigang, and Srikanthan Thambipillai

Centre for High Performance Embedded Systems,
School of Computer Engineering, Nanyang Technological University,
Singapore,639798, Republic of Singapore
{PA8760452, asjgwu, astsrikan}@ntu.edu.sg

Abstract. Hardware software partitioning is one of the most significant problem in embedded system design. The size of the total solution space for this problem is typically quite large. This problem has been investigated extensively. This paper is the first work to model the problem into a knapsack problem. We present a way to split the problem into standard 0-1 knapsack problems, so that most of the classical approaches for 0-1 knapsack problems can be directly applied. We use tight lower bound and tight upper bound on each of these knapsack problems to eliminate sub-problems, which are guaranteed not to give optimal results.

1   Introduction

Hardware software partitioning (HSP) problem is the problem of deciding for each subsystem, whether the required functionality is to be implemented in hardware or software to get the desired performance, while maintaining least cost. At the same time, hardware area minimization and latency constraints present contradictory objectives to be achieved through hardware-software partitioning.

Most of the existing approaches to HSP are based on either hardware oriented partitioning or software oriented partitioning. A *software-oriented* approach means that initially the whole application is allotted to software and during partitioning system parts are moved to hardware until constraints are met. In a *hardware-oriented* approach on the other hand, the whole application is implemented in hardware and during partitioning the parts are moved to software until constraints are violated. A software oriented approach has been proposed by Ernst *et al* [3], Vahid *et al* [4]. Hardware oriented approach has been proposed in Gupta *et al* [5], Niemann *et al* [6]. In [9], the authors proposed a flexible granularity approach for hardware software partitioning. Karam *et al* [7], proposes partitioning schemes for transformative applications i.e., multimedia and digital signal processing applications. The authors try to optimize the number of pipeline stages and memory required for pipelining. The partitioning is done in an iterative manner. Rakhmatov *et al* [8] modeled the hardware software partitioning as a unconstrained bipartitioning problem.

In this paper, we model the partitioning problem into some standard knapsack problems, which can be solved independently to arrive at the solution. Also

M. Bubak et al. (Eds.): ICCS 2004, LNCS 3036, pp. 200–205, 2004.

for every subproblem we calculate the lower bound and its upper bound, this helps in rejecting subproblems, which are not expected to give optimal results. The advantage of our work is that it provides the optimal solution. Moreover, many problems are rejected based on their lower bound and upper bound, and this reduces the number of subproblems that need to be solved; hence the algorithm is quite fast.

2   Model of the Physical Problem

We consider a basic case which can be later extended. In our case the application can be broken down into parts such that each of them can be run simultaneously or in other words the parts do not have any sort of data dependency between them. So we have a set of items $S = \{p_1, p_2, \ldots, p_n\}$ to be partitioned into hardware and software. Let h_i and s_i be the time required for the part p_i to be run in hardware and software respectively. Also let a_i be the area required for hardware implementation of part p_i. And let A be the total area available for hardware implementation. Our goal is to allot each part into hardware and software so that the combined running time of the whole application is minimized in such a way that the area constraint is satisfied. Let us denote the solution of the problems as a vector $X = [x_1, x_2, \ldots, x_n]$ such that $x_i \in \{0, 1\}$, where $x_i = 0\,(1)$ implies that the part p_i is implemented in software (hardware). Since the hardware and software can be run in parallel, the total running time of the application is given by

$$T(X) = max\{H(X), S(X)\} \tag{1}$$

where $H(X)$ is the total running time of the parts running in hardware and $S(X)$ is the total running time of the parts in software. Since all the parts that are implemented in hardware can be run in parallel to each other and all the software parts has to be run in serial, we have

$$H(X) = \max_{1 \leq i \leq n} \{x_i \cdot h_i\} \quad and \quad S(X) = \sum_{i=1}^{n} (1 - x_i) \cdot s_i.$$

Hence, the problem discussed in this paper can be modeled into

$$\mathcal{P} \left\{ \begin{array}{l} \text{minimize } T(X) \\ \text{subject to } \sum_{i=1}^{n} x_i \cdot a_i \leq A \end{array} \right. \tag{2}$$

3   Problem Splitting and Algorithm

Given a knapsack capacity C and set of items $S = \{1, \ldots, n\}$, where each item has a weight w_i and a benefit b_i. The problem is to find a subset $S' \subset S$, that maximizes the total profit $\sum_{i \in S'} b_i$ under the constraint that $\sum_{i \in S'} w_i \leq C$. This is the knapsack problem. Mathematically, it can be described as follows

$$0\text{-}1 \text{ KP} \begin{cases} \text{maximize } \sum_{i=1}^{n} p_i \cdot x_i \\ \text{subject to } \sum_{i=1}^{n} w_i \cdot x_i \leq C, \\ \qquad\qquad x_i \in \{0, 1\}, i = 1, \ldots n \end{cases} \tag{3}$$

where x_i is a binary variable equalling 1 if item i should be included in the knapsack and 0 otherwise. It is well known that this problem is NP-complete [2, 1, 10].

Let's assume that the items are ordered by their efficiencies in a non-increasing manner, where the efficiency is defined as

$$e_j = \frac{b_j}{w_j} \tag{4}$$

Let

$$\overline{b}_j = \sum_{i=1}^{j} b_i \quad and \quad \overline{w}_j = \sum_{i=1}^{j} w_i \quad j = 1, 2, \ldots, n \tag{5}$$

The residual capacity r is defined as

$$r = C - \overline{w}_{t-1} \tag{6}$$

By linear relaxation, [2] showed that an upper bound on the total benefit of 0-1 KP is

$$u = \lceil \overline{b}_{t-1} + r \cdot \frac{b_t}{w_t} \rceil \tag{7}$$

and the lower bound is given by,

$$l = \overline{b}_{t-1} \tag{8}$$

In HSP problem, sort all the items $p_1, p_2, \ldots, p_n$ in decreasing order of their hardware running time, so that after sorting we have the items ordered as $p'_1, p'_1, \ldots, p'_n$ and the following condition is satisfied

$$h'_i \geq h'_j \qquad \text{for all} \quad i \leq j \quad \text{and} \quad 0 \leq i, j \leq n. \tag{9}$$

Let us define $S_T = \sum_{i=1}^{n} s'_i$ and $R_i = S_T - s'_i$. Now we split the problem $\mathcal{P}$ into the following n subproblems $\mathcal{P}_1, \mathcal{P}_2, \cdots, \mathcal{P}_n$.

3.1   Subproblem $\mathcal{P}_k$:

Let $k \geq 1$. We fix p'_k to be implemented in hardware i.e., $x_k = 1$, and all the items $1, 2, \ldots, k-1$ are in software, because if any of them, say j is in hardware then any subproblem l such that $l > j$ is a subset of subproblem j. That is we have $x_1 = 0, x_2 = 0, \cdots, x_{k-1} = 0$. The total time is

$$T(X) = max \left\{ h'_k, R_k - \sum_{i=k+1}^{n} x_i \cdot s'_i \right\}$$

We have to minimize the total running time $T(X)$, i.e.,

$$minimize \quad T(X) \Longleftrightarrow minimize \left\{ R_k - \sum_{i=k+1}^{n} x_i \cdot s'_i \right\}$$

$$\Longleftrightarrow maximize \left\{ \sum_{i=k+1}^{n} x_i \cdot s'_i \right\}$$

subject to the constraint $x_i \in \{0, 1\}$ and the area constraint

$$\sum_{i=k+1}^{n} x_i \cdot a_i \leq A - a_k \tag{10}$$

Formally, the subproblem k is described as

$$\mathcal{P}_k \begin{cases} maximize \ \sum_{i=k+1}^{n} x_i \cdot s'_i \\ subject\ to \ \sum_{i=k+1}^{n} x_i \cdot a_i \leq A - a_k \end{cases} \tag{11}$$

The bounded interval of subproblem $\mathcal{P}_k$ are

$$L_k = \max\{h'_k, R_k - u_k\},$$
$$U_k = \max\{h'_k, R_k - l_k\},$$

and the optimal solution of $\mathcal{P}_k$ would lie in the range $[L_k, U_k]$ where l_k and u_k are the lower and upper bound of total benefit of $\mathcal{P}_k$, respectively.

If after creating subproblem i, we find that

$$\sum_{j=i+1}^{n} a_j \leq A - a_i \tag{12}$$

This means that after we have fixed items a_i to be implemented in hardware and $a_1, a_2, \ldots, a_{i-1}$ into software, the rest of the items left to be partitioned can be implemented easily in hardware as there are enough hardware space left. Hence, we can stop creating more subproblems as soon as eq[12] is satisfied.

A point to be noted is that all subproblems are not of the same size. This is because for subproblem P_i, we fix item i to be implemented in hardware and all the items $1, 2, \ldots, i-1$ are in software. This is because if in subproblem P_i, item $k, \{k < i\}$ is implemented in hardware then it becomes a subset of subproblem P_k. Therefore the size of the problem decreases from subproblem 1 to n.

3.2 Algorithm

The outline of the algorithm for solving the HSP problem is given below:

```
BOUND := 0;
sort all the items to be partitioned in decreasing order of
their hardware running time;
form the subproblems P(i), i=1, 2,...,n;
for(i:=1 ; i <= n ; i++ ){
    calculate the upper bound U(i)
        and the lower bound L(i) for P(i);
    if( L(i) > BOUND){
        BOUND := L(i);
    }
}
while( there are subproblems left to be solved){
    select the subproblem with the highest lower bound;
    if( U(i) < BOUND ){
        reject this subproblem;
    }else {
        solve this subproblem;
        B(i): = benifit of the above solution;
        if ( B(i) > BOUND ){
            BOUND := B(i);
        }
    }
}
```

4   Experimental Works

The proposed algorithm was implemented on a Pentium, 500MHz system, running on Linux. We used random data for partitioning. For solving the individual 0-1 knapsack problems we used the algorithm for 0-1 knapsack problem, given in [1]. The experiment was performed for different problem sizes and area constraints. Table 1 gives a count of the number of subproblems that needed to be solved to arrive at the optimal solution for the whole problem.

5   Conclusion

We have proposed a algorithm for hardware/software partitioning problem. The proposed algorithm models the problem into the Knapsack problem, which is a known NP-complete problem, and then splits the whole problem into some independent sub-problems. Upper and lower bounds of each subproblem is used to reject some sub-problems. As a result, fewer subproblems needs to be solved.

Table 1. Number of subproblems solved for different problem size and area constraints.

size	fraction of area put as constraint								
n	0.1	0.2	0.3	0.4	0.5	0.6	0.7	0.8	0.9
30	1	3	1	1	1	1	1	1	1
60	1	1	1	1	1	1	1	1	1
90	1	1	1	1	1	1	1	1	1
300	1	1	1	1	1	1	1	1	1
500	1	1	1	1	1	1	1	1	1
700	1	1	1	1	1	1	1	1	1
1000	1	1	1	1	1	1	1	1	1
2000	1	1	1	1	1	1	1	1	1
3000	1	1	1	1	1	1	1	1	1

Also, since the subproblems can be solved in parallel, this approach can be effectively used in a distributed computing environment. We are currently extending our knapsack model to consider cases when the items to be partitioned are not independent and hence communication between the items to be partitioned becomes an issue.

References

1. D. Pisinger,*A minimal algorithm for the 0-1 knapsack problem*, Operations Research, 1997,Page(s): 758-767.
2. D. Pisinger, *Algorithms for knapsack problems*, Ph.D. Thesis, 1995, Page(s):1-200.
3. R. Ernst, J.Henkel and T. Benner,*Hardware-Software Cosynthesis for Microcontrollers*,IEEE Design and Test of Computers, 1993, Page(s):64-75.
4. F.Vahid, D.D. Gajski and J. Jong, *A binary-constraint search algorithm for minimizing hardware during hardware/software partitioning*, IEEE/ACM Proceedings European Conference on Design Automation(EuroDAC), 1994, Page(s):214-219.
5. R.K. Gupta and G.D. Micheli, *System-level synthesis using reprogrammable components*, Procccdings. [3rd] European Conference on Design Automation, 1992, Page(s):2-7.
6. R. Niemann and P. Marwedel, *Hardware/software partitioning using integer programming*, Proceedings European Design and Test Conference, 1996. ED&TC 96, Page(s):473-479.
7. K.S. Chatha and R. Vemuri, *Hardware-software partitioning and pipelined scheduling of transformative applications*, Very Large Scale Integration (VLSI) Systems, IEEE Transactions on , Volume: 10 Issue: 3 , Jun 2002, Page(s):193-208.
8. D.N. Rakhmatov and S.B.K. Vrudhula, *Hardware-software bipartitioning for dynamically reconfigurable systems*, Hardware/Software Codesign, 2002. CODES 2002. Proceedings of the Tenth International Symposium on, Page(s):145-150.
9. Jorg Henkel and Rold Ernst, *An approach to automated hardware/software partitioning using a flexible granularity that is driven by high-level estimation technique*, Very Large Scale Integration (VLSI) Systems, IEEE Transactions on , Volume: 9 Issue: 2 , Apr 2001, Page(s):273-289.
10. W. Jigang, L. Yunfei and H. Schroeder, *A minimal reduction approach for the collapsing knapsack problem*,Computing and Informatics, Volume: 20, 2001,Page(s): 359-369.

A Simulated Annealing Algorithm for the Circles Packing Problem

Defu Zhang[1] and Wenqi Huang[2]

[1] Department of Computer Science, Xiamen University, 361005, China
dfzhang@xmu.edu.cn
[2] School of Computer Science, Huazhong University of Science and Technology,
Wuhan 430074,China

Abstract. We present a heuristic simulated annealing algorithm to solve the circles packing problem. For constructing a special neighborhood and jumping out of the local minimum trap, some effective heuristic strategies are presented. These strategies are from nature and can allow the iterative process to converge fast. The HSA algorithm inherits the merit of the simulated annealing algorithm, and can avoid the disadvantage of blind search in the simulated annealing algorithm to some extent according to the special neighborhood. The computational results show that the performance of the presented algorithm outperforms that of the quasi-physical quasi-human algorithm.

1 Introduction

Given n objects, each with given shape and size, and a bounded space, the packing problem is how to pack the objects into the bounded space without overlapping. In this paper, we consider the circles packing problem, its aim is to pack different-sized circles into a larger circle. This problem has been shown to be NP-hard [1], investigations from the 70s of last century up to now, however, show that for NP-hard problems, there does not exist an algorithm that is both rigorous and fast [1,2]. Hence people turn to nature for wisdom, to obtain heuristic approximate algorithm that is not absolutely rigorous, but is of high speed, high reliability and high efficiency.

The packing problem arises in many scientific and engineering fields, and finds many significant practical applications. The packing problem has a long history in the literature [3,4,5,6]. But the history of the circles packing problem is very short [5,6], and some results can be found in [6]. For the simpler cases of the circles packing problem, it is possible to obtain optimal solutions based on lattice patterns. However, for more complex cases, we have to depend on fast heuristic algorithms to generate approximate solutions. At present, some heuristic algorithms have been presented. A fast quasi-physical algorithm has been presented in [7], its work path is to find natural phenomena in the physical world equivalent to the original mathematical problem, and then to observe the evolution of matter in it. However, due to the work path, this algorithm is too determinate, it is often easy to get stuck in local minimum. A very

M. Bubak et al. (Eds.): ICCS 2004, LNCS 3036, pp. 206–214, 2004.

clever polynomial time approximate algorithm has been reported in [2]. This algorithm is of great theoretical significance, but it is of less practical value. Several heuristic procedures have been developed in [9]. The heuristics are based on a variety of solution building rules that emulate a certain process of packing circles into a rectangle container. Unfortunately, the running time of this algorithm is not reported. Recently, an improved quasi-physical quasi-human (QPQH) algorithm has been given in [8]. This algorithm, which combines the quasi-physical approach and the quasi-human strategy, is one of the best current heuristic methods. However, for more complex problems, it is still difficult to jump out of local minimum and arrive at a nice place.

Inspired by biology evolution, physical process, social life etc., human beings have found many good algorithms, for example, genetic algorithm and simulated annealing algorithm (SA)[10]. Especially, SA is a general stochastic search algorithm for combinatorial optimization problems. In contrast to other local search algorithms, it provides more opportunities to escape from local minimum. However, it often costs too much time for finding a solution, this situation prevents it from being applied to many practical problems. What is more, for enhancing the performance of SA, some effective heuristic strategies are presented. These strategies are from nature, and can allow the iterative process to converge fast. Based on SA and heuristics presented by this paper, we present a heuristic simulated annealing algorithm (HSA) for the circles packing problem. HSA inherits the merit of SA, and can avoid the disadvantage of blind search in SA to some extent according to the special neighborhood. The computational results show that the performance of HSA outperforms that of QPQH.

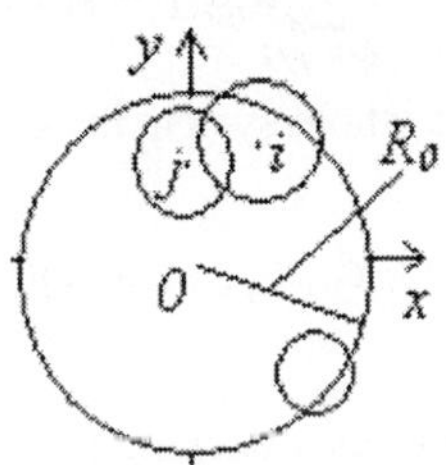

Fig. 1.

2 Mathematical Formulation of the Problem

Given an empty round plate and n circles of different sizes, here, n is a positive integer, we shall ask if these circles can be packed into the empty round plate without overlapping one another. This problem is stated more formally as follows [2,7,8]:

Take the origin of two-dimensional Cartesian coordinate system (See Fig. 1) at the central point of the round plate with the radius of R_0. The coordinates of the center of circle i is denoted by (x_i, y_i), the radius of circle i is R_i, we ask if there exist a set of real numbers or a configuration $(x_1, y_1, \cdots, x_i, y_i, \cdots, x_n, y_n)$, such that

$$\begin{cases} \sqrt{x_i^2 + y_i^2} \le R_0 - R_i \\ \sqrt{(x_i - x_j)^2 + (y_i - y_j)^2} \ge R_i + R_j \end{cases}, i, j = 1, 2, \cdots, n, i \ne j.$$

If there exist such real numbers, then please give them.

According to elasticity mechanism, the extrusion elastic potential energy (u_{ij}) between two smooth elastic objects is proportional to the square of their mutual embedding depth

$$u_{ij} = kd_{ij}^2, i, j = 0, 1, \cdots, n, i \ne j. \tag{1}$$

Here, k is a proportional coefficient, $k > 0$. In our computational program, we set $k = 1$. d_{ij} can be calculated as follows:

$$d_{0i} = \begin{cases} \sqrt{x_i^2 + y_i^2} + R_i - R_0 & if \sqrt{x_i^2 + y_i^2} + R_i > R_0, \\ 0, & else, \end{cases}$$

$$d_{ij} = \begin{cases} R_i + R_j - \sqrt{(x_i - x_j)^2 + (y_i - y_j)^2} & if \sqrt{(x_i - x_j)^2 + (y_i - y_j)^2} < R_i + R_j, \\ 0, & else. \end{cases}$$

According to (**1**), the extrusion elastic potential energy (U_j) possessed by circle j is

$$U_j = \sum_{i=0, j \ne i}^{n} u_{ij}, j = 1, 2, \cdots, n.$$

Thus, the potential energy of the whole system is

$$U(X) = U(x_1, y_1, x_2, y_2, \cdots, x_n, y_n) = \sum_{j=1}^{n} U_j. \tag{2}$$

Clearly, $U(X) \ge 0$. Hence the circles packing problem has been transformed into an optimization problem for the known potential energy function (2), that is to say, the minimum configuration $X^* = (x_1^*, y_1^*, \cdots, x_n^*, y_n^*)$ of the potential energy function should be found. Therefore, if $U(X^*) = 0$, then X^* is one solution for the circles packing problem, whereas if $U(X^*) > 0$, then the problem has no solution.

3 The Proposed HSA Algorithm

3.1 Neighborhood Structure

Definition. For any given configuration $X = (x_1, y_1, \cdots, x_i, y_i, \cdots, x_n, y_n)$, we call $X' = (x_1, y_1, \cdots, x_i', y_i', \cdots, x_n, y_n)$ as a neighboring configuration of X, where the position of circle i in the configuration X' is randomly generated and different from the position of circle i in the configuration X, whereas the positions of other circles in X' is the same as X. The set of all these neighboring configurations of X is called the neighborhood of X, denoted by $N(X)$.

By Definition, there exist many configurations in $N(X)$ because circle i in X' is selected randomly and its position is generated randomly. Clearly, the range of search is too large, it can easily result to blind search and waste too much time, so the efficiency of computation is very low. According to these analyses, in order to make iterative process converge fast, it is necessary to reduce search space. Therefore, it is crucial to know how to generate X' for further constructing the neighborhood. In order to enhance the efficiency of the annealing process, according to the characteristics of the circles packing problem, we present the following heuristic strategy for generating X' by simulating circle's physical moving process: For any given $X = (x_1, y_1, \cdots, x_i, y_i, \cdots, x_n, y_n)$, $X' = (x_1, y_1, \cdots, x_i', y_i', \cdots, x_n, y_n)$ can be determined by the vector sum of embedding depth of circle i. Here, we look on the embedding depth as a vector. Thus, the displacement of circle i is the module of the vector sum of its embedding depth. The moving direction of circle i is the direction of the vector sum of the extrusion elastic forces acted on it. In Fig. 2, the way of generating X' is as follows.

For given $X = (x_1, y_1, \cdots, x_i, y_i, \cdots, x_n, y_n)$, we consider the change of circle i's position under the extrusion elastic forces. With the help of Fig. 2, we have

$$\frac{x_i - x_j}{dx_{ij}} = \frac{D_{ij}}{d_{ij}}, \frac{y_i - y_j}{dy_{ij}} = \frac{D_{ij}}{d_{ij}}, \tag{3}$$

where, D_{ij} denotes the distance from the center of circle i to that of circle j, d_{ij} is the same as previous definition, dx_{ij} is the projection of d_{ij} in the horizontal axis x, dy_{ij} is the projection of d_{ij} in the vertical axis y. It is noted that dx_{ij}, dy_{ij} is not distance, so they may be positive or negative. From (3), we have

$$dx_{ij} = \frac{x_i - x_j}{D_{ij}} d_{ij},$$

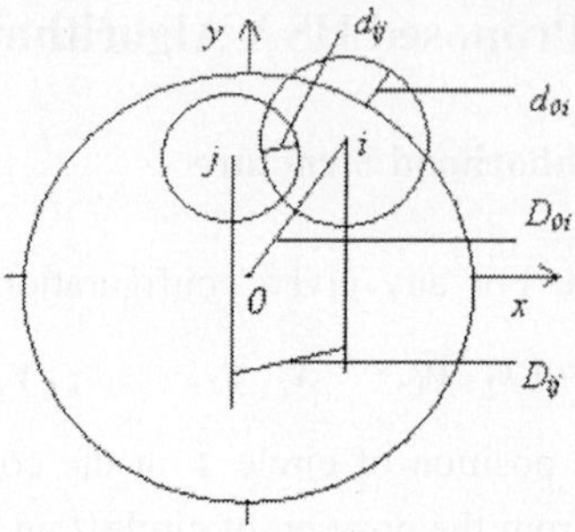

$$dy_{ij} = \frac{y_i - y_j}{D_{ij}} d_{ij}.$$

Specially, we have

$$dx_{i0} = \frac{-x_i}{D_{0i}} d_{0i},$$

$$dy_{i0} = \frac{-y_i}{D_{0i}} d_{0i},$$

Fig. 2.

where, D_{0i} denotes the distance from the center of the plate to the center of circle i, d_{0i} is the same as previous definition, dx_{i0} is the projection of d_{0i} in the horizontal axis, dy_{i0} is the projection of d_{0i} in the vertical axis. Therefore, the next position of circle i is

$$\begin{cases} x_i' = x_i + dx_{ij} + dx_{i0} \\ y_i' = y_i + dy_{ij} + dy_{i0} \end{cases}, i = 1, 2, \cdots, n.$$

With the help of the configuration $X = (x_1, y_1, \cdots, x_i, y_i, \cdots, x_n, y_n)$, we obtain a configuration $X' = (x_1', y_1', \cdots, x_i', y_i', \cdots, x_n', y_n')$ by simulating the physics moving process of circle i. It is noted that when we calculate X', we only consider those circles or the plate that embed with circle i. Similarly, we can construct other neighboring configurations of X. Clearly, the number of the configurations in $N(X)$ is n. Since the way of generating X' is determinate, the range of search is significantly reduced. Therefore, we believe this strategy enlightened by physical process may avoid blind search and allow iterative process to converge fast.

3.2   Strategies of "Jumping out of the Trap"

For SA , when the temperature T tends to zero at the end of the process, the probability of accepting worse neighboring configurations is approximately zero. In that case, the SA algorithm loses its feature to accept worse configurations, thereby becoming identical to other local search algorithms. In this way the optimization process may get stuck in a local optimum. In addition, the way of generating X' may often lead the SA algorithm to get stuck in a local optimum during the course of execution. Under this circumstance, the promising approach is to put forward some good heuristic strategies for jumping out of the local minimum trap by taking the calculating point out of the local minimum and place it in a position with better prospects. Then new SA process can be carried out over again.

This strategy of "jumping out of the trap" can be obtained by observing and learning from the social and nature phenomena and is therefore called heuristic strategy. In daily life, when we pack some objects into a trunk, we always pack large ones first, and then pack small ones. Otherwise, it is possible that the large objects cannot be packed into the trunk, because small ones have occupied the needed space of the large objects. Under this circumstance, we have to adjust the positions of small ones to find better layout. Namely, move away one small object from among small objects huddled together. What is more, the nature phenomena of "things of the same character repulse each other, things of the different character attract each other" are familiar. In addition, in order to decrease conflicts, the circle with maximum potential energy should be considered [8]. All of these inspire us to obtain the following heuristic strategies.

When getting stuck, we classify the circles by size. The circles of the same radius form one group, then we judge whether circles in the same group embed each other from the smallest group to the largest group in order. If circles in the same group embed each other, then randomly select an embedded circle and randomly determine its new position in the plate, and proceed with the SA process again. If circles in each group do not embed each other, then the circle with maximum potential energy can be picked out and randomly placed in the plate. These strategies are generally straightforward and intuitive, and can help the search process jump out of local minimum.

3.3 Statement of HSA

Suppose the circles have been sorted as M groups, from $i = 1$ to M, the radius increases in order. Integrating previous heuristic strategies into SA, HSA for the circles packing problem can be given as follows (See Fig. 3):

```
program HSA
   Begin
      Step one: Running the SA procedure: SA( );
      Step two: If U > 10⁻⁶ then
      Begin
         i := 1;
         While i ≤ M
         begin
            If circles in the i-th group embed each other;
            Begin;
               Randomly select an embedded circle in the i-
   th    group and randomly determine its new position in
   the plate, go to Step one;
            End
            i := i + 1;
         End
```

```
        The circle with maximum potential energy is
  picked out and randomly placed in the plate, go to Step
  one;
      End
      Step three. Stop.
  end.
```

Fig. 3. The HSA algorithm

In Step two, we make use of strategies of "jump out of the trap": dropping an overlapping circle in the group of small circles huddled together is to free space occupied by it and find better packing. In addition, dropping a circle with maximum potential energy is to decrease conflict. From statement of HSA, we understand that HSA includes SA, thus inherits the merit of SA.

4 Computational Results

We compare HSA with QPQH, both algorithms have been implemented with C language on an IBM PC-586 to perform large amounts of calculation with problem instances taken from [8] or generated for the purpose of this experiment. For the simpler cases of this problem, the calculations are carried out very smoothly and rapidly without exception. So what are chosen here are more complex cases. The 4 problem instances below, which include 1 instances of packing equal circles and 3 instances of packing unequal circles, are typical representatives. For each instance, 5 times of trial calculation have been executed with each algorithm. The average execution time, the times of reaching the solutions and the geometric morphologies of the solutions are shown in Table 1. Here, P denotes the problem instance, $\bar{t}$ is used to denote the average execution time, other symbols is as before.

From the results, as shown in Table 1, we observed the speed of HSA is about eight times that of QPQH. For rather difficult problem instances, for example, instance 4, such an increase in speed is especially notable.

5 Conclusions

The heuristic strategies are from nature, so they are very intuitive, and can allow the iterative process to converge fast. HSA inherits the merit of SA, and can avoid the disadvantage of blind search in SA to some extent. The computational results show that the performance of HSA outperforms that of QPQH.

Owing to the complexity and intractability of the circles packing problem, the Hochbaum-Maass algorithm is impractical. This has been thoroughly analyzed by them [2]. Our algorithm is a direct answer to the question asked by Dorit S. Hochbaum and Wolfgang Maass. For the case of equal circles, some results can be found in [6,9,18], but the execution time is not reported. So HSA may remedy this deficiency and is a significant supplement to this problem. Moreover, the global minimum configurations obtained by us have the same geometric morphologies as that obtained by [6,9,18].

Table 1. Comparisons between HSA and QPQH

P	$R_0, n, R_1, \cdots, R_n$	HSA $\bar{t}$	QPQH $\bar{t}$	The geometric morphologies of the solutions
1	$R_0 = 215.47, n = 12,$ $R_1 = \cdots = R_6 = 23.72,$ $R_7 = \cdots = R_9 = 48.26,$ $R_{10} = R_{11} = R_{12} = 100$	9.412	422.114	
2	$R_0 = 39.37,$ $n = 15, R_1 = 1,$ $R_{i+1} = R_i + 1,$ $i = 1, 2, \cdots, 14$	15.93	57.62	
3	$R_0 = 2.4143,$ $n = 17, R_{1\sim4} = 1,$ $R_{5\sim9} = 0.41415,$ $R_{10\sim17} = 0.2$	42.44	526.11	
4	$R_0 = 159.32,$ $n = 50,$ $R_{1\sim50} = 20$	168.27	439.72	

In addition, HSA is easily extendable to pack circles in other bounded space (for example, rectangle or triangle). The HSA algorithm may be of practical value to the rational layout of the round objects in the engineering fields, such as the optic-fiber communication and the transportation of the steel pipes in shipping containers. We hope to find highly efficient algorithm for other packing problem of even greater practical significance in the near future.

References

1. M.R. Garey, D.S. Johnson: Computers and intractability: A Guide to the Theory of NP-Completeness. San Francisco: Freeman (1979)
2. Dorit S. Hochbaum, Wolfgang Maass: Approximation schemes for covering and packing problems in image processing and VLSI, Journal of the Association for Computing Machinery 1(32) (1985) 130–136
3. Haessler. R.W. and Sweeney. P.E.: Cutting stock problems and solution procedures. European Journal of Operational Research 54 (1) (1991) 141–150
4. Dowsland. K.A., Dowsland. W.B.: Packing problems. European Journal of Operational Research 56 (1) (1992) 2–14
5. G. R. Raidl, G. Kodydek: Genetic Algorithms for the Multiple Container Packing Problem. in Proc. of the 5th Int. Conference on Parallel Problem Solving from Nature V, Amsterdam, The Netherlands, Springer LNCS (1998) 875–884
6. D. Lubachevsky, R.L. Graham: Curved hexagonal packing of equal disks in a circle. Discrete & Computational Geometry (1997) 179–194
7. Huang Wenqi, Zhan Shuhao: A quasi-physical method for solving packing problems, Acta Mathematicae Applagatae Sinica 2 (2) (1979) 176–180
8. Huaiqing Wang, Wenqi Huang, Quan Zhang, Dongming Xu: An improved algorithm for the packing of unequal circles within a larger containing circle. European Journal of Operational Research 141 (2002) 440–453
9. John A.George, Jennifer M.George, Bruce W.Lammar: Packing different-sized circles into a rectangular container, European Journal of Operational Research 84 (1995) 693–712
10. S. Kirkpatrick, C. D. Gelatt, M.P. Vecchi: Optimization by simulated annealing. Science 220 (1983) 671–689

Parallel Genetic Algorithm for Graph Coloring Problem

Zbigniew Kokosiński, Marcin Kołodziej, and Krzysztof Kwarciany

Faculty of Electrical & Computer Eng., Cracow University of Technology,
ul. Warszawska 24, 31-155 Kraków, Poland
zk@pk.edu.pl

Abstract. In this paper a new parallel genetic algorithm for coloring graph vertices is presented. In the algorithm we apply a migration model of parallelism and define two new recombination operators: SPPX and CEX. For comparison two recently proposed crossover operators: UISX and GPX are selected. The performance of the algorithm is verified by computer experiments on a set of standard graph coloring instances.

1   Introduction

Graph coloring problem (GCP) belongs to the class of NP–hard combinatorial optimizations problems. GCP is defined for an undirected graph as a problem of assignment of available colors to graph vertices providing that adjacent vertices are assigned different colors and the number of colors is minimal. There are many variants of the problem when some additional assumptions are made [6,10]. Intensive research conducted in this area resulted in a large number of exact and approximate algorithms, heuristics and metaheuristics [9]. GCP was the subject of Second DIMACS Implementation Challenge in 1993 and Computational Symposium on Graph Coloring and Generalizations in 2002. A collection of graph coloring instances in DIMACS format and summary of results are available at [11,12,13].

Genetic algorithms (GA) are metaheuristics often used for GCP [3,4,5,8]. Recently a number of parallel versions of GA were studied. This approach is based on co–evolution of a number of populations that exchange genetic information during the evolution process according to a communication pattern [1,2].

In this paper we present results of our experiments with parallel genetic algorithms (PGA) for graph coloring problem.

In the paper two new recombination operators for coloring chromosomes are proposed: SPPX (Sum–Product Partition Crossover) in which simple set operations and random mechanisms are implemented, and CEX (Conflict Elimination Crossover) that is focused on the offspring quality.

In computer simulations of PGA we used DIMACS benchmarks. The obtained results are very promising and encourage future research focused on PGA and new genetic operators for graph coloring problems.

M. Bubak et al. (Eds.): ICCS 2004, LNCS 3036, pp. 215–222, 2004.

2   Migration Model of Parallel Genetic Algorithm

There are many models of parallelism in evolutionary algorithms: master–slave PGA, migration based PGA, diffusion based PGA, PGA with overlaping subpopulations, population learning algorithm, hybrid models etc.

Migration models of PGAs consist of a finite number of subpopulations that evolve in parallel on their "islands" and only occasionaly exchange the genetic information under the control of a migration operator. Co–evolving subpopulations are built of individuals of the same type and are ruled by one adaptation function. The selection process is decentralized.

In our model the migration is performed on a regular basis. During the migration phase every island sends its representatives (emigrants) to all other islands and receives their representatives (immigrants) from all co–evolving subpopulations. This topology of migration reflects so called "pure" island model.

The migration process is fully characterized by migration size, distance betweeen populations and migration scheme. Migration size determines the emigrant fraction of each population. The distance between migrations determines how often the migration phase of the algorithm occurs. Two migration schemes are applied: migration of best individuals of the subpopulation or migration of individuals randomly selected.

In our algorithm we applied a specific model of migration in which islands use two copies of genetic information: migrating individuals still remain members of their original subpopulation. In other words they receive new "citizenship" without losing the former one. Incoming individuals replace the chromosomes of host subpopulation at random. Then, a selection process is performed. The rationale behind such a model is as follows. Even if the best chromosomes of host subpopulation are eliminated they shall survive on other islands where their copies were sent. On the other hand any eliticist scheme or preselection applied to the replacement phase leads to premature elimination of worse individuals and lowers the overall diversity of subpopulation.

3   Genetic Operators for GCP

In graph coloring problem k–colorings of graph vertices are encoded in chromosomes representing set partitions with exactly k blocks. In partition representation each block of partition does correspond to a single color. In assignement representation available colors are assigned to an ordered sequence of graph vertices. In this section we introduce a collection of genetic crossover, mutation and selection operators that are used in our PGA.

3.1   Sum–Product Partition Crossover

The first recombination operator called Sum–Product Partition Crossover (SPPX) employs for offspring generation simple set sum and set product operations on block of partitions and a random mechanism of operand selection

```
procedure: SPPX (p1,p2,r1,r2,s1,s2,t1,t2,Prob(PRODUCT),Prob(SUM))
begin
  s1=t1=s2=t2=∅;
  generate random numbers rand1, rand2 : 0 ≤rand1,rand2≤1;
  if rand1 ≤ Prob(PRODUCT) then PRODUCT(p1,r1,s1,t1);
  if rand2 ≤ Prob(SUM) then SUM(p2,r2,s2,t2);
end SPPX;
PRODUCT(p,r,s,t)
begin
  select random h (1≤h≤k) and j (1≤j≤l);
```
$$V_1^s = V_1^t = (V_h^p \cap V_j^r);$$
```
  for i = 1 to k do
    if i ≠ h do if (V_i^p \ V_1^s) nonempty then
      add next block V_i^p \ V_1^s to s;
  for i = 1 to l do
    if i ≠ j do if (V_i^r \ V_1^t) nonempty then
      add next block V_i^r \ V_1^t to t;
end PRODUCT;
SUM (p,r,s,t)
begin
  select random h (1≤h≤k) and j (1≤j≤l);
```
$$V_1^s = V_1^t = (V_h^p \cup V_j^r);$$
```
  for i = 1 to k do
    if i ≠ h do if (V_i^p \ V_j^r) nonempty then
      add next block V_i^p \ V_j^r to s;
  for i = 1 to l do
    if i ≠ j do if (V_i^r \ V_h^p) nonempty then
      add next block V_i^r \ V_h^p to t;
end SUM;
```

Fig. 1. The recombination operator SPPX.

from randomly determined 4 parental chromosomes. SPPX is composed of two procedures PRODUCT and SUM which are applied to the pair of chromosomes $p=\{V_1^p, \ldots, V_k^p\}$, $r=\{V_1^r, \ldots, V_l^r\}$ and produce a pair of chromosomes $s=\{V_1^s, \ldots, V_m^s\}$ and $t=\{V_1^t, \ldots, V_n^t\}$ with probabilities of elementary operations satisfying: $0 \leq \text{Prob(PRODUCT)} < \text{Prob(SUM)} \leq 1$. A pseudocode of the procedure SPPX is presented in Fig.1.

Example1

Four parents represent different 3–colorings of a graph with 10 vertices: p1={ABC,DEFG,HIJ}, r1={CDEG,AFI,BHJ}, p2={CDG,BEHJ,AFI} and r2={ACGH,BDFI,EJ}. Let us assume Prob(PRODUCT)=0.5, Prob(SUM)=0.7. Let rand1=0.4 and PRODUCT is computed with h=3, j=2. Thus, V_3^p={HIJ} and V_2^r={AFI}. We obtain $V_1^s = V_1^t = $ {I} and s1={I,ABC,DEFG,HJ} and t1={I,CDEG,AF,BHJ}. Let rand2=0.3 and SUM is computed with h=2, j=1. Thus, V_2^p={BEHJ} and V_1^r={ACGH}. We obtain $V_1^s = V_1^t = $ {ABCEGHJ} and then s2={ABCEGHJ,D,FI} and t2={ABCEGHJ,DFI}. As a result of the crossover we obtain four children: s1, t1, s2 and t2 representing 2,3 and 4–colorings of the given graph.

Let us notice that operation PRODUCT may increase the initial number of colors while the operation SUM may reduce this number. The probability of PRODUCT should be lower then the probability of SUM. The recombination operator SPPX which can be used as a versatile operator in evolutionary algorithms for many other partition problems.

3.2   Conflict Elimination Crossover

In conflict–based crossovers for GCP an assignement representation of colorings is used and the offspring try to copy conflict–free colors from their parents. The next recombination operator called Conflict Elimination Crossover (CEX) reveals some similarity to the classical crossover. Each parental chromosome is partitioned into two blocks. The first block consists of conflict–free nodes while the second block is built of the remaining nodes that break the coloring rules. The last block in both chromosomes is then replaced by corresponding colors taken from the other parent. This recombination scheme provides inheritance of all good properties of one parent and gives the second parent a chance to reduce the number of existing conflicts. However, if a chomosome represents a feasible coloring the recombination mechanism is not working. Therefore, the recombination must be combined with an efficient mutation mechanism. The operator CEX is almost as simple and easy to implement as the classical crossover (see Fig.2).

```
procedure: CEX (p,r,s,t)
begin
  s = r;
  t = p;
  copy conflict-free vertices V_cf^p from p to s;
  copy conflict-free vertices V_cf^r from r to t;
end
```

Fig. 2. The recombination operator CEX.

Example2

Two parents represent different 5–colorings of a graph with 10 vertices i.e. sequences: p=<5,2,**3**,**1**,1,4,3,5,1,**2**> and r=<1,4,**5**,2,3,3,**2**,4,2,1>. Vertices with conflict colors are marked by bold fonts.

Replacing the vertices with color conflicts by vertices taken from the other parent we obtain the following two chromosomes: s=<5,2,**5**,2,1,3,3,5,1,**1**> and t=<1,4,**3**,2,3,3,**3**,4,2,**2**>.

We can observe that the obtained chromosomes represent now two different 4–colorings of the given graph (reduction by 1 with respect to initial coloring) and the number of color conflicts is now reduced to 2 in each chromosome.

3.3   Union Independent Set Crossover

The greedy operator proposed by Dorne and Hao [4] and called Union Independent Sets (UISX) works on pairs of independent sets taken from two parent colorings. In any feasible graph coloring all graph vertices are partitioned into blocks that are disjoint independent sets (IS). A coloring is not feasible if it contains at least one block which is a non–independent set. Each block of a partition is assigned one color. "If we try to maximize the size of each IS by a combination mechanism, we will reduce the sizes of non–independent sets, which in turn helps to push these sets into independent sets" [4].

In the initial step disjoint ISs in both parents are determined. Then we compute coloring for the first child. At first, we select the maximum IS from the first parent and compute set intersections with ISs from the second parent. The union of a pair of ISs with maximum intersection is colored in the offspring with the IS color from the first parent. In the case of a tie a random IS is always chosen. Then the colored vertices are removed from the both parents and the coloring procedure is repeated as long as posible. The vertices without any color are assigned the original color from the first parent. The coloring for the second child is computed with reversed roles of both parents.

3.4   Greedy Partition Crossover

The method called Greedy Partition Crossover (GPX) was designed by Galinier and Hao [5] for recombination of colorings or partial colorings in partition representation. It is assumed that both parents are randomly selected partitions with exactly k blocks that are independent sets. The result is a single offspring (a coloring or partial coloring) that is built successively in a greedy way. In each odd step we select the maximum block from the first parent. Then, we add the block to the result and remove all its nodes from the both parents. In each even step we select the maximum block from the second parent. Then, we add the block to the result and remove all its nodes from the both parents. The procedure is repeated at most k times since, in some cases, the offspring has less blocks then the parents (see Example 3). This possibility is not considered in the original paper [5]. Finally, unassigned vertices (if they exist) are assigned at random to existing blocks of partition. A corrected version of GPX is shown in Fig.3.

The first parent is replaced by the offspring while the second parent returns to population and can be recombined again in the same generation. GPX crossover is performed with a constant probability.

Example 3

Two parents represent different 3–colorings of a graph with 10 vertices, i.e. partitions: p0={ABFGI,CDE,HJ}, p1={ABF,CDEGHJ,I}.

```
procedure: GPX (p0,p1,s)
begin
  s = ∅;
  i = 1;
  repeat
    select block V with maximum cardinality from the partition p(i mod2);
    s = s∪V -- add the block V to partition s;
    remove all vertices of V from p0 and p1;
    i = i + 1;
  until (i > k) or (all blocks of p1 and p2 empty);
  assign randomly all unassigned vertices to blocks of s;
end
```

Fig. 3. The modified recombination operator GPX.

For i=1 the maximum block {CDEGHJ} is selected from p1 and is added to s. After removing the block vertices from the parents we obtain p0={ABFI}, p1={ABF,I}. For i=2 the maximum block {ABFI} is selected from p0 and is added to s. Termination condition is satisfied and we obtain result partition s={ABFI,CDEGHJ} that is a valid 2–coloring.

3.5   Mutation Operators

Transposition is a classical type of mutation that exchange colors of two randomly selected vertices in the assignment representation. The second mutation operation called First Fit is designed for colorings in partition representationand and is well suited for GCP. In the mutation First Fit one block of the partition is selected at random and we try to make a conflict–free assignment of its vertices to other blocks using the heuristic First Fit. Vertices with no conflict–free assignment remain in the original block. Thus, as a result of the mutation First Fit the color assignment is partially rearranged and the number of partition blocks is often reduced by one.

3.6   Selection Operator

The quality of a solution is measured by the following cost function:
$$f(p) = \sum_{(u,v)\in E} q(u,v) + d + C \text{ , where:}$$
p – is a graph coloring,
q – is a penalty function for pairs of vertices connected by an edge $(u,v) \in$ E:
    $q(u,v) = 2$ when $c(u) = c(v)$, and $q(u,v) = 0$, otherwise,
d – is a general penalty function applied to graph colorings:
    $d = 1$, when $\sum_{(u,v)\in E} q(u,v) > 0$, and $d = 0$ when $\sum_{(u,v)\in E} q(u,v) = 0$,
C – is the number of colors used.

The proportional selection is performed in our PGA with the fitness function $1/f(p)$.

Table 1. Performance of the migration–based PGA with various crossover operators

no	graph	vertices	edges	colors	Crossover operator							
					UISX		GPX		SPPX		CEX	
					n	t[s]	n	t[s]	n	t[s]	n	t[s]
1	anna	138	493	11	19	2.0	21	6.0	61	3,2	15	1.3
2	david	87	406	11	22	2.0	24	3,9	74	3,6	20	1.0
3	huck	74	301	11	8	0,8	7	1.0	29	0,5	12	0.7
4	miles500	128	1170	20	95	9,5	59	40	152	38	100	3.0
5	myciel7	191	2360	8	18	2.0	21	7.9	76	7.6	20	1.0
6	mulsol$_1$	197	3925	49	90	31	60	35	180	34	58	2.0

4   Experimental Verification

For computer experiments we used graph coloring instances available in the
web archive [11]. This is a collection of graphs in DIMACS format with known
parameters, including graph chromatic numbers.

In our computer program *PGA for GCP* two basic models of PGA: migration and master–slave can be simulated. It is possible to set up most parameters
of evolution, monitor evolution process and measure both the number of generations and time of computations. In the preprocessing phase we converted
list of edges representation into adjacency matrix representation. The program
generates detailed reports and basic statistics.

We tested the influence of migration scheme on the PGA efficiency measured
by the number of generations n needed to obtain an optimal coloring when
the chromatic number is known. Computations were performed on 5 graphs
with the following parameters: *population size = 60 , number of islands = 5,
migration rate = 5, crossover = SPPX, mutation =* First Fit, *and
mutation probability = 0.1* . All experiments were repeated 30 times. For all
graphs migration of best individuals always gives the best results. Migration of
random individuals is almost useless except huge graphs like mulsol.i.4 where
random migration is also efficient.

The experiments confirmed that the mutation First Fit is superior to the
Transposition mutation for graph coloring problems. It works particularly well
with CEX crossover.

In the main experiment the efficiency of all 4 crossover operators was tested
in the migration model. Computations were performed on 6 graphs with the
parameters: *population size = 60 , number of islands = 3, migration rate =
5, migration size = 5, mutation =* First Fit, *and mutation probability = 0.1* .
All experiments were repeated 30 times. The results are presented in Table 1.

We can observe that the proposed crossover operators are efficient in terms
of computation time. SPPX requires more generations then GPX in order to
find an optimal coloring but it is simpler and therefore a bit faster then GPX.
In some cases the operator UISX requires less generations then GPX but it
always produces an optimal solution faster then two previous operators. The

most efficient operator in the experiment is CEX which dominates all others operators under the both criteria (except the smallest graph instance).

All computer experiments were performed on a computer with AMD Athlon 1700+ processor (1472 MHz) and 256 MB RAM.

5   Conclusions

In the paper we proved by computer simulation that parallel genetic algorithms can be efficiently used for the class of graph coloring problems. In island model of PGA the searched space is significantly enlarged and the migration between co–evolving subpopulations improves the overall convergence of the algorithm.

PGA is particularly efficient for large scale problems like GCP. The results presented in this paper encourage further research in this area. The authors intend to continue their work. One obvious direction is to extend the experiments on other DIMACS benchmarks including a class of random graphs. It is also worth to consider some variants of SPPX operator that will make it more problem–oriented. The search for new efficient genetic operators for GCP still remains an open question.

References

1. Alba E., Tomasini M.: Parallelism and evolutionary algorithms, IEEE Trans. Evol. Comput. Vol.6 (2002) No.5, 443–462
2. Bäck T.: Evolutionary algorithms in theory and practice, Oxford U. Press (1996)
3. Croitoriu C., Luchian H., Gheorghies O., Apetrei A.: A new genetic graph coloring heuristic, Computational Symposium on Graph Coloring and Generalizations COLOR'02, [in:] Proc. Int. Conf. Constraint Programming CP'02 (2002)
4. Dorne R., Hao J-K.: A new genetic local search for graph coloring, Parallel Problem Solving from Nature 1998, LNCS 1498 (1998) 745–754
5. Galinier P., Hao J-K.: Hybrid evolutionary algorithms for graph coloring, J. Combinatorial Optimization (1999) 374–397
6. Jensen T.R., Toft B.: Graph coloring problems, Wiley Interscience (1995)
7. Johnson D.S., Trick M.A.: Cliques, coloring and satisfiability: 2nd DIMACS Impl. Challenge, DIMACS Series in Discr. Math. and Theor. Comp. Sc. Vol.26 (1996)
8. Khuri S., Walters T. Sugono Y.: Grouping genetic algorithm for coloring edges of graph, Proc. 2000 ACM Symposium on Applied Computing (2000) 422–427
9. Kubale M.: Introduction to computational complexity and algorithmic graph coloring, GTN, Gdańsk (1998)
10. Kubale M. (ed): Discrete optimization. Models and methods for graph coloring, WNT, Warszawa (2002) (in Polish)
11. http://mat.gsia.cmu.edu/COLOR/instances.html
12. ftp://dimacs.rutgers.edu/pub/challenge/graph/benchmarks/
13. http://mat.gsia.cmu.edu/COLORING03/

Characterization of Efficiently Parallel Solvable Problems on a Class of Decomposable Graphs

Sun-Yuan Hsieh

Department of Computer Science and Information Engineering
National Cheng Kung University
No 1. University Road, Tainan 701, Taiwan
hsiehsy@mail.ncku.edu.tw

Abstract. In this paper, we sketch characteristics of those problems which can be systematically solved on decomposable graphs. Trees, series-parallel graphs, outerplanar graphs, and bandwidth-k graphs all belong to decomposable graphs. Let $T_d(|V|, |E|)$ and $P_d(|V|, |E|)$ denote the time complexity and processor complexity required to construct a parse tree representation T_G for a decomposable $G = (V, E)$ on a PRAM model M_d. We define a general problem-solving paradigm to solve a wide class of subgraph optimization problems on decomposable graphs in $O(T_d(|V|, |E|) + \log |V(T_G)|)$ time using $O(P_d(|V|, |E|) + |V(T_G)|/\log |V(T_G)|)$ processors on M_d. By using our paradigm, we show the following parallel complexities: (a) The maximum independent set problem on trees can be solved in $O(\log |V|)$ time using $O(|V|/\log |V|)$ processors on an EREW PRAM. (b) The maximum matching problem on series-parallel graphs can be solved in $O(\log |E|)$ time using $O(|E|/\log |E|)$ processors on an EREW PRAM. (c) The efficient domination problem on series-parallel graphs can be solved in $O(\log |E|)$ time using $O(|E|/\log |E|)$ processors on an EREW PRAM.

1   Introduction

A class of graphs is *recursive* if every graph of the class can be constructed by a finite number of applications of *composition operations* starting with a finite set of *basis* graphs. The recursive class Γ of graphs is said to be *decomposable* if each graph in Γ has a set of some specified vertices called *terminals*, and each composition operation is defined in terms of certain primitive operations on terminals. Trees, series-parallel graphs, outerplanar graphs, protoHalin graphs, and bandwidth-k graphs are all decomposable graphs [3]. Also, every decomposable graph has a fixed upper bound on the *treewidth* of the graphs in the class, and graphs with treewidth at most k for fixed k are partial k-trees [8].

Properties of decomposable graphs are studied by many researchers [2,3,7,8, 9,11,12] which resulted in sequential algorithms to solve quite a few interesting graph-theoretical problems on this special class of graphs. However, there are few results in the viewpoint of parallel computation. Given a graph problem, we say it belongs to the class of *subgraph optimization* problem if the object of this

M. Bubak et al. (Eds.): ICCS 2004, LNCS 3036, pp. 223–230, 2004.
© Springer-Verlag Berlin Heidelberg 2004

problem is to find a subgraph of the input graph to satisfy the given properties which includes an optimization constraint. For example, the problem of finding a maximum independent set is a subgraph optimization problem.

In this paper, we propose a different parallel strategy on the deterministic parallel random access machine (PRAM) [6]. Given a decomposable graph represented by its parse tree form, we define a class of subgraph optimization problems, called the (k, Θ)-*regular problem*, and show such a class of problems can be efficiently parallelized by applying the binary tree contraction technique to the given parse tree. Let $T_d(|V|, |E|)$ and $P_d(|V|, |E|)$ denote the time complexity and processor complexity required to construct a parse tree T_G of a decomposable graph $G = (V, E)$ on a PRAM model M_d. We show that a (k, Θ)-regular problem can be solved in $O(T_d(|V|, |E|) + \log |V(T_G)|)$ time using $O(P_d(|V|, |E|) + |V(T_G)| / \log |V(T_G)|)$ processors on M_d. Moreover, each (k, Θ)-regular problem can be solved in $O(\log |V(T_G)|)$ time using $O(|V(T_G)| / \log |V(T_G)|)$ processors on an EREW PRAM if T_G is given to be an input instance. Based on the technique, we obtain the following results: (a) The maximum independent set problem on trees can be solved in $O(\log |V|)$ time using $O(|V| / \log |V|)$ processors on an EREW PRAM, (b) The maximum matching problem can be solved in $O(\log |E| \log^* |E|)$ time using $O(|E| / \log |E| \log^* |E|)$ processors on an EREW PRAM, and (c) The efficient domination problem on series-parallel graphs can be solved in $O(\log |E| \log^* |E|)$ time using $O(|E| / \log |E| \log^* |E|)$ processors on an EREW PRAM. Given a parse tree of a series-parallel graph, the problems in (b) and (c) can be optimally solved in $O(\log |E|)$ time using $O(|E| / \log |E|)$ processors on an EREW PRAM. To our knowledge, no NC algorithm exists for solving the problem in (c) in the literature.

2   Preliminaries

This paper considers finite, simple[1] and undirected graphs $G = (V, E)$, where V and E are the vertex and edge sets of G, respectively. Let $n = |V|$ and $m = |E|$. For two graphs $G_1 = (V_1, E_1)$ and $G_2 = (V_2, E_2)$, the *union of G_1 and G_2*, denoted by $G_1 \cup G_2$, is the graph $(V_1 \cup V_2, E_1 \cup E_2)$. We say that a graph $G' = (V', E')$ is a *subgraph* of $G = (V, E)$ if $V' \subseteq V$ and $E' \subseteq E$. Given a set $V' \subseteq V$, the subgraph of G *induced* by V' is the graph $G' = (V', E')$, where $E' = \{(u, v) \in E | \ u, v \in V'\}$. Let $G[X]$ denote the subgraph of G induced by $X \subseteq V$. For a vertex $v \in V$ of a graph $G = (V, E)$, the *neighborhood of v* is $N_G(v) = \{u \in V | \ (u, v) \in E\}$ and the *closed neighborhood of v* is $N_G[v] = N_G(v) \cup \{v\}$. The subscript G in the notations used in this paper can be omitted when no ambiguity arises. Given a node v in a rooted tree T, let $T(v)$ be a subtree of T rooted at v. For graph-theoretic terminologies and notations not mentioned here, see [5].

We follow the notations used in [8] to define the class of decomposable graphs.

---

[1] We only consider simple graphs in this paper although some of the results also apply to multigraphs.

Definition 1. Let $G = (V, E, S)$ be a graph with vertex V, edge set E, and an ordered list S of t *terminals* chosen from V for some fixed integer t. We note that the elements of S are not necessary distinct.

(1) Let $B = \{B_1, B_2, \ldots, B_l\}$ be a finite set of *basis graphs*, where each B_i is a finite graph having an ordered list of t (not necessary distinct) terminals.

(2) Let $O = \{*_1, *_2, \ldots, *_q\}$ be a finite set of binary rules of *composition*, whereby two graphs $G_i = (V_i, E_i, S_i)$ and $G_j = (V_j, E_j, S_j)$ can be combined to produce new graphs $G_i *_c G_j$, $1 \leq c \leq q$. Each rule of composition $*_c$ consists of three suboperations on the terminals S_i and S_j:

    (i) Choose a subset $S_i{}'$ of distinct terminals from the list S_i and identify each $x \in S_i{}'$ with a unique $y \in S_j$. Let $S_j{}'$ denote the subset of affected terminals from the list S_j.

    (ii) Add any subset of the edges $\{(x, y) | x \in \overline{S_i{}'}, y \in \overline{S_j{}'}\}$ to $G_i *_c G_j$, where $\overline{S_i{}'}$ is the subset of terminals in the list S_i but not in $S_i{}'$, and $\overline{S_j{}'}$ is defined similarly.

    (iii) Select an ordered list of t (not necessarily distinct) terminals from the list S_i and S_j to the terminals of $G_i *_c G_j$.

(3) The class Γ of decomposable graphs is recursively defined as follows:

    (i) Any $B_i \in B$ is in Γ.

    (ii) If G_i and G_j are in Γ and $*_c$ is an operation in O, then the graph $G_i *_c G_j$ is also in Γ.

Definition 2. Let Γ be the class of decomposable graphs. The *parse tree* T_G of a graph $G \in \Gamma$ is a tree in which the leaves correspond to the basis graphs from which G is constructed, and each internal node represents the result of applying a composition operation to the graphs represented by the subtrees rooted at its children. Let G_v be the subgraph of G corresponding to a node v of a parse tree. Note that $T_G(v)$ is a parse tree of G_v.

3 A General Problem-Solving Paradigm

3.1 The (k, Θ)-Parse Tree

Given a graph G, let $\mathcal{U}_{V(G)}$ (respectively, $\mathcal{U}_{E(G)}$) be the set consisting of all subsets of $V(G)$ (respectively, $E(G)$). Given $\mathcal{Q} = \{Q_1, Q_2, \ldots, Q_l\}$, where $Q_i \in \mathcal{U}_{V(G)}$ (respectively, $Q_i \in \mathcal{U}_{E(G)}$), we define MIN_v (respectively, MIN_e) to be an operator on $\mathcal{Q}$ that returns a minimum-cardinality set Q_j for some $1 \leq j \leq l$. The operators MAX_v (respectively, MAX_e) can be defined similarly. For two lists $L_1 = \langle l_1, l_2, \ldots, l_i \rangle$ and $L_1{}' = \langle l_1{}', l_2{}', \ldots, l_j{}' \rangle$, we define the *concatenation of L_1 and $L_1{}'$*, denoted by $L_1 \bullet L_1{}'$, to be the list $\langle l_1, l_2, \ldots, l_i, l_1{}', l_2{}', \ldots, l_j{}' \rangle$.

Definition 3. Let $G = (V, E)$ be a decomposable graph and let T_G be a parse tree of G. Given a positive integer k, and an operator $\Theta \in \{\mathrm{MIN}_v, \mathrm{MIN}_e, \mathrm{MAX}_v, \mathrm{MAX}_e\}$, T_G is a (k, Θ)-*parse tree of G* if the following conditions hold. Let v be a node of T_G and let N_i be the set of integers from 1 to i.

(1) If v is an internal node, then it is associated with k integers $a_{v,1}, a_{v,2}, \ldots, a_{v,k}$

from N_k, and the following $2k$ functions $f_i : \{v\} \times N_{a_{v,i}} \mapsto N_k$ and $g_i : \{v\} \times N_{a_{v,i}} \mapsto N_k$, $1 \le i \le k$.

(2) Node v is also associated with a list of k subgraphs[2] $R_v = \langle R_{v,1}, R_{v,2}, \ldots, R_{v,k} \rangle$, called the *target subgraphs of v*, which are defined as follows.

CASE 1: v is a leaf. R_v is a list of k subgraphs selected from $\mathcal{U}_{V(G_v)}$ (respectively, $\mathcal{U}_{E(G_v)}$) if $\Theta \in \{\text{MIN}_v, \text{MAX}_v\}$ (respectively, $\Theta \in \{\text{MIN}_e, \text{MAX}_e\}$).

CASE 2: v is an internal node. Let u and w be two children of v. Then, $R_{v,i} = \Theta\{R_{u,f_i(u,1)} \cup R_{w,g_i(w,1)}, R_{u,f_i(u,2)} \cup R_{w,g_i(w,2)}, \ldots, R_{u,f_i(u,a_{v,i})} \cup R_{w,g_i(w,a_{v,i})}\}$, where $1 \le i \le k$.

Definition 4. Let T_G be a (k, Θ)-parse tree. The (k, Θ)-*parse tree problem* is the problem to find the k target subgraphs of the root of T_G.

Lemma 1. *The (k, Θ)-parse tree problem can be solved in $O(k^2 n)$ time, where n is the number of vertices of the given tree.*

3.2   Parallel Complexities of the (k, Θ)-Regular Problem

In this section, we apply the binary tree contraction technique described in [1] to parallelize the (k, Θ)-regular problem. This technique recursively applies two operations, *prune* and *bypass*, to a given binary tree. *Prune(u)* is an operation which removes a leaf node u from the current tree, and *bypass(v)* is an operation (following a prune operation) that removes a node v with exactly one child w and then lets the parent of v become the new parent of w. We define a *contraction phase* to be the consecutively execution of prune and bypass operations.

Let T be an n-leave binary tree with the root r. Given a Euler tour starting from r of T, the algorithm initially numbers the leaves from 1 to n according to the order of their appearances in the tour. Then, the algorithm repeats the following steps. In each step, *prune* and *bypass* work only on the leaves with odd index and their parents. Hence, these two operations can be performed independently and delete $\lfloor \frac{l}{2} \rfloor$ leaves together with their parents on the binary tree in each step, where l is the number of the current leaves. Therefore, the tree will be reduced to a three-node tree after repeating the steps in $\lceil \log n \rceil$ times.

Lemma 2. *[1] If the prune operation and bypass operation can be performed by one processor in constant time, the binary tree contraction algorithm can be implemented in $O(\log n)$ time using $O(n/\log n)$ processors on an EREW PRAM, where n is the number of nodes in an input binary tree.*

Consider a node x in a rooted tree T. Any node y on the unique path from x to the root is called an *ancestor* of x. If y is an ancestor of x, then x is a *descendant* of y. Further, x is a *proper descendant* of y when $x \ne y$. Note that every node is both an ancestor and a descendant of itself. For convenience, we allow $\mathcal{U}_G$ to represent one of $\mathcal{U}_{V(G)}$ and $\mathcal{U}_{E(G)}$ if it is not particularly specified.

---

[2] In this paper, a subgraph H of G is represented by a set Q: If $Q \in \mathcal{U}_{V(G)}$, then $H = (Q, \emptyset)$; If $Q \in \mathcal{U}_{E(G)}$, then $H = (\{x \mid x \text{ is an endpoint of an edge in } Q\}, Q)$.

Definition 5. Let u and v be two nodes of a (k, Θ)-parse tree T such that u is a descendant of v. A k-ary function $h : \mathcal{U}_{G_u}{}^k \mapsto \mathcal{U}_{G_v}$ possesses the *canonical form*, if $h(X_1, \ldots, X_k) = \Theta\{X_{b_1} \cup C_1, X_{b_2} \cup C_2, \ldots, X_{b_a} \cup C_a\}$, where $b_i \neq b_j$ for two distinct $1 \leq i, j \leq a$, and $C_i \in (\mathcal{U}_{G_v} \setminus \mathcal{U}_{G_u})$.

The following lemma can be shown by the set theory and properties of the function composition.

Lemma 3. *Let $\Theta \in \{\mathrm{MIN}_v, \mathrm{MIN}_e \mathrm{MAX}_v, \mathrm{MAX}_e\}$, and let $h_0 : \mathcal{U}_{G_u}{}^k \mapsto \mathcal{U}_{G_v}$ be a function with the canonical form, where u is a descendant of v. If k functions $h_i : \mathcal{U}_{G_w}{}^k \mapsto \mathcal{U}_{G_u}$ possess the canonical form, where $1 \leq i \leq k$ and w is a descendant of u, then the function obtained from the composition $h_0 \circ (h_1, h_2, \ldots, h_k) : \mathcal{U}_{G_w}{}^k \mapsto \mathcal{U}_{G_v}$ possesses the canonical form.*

We next develop a parallel algorithm for the (k, Θ)-parse tree problem. For a node x in the current tree H, let $par_H(x)$ (respectively, $child_H(x)$) denote the *parent* (*children*) *of* x, and let $sib_H(x)$ denote the *sibling of* x. The subscript H can be omitted if no ambiguity arises. Recall that $H(x)$ be the subtree of H rooted at x, and $R_x = \langle R_{x,1}, \ldots, R_{x,k} \rangle$ is the list of the target subgraphs associated with x.

During the process of executing the tree contraction, we aim at constructing k k-ary functions $h_{x,1}, h_{x,2}, \ldots, h_{x,k}$ associated with each node x of the current tree such that $h_{x,i}$'s possess the canonical form and satisfy the condition described below. Let v be an internal node in the current tree whose left child and right child are u and w, respectively. Also let u' be the left child and w' be the right child of v in the original tree. For the remainder of this section, we call u' and w' *replacing ancestors of u and w with respect to v*, respectively. Once $R_{u,i}$ and $R_{w,i}$, $1 \leq i \leq k$, are provided as the inputs of $h_{u,i}$ and $h_{w,i}$, respectively, the target subgraphs of v can be obtained from $R_{u'} = \langle R_{u',1}, \ldots, R_{u',k} \rangle = \langle h_{u,1}(R_{u,1}, \ldots, R_{u,k}), \ldots, h_{u,k}(R_{u,1}, \ldots, R_{u,k}) \rangle$, and $R_{w'} = \langle R_{w',1}, \ldots, R_{w',k} \rangle = \langle h_{w,1}(R_{w,1}, \ldots, R_{w,k}), \ldots, h_{w,k}(R_{w,1}, \ldots, R_{w,k}) \rangle$, using the formula

$$R_{v,i} = \Theta\{R_{u',f_i(u',1)} \cup R_{w',g_i(w',1)}, R_{u',f_i(u',2)} \cup R_{w',g_i(w',2)}, \ldots, R_{u',f_i(u',a_{v,i})} \cup$$
$$R_{w',g_i(w',a_{v,i})}\}. \tag{1}$$

where, $R_{u',f_i(u',j)} = h_{u,f_i(u',j)}(R_{u,1}, \ldots, R_{u,k})$ and $R_{w',g_i(w',j)} = h_{w,g_i(w',j)}(R_{w,1}, \ldots, R_{w,k})$ for $1 \leq j \leq a_{v,i}$.

We call the functions $h_{x,i}$, $1 \leq i \leq k$, computed for each node x in the current tree *the crucial functions of x*.

We next describe the details of our algorithm. Initially, for each node v in the given tree we construct k functions $h_{v,i}(X_1, \ldots, X_k) = \Theta\{X_i \cup \emptyset\}$, $1 \leq i \leq k$. Clearly, these functions are crucial functions.

In the execution of the tree contraction, assume that $prune(u)$ and $bypass(par(u))$ are performed consecutively. Let $par(u) = v$ and $sib(u) = w$ in the current tree. Let u' and w' be the replacing ancestors of u and w with respect to v, respectively. Assume that $h_{u,i}$ and $h_{w,i}$, $1 \leq i \leq k$, are crucial functions of u

and w in the current tree. Thus $R_{u'} = \langle h_{u,1}(R_{u,1}, \ldots, R_{u,k}), \ldots, h_{u,k}(R_{u,1}, \ldots, R_{u,k}) \rangle$ and $R_{w'} = \langle h_{w,1}(R_{w,1}, \ldots, R_{w,k}), \ldots, h_{w,k}(R_{w,1}, \ldots, R_{w,k}) \rangle$. Since u is a leaf, $R_{u,i}$'s are associated with u before executing the tree contraction algorithm. Therefore, the above k target subgraphs $R_{u'}$ can be obtained through function evaluation. On the other hand, since w is not a leaf in the current tree, $R_{w,i}$, $1 \leq i \leq k$, is an indeterminate value represented by variable X_i. Hence. $R_{w'}$ can be represented by $\langle h_{w,1}(X_1, \ldots, X_k), \ldots, h_{w,k}(X_1, \ldots, X_k) \rangle$. By Equation 1, we construct k intermediate functions representing k target subgraphs R_v from $R_{u'}$ and $R_{w'}$ by:

$$R_{v,i} = \Theta\{R_{u',f_i(u',1)} \cup R_{w',g_i(w',1)}, R_{u',f_i(u',2)} \cup R_{w',g_i(w',2)}, \ldots, R_{u',f_i(u',a_{v,i})} \cup R_{w',g_i(w',a_{v,i})}\}, \tag{2}$$

where $R_{w',g_i(w',j)} = h_{w,g_i(w',j)}(X_1, \ldots, X_k)$, $1 \leq j \leq a_{v,i}$

As with the proof similar to that of Lemma 3, Equation 2 can be further simplified as

$$R_{v,i} = \Theta\{X_{b_1} \cup C_1, X_{b_2} \cup C_2, \ldots, X_{b_a} \cup C_a\}, \tag{3}$$

where $b_i \neq b_j$ for two distinct $1 \leq i, j \leq a$, X_{b_i} are variables drawn from $\mathcal{U}_w$, and $C_i \in (\mathcal{U}_{G_v} \setminus \mathcal{U}_{G_w})$.

Therefore, the above functions (constructed after executing $prune(u)$) possess the canonical form. Given those functions $R_{v,i}$'s, the contribution to the k target subgraphs of $par(v)$ is obtained by function composition $h_{v,i}(R_{v,1}, \ldots, R_{v,k})$ for all $1 \leq i \leq k$. These functions are constructed for w after executing $bypass(par(v))$. By Lemma 3, $h_{v,i}(R_{v,1}, \ldots, R_{v,k})$, $1 \leq i \leq k$, possesses the canonical form. Hence, we have the following lemma.

Lemma 4. *During the process of executing the binary tree contraction on a (k, Θ)-parse tree to remove some nodes, the crucial functions of the remaining nodes of the current tree can be constructed in $O(k^3)$ time using one processor.*

Theorem 1. *The (k, Θ)-parse tree problem can be solved in $O(k^3 \log n)$ time using $O(n/\log n)$ processors on an EREW PRAM, where n is the number of nodes of the input tree.*

Definition 6. Let G be a decomposable graph and let T_G be a parse tree. A problem $\mathcal{P}$ is said to be a (k, Θ)-*regular problem* on G if $\mathcal{P}$ can be reduced to a (k, Θ)-parse tree problem $\mathcal{B}$ on T_G such that the solution of $\mathcal{B}$ is exactly the solution of $\mathcal{P}$. Moreover, the reduction scheme takes $O(k^3 \log |V(T_G)|)$ time using $O(|V(T_G)|/\log |V(T_G)|)$ processors on an EREW PRAM.

Note that each (k, Θ)-regular problem corresponds to a (k, Θ)-parse tree. This tree is obtained from a parse tree T_G in which some additional data structures are associated with $V(T_G)$ (refer to Definition 3). In Section 4, we assume that a parse tree is given for solving a (k, Θ)-regular problem on a decomposable graph.

The following result directly follows from Definition 6 and Theorem 1.

Theorem 2. *Given a parse tree of a decomposable graph G, a (k, Θ)-regular problem on G can be solved in $O(k^3 \log |V(T_G)|)$ time using $O(|V(T_G)| / \log |V(T_G)|)$ processors on an EREW PRAM.*

Corollary 1. *A (k, Θ)-regular problem of a decomposable graph $G = (V, E)$ can be solved in $O(T_d(|V|, |E|) + \log |V(T_G)|)$ time using $O(P_d(|V|, |E|) + |V(T_G)| / \log |V(T_G)|)$ processors on M_d.*

4 (k, Θ)-Regular Problems

Given a problem $\mathcal{P}$, a graph G_1, a subgraph G_2 of G_1, and a subset Q of vertices in G_2, $\mathcal{P}_Q(G_1, G_2)$ is a solution to the input graph G_1 such that this solution contains all vertices in Q and is in G_2. For the case of $Q = \emptyset$, i.e., $\mathcal{P}_\emptyset(G_1, G_2)$, the notation represents a solution to G_1 and this solution is contained in G_2. For brevity, let $\mathcal{P}_Q(G, G) = \mathcal{P}_Q(G)$.

An *independent set* of a graph is a subset of its vertices such that no two vertices in the subset are adjacent. The *maximum independent set problem $\mathcal{I}$* is the problem of finding a maximum-cardinality independent set in the input graph. Using our notation, given an input graph G, a solution is $\mathcal{I}_\emptyset(G)$. For a basis rooted tree $G = (\{r\}, \{\}, (r))$, $\mathcal{I}_\emptyset(G)$ and $\mathcal{I}_{\{r\}}(G)$ are both equal to $\{r\}$, and $\mathcal{I}_\emptyset(G[V \setminus \{r\}]) = \emptyset$.

Lemma 5.
Assume $G = (V_1 \cup V_2, E_1 \cup E_2 \cup \{(r_1, r_2)\}, (r_1))$ is obtained from $G_1 = (V_1, E_1, (r_1))$ and $G_2 = (V_2, E_2, (r_2))$.

(1) $\mathcal{I}_\emptyset(G) = \mathrm{MAX}_v \{ \mathcal{I}_{\{r_1\}}(G_1) \cup \mathcal{I}_\emptyset(G_2[V_2 \setminus \{r_2\}]),$
    $\mathcal{I}_\emptyset(G_1[V_1 \setminus \{r_1\}]) \cup \mathcal{I}_{\{r_2\}}(G_2), \mathcal{I}_\emptyset(G_1[V_1 \setminus \{r_1\}]) \cup \mathcal{I}_\emptyset(G_2[V_2 \setminus \{r_2\}]) \};$
(2) $\mathcal{I}_{\{r\}}(G) = \mathcal{I}_{\{r_1\}}(G_1) \cup \mathcal{I}_\emptyset(G_2[V_2 \setminus \{r_2\}]);$
(3) $\mathcal{I}_\emptyset(G[V \setminus \{r\}]) = \mathcal{I}_\emptyset(G_1[V_1 \setminus \{r_1\}]) \cup \mathcal{I}_\emptyset(G_2).$

Proof. Straightforward.   □
By the above result, it is not difficult to obtain the following two theorems.

Theorem 3. *The maximum independent set problem is a $(3, \mathrm{MAX}_v)$-regular problem on trees.*

Theorem 4. *The maximum independent set problem on trees can be solved in $O(\log n)$ time using $O(n / \log n)$ processors on an EREW PRAM, where n is the number of vertices of the input graph.*

Given an undirected graph $G = (V, E)$, a *matching* is a subset of edges $M \subseteq E$ such that for all vertices $v \in V$, at most one edge of M is incident on v. The *maximum matching problem $\mathcal{M}$* is the problem of finding a matching of maximum cardinality. For a basis series-parallel graph $G = (\{l, r\}, \{(l, r)\}, (l, r))$, $\mathcal{M}_\emptyset(G) = \{(l, r)\}, \mathcal{M}_{\{l\}}(G[V \setminus \{r\}]) = \emptyset, \mathcal{M}_{\{r\}}(G[V \setminus \{l\}]) = \emptyset, \mathcal{M}_{\{l,r\}}(G) = \{(l, r)\}, \mathcal{M}_\emptyset(G[V \setminus \{l, r\}]) = \emptyset$. We can further show that the maximum matching problem is a $(5, \mathrm{MAX}_e)$-regular problem on series-parallel graphs.

By the methods described in [4,10] to construct parse trees of series-parallel graphs, we have the following theorem.

Theorem 5. *The maximum matching problem on series-parallel graphs can be solved in sequential $O(n+m)$ time, and in parallel in $O(\log m \log^* m)$ time using $O(m/\log m \log^* m)$ processors on an EREW PRAM.*

Given a simple graph $G = (V, E)$, a vertex $v \in V$ is said to *dominate* itself and all vertices adjacent to v. A subset D of V is called an *efficient dominating set* of G if every vertex in V is dominated by exactly one vertex in D. Note that not all graphs have efficient dominating sets. Moreover, if a graph possesses an efficient dominating set, then all these sets have the same cardinality. The *efficient domination problem* $\mathcal{D}$ is the problem to find an efficient dominating set of a given graph if such a set exists. Using our paradigm, we can also show the following result.

Theorem 6. *The efficient domination problem on series-parallel graphs can be solved in linear $O(n + m)$ time, and in parallel in $O(\log m \log^* m)$ time using $O(m/\log m \log^* m)$ processors on an EREW PRAM.*

References

1. K. Abrahamson, N. Dadoun, D. G. Kirkpatrick, and T. Przytycka, A simple parallel tree contraction algorithm, *Journal of Algorithms*, 10, pp. 287-302, 1989.
2. S. Arnborg and A. Proskurowski, Linear time algorithms for NP-hard problems restricted to partial k-trees, *Discrete Applied Mathematics*, 23, pp. 11-24, 1989.
3. M. W. Bern, E. L. Lawler, and A. L. Wong, Linear-time computation of optimal subgraphs of decomposable graphs, *Journal of Algorithms*, 8:216-235, 1987.
4. H. L. Bodlaender and B. van Antwerpen-de Fluiter, Parallel algorithms for series parallel graphs and graphs with treewidth two, *Algorithmica*, 29(4):534-559, 2001.
5. M. C. Golumbic, *Algorithmic graph theory and perfect graphs*, Academic press, New York, 1980.
6. R. M. Karp and V. Ramachandran, Parallel algorithms for shared memory machines, *Handbook of Theoretical Computer Science*, North-Holland, Amsterdan, pp. 869-941, 1990.
7. S. Mahajan and J. G. Peters, Algorithms for regular properties in recursive graphs, in: *Proceedings of the 25th Allerton Conference on Communication, Control, and Computing*, pp. 14-23, 1987.
8. S. Mahajan and J. G. Peters, Regularity and locality in k-terminal graphs, *Discrete Applied Mathematics*, 54:229-250, 1994.
9. K. Takamizawa, T. Nishizeki, and N. Saito, Linear-time computability of combinatorial problems on series-parallel graphs, *Journal of the ACM*, 29:623-641, 1982.
10. J. Valdes, R. E. Tarjan, and E. L. Lawler, The recognition of series-parallel digraphs, *SIAM Journal on Computing*, 11:298-313, 1982.
11. T. V. Wimer, Linear algorithms on k-terminal graphs, Ph.D. Thesis, Clemson University, Clemson, SC, 1987.
12. T. V. Wimer and S. T. Hedetniemi, K-terminal recursive families of graphs, *Congressus Numerantium*, 63:161-176, 1988.

The Computational Complexity of Orientation Search in Cryo-Electron Microscopy[*]

Taneli Mielikäinen[1], Janne Ravantti[2], and Esko Ukkonen[1]

[1] Department of Computer Science
[2] Institute of Biotechnology and Faculty of Biosciences
University of Helsinki, Finland
{tmielika,ravantti,ukkonen}@cs.Helsinki.FI

Abstract. In this paper we study the problem of determining three-dimensional orientations for noisy projections of randomly oriented identical particles. The problem is of central importance in the tomographic reconstruction of the density map of macromolecular complexes from electron microscope images and it has been studied intensively for more than 30 years.
We analyze the computational complexity of the problem and show that while several variants of the problem are NP-hard and inapproximable, some restrictions are polynomial-time approximable within a constant factor or even solvable in logarithmic space. The negative complexity results give a partial justification for the heuristic methods used in the orientation search, and the positive complexity results have some positive implications also to a different problem of finding functionally analogous genes.

1   Introduction

Structural biology studies how biological systems are built. Especially, determining three-dimensional electron density maps of macromolecular complexes, such as proteins or viruses, is one of the most important tasks in structural biology [1].

Standard techniques to obtain three-dimensional density maps of such particles (at atomic resolution) are by X-ray diffraction (crystallography) and nuclear magnetic resonance (NMR) studies. However, X-ray diffraction requires that the particles can form three-dimensional crystals and the applicability of NMR is limited to relatively small particles [2]. For example, there are many well-known viruses that do not seem to crystallize and are too large for NMR techniques.

A more flexible way to reconstruct density maps is offered by cryo-electron microscopy [1,3]. Currently the resolution of the cryo-electron microscopy reconstruction is not quite as high as resolutions obtainable by crystallography or NMR but it is improving steadily.

Reconstruction of density maps by cryo-electron microscopy consists of the following subtasks [1]:

---

[*] A work supported by the Academy of Finland.

M. Bubak et al. (Eds.): ICCS 2004, LNCS 3036, pp. 231–238, 2004.
© Springer-Verlag Berlin Heidelberg 2004

Specimen preparation. A thin layer of water containing a large number of identical particles of interest is rapidly plunged into liquid ethane to freeze the specimen very quickly. Quick cooling prevents water from forming regular structures. Moreover, the particles get frozen in random orientations in the iced specimen.

Electron microscopy. The electron microscope produces an image representing a two-dimensional projection of the mass distribution of the iced specimen. This image is called a *micrograph*. Unfortunately the electron beam of the microscope rapidly destroys the specimen so getting accurate images from it is not possible.

Particle picking. Individual projections of particles are extracted from the micrograph. The number of projections obtained may be thousands or even more.

Orientation search. The orientations (i.e., the projection directions for each extracted particle) for the projections are determined. There are a few heuristic approaches for finding the orientations.

Reconstruction. If the orientations for the projections are known then quite standard tomography techniques can be applied to construct the three-dimensional electron density map from the projections.

In this paper we study the computational complexity of the orientation search problem which is currently the major bottleneck in the reconstruction process. On one hand we show that several variants of the task are computationally very difficult. This justifies (to some extent) the heuristic approaches used in practice. On the other hand we give exact and approximate polynomial-time algorithms for some special cases of the task that are applicable e.g. to the seemingly different task of finding functionally analogous genes [4].

The rest of this paper is organized as follows. In Section 2 the orientation search problem is described. Section 3 analyzes the computational complexity of the orientation search problem. The paper is concluded in Section 4.

Due to the page limitations the proofs of theorems and further details appear in the full version [5].

2   The Orientation Search Problem

A *density map* is a mapping $D : \mathbb{R}^3 \to \mathbb{R}$ with a compact support. An *orientation* o is a rotation of the three-dimensional space and it can be described e.g. by a three-dimensional rotation matrix. A *projection* p of a three-dimensional density map D to orientation o is the integral

$$p(x, y) = \int_{-\infty}^{\infty} D\left(R_o\, [x, y, z]^T\right) dz$$

where R_o is a rotation matrix, i.e., the mass of D is projected on a plane passing through the origin and determined by the orientation o.

Based on the above definitions, the orientation search task is, given projections $p_1, \ldots, p_n$ of the same underlying but unknown density map D to find good orientations $o_1, \ldots, o_n$ for them. There are several heuristic definitions of what are the good orientations for the projections.

One possibility is to choose those orientations that determine a good density map although it might not be obvious what a good density map is nor how it should be constructed from oriented projections. A standard solution is to compare how well the given projections fit to the projections of the reconstructed density map. This kind of definition of good orientations suggests an Expectation Maximization-type procedure of repeatedly finding the best model for fixed orientations and the best orientations for a fixed model, see e.g. [6,7,8]. Due to the strong dependency on the reconstruction method, it is not easy to say analytically much (even whether it converges) about this approach in general. In practice, this approach to orientation search works successfully if there is an approximate density map of the particle available to be used as an initial model.

The orientations can be determined also by *common lines* [9]: Let p_i and p_j be projections of a density map D onto planes corresponding to orientations o_i and o_j, respectively. All one-dimensional projections of D onto a line passing through the origin in the plane corresponding to the orientation o_i (o_j) can be computed from the projection p_i (p_j); this collection of projections of p_i (p_j) is also called the *sinogram* of p_i (p_j). As the two planes intersect, there is a line for which the projections of p_i and p_j agree. This line (which actually is a vector since the one dimensional projections are oriented, too) is called the common line of p_i and p_j.

If the projections are noiseless then already the pairwise common lines of three projections determine the relative orientations of the projections in three-dimensional space uniquely (except for the handedness) provided that the possible symmetries of the particle are taken into account. Furthermore, this can be computed by only few arithmetic and trigonometric operations [10].

However, the projections produced by the electron microscope are extremely noisy and so it is highly unlikely that two projections have one-dimensional projections that are equal. In this case it would be natural to try to find the best possible approximate common lines, i.e., a pair of approximately equal rows from the sinograms of the two projections. Several heuristics for the problem have been proposed [3,10,11,12,13]. However, they usually assume that the density map under reconstruction is highly symmetric which radically improves the signal-to-noise ratio. In the next section we partially justify the use of heuristics by showing that many variants of the orientation search problem are computationally very difficult.

3 The Computational Complexity of Orientation Search Problem

In this section we show that finding good orientations using common lines is computationally very difficult in general but it has some efficiently solvable special cases. First, we consider the decision versions of the orientation search problem. Second, we study the approximability of several optimization variants.

We would like to point out that some of the results are partially similar to the results of Hallett and Lagergren [4] for their problem CORE-CLIQUE that models the problem of finding functionally analogous genes. However, our problem of finding good orientations based on common lines differs from the problem of finding functionally analogous genes, e.g., by its geometric nature and by its very different application domain. Furthermore, we provide relevant positive results for finding functionally analogous genes: we describe an approximation algorithm with guaranteed approximation ratio of $\beta(2 - o(1))$, if the distances between genes adhere to the triangle inequality within a factor β.

3.1 Decision Complexity

As mentioned in Section 2, the pairwise common lines cannot be detected reliably when the projections are very noisy. A natural relaxation is to allow several common line candidates for each pair of projections. In this section we study the problem of deciding whether there exist common lines in given sets of pairwise common lines that determine consistent orientations. We show that some formulations are NP-complete in general but there are nontrivial special cases that are solvable in nondeterministic logarithmic space. Due to the page limitations the proofs and further details appear in the full version [5].

The common lines-based orientation search problem can be modeled at a high level as the problem of finding an n-clique from an n, m-partite graph $G = (V_1, \ldots, V_n, E)$, i.e., a graph consisting independent sets $V_1, \ldots, V_n$ of size m.

Problem 1 (n-clique in an n, m-partite graph). Given an n, m-partite graph $G = (V_1, \ldots, V_n, E)$, decide whether there is an n-clique in G.

Problem 1 can be interpreted as the orientation search problem in the following way: each group V_i describes the possible orientations of the projection p_i and each edge connecting two oriented projections says that the projections in the corresponding orientations are consistent with each other.

On one hand already three different orientations for each projection can make the problem NP-complete:

Theorem 1. *Problem 1 is NP-complete if $m \geq 3$.*

On the other hand the problem can be solved in nondeterministic logarithmic space if the number of orientations for each projection is at most two:

Theorem 2. *Problem 1 is NL-complete if $m = 2$.*

The formulation of the orientation search problem as Problem 1 seems to miss some of the geometric nature of the problem. As a first step toward the final formulation, let us consider the problem of finding a constrained line arrangement, the constraint being that any two lines of the arrangement are allowed to intersect only at a given set of points, each such set being of size $\leq l$:

Problem 2 (l-constrained line arrangement). Given sets $P_{ij} \subset \mathbb{R}^2, |P_{ij}| \leq l, 1 \leq i < j \leq n$, decide whether there exist lines $L_1, \ldots, L_n$ in $\mathbb{R}^2$ such that L_i and L_j intersect only at some $p \in P_{ij}$ for all $1 \leq i < j \leq n$.

This problem has some interest of its own since line arrangements are one of the central concepts in computational and discrete geometry [14]. If we require that the lines are in general position, i.e., that they are not parallel nor they intersect in same points, then we get the following hardness result:

Theorem 3. *Problem 2 for lines L_i in general position is NP-complete if $l \geq 9$.*

The result can be slightly improved if we allow also parallel lines in the arrangement:

Theorem 4. *Problem 2 is NP-complete if $l \geq 6$.*

However, the orientation search is not about arranging lines on the plane but great circles on the (unit) sphere $S = \{(x, y, z) \in \mathbb{R}^3 : x^2 + y^2 + z^2 = 1\}$ as the orientations and the great circles are obviously in one-to-one correspondence. Thus, we should study the great circle arrangements:

Problem 3 (l-constrained great circle arrangement). Given sets $P_{ij} \subset S_+ = \{(x, y, z) \in S : z \geq 0\}, |P_{ij}| \leq l, 1 \leq i < j \leq n$, decide whether there exist great circles $C_1, \ldots, C_n$ on S such that C_i and C_j intersect on S_+ only at some $p \in P_{ij}$ for all $1 \leq i < j \leq n$.

It can be shown that the line arrangements and great circle arrangements are equivalent through the stereographic projection [14]:

Theorem 5. *Problem 3 is as difficult as Problem 2.*

Still, our problem formulation is lacking some of the important ingredients of the orientation search problem: it is not possible to express at this stage the common line candidates by giving the allowed pairwise intersection points on the sphere S. Rather, one can represent a common line only in the internal coordinates of the two great circles that correspond to the two projections intersecting. Each coordinate is in fact an angle giving the rotation angle of the common line on the projection. Hence the representation is a pair of angles:

Problem 4 (locally l-constrained great circle arrangement on sphere). Given sets $P_{ij} \subset [0, 2\pi) \times [0, 2\pi), |P_{ij}| \leq l, 1 \leq i < j \leq n$, decide whether there exist great circles $C_1, \ldots, C_n$ on S such that C_i and C_j intersect only at some $p \in P_{ij}$ for all $1 \leq i < j \leq n$, where p defines the angles of the common line on C_i and C_j.

Also this problem can be shown to be equally difficult to decide:

Theorem 6. *Problem 4 is NP-complete.*

Thus, deciding whether there exist consistent orientations seems to be difficult in general.

3.2   Approximability

As finding a consistent orientation for the projections is by the results of Section 3.1 difficult, we should consider also orientations that may cover only a large subset of the projections or resort to common lines that are as good as possible.

A simple approach to allow errors in solutions is look for large cliques in the n, m-partite graph $G = (V_1, \ldots, V_n, E)$ instead of exact n-cliques. In the world of orientations this means that instead of finding consistent orientations for all projections we look for consistent orientations for as many projections as we are able to and neglect the other projections.

Containing a clique is just one example of a property a graph can have. Also other graph properties might be useful. Thus, we can formulate the problem in a rather general form as follows:

Problem 5 (Maximum subgraph with property P in an n, m-partite graph). Given an n, m-partite graph $G = (V_1, \ldots, V_n, E)$, find the largest $V' \subset V_1 \cup \ldots \cup V_n$ such that the induced subgraph satisfies the property P and $|V' \cap V_i| \leq 1$ for all $1 \leq i \leq n$.

This resembles the following fundamental graph problem in combinatorial optimization and approximation algorithms:

Problem 6 (Maximum subgraph with property P [15]). Given a graph $G = (V, E)$, find the largest $V' \subseteq V$ such that the induced subgraph satisfies the property P.

It is not very difficult to see that the two problems are equivalent:

Theorem 7. *Problem 5 is as difficult as Problem 6.*

Thus, Problem 5 is very difficult w.r.t. several properties. For example, due to Theorem 7, finding the maximum clique from the n, m-partite graph cannot be approximated within ratio $n^{1-\epsilon}$ for any fixed $\epsilon > 0$ [16]. Note that the approximation ratio n can be achieved trivially by choosing any of the vertices in G which is always a clique of size 1.

In practice the techniques for finding common lines or common line candidates actually produce distances between all possible intersections of two projections. Thus, we could assume that there is always at least one feasible solution and study the following problem:

Problem 7 (Minimum weight n-clique in a complete n, m-partite graph). Given a complete n, m-partite graph $G = (V_1, \ldots, V_n, E)$ and a weight function $w : E \to \mathbb{N}$, find $V' \subset V_1 \cup \ldots V_n$ such that the weight $\sum_{u,v \in V', u \neq v} w(\{u, v\})$ is minimized and $|V' \cap V_i| \leq 1$ for all $1 \leq i \leq n$.

Unfortunately, it turns out that in this case the situation is extremely bad:

Theorem 8. *Problem 7 with $m \geq 3$ is not polynomial-time approximable within 2^{n^k} for any fixed $k > 0$ if $P \neq NP$.*

When there are only two vertices in each group the problem admits a constant factor approximation ratio but no better:

Theorem 9. *Problem 7 is APX-complete if $m = 2$.*

An easier variant of Problem 7 is the case where the edge weights admit triangle inequality within a factor β, i.e., for all edges $\{t, u\}$, $\{t, v\}$ and $\{u, v\}$ it holds

$$w(\{t, u\}) \leq \beta(w(\{t, v\}) + w(\{u, v\})).$$

A good approximation of the lightest clique can be found by finding the minimum weight star that contains one vertex from each group V_i. (The algorithm is described in the full version of this paper [5].) This gives constant-factor approximation guarantees and the approximation is stable (for details on approximation stability, see [17]):

Theorem 10. *Problem 7 is polynomial-time approximable within $\beta (2 - o(1))$ if the edge weights satisfy triangle inequality within factor β.*

This algorithm might not be applicable in orientation search as there seems to be little hope of finding distance functions (used in selecting the best common lines) satisfying even the relaxed triangle inequality for the noisy projections. However, in the case of finding functionally analogous genes this is possible since many distance functions between sequences are metric. Thus, the algorithm seems to be very promising for that task.

Another very natural relaxation of the original problem is to allow small changes to common line candidates to make the orientations consistent:

Problem 8 (Minimum error l-constrained line arrangement). Given sets $P_{ij} \subset \mathbb{R}^2, |P_{ij}| \leq l, 1 \leq i < j \leq n$, find lines $L_1, \ldots, L_n$ in $\mathbb{R}^2$ that minimize the sum of distances $\min_{p_{ij} \in P_{ij}} |p_{ij} - \hat{p}_{ij}|^q$ where $\hat{p}_{ij}$ is the actual intersection point of lines L_i and L_j and $q > 0$.

Theorem 11. *Problem 8 with $l \geq 6$ is not polynomial-time approximable within 2^{n^k} for any fixed $k > 0$ if $P \neq NP$.*

4   Conclusions

In this paper we have shown that some approaches for determining orientations for noisy projections of identical particles are computationally very difficult, namely NP-complete and inapproximable. These results justify (to some extent) the heuristic approaches widely used in practice.

On the bright side, we have been able to detect some polynomial-time solvable special cases. Also, we have described an approximation algorithm that achieves the approximation ratio $\beta (2 - o(1))$ if the instance admits the triangle inequality within a factor β. It has promising applications in search for functionally analogous genes. As a future work we wish to study the usability of current state of art in heuristic search to find reasonable orientations in practice. This is very challenging due to the enormous size of the search space. Another goal is to analyze the complexity of other approaches for determining the orientations for the projections.

References

1. J. Frank, Three-Dimensional Electron Microscopy of Macromolecular Assemblies, Academic Press, 1996.

2. J. M. Carazo, C. O. Sorzano, E. Rietzel, R. Schröder, R. Marabini, Discrete tomography in electron microscopy, in: G. T. Herman, A. Kuba (Eds.), Discrete Tomography: Foundations, Algorithms, and Applications, Applied and Numerical Harmonic Analysis, Birkhäuser, 1999, Ch. 18, pp. 405–416.
3. R. Crowther, D. DeRosier, A. Klug, The reconstruction of a three-dimensional structure from projections and its application to electron microscopy, Proceedings of the Royal Society of London A 317 (1970) 319–340.
4. M. T. Hallett, J. Lagergren, Hunting for functionally analogous genes, in: S. Kapoor, S. Prasad (Eds.), Foundations of Software Technology and Theoretical Computer Science, Vol. 1974 of Lecture Notes in Computer Science, Springer-Verlag, 2000, pp. 465–476.
5. T. Mielikäinen, J. Ravantti, E. Ukkonen, The computational complexity of orientation search problems in cryo-electron microscopy, Report C-2004-3, Department of Computer Science, University of Helsinki (2004).
6. P. C. Doerschuk, J. E. Johnson, *Ab initio* reconstruction and experimental design for cryo electron microscopy, IEEE Transactions on Information Theory 46 (5) (2000) 1714–1729.
7. Y. Ji, D. C. Marinescu, W. Chang, T. S. Baker, Orientation refinement of virus structures with unknown symmetry, in: Proceedings of the International Parallel and Distributed Processing Symposium, IEEE Computer Society, 2003, pp. 49–56.
8. C. J. Lanczycki, C. A. Johnson, B. L. Trus, J. F. Conway, A. C. Steven, R. L. Martino, Parallel computing strategies for determining viral capsid structure by cryo-electron microscopy, IEEE Computational Science & Engineering 5 (1998) 76–91.
9. T. S. Baker, N. H. Olson, S. D. Fuller, Adding the third dimension to virus life cycles: Three-dimensional reconstruction of icosahedral, Microbiology and Molecular Biology Reviews 63 (4) (1999) 862–922.
10. M. van Heel, Angular reconstitution: a posteriori assignment of projection directions for 3D reconstruction, Ultramicroscopy 21 (1987) 11–124.
11. P. L. Bellon, F. Cantele, S. Lanzavecchia, Correspondence analysis of sinogram lines. Sinogram trajectories in factor space replace raw images in the orientation of projections of macromolecular assemblies, Ultramicroscopy 87 (2001) 187–197.
12. P. A. Penczek, J. Zhu, J. Frank, A common-lines based method for determining orientations for $N > 3$ particle projections simultaneously, Ultramicroscopy 63 (1996) 205–218.
13. P. A. Thuman-Commike, W. Chiu, Improved common line-based icosahedral particle image orientation estimation algorithms, Ultramicroscopy 68 (1997) 231–255.
14. H. Edelsbrunner, Algorithms in Combinatorial Geometry, Vol. 10 of EATCS Monographs on Theoretical Computer Science, Springer-Verlag, 1987.
15. G. Ausiello, P. Crescenzi, V. Kann, A. Marchetti-Spaccamela, M. Protasi, Complexity and Approximation: Combinatorial Optimization Problems and Their Approximability Properties, Springer-Verlag, 1999.
16. J. Håstad, Clique is hard to approximate within $n^{1-\epsilon}$, Acta Mathematica 182 (1999) 105–142.
17. H.-J. Böckenhauer, J. Hromkovič, R. Klasing, S. Seibert, W. Unger, Towards the notion of stability of approximation for hard optimization tasks and the traveling salesman problem, Theoretical Computer Science 185 (1) (2002) 3–24.

Advanced High Performance Algorithms
for Data Processing

Alexander V. Bogdanov and Alexander V. Boukhanovsky

Institute for High Performance Computing and Information Systems,
St. Petersburg, Russia
{bogdanov, avb}@csa.ru, http://www.csa.ru

Abstract. We analyze the problem of processing of very large datasets on parallel systems and find that the natural approaches to parallelization fail for two reasons. One is connected to long-range correlations between data and the other comes from nonscalar nature of the data. To overcome those difficulties the new paradigm of the data processing is proposed, based on a statistical simulation of the datasets, which in its turn for different types of data is realized on three approaches - decomposition of the statistical ensemble, decomposition on the base of principle of mixing and decomposition over the indexing variable. Some examples of proposed approach show its very effective scaling.

1 Introduction

The amount of data, generated by scientific and technological activity of humanity is increasing by the order of magnitude every couple of years. The new features, which become evident in the last years are the nonhomogeneous types of the data and the need of determination of the detailed characteristics of the process, including higher order moments of it. Presence of long-range correlations and nonscalar nature of data make it very difficult to use large parallel computer systems for their processing. Even when it becomes possible, the effect of parallelization can be very small, because of the bad load balancing. To build the effective algorithms several steps should be taken, that make it possible to decrease the dimension of the problem, to find the variable, that make it possible to make uniform indexing of data through all set and finally to simulate the initial process by the relevant procedures, that can be mapped effectively onto the large computer system.

Due to the complexity of the natural and technological phenomena it is possible to formulate several approaches within proposed paradigm for solution of pertinent problem. That is why we give also the algorithms for determination of the optimal approach for given multiprocessor system. The same technology can be used for control of the load balancing.

M. Bubak et al. (Eds.): ICCS 2004, LNCS 3036, pp. 239–246, 2004.

2   Challenges in Parallel Data Processing

The problem of the processing of very large datasets can be illuminated for the simplest probabilistic models, e.g. – model of random value (RV) or multivariate random value (MRV). These models allow the use of the classical statistical approaches for the parallel processing of independent data flows on the transputers or multicomputer farms. The development of the parallel algorithms for processing of a more complicated data models, e.g. sets of time series (TS) $\zeta(t)$ or spatiotemporal random fields $\Xi(\mathbf{r},t)$ (STRF), is not so easy, as seems.

The main source of difficulties of the parallelization of the statistical algorithms is the correlations in the multivariate data. It is the result of the multiscale variability, nonstationarity and inhomogeneity for both the natural and technical complex systems. The two types of such correlations could be considered [4]:

– Spatiotemporal dependence of the data in different points $\mathbf{r}$ at the time t, as result of the non-local effects (non-stationary behavior and spatial inhomogeneity).

– Intra-element dependence, as the result of the axiomatic representation of the multivariate random data (as the system of scalar values, Euclidean of affine vectors, functions etc.).

Multivariate statistical analysis (MSA) is the traditional tool for the development of the statistical models of the correlated data (e.g. [2]). Its goals are the reduction of the dimensionality, determination of correlations and description of inhomogeneity of the multivariate statistical sample. Nowadays classical MSA is proposed for model of MRV, but for TS and STRF the principal approaches were not generated yet, in spite of some specific approaches for certain classes of the data [9,19]. Hence, the more general approach on the base of the functional analysis is needed for formalization of both types of the dependence.

Moreover, the generalization of MSA procedures for more general models is associated with the complicate statistical inference tools. It crucially restricts the possibility to obtain the simple and transparent analytical expressions even for the simplest statistical estimates. Hence, the computational tools of the statistics on the base of Monte-Carlo simulation must be widely used [24].

Thus, the principal problem of the development of high-performance statistical algorithms is not in extensive code optimization only. The development of the adequate parallel models for statistical description of the multivariate data is of the prime importance. This approach must take into account both the spatiotemporal and intra-element variability of the data. Only such approach allows to achieve the direct *intrinsic* mapping to architecture of the parallel computer system.

3   Regenerative Paradigm of Multivariate Statistics

The problem of the development of the parallel statistical models could be solved in the frame of regenerative paradigm of the computational statistics [23]. It means, that the result of any data processing could be considered as the imitation model (algorithm for Monte-Carlo simulation) for the initial dataset. It allows simulate the large-size ensemble of the data realizations for the numerical studies of different features of the data, especially – non-observable events etc. [5].

This paradigm leads to the promising new possibilities for the development of the parallel algorithms for both the statistical analysis and synthesis. It allows constructing the *intrinsic* parallel models for the dependent data, when the parallelization of the classical statistical procedures is impossible. Thus, the problem of parallel decomposition may be solved on the level of the imitative model.

The development of the parallel statistical models for TS and STRF variability is possible to represent as the next four stages.

Reduction of the dimensionality for the initial data $\Xi(\mathbf{r},t) \in H$ in the linear space H. The goal of this stage is the construction of the set of most informative indexes, characterizing the sample variability. This gives the system of the linear operators $I_k : H \to X$, where $\dim(X) \le \dim(H)$, and allows to project the initial data set on a subspace. The sequential (in order p) application of the hierarchy of the operators

$$I_k^{(p)} : H_p \times H_{p+1} \to R \times H_{p+1}, \ p = 1,2,\dots, \tag{1}$$

allows not only to "fold" the multivariate data space to R, but also simplify the probabilistic data model. E.g., in accordance with Eqn. (1) the representation of non-scalar STFR reduced to analysis of the set of TS, and further – to MRV. This principal step makes it possible the complete use of the traditional techniques MSA MRV.

Identification of the model. The Eqn. (1) allows to express the dependence between non-scalar components of $\Xi(\mathbf{r},t) \in H$ in terms of the system of scalar indexes $Z = \{z_k(t)\}$. These indexes may be treated as the MRV or system of TS. For the quantitative description of the temporal (t) and intra-element (k) dependencies of these data the model of linear stochastic dynamic system has been considered [1]:

$$LZ = RE + BH. \tag{2}$$

Here L, R, B – are the linear differential operators, E – is the multivariate white noise (independent realizations of random value), and H is the set of driving stochastic factors (predictors). The objects $Z(t), E(t), H(t)$ are multivariate, and possible $\dim(\Xi) \ne \dim(H)$. The Eqn. (2) is the generalization for different regression models for TS, e.g. ARMA [13], dynamic [22] and spectral [12] regressions. This way also allows intrinsically extend the qualitative correlation theory of the RV on the TS and STRF, because Eqn. (2) is justified the terms of *functions* of partial, multiple and canonical correlations. These characteristics reflect the non-local dependencies of TS on the whole time interval. The results of the qualitative analysis are used for the identification of the set of model parameters ϑ (coefficients of the L, R, B) on the initial data sample.

Statistical synthesis. The Eqn. (2) may be treated as the algorithm for Monte-Carlo simulation of multivariate TS $Z = \{z_k(t)\}$, using the advanced numerical techniques [20]. Hence it is considered as the milestone for the construction of the hierarchy of the stochastic operators $J^{(p)}$ of Monte-Carlo procedure opposite to Eqn. (1). It allows synthesizing the large-size ensemble $\Xi(\vec{r},t) \in H$ on the base of the estimated parameters ϑ.

Verification, scenarios and forecast. The procedure of verification (error analysis) is proposed as the technique for qualitative control of the statistical model. It allows to establish the degree of the model adequacy to initial data. The verification based on the statistical comparison of the simulated and sample characteristics has not been used for the identification of the parameters ϑ. The elements of the verified simulated ensemble may be treated as the statistical scenarios of the non-observable events, in respect to the probability of its occurrence [5]. On the base of statistical scenarios the different statistical problems for the inferences, control, monitoring and statistical forecast may be solved.

In view of parallel processing, the principal feature of $I^{(p)}, J^{(p)}$ construction is the intrinsic formalization of the parallel algorithm, using the possibility of the elimination of the correlations between data in computational procedure.

4 Principles of Intrinsic Parallelization

Generally, there is no unique way to parallel formalization of all the types of statistical models, due to complexity of the mathematical tools. But the mapping of the statistical algorithms on the parallel architecture may be based on the three principles [7]. These principles allow classifying the methods of statistical processing and Monte-Carlo simulation by means of the natural way of intrinsic parallelization.

Decomposition of the statistical ensemble. This principle reflects the postulate of the independence of sample elements. It allows dividing the sample on the independent fragments and process these data in parallel. The resulting computational algorithm is rather homogenic. Hence, the most statistical procedures for both RV and MRV models may be, in principle, adopted for parallel architecture. The main problem of the ensemble decomposition is the further integration of the estimates, obtained on the different processors. If each parallel estimate is treated as the realization of RV, the theory of small sample may be adopted for the formalization of the results of parallel processing.

Decomposition on the base of principle of mixing. This is the modification of statistical ensemble decomposition for the model of TS with local dependence between data. The idea of mixing principle [15] is the possibility to consider the values $\Xi(t)$ and $\Xi(s)$ independently, when $|t-s| \gg 1$. Hence, the realization of TS may be divided on the set of uncrossed fragments. Each fragment simulated by Eqn. (2) in parallel. After that, the matching of the parallel fragments may be organized as binary tree algorithm, when Eqn. (2) is considered as the boundary problem, where boundary conditions are the values, obtained on the previous step.

Decomposition of the indexing variable. This principle corresponds to alternative way for dependence elimination in stochastic model. It is important for the multivariate data, e.g. – for model of inhomogeneous STRF $\Xi(\mathbf{r},t)$, where the mixing principle is out of consideration. The general approach is based on the specific construction of the operators $I^{(p)}, J^{(p)}$ of data transformation, thus the values of the transformed data for different values of index variable (e.g. - $\mathbf{r}$) could be computed independently. If the operators in Eqn. (1) may be expressed by means of orthogonal expansions technique [3,16]:

$$\Xi(\mathbf{r},t) = \sum_k z_k(t)\phi_k(\mathbf{r},t)\,, \qquad I_k \overset{def}{\equiv} z_k(t) = (\Xi,\phi_k)_{\mathbf{r}}\,, \qquad (3)$$

the values of $\Xi(\mathbf{r},t)$ for different $\mathbf{r}$ may be expressed in parallel. Hence, the spatial domain $\mathbf{r} \in \mathfrak{R}$ allows the intrinsic decomposition on the fragments has been processed in parallel.

Let us note, that the development of rather complicated models of computational multivariate statistics for the data with both the intra-element, spatial and temporal (spatiotemporal) dependence, not allows using only one principle of the parallel decomposition. Usually some combinations are used for multivariate data in respect to features of the each type of variability.

5 Performance Analysis and Load Balancing Optimization

Non uniqueness of the principles of intrinsic parallel decomposition, agglomeration and communications require the use of the specific techniques [11] for the preliminary quantitative analysis of the parallel performance for statistical algorithms. The main object of the analysis is the parallel speed-up S. Let us consider, that the initial dataset is characterized by means of set of parameters χ (sample size, dimension of the data, number of model coefficients etc.), and the architecture of the parallel computer system is characterized by parameters (t_s, t_w, t_c) (latency, communication time, and computation time respectively) [10]. It allows developing of the analytical model for comparison of the two (or more) statistical algorithms for the parallel processing. The isoefficiency surface in terms of S_p for two algorithms ("A" and "B") may be expressed in implicit form:

$$S_A(t_s, t_c, t_w, p, \chi) = S_B(t_s, t_c, t_w, p, \chi)\,. \qquad (4)$$

When (t_s, t_w, t_c) is fixed (for concrete architecture of computer system), the Eqn. (4) describe the surface in space (p, χ). This surface may be treated as the barrier for the success domains for each algorithm.

The Eqn. (4) allows formulating of the so call "concurrence principle", as the criterion of the selection of most effective algorithm. The realization of this principle is the intellectual technology of mapping that allows to take into account both features of the initial data and specific features of the parallel architecture. Hence, it provides the possible scalability of the algorithm, and the code may be effectively ported for different parallel systems.

The use of the theoretical performance model (Eqn. (4)) is possible only for computational systems with rather simple topology of the network, and low number of processors (not more 256). Generally, these models do not take into account the technical features of MPP-systems, e.g. the possibility of communications via more, than one router. The different ways for model improvement are rather specific [21]. Hence, the problem of practical scalability for large number of processors requires the special techniques of parallel load balancing, instead of analytical models (4). For the development of algorithms for scheduling and load balancing the same principles of decomposition, as for parallel statistical models, may be used.

6 Applications

Statistical analysis and simulation of different natural and technical complex events in frame of regenerative approach requires the development of the different models associated with specific parallel representation. Let us consider two crucial computational problems.

6.1 Estimation of the Extreme Waves in the Storm once T-years

The stochastic model of multiscale (synoptic, annual, year-to-year) spatiotemporal variability of sea wave fields is considered in [6]. Using of this model for estimation of T-years waves require a lot of computational resources. E.g., even for Barents Sea on the gridpoint ($0.5^0 \times 1.5^0$) for statistical estimation of 100-years waves the computations of 10^{11} values are required. The principle of the decomposition on indexing variable is the best (in accordance with Eqn. (4)) for these data. In the fig. 1(a,b) the results of the computation of 10- and 100-years significant wave height extremes in different points of the sea are shown. Let us note, that the estimates of field extremes (e.g. – simultaneous values in the few points) may be obtained by means of regenerative approach only, because all other methods are extensively oriented on rare events in a fixed point [17].

6.2 Simulation of Multivariate ECG Signal Variability

Another application of the regenerative approach is the stochastic simulation of ECG signal for criterion formulation of the pathological cardio-dynamics [8]. The most known models of ECG have been developed for unichannel signal for the purposes of the classification and discrimination [14]. But arising of the new multichannel cardiomonitors with extremely high resolutions require using parallel computations for real-time simulations. The most crucial elements of the model are the estimation of the basic functions $\{\phi_k\}$ in Eqn. (3) on the sample data, and the direct Monte-Carlo simulation. In the fig. 1(c) the absolute computational time for different stages of modeling are shown for the different number of the processors. In the fig. 1(d) the result of dynamic balancing of the computational algorithm (in term of the parallel efficiency E_p) are shown in dependence of value m - number of sub-samples that geometrically distributed on all the processors.

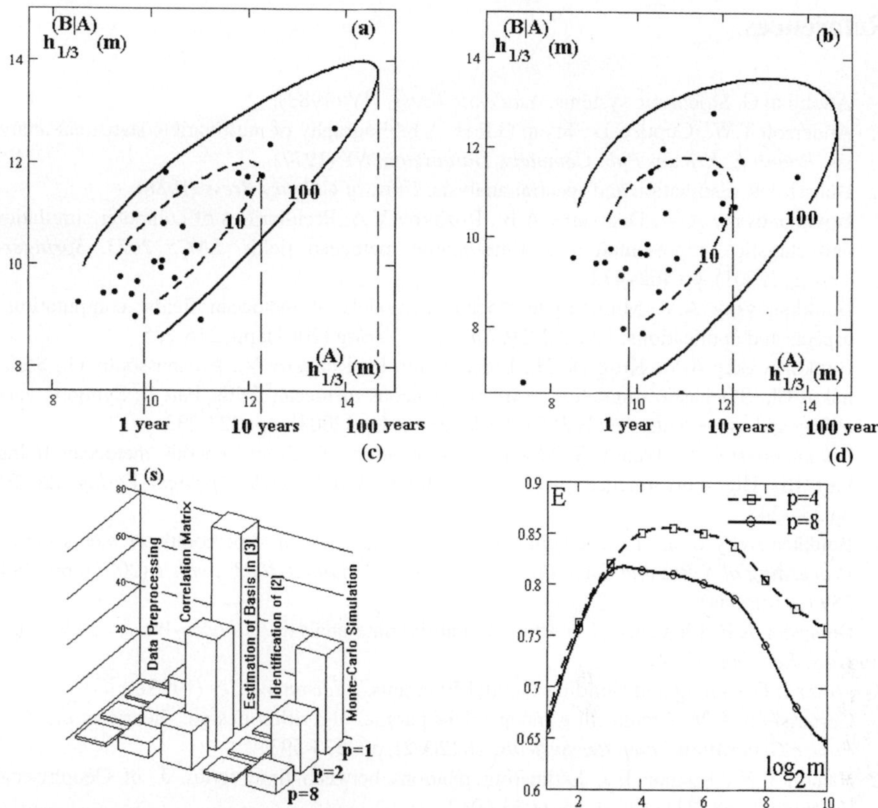

Fig. 1. Result of the parallel stochastic models applications; (a,b) – estimation of combinations of significant wave heights once 10- and 100-years in the points A and B in a Barents sea: (74N,30E-74N,35E) (a) and (74N,30E-74N,40E) (b). Points are the annual extremes on hydrodynamic modeling [18]; (c,d) – performance of the stochastic model of multilead ECG: Total computational time of model stages (c), and parallel efficiency for different regimes of the dynamic load balancing (d).

7 Conclusions

The new paradigm of the data processing is proposed, based on a statistical simulation of the datasets, which in its turn for different types of data is realized on three approaches – decomposition of the statistical ensemble, decomposition on the base of principle of mixing and decomposition over the indexing variable. In practical realization the combination of only one or two of those approaches can be used, that open effective possibilities for processing of very large data on multiprocessors computer system. The concurrence principle is proposed for choosing the most effective algorithm and to make load balancing more easily. Some examples of proposed approach show, that even for most complex natural processes it is possible to find the approach for determination most sensitive characteristics of the phenomenon.

References

1. Adomian G. Stochastic systems. *Academic Press*, NY (1983).
2. Anderson T.W., Gupta S.D., Styan G.P.H. A bibliography of multivariate statistical analysis. *Robert E. Krieder Pub. Company, Hantington, NY* (1977)
3. Blais J.A.R. Estimation and spectral analysis. *Univ. of Calgary Press* (1988)
4. Boukhanovsky A.V., Degtyarev A.B., Rozhkov V.A. Peculiarities of computer simulation and statistical representation of time–spatial metocean fields. *LNCS 2073, Springer–Verlag,* (2001), pp.463–472.
5. Boukhanovsky A.V. Multivariate stochastic models of metocean fields: computational aspects and applications. *LNCS 2329, Springer–Verlag* (2002), pp. 216-225
6. Boukhanovsky A.V., Krogstad H., Lopatoukhin L., Rozhkov V., Athanassoulis G., Stefanakos Ch. Stochastic simulation of inhomogeneous metocean fields. Part II: Synoptic variability and rare events. *LNCS 2658, Springer-Verlag* (2003), pp. 223-233.
7. Boukhanovsky A., Ivanov S. Stochastic simulation of inhomogeneous metocean fields. Part III: High-performance parallel algorithms. *LNCS 2658, Springer-Verlag* (2003), pp.234-243.
8. Boukhanovsky et al. Telemedicine complex on the base of supercomputer technologies. *Proceeding of X Russian Scientific Conference "Telematica-2003", vol. 1* (2003), pp. 288-289 (in Russian).
9. Dempster A.P. Elements of continuous multivariate analysis. *Addison-Wesley Pub. Company, Reading* (1969)
10. Foster J. Designing and Building Parallel Programs. *Addison-Wesley* (1995).
11. Gerbessiotis A.V. Architecture independent parallel algorithm design: theory vs practice. *Future Generation Computer Systems*, 18 (2002), pp. 573-593.
12. Hamon B.V., Hannan E.J. Estimating relations between time series. J. of Geophysical Research, v. 68 (21) (1963), pp. 6033-6042.
13. Jenkins G.M., Watts D.G. Spectral analysis and its application. *Holden-Day, San-Francisco* (1969).
14. Koski A. Modelling ECG signals with hidden Markov models. *Artificial intelligence in medicine (8)* (1996), pp. 453-471.
15. Leadbetter M., Lindgren G., Rootzen H. Extremes and related properties of random sequences and processes. *Springer-Verlag, NY,* (1986).
16. Loeve M. Fonctions aleatories de second odre. *C.R. Acad. Sci. 220,* (1945).
17. Lopatoukhin L.J., Rozhkov V.A., Ryabinin V.E., Swail V.R., Boukhanovsky A.V., Degtyarev A.B. Estimation of extreme wave heights. *JCOMM Technical Report, WMO/TD #1041* (2000).
18. Lopatoukhin L.J. et al. The spectral wave climate in the Barents sea. *Proceedings of Int. Conf OMAE'02, Oslo, Norway, June 23-28* (2002) (CD-version).
19. Lutkepohl H. Introduction to multivariate time series analysis. *Springer-Verlag* (1991)
20. Ogorodnikov V.A., Prigarin S.M. Numerical modelling of random processes and fields: algorithms and applications. *VSP, Utrecht, the Netherlands* (1996).
21. Pandle S., Agrawal D.P. (Eds.) Compiler Optimization for Scalable PC, *LNCS 1808, Springer-Verlag* (2001).
22. Pesaran M.H., Slater L.J. Dynamic regression: theory and algorithms. *Ellis Horwood Limited, NY* (1980).
23. Rubinstein R.Y. Simulation and the Monte-Carlo method. *John Wiley & Sons* (1981)
24. Yakowitz S.J. Computational Probability and Simulation. *Addison-Wesley* (1977)

Ontology-Based Partitioning of Data Steam for Web Mining: A Case Study of Web Logs

Jason J. Jung

School of Computer and Information Engineering, Inha University,
253 Yonghyun-dong, Incheon, Korea 402-751
`j2jung@intelligent.pe.kr`

Abstract. This paper presents a nevel method partitioning steaming data based on ontology. Web directory service is applied to enrich semantics to web logs, as categorizing them to all possible hierarchical paths. In order to detect the candidate set of session identifiers, semantic factors like semantic mean, deviation, and distance matrix are established. Eventually, each semantic session is obtained based on nested repetition of top-down partitioning and evaluation process. For experiment, we applied this ontology-oriented heuristics to sessionize the access log files for one week from IRCache. Compared with time-oriented heuristics, more than 48% of sessions were additionally detected by semantic outlier analysis.

1 Introduction

As the concern for searching relevant information from the web has been exponentially increasing, the very large amount of log data have been generated in web servers. Thus, many applications have been focusing on various ways to analyze them in order to recognize the usage patterns of users and discover other meaningful patterns [1], [5]. Among the whole steps of web user profiling mentioned in [2], we have taken the session identification for segmenting web log data in consideration. For partitioning each user activity into sequences of entries corresponding to each user visit, mainly two kinds of sessionization heuristics, which are time-oriented heuristics [3] and navigation-oriented heuristics [4] have been introduced. However, knowledge extractable from sessions identified by those heuristics is limited like frequent and sequential patterns represented by URLs. It means that web logs has to be sessionized with semantic enrichment based on ontology in order to find out more potential and meaningful information like a user's preference and intention. More importantly, web caching(or proxy) servers have to track streaming URL requests from multiple clients, because they have to increase predictability for prefetching web content that is expected in next request. Enriching web logs with their corresponding semantic information has been attempted in some studies [6], [10] such as mapping URLs to set of concepts as a feature vector and a specific value, respectively. We present conceptualizing an URL information itself by using web directory and introduce representing conceptualized URLs as tree-like information.

M. Bubak et al. (Eds.): ICCS 2004, LNCS 3036, pp. 247–254, 2004.

2   Data Model of Web Log and Problem Statement

Several standard data models of web logs, generally, have some problems to analyze these web logs such as their anonymity, rotating IP addresses connections through dynamic assignment of ISPs, missing references due to caching, and inability of servers to distinguish among different visits. Therefore, we note the problem statements concentrated for semantic sessionization in this paper, as follows.

- **Weakness of IP address field as session identifier.** The same IP address field in a web logs (within the time window or not) can not guarantee that those requests are caused by only one user, and reversely, requests from the different IPs can be generated by a particular user.
- **Simultaneous user requests based on multiple intention.** It means we have to consider multiple intention of users by classifying mixed logs according to the corresponding semantics.

Each request consists of timestamp, IP address, and URL fields. URL field is divided into base url and reminder, which are the host name of web server and the rest part of full URL, respectively. Then, we assume that each URL is semantically characterized by its base URL. For example, we are given a web log composed of eight requests ordered by timestamps from t_1 to t_8. We denote the URL set of sequential requests by $< ..., b\_url_i+r_j, ... >$ mapped to the timestamps $< ..., t_i, ... >$. These logs are partitioned with respect to an IP address ip_i. After partitioning, we compare semantic distance between base URLs in a set of requests, because we regard a semantic session as the sequence of URL having similar semantics. In other words, we investigate if an user's intention is retained or not.

3   Ontology-Oriented Heuristics for Sessionization

An ontology, a so-called semantic categorizer, is an explicit specification of a conceptualization. It means that ontologies can play a role of enriching semantic or structural information to unlabeled data. Web directories like Yahoo and Cora can be used to describe the content of a document in a standard and universal way as ontology [7]. Besides, web directory is organized as a topic hierarchical structure which is an efficient way to organize, view, and explore large quantities of information that would, otherwise, be cumbersome [9].

In this paper we assume that all URLs can be categorized by a well-organized web directory service. There are, however, some practical obstacles to do that, because most of web directories are forced to manage a non-generic tree structure in order to avoid a waste of memory space caused by redundant information [8]. We briefly note that problems with categorizing an URL with web directory as an ontology are the following:

- **The multi-attributes of an URL.** An URL can be involved in more than a category. The causal relationships between categories makes their hierarchical structure more complicated. As shown in Fig. 1 (1), an URL can be included in some other categories, named as A or B.

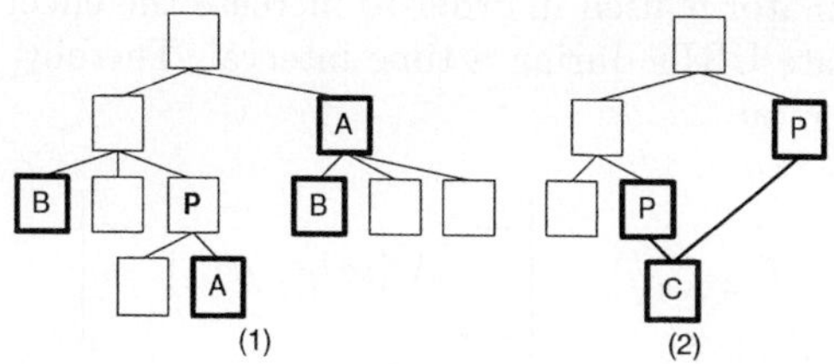

Fig. 1. (1) The multi-attribute of URLs; (2) The subordinate relationship between two categories

- **The relationship between categories.** A category can have more than a path from root node. As shown in Fig. 1 (2), the category C can be a subcategory of more than one like P. Furthermore, some categories can be semantically identical, even if they have different labels.
  - Redundancy between semantically identical categories
  - Subordination between semantically dependent categories

In order to simply handle these problems, we categorize each URL to all possible categories causally related with itself. Therefore, an URL url_i is categorized to a category set $Category(url_i)$, and the size of this category set depend on the web directory. Each element of a category set is represented as a path from the root to the corresponding category on web directory. Let the base URLs {b_url$_1$, b_url$_2$, b_url$_3$} semantically enriched to {<a:b:d, a:b:f:k>, <a:c:h>, <a:b:f:j>}. The leftmost concept "a" is indicating the root of web directory and these base URLs are categorized to <d, k>, <h>, and <j>, respectively. In particular, due to multi-attribute of base URL b_url$_1$, $Category$(b_url$_1$) is composed of two different concepts.

We define semantic factors measuring the relationship between two log data. All possible categorical and ordered paths for the requested URL, above all, are obtained, after conceptualizing this URL by web directory. Firstly, the semantic distance is formulated for measuring the semantic difference between two URLs. Let an URL url_i categorized to the sets $\{path_i | path_i^m \in Category(url_i), m \in [1,\ldots,M]\}$ where M is the number of total categorical paths. As simply extending Levenshtein edit distance, the semantic distance $\Delta^\diamond$ between two URLs url_i and url_j is given by

$$\Delta^\diamond[url_i, url_j] = \arg\min_{m=1,n=1}^{M,N} \frac{\min\left((L_i^m - L_C^{(m,n)}), (L_j^n - L_C^{(m,n)})\right)}{\exp(L_C^{(m,n)})} \tag{1}$$

where L_i^m, L_j^n, and $L_C^{(m,n)}$ are the lengths of $path_i^m$, $path_j^n$, and common part of both of them, respectively. As marking paths representing conceptualized URLs on trees, we can easily get this common part overlapping each other. $\Delta^\diamond$ compares all combination of two sets ($|path_i| \times |path_j|$) and returns the minimum among values in the interval $[0, 1]$, where 0 stands for complete matching. Exponent function in denominator is used in order to increase the effect of $L_C^{(m,n)}$. Second factor is to aggregate URLs during a time interval. Thereby, semantic distance matrix $D_{\Delta^\diamond}$ is given by

$$D_{\Delta^\diamond}(i,j) = \begin{bmatrix} \cdots & \cdots & \cdots \\ \cdots & \Delta^\diamond[url_{t_i}, url_{t_j}] & \cdots \\ \cdots & \cdots & \cdots \end{bmatrix} \tag{2}$$

where the predefined time interval T is the size of matrix and diagonal elements are all zero. Based on $D_{\Delta^\diamond}$, the semantic mean $\mu^\diamond$ is given by

$$\mu^\diamond(t_1, \ldots, t_T) = \frac{2 \sum_{i=1}^{T} \sum_{j=i}^{T} D_{\Delta^\diamond}(i,j)}{T(T-1)} \tag{3}$$

where $D_{\Delta^\diamond}(i,j)$ is the (i,j)-th element of distance matrix. This is the mean value of upper triangular elements except diagonals. Then, with respect to the given time interval T, the semantic deviation $\sigma^\diamond$ is derived as shown by

$$\sigma^\diamond(t_1, \ldots, t_T) = \sqrt{\frac{2 \sum_{i=1}^{T} \sum_{j=i}^{T} (D_{\Delta^\diamond}(i,j) - \mu^\diamond(t_1, \ldots, t_T))^2}{T(T-1)}} \tag{4}$$

These factors are exploited to quantify the semantic distance between two random logs and statistically discriminate semantic outliers such as the most distinct or the N distinct data from the rest in the range of over pre-fixed threshold, with respect to given time interval.

When we try to segment web log dataset, log entries are generally time-varying, more properly, streaming. In case of streaming dataset, not only semantic factors in a given interval but also the distribution of the semantic mean $\mu^\diamond$ is needed for sessionization. This will be described in the Sect. 4. We, hence, simply assume that a given dataset is time-invariant and its size is fixed in this section.

In order to analyze semantic outlier for sessionization, we regard the minimize the sum of partial semantic deviation $\mu^\diamond$ for each session as the most optimal partitioning of given dataset. Thereby, the principle session identifiers $PSI = \{psi_a | a \in [1, \ldots, S-1], psi_a \in [1, \ldots, T-1]\}$ is defined as the set of boundary positions, where the variables S and T are the required number of sessions and the time interval, respectively.

The semantic outlier analysis for sessionizing static logs SOA_S as objective function with respect to PSI is given by

$$SOA_S(PSI) = \sum_{i=1}^{S} \mu_i^\diamond \tag{5}$$

where $\mu_i^\diamond$ means partial semantic deviation of ith segment. In order to minimize this objective function, we scan the most distinct pairs, in other words, the largest value in the semantic distance matrix $D_{\Delta^\diamond}$, as follows:

$$\Delta_{MAX}^\diamond[T_a, T_b] = \arg \max_{i=1, j=1}^{T} D_{\Delta^\diamond}(i, j) \tag{6}$$

where $\arg \max_{i=1}^{T}$ is the function returning the maximum values during a given time interval $[T_a, T_b]$. When we obtain $D_{\Delta^\diamond}(p, q)$ as the maximum semantic distance, we assume there must be at least a principle session identifier between p^{th} and q^{th} URLs. Then, the initial time interval $[T_a, T_b]$ is replaced by $[T_p, T_q]$, and the maximum semantic distance in reduced time interval is scanned, recursively. Finally, when two adjacent elements are acquired, we evaluate this candidate psi by using $SOA_S(psi)$. If this value is less than $\sigma^\diamond$, this candidate psi is inserted in PSI. Otherwise, this partition by this candidate psi is cancelled. This sessionization process is top-down approaching, until the required number of sessions S is found. Furthermore, we can also be notified the oversessionization, which is a failure caused by overfitting sessionization, detected by the evaluation process $SOA_S(PSI)$.

4 Session Identification from Streaming Web Logs

Actually, on-line web logs are continuously changing. It is impossible to consider not only the existing whole data but also streaming data. We define the time window W as the pre-determined size of considerable entry from the most recent one. Every time new URL is requested, this time window have to be shifted. In order to semantic outlier analysis of streaming logs, we focus on not only basic semantic factors but also the distribution of the semantic mean with respect to time window, $\mu^\diamond(W^{(T)})$.

As extending SOA_S, the objective function for analyzing semantic outlier of dynamic logs SOA_D is given by

$$SOA_D^{W^{(i)}}(PSI) = \sum_{k=1}^{S} \mu_k^\diamond|_{W^{(i)}} \tag{7}$$

where the $W^{(i)}$ means that the time window from i^{th} URL is applied. We want to minimize this $SOA_D(PSI)$ by finding the most proper set of principle session identifiers. The candidate psi_i is estimated by the difference between the semantic means of contiguous time windows and predefined threshold ε, as shown by

$$\left| \mu^\diamond(W^{(i)}) - \mu^\diamond(W^{(i-\tau)}) \right| \geq \varepsilon \tag{8}$$

where τ is the distance between both time windows and assumed to be less than the size of time window $|W|$. Similar to the evaluation process of SOA_S, once a candidate psi_i is obtained, we evaluate it by comparing $SOA_D^{W^{(i)}}$ and

$SOA_D^{W^{(i-1)}}$. Finally, we can retrieve PSI to sessionize streaming web logs. In case of streaming logs, more particularly, a candidate psi meeting the evaluation process can be appended into unlimited size of PSI.

5   Experiments and Discussion

For experiments, we collected the sanitized access logs from sv.us.ircache.net, one of web cache servers of IRCache. These raw files, generated from 20 March 2003 to 26 March 2003, consist of 11 attributes and about 9193000 entries. We verified sessionizing process proposed in this paper on a PC with a 1.2 GHz CPU clock rate, 256 MB main memory, and running FreeBSD 5.0. During data cleansing, logs whose URL field is ambiguous (wrong spelling or IP address) are removed, as referring to web directory.

Table 1. The number of sessions by time-oriented heuristics and ontology-oriented heuristics (static and dynamic logs) from logs for seven days (20-36 March 2003).

	1	2	3	4	5	6	7
Time-oriented	1563	1359	1116	877	1467	1424	1384
Ontology-oriented	907	923	692	421	807	783	844
(Static logs, SOA_S)	(58%)	(68%)	(62%)	(48%)	(55%)	(55%)	(61%)
Ontology-oriented	983	1051	939	683	1118	827	1105
(Dynamic logs, SOA_D)	(63%)	(77%)	(84%)	(78%)	(76%)	(58%)	(80%)
Common Session Boundary	47%	51%	49%	48%	57%	32%	74%

We compared two sessionizations based on time oriented and ontology oriented heuristics, with respect to the number of segmented sessions and the reasonability of association rules extracted from them. In case of ontology-oriented sessionization, fields related with time such as "Timestamp" and "Elapsed Time" were filtered. Time-oriented heuristics simply sessionized log entries between two sequential requests whose difference of field "Timestamp" is more than 20 milliseconds with respect to the same IP address. On the other hand, for ontology-oriented heuristics, the size of time window W was predefined as 50. The numbers of sessions generated in both cases are shown in Table 1. Time-oriented heuristics estimate denser sessionization than two ontology-oriented approaches. It means that ontology-oriented heuristics based on SOA_S or SOA_D, generally, can make URLs requested over time gap semantically connected each other. They, SOA_S or SOA_D, decreased the number of sessions to, overall, 58.14% and 73.71%, respectively, compared to time-oriented heuristics. Even though ontology-oriented heuristics searched fewer sessions, the rate of common session boundaries (the number of common sessions matched with time-oriented heuristics over the number of sessions of SOA_D) is average 51.1%. It shows that more than 48% of sessions not segmented by time-oriented heuristics can be detected by semantic outlier analysis. While time oriented sessionization is impossible to recognize patterns of users who is easily changing their preferences or simultaneously trying to search various kinds of information on the web, ontology-oriented method can discriminate these complicated patterns.

Table 2. Evaluation of the reasonability of the extracted ruleset (hit ratio (%))

	1	2	3	4	5	6	7
Time-oriented	0.06	0.32	0.46	0.41	0.51	**0.52**	0.49
Static logs, SOA_S	0.05	0.45	0.66	0.72	**0.76**	0.74	0.75
Dynamic logs, SOA_D	0.05	0.46	0.52	0.67	0.70	**0.75**	0.72

We also evaluated the reasonability of the rules extracted from three kinds of session sequences. According to the standard *least recently used* (LRU), we organized the expected set of URLs, which means the set of objects that cache server has to prefetch. The size of this set is constantly 100. As shown in Table 2, we measured the two hit ratios by both of their sessionizations for seven days. The maximum hit ratios in three sequences were obtained 0.52, 0.76, and 0.75, respectively. Ontology-oriented sessionization SOA_S acquired about 24.5% improvement of prefetching performance, compared with time-oriented. Moreover, we want to note that the difference between SOA_S and SOA_D. For the first three days, the hit ratio of SOA_S was higher than that of SOA_D by over 5%. Because of streaming data, SOA_D showed the difficulty in initializing the ruleset. After initialization step, however, the performances of SOA_S and SOA_D were converged into a same level.

6   Conclusions and Future Work

In order to mine useful and significant association rules from web logs, many kinds of well-known association discoverying methods have been developed. Due to the domain specific properties of web logs, sessionization process of log entries is the most important in a whole step. We have proposed ontology-oriented heuristics for sessionizing web logs. In order to provide each requested URL with the corresponding semantics, web directory service as ontology have been applied to categorize this URL. Especially, we mentioned three practical problems for using real non-generic tree structured web directories like Yahoo. After conceptualizing URLs, we measured the semantic distance matrix indicating the relationships between URLs within the predefined time interval. Additionally, factors like semantic mean and semantic deviation were formulated for easier computation. We considered two kinds of web logs which are stationary and streaming. Therefore, two semantic outlier analysis approaches SOA_S and SOA_D were introduced based on semantic factors. Through the evaluation process, the detected cadidate semantic outliers were tested whether their sessionization is reasonable or not. According to results of our experiments, investigating semantic relationships between web logs is very important to sessionize them. Classifying semantic sessions, 48% of total sessions, brought about 25% higher prefetching performance, compared with time-oriented sessionization. Complex web usage patterns seemed to be meaninglessly mixed along with "time" can be analyzed by ontology.

References

1. Cooley, R., Srivastava, J., Mobasher, B.: Web Mining: Information and Pattern Discovery on the World Wide Web. Proc. of the 9th IEEE Int. Conf. on Tools with Artificial Intelligence (1997)
2. Mobasher, B., Cooley, R., Srivastava, J.: Automatic personalization based on Web usage mining. Comm. of the ACM **43**(8) (2000)
3. Berendt, B., Mobasher, B., Nakagawa, M., Spiliopoulou, M.: The Impact of Site Structure and User Environment on Session Reconstruction in Web Usage Analysis. Proc. of the 4th WebKDD Workshop at the ACM-SIGKDD Conf. on Knowledge Discovery in Databases (2002)
4. Chen, Z., Tao, L., Wang, J., Wenyin, L., Ma, W.-Y.: A Unified Framework for Web Link Analysis. Proc. of the 3rd Int. Conf. on Web Information Systems Engineering (2002) 63–72
5. Cooley, R., Mobasher, B., Srivastava, J.: Data Preparation for Mining World Wide Web Browsing Patterns. Knowledge and Information Systems **1**(1) (1999) 5–32
6. Dai, H., Mobasher, B.: Using ontologies to discover domain-level web usage profiles. Proc. of the 2nd Semantic Web Mining Workshop at the PKDD 2002 (2002)
7. Labrou, Y., Finin, T.: Yahoo! as an Ontology: Using Yahoo! Categories to Describe Documents. Proc. of the 8th Int. Conf. on Information Knowledge Management (1999) 180–187
8. Jung, J.J., Yoon, J.-S., Jo, G.-S.: Collaborative Information Filtering by Using Categorized Bookmarks on the Web. Proc. of the 14th Int. Conf. on Applications of Prolog (2001) 343–357
9. McCallum, A., Nigam, K., Rennie, J., Seymore, K.: Building Domain-Specific Search Engines with Machine Learning Techniques. AAAI Spring Symp. (1999)
10. Berendt, B., Spiliopoulou, M.: Analysing navigation behaviour in web sites integrating multiple information systems. The VLDB Journal **9**(1) (2000) 56–75

Single Trial Discrimination between Right and Left Hand Movement-Related EEG Activity[*]

Sunyoung Cho[1], Jung Ae Kim[2], Dong-Uk Hwang[2], and Seung Kee Han[1,2]

[1]Basic Science Research Institute, Chungbuk National University, Cheongju, Korea,
sycho@chungbuk.ac.kr
[2]Department of Physics, Chungbuk National University, Cheongju, Korea,

Abstract. We propose an EEG-based discrimination method for the right/left hand movement in a single trial. The EEG was recorded during the voluntary movement and imagination of the hand movement. We made a feature vector for every second that represents the characteristics to reflect the process of the right/left movement. It was composed of the ERD, ERS patterns of the mu and beta rhythm and the coefficients of the autoregressive model best fitting for the data of the given period. Linear discrimination of their distributions in the vector space classified the right/left hand movement-related EEG activity efficiently.

1 Introduction

The ongoing EEG (electroencephalogram) signals are including useful information to reflect the neuronal processing for the specific mental and/or physical functions. There is a plenty of evidence indicating the frequency-specific changes of EEG may correlate to the sensory, motor and cognitive processing [1, 2, 3]. With high temporal resolution and a low cost, EEG is widely used in assessing brain processes. This EEG signals could be applied for the communication between the brain and an electronic system like a computer – a Brain-Computer Interface (BCI) [4].

The EEG changes reflecting the human intention related the limb movements or the imagination of the movements have been researched extensively and applied to BCI [5, 6]. During the preparation or imagination of the movements, the EEG signals show frequency-specific changes time-locked to the event. These event-related changes consist of decrease or increase of the power in given frequency bands, which might be due to decrease or increase in synchronous activities of the underlying neuronal populations. These are called event-related desynchronization (ERD) and event-related synchronization (ERS) respectively [7].

---

* This work was supported by Korean Research Foundation, KRF 2002-075-H0007 to S.Y. Cho, and a grant(M103KV010011 03K2201 01130) from Brain Research Center of the 21st Century Frontier Research Program funded by the Ministry of Science and Technology of Republic of Korea to S.K. Han.

M. Bubak et al. (Eds.): ICCS 2004, LNCS 3036, pp. 255–262, 2004.

In this study, the EEG signals were recorded during the performance and imagination of the hand movement and analyzed to generate feature vectors for every second EEG data. A feature vector was composed of the ERD, ERS patterns of the mu and beta rhythm and the coefficients of the autoregressive model best fitting for the data of the given period. Linear discrimination of their distribution in the vector space divided the right and left hand movement efficiently.

2   Method

2.1   EEG Data Acquisition

Thirty-five subjects aged 19 to 25 years participated in the study. All subjects were right-handed and free of neurological disorders. The EEG was recorded from the whole scalp with 32 Ag/AgCl electrodes placed according to the international 10-20 system (Neuroscan amplifier, sampling rate 1000Hz, bandwidth filtering 1.5~100Hz). Three kinds of experimental paradigms were used; *self-paced hand movement* in which subjects push a button with the index finger on their own pace in 12-18 s intervals, *tone-triggered hand movement* in which subjects perform the movements after the presentation of tone (1kHz, duration 100ms), and *tone-trigger imagination of hand movement* in which subjects were instructed to imagine performing the movement after the tone stimulation. The EEG was recorded continuously to be selected 12s epoch in each trial, time-locked with the movement-onset or tone stimulation.

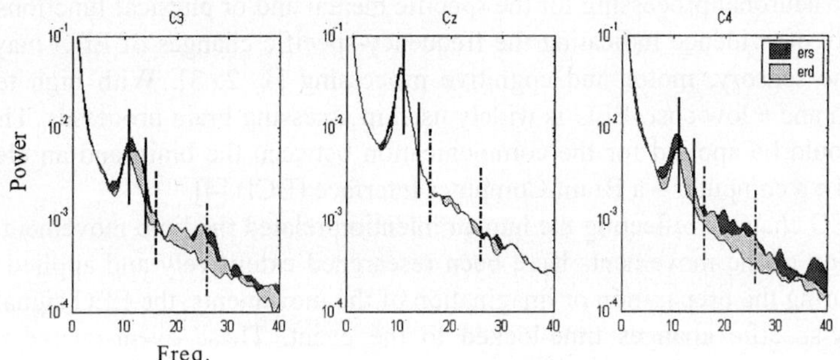

Fig. 1. Examples of 1sec power spectra from EEG data of C3, Cz, and C4 recorded during pre-movement reference period (center line), movement period (lower line), and post-movement period (upper line) while tone-triggered left hand movement. The frequency ranges displaying significant power decrease or increase are marked with a pair of vertical solid lines of 11~14Hz (μ) and a pair of dash-dot lines of 16~22Hz (β)

2.2 Power Spectrum and ERD/ERS Computation

To select the most reactive frequency components to reveal the ERD/ERS patterns related to the hand movement, the power spectra for three periods were compared. For the data from C3, Cz, and C4 electrodes, the power spectra of 1-s pre-movement period as a reference, 1-s movement period around the movement onset, and 1-s post-movement period after movement offset were calculated. Examples from a subject are presented in Fig. 1. In these examples, similar to the formal studies [8, 9], they showed different in the frequency band between 11-14Hz (mu) and 16-22 Hz (beta).

The ERD/ERS time curves were calculated for the selected frequency bands. This procedure involved band pass filtering, squaring of amplitude to obtain power values, averaging of power over all trials, normalizing, and computing of percentages with respect to the reference interval.

$$ERD/ERS(\%) = (P_{segment} - P_{reference})/P_{reference} \times 100$$

2.3 Coefficients of Autoregressive Model

We adopted the coefficients of the autoregressive model as useful indices to discriminate right/left hand movement. In each trial, the EEG signals for 12-s epoch were divided into 1-s window segments with 500ms overlap. The coefficients of the autoregressive model best fitting for the data of each segment were calculated using the following model, in which delay time d=5, and mode order k=6.

$$x_n = a_1 x_{n-d} + a_2 x_{n-2d} + \cdots + a_k x_{n-kd}$$

Fig. 2 illustrates the time curve of the one coefficient (a1) piled up across all trials. They showed the different patterns for the right/left directions (the left hand movement in this figure) and time locked to the movement onset and offset. Therefore, these coefficients were included in our feature vector.

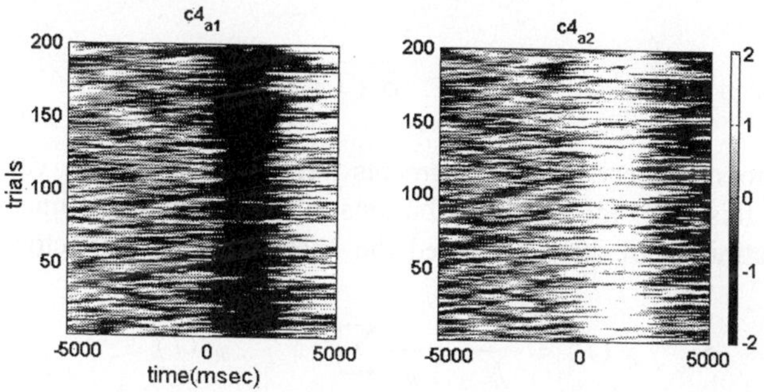

Fig. 2. The change of a coefficient of the autoregressive model in each trial while tone-triggered left hand movement. The x axis indicates the time (msec) from the movement onset, and y axis indicates the trial number

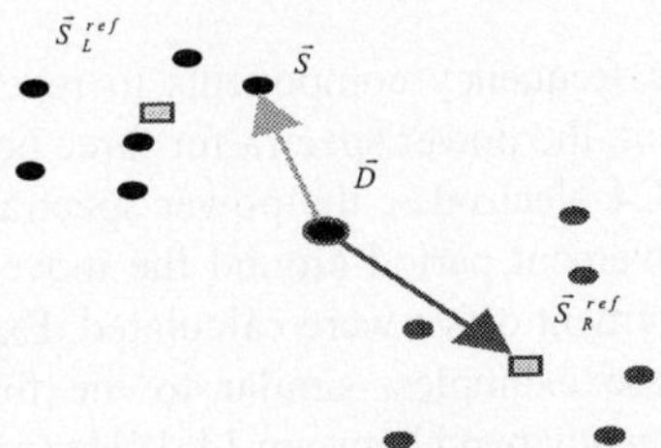

Fig. 3. Schematic diagram of the distribution of feature vectors in a vector space. If dark and thin circles represent the feature vectors of right and left movement, two rectangles ($\vec{s}_L^{\,ref}$, $\vec{s}_R^{\,ref}$) indicate the middle points of each group. The center point $\vec{D}(t)$ is an average of two rectangles. The value of d(t) is inner product of ($\vec{s}(t) - \vec{D}(t)$) and ($\vec{s}_L^{\,ref}(t) - \vec{D}(t)$), which is qualifying the position of the vector compared to the middle point of one group, $\vec{s}_L^{\,ref}(t)$ in this equation. If d(t) is positive value, therefore, it means the vector is included in the group. If negative, the vector is included in the other group

2.4 Feature Vector and Liner Discrimination

The feature vectors were composed with the characteristics proven to be useful for the right/left discrimination in our analysis. We made feature vectors for every 1-s window segments using the EEG signals from C3 and C4 sites. A feature vector includes 6 coefficients of autoregressive model for C3, 6 coefficients for C4, and the ratios of the power change in mu and beta bands for C3 and C4 (that is, ERD/ERS ratio in the period). To compare across trials, the values were standardized in each trial to the reference period of the formal 6 s before the tone onset.

$$\vec{S}(t) =$$

$$(\bar{a}_1^{C3}, \bar{a}_2^{C3}, \ldots, \bar{a}_6^{C3}, \bar{a}_1^{C4}, \bar{a}_2^{C4}, \ldots, \bar{a}_6^{C4}, \overline{P}_{hi-\alpha}^{C3}, \overline{P}_{\beta}^{C3}, \overline{P}_{hi-\alpha}^{C4}, \overline{P}_{\beta}^{C4})$$

$$\overline{x} = \frac{x - \langle x \rangle_{ref}}{\sigma(x)_{ref}}$$

Feature vectors of every window segments were projected to the vector space (16 dimensions in this case) for their distributions to be discriminated linearly. Fig 3 explained the definition of d(t), quantified the position of each vector in the vector space.

$$\vec{S}_{L/R}^{\,ref}(t) = \frac{1}{N_{L/R}} \sum \vec{S}_{L/R}(t)$$

$$\vec{D}(t) = (\vec{S}_L^{\,ef}(t) + \vec{S}_R^{\,ref}(t)) / 2$$

$$d(t) = (\vec{S}(t) - \vec{D}(t)) \cdot (\vec{S}_R^{\,ref}(t) - \vec{D}(t))$$

$$if \; d \; > 0 \; \rightarrow \; Right$$

$$if \; d \; < 0 \; \rightarrow \; Lelf$$

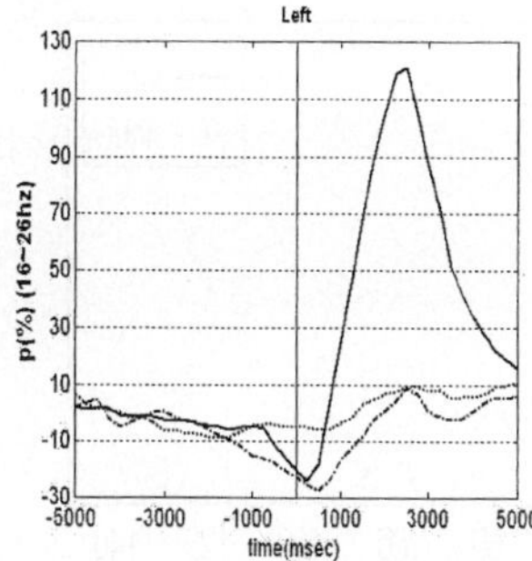
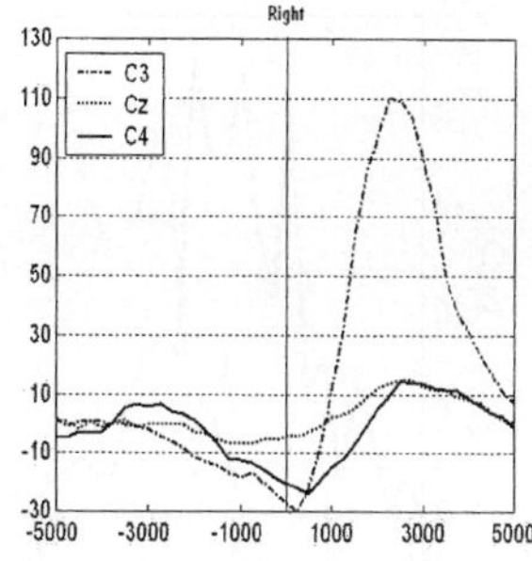

Fig. 4. Grand average time courses of β rhythm in C3 (dash-dot line) Cz (thin line), and C4 (solid line) while tone-triggered movement of left (left box) and right (right box) hand. The x axis indicates the time (msec) from the onset of the tone (the vertical bar). The y axis indicates the percentage of the relative power change, to show ERD and ERS specifically dominant in the contralateral somatomotor area

3   Results

Fig. 4 displays the grand average ERD/ERS time courses of the beta band activity from C3, Cz, and C4 data. For each side movement, it is seen that the post-movement power increases (ERS) are larger in the contralateral hemisphere than the ipsilateral hemisphere. In case of the mu band, a prominent ERD was found in the contralateral hemisphere, followed by ERS.

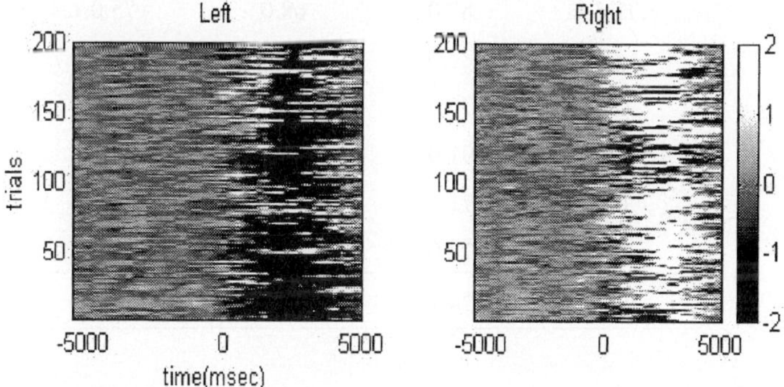

Fig. 5. The time course of d(t) value of the feature vector while tone-triggered movement of left (left box) and right (right box) hand. The x axis indicates the time (msec) from the movement onset, and y axis indicates the trial number

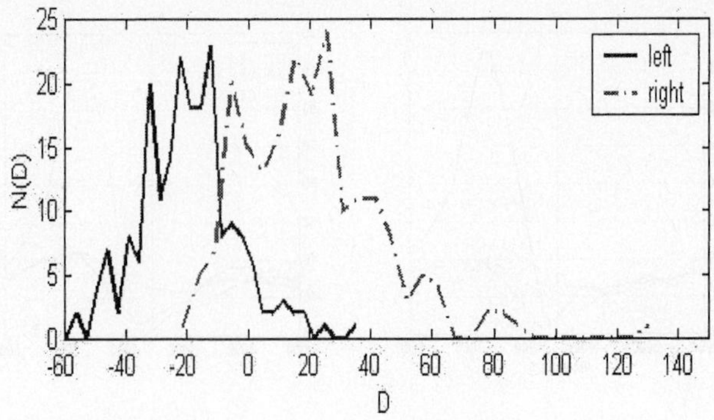

Fig. 6. Histogram of d(t) value of the feature vector while tone-triggered movement of left (solid line) and right (dot line) hand. The x axis indicates the value of d(t) and the y axis indicates the number of the feature vectors that have the value. In this subject, the recognition rates of left and right are 91% and 87%

Fig. 5 illustrates the time curve of the d(t) value piled up across all trials. They are changed consistently after movement onset by the right/left movement across the trials. We made a histogram accumulated by the value of d(t) for every right/left trials in each subject. As shown in Fig. 6, the distributions for the right and left movement could be discriminated well. Table 1 presents the recognition ratios for right/left movement using the linear discrimination of the feature vectors for 6 subjects.

Table 1. The recognition rate while tone-triggered hand movement

subject	Left(%)	Right(%)	Total(%)
HMA	82.0	68.0	75.0
JJH	81.0	76.5	78.7
CSY	75.5	66.5	71.0
KSM	91.0	80.0	85.5
JWR	71.5	79.3	75.5
PMJ	72.0	80.0	76.0
Total	78.8	75.1	77.1

4   Discussion

For the application to the BCI system, it is necessary that the EEG features related to the human intent were analyzed with the EEG signals in a single trial. The present study determined the features that could reveal the intention and performance of the right/left hand movement and proposed the discrimination method using the features in a single trial.

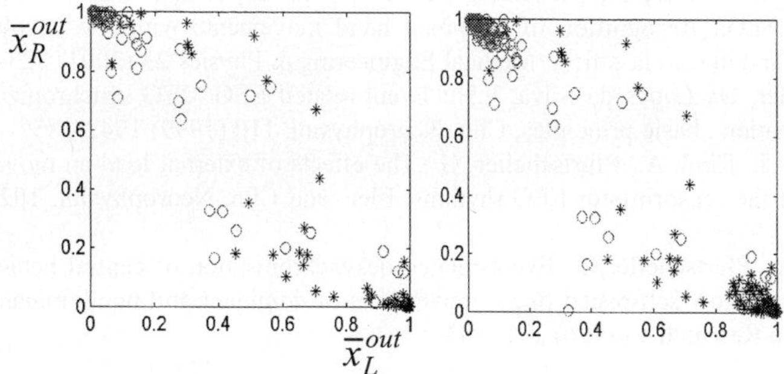

Fig. 7. The performance of the artificial neural network using the units of the feature vector as input nodes while tone-triggered hand movement. We set two output nodes for right and left, indicated in the y and x axis. Two kinds of test data set were applied to generate recognition rate 87% and 93%

As units of our feature vector, we used the single-trail ERD/ERS patterns that were well known as grand averaged ones. And the coefficients of the autoregressive model were used as other units, which showed consistent time-course changes across the trials and differences in the right/left movement.

We further tried to use an artificial neural network for the discrimination of the right/left movement. The units of our feature vectors after the movement onset in each trial were used as the value of the input nodes (multi-layered perception model, input node 80, one-layer hidden node 10, output node 2, and learning rule: feed-forward backpropagation). The preliminary result showed a similar recognition rate with the linear discrimination method (Fig. 7).

References

1. Bastiaansen, M.C.M., Bocker, K.B.E., Brunia, C.H.M., Munch, J.C., Spekreijse, H.: Event-related desynchronization during anticipatory attention for an upcoming stimulus: a comparative EEG/EMG study. Clin. Neurophysiol. 112 (2001) 393-403
2. Leocani, L., Toro, C., Zhuang, P., Gerlff, C. Hallett, M.: Event-related desynchronization in reaction time paradigms: a comparison with event-related potential and cortical excitability, Clin. Neurophysiol. 112 (2001) 923-930
3. Singer, W.: Synchronization of cortical activity and its putative role in informstion processing and learning, Annual Review of physiology 55 (1993) 349-374
4. Wolpaw, J. R., Birbaumer, N., McFarland, D.J., Pfurtscheller, G., Vaughan, T.M.: Brain-computer interfaces for communication and control, Clin. Neurophysiol. 113 (2002) 767-791
5. Pfurtscheller, G., Neuper, C., Guger, C., Harkam, W., Ramoser, H., Schlögl, A.,Obermaier, B., Pregenzer, M.: Current Trends in Graz Brain–Computer Interface (BCI) Research, IEEE Trans. Rehabil. Engineering 8 (2000) 216-219

6. Babiloni, F., Cincotti, F., Bianchi, L., Pirri, G., Millan, J.R., Mourin, O.J., Salinari, S., Marciani, M.G.: Recognition of imagined hand movements with low resolution surface Laplacian and linear classifiers, Medical Engineering & Physics 23 (2001) 323–328
7. Pfurtscheller, G., Lopes da Silva, F.H.: Event-related EEG/MEG synchronization and desynchronization : basic principles, Clin. Neurophysiol. 110 (1999) 1942-1857
8. Stancak, A.J., Riml, A., Pfurtscheller, G.: The effects of external load an movement-related changes of the sensorimotor EEG rhythms, Elec. and Clin. Neurophysiol. 102 (1997) 495-504
9. Stancak, A., Pfurtscheller,G.: Event-related desynchronisation of central beta-rhythms during brisk and slow self-paced finger movements of dominant and nondominant hand, Cognitive Brain Research 4 (1996) 171–183

WINGS: A Parallel Indexer for Web Contents

Fabrizio Silvestri[1,2], Salvatore Orlando[3], and Raffaele Perego[1]

[1] Istituto di Scienze e Tecnologie dell'Informazione - ISTI–CNR, Pisa, Italy
[2] Dipartimento di Informatica, Università di Pisa, Italy
[3] Dipartimento di Informatica, Università Ca' Foscari di Venezia, Italy

Abstract. In this paper we discuss the design of a parallel indexer for Web documents. By exploiting both data and pipeline parallelism, our prototype indexer efficiently builds a partitioned inverted compressed index, a suitable data structure commonly utilized by modern Web Search Engines. We discuss implementation issues and report the results of preliminary tests conducted on a SMP PCs.

1 Introduction

Nowadays, Web Search Engines (WSEs) [1,2,3,4] index hundreds of millions of documents retrieved from the Web. Parallel processing techniques can be exploited at various levels in order to efficiently manage this enormous amount of information. In particular it is important to make a WSE scalable with respect to the size of the data and the number of requests managed concurrently.

In a WSE we can identify three principal modules: the *Spider*, the *Indexer*, and the *Query Analyzer*. We can exploit parallelism in all the three modules. For the Spider we can use a set of parallel agents which visit the Web and gather all the documents of interest. Furthermore, parallelism can be exploited to enhance the performance of the Indexer, which is responsible for building an index data structure from the collection of gathered documents to support efficient search and retrieval over them. Finally, parallelism and distribution is crucial to improve the throughput of the Query Analyzer (see [4]), which is responsible for accepting user queries, searching the index for documents matching the query, and returning the most relevant references to these documents in an understandable form.

In this paper we will analyze in depth the design of a parallel Indexer, discussing the realization and the performance of our WINGS (Web INdexinG System) prototype. While the design of parallel Spiders and parallel Query Analyzers has been studied in depth, only a few papers discuss the parallel/distributed implementation of a Web Indexer [5,6].

Several sequential algorithms have been proposed, which try to well balance the use of core and out-of-core memory in order to deal with the large amount of input/output data involved. The *Inverted File* (*IF*) index [7] is the data structure typically adopted for indexing the Web. This is mainly due to two main reasons. First it allows the efficient resolution of queries on huge collections of Web pages, second it can be easily compressed to reduce the space occupancy in order to obtain a better exploitation of the memory hierarchy [8].

M. Bubak et al. (Eds.): ICCS 2004, LNCS 3036, pp. 263–270, 2004.

An IF index on a collection of Web pages consists of several interlinked components. The principal ones are: the *lexicon*, i.e. the list of all the *index terms* appearing in the collection, and the corresponding set of *inverted lists*, where each list is associated with a distinct term of the lexicon. Each inverted list contains, in turn, a set of *postings*. Each posting collects information about the *occurrences* of the corresponding term in the collection's documents. For the sake of simplicity, in the following discussion we will consider that each posting only includes the identifier of the document (*DocID*) where the term appears, even if postings actually store other information used for document ranking purposes.

Another important feature of the IF indexes is that they can be easily partitioned. In fact, let us consider a typical parallel Query Analyzer module: the index can be distributed across the different nodes of the underlying architecture in order to enhance the overall system's throughput (i.e. the number of queries answered per each second). For this purpose, two different partitioning strategies can be devised. The first approach requires to horizontally partition the whole inverted index with respect to the lexicon, so that each query server stores the inverted lists associated with only a subset of the index terms. This method is also known as *term partitioning* or *global inverted files*. The other approach, known as *document partitioning* or *local inverted files*, requires that each query server becomes responsible for a disjoint subset of the whole document collection (vertical partitioning of the inverted index). Following this last approach the construction of an IF index become a two-staged process. In the first stage each index partition is built locally and independently from a partition of the whole collection. The second phase is instead very simple, and is needed only to collect global statistics computed over the whole IF index.

Since the document partitioning approach provides better performance figures for processing typical Web queries than the term partitioning one, we adopted it in our Indexer prototype. Figure 1 illustrates this choice, where the document collection is here represented as a set of html pages.

Note that in our previous work [4] we conducted experiments where we pointed out that the huge sizes of the Web make the global statistics useless. For this reason, we did not consider this phase in the design of our indexer.

The paper is organized as follow. Section 2 motivates the choices made in the design of WINGS and discusses parallelism exploitation and implementation issues. Some encouraging experimental results obtained running WINGS on Linux SMP PCs are instead presented and discussed in Section 3. Finally, Section 4 draws some conclusions and outlines future work.

2   The Design of WINGS

The design of a parallel Indexer for a WSE adopting the document partition approach (see Figure 1, can easily exploit *data parallelism*, thus independently indexing disjoint sub-collections of documents in parallel. Besides this natural form of parallelism, in this paper we want to study in depth the parallelization

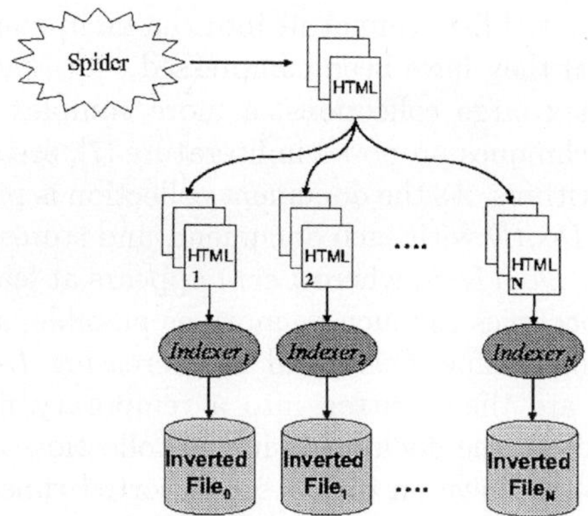

Fig. 1. Construction of a distributed index based on the document partition paradigm, according to which each local inverted index only refers to a partition of the whole document collection.

opportunities within each instance of the Indexer, say $Indexer_i$, which accomplishes its indexing task on a disjoint partition i of the whole collection.

Table 1. A toy text collection (a), where each row corresponds to a distinct document, and (b), the corresponding inverted index, where the first column represents the lexicon, while the last column contains the inverted lists associated with the index terms.

Document Id	Document Text
1	Pease porridge hot, pease porridge cold.
2	Pease porridge in the pot.
3	Nine days old.
4	Some like hot, some lite it cold.
5	Some like in the pot.
6	Nine days old.

(a)

Term	Postings list
cold	1, 1
days	3, 6
hot	1, 4
in	2, 5
it	4, 5
like	4, 5
nine	3, 6
old	3
. . .	. . .

(b)

The job performed by each $Indexer_i$ to produce a local inverted index is apparently simple. If we consider the collection of documents as modeled by a matrix (see Table 1.(a)), building the inverted index simply corresponds to transpose the matrix (see Table 1.(b)). This matrix transposition or inversion can be easily accomplished in-memory for small collections. Unfortunately, a naive in-core algorithm becomes rapidly unusable as the size of the document collection grows. Note that, with respect to a collection of some GBs of data, the size of the final lexicon is usually a few MBs so that it can be maintained in-core, while the inverted lists cannot fit into the main memory and have to be stored on disk, even if they have been compressed.

To efficiently index large collections, a more complex process is required. The most efficient techniques proposed in literature [7], are all based on external memory sorting algorithms. As the document collection is processed, the Indexer associates a distinct $DocID$ with each document, and stores into a in-core buffer all the pairs $< Term, DocID >$, where $Term$ appears at least once in document $DocID$. Buffer size occupies as much memory as possible, and when it becomes full, it is sorted by increasing $Term$ and by increasing $DocID$. The resulting *sortedruns* of pairs are then written into a temporary file on disk, and the process repeated until all the documents in the collection are processed. At the end of this first step, we have on disk a set of sorted runs stored into distinct files. We can thus perform a multi-way merge of all the sorted runs in order to materialize the final inverted index.

According to this approach, each $Indexer_i$ works as follows: it receives a stream of documents, subdivides them in blocks, and produce several disk-stored sorted runs, one for each block. We call this phase $Indexer_i^{pre}$. Once $Indexer_i^{pre}$ is completed, i.e. the stream of pages has been completely read, we can start the second phase $Indexer_i^{post}$, which performs the multi-way merge of all the sorted runs.

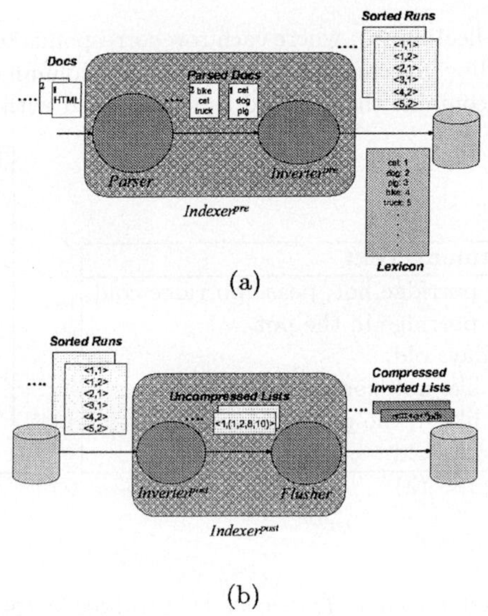

(a)

(b)

Fig. 2. Forms of parallelism exploited in the design of a generic *Indexer*, in particular of the first module (a) $Indexer_i^{pre}$, and the (b) $Indexer_i^{post}$ one.

The main activities we have identified in $Indexer^{pre}$ and $Indexer^{post}$ are illustrated in Figure 2.(a) and Figure 2.(b), respectively.

Indexer^{pre} can be in turn modeled as a two stage pipeline, *Parser* and *Inverter^{pre}*. The former recognizes the syntactical structure of each document (html, xml, pdf, etc.), and generates a stream of the terms identified. The *Parser* job is indeed more complex than the one illustrated in figure. It has to determine the local frequencies of terms, remove stop words, perform stemming, store information about the position and context of each occurrence of a term to allows phase searching, and, finally, collect information on the linking structure of the parsed documents. Note that information about term contexts and frequencies are actually forwarded to the following stages, since they must be stored in the final inverted files in order to allow to rank the results of each WSE query. For the sake of clarity, we will omit all these details in the following discussion.

The latter module of the first pipeline, *Inverter^{pre}*, thus receives from the *Parser* stage a stream of terms associated with distinct *DocID*s, incrementally builds a lexicon by associating a *TermID*s with each distinct term, and stores on disk large sorted runs of pairs $< TermID, DocID >$. Before storing the run, it has to order them first by *TermID*, and then by *DocID*. Note that the use of integer *TermID* and *DocID* not only reduces the size of each run, but also makes faster comparisons and thus runs' sorting. *Inverter^{pre}* has a sophisticated main memory management, since the lexicon has to be kept in-core, each run has to be as large as possible before flushing to disk, and we have to avoid memory swapping.

When the first pipeline *Indexer^{pre}* ends processing a given document partition, the second pipeline *Indexer^{post}* can start its work. The input to this second pipeline is exactly the output of *Indexer^{pre}*, i.e., the set of sorted runs and the lexicon relative to the document partition. The first stage of the second pipeline is the *Inverter^{post}* module, whose job is to produce a single sorted run starting from the various disk-stored sorted runs. The sorting algorithm is very simple: it is a in-core multi-way merge of the n runs, obtained by reading into the main memory the first block b of each run, where the size of the block is carefully chosen on the basis of the memory available. The top pairs of all the blocks are then inserted in a (min-)heap data structure, so that the top of the heap contains the smallest *TermID*, in turn associated with the smallest *DocID*. As soon as the lowest pair p is extracted from the heap, another pair, coming from the same sorted run containing p (i.e., form the corresponding block), is inserted into the heap. Finally, when an in-core block b is completely emptied, another block is loaded from the same disk-stored run. The process clearly ends when all the disk-stored sorted runs have been enterely processed, and all the pairs extracted from the top position of the heap.

Note that *Indexer^{post}* as soon as extracts each ordered pair from the heap, forwards it to the *Flusher*, i.e. the latter stage of the second pipeline. This stage receives in order all the postings associated with each *TermID*, compresses them by using the usual techniques based on the representation of each inverted list as a sequence of gaps between sorted *DocID*s, and stores each compressed list on disk. A scratch of the inverted index produced is shown in Figure 3. Note that the various lists are stored in the inverted file according to the ordering given

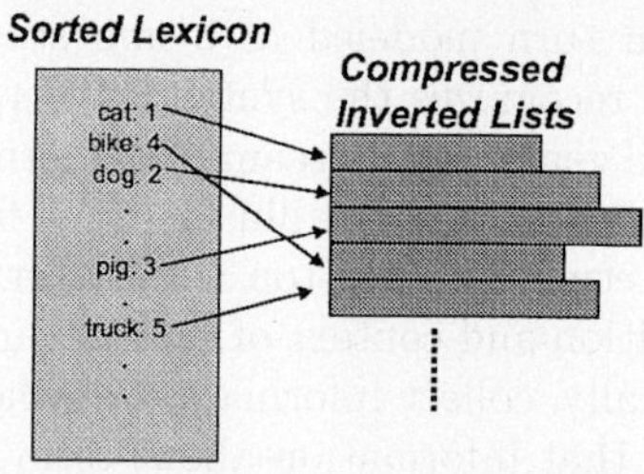

Fig. 3. Inverted file produced by the *Flusher*.

by the *TermID* identifiers, while the lexicon shown in figure has to be finally sorted according to lexicographic order given by the corresponding terms.

3   Experimental Results

All the tests were conducted on a 2-way SMP equipped with two Intel 2.0 GHz Xeon processors, one GB of main memory, and one 80 GB IDE disk.

For the communication between the various stages of the *Indexer* pipelines, we used an abstract communication layer that can be derived in order to exploit several mechanisms. In the tests we performed, since the pipeline stages are mapped on the same node, we exploited the System V message queue IPC.

In order to evaluate the opportunity offered by the pipelined parallelization scheme illustrated above, we have first evaluated the computational cost of the *Parser*, the *Inverter*, and the *Flusher* module, where each of them uses the disk for input/output.

Table 2. Sequential execution times for different sizes of the document collection.

Collection Size (GB)	Time (s) *Parser*	Time (s) *Inverter*	Time (s) *Flusher*	Tot. Throughput (GB/h)
1	3610	1262	144	0.71
2	7044	3188	285	0.68
3	11355	4286	429	0.67
4	14438	5714	571	0.69
5	18047	6345	725	0.71

As you can observe from the execution times reported in Table 2, the most expensive module is the *Parser*, while the execution time of the *Flusher* is, on average, one order of magnitude smaller than whose of the *Inverter*. Moreover, we have to consider that, in the pipeline version, the *Inverter* module will be split into two smaller ones, $Inverter^{pre}$ and $Inverter^{post}$, whose cost is smaller than the whole *Inverter*. From the above considerations, we can conclude that

the pipeline implementation will result in an *unbalanced computation*, so that
if we execute a single pipelined instance $Indexer_i$ on a 2-way multiprocessors,
processors, it should result in a under-utilization of the workstation. In particular, the $Inverted^{pre}$ should waste most of its time waiting for data coming from
the *Parser*.

When a single pipelined $Indexer_i$ is executed on a multiprocessor, a way to
increase the utilization of the platform is to try to balance the throughput of
the various stages. From the previous remarks, in order to balance the load we
could increase the throughput of the *Parser*, i.e. the most expensive pipeline
stage, for example by using a multi-thread implementation where each thread
independently parses a distinct document[1]. The other way to improve the multiprocessor utilization is to map multiple pipelined instances of the indexer, each
producing a local inverted index from a distinct document partition.

Table 3. Total throughput (GB/s) when multiple instances of the pipelined $Indexer_i$
are executed on the same 2-way multiprocessor.

Coll. Size	No. of instances			
(GB)	1	2	3	4
1	1.50	2.01	2.46	2.57
2	1.33	2.04	2.44	2.58
3	1.38	1.99	2.38	2.61
4	1.34	2.04	2.45	2.64
5	1.41	2.05	2.45	2.69

In this paper we will evaluate the latter alternative, while the former one will
be the subject of a future work. Note that when we execute multiple pipeline
instances of $Indexer_i$, we have to carefully evaluate the impact on the shared
multiprocessor resources, in particular the disk and the main memory. As regards
the disk, we have evaluated that the most expensive stage, i.e. the *Parser*, is
compute-bound, so that the single disk suffices to serve requests coming from
multiple *Parser* instances. As regards the main memory, we have tuned the
memory management of the stages $Inverter^{pre}$ and $Inverter^{post}$, which in principle could need the largest amount of main memory to create and store the
lexicon, store and sort the runs before flushing to the disk, and to perform the
multi-way merge from the sorted runs.

In particular, we have observed that we can profitably map up to 4 instances
of distinct pipelined $Indexers_i$ on the same 2-way processor, achieving a maximum throughput of 2.64 GB/hour. The results of these tests are illustrated in
Table 3. Note that, when a single pipelined $Indexer_i$ is executed, we were however able to obtain a optimal speedup over the sequential version ($\simeq 1.4\ GB/h$

---

[1] The *Parser* is the only pipeline stage that can be further parallelized by adopting
a simple data parallel scheme.

vs. $\simeq 0.7\ GB/h$), even if the pipeline stages are not balanced. This is due to the less expensive in-core pipelined data transfer (based on message queues) of the pipeline version, while the sequential version must exploit the disk to save intermediate results.

4   Conclusion

We have discussed the design a WINGS, a parallel indexer for Web contents that produces local inverted indexes, the most commonly adopted organization for the index of a lrge-scale parallel/distributed WSE. WINGS exploits two different levels of parallelism. Data parallelism, due to the possibility of independently building separate inverted indexes from disjoint document partitions. This is possible because WSEs can efficiently process queries by broadcasting them to several searchers, each associated with a distinct local index, and by merging the results. In addition, we have also shown how a limited pipeline parallelism can be exploited within each instance of the indexer, and that a low-cost 2-way workstation equipped with an inexpensive IDE disk is able to achieve a throughput of about 2.7 $GB/hour$, when processing four document collections to produce distinct local inverted indexes. Further work is required to assess the performance of our indexing system on larger collections of documents, and to fully integrate it within our parallel and distributed WSE prototype [4]. Moreover, we plan to study how WINGS can be extended in order to exploit the inverted lists active compression strategy discussed in [8].

References

1. Baeza–Yates, R., Ribiero–Neto, B.: Modern Information Retrieval. Addison Wesley (1998)
2. Witten, I.H., Moffat, A., Bell, T.C.: Managing Gigabytes – Compressing and Indexing Documents and Images. second edition edn. Morgan Kaufmann Publishing, San Francisco (1999)
3. Brin, S., Page, L.: The anatomy of a large-scale hypertextual Web search engine. Computer Networks and ISDN Systems **30** (1998) 107–117
4. Orlando, S., Perego, R., Silvestri, F.: Design of a Parallel and Distributed WEB Search Engine. In: Proceedings of Parallel Computing (ParCo) 2001 conference, Imperial College Press (2001) 197–204
5. Jeong, B., Omiecinski, E.: Inverted File Partitioning Schemes in Multiple Disk Systems. IEEE Transactions on Parallel and Distributed Systems (1995)
6. Melnik, S., Raghavan, S., Yang, B., Garcia-Molina, H.: Building a Distributed Full–Text Index for the Web. In: World Wide Web. (2001) 396–406
7. Van Rijsbergen, C.: Information Retrieval. Butterworths (1979) Available at http://www.dcs.gla.ac.uk/Keith/Preface.html.
8. Silvestri, F., Perego, R., Orlando, S.: Assigning document identifiers to enhance compressibility of web search. In: Proceedings of the Symposium on Applied Computing (SAC) - Special Track on Data Mining (DM), Nicosia, Cyprus, ACM (2004)

A Database Server for Predicting Protein-Protein Interactions*

Kyungsook Han** and Byungkyu Park

School of Computer Science and Engineering, Inha University, Inchon 402-751, Korea

Abstract. Large-scale protein interactions are known for several species due to the recent improvements in experimental methods for detecting protein interactions. However, direct determination of all the interactions between the human proteins is difficult even with current high-throughput methods. This paper describes a database server called HPID (http://www.hpid.org) that (1) provides structural interactions between human proteins precomputed from existing structural and experimental data and (2) predicts structural interactions between proteins submitted by users. The structural interactions were obtained by finding known structural interactions of PDB in SCOP domains and then by finding homologs of the domains in target proteins. Based on the structural interactions, we constructed two protein interaction maps, one for human and another for yeast. We believe this is the first attempt to map a whole human interactome at the superfamily level and to compare a human protein interaction map with other species' interaction map.

1   Introduction

One of today's challenges in bioinformatics is to identify all the interactions of human proteins. Large-scale protein interactions are known for several organisms due to the development of high-throughput methods for detecting protein interactions, such as two-hybrid method and mass spectrometry. However, determination of genome-wide protein interactions by experimental methods is limited to low-order organisms such as yeast and Helicobacter pylori [1, 2]. The genes of the human genome are known, but direct determination of all the interactions between the human proteins is still difficult even with high-throughput methods. An intrinsic problem with high-throughput methods is that protein interactions detected by the methods include many false positives. In fact, more than half of current high-throughput data are estimated to be spurious [3]. Considering these constraints, it is important to develop computational methods that can predict protein interactions, and compare different sets of predicted or experimental data of protein interactions.

---

* This work was supported by the Ministry of Information and Communication of Korea under grant IMT2000-C3-4. We would like to thank Hyongguen Kim and Jinsun Hong for providing the sub-cellular localization information of human proteins.

** To whom correspondence should be addressed. Email: khan@inha.ac.kr

M. Bubak et al. (Eds.): ICCS 2004, LNCS 3036, pp. 271–278, 2004.

It has been widely conjectured that core structural protein interactions are conserved among different organisms. The number of distinct protein domains known so far is around 1,000 (http://scop.mrc-lmb.cam.ac.uk/, SCOP version used here is 1.57 unless otherwise stated) [4]. Therefore, it is inevitable that the same kind of protein domains are involved in diverse types of protein-protein interactions [5]. We have previously predicted protein interactions in human by homologous interactions in yeast, and compared them [6]. However, the predicted interactions between human proteins are estimated to contain many false positives partly because they are derived from the experimental data of yeast protein interactions and the experimental data themselves contain many false positives.

As an improvement of our previous study [6], we attempted to predict structural protein interactions of human and yeast and to compare them. The structural protein interactions are expected to be more accurate than the interactions obtained from the previous study for the following reason. Protein interactions are predicted by finding known structural interactions of PDB [7] in SCOP domains [4] and then by finding homologs of the domains in human proteins. X-ray crystallography and NMR (Nuclear Magnetic Resonance) were main methods for determining the structure data of PDB, and they are more precise experimental techniques than the high-throughput methods for detecting protein interactions.

The aim of this comparative study is to estimate the extent of protein superfamilies in human structural interactome and examine how much overlap exists between the two very diverse eukaryotes. The overall procedure from assigning protein folds to the whole predicted proteome of the complete genomes to visualizing the large network of interactions forms a systematic methodology that can be applied to other genomes. This pilot study can provide the major problems associated in such a bioinformatics analysis and a rough insight on the comparative structural interactomics.

2 Prediction Method

Structural interactions were determined based on the Protein Structural Interaction Map (PSIMAP) [8], which classifies interactions between all known structural protein superfamilies. Structures were assigned to the whole genome (predicted coding regions in the genome) by homology search. The level of homology interaction applied is at the SCOP superfamily. This means that the estimated structural interactome of human does not describe the protein-protein or domain-domain interactions at the molecular level, but at the protein family level. This section starts with definitions of the main concepts in the work.

Definition 1. Given a set P of proteins and a set S of protein structures in an organism, a set B of pairs of the form (protein, structure) shows the structure assignment to proteins in the organism.

$$B = \{(p,s) \mid p \in P, s \in S\} \tag{1}$$

Fig. 1 shows the data schema of the species_proteins_node table, species_proteins table, and protein_structures table in HPID. The species_proteins_node table represents the set B, and the protein_structures table is constructed with the data from the SCOP [4] and Pfam [13] databases.

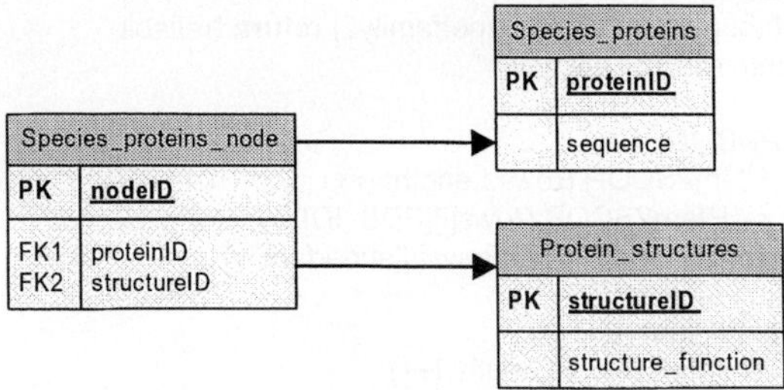

Fig. 1. Data schema for the species_proteins_node table, species_proteins table, and protein_structures table.

Definition 2. If structures s_1 and s_2 interact to each other, and there exist structure assignments $(p_1, s_1) \in B$, and $(p_2, s_2) \in B$ such that $p_1, p_2 \in P$ and $s_1, s_2 \in S$, then proteins p_1 and p_2 interact to each other.

The edge table in Fig. 2 represents the data schema for protein-protein interactions, obtained by definition 2. The global framework of HPID is shown in Fig. 3.

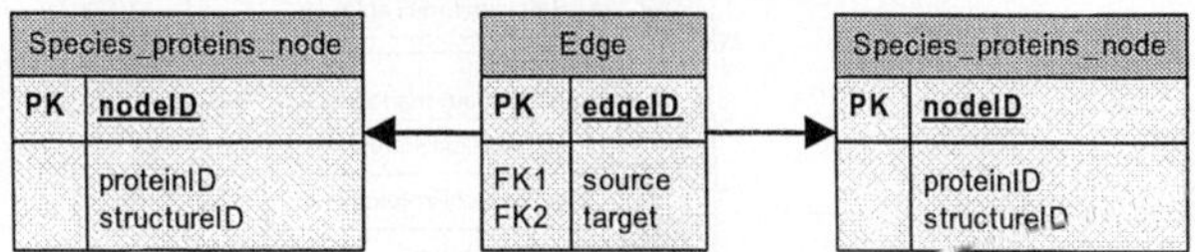

Fig. 2. Data schema for protein-protein interactions.

For the homology search, we constructed a composite database with 27,049 human proteins (http://www.ensembl.org/, v7.29.1), 3,877 yeast proteins, SCOP proteins [4] and NRDB90 [9]. PSI-BLAST [10] was run on the composite DB with SCOP domain sequences as query sequences, the e-value threshold of 0.0005, and the maximum number of profile search rounds of 10 (iterations above 5 does not bring a very high number of distant homologs). The output of PSI-BLAST was parsed by our MS C# program to extract human proteins and yeast proteins matched to SCOP domain sequences and the start and end positions of the matched parts.

Following algorithms describe the procedure for determining the reliability of the predicted protein interactions. 42% of the predicted interactions between human proteins were 'reliable' and the rest 58% were 'unknown'. Fig. 4 shows the data objects used by the algorithms.

```
Check-Reliable-Assignment (protein_ID)
1    superfamily1 ← Online_prediction_result.Rows[protein_ID][superfamily]
2    for (int i=0; i < EnsMart.Rows.Length; i++)
3       if (EnsMart.Rows[i][protein_ID] == protein_ID)
4          superfamily2←Get-Superfamily (Get-PDB_ID-in-Pfam (EnsMart.Rows[i][Pfam_an]))
5          If (superfamily2 != null)
6                if (superfamily1 == superfamily2) return "reliable"
7                else return "unknown"


Get-Superfamily (PDB_ID)
1    for (int i=0; i < Pfam2SCOP.Rows.Length; i++)
2       if (PDB_ID == Pfam2SCOP.Rows[i][PDB_ID])
3                return Pfam2SCOP.Rows[i][superfamily]


Get-PDB_ID-in-Pfam (Pfam_an)
1    for (int i=0; i < EnsMart.Rows.Length; i++)
2       if (EnsMart.Rows[i][Pfam_an] == Pfam_an)
3                return EnsMart.Rows[i][PDB_ID]
```

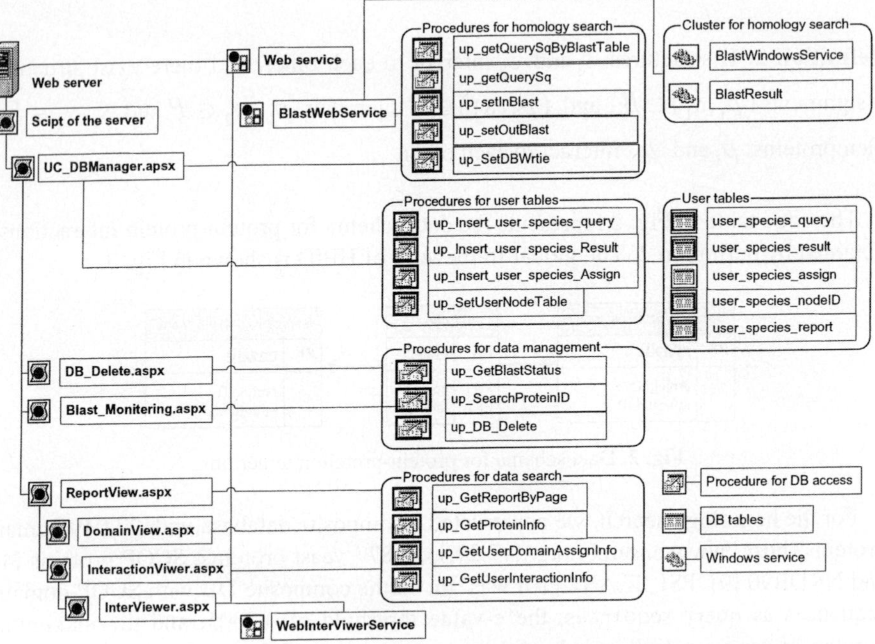

Fig. 3. The framework of the database server.

EnsMart	
PK	**EnsMart_pk**
	protein_ID Pfam_an

Pfam2SCOP	
PK	**Pfam2SCOP_pk**
	Pfam_an PDB_ID superfamily

Online_prediction_result	
PK	**result_pk**
	protein_ID superfamily

Fig. 4. The data objects of EnsMart, Pfam2SCOP, and Online_prediction_result.

The database allows the user to infer potential interactions between proteins submitted by the user. Registration is required to use the online prediction service since prediction results are maintained for individual users. The only required information for registration is the email address, and a user ID and password for the user to use when logging onto the database to view prediction results. When a registered user logs onto to the database server, the status of the user's previous job is displayed regarding whether there is an error in the submitted protein sequences, and whether homology search is complete, in progress, or has not been started yet.

3 Results and Discussion

A protein superfamily was assigned to a human protein (or yeast protein), in a conservative manner, when the matched part is 70% or longer of the original protein superfamily. When multiple superfamilies were matched to a same location of a protein, a superfamily with the highest matching score was assigned to that location and overlap of superfamilies was not allowed. 46% (12,550 proteins) of the total 27,049 human proteins were assigned one or more superfamilies. One human protein was assigned 152 superfamilies and others were assigned 52 or fewer superfamilies (Fig. 5A). 39% (1,509 proteins) of the total 3,877 yeast proteins were assigned at least one superfamily but no more than 6 superfamilies (Fig. 5B).

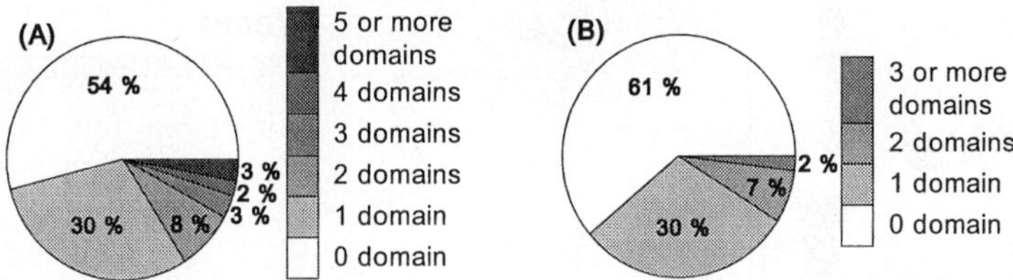

Fig. 5. (A) Superfamily assignment to 27,049 human proteins. (B) Superfamily assignment to 3,877 yeast proteins.

In order to assess the reliability of predicted interactions, we scored interactions based on the identity value of matched parts. In human protein interactions the average identity (X) was 34% with standard deviation (σ) of 23% whereas the average identity was 32% with standard deviation of 25% in yeast protein interactions. An interaction (p1, p2) between proteins p1 and p2 was declared to have a high identity score when both proteins p1 and p2 were assigned a superfamily with an identity $\geq$ X+σ. As shown in Table 1, 220,066 interactions between 2,424 human proteins had a high identity ($\geq$57%=34%+23%), whereas 1,127 interactions between 184 yeast proteins had a high identity ($\geq$57%=32%+25%). The 220,066 human protein interactions with a high identity correspond to 617 interactions at the superfamily level (i.e., at the PSIMAP data) and the 1,127 yeast protein interactions with a high identity correspond to 157 interactions at the superfamily level. Yeast and human are evolutionarily distant species but 74.5% (117 interactions) of the 157 yeast interactions at the superfamily-level were also found in human protein interactions.

Table 1. Protein interactions (including self-loops) in human and yeast. 74.5% (117 interactions) of the 157 yeast interactions at the superfamily-level were also found in human protein interactions.

	human		yeast	
	# proteins (# superfamilies)	# interactions	# proteins (# superfamilies)	# interactions
total interactions	12,550	7,000,943	1,509	88,855
interactions w/high identity at protein level	2,424	220,066	184	1,127
interactions w/high identity at superfamily level	358	617	102	157

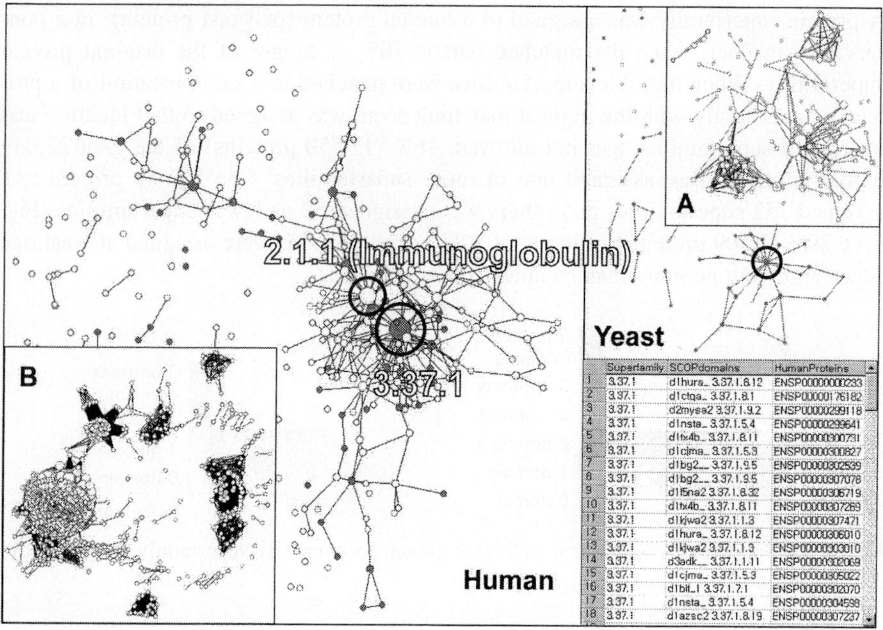

Fig. 6. The large maps visualize protein interactions at the superfamily level with a high identity in human and yeast (the last row of Table 1). Interaction maps at the protein level are shown in boxes A and B for human and yeast, respectively (the second row of Table 1). Red nodes represent superfamilies shared by human and yeast. A box at the lower right corner lists proteins with superfamily '3.37.1' assigned.

Fig. 6 shows protein interactions with a high identity both at the superfamily level (the last row of Table 1) and at the protein level (the second row of Table 1), visualized by WebInterViewer [11]. Node '3.37.1' has the largest number of interacting superfamilies in both human and yeast. 3.37.1 is a p-loop containing hydrolase. It is one of the most important protein structures occurring in all 4 superkingdoms of life. Its functions are tightly related to energy metabolism such as ATP synthase and signal transduction such as G-protein containing pathways. Therefore, it is not surprising that it has diverse superfamily level structural interactions. Proteins with superfamily

'3.37.1' assigned were also listed in the lower right corner of Fig. 6. Node '2.1.1' of Fig. 6 (IG superfamily. b.1.1 in original SCOP versions) represents the immuno-globulin superfamily, which was found in human, but was missing in yeast assign-ment. This corroborates a well-known fact that immunoglobulin exists more pre-dominantly in the high-order species only.

Fig. 7 shows many homologs of human proteins that have known homologs in PDB as structural interaction pairs. One of the points of it is that the seemingly com-plicated interaction patterns in human can be reduced dramatically to a simple basic backbone of family-family interactions. In smaller genomes, the same interaction patterns are found but with much fewer homologs associated with the graph. It fol-lows a scale-free network [12] in parameters. However, the result is a theoretical estimation of human structural interactome. It is analogous to genome draft showing the magnitude of the problem and a rough map for higher resolution mapping of protein-protein interaction.

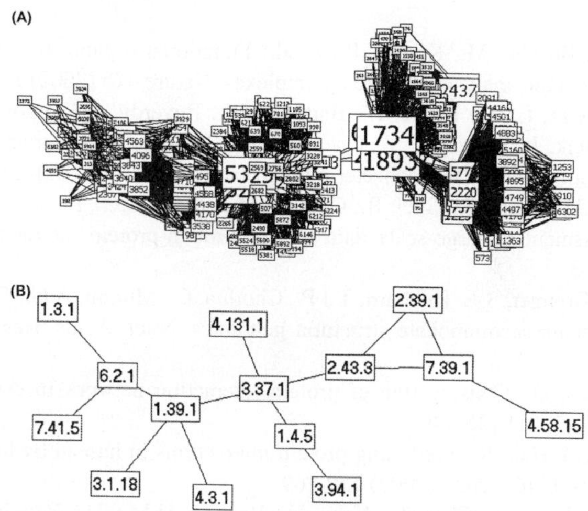

Fig. 7. (A) The second largest connected component in the human protein interaction network. (B) When grouping proteins directly interacting, the apparently complex network in (A) is reduced to a simple network in (B). The simplified network corresponds to a subnetwork of PSIMAP, consisting of superfamilies assigned to the human proteins in (A)

4 Conclusion

We constructed a database server for (1) providing structural interactions between human proteins precomputed from existing structural and experimental data and for (2) predicting structural interactions between proteins submitted by users. The struc-tural interactions between human proteins were compared with those between yeast proteins. They revealed a significant overlap in structural interactome (75%), showing the high rate of conservation of structural interaction at the protein superfamily level. Even though the portion of genes assigned to the whole genomes at present, this indicates that the functional diversification of humans, on the large, is not correlated from different or new interactions derived from new superfamilies. It is more likely that the core structural interactions in life are tightly conserved and the functional diversification and species differentiation are more associated with complex regulatory differentiation. It is possibly related to subtle differentiation in interactions with a basic set of protein structures and their interaction types. As illustrated in Fig. 7, a complicated network of interacting proteins with many homologs can be dramatically reduced to a single backbone network using the family-family interaction concept. The methodology applied here covers many different computational steps and forms a pipeline of structural interactome analysis. Albeit partial, we believe this is the first bioinformatics attempt to map a whole human interactome and to compare a human protein interaction map with other species' interaction map.

References

1. Gavin, A.-C., Bosche, M., Krause, R., et al.: Functional organization of the yeast proteome by systematic analysis of protein complexes. Nature 415 (2002) 141-147
2. Rain, J.-C., Selig, L., De Reuse, H., Battaglia, V., Reverdy, C., Simon, S., Lenzen, G., Petel, F., Wojcik, J., Schachter, V., Chemama, Y., Labigne, A., Legrain, P.: The protein-protein interaction map of Helicobacter pylori. Nature 409 (2001) 211-215
3. von Mering, C., Krause, R., Snel, B., Cornell, M., Oliver, S.G., Field, S., Bork, P.: Comparative assessment of large-scale data sets of protein-protein interactions. Nature 417 (2002) 399-403
4. Lo Conte, L. Brenner, S.E. Hubbard, T.J.P., Chothia, C., Murzin, A.G.: SCOP database in 2002: refinements accommodate structural genomics. Nucl. Acids. Res. 30 (2002) 264-267
5. Park, J., Bolser, D.: Conservation of protein interaction network in evolution. Genome Informatics 12 (2001) 135-140
6. Kim, H., Park, J. Han, K.: Predicting protein interactions in human by homologous interactions in yeast. LNCS 2637 (2003) 159-169
7. Westbrook, J., Feng, Z., Chen, L., Yang, H., Berman, H.M.: The Protein Data Bank and structural genomic. Nucl. Acids. Res. 31 (2003) 489-491
8. Park, J., Lappe, M., Teichmann, S.: Mapping protein family interactions: intramolecular and intermolecular protein family interaction repertoires in the pdb and yeast. J. Mol. Biol. 307 (2001) 929-938
9. Lappe, M., Park, J., Niggemann, O., Holm, L.: Generating protein interaction maps from incomplete data: application to fold assignment. Bioinformatics 17 (2001) 149-156
10. Altschul, S.F., Madden, T.L., Schäffer, A.A., Zhang, J., Zhang, Z., Miller, W., Lipman, D.J.: Gapped BLAST and PSI-BLAST: a new generation of protein database search programs. Nucl. Acids. Res. 25 (1997) 3389-3402
11. Han, K., Ju, B.-H.: fast layout algorithm for protein interaction networks. Bioinformatics 19 (2003) 1882-1887
12. Jeong, H., Tombor, B., Albert, R., Oltvai, Z.N., Barabasi, A.L.: The large-scale organization of metabolic networks. Nature 407 (2000) 651-654
13. Bateman, A., Birney, E., Cerruti, L., Durbin, R., Etwiller. L., Eddy, S.R., Griffiths-Jones. S., Howe. K.L., Marshall, M., Sonnhammer. E.L.: The Pfam Protein Families Database. Nucleic Acids Research 30 (2002) 276-280

PairAnalyzer: Extracting and Visualizing RNA Structure Elements Formed by Base Pairing

Daeho Lim and Kyungsook Han[*]

School of Computer Science and Engineering, Inha University, Inchon 402-751, Korea
`Kingtiger1@hanmail.net, khan@inha.ac.kr`

Abstract. Most currently known molecular structures were determined by X-ray crystallography or Nuclear Magnetic Resonance (NMR). These methods generate a large amount of structure data, even for small molecules, and consist mainly of three-dimensional atomic coordinates. These are useful for analyzing molecular structure, but structure elements at higher level are also needed for a complete understanding of structure, and especially for structure prediction. Computational approaches exist for identifying secondary structural elements in proteins from atomic coordinates. However, similar methods have not been developed for RNA, due in part to the very small amount of structure data so far available, and extracting the structural elements of RNA requires substantial manual work. Since the number of three-dimensional RNA structures is increasing, a more systematic and automated method is needed. We have developed a set of algorithms for recognizing secondary and tertiary structural elements in RNA molecules and in the protein-RNA structures in protein data banks (PDB). The algorithms were implemented in to a web-based program called PairAnalyzer. The present work represents the first attempt at extracting RNA structure elements from atomic coordinates in structure databases. The regularities in the structure elements revealed by the algorithms should provide useful information for predicting the structure of RNA molecules bound to proteins. PairAnalyzer is accessible at http://wilab.inha.ac.kr/PairAnalyzer/.

1 Introduction

Mining biological data in databases has become the focus of increasing interest over the past several years. However, most data mining in bioinformatics is limited to sequence data. The structure of a molecule is much more complex, but it is important as it determines the biological function of the molecule. It is therefore not enough just to analyze sequence data if one wishes to understand the structure of a molecule more completely.

We have developed a set of algorithms and a program called PairAnalyzer that recognize secondary and tertiary RNA structure elements from the three-dimensional atomic coordinates of protein-RNA complexes obtained from protein data bank (PDB), which provides a rich source of structural data [1]. The structure data were first cleaned up to make all the atoms accurately named and ordered, and no atoms

---

[*] To whom correspondence should be addressed. Email: khan@inha.ac.kr

M. Bubak et al. (Eds.): ICCS 2004, LNCS 3036, pp. 279–286, 2004.
© Springer-Verlag Berlin Heidelberg 2004

have alternate locations. PairAnalyzer identifies hydrogen bonds and base pairs, and classifies the base pairs into one of 28 types [2]. These base pairs include non-canonical pairs such as purine-purine pairs and pyrimidine-pyrimidine pairs as well as canonical pairs such as Watson-Crick pairs and wobble pairs. PairAnalyzer also extracts RNA sequences to integrate them with the data of base pairs. Secondary or tertiary structural elements consisting of base pairs are then visualized for user scrutiny. To the best of our knowledge, this is the first attempt to extracting RNA structural elements from the atomic coordinates in structure databases. PairAnalyzer is intended for analyzing RNA structures. However, it can also be used for analyzing DNA structures since DNA is similar to RNA in hydrogen bonding between complementary bases.

2   Background of Base Pairs and Base Pairing Rules

An RNA nucleotide consists of a molecule of sugar, a molecule of phosphoric acid, and a molecule called a base. A base pair is formed when one base is paired with another base by hydrogen bonds. Base pairs can be classified into canonical base pairs (Watson-Crick base pairs) and non-canonical base pairs. We consider base pairs of 28 types [2] comprising both canonical and non-canonical base pairs. Fig. 1 shows four base pairs.

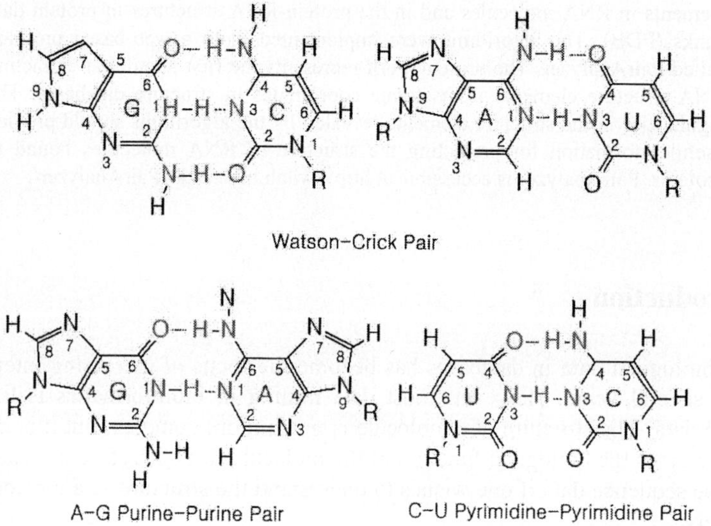

Watson–Crick Pair

A–G Purine–Purine Pair   C–U Pyrimidine–Pyrimidine Pair

Fig. 1. G-C and A-U Watson-Crick pairs, G-A purine-purine pair, and C-U pyrimidine-pyrimidine pair

A base consists of a fixed number of atoms (see Fig 1). These fixed numbers provide important clues for extracting base pair data and classifying the data into types of base pairs. Base pairs are formed by hydrogen bonding between atoms of base. For example, the Watson-Crick A-U pair has two hydrogen bonds: between N1 of adenine (A) and N3 of uracil (U), and between N6 of A and O4 of U. Thus we can define the hydrogen bonds that generate base pairs and classify the base pairs. In this study we define base pair rules to classify base pairs, and divide them into 28 types by means of these base pair rules.

3 Algorithms

Our algorithm is divided into two parts. The first part extracts information about secondary and tertiary structure elements of RNA by analyzing data in a PDB file [1]. We use HB-plus [3] to obtain data on all the hydrogen bonds that are present from the PDB file, and this data is used to generate base pair data. We can then obtain insight into the secondary or tertiary structure elements of the RNA by analyzing this data and integrating it with sequence data. The second part derives a visual representation of the structure of the RNA by integrating the information about structure elements obtained in the first part with knowledge of the coordinates of the nucleotides. Fig. 2 and 3 show the framework of the first part and second part, respectively. This section describes the algorithms for the two parts.

3.1 First Part: Extracting Structure Elements of RNA

The first part consists of 5 steps, and the final output is information about the secondary and tertiary structure elements of the RNA.

Step 1: From a PDB file, extract data on all the hydrogen bonds by using HB-plus [3], and record this data in Hydrogen Bonds.

Step 2: Extract the RNA sequence data by analyzing the PDB file, and record it in RNA-SEQ.

Step 3: Extract only those hydrogen bonds that bond one base to another from the hydrogen bond data obtained in Step 1. Record these hydrogen bonds in the Base-Base List.

Step 4: Extract those hydrogen bonds that are involved in base pairing, and classify them into the 28 types by means of the Base pair rules. Record these hydrogen bonds separately in the Base-Pair List.

Step 5: Integrate the sequence data in RNA-SEQ with the base pairs data in the Base-Pair List. Match all the nucleotides in RNA-SEQ to the nucleotides in the Base-Pair List to determine the hydrogen bonding relationships of each nucleotide.

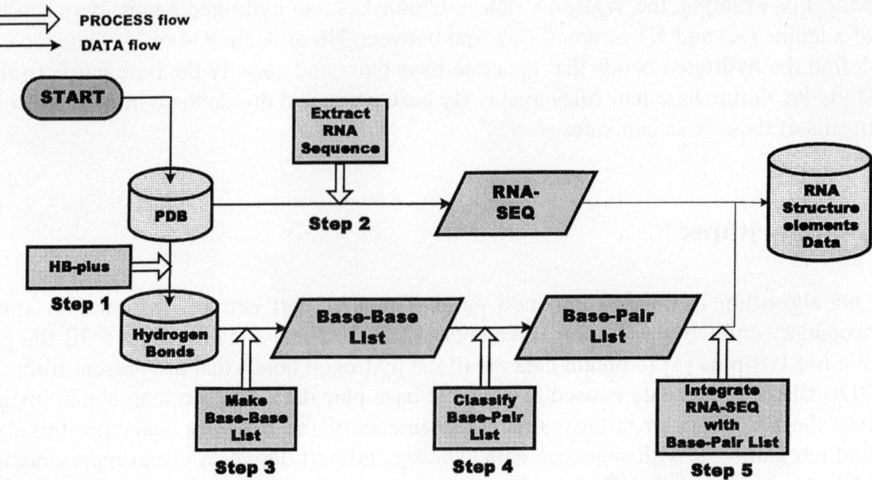

Fig. 2. Framework for extracting base pairs of 28 types and structure elements of RNA from PDB

3.2 Second Part: Visualizing RNA Structure

The 3 steps of the second part are outlined as follows.

Step 1: Obtain the 3D coordinate of every nucleotide by computing the average coordinate values of all the atoms of the nucleotide.

Step 2: Integrate the 3D coordinates of nucleotides with structure elements, and derive the connectivity relation among nucleotides.

Step 3: Represent the structure of the RNA visually by combining the information about structure elements with the coordinate values of the nucleotides.

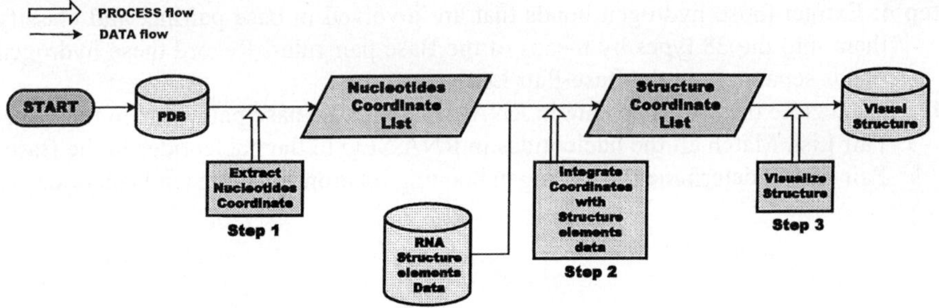

Fig. 3. Framework for visualizing the RNA structure

4   Experimental Results

PairAnalyzer is written in Microsoft Visual C#, and is executable within a Web browser on any PC with Windows 2000/XP/Me/98/NT 4.0 as its operating system. PairAnalyzer takes as input PDB file and HB2 file. As output, PairAnalyzer produces a drawing of RNA structure elements and information about base pairs. Fig. 4 displays the tertiary structure of tRNA (PDB identifier: 1EHZ) derived by PairAnalyzer. Nodes of the drawing indicate nucleotides of RNA and the blue lines indicate that nucleotides are connected in the RNA backbone. In addition, the red dotted lines indicate that two bases are hydrogen bonded. The text window in the left shows the information about the nucleotide sequence, base pairs and their types of tRNA.

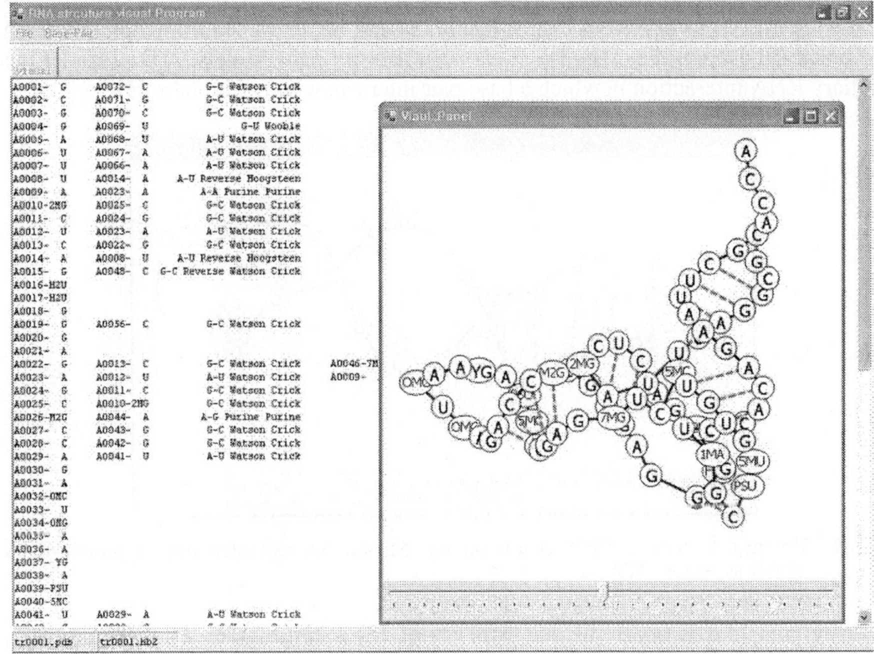

Fig. 4. Interface of PairAnalyzer with the main window and visual panel

Although PairAnalyzer is intended for analyzing RNA structures, it can be used for analyzing DNA structures since DNA is similar to RNA in hydrogen bonding between complementary bases. Fig. 5 shows a small structure of Z-DNA (PDB identifier: 249D) obtained by PairAnalyzer. Z-DNA is a left-handed structure [4]. PairAnalyzer can extract structure elements of left-handed nucleic acids as well as regular nucleic acids because its structure analysis is based on extracting base pairs formed by hydrogen bonds.

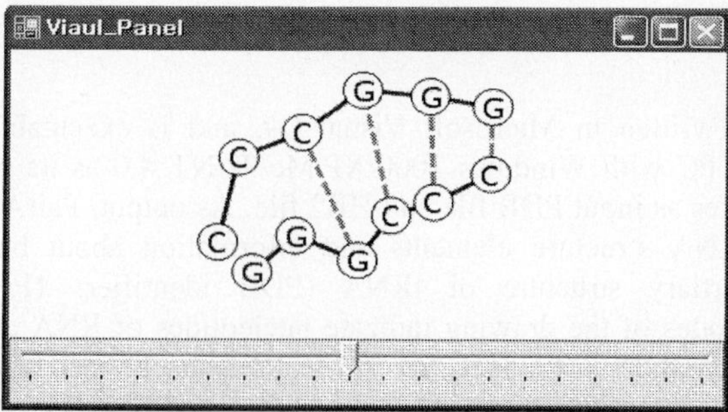

Fig. 5. Z-DNA structure (PDB identifier: 239D) extracted by PairAnalyzer

PairAnalyzer can extract a structure involving multiple RNA stands. The structure shown in Fig. 6 has two RNA chains (chains M and N), extracted from a protein-RNA complex (PDB identifier: 1DFU). It can also identify base-triplets. A base-triplet is a tertiary RNA interaction in which a base pair interacts with a third base [5].

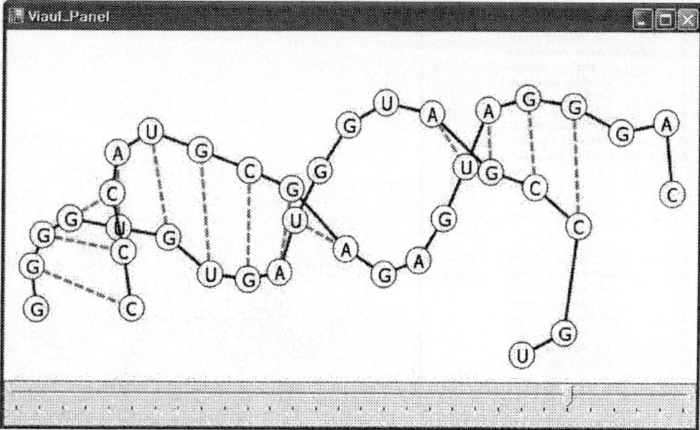

Fig. 6. Structure to have 2 RNA chains (chains M and N), extracted from a protein-RNA complex (PDB identifier: 1DFU)

Programs like Rasmol [6] and Mol-Script [7] can generate the structure of a molecule from the three-dimensional coordinates of its atoms. There are also programs that represent secondary or tertiary structure elements in a plane. However with programs like Rasmol and Mol-Script one cannot easily obtain information about each nucleotide in the RNA and the binding relations between the nucleotides, because these programs represent the structures of molecules at the atomic level. In addition programs that visualize structure elements in a plane have difficulty representing tertiary structure elements. On the other hand, our algorithm uses the three-dimensional coordinates of the nucleotides to generate secondary and tertiary structures. Hence it produces stereoscopic RNA structures. Moreover it provides not only the configuration of a given RNA molecule but also the bonding relations and types of base pairs between the nucleotides.

Fig. 7 shows the tertiary structure of domain V of 23S ribosomal RNA (PDB identifier: 1FFZ) drawn by PairAnalyzer. This structure consists of many nucleotides and has complex structure elements. If input data in PDB file is given, PairAnalyzer can analyze any structure and extract information about structure elements.

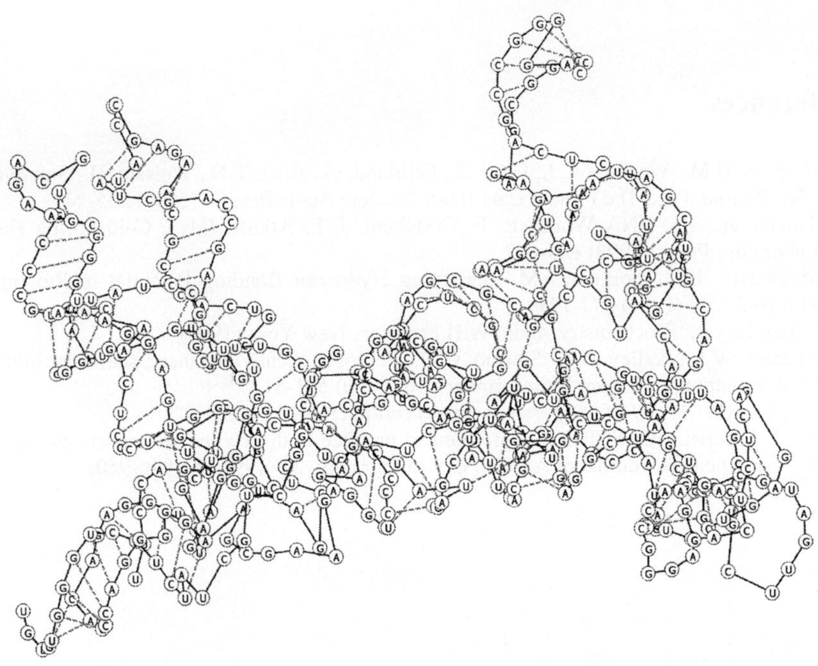

Fig. 7. Tertiary structure of domain V of 23S ribosomal RNA (PDB identifier: 1FFZ) extracted and visualized by PairAnalyzer

5   Conclusion

Up to now, extracting secondary and tertiary structure elements of RNA from the three-dimensional atomic coordinates has relied upon a substantial amount of manual work. In this study we have developed a set of algorithms for recognizing secondary or tertiary structure elements of RNA in protein-RNA complexes obtained from PDB. Experimental tests showed that our algorithm is easily capable of automatically extracting base-triplet structures and all secondary or tertiary structure elements formed by hydrogen bonding. To the best of our knowledge, this is the first attempt to extract and visualize RNA structure elements from the atomic coordinates in structure databases. We expect it to help research on RNA structures, and the regularities in the structure elements discovered should provide useful information for predicting the structure of RNA molecules bound to proteins.

Acknowledgements. This work was supported by the Ministry of Information and Communication of Korea under grant 01-PJ11-PG9-01BT00B-0012.

References

1. Berman, H.M., Westbrook, J., Feng, Z., Gilliland, G., Bhat, T.N., Weissig, H., Shindyalov, I.N., Bourne, P.E.: The Protein Data Bank. Nucleic Acids Res. 28 (2000) 235-242
2. Tinoco, Jr.: The RNA World (R. F. Gesteland, J. F. Atkins, Eds.), Cold Spring Harbor Laboratory Press, (1993) 603-607
3. McDonald, I.K. Thornton, J.M.: Satisfying Hydrogen Bonding Potential in Proteins. J. Mol.Biol. 238 (1994) 777-793
4. Lubert Stryer.: Biochemstry. 4edn. W.H.Freeman, New York (1995)
5. Akmaev, V.R., Kelley, S.T., Stormo, G.D.: Phylogenetically enhanced statistical tools for RNA structure prediction. Bioinformatics 16 (2000) 501-512
6. Roger Sayle : *RASMOL*, http://www.umass.edu/microbio/rasmol/
7. Per J. Kraulis: MOLSCRIPT: a program to produce both detailed and schematic plots of protein structures, Journal of Applied Crystallography. 24 (1991), pp 946-950.

A Parallel Crawling Schema Using Dynamic Partition

Shoubin Dong, Xiaofeng Lu, and Ling Zhang

Network Research Center, South China University of Technology,
510640 Guangzhou, China
`{sbdong,xflv,ling}@scut.edu.cn`
Tel: (8620)87110014
Fax: (8620)87110019

Abstract. Parallel crawling is a key issue for search engine. In this paper we propose a parallel crawling schema based on dynamic partition, in order to fully utilize the available resources and achieve the best of load balance. The crawling schema is evaluated based on parallel metrics and performance of load balance. A prototype system built on Grid middleware has been constructed to demonstrate its efficiency and flexibility.

1   Introduction

As the size of the web is growing explosively, web search engines are becoming increasingly important as the primary means to retrieve information on the Internet. Most search engines use parallel web crawlers to retrieve large collections of web pages, in order to achieve a maximized download rate. However the competition among parallel crawling may result in redundant crawling and wasted resources.

Several researches have been conducted on parallel and distributed crawlers [1-7]. Some features of Google have been introduced in [1], where the crawling mechanism is described as a two-stage procedure. First, a URL server sends URLs to several web crawlers, where pages are fetched in parallel. Second, the downloaded pages are sent to the a central indexer, in which new URLs are parsed out for PageRank computing, and then forwarded to the URL server for next crawling. UbiCrawler[2] is a scalable fully distributed web crawler, in which each agent/robot is responsible for approximately the same number of URLs using Identifier–Seeded Consistent Hashing to achieve load balancing. WebRace[3] is a java-implemented distributed crawler, to collect, annotate and disseminate information from heterogeneous Internet source and protocols. Shkapenyuk and Suel [4] gave a detailed description of the architecture of a distributed crawler. It primarily discusses about the I/O and network efficiency aspects of a crawling system and the scalability issues in terms of crawling speed and number of participating nodes. Hash function is engaged to partition the space of all possible web URLs. Walker [5] used MPI and genetic programming to simulate the results for parallel pseudo-search engine. Travatore [6] is a kind of platform-independent distributed crawler, of which an agent is consisted of three elements: store, frontier and controllers. In [7], Cho proposed a general parallel crawling structure, and evaluated some crawling schema based on static partition according to the parallel metrics.

M. Bubak et al. (Eds.): ICCS 2004, LNCS 3036, pp. 287–294, 2004.

In our work, we propose a new parallel crawling schema based on dynamic partition mechanism. Based on the dynamic partition, we are able to extend the parallel crawler to run on grid nodes, thus to construct a high performance, fault tolerant and scalable crawler in grid environment.

2 Parallel Crawling Architecture

The parallel crawling architecture, shown in figure 1, is composed of agents and coordinators. An **agent** is a grid node, it performs its download task by running several threads, each of which is called **C-proc**. Each C-proc is dedicated to the visit of a single site. We make sure that different C-proc visit different sites at the same time, so that each site is not overloaded by too many requests. A **coordinator** is to coordinate the behavior of the agents. It may be also an agent.

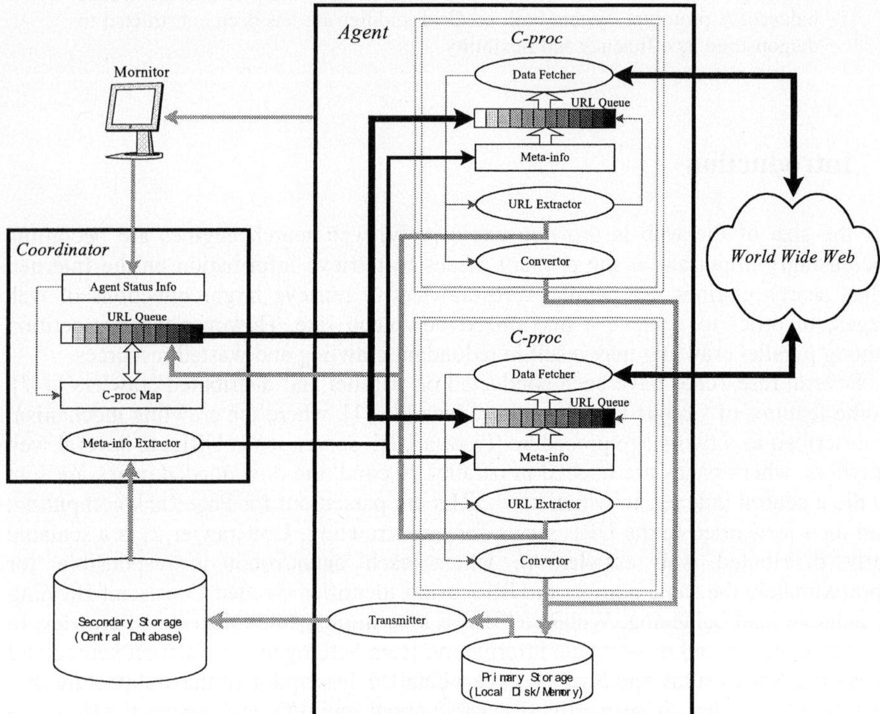

Fig. 1. The Parallel Crawling Architecture

Before the crawler starts the crawling procedure, a number of seed URLs, which are grouped by site names and ordered by prior knowledge, if available, have to be set to the "URL Queue" in the coordinator. The monitor collects the status (such as CPU usage, memory usage, disk size, network bandwidth, etc.) of every agent, and save the information in the "Agent Status Info" of coordinator. When the crawler gets started, the coordinator assigns the number of available C-proc's groups of seed URLs to appropriate C-proc's, based on agent status information and our dynamic partition algorithm, and uses "C-proc Map" to keep the information about which site each C-proc is responsible for at present. Meanwhile, the "Meta-Info Extractor" collects statistics of the formerly downloaded objects in the central database, and distributes them to the corresponding C-proc's, so that C-proc can determine whether pages within the site need to be downloaded. C-proc keeps this statistics in its "Meta-Info".

Objects are retrieved by "Data Fetcher" and then sent to "Convertor", where objects including web pages and objects of other formats are converted into XML formats and saved to the primary storage – agent's local memory or disk, which is supervised by a "Transmitter" located in the same agent. When the number of documents in the primary storage grows to certain degree, all data in the primary storage would be packed and sent to the central database through the transmitter. Meanwhile "URL extractor" extracts new URLs from the downloaded objects and sends those that are local to C-proc's assigned site to the "URL queue" of C-proc, while sends others back to the "URL queue" in the coordinator. Meanwhile "URL extractor" extracts new URLs from the downloaded objects and sends those that are local to C-proc's assigned site to the URL queue of C-proc, while sends others back to the "URL queue" in the coordinator. When some C-proc's URL queue is empty, it asks the coordinator for new assignment. These steps would repeat until there is no URL in the URL-queues of both coordinator and C-proc's.

3 Dynamic Partition Algorithm

The web can be partitioned in several ways; in particular, the partition can be obtained from URL-based hash, site-based hash or hierarchically by domain name. The key feature of our model is to conduct dynamic task assignment. This is currently done by means of partition algorithm based on two parameters: **size** of a site and **capacity** of a C-proc. Size of a site may be measured by the number of objects the site holds or the time elapse of last full download of all objects in this site. The size is estimated after the first download, and will be re-estimated after each download. Capacity of a C-proc is proportional to the available resources of an agent. The crawling related resources include CPU, memory, and storage and network bandwidth. According to our design, the partition of URL queues is implemented according to algorithm 3.1.

Algorithm 3.1 Dynamic partition algorithm

Input: S is an ordered set of sites according to their sizes
Input: P is an ordered set of C-proc's according to their capacities
Required: M is the mapping list of a pair set (site, C-proc)
Required: A is the list of all agents
Required: U is a list of global URL-queue group by sites, managed by coordinator
Required: URL-queue for each C-proc, maintained by each C-proc
1: Initial assignment: $p_0 \leftarrow s_0$ {assign s_0 to p_0}, $p_1 \leftarrow s_1$,....,
2: add pair $(s_0,p_0),(s_1,p_1)$... to the mapping list M;
3: delete the assigned sites from S
4: while $|S| > 0$ do

```
 5:   for all p_i □ P, do      /* Check for the tasks of C-proc's
 6:       if the task of this C-proc p_i has been finished, do
 7:           delete pair (s_k, p_i) from the mapping list M
 8:           assign p_i ← s_0, delete s_0 from queue S, add pair (s_0,p_i) to list M
 9:       end if
10:       if the C-proc p_i found some new URLs not local to its given site s_j, do
11:           send the URLs to URL-queue of coordinator U
12:       end if
13:   end for
14:   for all a_i □ A, do
15:       if the agent a_i has failed, do
16:           for all C-proc p_i running on this agent, do
17:               add s_k to the head of list S , where (s_k, p_i) is a pair value in M
18:               delete pair (s_k, p_i) from the mapping list M
19:           end for
20:       end if
21:   end for
22:   if some new agent is available, do
23:       start several C-proc on this agent
24:       add these C-proc to list P, re-order P according to their capacities
25:   end if
26:   Order the list of global URL-queue U according to their sites
27:   for all new URLs u_i □ U, suppose the site of u_i is s_k, do
28:       if (s_k, p_i) exists in the lists of M, do
29:           add new URLs u_i to the URL-queue of C-proc p_i
30:       else if s_k is a new discovered site, thus not belonging to S, do
31:           add s_k to the tail of list S
32:       end if
33:   end for
34: end while
```

It should be mentioned that (1) The assignment of $s_k \rightarrow p_i$ means that all URLs of the site s_k are put into URL-queue of C-proc p_i; (2) Step 10-12 indicates that the hyperlinks that are not local to the given site of C-proc are sent to coordinator, then dispatched to the right C-proc by coordinator (step 28-29) or saved in the URL-queue of coordinator for new assignment; (3) The agents is allowed to be dynamically changed in the running. New agents may participate in tasks, while crash agents may be left out.

It is obvious that this kind of assignment may achieve the optimized results because it fully utilizes the available resources. We will discuss this in detail by experiment in section 5.

4   Implementation Issues

To achieve our design goals, we propose a four layers' implementation architecture as shown in figure 2:

- Storage layer: to perform the distributed and parallel file management;
- Scheduling layer: to act as the status monitor and task scheduler;
- Communication layer: to coordinate the data transfer from primary storage to secondary storage and the communication between agents and coordinator;
- Application layer: to perform the tasks of data gathering.

We develop our system based on Globus Toolkit (http://www.globus.org/) and Sun Grid Engine (http://wwws.sun.com/software/gridware/sge.html). Agents and coordinator exchange URLs and control information by MPI (Message Passing Interface). The GridFTP of Globus is used to transfer the large amount of data, such as the data transfer from primary storage to secondary storage. Next we discuss two main issues.

Task Scheduling. Sun Grid Engine (SGE) is used to collect the status of grid nodes, and schedule the tasks running on the nodes. Each "Execution Host" acts as an agent, while "Master Host" acts as a coordinator. When data gathering is started, the ordered tasks according to prior knowledge are submitted to "Submit Host". Then the Master Host begins to schedule the tasks according to the resources of all Execution Hosts and SGE's policies that may be configured by the system administrator. When the tasks finish, the Execution Host may release the resources and Master Host will automatically assign new tasks to it. In the running, an agent may join or leave freely. Administration Host manages all register information of nodes, thus its role is the "Monitor". Master Host may re-schedule the tasks according to the register information and status of task execution. The system may also query SGE for the status of all grid nodes, and save them in "Status Info" for further use. Thus, the employee of SGE simplifies the implementation of the system.

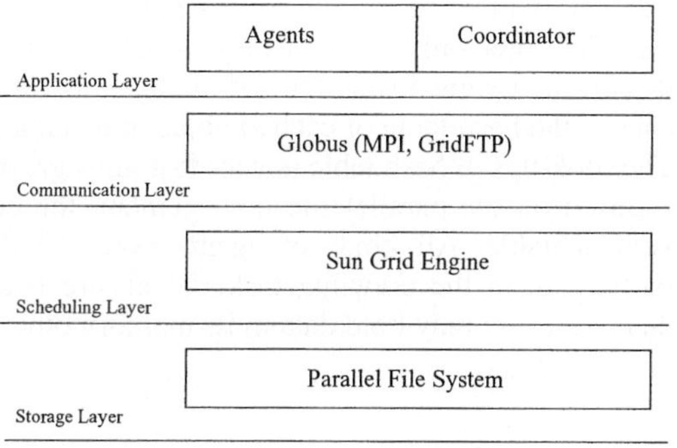

Fig. 2. The Implementation Architecture

Parallel File Management. Web crawler is a heavy I/O system, thus the file management is a key factor that greatly affects its performance. To support dynamic partition technique, we use a virtual central database to hold the data. This central database can be disk array, SAN (Storage Area Network), or Data Grid [8].

At first the data is downloaded into the primary storage. After some time (depend on the amount of data and available storage resources), the downloaded objects may be compressed and transferred back to the secondary storage, namely, central database. This kind of batch communication greatly improves the performance of I/O.

To avoid the overhead of the repeated downloading and analysis of documents that have not been modified, the Data Fetcher uses cached object metadata stored in "Meta-Info" to decide whether to download the documents that are already in the cache. The content of this metadata is represented as <URL, Size, Last-modified>. This metadata has been built into a hash table to facilitate the fast look-ups. According to the partition algorithm 3.1, the assignment of task to C-proc is site based. When one C-proc starts, the coordinator constructs a hash table of cache objects' metadata in the site assigned to the C-proc, and sends this hash table to the C-proc.

The "Data Fetcher" retrieves a URL from the URL-queue, make the HTTP connection, retrieve the URL and analyze the HTTP-header of the response message, and then access the hash table of cached object metadata and check if the URL retrieved from the URL-queue is corresponding to the cached object metadata. If the object is not in the cache or has been modified since last fetching, download the body of the object and store it in primary storage, and send the object to "URL Extractor" and "Convertor" for further processing.

Each C-proc only maintains the metadata of those cached objects, which are local to its current assigned site. Instead of checking for the files on the disk, the Data Fetcher checks the hash table of cached object metadata for the update information of objects. The look-ups of hash table is very fast, thus greatly improves the performance.

Based on the parallel file management, the construction of the parallel crawler is very scalable. All kinds of agents even the diskless workstations are allowed to participate in the crawling task. To ensure batch communication, the C-proc on a diskless agent may hold data in its memory other than local disk.

5 Evaluation

Evaluation model of parallel crawling schemes is described in details in [7]. Here we discuss the parallel properties based on these four key metrics of parallel crawlers.

Overlap. Overlap of downloaded pages is defined as $(N - I) / I$. Here, N represents the total number of pages downloaded by the overall crawler, and I represents the number of unique pages downloaded by the overall crawler. Even in the presence of faults, our model achieves overlap 0, which is optimal, because all C-proc's download the data according to the hash tables of cached objects, and the hash tables are constructed from the updated central database.

Coverage. Coverage of all pages that ought to be downloaded is defined as I / U, where U represents the total number of pages that the overall crawler has to download, and I is the same as in the definition of Overlap. Theoretically, our model achieves coverage 1, which is optimal, even when fault occurs.

If the objects stored locally on a crashed agent have not been transferred back into central database, they will be fetched by the new C-proc on other agents responsible for them. If the objects stored locally on a crashed agent have been transferred back into central database, they will not be fetched by any new C-proc. Because when coordinator starts to assign new task to a C-proc, it will send to the C-proc a hash table of cached objects that belong to the site that the C-proc will be responsible for, as stated in section 4.

Quality. Quality of downloaded pages is defined as $/ A_N \cap P_N / / / P_N /$, where A_N stands for the set of the most important N pages that an actual crawler would download, and P_N represents the set of the most important N pages that an ideal crawler would download. Though our crawler uses a parallel per-site breadth-first visit, without dealing with ranking and quality of page issues, it is proved that a breadth-first single-process visit tends to visit high-quality pages first [2,9]. Thus this kind of crawler tends to have a very good performance in quality.

Communication overhead. Communication overhead is defined as L/N, where L represents the total number of inter-partition URLs exchanged by the overall crawler, and N is the same as in the definition of Overlap, represents the total number of pages downloaded by the overall crawler. As it is stated in [2] that on the average every page contains just one link to another site, we have that n crawled pages will give rise to n URLs that must be potentially communicated to coordinator and other agents. By the definition, the communication overhead is thus no more than 1, which means that the crawler consumes very small network bandwidth for URL exchanges.

To evaluate the performance, we conducted an experiment to demonstrate the advantage of dynamic partition methods. In this experiment, six C-proc's were managed to run on three grid nodes to do the data gathering tasks. They downloaded totally 98585 pages from 30 different sites. We tested four partition methods: "Static-Random" is to partition tasks by static partition, each C-proc was responsible for nearly same number of sites; "Static-Size" is to partition sites according to their objects numbers by static partition, each C-proc tries to be responsible for nearly the same number of objects; "Dynamic-Random" is to partition tasks by dynamic partition algorithm (Algorithm 3.1), the difference is that set of sites S is a random set; "Dynamic-Size" is also to partition sites by dynamic partition algorithm, and S is an ordered set of sites by their sizes. All C-proc's cooperate together to finish exactly the same tasks each time for different partitions. The workload contributed by one C-proc, called "normalized workload", is measured by the download time of this C-proc divided by the sum of download time of all C-proc's. We use normalized workload rather than the crawling time, because the former is not affected by actual network conditions and server performance of every site while the latter does. Thus it is a good measure for the load balance.

Table 1. The normalization workload of C-proc's under different partition methods

Partition C-Proc	1	2	3	4	5	6	stdev
Static-Randam	0.27	0.20	0.09	0.14	0.16	0.14	6.19%
Static-Size	0.15	0.23	0.14	0.16	0.14	0.18	3.44%
Dynamic-Random	0.17	0.14	0.20	0.19	0.16	0.14	2.50%
Dynamic-Size	0.17	0.16	0.17	0.18	0.16	0.16	0.82%

The experimental result is shown in Table 1. The standard variation (stdev) means that the difference of normalized workload distributed in C-proc's. It is observed that the dynamic partition which achieve the smaller standard variation value, is very effective to shorten the difference of download time among C-proc's, thus provides the best of load balance. And the results also demonstrated that the system would achieve better load balance if there exists more prior knowledge.

6 Conclusion

In our work, we address the challenge of designing and implementing a parallel crawler in the context of Grid middleware. We introduce a parallel crawling schema designed using dynamic partition mechanisms to achieve high performance, fault-tolerance and scalability, and evaluate our model with the criterions of parallel crawler and performance of load balance. Further work will be included in the near future, such as the employment of page ranking technique to improve the crawling quality of the crawler.

References

1. S. Brin and L. Page: The anatomy of a large-scale hypertextual web search engine. Computer Networks (1998) 107-117
2. P. Boldi, B. Codenotti, M. Santini, and S. Vigna: Ubicrawler: A scalable fully distributed web crawler. In: Proc. AusWeb02. The Eighth Australian World Wide Web Conference (2002)
3. D. Zeinalipour-Yazti, M. Dikaiakos: Design and Implementation of a Distributed Crawler and Filtering Processor. In: A. Halevy, A. Gal (Eds.): Proceedings of the Fifth International Workshop on Next Generation Information Technologies and Systems (NGITS'2002). Lecture Notes in Computer Science, vol. 2382. Springer (2002) 58-74
4. V. Shkapenyuk and T. Suel: Design and implementation of a high-performance distributed Web crawler. In: Proceedings of the 18th International Conference on Data Engineering (ICDE'02). San Jose, CA (2002) 357-368
5. R. L. Walker: Dynamic load balancing model: Preliminary results for parallel pseudo-search engine indexers/crawler mechanisms using MPI and genetic programming. VECPAR 2000. Porto, Portugal (2000) 61-74
6. P. Boldi, B. Codenotti, M. Santini, and S. Vigna: Trovatore: Towards a highly scalable distributed web crawler. In: Proc. of 10th International World Wide Web Conference. Hong Kong, China (2001)
7. J. Cho and H. Garcia-Molina: Parallel crawlers. In: Proc. of the 11th International World–Wide Web Conference (2002)
8. P. Andrews, T. Sherwin and B. Banister: A centralized data access model for grid computing. In: Proceeding of the 20th IEEE/11th NASA Goddard Conference on Mass Storage Systems and Technologies (MSS'03). San Diego, California (2003)
9. Marc Najork and Janet L. Wiener: Breadth-first search crawling yields high quality pages. In: Proc. of 10th International World Wide Web Conference. Hong Kong, China (2001)

Hybrid Collaborative Filtering and Content-Based Filtering for Improved Recommender System

Kyung-Yong Jung[1], Dong-Hyun Park[2], and Jung-Hyun Lee[3]

[1] HCI Lab., Department of Computer Science & Engineering,
[2] Department of Industrial Engineering,
[3] Department of Computer Science & Engineering,
Inha University, Korea
`kyjung@gcgc.ac.kr, {dhpark,jhlee}@inha.ac.kr`

Abstract. The growth of the Internet has resulted in an increasing need for personalized information systems. The paper describes an autonomous agent, WebBot: Web Robot Agent, which integrates with the web and acts as a personal recommender system that cooperates with the user on identifying interesting pages. Hybrid components from collaborative filtering and content-based filtering, a hybrid recommender system can overcome traditional shortcomings. In this paper, we present an effective hybrid collaborative filtering and content-based filtering for improved recommender system. Experimental results indicate the hybrid collaborative filtering and content-based filtering better than collaborative, content-based, and combined filtering approach.

1 Introduction

The world wild web hypertext system is a very large distributed digital information space. Some estimates suggested that the web included about 160 million pages and this number double every four months. As more information becomes available, it becomes increasingly difficult to search for information without specialized aides.

Recommender systems are designed to predict user preferences using features of the items and ratings given by other users. To be effective, a recommender system must deal with well with two fundamental problems. First, the sparse rating problem; the number of ratings already obtained is very small compared to the number of ratings that need to be predicted. Effective generation from a small number of examples is thus important. This problem is particularly severe during the startup phase of the system when the number of users is small. Second, the first-rater problem; an item cannot be recommended unless a user has rated it before. This problem applies to new items and also obscure items and is particularly detrimental to users with eclectic tastes. [1,2,4,5,7,9,11] are presented as solutions to the aforementioned problems by using both collaborative filtering and content-based filtering method. LSI [11] and SVD [2] classification are used to decrease the number of dimensions in the matrix to solve the sparse rating problem in collaborative filtering, yet they fail to fix the first-

M. Bubak et al. (Eds.): ICCS 2004, LNCS 3036, pp. 295–302, 2004.

rater problem. [1,4,5] solve the first-rater problem, yet fail to fix the sparse rating problem. In an attempt to find a solution to both the sparse rating problem and the first-rater problem, method [9] was implemented. We overcome these drawbacks of collaborative filtering systems, by exploiting content information of the items already rated. Our basic approach uses content-based filtering to convert a sparse user ratings matrix into a full ratings matrix; and then uses collaborative filtering to provide recommendations. We present the framework for hybrid filtering. We apply this framework in the domain of movie recommendation and show that our approach performs significantly better than both collaborative and content-based filtering.

2   Collaborative Filtering and Content Based Filtering

2.1   Collaborative Filtering

Collaborative filtering systems recommend objects for a target user based on the opinions of other users by considering how much the target user and another users have agreed on other objects in the past [4,5]. Collaborative filtering technique predicts the rating of a particular user u for an item i. And it compares the predicted rating with the rating of all other users who have rated the item i. Then a weighted average of the other users rating is used as a prediction. If I_u is set of items that a user u has rated then we can define the mean rating of user u by Equation (1).

$$w(u,a) = \frac{\sum_{i \in I_u \cap I_a}(r_{u,i} - \overline{r_u})(r_{a,i} - \overline{r_a})}{\sqrt{\sum_{i \in I_u \cap I_a}(r_{u,i} - \overline{r_u})^2 \bullet \sum_{i \in I_u \cap I_a}(r_{a,i} - \overline{r_a})^2}} \qquad \overline{r_u} = \frac{1}{|I_u|}\sum_{i \in I_u} r_{u,i} \qquad (1)$$

Collaborative filtering algorithms predict the rating based on the rating of similar users. When Pearson correlation coefficient is used, similarity is determined from the correlation of the rating vectors of user u and another users a. It can be noted that $w \in [-1, +1]$. The value of w measures the similarity between the two users' rating vectors. A high value close to $+1$ signifies high similarity and a low value close to 0 signifies low correlation (not much can be deduced) and a value close to -1 signifies that users are often of opposite opinion. The general prediction formula is based on the assumption that the prediction is a weighted average of the other users' rating. The weights refer to the amount of similarity between the user u and the other users by Equation (2). U_i represents the users who rated item i.

$$p^{collab}(u,i) = \overline{r_u} + \frac{1}{\sum_{a \in U_i} w(u,a)} \sum_{a \in U_i} w(u,a)(r_{a,i} - \overline{r_a}) \qquad (2)$$

It is common for the active user to have highly correlated neighbors that are based on very few co-rated item (overlapping; $I_u \cap I_a$). These neighbors based on a small number of overlapping item tend to be bad predictor. To devalue the correlation based on few co-rated items, we multiply the correlation by significance weighting factor. If two users have less than 45 co-rated items, we multiply their correlation by a factor

$sg_{a,u}=n/45$, where n is the number of co-rated items. If the number of overlapping items in greater than 45, then we leave the correlation unchanged i.e. $sg_{a,u}=1$.

2.2 Content-Based Filtering

We use a multinomial text model, in which a document is modeled as an ordered sequence of ordered sequence of word events drawn from the same vocabulary, V. The naïve Bayes assumption states that the probability of each word event is dependent on the document class but independent of the word's context and position. For each class c_j, and word, $w_k \in V$, the probability, $P(c_j)$ and $P(w_k|c_j)$ must be estimated from training data. Then the posterior probability of each class given a document D, is computed using naïve Bayesian classifier [8] by Equation (3).

$$p(c_j \mid D) = \frac{P(c_j)}{P(D)} \prod_{i=1}^{|D|} P(a_i \mid c_j) \tag{3}$$

Where a_i is the ith word in the document, and $|D|$ is the length of the document in words. Since for any given document, the prior $P(D)$ is a constant, this factor can be ignored if all that is desired is a rating rather than a probability estimate. A ranking is produced by sorting documents by their odds ratio, $P(c_1|D)/P(c_0|D)$, where c_1 represents the positive class and c_0 represents the negative class. An example is classified as positive if the odds are greater than 1, and negative otherwise. In our case, since movies are represented as a vector of "documents", d_m, one for each feature (where f_m denotes the mth feature), the probability of each word given the category and the feature $P(w_k|c_j, f_m)$, must be estimated and the posterior category probability for a film, F, computed using Equation (4).

$$p(c_j \mid F) = \frac{P(c_j)}{P(F)} \prod_{m=1}^{|f|} \prod_{i=1}^{|d_m|} P(a_{m,i} \mid c_j, f_m) \tag{4}$$

Where $|f|$ is the number of features and $a_{m,i}$ is the ith word in the mth feature. The class with the highest posterior probability determines the predicted rating. The *Laplace* smoothing is used to avoid zero probability estimates [8].

3 An Approach Hybrid Collaborative Filtering and Content-Based Filtering

The proposed hybrid filtering transparently creates and maintains user preferences. It assists users by providing both collaborative filtering and content-based filtering, which are updated in real time whenever the user changes his/her current page using any navigation technique. The WebBot uses the URLs provided in the EachMovie dataset to download movie content from IMDb [6]. WebBot keeps track of each individual user and provides that a user online assistance. The assistance includes two lists of recommendations based on two different filtering paradigms: collaborative filtering and content-based filtering. WebBot updates the list each time the user changes his/her current page. Content-based filtering is based on the correlation between the content of the pages and the user preferences. The collaborative filtering is based on a comparison between the user path of navigation and the access patterns of past users. Hybrid filtering may eliminate the shortcomings in each approach. By making collaborative filtering, we can deal with any kind of content and explore new domains to find something interesting to the user. By making content-based filtering, we can deal with pages un-seen by others. Fig. 1 is the system overview for hybrid filtering.

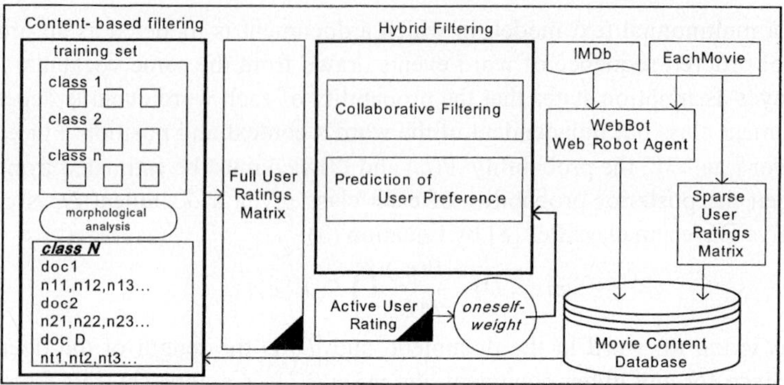

Fig. 1. System overview for hybrid collaborative filtering and content-based filtering

To overcome the problem of stateless connection in HTTP, WebBot follows users through tracking their IP address. To track user presence, a timeout mechanism is used to delete user's session information after a predetermined amount of idle time. So that, a connection after the specified period having the same IP is identified as a new user. This method is fairly easy to implement. Consequently, the IP of a proxy server may represent two or more people who are accessing the same web site simultaneously in their browsing sessions, causing an obvious conflict. However, the reality is that many large sites use this method and have not any clashes. The EachMovie dataset also provides the user-ratings matrix; which is a matrix of users versus items, where each cell is the rating given by a user to an item. We will refer to each row of this matrix as a user-rating vector. The user-ratings matrix is very sparse, because most users have not rated most items. The content-based predictor is trained on each user-ratings vector and a pseudo user-ratings vector is created. A pseudo user-ratings vector contains the user's actual ratings and content-based filtering for the un-rated items. All pseudo user-ratings vector put together from the pseudo ratings matrix, which is a full matrix. Now given an active user's ratings, filtering predictions are made for a new item using collaborative filtering on the full pseudo ratings matrix.

3.1 Extracting Information from Web Robot Agent and Building a Database

Our current prototype system, WebBot: Web Robot Agent uses a database of movie content information extracted from web page at IMDb (www.imdb.com). Therefore, the system's current content information about titles consists of textual metadata rather than the actual text of the items themselves. An IMDb subject search is performed to obtain a list of movie-description URLs of broadly relevant titles. WebBot then downloads each of these pages and uses a simple pattern based information extraction system to extract data about each title. Information extraction is the task of locating specific pieces of information from a document, thereby obtaining useful structured data from unstructured text. A WebBot follows the IMDb link provided for every movie in the EachMovie dataset [6] and collects information from the various links off the main URL. We represent the content information of every movie as a set of features. Each feature is represented simply as a bag of words. IMDb produces the information about related directors and movie titles using collaborative filtering: however, WebBot treats them as additional content about the movie. The text in each feature is then processed into an un-ordered bag of words and the examples represented as a vector of bags of words. Fig. 2 shows WebBot for extracting information, example web page.

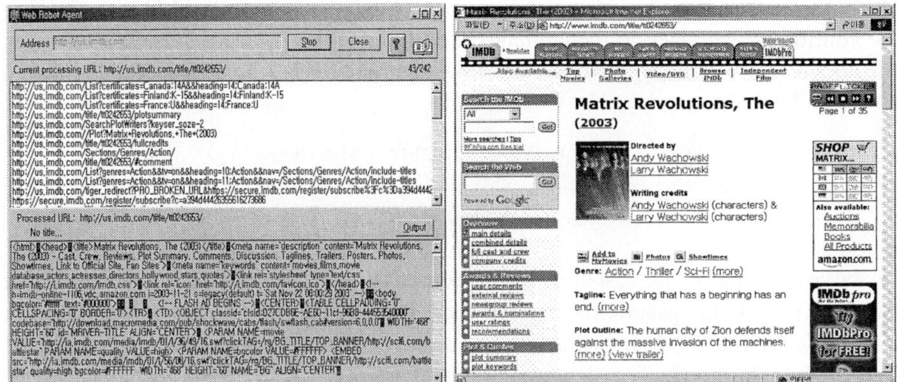

Fig. 2. WebBot: Web Robot Agent, example web page

3.2 Deriving Pseudo User-Ratings Vector from User-Item Matrix

We present now another approach which does not change the collaborative filtering algorithm but instead alters the rating database based on content-based criteria and ratings of real users. This approach was inspired by Sarwar's *rating-bots* approach for the Gouplens news filtering project: for that project software agents which used content-based criteria (spelling and article length) to rate news articles automatically and to increase the amount of ratings in the database [10]. The pseudo user-ratings are content-based filtering for that particular user. This means that some un-rated items are assigned a predicted rating, based on similarity between the rated items and the item for which the rating is missing [7]. We first create a pseudo user-ratings vector for every user u in the database. The pseudo user-ratings vector, v_u, consists of the item ratings provided by the user u, where available, and those predicted by the content-based filtering otherwise. The pseudo user-ratings vectors of all users put together gives the dense pseudo rating matrix V. We now perform collaborative filtering using this dense matrix. The similarity between the active user a and another user u is computed using Pearson correlation described in Equation (1). Instead of the original user votes, we substitute the votes provided by the pseudo user-ratings vectors v_a and v_u.

The accuracy of a pseudo user-ratings vector computed for a user depends on the number of movies he/she has rated. If the user rated many items, the content-based filtering is good and hence his pseudo user-ratings vector is fairly accurate. On the other hand, if the user rated only a few items, the pseudo user-ratings vector will not be as accurate. We found that inaccuracies in pseudo user-ratings vector often yielded misleadingly high correlations between the active user and other users.

3.3 Prediction of User Preference through Hybrid Filtering

The prediction for the active user is computed as a weighted sum of the mean-centered votes of the *best-n-neighbors* of the user. In our approach, we also add the pseudo active user to the neighborhood. However, we may want to give the pseudo active user more importance than the other neighbors. In other words, we would like to increase the confidence we place in the content-based filtering for the active user. Combining the above two weighting schemes, the final hybrid filtering prediction for the active user a and item i is produced by Equation (5).

$$p_{a.i} = \overline{v_a} + \frac{(c_{a,i} - \overline{v_a}) + \sum_{i=1,u\neq a}^{n} Hybridw_{a,u} w(a,i)(v_{u,i} - \overline{v_u})}{\sum_{i=1,u\neq a}^{n} Hybridw_{a,u} w(a,i)} \tag{5}$$

In above equation $c_{a,i}$ corresponds to the content-based filtering for the active user a and item i. $v_{u,i}$ is the pseudo user-rating for a user u and item i. And v_u is the mean over all items for that user. $w(a,i)$ are as shown in Equation (1) respectively; n is the size of neighborhood. The denominator is a normalization factor that ensures all weights sum to one.

4 Performance Evaluation

We used a subset of the EachMovie dataset [6]. This dataset contains 7,291 randomly selected users and 1,628 movies for which content was available from IMDb. To evaluate various approaches of filtering, we divided the rating dataset in test-set and training-set. The rating database is used a subset of the ratings data from the Eachmovie dataset. The training-set is used to predict ratings in the test-set using a commonly used error measure. The metrics for evaluating the accuracy of a prediction algorithm are used mean absolute error(MAE) and rank scoring measure(RSM) [3].

For evaluation, this paper uses the following methods: The proposed hybrid collaborative filtering and content-based filtering (HMW_HF), a collaborative filtering (P_Corr), the recommendation method using only the content-based filtering (Content), and a naïve combined approach (N_Com). The naïve combined approach takes the average of the ratings generated by the collaborative filtering and the content-based filtering. The various methods were used to compare performance by changing the number of clustering users. Also, the proposed method was compared with the previous methods in section 1 that use both collaborative filtering and content-based filtering method by changing the number of user evaluations on items. The aforementioned previous method includes the Soboroff method [11] that solved the sparse rating problem, the Fab method [1] that solved the first-rater problem, and the Pazzani method [9] that solved both the sparse rating problem and the first-rater problem.

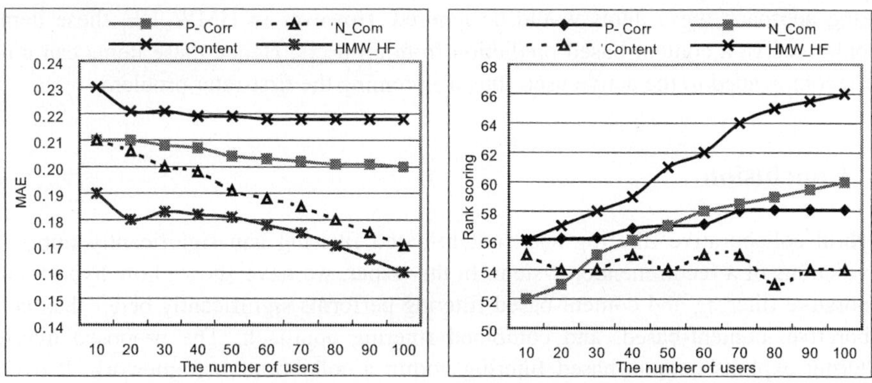

Fig. 3. MAE, Ranking scoring measure at varying the number of users

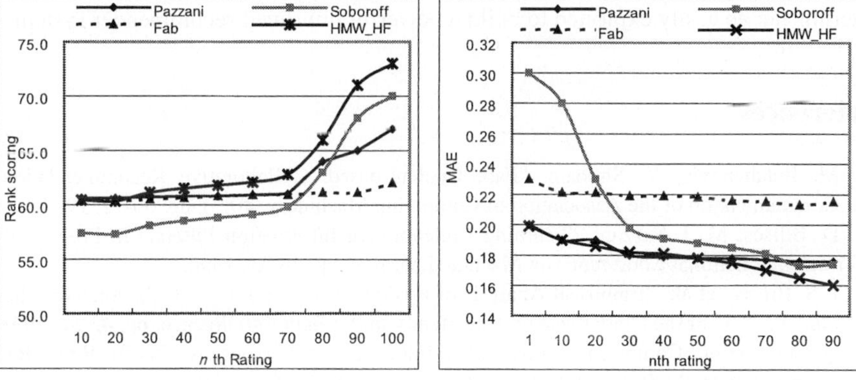

Fig. 4. MAE, Ranking scoring measure at nth rating

Fig. 3 shows the MAE and RSM of varying the number of users. Fig. 3, as the number of users increases, the performance of the HMW_HF, and the P_Corr also increases, whereas the method using content shows no notable change in performance. In terms of accuracy of prediction, it is evident that method HMW_HF, which uses both collaborative filtering and content-based filtering, is more superior to method N_Com. Fig. 4 is used to show the MAE and RSM when the number of user's evaluations is increased. In Fig. 4, the Soboroff method, which has the first-rater problem, shows low performance when there are few evaluations; the other methods outperform the Soboroff method. Although the Pazzani method, which solved both the sparse rating problem and the first-rater problem, along with the HMW_HF show high rates of accuracy, the HMW_HF shows the highest accuracy of all methods.

Since we use a pseudo ratings matrix, which is a full matrix, we eliminate the root of the sparse rating problem and the first-rater problem. Pseudo user-ratings vectors contain ratings for all items; and hence all users will be considered as potential neighbors. This increases the chances of finding similar users. The original user-ratings matrix may contain items that have not been rated by any user. In a collaborative filtering approach these items would be ignored. However in HMW_HF, these items would receive a content-based prediction from all users. Hence these items can now be recommended to the active user, thus overcoming the first-rater problem.

5   Conclusion

Hybrid collaborative filtering and content-based filtering can significantly improve predictions of a recommender system. In this paper, we have shown how hybrid collaborative filtering and content-based filtering performs significantly better than collaborative, content-based, and combined filtering approach. The proposed hybrid filtering exploits content-based filtering within a collaborative framework. It overcomes the disadvantages of both collaborative filtering and content-based filtering, by bolstering collaborative filtering with content and vice versa. Further, due to the nature of the approach, any improvements in collaborative filtering or content-based filtering can be easily exploited to build a powerful improved recommender system.

References

1. M. Balabanovic, Y. Shoham, "Fab: Content-based, Collaborative Recommendation," Communication of the Association of Computing Machinery, 40(3), pp. 66-72, 1997.
2. D. Billsus, M. J. Pazzani, "Learning Collaborative Information Filters," In Proc. of the 15th International Conference on Machine Learning, pp. 46-54, 1998.
3. J. S. Breese, et al., "Empirical Analysis of Predictive Algorithms for Collaborative Filtering," In Proc. of the Conference on Uncertainty in Artificial Intelligence, pp. 43-52, 1998.
4. N. Good, et al., Combining Collaborative Filtering with Personal Agents for Better Recommendations, In Proc. of National Conference on Artificial Intelligence, pp439-446, 1999.
5. W. S. Lee, "Collaborative Learning for Recommender Systems," In Proc. of the 18th International Conference on Machine Learning, pp. 314-321, 1997.
6. P. McJones, EachMovie dataset, URL: http://www.research.digital.com/SRC/eachmovie
7. P. Melville, et al., "Content-Boosted Collaborative Filtering for Improved Recommendations," In Proc. of the National Conference on Artificial Intelligence, pp. 187-192, 2002.
8. T. Mitchell, Machine Learning, McGraw-hill, New York, pp. 154-200, 1997.
9. M. J. Pazzani, "A Framework for Collaborative, Content-based and Demographic Filtering," Artificial Intelligence Review, 13(5-6), pp. 393-408, 1999.
10. B. M. Sarwar, et al., "Using Filtering Agents to Improve Prediction Quality in the GroupLens Research Collaborative Filtering System," In Proc. of the Conference on Computer Supported Cooperative Work, pp. 345-354, 1998.
11. Soboroff, C. Nicholas, "Combining Content and Collaboration in Text Filtering," In Proc. of the IJCAI'99 Workshop on Machine Learning in Information Filtering, pp. 86-91, 1999.

Object-Oriented Database Mining: Use of Object Oriented Concepts for Improving Data Classification Technique

Kitsana Waiyamai, Chidchanok Songsiri, and Thanawin Rakthanmanon

Computer Engineering Department, Kasetsart University, Thailand
Fengknw, fengtwr@ku.ac.th, schidchanok@kdl.cpe.ku.ac.th

Abstract. Complex objects are organized into class/subclass hierarchy where each object attribute may be composed of other complex objects. Almost of the existing works on complex data classification start by generalizing objects in appropriate abstraction level before the classification process. Generalization prior to classification produces less accurate result than integrating generalization into the classification process. This paper proposes CO4.5, an approach for generating decision trees for complex objects. CO4.5 classifies complex objects directly through the use of inheritance and composition relationships stored in object-oriented databases. Experimental results, using large complex datasets, showed that CO4.5 yielded better accuracy compared to traditional data classification techniques.

1   Introduction

Data classification, one important task of data mining, is the process of finding the common properties among a set of objects in a database and classifies them into different classes [2]. One of well-accepted classification method is the induction of decision trees. However, traditional decision tree based-algorithms [7], [8], [9] have failed to classify complex objects. Complex objects are organized into class/subclass hierarchy where each object attribute may be composed of other complex objects. The classification is usually performed at low-level concepts resulting in very bushy decision tress with meaningless results.

Several methods for generating decision trees from complex objects have been proposed. Wang et al. [3], [11] have integrated ID3 decision tree classification methods with AOI (*Attribute-Oriented Induction*) [4], [5] to predict object distribution over different classes. Han et al. [6] have developed a level-adjustment process to improve the classification accuracy in large databases. These methods require users to provide an appropriate concept hierarchy before the classification process. This is a difficult task even for experts to identify constraint such as balanced concept hierarchies. Further, generalization prior to classification using domain knowledge at the pre-processing stage produces less accurate result than integrating generalization/specialization into the classification process.

This paper proposes CO4.5, an approach for generating decision trees from complex objects. CO4.5 is based on inheritance relationships and composition relation-

M. Bubak et al. (Eds.): ICCS 2004, LNCS 3036, pp. 303–309, 2004.

ships stored in the object-oriented database. In the classification process, the method utilizes an object's attributes in different levels of the hierarchy by distinguishing object attributes and non-object attributes. For object attributes, they are reused several times in the classification process by generalizing them to their appropriate level through the use of inheritance relationships. By integrating generalization in the classification process, the method produces more efficient and accurate result compared to those methods using generalization prior to the classification process. Experimental results, using large complex datasets, showed that CO4.5 yielded better accuracy compared to the traditional data classification techniques.

The rest of the paper is organized as follows. Sect. 2 contains description of CO4.5 algorithm that generates decision trees through the use of inheritance and composition relationships. Sect. 3 analyzes the performance of the proposed algorithm. Finally, Sect. 4 contains conclusions and future work.

2   Using Object-Oriented Concepts for Complex Data Classification

Complex objects in object-oriented databases are objects that have inheritance and composition relationships between them. Inheritance relationship describes how a class reuses features from one another. Composition relationship describes how an object is composed of other objects. In this section, we present an algorithm CO4.5 for constructing decision trees. Object attributes and non- object attributes can be determined using composition relationships. Compared to non-object attributes, object attributes have more semantics; they are reused several times in the classification process by generalizing them to their appropriate level through the use of inheritance relationships. Once decision trees, of a given level, are obtained, the overall decision tree is obtained by combining all the level-based decision trees through the inheritance relationship. Level-based decision tree has *target class* called *determinant class*, which is lower-level class in the hierarchy, while the overall decision tree has traditional target class at the leaf nodes.

2.1   CO4.5 Algorithm

```
(1)     algorithm CO4.5(Node n)
(2)        begin
(3)            FindDTree(root_n, generalize-all-att-to-
        appropriate-level(Data_n), target_n)
(4)            if n ∉ Leaf then
(5)               for c ∈ child-of-n do begin
(6)                  CO4.5(c)
(7)               end for
(8)            end if
(9)     end.
```

Input of CO4.5 algorithm is set of complex objects stored in object-oriented databases. Notice that inheritance relationships can be inferred implicitly from the object-oriented database schema, or can be specified by the user. Output of CO4.5 is the overall aggregated decision trees. It is a greedy tree-growing algorithm, which constructs decision trees using top-down recursive divide-and-conquer strategy.

CO4.5 starts by building decision tree, using *FindDTree ()* function (described later in this section). Object-attributes are generalized into appropriate level by calling *generalize-all-att-to-appropriate-level ()* function. In the case that user specifies the concept level, CO4.5 then generalizes object-attributes to that specific level. If concept level is not specified, the attributes are automatically generalized into the most general level. The tree starts with single node (node *n*) containing the training samples in the root node (*root$_n$*). If the samples (*Data$_n$*: set of tuples with value and target class of node *n*) are all of the same class, then the node becomes a leaf node and is labeled with that determinant class. Otherwise, hierarchical and composition relationships are used to select "test-attribute".

A branch is created for each value of the test-attribute, and the samples are partitioned accordingly. The algorithm uses the same process recursively to form a decision tree for each partition. The recursive partitioning stops only when all samples at a given node belong to the same class, or when there are no remaining attributes on which the samples may be further partitioned. After level-based decision trees with determinant classes are obtained, CO4.5 utilizes inheritance relationships recursively to form decision tree of the descending node (*child-of-node-n*). The recursive partitioning stops when node n becomes leaf node of the hierarchy.

Execution Example of CO4.5

CO4.5 starts by constructing the decision of the top level of hierarchy, where determinant classes are the classes in lower level of the hierarchy. Fig. 1a shows the determinant classes of A which are class B and class C.

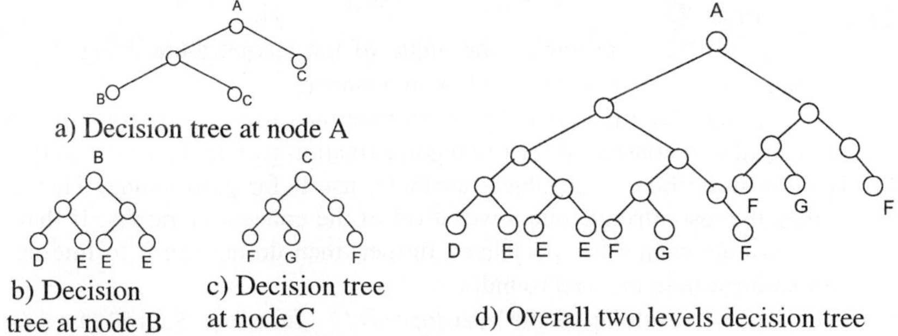

a) Decision tree at node A

b) Decision tree at node B

c) Decision tree at node C

d) Overall two levels decision tree

Fig. 1. Hierarchical decision tree

Then, decision trees of the lower level are constructed. Fig. 1b and 1c show, respectively, determinant classes of B which are D and E and determinant classes of C which are F and G. The overall decision tree in these two levels is shown in Fig. 1d. These steps are repeated until determinant classes of the bottom-level decision tree become user specified-target classes are obtained.

2.2 FindDTree Algorithm

Similar to C4.5, `FindDTree` is an algorithm for constructing a decision tree from data containing generalized complex attributes. All attributes (`att-of-Data`), object attributes and non-object attributes, are used to calculate the information gain in order to find the test-attribute (`testAtt`) (line 5-10).

```
(1)     algorithm FindDTree(DTree d, ComplexData Data,
Target target)
(2)        begin
(3)           if all-prop>κ-in-one-class or no-more-att or
percent-object-subclass < ε then
(4)              d = mainDist(target)
(5)           end if
(6)           for each Att a ∈ att-of-Data do begin
(7)              g=Gain(Data,a)
(8)              if max<g
(9)                 max=g
(10)                testAtt=a
(11)             end if
(12)          end for
(13)          for x ∈ value-of-testAtt do begin
(14)             NewData=newRound(Data_testAtt=x, testAtt)
(15)             d.value[i] = x
(16)             d.branch[i] = new DTree()
(17)             FindDTree(d.branch[i],NewData,target)
(18)          end for
(19)       end
```

Dataset is partitioned according to the value of the test-attribute using the *newRound()* function that works in the following manner:

a)   If the test-attribute is a non-object attribute then use it for partitioning. It is not useful to reconsider it for finding information gain in the next round.

b)   If the test-attribute is an object attribute, use it for partitioning. Then specialize the test-attribute one lower level of the concept hierarchy. If that object attribute cannot be specialized further, then do not use it to finding information gain in the next round.

For each dataset resulting of the *newRound()* function, *FindDTree()* is called recursively (line 14-17) until one of the following conditions is reached:

(1) If the portion of dataset in a given partition is less than the exception threshold ε, further partitioning of the dataset is halted

(2) Further partitioning of the dataset at a given node is terminated if the percentage of dataset belonging to any given class at that node exceeds the classification threshold κ [Kamber 1997] or less than exception threshold ε [Kamber 1997]. In this case the function `mainDist()` is called for determining target class with the main distribution

(3) No more attributes can be used for further classification (`no-more-att`)

3   Experimental Results

This section presents detailed evaluation of CO4.5 compared with AOI-based ID3 and the well-known C4.5 algorithms. The primary metric for evaluating classifier performance is classification accuracy. The comparison is made from different parameters such as number of data records, number of attributes, number of child/parents, number of target classes, and number of levels in the hierarchy.

In order to compare the three algorithms, two synthetic datasets are generated using randomizing functions with various parameters. The first one contains complete data with low noise, while the second one almost contains noisy data. Experiments have been realized on a Celeron CPU at clock rate of 1.2 GHz, 128 MB of main memory. In each experiment, results are the average percentage of accuracy using different data patterns. Number of training examples is fixed to 20,000, number of attributes: 11 (with 5 non-object attributes, 5 object attributes and 1 predictive attribute). For each attribute, number of distinct values is as follows: 3 for the non-object attributes, 3 for each level of the object attributes. Each object attribute has 5 levels. Predictive attribute has 2 target classes.

Fig. 2 shows how the total number of records affects the accuracy of the algorithms. Number of records is varying from 10000 to 80000. The experimental result demonstrates that all the three algorithms scale-up reasonably well with increasing number of records. Compared to AOI-based ID3 and C4.5, CO4.5, has the highest accuracy. Next experiment studies the accuracy of CO4.5, AOI-based ID3 and C4.5 as the number of attributes (object attributes for CO4.5 and AOI-based ID3 and non-object attributes for C4.5) increases. Fig. 3 shows that, with the use of semantics in object attributes, CO4.5 has better accuracy than AOI-based ID3 and C4.5.

Fig. 4 considers different levels of object attributes increasing from 3 to 6. Experimental result shows that accuracy of CO4.5 is increased with the number of object attribute levels. However, the accuracy continues to increase until a certain number of object attribute levels, and then it becomes steady. For AOI-based ID3, its accuracy depends on object attribute levels less than the threshold value, so the accuracy of AOI-based ID3 is rather steady. For C4.5, we notice that the number of object attribute levels does not affect its accuracy, which is rather steady as well. Fig. 5 shows the accuracy of the three algorithms as the number of child/parents increases. Through the semantics hidden in objects with large number of child/parents, CO4.5 provides better accuracy when the number of child/parents is increased.

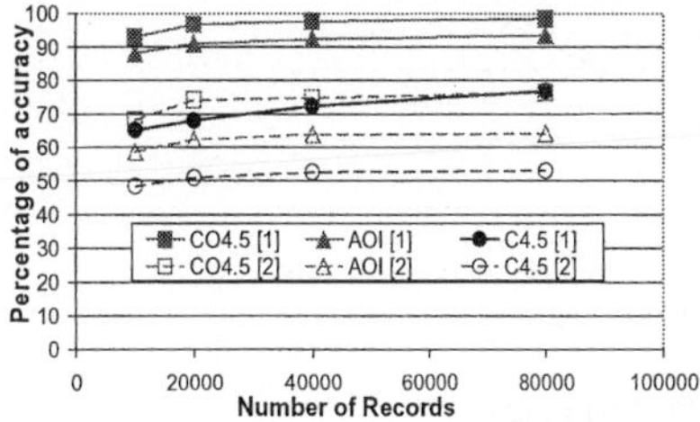
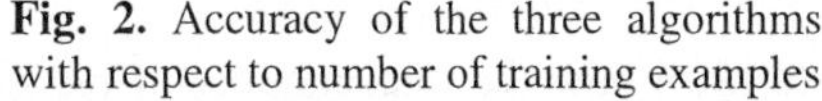
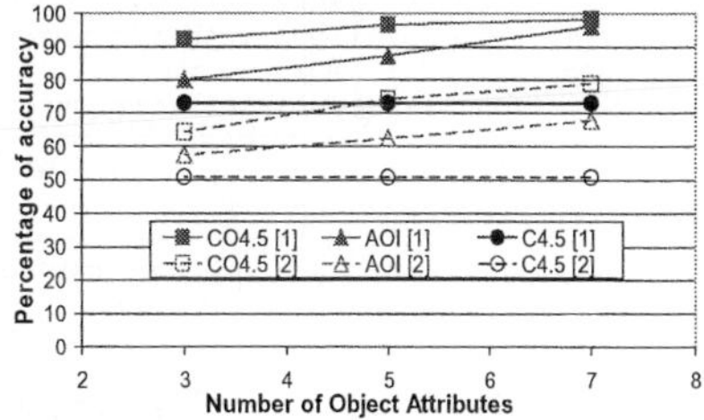

Fig. 2. Accuracy of the three algorithms with respect to number of training examples

Fig. 3. Accuracy of the three algorithms with respect to number of object attributes

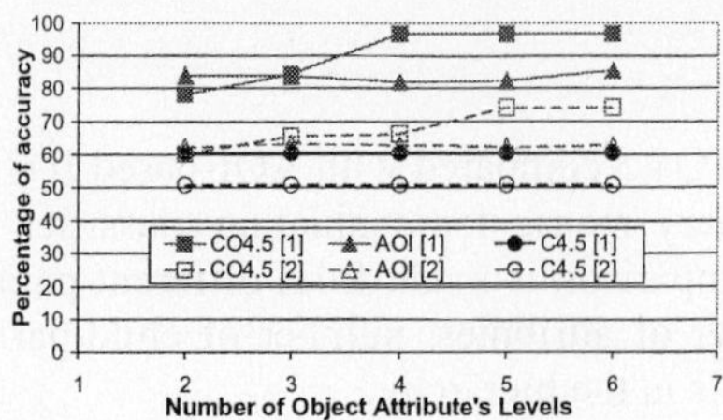

Fig. 4. Accuracy of the three algorithms with respect to number of object attribute levels

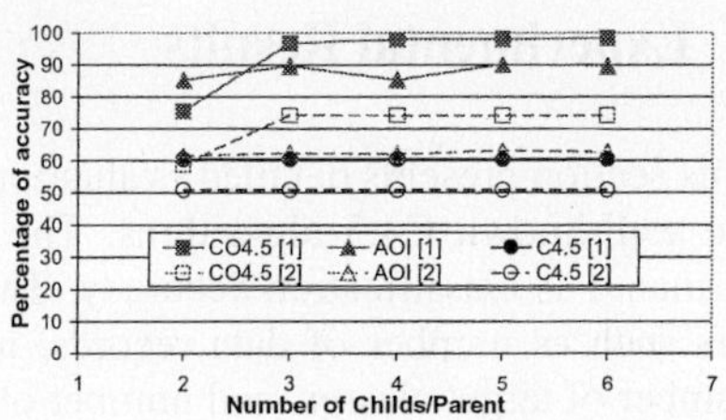

Fig. 5. Accuracy of the three algorithms with respect to number of child/parents

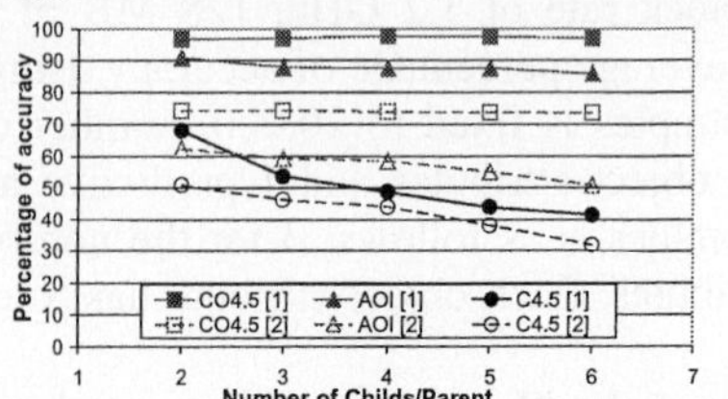

Fig. 6. Accuracy of the three algorithms with respect to number of target classes

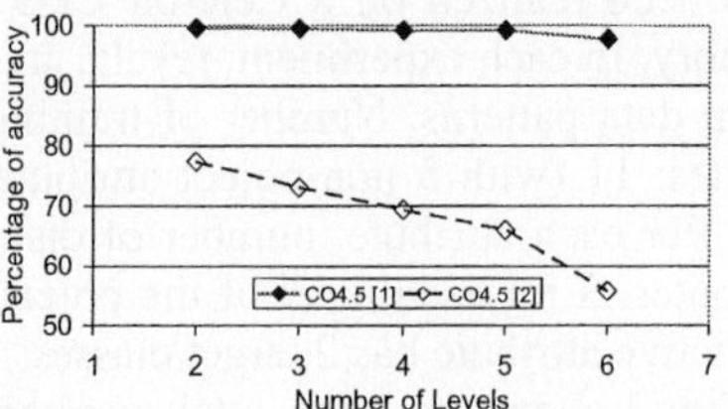

Fig. 7. Accuracy of CO4.5 with respect to number of levels in the hierarchy

Fig. 6 shows accuracy of the three algorithms with respect to number of target classes. Since CO4.5 uses the same object attribute several times, CO4.5 can classify data efficiently in any number of target classes. Unlike CO4.5, AOI-based ID3 and C4.5 have less accuracy when the number of target classes is increased. This can be explained by the fact that increasing target classes will decrease number of data per pattern, which leads to low accuracy. Fig. 7 shows accuracy of CO4.5 with respect to the number of levels in the hierarchy. Objective is to learn how the total number of hierarchical levels affects the accuracy of our algorithm. The number of hierarchical levels is varying from 3 to 6 with each node is divided to two nodes. This experiment shows that CO4.5 algorithm has less accuracy when the number of hierarchical levels is increased. This can be explained by the fact that there are errors in decision trees in each level. The more the number of levels, the more the overall error of the combining decision tree.

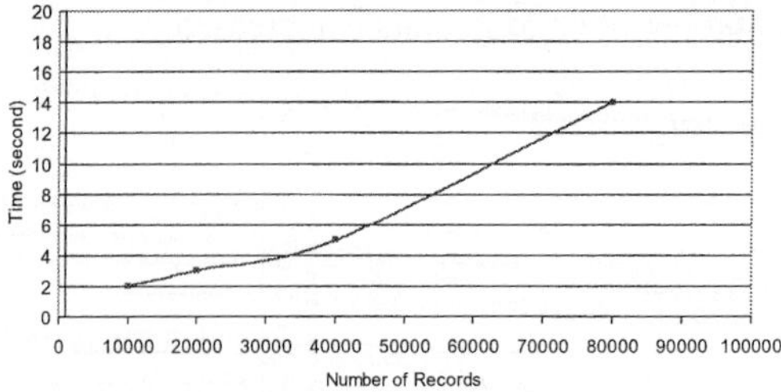

Fig. 8. Scale-Up Performance

In the last experiments, our objective is to learn how the total number of records affects the performance of CO4.5. In this experiment, only performance on the first dataset is shown since the two datasets produce the same execution time. Experimental result is shown in Fig. 8, where CO4.5 achieves nearly linear execution times. This explains the fact that the total classification time is dominated by the time used for specifying each object attribute.

4 Conclusion and Future Work

This paper proposes an approach for classification complex database. The main contribution is to use object's attributes in different levels of the hierarchy for the classification task. While almost of the existing works on complex data classification start by generalizing objects in appropriate abstraction level, the algorithm classifies complex objects directly through the use of inheritance relationships and composition relationship stored in object-oriented database. The proposed method solves the limitation of traditional algorithm in discovering complex object and in predicting multi-target class. The proposed algorithm can be further developed in the several ways such as increasing algorithm performance to support the complex hierarchical target class, apply the interesting object oriented database with this algorithm and use this new concept to develop on the current efficient data classification algorithm such as SLIQ, SPRINT.

References

1. Chen, M., Han, J., Yu, S. Data Mining: An Overview from Database Perspective. In IEEE Transactions on Knowledge and Data Engineering (1996)
2. Han, J., Kamber, M.: Data Mining Concepts and Techniques. Morgan Kaufmann Publishers (2001)
3. Han, J., Nishio, S., Kawano, H., Wang, W.: Generalization-based data mining in object-oriented databases using an object-cube model. In Data and Knowledge Engineering. 25 (1998) 55-97
4. Han, J., Fu, Y.: Exploration of the power of attribute-oriented induction in data mining. In: Fayyad, U.M., Piatetsky-Shapiro, G., Smyth, P., Uthurusamy, R. (eds.): Advances in Knowledge Discovery and Data Mining. AAAI/MIT Press (1996) 399-421
5. Han, J., Cai, Y., Cercone, N.: Data driven discovery of quantitative rules in relational databases. In IEEE Trans.Knowledge and Data Engineering. 5. (1993) 29–40
6. Kamber M., Winstone, L., Gong, W., Cheng, S., Han, J.: Generalization and decision tree induction: Efficient classification in data mining. Int. Workshop on Research Issues on Data Engineering , Birmingham, England (1997) 111-120
7. Mehta, M. Agrawal, R. Rissanen, J.: SLIQ: A Fast Scalable Classifier for Data Mining, Int Extending Database Technology. Avignon, France. (1996)
8. Quinlan, J. R.: Induction of decision trees. Machine Learning. 1. (1986) 81-106
9. Quinlan, J. R.: C4.5: Programs for Machine Learning. Morgan Kaufman (1993)
10. Songsiri, C., Waiyamai, K, Rakthanmanonn, T.: An Object-oriented Data Classification Technique (in Thai)". In Proc. The National Computer Science and Engineering Conference (2002)
11. Wang, W.: Predictive Modeling Based on Classification and Pattern Matching Methods. M.Sc. thesis. Computing Science, Simon Fraser University. (1999)

Data-Mining Based Skin-Color Modeling Using the ECL Skin-Color Images Database

Mohamed Hammami, Dzmitry Tsishkou, and Liming Chen

LIRIS, FRE 2672 CNRS, Ecole Centrale de Lyon
36, Avenue Guy de Collongue
69131 Ecully, France
{mohamed.hammami,dzmitry.tsishkou,liming.chen}@ec-lyon.fr

Abstract. Many human image processing techniques use skin detection as a first stage in subsequent feature extraction. In this paper we describe methods of skin detection using a data-mining technique. We also show the importance of the choice of the base simple to the performance of our skin analysis techniques. We present the details and the process of construction of our database which we have called "the ECL Skin-color Images Database from video ". We will show that the use of a database derived from live video gives better results than one derived from internet images for face detection in video application.

1   Introduction

Skin detection can be defined as the process of selecting which pixels of a given image correspond to human skin. Skin is arguably the most widely used primitive in human image processing research, with applications ranging from face detection[10] and person tracking to pornography filtering[5]. Skin detection techniques can be both simple and accurate, and so can be found in many commercial applications, for example the driver eye tracker developed by Ford UK [9].

Most potential applications of skin-color model require robustness to detect significant variations in races, differing lighting conditions, textures and other factors. This means large databases composed of tens of millions of pixels are necessary in the training stage in order to construct an effective skin color-model. It is a major challenge to manage such a complex structure. One of the driving applications for skin-color model construction of large datasets is data-mining.

Construction of a skin-color database may be done in a different way, depending on an application. Perhaps the four most significant factors are: (1) the size of the database; i.e. it should contain more than 10 million of skin-color pixels. (2) the variety in the image content, so that it should be representative for changes in race, sex, and environmental conditions. (3) the database source, that is particular depends on the application, such as www adult content filtering, face detection for security systems or face detection for video. Although there is a comprehensive database of skin-color images with a large number of subjects, and with significant race, sex and environmental conditions variations (Jones et al., 1998).

M. Bubak et al. (Eds.): ICCS 2004, LNCS 3036, pp. 310–317, 2004.

In this paper we propose a data-mining method for skin-color modeling. Our system uses the Ecl Skin color data base and SIPINA method [13] to automatically extract the most efficient parameters of the skin-color model.

The remainder of this paper is organized as follows. The Construction of the ECL skin color image database is presented in Section 2. The data-mining based skin-color modeling is presented in Section 3. The face detection system is presented in Section 4. Experimental evaluation of the proposed skin-color model is discussed in section 5. Section 6 draws conclusions on our work.

2   Data Collection: The ECL Skin-Color Database

Between April 2002 and June 2002 we collected a database of over 1110 skin-color images of more than 1110 people. (the total size of database is about 91 MB). We call this database the ECL Skin-color Images from Video (SCIV) database (Fig.1). To obtain wide variations in race, sex and environmental conditions we proceed more than 30 hours of TV news (Euronews, TF1, France2). For capturing Euronews TV stream we used Pinnacle TVSat hardware, for TF1 and France2 internal TV receiver was used together with Hauppage WinTV USB video tuner. All video were stored in MPEG4 compatible format using 320x240 resolution on image size, no audio, and medium compression settings. To obtain binary masks we used skin-color segmentation tool similar to one constructed by (Jones et al.,1998). We build a similar histogram based skin-color model and implement a software for skin-color images segmentation. Finally we manually corrected binary masks in Adobe Photoshop 7 software. Besides the two major partitions of the database (skin-color images and corresponding binary masks). The ECL SCIV database therefore consists of two major partitions, the first skin-color images, the second corresponding binary masks.

Fig. 1. Sample skin-color images from ECL SCIV database

2.1 Capture Apparatus

Obtaining images of skin-color from video requires special capturing hardware. Image quality depends on the video image type. There are two sources for ECL SCIV database: (1) video images captured from a satellite using Pinnacle TVSat hardware, which enables to receive and store MPEG2 stream on the hard disk, and (2) video images captured from an internal receiver with Hauppage WinTV USB hardware. The first solution represents digital TV standards, using PCI bus transfer it makes possible to store high quality video images. Meanwhile USB based tuner accommodates analogue TV so that medium image quality within low bandwidth transfer. Finally all video, with the total duration of 30 hours, was converted to MPEG4 compatible format using DivX 5.1 codec with no audio option.

2.2 Capture Procedure

To obtain significant variation in face expressions, pose, position, race, sex and environmental we did manual shot segmentation and extracted one frame per each shot, containing skin-color with total area size more than 5% of the whole image size. So that we extracted 1100 skin-color images from 30 hours of the video. The primary subject of our research activity is the skin detection in video (particulary TV news) application; therefore we captured 15 hours of Euronews with a high quality. The TF1 and France2 were captured during of 15 hours, covering the whole range of TV programs types, including TV news, sitcoms, movies, sport (World cup 2002), commercials and other.

2.3 Skin-Color Segmentation

Each skin-color image in ECL SCIV was proceed in the following manner: first skin-color regions were manually labelled using software tool shown on figure 2. This tool allows user to interactively segment skin-color regions by controlling the threshold of histogram based skin-color model pixel classifier.

The threshold slider controls the accuracy of segmenting the region of interest. In labelling we were careful to exclude the eyes, hair and mouth opening.

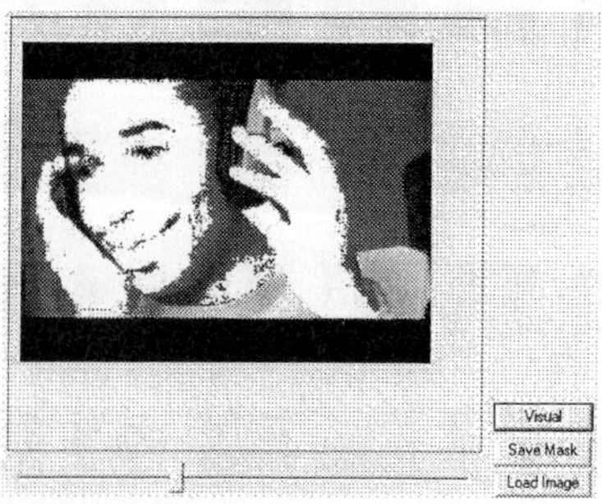

Fig. 2. Skin-Color segmentation tool

Second each binary mask was manually corrected in Adobe Photoshop 7 image editor.

Finally all segmented images also called as masks were stored in independent files with name corresponding to its origin images with white pixels representing skin-color pixels and black pixels –non skin color (Fig.4). Hence we are able to extract only skin color or non skin color pixels from image using mask automatically.

Fig. 3. Skin-color image from ECL SCIV database

Fig. 4. Binary mask, corresponding to the skin-color image on Fig.3

2.4 Database Organization

On average the capture, preprocessing and indexing took about 10 minutes per each image. The images are 320x240 color images 24BPP. Each image is compressed using the JPEG codec with high quality. The database is organized in 3 partitions, the first consisting of the skin-color images (manually extracted from video), the second consisting the corresponding binary masks (automatically segmented and manually corrected from corresponding skin-color images) and the third one – meta-data index (manually done skin-color images classification).

2.5 Meta-Data

Regarding potential uses of the database, besides the two major portions of the database, we also collected a set of meta-data to aid further ECL SCIV database analysis:

Race: We classify each image into European, Asian, Africa, Latin and Arabic. The race distribution graph shows that the database consists of 82% Europeans, 8% Africans, 5% Latin, 3% Asian and 2% Arabic people (Fig 5).

Gender: We manually did men/woman classification on ECL SCIV database. The results demonstrate 72% of men and 28% of women populating the database (Fig 6).

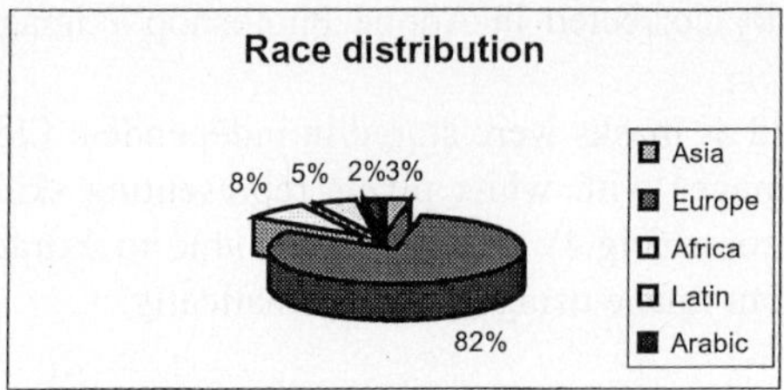

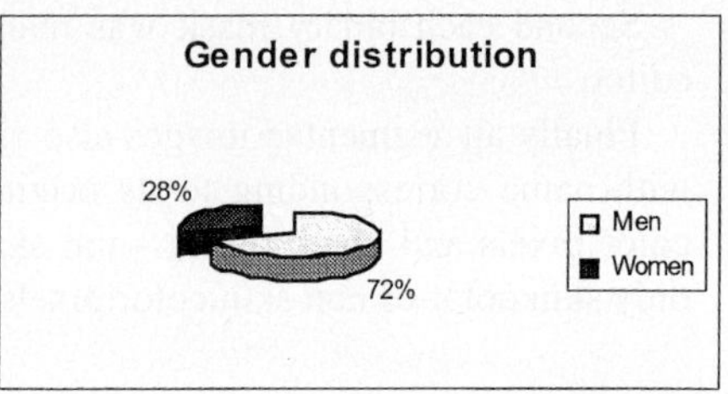

Fig. 5. ECL SCIV database: race distribution   **Fig. 6.** ECL database: Gender distribution

3 Data-Mining Based Skin-Color Modeling

3.1 Data Preparation

Large databases composed of tens of millions of pixels are necessary to explore various types of lighting conditions, races, etc. in order to construct an effective skin color-model. Several color spaces have been proposed in the literature for skin detection applications. YCbCr has been widely used since the skin pixels form a compact cluster in the Cb-Cr plane. As YCbCr is also used in video coding and then no transcoding is needed, this color space has been used in skin detection applications where the video sequence is compressed [1, 10]. In [11] two components of the normalized RGB color space (rg) have been proposed to minimize luminance dependence. And finally CIE Lu*v* has been used in [12]. However, it is still not clear which is the color space where the skin detection performance is the best.

To create skin-color model using color information we use ECL SCIV database; all intensity information on color pixels was extracted using binary masks.

For each pixel we compute its representation in following normalized colorspaces: RGB, HSV, YIQ, YCbCr, CMY in order to find the most discriminative set of color axes. To store the dataset we include information on pixel values in different color-spaces with corresponding binary masks values indicating the presence of either skin-color or non-skin color pixels[4].

3.2 Skin-Color Modeling: SIPINA

A number of classification techniques from the statistics and machine learning communities have been proposed [3, 7, 8]. A well-accepted method of classification is the induction of decision trees [2, 7]. A decision tree is a flow-chart-like structure consisting of internal nodes, leaf nodes, and branches. Each internal node represents a decision, or test, on a data attribute, and each outgoing branch corresponds to a possible outcome of the test. Each leaf node represents a class. In order to classify an unlabeled data sample, the classifier tests the attribute values of the sample against the decision tree. A path is traced from the root to a leaf node which holds the class predication for that sample. Decision trees can easily be converted into IF THEN rules and used for decision-making.

SIPINA [13] is a widely used technique for data-mining. The effectiveness of SIPINA is superior to classical methods such as ID3 and C4.5[8], because the distribution equivalency can be considered in a population-wise manner, irrespective of any fixed solutions proposed previously. This accounting mechanism accurately charts usage distribution and leads to the highest performance among other methods, particularly for skin-color modelling. Let the set of pixels   be extracted and pre-processed automatically from training images and corresponding binary masks. Each pixel W is associated with its C(w) class (skin-color, non skin-color).

$$C : \Omega \rightarrow \cent = \{\text{skin-color, non skin-color}\}. \tag{1}$$

$$W \rightarrow C(w). \tag{2}$$

The observation of C(w) is not easy, because of lighting conditions, race differences and other factors. All these factors complicate the skin-color classification process. Therefore we are looking for mean value   to describe class C of each element in different color spaces.

As a result of applying this method to a training set, a hierarchical structure of classifying rules of the type "IF...THEN..." is created.

Using the ECL data base of skin-color and non-skin color images and a data-mining technique we discover that HSV is the most discriminative colorspace. Detailed values on skin-color model decision rules and experimental results on the training process will be published in our extended technical report, however we will present experimental results on evaluation of this model for face detection in video.

4 Face Detection in Video Analysis and Indexing

Rapid growth of telecommunication data is caused by tremendous progress in information technology development and its applications. Modern figures of the volume of video data are ranged from hundred hours of home video for personal use to a million-hour TV company's archive. The condition of a successful practical use of this data is the stringent enforcement in the following areas: storage, transmission, preprocessing, analysis, indexing. Despite the fact that information storage and transmission systems are capable to supply video data to user there is still a developmental gap between the storage and the indexing. A problem of manual video data preprocessing, analysis and indexing has no practical solution in terms of human perception and physiological abilities. Therefore the construction of the fully automatic video indexing system is the current research subject. High level semantic analysis and indexing systems generally exploit some known specific properties of the video, e.g., that the video is a hierarchical multi-layer combination of background/foreground objects. One can decompose the video into the set of such objects and the structure. That requires video structure understanding and object recognition solutions. Examples of such object recognition solutions include face detection and recognition among others. There are several reasons why one is interested in a use of facial information for video analysis and indexing. First of all, visual facial information serves for clue up on a personal identity. That is one of the most informative tags for a particular video invent. The second reason why the face detection and recognition is important is that it has been given a great attention from the scientific society until present. It's a strong advantage, taking into account the fact that the number of potential objects to be recognized by the video analysis and indexing system is nearly infinite. Finally, the most important reason why facial information is important is that it enables us to dramatically increase the number of applications of the video analysis and indexing.

5 Experimental Evaluation

In this section we will deal with practical implementation of the skin-color model for face detection application. We performed two experiments and present the results here. All experiments were performed with video captured to files, so that pure performance (faster than real time) could be evaluated. The skin-color models used in experiments were: data-mining based using the ECL SCIV derived from live video (A), data-mining based using the data base derived from internet images [6] (B). We note that the face detection in video system was tuned for fast performance and some features were disabled. Meanwhile skin-color preprocessing takes less than 1% of total computational complexity; therefore the only parameter evaluated is a total number of detected persons per hour of video (table 1).

Table 1. Face detection in video system: parameters evaluation

Experimental conditions:	Full time French public TV (TF1, France2, France3, M6)
Duration	30 hours
Time interval	15 days
Skin-color pixels (per hour)	A-24770831 B-21615935
Skin-color regions (per hour)	A-1228209 B-1050561
Correct face images (detected per hour)	A – 2169 B – 2021
Performance	Average fps – 78 (Intel Pentium 2.0 GH)
Minimum size	20x20 pixels
Maximum size	infinite
In plane Rotation freedom	Up to 25 degrees

6 Conclusions

We have introduced in this paper a new skin-color modeling approach – data-mining based skin-color model. We presented the details and the process of construction of the "ECL Skin-color Images from Video Database", which was used a training dataset. Experimental evaluation has shown that data mining-based skin-color model could be integrated into the face detection in video system, providing acceptable detection results. Usage of a special training dataset slightly improves total performance of face detection in video system. Further research activities would be concentrated on a real-time data mining-based skin-color model adaptation to the lighting conditions.

References

1. A. Albiol, L. Torres, C.A. Bouman, and E. J. Delp, "A simple and efficient face detection algorithm for video database applications," in *Proceedings of the IEEE International Conference on Image Processing*, Vacouver,Canada, September 2000, vol. 2, pp. 239–242.
2. L. Breiman, J. Friedman, R. Olshen, and C. Stone. *Classification of Regression Trees.* Wadsworth, 1984.
3. U. M. Fayyad, S. G. Djorgovski, and N. Weir. Automating the analysis and cataloging of sky surveys. In U. Fayyad, G. PiatetskyShapiro, P. Smyth, and R. Uthurusamy, editors, *Advances in Knowledge Discovery and Data Mining*, pages 471–493. AAAI/MIT Press, 1996.
4. M. Hammami, Y. Chahir, L. Chen, D. Zighed, "Détection des régions de couleur de peau dans l'image" revuc RIA-ECA vol 17, Ed.Hermès, ISBN 2-7462-0631-5, Janvier 2003, pp.219-231.
5. M. Hammami, Y. Chahir, L. Chen, "Combining Text and Image Analysis in The Web Filtering System: WebGuard", *IADIS International Conference: WWW/Internet 2003*, ISBN 972-98947-1-X, Algarve, Portugal, November 5-8, pp.611-618
6. M.J. Jones, J..M. Regh, "Statistical Color Models with application to Skin Detection", Cambridge Research Laboratory, CRL 98/11, 1998.
7. J. R. Quinlan. Induction of decision trees. *Machine Learning*, 1:81–106, 1986.
8. J. R. Quinlan. C4.5: Programs for Machine Learning. Morgan Kaufmann, 1993.
9. D. Tock and I. Craw. Tracking and measuring drivers' eyes. Image and Vision Computing,14:541–548, 1996.
10. H. Wang and S-F. Chang, "A highly efficient system for automatic face region detection in mpeg video," IEEE Transactions on circuits and system for video technology, vol. 7, no. 4, pp. 615–628, August 1997.
11. J. G. Wang and E. Sung, "Frontal-view face detection and facial feature extraction using color adn morphological operators," Pattern recognition letters, vol. 20, no. 10, pp. 1053–1068, October 1999.
12. Ming-Hsuan Yang and Narendra Ahuja, "Detecting human faces in color images," in Proceedings of the International Conference on Image Processing, Chicago, IL, October 4-7 1998, pp. 127–130.
13. D.A. Zighed and R. Rakotomala. A method for non arborescent induction graphs. Technical report, Laboratory ERIC, University of Lyon 2, 1996.

Maximum Likelihood Based Quantum Set Separation

Sándor Imre and Ferenc Balázs*

Mobile Communications & Computing Laboratory
Department of Telecommunications
Budapest University of Technology and Economics
1117 Budapest, Magyar Tudósok krt. 2, Hungary
{imre,balazsf}@hit.bme.hu

Abstract. In this paper we introduce a method, which is used for set separation based on quantum computation. In case of no a-priori knowledge about the source signal distribution, it is a challenging task to find an optimal decision rule which could be implemented in the separating algorithm. We lean on the Maximum Likelihood approach and build a bridge between this method and quantum counting. The proposed method is also able to distinguish between disjunct sets and intersection sets.

1 Introduction

In the course of signal and/or data processing fast classification of the input data is often helpful as a preprocessing step for decision preparation. Assuming that the to be classified data $\mu \in M$ is well defined and it came under a given number of classes or sets, $A := \{\mu \in M : \mathcal{A}(\mu)\}, B := \{\mu \in M : \mathcal{B}(\mu)\}, \ldots, Z := \{\mu \in M : \mathcal{Z}(\mu)\}$. To perform the classification is in such a way equivalent to a set separation task.

The problem of separation could be manifold: sparsely distributed input data makes the determination of the decision lines between the classes to a hard (often nonlinear) task, or even the probability distribution of the input data is not known a-priori which is resulted in an unsupervised classification problem also known as clustering [1]. Further "open question" is to classify input sequences in the case of only the original measurement/information data is known almost sure, but the observed system adds a stochastically changing behavior to it, in this manner the classification becomes a statistical decision problem, which could be extremely hard to solve if the number of "possibilities" is increasing. Due to this fact to find an optimal solution is time consuming and yields broad ground to suboptimal ones. With assistance of quantum computation we introduce an optimal solution whose computational complexity is much lower contrary to the classical cases.

---

* The research project was supported by OTKA, id. Nr.: F042590

M. Bubak et al. (Eds.): ICCS 2004, LNCS 3036, pp. 318–325, 2004.

This paper is organized as follows. In Sect. 2. the set separation related quantum computation basics are highlighted. The system model is described in Sect. 3. together with the proposed set separation algorithm in Sect. 4. The main achievements are revised in Sect. 5.

2   Quantum Computation

In this section we give a brief overview about quantum computation which is relevant to this paper. For more detailed description, please, refer to [2,3,4,5].

In the classical information theory the smallest information conveying unit is the *bit*. The counterpart unit in quantum information is called the *"quantum bit"*, the qubit. Its state can be described by means of the state $|\varphi\rangle$, $|\varphi\rangle = \alpha|0\rangle + \beta|1\rangle$, where $\alpha, \beta \in \mathbb{C}$ refers to the complex probability amplitudes and $|\alpha|^2 + |\beta|^2 = 1$ [2,3]. The expression $|\alpha|^2$ denotes the probability that after measuring the qubit it can be found in computational base $|0\rangle$, and $|\beta|^2$ shows the probability to be in computational base $|1\rangle$. In more general description an N-bit *"quantum register"* (qregister) $|\varphi\rangle$ is set up from qubits spanned by $|x\rangle$ $x = 0 \ldots (N-1)$ computational bases, where $N = 2^n$ states can be stored in the qregisters at the same time [6]

$$|\varphi\rangle = \sum_{x=0}^{N-1} \varphi_x|x\rangle; \quad \varphi_x \in \mathbb{C}, \tag{1}$$

where N denotes the number of states and $\forall x \neq j$, $\langle x|j\rangle = 0$, $\langle x|x\rangle = 1$, $\sum |\varphi_x|^2 = 1$, respectively. It is worth mentioning, that a transformation U on a qregister is executed parallel on all N stored states, which is called *quantum parallelizm*. To provide irreversibility of transformation, U must be unitary $U^{-1} = U^\dagger$, where the superscript (†) refers to the Hermitian conjugate or adjoint of U. The quantum registers can be set in a general state using quantum gates [4,5] which can be represented by means of a unitary operation, described by a quadratic matrix.

3   System Model

For the sake of simplicity a 2-dimensional set separation is assumed, where the original source data can take the values $\mu \in \mathbf{M}^{[0,1]}$ and was chosen from the sets $s = 0$ and $s = 1$. Additional information on the source is not available, e.g. also nothing about the probability density function (*pfd*). The general set separation system is depicted in Fig. 1. The observed signal r, disturbed by the system A, becomes the input data which will be separated into the two sets ($s = 0$ and $s = 1$) again.

In the set separator a quantum register $|\varphi\rangle$ –as described by equation (1) and shown in Fig. 2.– is used to store all the parameters, e.g. delay, heat, velocity, etc. values of the possible system disturbance in a specially given quantization[1][2].

---

[1] Quantization is NOT a quantum computation operation!

[2] The quantization method, i.e. linear or nonlinear is out of the scope of this paper.

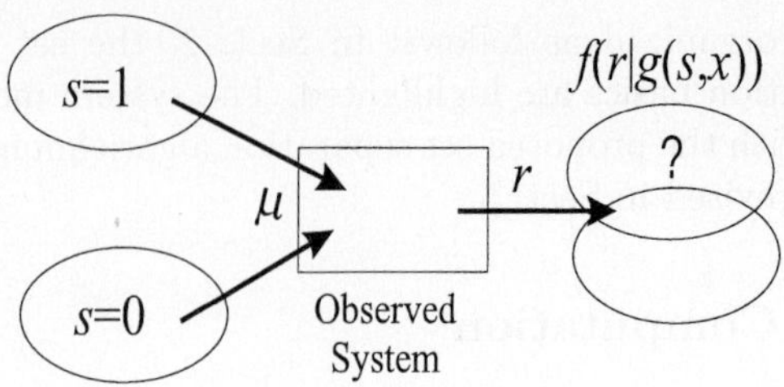

Fig. 1. General set separation system

As an example: in the qregister $|\varphi\rangle$, the properly prepared, quantized delay and velocity values are stored, e.g. the values $1.0 \cdot 10^{-1}, 1.1 \cdot 10^{-1}, \ldots, 1.0 \cdot 10^{-10}$ and $1.0\ m/s, 1.1\ m/s, \ldots, 100\ m/s$. This information is not utilizable so far but the combination of this effects, i.e. this values, whose extent could blast any database. To handle the large amount of data to be processed a virtual database should be introduced.

Definition 1 *To build up a virtual database a function*

$$y = g(s, \underline{x}),\tag{2}$$

is defined, where $s \in S$ identifies the sets and $\underline{x}$ denotes the index of the qregister $|\varphi\rangle$, respectively. The function $y_i = g(s, x_i)$ points to an record in the virtual database [7].

3.1 Properties of the Function $g(\cdot)$

The function $g(s, \underline{x})$ is not obligingly mutual unambiguous consequently, it is not reversible, except for several special cases, when the virtual database contains $\widehat{r} = g(s, \underline{x})$ only once. In this case the parameter settings of the system A are easy to determine. Nevertheless, the fact to have an entry only once in the virtual database described by the equation $g(s_i, \underline{x})$ does not exclude to have the same entry in other virtual databases generated by $g(s_j, \underline{x})$, where $i \neq j$, which makes a trivial decision impossible. Henceforth the fact should be kept in mind that $g(s, \underline{x})$ is in almost every case a so called one way function which is easy to evaluate in one direction, but to estimate the inverse is rather hard.

The function $g(.)$ generates all the possible disturbances additional to the considered input value μ belonging to the set $s = 0$ or $s = 1$ of the system. This is of course a large amount of information, $2N = 2^{n+1}$, where n is the length of the qregister $|\varphi\rangle$. For an example let us assume a 15-qbit qregister. The function $g(\cdot)$ in (2) generates $2^{15} = 32.768$ output values at the same time for $s = 0$ and the same number of outputs for $s = 1$. Taking into account the large number of possible points in the set surface the optimal classification in a classical way becomes difficult.

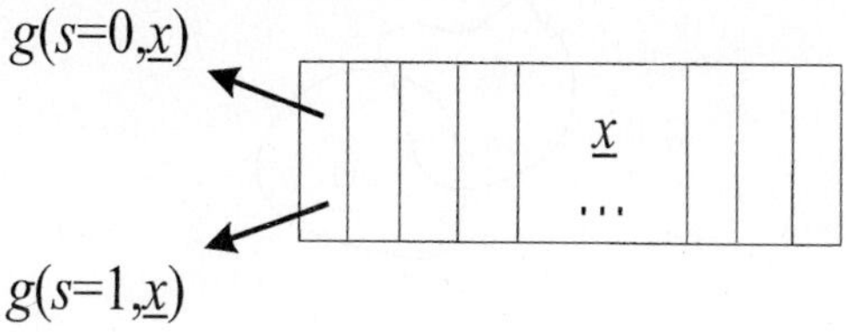

Fig. 2. Quantum register $|\varphi\rangle$

At the first glace this problem looks more difficult to solve, however, with exploiting the enormous computational power of quantum computation, in this case the Deutsch-Jozsa [8] quantum parallelization algorithm, an arbitrary unitary operation can be executed on all the prepared states contemporaneously.

3.2   Quantum Search in Qregister $|\varphi\rangle$

Roughly speaking the task is to find the entry (entries) in the virtual databases which is (are) equal to the observed data r. To accomplish the database search the Grover database search algorithm should be invoked [9]. In Sect. 2. we proposed to set up an qregister, which has to be built up only one time at all. It is obvious to choose a suitable database searching algorithm, to see which function $g_{0,1}(s_{0,1}, \underline{x})$ picking the vector $\underline{x}$ form qregister $|\varphi\rangle$ contains the searched *bit*, if any at all. We apply the optimal quantum search algorithm $\mathcal{G}$, as depicted in Fig. 3. proposed by Grover [10,11]. We feed the received signal $r(t)$ to the oracle (O), where the function $f(r, g(s, \underline{x}))$ is evaluated such that

$$f(a, b) = \begin{cases} 1 \text{ if } a = b \\ 0 \text{ otherwise.} \end{cases} \tag{3}$$

Assuming, there is again M solutions for the search in qregister $|\varphi\rangle$,

$$|\varphi\rangle = \sqrt{\frac{N_s - M}{N_s}} |\alpha\rangle + \sqrt{\frac{M}{N_s}} |\beta\rangle, \tag{4}$$

where $|\alpha\rangle$ consists of such configurations of $|x\rangle$, which does not results $\widehat{\mu} = r$, while $|\beta\rangle$ does.

Because of the fact of tight bound, in real application less iterations would be also appropriate [12].

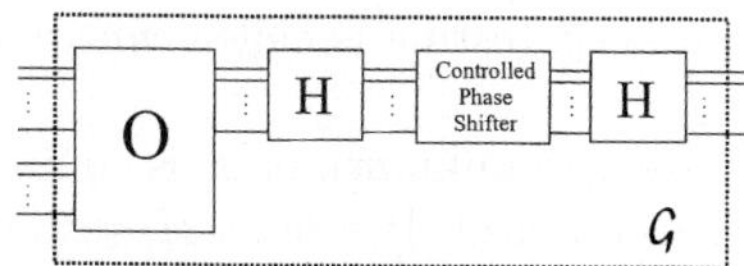

Fig. 3. The Grover database search circuit

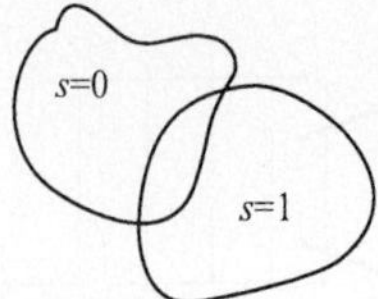

Fig. 4. Sets with intersection

4   Set Separation

Let us turn our interest back to the separation of the observed data r from the predefined sets.

Assuming the special case where only one of the virtual database descriptor functions, either $g(s_0, \underline{x})$ or $g(s_1, \underline{x})$ contains the entry identical to the observed data r a set separation can be performed easily.

A more realistic case is to have an intersection part of the two sets as shown in Fig. 4. Even so, due to passing the observed system, overlapping of the sets can be occurred due to disturbances. After evaluating the functions $g_{0,1}(s_{0,1}, \underline{x})$ it could happen that the same records are multiple present, which shows the irreversibility behavior of the function (2). Originally, the input signal was chosen from well defined disjunct sets without a-priori known probability distributions. The process, to put r to a set either to s_0 or to s_1 should be based on *Maximum Likelihood* decision.

Let us assume that we have a random variable r. Its measured value depends on a selected element x_l from a finite set $(l = 1, \ldots, L)$ and a process which can be characterized by means of a conditional pdf $f(r|x_l)$ belonging to the given element. Our task is to decide which x_l was selected if a certain r has been measured. Each guess H_l for x_l can be regarded as a hypothesis. Therefore decision theory is dealing with design and analysis of suitable rules building connections between the set of observations and hypotheses.

If we are familiar with the unconditional (a priori) probabilities $P(x_l)$ then the Bayes formula helps us to compute the conditional (a posteriori) probabilities $P(H_i|r)$ in the following way

$$P(H_l|r) = \frac{f(r|x_l)P(x_l)}{\sum_{i=1}^{L} f(r|x_i)P(x_i)}.$$

Obviously the most pragmatic solution if one chooses H_l belonging to the largest $P(H_l|r)$. This type of hypothesis testing is called *maximum a posteriori* (MAP) decision.

If the a priori probabilities are unknown or x_l is equiprobable then *maximum likelihood* (ML) decision can be used. It selects H_l resulting the largest $f(r|x_l)$ when the observed r is substituted in order to minimize the probability of error

$$\max_l L(r, x_l).$$

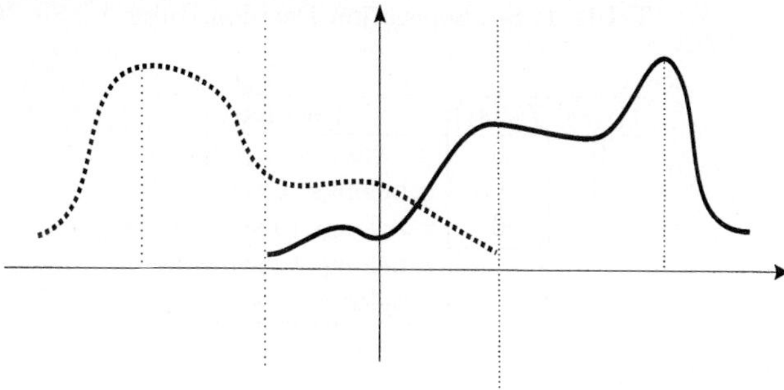

Fig. 5. The two density functions $f(r|s=0)$ and $f(r|s=1)$

The Maximum Likelihood estimator requires to know the probability density function of the observed signal. Employing the Grover database search algorithm we are able to find the entries in the virtual databases, however, it is not needed to perform a complete search because the search result –the exact index (indices) of the searched item(s)– is (are) not interesting but the number how often a given configuration is involved in $g(s, \underline{x})$ or not. For that purpose a new function $f(\cdot)$ is defined.

Definition 2 *The function*

$$f(r|s) = \frac{\sharp\left(x : r = g(s, x)\right)}{\sharp(x)}, \tag{5}$$

counts the number of similar entries in the virtual database, which corresponds to the conditional probability density function r to be in the set s.

For that reason it is worth stepping forward to quantum counting [13] based on Grover iteration.

4.1 Set Separation Method

The both curves in Fig. 5. represent the number of the same entries in the virtual databases, i.e. the pdf's, according to $f(r|s=0)$ and $f(r|s=1)$, respectively. In case of having entry(entries) only in y_i but not in y_j of function $g_{0,1}(s_{0,1}, \underline{x})$, where $i, j \in [0, 1]$, and $i \neq j$, means a 100 percent sure decision, following the decision rules in Table 1. This areas are the non-overlapping parts of the sets in Fig. 4. and the outer parts (until the vertical dashed black lines) in Fig. 5. However, in the case of non zero $f(r|s=0)$ and $f(r|s=1)$ values an accurate prediction can be given relating to the Maximum Likelihood decision rule.

   All the possible states from the qregister $|\varphi\rangle$ will be evaluated by the function (2) for $s = 0$ and also for $s = 1$, simultaneously, which will be collated with the system output r. If at least one output y_0 or y_1 with the parameter settings x

Table 1. Set Separation Decision Rules

$f(r\|s_0)$	$f(r\|s_1)$	Decision
0	0	$\|\varphi\rangle$ was badly prepared
0	$\neq 0$	r belongs to set $s = 1$
$\neq 0$	0	r belongs to set $s = 0$
$>$		r belongs to set $s = 0$
$<$		r belongs to set $s = 1$

is matched to the system output r, it will be put to the set $s = 0$ or $s = 1$, respectively. In a more exciting case at least one similarity of y_0 and also at least one of y_1 to r is given, the system output could be classified to the both sets, an intersection is drawn up. This result in a not certainty prediction, which piques our interest and sets our focus not this juncture.

We assume no a-priori knowledge on the probability distribution of the input sequence μ, so it is assumed to be equally distributed. Henceforward we suppose that after counting the evaluated values $f(r|s = i)$ the number of similarity to the system output r is higher than in case of $f(r|s = j)$, where $i, j \in [0, 1]$. In pursuance of the decision rule in Table 1., r belongs rather to set $s = i$ than to set $s = j$.

The Method. To perform a set separation nothing else is required as

1. Prepare the qregister $|\varphi\rangle$,
2. Evaluate the functions $y_i = g_i = (s = i, \underline{x})$, where $i \in [0, 1]$ in 2-dimensional case,
3. Count the identical entries in the virtual databases which are equal to the observed data r, $f(r|s)$, (see Fig. 5).,
4. Use the decision table Table 1 to assign r to the sets $s = 0$ or $s = 1$.

5 Concluding Remarks

In this paper we showed a connection between *Maximum Likelihood* hypothesis testing and Quantum Counting used for quantum set separation. We introduced a set separation algorithm based on quantum counting which was employed to estimate the conditional probability density function of the observed data in consideration to the belonging sets. In our case the *pdf*'s are estimated fully at a single point by invoking the quantum counting operation only once, that makes the decision facile and sure. In addition one should keep in mind that the qregister $|\varphi\rangle$ have to be set up only once before the separation. The virtual databases are generated once and directly leaded to the Oracle of the Grover block in the quantum counting circuite, which reduce the computational complexity, substantially.

References

1. K. Fukunaga, *Introduction to Statistical Pattern Recognition*, 2nd ed., ser. Electrical Science, H.G. Booker & N. DeClaris, Ed. New York, London: Academia Press, INC., 1972.
2. P. Shor, "Quantum computing," *Documenta Mathematica*, vol. 1-1000, 1998, extra Volume ICM.
3. D. Deutsch, "Quantum theory of probability and decisions," *Proc. R. Soc. London, Ser. A*, 2000.
4. M.A. Nielsen, I.L. Chuang, *Quantum Computation and Quantum Information.* Cambridge University Press, 2000.
5. A. Ekert, P. Hayden, H. Inamori, "Basic concepts in quantum computation", 16 January 2000.
6. S. Imre, F. Balázs, "Quantum multi-user detection," *Proc. 1st. Workshop on Wireless Services & Applications, Paris-Evry, France*, pp. 147–154, July 2001, ISBN: 2-7462-0305-7.
7. ——, "Non-coherent multi-user detection based on quantum search," *IEEE International Conference on Communication (ICC)*, 2002.
8. D. Deutsch, R. Jozsa, "Rapid solution of problems by quantum computation," *Proc. R. Soc. London, Ser. A*, pp. 439,553, 1992.
9. L. Grover, "A fast quantum mechanical algorithm for database search," *Proceedings, 28th Annual ACM Symposium on the Theory of Computing*, pp. 212–219, May 1996, e-print quant-ph/9605043.
10. ——, "How fast can a quantum computer search?" April 1999.
11. C. Zalka, "Grover's quantum searching algorithm is optimal, e-print quant-ph/9711070v2," December 1999.
12. S. Imre, F. Balázs, "The generalized quantum database search algorithm," *Submitted to Computing Journal*, 2003.
13. G. Brassard, P. Hoyer, A. Tapp, "Quantum counting," *Lecture Notes in Computer Science*, vol. 1443, pp. 820+, 1998. [Online]. Available: http://xxx.lanl.gov/archives/9805082

Chunking-Coordinated-Synthetic Approaches to Large-Scale Kernel Machines

Francisco J. González-Castaño[1] and Robert R. Meyer[2]

[1] Departamento de Ingeniería Telemática, Universidad de Vigo, Spain
ETSI Telecomunicación, Campus, 36200 Vigo, Spain
[2] Computer Sciences Department, University of Wisconsin-Madison, USA
javier@det.uvigo.es, rrm@cs.wisc.edu

Abstract. We consider a kernel-based approach to nonlinear classification that coordinates the generation of "synthetic" points (to be used in the kernel) with "chunking" (working with subsets of the data) in order to significantly reduce the size of the optimization problems required to construct classifiers for massive datasets. Rather than solving a single massive classification problem involving all points in the training set, we employ a series of problems that gradually increase in size and which consider kernels based on small numbers of synthetic points. These synthetic points are generated by solving and combining the results of relatively small nonlinear unconstrained optimization problems. In addition to greatly reducing optimization problem size, the procedure that we describe also has the advantage of being easily parallelized. Computational results show that our method efficiently generates high-performance simple classifiers on a problem involving a realistic dataset.

1 Introduction

Suppose that the following classification problem is given:

A set of m training points in a n-dimensional space is given by the $m \times n$ matrix Y. These points are divided into two classes, types 1 and -1. A classifier is to be constructed using this data and a nonlinear kernel function K that is a mapping from $R^n \times R^n$ to R. We assume that construction of the corresponding classifier involves generating a function $g(x) = wK(C, x)$, where w is a row vector of size s and C is a set of s points in R^n and it is understood that, for such a set of points C, $K(C, x)$ is a column vector of size s whose entries are given by $K(c_i, x)$ for the rows c_i of C (in a further extension of this notation below, we will assume that for any two matrices A and B each of which has n columns, that $K(A, B)$ is a matrix whose (i,j) entry is $K(A_i, B_j)$ where A_i and B_j are the rows corresponding to indices i and j. In addition, we will use a juxtaposition of vectors and matrices to indicate a product operation, without introducing an extra symbol to indicate the transposition of the vector that may be required). The points used in a set C will be termed classifier points. While it is customary to set $C = Y$, one of the goals of this paper is to investigate alternative approaches to constructing C, particularly when Y is a large dataset.

M. Bubak et al. (Eds.): ICCS 2004, LNCS 3036, pp. 326–333, 2004.

In conjunction with the determination of the weights w described above, a constant γ is also computed so that the resulting classifier designates (as correctly as possible) a point as type 1 if $g(x) > \gamma$ and a point as type -1 if $g(x) < \gamma$.

We assume the kernel *Mercer inner product property* [2] $K(y,x) = f(y)f(x)$, where f is a mapping from R^n to another space R^k, and $f(y)f(x)$ represents an inner product (for a simple example, suppose that $n = 2$ and $K(y,x) = (yx)^2$. Then by taking $f(z) = (z_1^2, 2^{1/2}z_1z_2, z_2^2)$, where z is a 2-vector, it is easy to verify that $K(y,x) = f(y)f(x)$ holds for any pair of vectors). The original space R^n is finite dimensional and is referred to as "input space" whereas R^k may have a much higher dimension (or even be infinite-dimensional) and is referred to as "feature space" (in the simple example given, $n = 2$ and $k = 3$, but note that even for this simple quadratic kernel the dimension of the feature space is proportional to the square of the dimension of the input space). Many commonly used nonlinear classifiers (such as Gaussians or homogeneous polynomials) have this Mercer property, and although evaluation of the corresponding function f itself is in general not computationally practical, the inner product representation $K(y,x) = f(y)f(x)$ allows classifier approximation problems to be formulated in an manner that is computationally tractable. In fact, by taking advantage of this Mercer property, we will be able to solve via a relatively small unconstrained nonlinear program (NLP), the problem of approximating a classifier with $C = Y$ (or subsets of Y) by classifiers based on very small numbers of "synthetic points" that are not necessarily in Y.

Classifiers are usually constructed by taking the set C to be the set Y of training points and then solving an optimization problem for the values of w and γ. However, for massive datasets this choice of C results in intractably large optimization problems. Moreover, it may be the case that even for medium-sized datasets, the classifiers using $C = Y$ can be outperformed by alternative choices of C. For example, if the type 1 points form a cluster about a center point z that is not in Y, and the type -1 points lie beyond this cluster, then an ideal classifier may be constructed using a Gaussian kernel involving the point z alone rather than any of the many points from Y. As observed in a noisy dataset below, the generalizability associated with alternative smaller choices of C may also be better than the choice $C = Y$, since the latter can lead to overtraining.

In the initialization step of our method we construct classifiers using C's that are small subsets of Y, and then in successor steps we approximate classifiers for successively larger subsets of the training set by using C's corresponding to even smaller sets of synthetic points (as opposed to using points from Y). We describe this procedure in the next section. It should be noted that this chunking strategy of considering successively larger subsets is one of the elements that differentiates this research from that of [2], which emphasizes synthetic point generation as an *a posteriori* procedure applied to simplify a classifier obtained in the standard manner from the full dataset. A second difference is our coordination of the results of parallel optimization problems for the synthetics.

The idea of simplifying classifiers by using synthetic points was introduced in [2]. Alternative approaches for generating synthetic points are described in [3,8].

Other approaches that seek to reduce kernel-classifier complexity are discussed in [7]. In the linear case, classifiers have also been simplified by means of feature selection via concave programming [1].

2 Chunking-Coordinated-Synthetic (CCS) Approaches

Our approaches are based on a sequence of classifier generating problems interspersed with a sequence of classifier approximation problems (that generate synthetic points). The former we refer to as classifier problems and the latter, as approximation problems. We now define the format of the problems used in the specific implementation presented below.

$LP(S, C)$ is the linear programming classifier generating problem:

$$\min_{u,\gamma,E} \quad \nu \parallel E \parallel_1 + \parallel u \parallel_1 \text{ s.t. } D(uK(C,S) - \gamma\mathbf{1}) + E \geq \mathbf{1}, \ E \geq \mathbf{0}, \quad (1)$$

where ν is a weighting factor, $uK(C, S)$ is the vector obtained by applying the kernel corresponding to $uK(C, \cdot)$ to each element of the subset S of the training set, D is a diagonal matrix of ±1's corresponding to the appropriate classification of the corresponding element of S, E is a vector of "errors" (with respect to the "soft" margin bounding surfaces corresponding to $uK(C, x) = \gamma\pm1$), $\mathbf{1}$ is a vector of 1's, and $\mathbf{0}$ is a vector of 0's. Note that the standard classification problem in this framework would be obtained by setting $C = S = Y$, but this leads to a problem with $O(m)$ constraints and variables, which is intractable if m is large. Other classifier problems (such as classifiers with quadratic objective terms) may be substituted for the LP in (1). In particular, in future research we will experiment with further reductions in problem size via the use of *unconstrained* classifier models.

Using the notation $f(C)$ to denote the array obtained by applying f to each of the s points of C, consider the classifier term $u^*f(C)$, where u^* is obtained by solving (1), and generate an approximation (using $t < s$ points) to this term by considering the problem $NLP(u^*, C, t)$, which is the unconstrained problem $\min_{w,Z} d(u^*f(C), wf(Z))$, where $d()$ provides a measure of distance. As a particular case, in this research we used the 2-norm:

$$\min_{w,Z} \parallel u^*f(C) - wf(Z) \parallel_2^2, \quad (2)$$

where w is a vector of size t, and Z is a set of t synthetic points, which may be distinct from points in C. Thus, the optimal solution $w^*f(Z^*)$ approximates the classifier term $u^*f(C)$. In order to avoid computation with f in feature space, this NLP is re-formulated by expanding the squared 2-norm and applying the Mercer inner product property to obtain an equivalent problem expressed in input space [2]. The expanded problem (2) is $\min_{w,Z} u^*f(C)f(C)u^* - 2u^*f(C)f(Z)w + wf(Z)f(Z)w$, so the corresponding problem in input space via the inner product property is $\min_{w,Z} u^*K(C,C)u^* - 2u^*K(C,Z)w + wK(Z,Z)w$. Note that a numerical value of $K(y, x)$ for a specific pair (y, x) is computed by operations in input space (for example, we may evaluate $K(y, x) = (yx)^p$ where p is a positive integer and yx represents the inner product in input space. Evaluation of $K(y, x)$ via the expression $f(y)f(x)$ would be impractical for $p > 1$ and large n).

The CCS strategy considered here allows different algorithmic formulations. We now describe the i-th stage of the multistage CCS algorithm that we used in this research. In our approach we perform p *independent runs of the i-th stage process*, using a different random subset of training points in each run; the results of these p runs are then combined as described below to provide the initial set of classifier points C_{i+1} for stage $i + 1$. In our results we use the settings $p = 10$, $t = 10$ (number of points used in classifier approximation problems).

At each successive stage, the size of the subset of the training set used to provide data for the classification problem is increased as described below until a classifier problem for the full training set is reached.

Stage i process (performed p times using p independent samples from the training set):

a) Let C_i be a set of s points in R^n, with $s = pt << m$ (C_0 is chosen as a subset of the training set; for $i > 0$ the C_i are sets of synthetic points obtained from the preceding stage).

b) Let Y_i be a randomly chosen subset of Y such that $size(Y_i) > size(Y_{i-1})$; solve the **classifier problem** $LP(Y_i, C_i)$.

c) Letting Z_i, w_i be a set of initial values for a nonlinear optimization procedure, solve the **approximation problem** $NLP(u_i^*, C_i, t)$ to obtain a set of synthetics Z_i^* .

d) Solve the **classifier problem**: $LP(Y_i, Z_i^*)$.

Coordination Step (coordinates the p i-th stage processes): for each run performed in stage $i + 1$, C_{i+1} is the **union** of all Z_i^* for the p runs of the i-th stage (yielding a total of $s = pt$ potential classifier points for each of the initial classifier problems of step (b) in stage $i + 1$).

It should be noted that step (d) of this chunking strategy serves to validate the choices made for the strategic parameters p and t in the sense that the testing set correctness should be similar for steps (b) and (d) and correctness should stabilize as the subsets of Y approach the size of the original training set. This behavior was observed in the computational results that we now present.

Important remark: alternative implementations of step (c) could employ other strategies to generate appropriate synthetic points. Some possible methods are described in [3,8,10].

3   Computational Results, Modified USPS Problem

In order to evaluate the CCS algorithm, we considered a variant of the US Postal Service problem [9]. The USPS problem is composed of 9,298 patterns, 256 features each, corresponding to scanned handwritten numbers (0-9). Thus, it is a multi-category problem. Roughly, each category has the same number of representatives. To generate a (harder) modified version, inspired by [8], we add noise to a sample of USPS points in order to simulate a noisier dataset with less separability. We use a USPS sample as centers. In order to balance the problem, the sample is composed of 90 random '8' patterns and a random 10-pattern subset for each of the other nine digits. The resulting 180-center set was used as input for the NDC data generator [6] (thus, NDC centers are real data, instead of random points). The NDC generator typically assigns labels to centers by means of a random separating hyperplane, but in our case, this assignment method is not necessary, because labels are known *a priori*: 1 for '8' centers and -1 for the other centers and (as is usual in NDC) all points that are generated from a specific center are assigned the center's label. The overall result is a training set with approximately 4000 points of category 1 and 4000 of category -1. A Gaussian kernel $K(y,x) = \exp(- \mid y - x \mid^2 /128)$ was used as in [8], where it provided good results. All our runs took place on a node of a Sun Enterprise machine with an UltraSPARC II processor at 250 MHz. Step (b) and any other linear problems in our tests were solved using CPLEX 6.0. The nonlinear approximation problem in step (c) was solved by invoking CONOPT 2 via GAMS 2.50A (we chose an "off-the-shelf" NLP solver for this initial implementation, and found it to be quite effective. Other NLP techniques could, of course, be more efficient for this particular class of applications. With CONOPT 2 we found that the solution process was accelerated by imposing box constraints on the variables w. In particular, we impose the constraints: $-\alpha|u_i^*| \leq w \leq \alpha|u_i^*|$ for some $\alpha > 0$. The CONOPT 2 option `rtredg`, reduced gradient tolerance for optimality, was set to 1e-4). Both CPLEX and GAMS were called from a C program that implemented the whole algorithm.

All settings were NDC defaults, except for the number of points (8000 used here), number of features (256) and expansion factor (25) (this factor controls the expansion of point clouds around centers, using a random covariance matrix for each center). The expansion factor was selected to produce a separability in the range of 70% for the Gaussian kernel $K(y,x) = \exp(-|y - x|^2/128)$ in a standard 1-norm classification problem. The value $\nu = 10.0$ was found to yield the best training correctness for this dataset. Four stages were used, employing the following *sizes for the randomly selected subsets of the data*: Y_{m0}: 1000 USPS/NDC points, (500 testing + 500 training), Y_{m1}: 2000 USPS/NDC points, (1000 testing + 1000 training), Y_{m2}: 4000 USPS/NDC points, (2000 testing + 2000 training) and Y_{m3}: 8000 USPS/NDC points, (4000 testing + 4000 training).

A point in Y_{mi} is termed a *support vector* if the dual variable of its classification constraint is nonzero at the solution, implying its classification constraint is active at the solution. Therefore, for similar testing correctness, the number of support vectors shown below is a measure of the robustness of the classifier in the sense that the support vectors include (in addition to "error" points) those points that are correctly classified, but lie in the margin and hence are not "widely separated" from the points of the other category.

The following implementation choices were made:

- α=1.25.
- The initial value set Z_0 for the initial approximation problem is a random subset of t Y_{m0} points with $u \neq 0$ after step (a) in stage 0. The w_0 are their u multipliers.

- For $i \geq 0$, the initial value set Z_{i+1} is the best Z_i^* obtained from an approximation problem in step (c), over p runs, in terms of testing set classification in step (c). The initial w_{i+1} are the corresponding w, also after step (c).

In order to evaluate the efficiency of the CCS algorithm, it is instructive to compare the quality of the solution at step (d) of stage i with the output of the **standard 1-norm classification problem** in which $C_i = Y_{mi}$:

$$\min_{u,\nu,E} \quad \nu\|E\|_1 + \|u\|_1 \text{ s.t. } D(uK(Y_{mi}, Y_{mi}) - \gamma\mathbf{1}) + E \geq \mathbf{1}, E \geq \mathbf{0}. \quad (3)$$

Note that, in the particular case of stage 0, the problem (3) is the same as step (b). However, relative to stages 1, 2 and 3, the number of classifier points that are allowed and the corresponding number of variables grow significantly in the case of problem (3).

The results for **stage 0** were:

- \# of classifier points, p runs: 96.8 (step (b), avg.), 10 (step (d), max.).
- Avg. training correctness: 84.3% (step (b)), 84.38% (step (d)).
- Avg. testing correctness: 72.12% (step (b)), 72.42% (step (d)).
- Avg. \# of support vectors out of 500: 362.5 (step (b)), 212.5 (step (d)).

We see that the number of support vectors in step (d) is considerably less than in the standard 1-norm classification problem. Also, using only a small number of synthetic points does not degrade testing correctness, but instead yields a small improvement (similar results are obtained for the larger subsets considered below).

Step (c) yields an approximation of the classifier obtained in step (b). In order to evaluate the need of step (d), we evaluated the average quality of the step (c) classifier over all runs in stage 0 ("pure" synthetic-point classifier, without the re-classification step (d)): we observed an average testing correctness of 56.14% and an average training correctness of 56.28% in step (c).

Consequently, the benefit of the re-classification in step (d) is evident. Another question that may arise is the quality of trivial alternatives to the NLP synthetic point generation procedure of step (c). Thus, we considered a simple choice of classifier points: given the subset of C_0 such that $u \neq 0$ after step (b), we took a t-point subset whose u multipliers had largest absolute value, and used them for re-classification of the training points in stage 0, yielding an average training correctness of 60.54% in modified step (d) and an average testing correctness of 58.16% using the approximate classifier from step (d).

We observe again the considerable advantage of using synthetic points. Similar results were obtained when t-point random samples of Y_{m0} were used.

In stages 1 and 2 the number of classifier points in step (b) is progressively lower. Specifically, in stage 2 we obtained the following results:

- \# of classifier points, p runs: 13.1 (step (b), avg.), 10 (step (d), max.).
- Avg. training correctness: 76.77% (step (b)), 76.71% (step (d)).
- Avg. testing correctness: 76.76% (step (b)), 76.62% (step (d)).
- Avg. \# of support vectors out of 2000: 1104 (step (d)).

Nevertheless, when we used the **standard 1-norm formulation** in problem (3) the results were much worse:

- Avg. # of classifier points, p runs: 122.2 (Y_{m1}), 180.6 (Y_{m2}).
- Avg. training correctness: 82.06% (Y_{m1}), 81.94% (Y_{m2}).
- Avg. testing correctness: 73.3% (Y_{m1}), 75.47% (Y_{m2}).
- Avg. # of support vectors: 672.5 out of 1000 (Y_{m1}), 1181 out of 2000 (Y_{m2}).

Note that while *training correctness* is improved by allowing all 2000 points in the stage 2 training subset to be used to construct the classifier, the resulting *testing correctness* of 75.47% is actually worse than that obtained by using the classifier points from the much smaller synthetic sets C_i (whose 100 points yield 76.76% classification) or Z_i^* (whose 10 points yield 76.62%). This apparently surprising result that we already observed in stage 0 is analogous to results in [4], where random 1%-5% subsets of the Adult Dataset [5] were allowed in the Gaussian kernel classifier. Observe that if the solution quality is measured by the number of support vectors, then by this measure as well, the smaller classifier sets provide better quality solutions than the full training set. Similar results are obtained when the full dataset is used in stage 3, except that the standard 1-norm LP could not be solved in 48 hours, illustrating the scalability difficulties associated with the standard approach.

We thus compared our results (as provided by the use of synthetic points in the classifier) with *random subset classifiers*. Let θ be the number of random points chosen in Y_{m2}. We observed the following:

- Avg. # of classifier points, 1-norm, p runs: 63.9 ($\theta=100$), 10 ($\theta = 10$).
- Avg. training correctness: 74.28% ($\theta=100$), 59.18% ($\theta=10$).
- Avg. testing correctness: 70.83% ($\theta=100$), 57.74% ($\theta=10$).
- Avg. # of support vectors: 1814.3 out of 2000 ($\theta=10$).

For this dataset, randomly chosen subsets of the training set produce average classifications that are significantly worse than those obtained with either the 1-norm classification with the full training set or our small sets of synthetic points.

4   Conclusions and Directions for Future Research

In this paper, we have analyzed the fusion of synthetic point generators (small nonlinear programs) with chunking algorithms to develop a chunking-coordinated-synthetic (CCS) approach that achieves good generalizability while greatly reducing the size of the optimization problems that are required to produce nonlinear classifiers. Our numerical results on a modified USPS dataset show that the classifiers obtained using very small numbers of synthetic points (*as few as 10*) not only yield good generalizability (in terms of good testing set classification in ten-fold cross-validation), but also, for the noisy data considered, actually yield classifiers with *better generalizability* than either various other choices of reduced sets of training points or even the full training set (the latter appears to result in over-training as was noted in [4] for other datasets and other reduced set approaches). Finally, since the CCS approach utilizes the solution of independent optimization problems at each stage, computation may be further accelerated by parallelization. One of our future directions for research will be the parallel implementation of CCS.

Finally, although the emphasis of this research is on nonlinear kernels, similar ideas could be applied to linear kernels. For them, problem (2) is trivial, since f is the identity function and (2) is solved by taking $w = 1$ and $Z = u^*C$. The interesting aspect of CCS in this case is the coordination step in which these single-point "ideal" classifiers for subsets are combined to produce a small classifier set for a larger subset. The key issue is whether these small classifier sets will produce good classifiers.

References

1. P.S. Bradley and O.L. Mangasarian. "Feature selection via concave minimization and support vector machines". In J. Shavlik, editor, Machine Learning Proceedings of the Fifteenth International Conference (ICML'98), pp. 82-90, San Francisco CA, 1998, Morgan Kaufmann.
2. C. J. C. Burges. "Simplified Support Vector Decision Rules". In L. Saitta, editor, Proceedings 13th Intl. Conf. on Machine Learning, pp. 71-77, San Mateo CA, 1996, Morgan Kaufmann.
3. C. J. C. Burges and B. Scholkopf. "Improving the Accuracy and Speed of Support Vector Machines". In M. Mozer, M. Jordan, and T. Petsche, editors, Advances in Neural Information Processing Systems 9, pages 375-381, Cambridge, MA, 1997. MIT Press.
4. Y.-J. Lee and O. L. Mangasarian, "RSVM: Reduced Support Vector Machines". Data Mining Institute technical report 00-07, July 2000.
5. P. M. Murphy and A. W. Aha. UCI repository of machine learning databases, 1992. www.ics.uci.edu/~mlearn/MLRepository.html.
6. D. R. Musicant. NDC: Normally Distributed Clustered datasets, 1998. www.cs.wisc.edu/~musicant/data/ndc.
7. E. Osuna and F. Girosi. "Reducing the Run-time complexity of Support Vector Machines". In Proceedings of the 14th International Conference on Pattern Recognition, Brisbane, Australia, 1998.
8. B. Scholkopf, S. Mika, C. J. C. Burges, P. Knirsch, K.-R. Muller, G. Ratsch and A. J. Smola. "Input Space vs. Feature Space in Kernel-Based Methods". IEEE Transactions on Neural Networks 10(5):1000-1017, 1999.
9. B. Scholkopf and A. J. Smola. Kernel machines page, 2000. www.kernel-machines.org.
10. D. DeCoste and B. Scholkopf. "Training Invariant Support Vector Machines". Machine Learning 46: 161-190, 2002.

Computational Identification of −1 Frameshift Signals

Sanghoon Moon, Yanga Byun, and Kyungsook Han[*]

School of Computer Science and Engineering, Inha University, Inchon 402-751, Korea
{jiap72,quaah}@hanmail.net, khan@inha.ac.kr

Abstract. Ribosomal frameshifts in the −1 direction are used frequently by RNA viruses to synthesize a single fusion protein from two or more overlapping open reading frames. The slippery heptamer sequence XXX YYY Z is the best recognized of the signals that promote −1 frameshifting. We have developed an algorithm that predicts plausible −1 frameshift signals in long DNA sequences. Our algorithm is implemented in a working program called FSFinder (Frameshift Signal Finder). We tested FSFinder on 72 genomic sequences from a number of organisms and found that FSFinder predicts −1 frameshift signals efficiently and with greater sensitivity and selectivity than existing approaches. Sensitivity is improved by considering all potentially relevant components of frameshift signals, and selectivity is increased by focusing on overlapping regions of open reading frames and by prioritizing candidate frameshift signals. FSFinder is useful for analyzing −1 frameshift signals as well as discovering unknown genes.

1 Introduction

Translation is the mechanism of protein synthesis in which RNA messages are transformed into the amino acid sequences of proteins. Two kinds of errors can alter the reading frame during translational elongation. One is spontaneous error that occurs at a frequency of less than 5×10^{-5} per codon in all species. The other is non-standard error (also called *programmed translational frameshift*) that occurs in some genes with a frequency close to 100% [1, 2].

Programmed frameshift occurs in genes of organisms ranging from bacteria to lower eukaryotes, as well as in animal and plant viruses. The analysis of programmed frameshift is important because it plays a significant role in viral particle morphogenesis, and in the genetic control of alternative enzymatic activities [2]. In this process the ribosome shifts a reading frame by one or a few nucleotides at a specific site in a messenger RNA. The most common of these events requires the ribosome to shift to a codon that overlaps a codon in the existing frame. The shift of a single step backwards in effect reassigns a single nucleotide (-1 frameshift), whereas a slip forwards skips a single nucleotide (+1 frameshift) [3]. The most common type of frameshift is a -1 shift. The most common elements causing eukaryotic frameshifts consist of a slippery site that promotes frameshifting mechanically, and a stimulatory structure that probably induces the ribosome to pause [4]. The slippery site consists of

---

[*] To whom correspondence should be addressed. Email: khan@inha.ac.kr

M. Bubak et al. (Eds.): ICCS 2004, LNCS 3036, pp. 334–341, 2004.
© Springer-Verlag Berlin Heidelberg 2004

a heptameric sequence of the form X-XXY-YYZ (in the incoming 0-frame), where X, Y and Z can be the same nucleotide [4]. The downstream stimulatory structure is usually a pseudoknot in which certain bases in a loop pair with complementary bases outside the loop, or it is a simple stem-loop. The slippery heptamer is separated from the stimulatory structure by a short sequence of 5 to 9 nucleotides, the so-called spacer [5, 6]. The length of the spacer is known to influence the probability of frameshifting. Typically viral frameshifts produce fusion proteins in which the amino- and carboxy-terminal domains are encoded by overlapping open reading frames [7], as shown in Fig. 1.

Many existing approaches to identifying frameshift signals either depend on comparing DNA sequences with protein sequences in databases [11, 12], or focus on detecting experimental errors [13]. We have developed a set of algorithms that consider both downstream pseudoknots and simple stem-loops as downstream stimulatory structures in the overlaps between open reading frames. We have implemented these algorithms in a program called FSFinder (Frameshift Signal Finder).

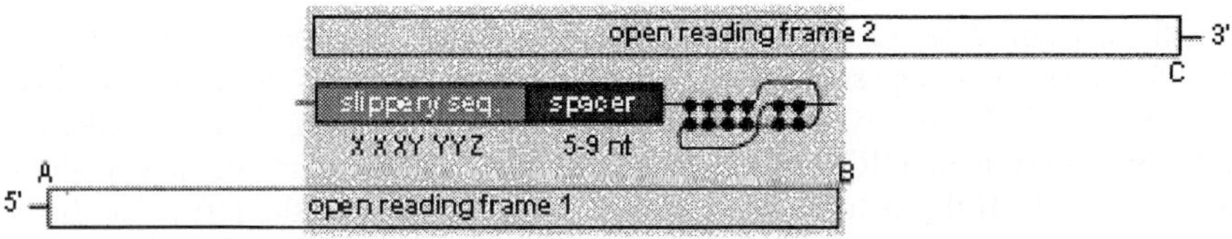

Fig. 1. Three components of −1 frameshift signals in the overlap between two open reading frames: slippery sequence, spacer, and pseudoknot (or stem-loop). When a frameshift takes place, protein synthesis terminates at C rather than at B

2 Computational Model

2.1 Components of Frameshift Signals

We extended the computational model for −1 frameshift signals of Hammell *et al.* [7] to improve its sensitivity and selectivity. Sequences of 3 codons (9 nucleotides) in a genomic sequence are first examined for possible slippery sequences X XXY YYZ. In X XXY YYZ, X and Z can be any nucleotide, and Y can be A or U (in Hammell's model, Z is either A, U, or C). If a slippery sequence is identified, FSFinder searches for a downstream structure by sliding along the spacer from one to 11 nucleotides. Fig. 2 (A) shows a programmed −1 frameshift signal with a pseudoknot as stimulatory structure. The pseudoknot is of the H-type, in which stem 1 has   13 base pairs, stem 2 has   6 base pairs, and both loops of the pseudoknot have   6 nucleotides. The first 4 base pairs of stem 1 must include at least 2 G-C pairs.

Some programmed −1 frameshift signals have a simple stem-loop as stimulatory structure. As explained in Fig. 2 (B), we examine the nucleotides in both directions from every pivot nucleotide for possible base pairing. The pivot nucleotide can be either included or excluded in the base pairing.

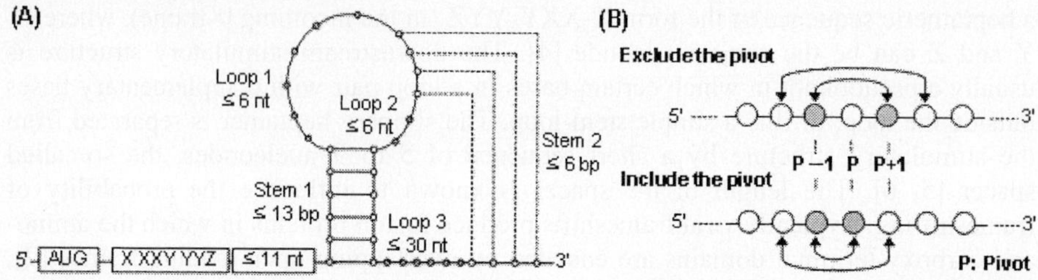

Fig. 2. (A) A programmed -1 ribosomal frameshift signal with an H-type pseudoknot. (B) The process of finding a simple stem-loop structure downstream from a slippery sequence. Nucleotides in both directions from each pivot nucleotide are examined for possible base pairing

2.2 Algorithms for Predicting Frameshift Signals

Algorithms 1 and 2 search for stem-loops and canonical base pairs, respectively. If a stem-loop crosses other stem-loops, they are considered to form a pseudoknot. Algorithm 3 finds an overlapping region of open reading frames (ORF). An overlapping region of ORFs is identified by first finding pairs of stop codons in frames −1 and 0. If the second stop codon of in frame −1 is to the left to the fist stop codon in frame 0, an overlapping region of the two frames is found. Overlapping frames with the largest ORF (light yellow) have the highest probability of containing frameshift signals, and overlapping frames with the second largest ORF (sky blue) have the second highest probability of having frameshift signals (see Fig. 3).

Algorithm 1 Find stems

```
Initialize boolean sequence array S[n,2], n is a length of the
sequence
  Set S according to each character of the sequence
   [false, true]: boolean type of 'A'
   [false, false]: boolean type of 'G'
   [true, true]: boolean type of 'C'
   [true, false]: boolean type of 'U'
  Find slippery sites
  for each slippery site do
    for i ← end index of a slippery site + 1 to n - 8 do
      Set ST be an empty list of stems
      checkCanonicalPair(S, false,
                    start index of a slippery site, i, ST)
    checkCanonicalPair(S, true,
                start index of a slippery site, i, ST)
    end for
    Set stems of slippery site as ST
    findPseudoknots(slippery site)
  end for
```

Algorithm 2
```
checkCanonicalPair(seqArray, includePivot, startIndex,
            pivotIndex, ST)
```

```
Set L be an empty list of base pairs
if startIndex = pivotIndex then
  return
end if
pairCount ← pivotIndex  2 + 1
insertLase ← false
if not checkMiddle then
  pivotIndex ← pivotIndex - 1
  pairCount ← pairCount - 1
end if
for i ← startIndex to pivotIndex do
  pairIndex ← pairCount - i
  if isCanonicalPair(seqArray[i], seqArray[pairIndex]) then
  if not insertLast then
    if size of L ≥ 4 then
      insert L to ST
    end if
   end if
  Set L be an empty list of pair
   insert pair(i, pairIndex) to L
    insertLast ← true
  else
  insertLast ← false
  end if
end for
if size of L ≥ 4 then
insert L to ST
end if
```

Algorithm 3 `FindOverlappingRegion`

```
Set F0 be N + 1 element array of regions, N is count of frame 0 stop
codons
Set FM1 be M + 1 element array of regions, M is count of frame −1
stop codons
Set each element of F0 and FM1 as a region between front stop codon
and next stop codon, order of frame0 and frame −1 stop codon index
Set O be an empty array of overlapping region
j ← −1
for i ← 0 to N do
  while j < M do
    j ← j + 1
   if end of F0[i] ≤ start of FM1[j] then
       break
     else if (start of F0[i] < start FM1[j]) and
        (end of F0[i] < end of FM1[j]) then
        Let R as a region (start of FM1[j] to end of F0[i])
        if any slippery site exists in R then
          insert R to O
     end if
     end if
 end while
end for
```

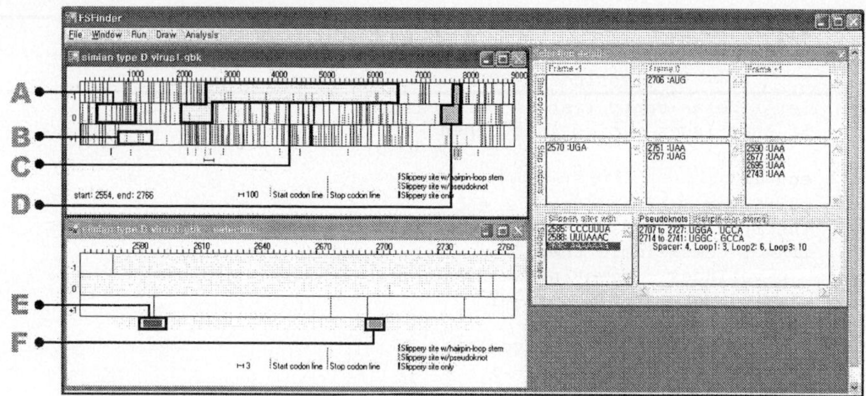

Fig. 3. Graphical user interface of FSFinder. **A.** Stop codons (long, blue lines). **B.** Start codons (short, red lines). **C.** Frameshift signal with the highest probability (light yellow). **D.** Frameshift signal with the second highest probability (sky blue). **E.** Frameshift signal with a stem-loop (green bar). **F.** Frameshift signal with a pseudoknot (pink bar)

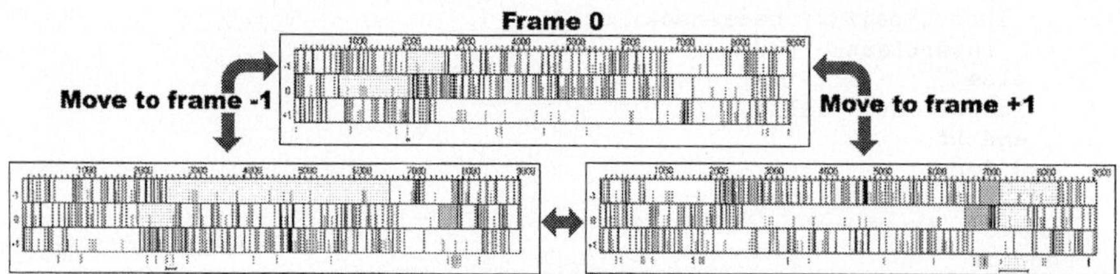

Fig. 4. Alternating frames

2.3   Implementation

FSFinder was implemented in Microsoft C#. It provides graphical views of -1, 0, and +1 frames, like DNA Strider [8]. The three frames (-1, 0 and +1 frames) are shown in the left upper window of Fig. 3. If a user specifies a region for detailed examination by the drag and drop operation in the left upper window, the specified region is enlarged in the lower left window. The right window displays the positions of start and stop codons, slippery sequences, pseudoknots and stem-loops found in the frames in the left window. Users can change the stem and loop sizes of a stem-loop or pseudoknot. They can also alternate frames to find frameshift signals in different overlapping frames. (see Fig. 4).

3   Results and Discussion

FSFinder was tested on 71 organisms with known programmed -1 frameshift mutations obtained from the databases PseudoBase [9] and RECODE [10]. PseudoBase contains 20 eukaryotic viruses and RECODE has 65 prokaryotes, eukaryotic viruses, bacteriophages, eukaryotic transposable elements and bacterial insertion sequences. The two databases share 14 frameshifts. Each of these organisms and elements has one or two authentic programmed -1 frameshift sites.

Hammell *et al.* [7] have attempted to identify frameshift signals in prokaryotic and eukaryotic DNA sequences [7], but the sensitivity of their approach is low. It misses many frameshift signals because it only considers pseudoknots as downstream stimulatory structures, the definition of pseudoknots is too restricting, and X XXY YYG is not considered a slippery sequence. For example, their approach does not locate the frameshift signals in Rous sarcoma virus (RSV), because loops 1 and 2 of the pseudoknots involved are larger than their approach permits. On the other hand, the selectivity of the computational model of Bekaert *et al.* [5] is low because it predicts too many false positives. Other computational models can identify potential frameshift signals only when they are given reference protein sequences along with DNA sequences [11, 12].

FSFinder identifies more frameshift signals than the approach of Hammell *et al.* because both pseudoknots and simple stem-loops are considered as downstream secondary structures and because conditions for slippery motifs and pseudoknots are relaxed. On the other hand, FSFinder finds less potential frameshift signals than the approach of Bekaert *et al.* because it searches for frameshift signals only in the overlapping regions of open reading frames, and prioritizes candidate frameshift signals.

A total of 26 frameshift signals in RECODE have simple stem-loops as downstream secondary structures, but 5 of these were excluded because PseudoBase assigns them different stimulatory structures. Seventeen of the remaining 21 frameshift signals were detected by FSFinder while 4 could not be found because their slippery sequences do not conform to the motif X XXY YYZ. It turns out that many frameshift signals have the slippery motif X XXY YYG. FSFinder identified 13 such sequences, and these can be classified into two types: A AAA AAG and G GGA AAG. The frameshift signals of RSV were also detected.

Table 1. Frameshift signals in RECODE with downstream stem-loops and X XXY YYG slippery sequences. * indicates a frameshift signal that was not identified by FSFinder because the slippery sequence does not conform to the motif X XXY YYZ.

ID	frameshift signals with X XXY YYZ (Z≠G) and a downstream stem	ID	frameshift signals with X XXY YYG and a downstream stem	ID	frameshift signals with X XXY YYG and other downstream structures
82	HIV type 1	71	Escherichia coli	104	Bacteriophage lambda
83	HIV type 2*	238	IS911	237	IS2
84	Human T-cell lympotrophic virus type 1	251	IS150		
85	Human T-cell lympotrophic virus type 2	252	IS1221A		
92	RCNMV *	360	Salmonella typhi		
97	Simian T-cell lymphosropic virus type 1	361	Salmonella typhimurium		
106	Drosophila buzzatii Ossvaldo retrotransposon	362	Vibrio cholerae		
257	Carrot mottle mimic virus*	363	Neisseria meningtidis		
258	Groundnut rosette virus	364	Neisseria gonorrhoeae		
260	PEMV RNA 2*	365	Neisseria meningitides		
		392	Yersinia pestis		

Searching for frameshift signals in the overlapping region of ORFs is effective in predicting strong frameshift signal candidates. For example, a total of 157 potential frameshift signals were found in the sequences of the test cases in PseudoBase. Only 33 of these were in overlapping ORFs, and 19 of 33 proved to be the only genuine frameshift signals. FSFinder also identifies frameshift signals in alternative frames. For example, simian type D virus 1 has two slippery sequences G GGA AAC and A AAU UUU in different frames at positions 2058 and 2585, respectively. FSFinder detects two different signals in each of 6 organisms in RECODE: human T-cell lymphotropic virus type 2, mouse mammary tumor virus, simian type D virus 1, simian retrovirus type 2, simian T-cell lymphotropic virus type 1, and visna virus. There was only one alternative signal (in mouse mammary tumor virus) that could not be identified as it has a different motif (G GAU UUA). Table 2 summarizes the predicted frameshift signals in PseudoBase.

Table 2. Predicted frameshift signals in PseudoBase. * indicates a frameshift signal that was not detected by FSFinder because the slippery sequence does not conform to the motif X XXY YYZ

PseudoBase numbers	Organisms	frameshift signals in the entire region	frameshift signals in the overlapping region
PKB1	Bovine Leukemia Virus	14	4
PKB2	Beet Western-Yellow Virus	7	4
PKB3	Equine Infectious Anemic Virus	12	2
PKB4	Feline Immunodeficiency Virus	14	1
PKB42	Potato Leafroll Virus-W	2	1
PKB43	Potato Leafroll Virus-S	2	2
PKB44	CABYV	4	1
PKB45	Pea Enation Mosaic Virus	6	3
PKB46	Barley Yellow Dwarf Virus	4	2
PKB80	Mouse Mammary Tumor Virus	12	1
PKB106	Infectious Bronchitis Virus	1	1
PKB107	Semian Retro Virus -1	9	2
PKB127	Equine Arteritis Virus*	-	-
PKB128	Berne Virus	11	1
PKB171	Human Corona Virus 229E	12	1
PKB174	Rous Sarcoma Virus	4	1
PKB217	LDV-C	1	1
PKB218	PRRSV-16244B	16	1
PKB233	PRRSV-LV	17	1
PKB240	Beet Chlorosis Virus	9	3
Total number of true positives		19	19
Total number of candidates		157	33

4 Conclusion

Identifying programmed -1 frameshifts is difficult because they are not uniform. However it is very important to achieve this identification in order to fully understand the underlying mechanisms and to discover new genes. Existing computational models predict too many false positives, or need reference protein sequences together with DNA sequence data from similar organisms.

We have developed an algorithm and a program called FSFinder for predicting plausible –1 frameshift signals in long DNA sequences. FSFinder was tested on 71 genomic sequences from different organisms and it predicted –1 frameshift signals more sensitively and selectivity than existing approaches. The procedure increases sensitivity by considering all potentially relevant components, and has increased selectivity because it focuses on the overlapping regions of open reading frames and prioritizes candidate signals. We believe FSFinder will be useful for analyzing –1 frameshift signals as well as for discovering novel genes.

Acknowledgement. This work was supported by the Korea Science and Engineering Foundation (KOSEF) under grant R01-2003-000-10461-0.

References

1. Vimaladithan, A., Farabaugh, P.J.: Identification and analysis of frameshift sites. Methods in Molecular Biology 77 (1998) 399-411
2. Farabaugh, P.J.: Programmed translational frameshifting. Microbiological Reviews 60 (1996) 103-134
3. Farabaugh, P.J.: Programmed translational frameshifting. Annual Review of Genetics 30 (1996) 507-528
4. Jacks, T., Varmus, H.E.: Expression of the Rous sarcoma virus pol gene by ribosomal frameshifting. Science 230 (1985) 1237-1242
5. Bekaert, M., Bidou, L., Denise, A., Duchateau-Nguyen, G., Forest, J., Froidevaux, C., Hatin, I., Rousset, J., Termier, M.: Towards a computational model for -1 eukaryotic frameshifting sites. Bioinformatics 19 (2003) 327-335
6. Dinman, J.D., Icho, T., Wickner, R.B.: A -1 ribosomal frameshift in a double-stranded RNA virus of yeast forms a gag-pol fusion protein. Proc. Natl Acad. Sci. USA 88 (1991) 174-178
7. Hammell, A.B., Taylor, R.C., Peltz, S.W., Dinman, J.D.: Identification of putative programmed -1 ribosomal frameshift signals in large DNA databases. Genome Res. 9 (1999) 417-427
8. Marck, C.: DNA Strider: a C program for the fast analysis of DNA and protein sequence on the Apple Macintosh family of computers. Nucleic Acids Research 16 (1988) 1829-1836
9. van Batenburg, F.H.D., Gultyaev, A.P., Pleij, C.W.A., Ng, J., Oliehoek, J.: PseudoBase: a database with RNA pseudoknots. Nucleic Acids Research 28 (2000) 201-204
10. Baranov, P., Gurvich, O.L., Hammer, A.W., Gesteland, R.F., Atkins, J.F.: RECODE Nucleic Acids Research 31 (2003) 87-89
11. Birney, E., Thompson, J.D., Gibson, T.J.: PairWise and SearchWise: finding the optimal alignment in a simultaneous comparison of a protein profile against all DNA translation frames. Nucleic Acids Research 24 (1996) 2730-2739
12. Halperin, E., Faigler, S., Gill-More, R.: FramePlus: aligning DNA to protein sequences. Bioinformatics 15 (1999) 867-873
13. Fichant, G.A., Quentin, Y.: A frameshift error detection algorithm for DNA sequencing projects. Nucleic Acids Research 23 (1995) 2900-2908

Mobility Management Scheme for Reducing Location Traffic Cost in Mobile Networks

Byoung-Muk Min[1], Jeong-Gyu Jee[2], and Hea Seok Oh[1]

[1] School of Computing Soongsil Univ., Seoul, Korea
bmmin@yahoo.com
[2] Korea Research Foundations, Seoul, Korea

Abstract. Even when users are moving, a major problem in such a mobile networks is how to locate Mobile Hosts (MHs). In this paper we propose mobility strategy that minimizes the costs of both operations, the location registration and the call tracking, simultaneously. In numerical results, the proposed method proves that it has more improved performance than the previous methods.

1 Introduction

To effectively monitor the movement of each MH, a large geographical region is partitioned into small Registration Areas (RAs). Figure 1 shows the architecture of a mobile system. Each RA has a Mobile Switch Center (MSC, also called a Base Station (BS)) which serves as the local processing center of the RA. The profiles of MH inside a RA are kept, in the MSC's Visitor Location Register (VLR). On top of several MSC/VLRs is a Local Signaling Transfer Point (LSTP) and on top of several LSTPs again is a Remote Signaling Transfer Point (RSTP). In this way, the whole system forms a hierarchy of station. The LSTP and the RSTP are routers for handling message transfer between stations. For one RSTP there is a Home Location Register (HLR). Each MH must register in a HLR. When a MSC needs to communicate with another MSC. MSC first sends a message to the LSTP on top of it. If another MSC is under the same LSTP as MSC, then the message is forwarded to another MSC without going through the RSTP. Otherwise, the message has to be through the RSTP and then down to a proper LSTP ad then to another MSC. In spite of many advantages available in wireless communication. It is not without difficulties to realize such systems. The first problem is how to locate a Mobile Host (MH) in a wireless environment. The IS-95 strategy is most often referred in resolving this problem. IS-95 used in the United States and GMS [6] used in Europe are examples of this strategy.

Many papers in the literature have demonstrated that the IS-95 strategy does not perform well. This is mainly because whenever a MH moves. The VLR of a Registration Area (RA) which detected the arrival of the host always reports to the HLR about the host's new location.

M. Bubak et al. (Eds.): ICCS 2004, LNCS 3036, pp. 342–348, 2004.

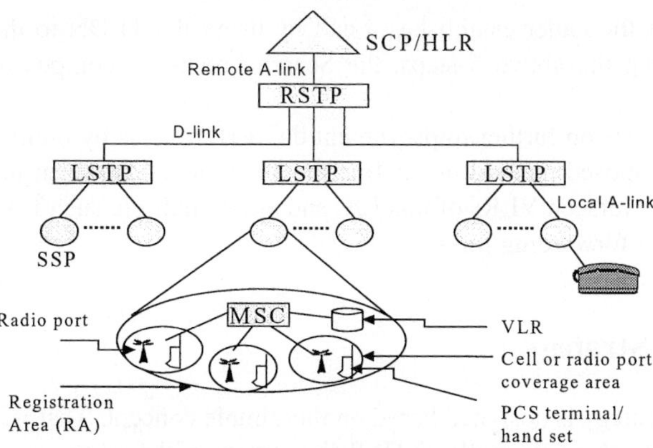

Fig. 1. Architecture of mobile networks

Among them, the Forwarding strategy [4, 5], the Local Anchor (LA) strategy, and the Caching strategy [3,7] are representatives of the old VLR to the new VLR. Update of the client's location to the HLR's database is not always needed to minimize communications to the HLR. To locate a callee however, some extra time is required to follow the forwarding link to locate the host. When the number of the forwarding links is high, the locating cost becomes significant.

In IS-95 scheme, the BS reserves only the resources corresponding to the minimum transmission rate to the mobile. According to the IS-95 strategy, the HLR always knows exactly the ID of the serving VLR of a mobile terminal. We outline the major steps of the IS-95 location registration scheme as follows [6]:

1. The mobile terminal sends a registration request (REGREQ) message to the new VLR.
2. The new VLR checks whether the terminal is already registered. If not, it sends a registration notification (REGNOT) message to the HLR
3. The HLR sends a registration cancellation (REGCANC) message to the old VLR.

The old VLR deletes the information of the terminal and the IS-95 call tracking scheme is outlined as follows:

1. The VLR of caller is queried for the information of callee. If the callee is registered to the VLR, the Search process is over and the call is established. If not, the VLR sends a location Request (LOCREQ) message to the HLR.
2. The HLR finds out to which VLR the callee is registered, and sends a routing request (ROUTREQ) message to the VLR serving the callee. The VLR finds out the location information of the callee.
3. The serving MSC assigns a temporary local directory numbers (TLDN) and returns the digits to the VLR which sends it to the HLR.
4. The HLR sends the TLDN to the MSC of the caller.

5. The MSC of the caller establishes a call by using the TLDN to the MSC of the callee. Among the above 5 steps, the Search process is composed of step1 and step2.

This paper proposes on further improvement the performance by minimizing location traffic. In the proposed method, we define the VLRs that have been linked to by the same LA as the overseen VLRs of this LA, and allow multiple such LAs to be linked together by using forwarding links.

2 Proposed Strategy

The proposed strategy is designed based on this simple concept. That is, when a call is made, instead of asking the callee's HLR the system will find the callee's VLR and from there the LA and then the callee. To accomplish this, the concept of the past LA strategy is adopted in this work. Depending on whether the caller's VLR statically or determine where to search for the callee. In both of these methods, the callee's profile needs to be kept in his visited VLRs. To serve for this purpose, a data structure named the MH table is defined to save some information of visited mobile users for each VLR. This table maintains for each visited MH the host's ID, a Type and a Pointer. The host's ID has the identifier of the host. Each host is assumed to have a different host's ID. Every VLR maintains a MH table records the information for each MH who has visited this VLR. Whether a VLR is the LA of a mobile client can be examined by using the value of Type in the MH table. The schema of the MH table is quite simple and easy to implement. The size of each record in this table can be as small as eight byte. The size of Type value is two bits and that of a Pointer value is also four bytes. It is easily manageable by any current DBMS. However the size of a MH table grows when more and more clients visited this. This problem can be easily resolved by removing obsolete records from the table when necessary. If from the table the system cannot locate the host, then the system simply asks the HLR of the callee about the current location of the callee.

2.1 Location Registration

We formally present the algorithms in the following subsections. Basically, take of location registration are to save a new record in visiting VLR, and to update location of the MH recorded in the old VLR. We provide for each algorithm to illustrate how the algorithm works.

1. The new VLR learns that the MH is inside its territory and informs the old VLR that MH is in its RA. The MH table of the new VLR is inserted a new record describing the coming MH. If the mobile client visited this new VLR in the past, then the system only updates the Type and the Pointer values.
2. The old VLR replies an acknowledgement to the new VLR.
3. The old VLR informs the LA that the MH has moved to the new VLR. Also the old VLR update its own MH table by replacing the MH's Type value with Visited VLR and the Pointer value with the LA's; location.

4. The LA replies a message to the old VLR, and updates its own MH table. The Type value of the mobile client is not changed. The Pointer value is modified to the new VLR's location.
5. End.

2.2 Call Tracking

We describe the call tracking operation. The algorithm of the call tracking is as follows.

1. When a VLR receives a request of locating a callee, it first checks whether its Mobile Host table has the callee's record. If yes, then sends the locating requite to the LA stated in this record. Otherwise, jump to Step 7.
2. / The caller is currently at a location where the callee visited before./
    if the record of the Mobile Host table of the LA stated in Step 1 says that this LA is a "Visited LA", then goto Step 3. If it says "Latest LA", goto Step 4.
3. The locating request is forwarded to this visited LA. While the request is forwarded to the next LA, the callee's record is again searched from this LA's Mobile Host table. Goto Step 2.
4. The latest LA finds the callee's record from the Mobile Host table. If the value of the Pointer field is NULL, then the callee is right in one of this LA's governing RAs. Hence, a message is forwarded to the caller's VLR to make the connection. Goto step 13. If the value of the Pointer field is not NULL, then it must be a VLR who is currently overseeing the callee. Hence, the call tracking request is sent to the latest VLR to which the Pointer field refers.
5. The latest VLR sends a message to the caller's VLR to make a connection.
6. Goto Step 13
7. / The caller is at a location where the callee has not visited before. Updates of the callee's new location in the LA, VLR, and HLR are associated with this call tracking operation /
8. The HLR forwards the request to the callee's LA.
9. The callee's latest LA forwards the request to the latest. Also, the callee's record in this VLR's Mobile Host table is updated by replacing its Type with "Latest LA" and Pointer with NULL.
10. The callee's VLR acknowledges the receipt of the message to the LA and the LA will then update the callee's record in its Mobile Host table by replacing type with "Visited LA" and Pointer with a pointer to the callee's current residing.
11. The callee's VLR sends a message to the HLR. The HLR updates the callee's new location to the new latest .
12. The HLR forwards the message about the current location (VLR) of the callee to the caller's VLR and the connection between the caller's VLR and the callee's VLR is built.
13. End.

3   Performance Model

We present the cost models that used to evaluate the performance of the proposed strategy. We list the parameters used in the models. Then, we derive the cost functions for the mobility strategies to be compared. The parameters used in our cost models are listed in Fig. 2. The costs of the IS-95strategy, was discussed in the literature [4]. But the environments and the details that were referenced in their derivations are different in many ways. In order to make a fair and reasonable comparison, we make some general assumptions and based on which we derive their cost functions in a uniform way. As the local database processing cost is insignificant comparing to the long communication time, we only consider communication cost in this derivation. Communication cost is dependent on the "distance" between two parties, and is classified into three levels: two parties are under different RSTPs, two parties are under the same RSTP but different LSTPs, and two parties under the same LSTP. Their costs are respectively C_1, C_2, and C_3, we also need to use probability to model the location distribution of two communicating parties. VLR, and two linked LAs. For simplicity, in all three sub-cases we assume that the two communicating parties are arbitrarily distributed.

Cost Function: The tasks of location management include managing location registration and call tracking. Hence, the location management cost is computed according to these two operations. As the ratio of the number of calls to mobility and defined as

$$Total\ \cos t = \frac{1}{CMR} \cdot Registration\ \cos t + Call\ tracking\ \cos t.$$

Cost of IS-95: The total cost of the IS-95 strategy can be represented as follows.

$$C_{IS-95}^{total} = (\frac{1}{CMR}) \cdot C_{IS-95}^{R} + C_{IS-95}^{T}$$

The registration cost and the call tracking cost of the IS-95 strategy is therefore.

$$C_{IS-95}^{R} = 2 \cdot C_1, \quad C_{IS-95}^{T} = 4 \cdot C_1.$$

Cost of Proposed Method: From the previous discussion, we understand that the difference of the registration operation between the proposed strategy and the Static LA strategy is that the MH record of a host is saved in the VLRs that the client has visited, whereas it's not in the LA strategy. Therefore, the registration cost of those two strategies should be the same. That is,

$$C_{proposed}^{R} = P_L \cdot (2 \cdot C_3) + P_R \cdot (2 \cdot C_2) + (1 - P_L - P_R) \cdot (2 \cdot C_1)$$

For the call tracking operation, two cases are involved: The caller is at a VLR that the callee has never visited. The caller is at a VLR that the callee visited before.

$$C_{prooposed}^{T} = (1 - \sum_{i=0}^{k} p_i) \cdot C_{LA}^{T} + \sum_{i=0}^{k} (p_i \cdot (i \cdot C_{proposed\ -link}) \cdot$$

Symbol	Meaning
C_1	The cost of sending a message from VLR to another VLR under a different RSTP
C_2	The cost of sending a message from VLR to another VLR under a different LSTP but the same RSTP
C_3	The cost of sending a message from VLR to another VLR under the same LSTP
P_L	The probability of a mobile client's moving into a new RA which is under the same LSTP as the last RA that the client just left
P_R	The probability of a mobile client's moving into a new RA which is under the same RSTP as the last RA that the client just left
CMR	The call-to-mobility ratio
P_i	The probability that a caller's request is issued from LA_i and its overseeing VLRs

Fig. 2. Symbols of the parameters

4 Performance Analysis

From the above discussion, we see two important factors that affect the performance of the proposed strategy: K and P_i. Both of these parameters help to indicate how many calls could be from a VLR that the callee has visited in the past. For such callers, the locating cost could be cheap. But the tradeoff is that a long LA link will increase the cost for traversing through the LAs. Hence, we study the effect of these two factors. Also, we vary the ratio C_1/C_3, which represents varying region size of a RSTP versus a LSTP. This is a general factor which affects all strategies. The default values of the parameters used in our evaluation are given ; C_1, C_2, and C_3 are 4, 2, 1, respectively, P_L is 0.7, P_R is 0.2, CMR is 0.5, Ks is 6, and P_i is 0.05. P_i is the probability that a caller places a call from a VLR that happens to be under one of the linked LAs of the callee. When this occurs, the call tracking cost is cheap. We vary P_i from 0.01 to 0.16. As the default K is 6, the total probability of a call from the VLR under a linked LAs is actually 0.06~0.96. The result given in Fig. 3 shows. However, for the proposed method a dramatic decrease of the cost when P_i increases. Although in general P_i may not be large for every kind of MHs, it could definitely be so for a certain type of users. Our performance result shows that the proposed strategy is especially good for managing MH of this kind. Figure 4 shows the result by varying length of this link K. A large K means that many VLRs that are under the linked LAs can locate a callee through the providing links of LAs, which helps to reduce the locating cost. Hence, the higher the K, the lower the cost of the proposed strategy. The improvement of the proposed strategy over the IS-95 strategies is very significant.

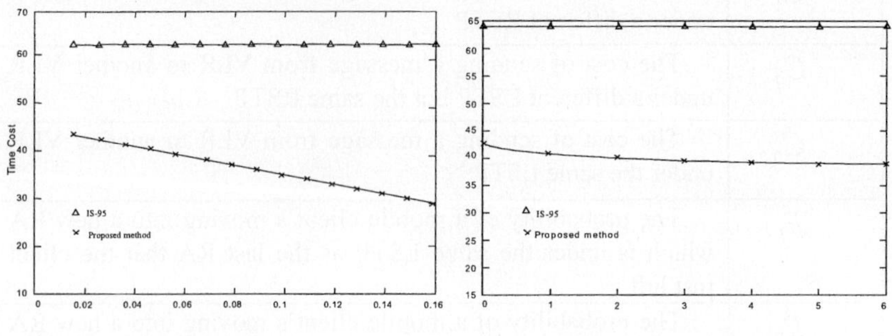

Fig. 3. Probability from LA_i Fig. 4. Length of link

5   Conclusions

In this paper the proposed strategy could be reduced location traffic cost. The proposed strategy avoids updating the host's location to the HLR when the client moves to a new VLR. The host's new VLR always updates the host's location to the LA. We also derived the cost models of the proposed strategies and several other methods. Our analysis results reveal that in most case the proposed strategy performs better than the IS-95 strategies.

References

1. Amotz Bar-Noy and Ilan Kessler, "Tracking Mobile Users in Wireless Communication Networks," IEEE Trans. on Information Theory, Vol. 39, 1993
2. Ing-Ray Chen, Tsong-Min Chen, "Modeling and Analysis of Forwarding and Resetting Strategies for Location Management in Mobile Environments," Proc. of ICS'96, 1996
3. Ing-Ray Chen, Tsong-Min Chen, and Chiang Lee, "Performance Characterization of Forwarding Strategies in Personal Communication Networks", Proc. of IEEE COMPSAC'97, 1997
4. Ing-Ray Chen, Tsong-Min Chen, "Performance Evaluation of Forwarding Strategies for Location Management in Mobile Networks", the Computer Journal, Vol. 41, No. 4, 1998,
5. Ing-Ray Chen, Tsong-Min Chen, and Chiang Lee, "Analysis and Comparison of Location Strategies for Reducing Registration Cost in CS Networks", Wireless Personal Communications journal, Vol. 12, No. 2, 2000, pp. 117-136.
6. EIA/TIA IS-41.3, "Cellular Radio Telecommunications Intersystem Operations", Technical Report (Revision B), July 1997.
7. Joseph S. M. Ho and Ian F. Akyildiz, "A Dynamic Mobility Tracking Policy for Wireless Personal Communications Networks", Prof. of GLOBECOM'95, 1995,

Performance Analysis of Active Queue Management Schemes for IP Network

Jahwan Koo[1], Seongjin Ahn[2], and Jinwook Chung[1]

[1] School of Information and Communications Engineering, Sungkyunkwan Univ.
Chunchun-dong 300, Jangan-gu, Suwon, Kyounggi-do, Korea
{jhkoo,jwchung}@songgang.skku.ac.kr
[2] Department of Computer Education, Sungkyunkwan Univ.
Myeongnyun-dong 3-ga 53, Jongno-gu, Seoul, Korea
sjahn@comedu.skku.ac.kr

Abstract. Active Queue Management schemes have evolved over time and continue to do so. In this paper, we present a comprehensive survey of AQM schemes for IP network. Its purpose is to identify the basic approaches that have been proposed and classify them according to the design goals and performance issues of AQM schemes. The results from several performance evaluation such as the link utilization, the average delay, and the loss rates have been provided. In particular, simulation-based comparisons of AQM schemes help to understand how they differ from in terms of fairness, global synchronization, performance guarantee, complexity and scalability.

1   Introduction

A QoS-enabled network is composed of various functions for providing different types of service to different packets such as rate controller, classifier, scheduler, and admission control. The scheduler function of them determines the order in which packets are processed at a node and/or transmitted over a link. The order in which the packets are to be processed is determined by the congestion avoidance and packet drop policy (also called *Active Queue Management*) at the node. Although there are many papers related the AQM algorithms, there were few discussed together in a single paper.

In the next two sections, we present a comprehensive survey of all possible AQM schemes. Its purpose is to identify the basic approaches that have been proposed and classify them according to the design goals and performance issues of AQM schemes. Next, we provide a description of the basic algorithms for IP network, including drop tail, random early detection (RED) [1], BLUE [2], Random Exponential Marking (REM) [3], Proportional Integral (PI) [4], and Joint Buffer management and Scheduling (JoBS) [5]. Section 4 discuses the performance evaluation of the AQM schemes via simulation. The final section offers some concluding remarks.

M. Bubak et al. (Eds.): ICCS 2004, LNCS 3036, pp. 349–356, 2004.

2   Basic Algorithms

2.1   Drop Tail

The drop tail algorithm maintains exactly simple FIFO queues. There is no methods, configuration parameter, or state variables that are specific to drop tail queues. Although simple and easy to implement, drop tail has two well-known drawbacks, the lock-out and full queue phenomena.

2.2   RED

The RED algorithm [1] was presented with the objective to minimize packet loss and queueing delay, avoid global synchronization of sources, maintain high link utilization, and remove biases against bursty sources. To achieve these goals, RED utilizes two thresholds, min_{th} and max_{th}, and a exponentially-weighted moving average (EWMA) formula to estimate the average queue length, $Q_{avg} = (1 - W_q) * Q_{avg} + W_q * Q$, where Q is the current queue length and W_q is a weight parameter, $0 \leq W_q \leq 1$. The two thresholds are used to establish three zones. If the average queue length is below the lower threshold (min_{th}), the algorithm is in the normal operation zone and all packets are accepted. On the other hand, if it is above the higher threshold (max_{th}), RED is in the congestion control region and all incoming packets are dropped. If the average queue length is between both thresholds, RED is in the congestion avoidance region and the packets are discarded with a certain probability P_a. This probability is increased by two factors. A counter is incremented every time a packet arrives at the router and is queued, and reset whenever a packet is dropped. As the counter increases, the dropping probability also increases. In addition, the dropping probability also increases as the average queue length approaches the higher threshold. In implementing this, the algorithm computes an intermediate probability P_b, whose maximal value given by P_{max} is reached when the average queue length is equal to T_{high}. For a constant average queue length, all incoming packets have the same probability to get dropped. As a result, RED drops packets in proportion to the connections' share of the bandwidth.

2.3   BLUE

The BLUE algorithm [2] uses different metrics to characterize the probability of dropping an arrival. This algorithm uses the current loss ratio and link utilization as input parameters. It maintains a single probability, P_m, which it uses to mark (or drop) packets when they are enqueued. If the queue is continually dropping packets due to queue overflow, BLUE increments P_m, thus increasing the rate at which it sends back congestion notification. Conversely, if the queue becomes empty or if the link is idle, BLUE decreases its marking probability. This effectively allows BLUE to "learn" the correct rate it needs to send back congestion notification. Besides the marking probability, BLUE uses two other parameters which control how quickly the marking probability changes over time.

The first is *freeze time*. This parameter determines the minimum time interval between two successive updates of P_m. This allows the changes in the marking probability to take effect before the value is updated again. The other parameters used, (d_1 and d_2), determine the amount by which P_m is incremented when the queue overflows or is decremented when the link is idle.

2.4   REM

The REM algorithm [3] is an AQM scheme that measures congestion not by a performance measure such as loss or delay, but by a quantity we call price. It treats the problem of marking (or dropping) arrivals as an optimization problem. The objective is to maximize a utility function subject to the constraint that the output link has a finite capacity. REM algorithm marks packets with a probability exponentially dependent on the cost of a link. The cost is directly proportional to the queue occupancy.

2.5   PI

The PI algorithm [4] uses a feedback-based model for TCP arrival rates to let the queue occupancy converge to a target value, but assumes a priori knowledge of the round-trip times and of the number of flows traversing the router. A feedback-based model consists of: 1) a *plant* which represents a combination of subsystems such as TCP sources, routers and TCP receivers, 2) the queue length at a router as a plant variable denoted by Q, 3) a desired queue length at a router (i.e., a *reference input*) denoted by Q_{ref}, 4) a *feedback signal* which is a sampled queue length used to obtain the error term, $Q_{ref} - Q$, 5) an *AQM controller* which controls the packet arrival rate to the router by generating a packet drop probability as a control signal. In [4], a simplified TCP flow dynamics model was developed. There, the open-loop transfer function (OLTF) of the plant was given by

$$P(s) = P_{TCP}(s) \cdot P_{QUEUE}(s) = \left(\frac{\frac{R_0 C^2}{2N^2}}{s + \frac{2N}{R_0^2 C}} \right) \cdot \left(\frac{\frac{N}{R_0}}{s + \frac{1}{R_0}} \right) \tag{1}$$

where N is a load factor (i.e., number of TCP connections), R_0 is the round-trip time, and C is the link capacity. Two main functions are used in the PI algorithm: one is the congestion indicator (to detect congestion) and the other is the congestion control function (to avoid and control congestion). The PI-controller has been designed based on (1) not only to improve responsiveness of the TCP flow dynamics but also to stabilize the router queue length around Q_{ref}. The latter can be achieved by means of integral (I)-control, while the former can be achieved by means of proportional (P)-control using the instantaneous queue length rather than using the EWMA queue length.

2.6 JoBS

The JoBS algorithm [5]) is capable of supporting a wide range of relative, as well as absolute, per-class guarantees for loss and delay, without assuming admission control or traffic policing.

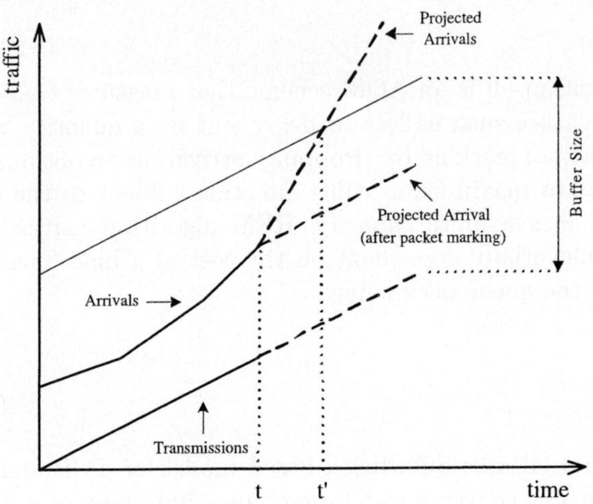

Fig. 1. The concepts of JoBS algorithm

The main idea of the JoBS algorithm is the following: At time t of a packet arrival, the router estimates the congestion window size and the round-trip time of the TCP flow. With these estimates, future traffic arrivals are projected, and impending buffer overflows are inferred, as illustrated in Figure 1. If a packet loss is projected, the algorithm reduces the congestion window size of the TCP source by marking packets with Explicit Congestion Notification (ECN). By reducing the congestion window size, the sending rate of the TCP source is reduced, and impending packet losses can be avoided. At any time t, the backlog at the router is equal to $R_{in}(t) - R_{out}(t)$. Hence, the JoBS makes an effort to meet the following requirement:

$$\forall t : R^{in}(t) - R^{out}(t) \leq B_{\lim}. \tag{2}$$

where B_{lim} is the size of the router's buffer, $R^{in}(t)$ is the total amount of traffic that has entered the router until time t, and $R^{out}(t)$ is the total amount of traffic that has left the router until time t. A unique feature of JoBS is that it considers scheduling and queue management (dropping) together in a single step.

3   Classification

AQM schemes can be classified in different ways. In [6], these were classified based on two dimensions : one is when decision on discarding packets is made and the other is what information is used to make packet discard decisions. In [7], a detailed classification of the various schemes has been proposed. It was based on the networking environment (ATM or IP networks), the type of congestion management mechanism implemented (congestion avoidance or congestion control and recovery), the number of thresholds used (none, global, or per-connection), decision information (global or per-connection), and the queue behavior (static or dynamic). In this paper, a wide variety of AQM schemes with

Table 1. Classification of AQM Schemes

Criteria	Congestion		Thresholds			State-Info	
Schemes	Control	Avoidance	None	Global	Per-Conn	Global	Per-Conn
DT	•		•			•	
RED [1]		•		•		•	
BLUE [2]		•		•		•	
REM [3]		•		•		•	
PI [4]		•			•		•
JoBS [5]		•			•		•

many different characteristics for IP network are considered, and hence a more extensive classification is proposed in Table 1. We refer to a survey article [7] for an exhaustive review of all possible AQM schemes. In summary, various kinds of AQM Schemes described section 2 have mainly focused on seven issues: 1) avoid congestion, 2) reduce the packet transfer delay, keeping the queue lengths at low levels, 3) avoid the TCP global synchronization problem, 4) achieve fairness among different traffic types, 5) deliver service guarantee (guaranteed or differentiated), 6) reduce the program complexity, and 7) increase the scalability. These issues are all inter-related.

Initial proposals for AQM for IP network [1], [2], and [3] were motivated by the need to improve TCP performance, without considering service differentiation. More recent research efforts [4] and [5] enhance these initial proposals in order to provide service differentiation.

4   Performance Analysis

We present an evaluation of our surveyed AQM schemes via simulation, using the *ns-2* network simulator [8].

4.1   Simulation Setup

We consider a bottleneck link with 45 Mbps bandwidth, 10 ms propagation delay, and 100,000 bytes queue size. The rest of the links (edge links) are all 100 Mbps with 200 packet queues. Their propagation delay is randomly in the range 5-20 ms. Each source is connected to the corresponding sink at the other side of the network, i.e., source S_i is connected to sink R_i. There are 3 TCP source/sinks and one UDP source/sink connected to each edge node. Each TCP source is an FTP application on top of NewReno TCP. The FTP packet size is 1,000 bytes. Each UDP source is an Exponential On-Off source with peak rate 1 Mbps, average rate 200 Kbps, average Off duration 80 ms, and packet size 100 bytes. The experiment lasts for 20 seconds of simulated time, and ECN is available in the entire network. We compare the performance of six different algorithms at the core router governing the bottleneck link.

- **Drop-Tail**. We use Drop-Tail to have an estimate of the loss rates encountered without AQM. With Drop-Tail, incoming packets are discarded only when the queue is full.
- **RED**. We use RED with a minimum threshold $min_{th} = 20,000$ bytes, and a maximum threshold $max_{th} = 80,000$ bytes. The parameter max_p is set to 1, and the weight used in the computation of the average queue size is set to $W_q = 0.002$.
- **BLUE**. We use the minimum time interval *freeze time* $= 100$ms. Other parameters $d1$ and $d2$ are set to 0.02 and 0.002, respectively.
- **REM**. The parameter values of REM are set to $\phi = 1.001$, $\alpha = 0.1$, $\gamma = 0.001$, and $b^* = 20$. Their behavior according to the probability determined by the link algorithm is described in Section 2.
- **PI**. We configure the PI algorithm with approximate RTTs and a tight upper bound on the round-trip times $R+ = 180$ ms, with a sampling frequency of 160 Hz, and get $a = 1.643e - 4$ and $b = 1.628e - 4$. The target queue length Q_{ref} is set to 70,000 bytes.
- **JoBS**. We use parameter settings of $K = 10$ and $\alpha = 0.9$ which are proper values in [5].

4.2   Simulation Results

For each algorithm, we monitor the link utilization, the average delay, and the loss rates at the core router, and present our results in Figure 2 (a), (b), and (c), respectively. Not surprisingly, the Drop-Tail is almost always full, which explains the relatively high loss rates. One of the biggest problems with TCP's congestion control algorithm over drop-tail queues is that the sources reduce their transmission rates only after detecting packet loss due to queue overflow. Figure 2 (c) tells us that, without AQM, one can expect loss rates in the order of 12%. RED manages to stabilize the queue length around $max_{th} = 80,000$ bytes. However, it is unable to detect incipient congestion caused by short-term traffic load changes. As a result, AQM parameter configuration has been a main design

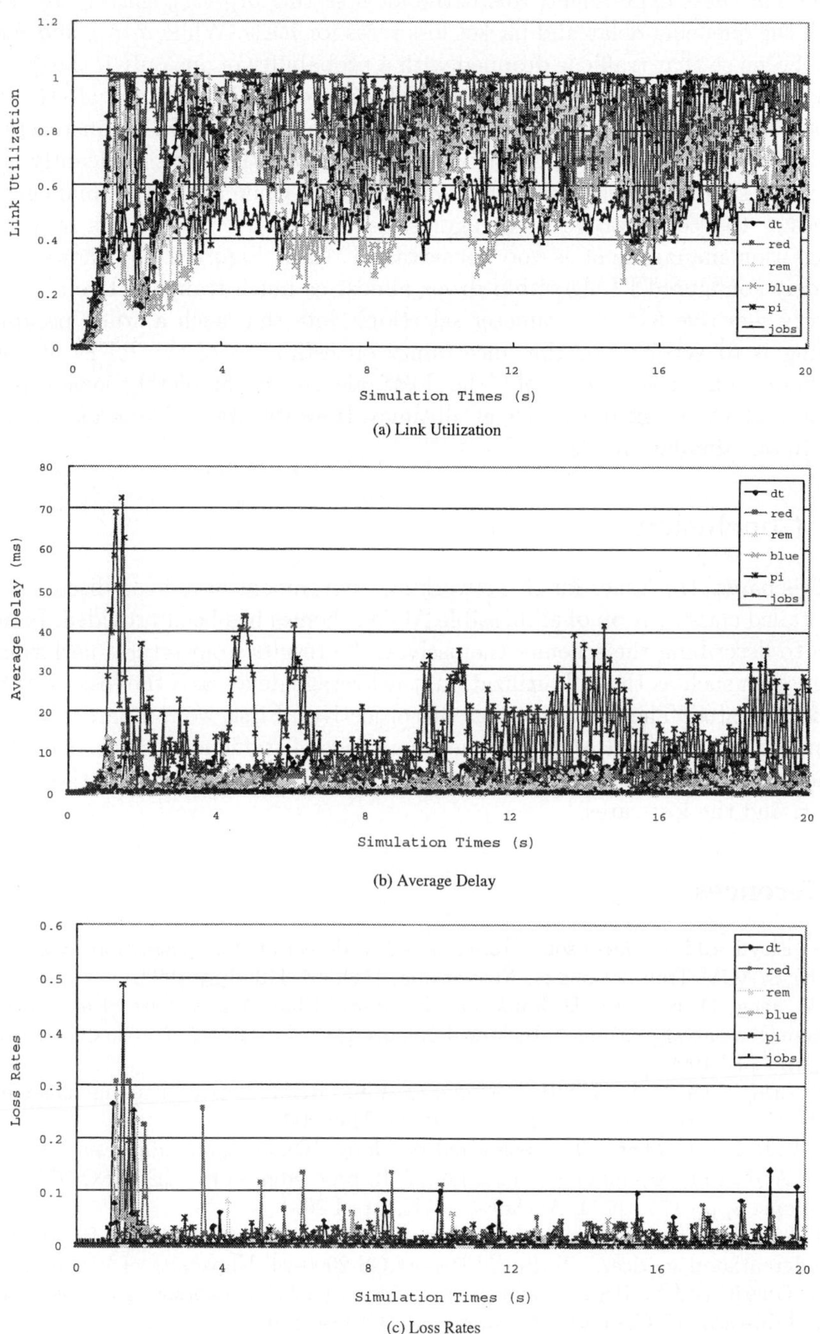

(a) Link Utilization

(b) Average Delay

(c) Loss Rates

Fig. 2. Simulation Results

issue. For these experiment, this is the ideal setting of max_p since it minimizes both the queueing delay and packet loss rates for RED. While min_{th} and max_{th} are chosen so that traffic is dropped with a probability of one only if the queue is full, other parameters are the default RED parameters in ns-2, and are therefore expected to cover a large range of operating conditions. RED is instructed to use ECN when needed. For the BLUE experiments, $d1$ is set significantly larger than $d2$. This is because link utilization can occur when congestion management is either too conservative or to aggressive, but packet loss occurs only when congestion management is too conservative. Also, Figure 2 (b) shows that a crudely configured PI algorithm drops almost as much traffic as Drop-Tail and is very sensitive to the parameter selection. Note that such a crude parameter tuning is to account for the uncertainty on estimates of the RTTs at router configuration time. Conversely, the JoBS algorithm completely avoids packet losses and lowers the queue size at all times. However, its utilization is relatively low in our simulation.

5   Conclusions

In this paper, the issues for IP networking environment have been discussed and a detailed classification of all possible AQM schemes has been provided. In addition to describing the schemes themselves, the results from several performance evaluation such as the link utilization, the average delay, and the loss rates have been presented. The evaluation has two objectives. First, we compare the performance of our surveyed AQM schemes in a range of traffic conditions. Second, we compare each algorithm's design goals such as the link utilization, the average delay, and the loss rates.

References

1. S. Floyd and V. Jacobson. "Random early detection for congestion avoidance," IEEE/ACM Transactions on Networking, 1(4):397-413, July 1993.
2. W. Feng, D. Kandlur, D. Saha, and K. Shin. "Blue: A new class of active queue management algorithms," Technical Report CSE-TR-387-99, University of Michigan, April 1999.
3. S. Athuraliya, V.H. Li, S.H. Low and Q. Yin. "REM: Active queue management," IEEE Network, Vol. 15, Issue. 3, pp 48-53, May 2001.
4. C.V. Hollot, V. Misra, D. Towsley and W. Gong. "On designing improved controllers for AQM routers supporting TCP flows," In proceedings of IEEE INFOCOM 2001, volume 3, pp 1726-1734, Anchorage, AK, April 2001.
5. J. Liebeherr and N. Christin. "Buffer management and scheduling for enhanced differentiated services," Technical Report CS-2000-24, University of Virginia, 2000.
6. R. Guerin and V. Peris. "Quality-of-service in packet networks: basic mechanisms and directions," Computer Networks, 31:169-189, 1999.
7. M. Labrador and S. Banerjee. "Packet dropping policies for ATM and IP networks," IEEE Communications Surveys, Vol. 2, No. 3, pp. 2-14, Third Quarter 1999.
8. ns-2 network simulator. http://www.isi.edu/nsnam/ns/.

A Real-Time Total Order Multicast Protocol

Kayhan Erciyes[1] and Ahmet Şahan[2]

[1] California State University, San Marcos,
Computer Sci. Dept., 333 S.Twin Oaks Valley Rd.,
San Marcos CA 92096, U.S.A.
`kerciyes@csusm.edu`
[2] Ege University International Computer Inst.
35100 Izmir, Turkey
`sahan@ube.ege.edu.tr`

Abstract. We describe, analyze and submit results of a real-time total order multicast protocol developed on a distributed real-time system architecture that consists of hierarchical rings with synchronous packet delivering characteristics. The protocol is structured on and closely interacts with the distributed clock synchronization and the real-time group management modules. The synchronous characteristics of the protocol makes it suitable for hard real-time applications where total ordering is required. The complexity analysis of the protocol is given and the performance results are shown for several scenarios. We show that the developed protocol is correct, scalable and real-time ...

1   Introduction

A group is a logical name for a set of computing elements whose membership may change with time. Replication using process groups for fault tolerance has attracted many researchers for many years [3][4]. There are several systems which provide fault tolerant group communication such as Horus [9] and Totem [2]. Moshe [6] extends these services to a WAN. The common goal of these projects is to provide a reliable multicast communication for process groups. Most of these systems are event-driven or asynchronous systems and their suitability to a distributed real-time system such as a process control or a flight control system should be considered with care. For distributed real-time systems, synchronous systems which include periodic functionalities, can yield better performance and are usually preferred for communication in such systems [8].

Total Order Multicast (TOM) is the basic paradigm to provide message ordering in fault tolerant systems that use active replication [7]. TOM ensures that no pair of messages are delivered to the members of a group in a different order. The fundamental properties of TOM are *Validity, Uniform Agreement, Uniform Integrity* and *Uniform Total Order* [5]. *Atomic broadcast* is a special case of total order multicast where a TOM message is delivered to all of the group members or none. Atomic Broadcast or Reliable TOM protocols can be symmetric or asymmetric depending on whether some privileged nodes exist in

M. Bubak et al. (Eds.): ICCS 2004, LNCS 3036, pp. 357–364, 2004.

the system or not. TOM has been studied extensively and many protocols have been proposed. A detailed survey is given in [5]. However, to our knowledge, there is not any significant research directed toward a real-time total ordering protocol.

The aim of this study is to revise and implement a distributed real-time system model that was designed previously [8] and then to design and implement a real-time total ordering protocol with distributed clock synchronization and group management modules. The model consists of hierarchical clusters of processing nodes which are connected by some communication medium. Tokens are delivered on hierarchical rings periodically enabling a synchronous communication. We also show that this model is suitable for any distributed real-time application that requires deadline guarantees and scalability. The paper is organized as follows. In Section 2, the base distributed real-time system model is described. The distributed clock synchronization and the total ordering protocol are described in sections 3 and 4. The analysis of the proposed architecture and the protocol are detailed in Section 5, the implementation results for regular cycle and total ordering protocol are given in Section 6 and the conclusions are outlined in the Conclusions section.

2   Distributed Real-Time System Architecture

The distributed real-time system model designed consists of hierarchical clusters of node members where each node represents a processor as shown in Fig. 1. Each cluster has a coordinator which is called the *Representative*. In the three layer design, two kinds of representatives provided are called as the *Sub-Representative* which is the coordinator for the lower ring and the central coordinator of rings that have Sub-Representatives members is the *Super-Representative*.

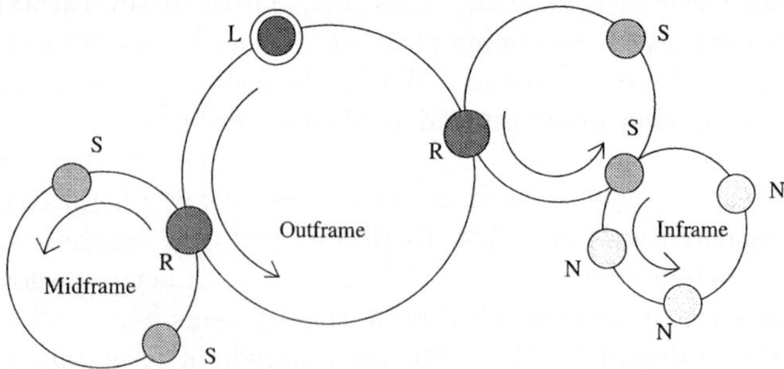

Fig. 1. The Distributed Real-time Model, L: Leader, R: Super-Representative, S: Sub-Representative, N: Node

At the highest level, the Leader is the coordinator and issues a token periodically called *outframe* on its ring, When Representatives receive the token, they also issue a token called *inframe* or *midframe* to collect data from their nodes.

Based on this protocol, the Total Order Multicast protocol is designed with the group management and the distributed clock synchronization modules as shown in Fig. 2. The Total Order protocol needs the clock synchronization service to sort the messages with respect to their valid timestamps and also the group management module to identify and manage atomicity of message delivery as detailed in sections 4 and 5.

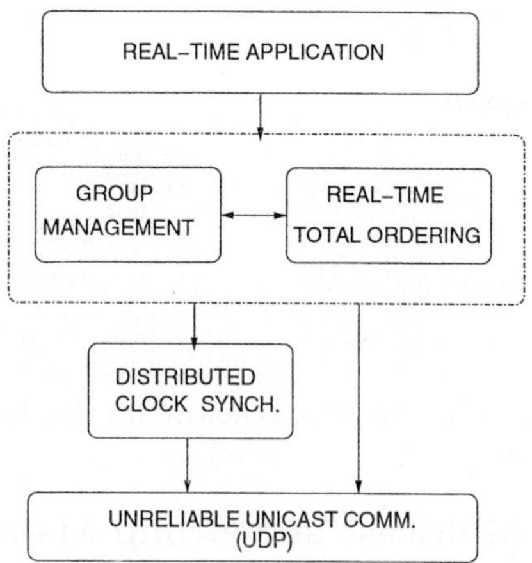

Fig. 2. The Real-time Total Order Architecture

3   Distributed Clock Synchronization

The clock synchronization is the base layer in many distributed systems where time services can be provided to any upper layer. The Total Order Multicast protocol proposed requires the clocks to be synchronized so that a global time frame is established. The three types of messages that can reach the nodes are COLLECT, SET or NONE. If NONE comes, it means increment your virtual clock value as frame period parameter and if COLLECT is received, the node writes its clock data into its slot in inframe. Finally, if SET is received, the node changes its clock value by adding a constant communication delay to the new coming value. When a COLLECT message comes from the Leader to a Representative, it collects values from the nodes, estimates an arithmetic average and sends it to the Leader. After the Leader issues a COLLECT message and receives all clock values, it also estimates an average value by adding the communication delay. The communication delay is related with frame issue and arrival time values. At the next period, it sends this new clock value to all by a new SET message. The finite state machine diagrams of the three processes for clock synchronization are depicted in Fig. 3.

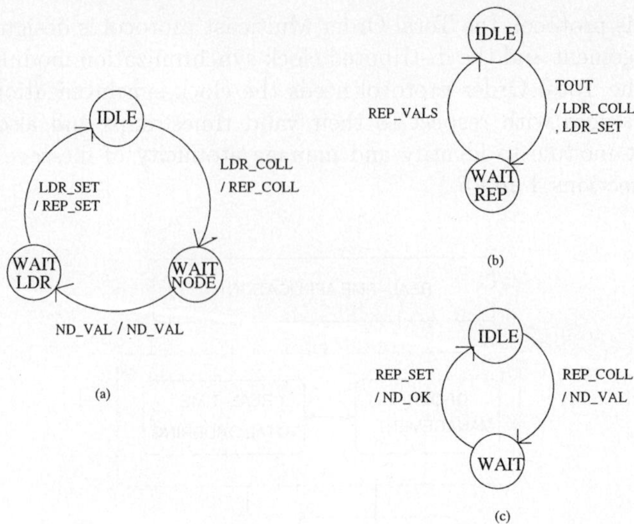

Fig. 3. FSM Diagrams of the Sub-Representative (a), the Leader (b) and the Node (c) for Clock Management

4   Total Order Multicast and Group Management

The Real-Time Total Order Multicast (RT-TOM) protocol is based on the normal operation of the protocol where the Leader functions as the central sequencer. It periodically gathers the requests from the Super-Representatives and sequences them with respect to their timestamps. The messages however are delivered to the nodes asynchronously as they are sent but only delivered to the application upon the received order sent by the Leader after sorting. The code for the Super or Sub-Representative is shown below.

The Total Order Representative

```
process TO_Rep
begin
   repeat
    msg=receive_msg();
    switch (msg.type):
        case NODE_MULTICAST : send_msg(node_msg, next_rep);
                              send_msg(node_msg, first_node);
                              insert_msg(node_msg, unsorted_list);
        case LDR_SOLICIT    : append(unsorted_list, ldr_msg);
                              send_msg(ldr_msg, next_rep);
        case LDR_ORDER      : send_msg(ldr_order, next_rep);
                              send_msg(ldr_order, first_node);
    until forever
end.
```

When a node sends a multicast message (NODE_MULTICAST), the representative broadcasts this message to its local ring and its upper ring. It also enqueues the identity of this message with its timestamp, to be sent to the Leader when the Leader request (LDR_SOLICIT) is received. The cycle is completed when the ordered identities of the messages are received from the leader (LDR_ORDER) in which case the order is broadcast in the local ring. The Total Order node process *TO_Node* delivers the message to the application with respect to the order sent by the central sequencer (Leader) only when the order is received. The actual delivery of the message to the *TO_Node* process however is performed independently as the message is circulated via the representatives without any interface from the Leader. Group Message Cycle is defined as atomic which provides that a group message is received by all or none of the members of the group. The atomicity control is managed by the Leader which checks the group member count with the coming acknowledgement message count from the nodes and if these do not match, the delivery is abandoned.

5   Analysis

The performance analysis of the message circulation for the *collect* operation of the three layer ring protocol should include the following

1. Distribution of the Leader message to the Super-Representatives : O_1
2. Distribution of Super-Representative message to Sub-Representatives : O_2
3. Distribution and the collection of the nodes information from the individual nodes at the Sub-Representatives: O_3
4. Collection of Sub-Representative information at Super-Representatives : O_4
5. Collection of the Super-Representatives information at the leader : O_5

Lemma 1. *Distribution and collecting of the leader message to/from the individual nodes (steps 1, 2, 3, 4 and 5 above) take $O_{ntime}(m)$ time where m is an upper bound on the number of nodes in the lowest cluster.*

Proof. Assume k, l and m are the upperbounds for the number of nodes in the outer, middle and inner rings respectively. The leader sends the outer ring message to the Super-Representatives in $O_1(k)$ steps as *all-to-all communication* in a unidirectional ring takes $O(n-1)$ where n is the number of nodes in the ring. Similarly, the Super-Representatives transfer this information to the Sub-Representatives in $O_2(l)$ steps in parallel with the other representatives. Finally, the Sub-Representatives will transfer and gather the information such as current clock values from the nodes in $O_3(m)$ steps. Gathering of this information at the Super and Sub-Representatives are similarly done in $O_4(m)$ and $O_5(l)$ steps. All of the five steps above take $T_{ntime}(2k+2l+m)$ time. For simplicity, let us consider that k, l and m are equal to each other. The total time complexity of the *collect* operation is then $O_{ntime}(m)$. It should also be noted that the minimum value for the frame period issued by the leader T_{lmin} should be approximately equal to $(k+l+m)t_m$ where t_m is the average message transfer time between a pair of consecutive nodes so that required data is available at the Super-Representative nodes when the next token is issued. The *collect* operation requires two frame periods ($2T_{lmin}$) and the *set* operation requires one frame period (T_{lmin}).

Lemma 2. *The total number of messages transferred during a normal ring operation is $O_{nmsg}(m^3)$ where m is an upper bound on the number of nodes in the inner ring.*

Proof. According to Lemma 1, the total number of messages transferred is $O_1(k)$ in the first step of data transfer. Each Super-Representative will then initiate transfer of l messages for a total of $O_2(kl)$ messages. In the third step, the total messages in transit are $O_3(klm)$. Similarly, the returning of the messages with valid information from the nodes require $O_4(kl)$ and $O_5(k)$ messages to reach the Leader. Assuming k, l and m are equal, the total complexity is then $O_{nmsg}(m^3)$.

Theorem 1. *The Speedup S obtained by the ring protocol with respect to a one level ring protocol is $O(m^2)$ where m is an upper bound in the number of nodes in the inner rings.*

Proof. If a single layer ring is constructed with the same number of nodes as a three layer ring, it would have klm or m^3 members. The total transfer time would then be $O_{otime}(m^3)$ for this one layer ring. The speedup obtained by the ring protocol is then :

$$S = O_{otime}(m^3)/O_{ntime}(m) = O(m^2) \tag{1}$$

Corollary 1. *Total time spent for message delivery by RT-TOM is $O_{rttom}(m)$ where m is an upperbound on the number of nodes in the inner ring.*

Proof. The total order multicast is performed by a *collect* operation $(2T_{lmin})$, followed by the *sort* operation and then the *set* operation (T_{lmin}) where the Leader sends the order to the individual nodes via the representatives. To perform atomicity a further step (T_{lmin}) to get and then a step to check acknowledgements followed by a *set* operation (T_{lmin}) is needed. Assuming sorting and acknowledgement checking are each done in T_{lmin} time in the worst case, the total time spent for RT-TOM is $T_{rttom}(7T_{lmin})$ and its complexity is $O_{rttom}(m)$.

6   Implementation and Results

The Clock Synchronization and RT-TOM in addition to the middle member representatives are implemented in the scope of this study, other parts are implemented in [1]. Nodes, Representatives and the Leader are implemented as POSIX threads on Sun OS v5.7 UNIX operating system running on DEC Alpha workstations where each thread can run on different workstations. The communication between member threads is performed by FIFO queues and UDP/IP sockets and the network environment was 100 Mbps FDDI backbone. The packet circulation test results during normal operation with the frame period T_l set to 170 ms are shown in Table. 1 for light load in the system. We can see that for the first case, the frame period is approximately equal to the sum of the individual

Table 1. Packet Circulation Tests (ms)

Total Ordinary Nodes	Configuration	Inframe	Midframe	Outframe
81	*Sup:3; Sub:3; Node:9*	114	35	27
54	*Sup:3;Sub:3;Node:6*	74	15	26

frame circulation values measured which means $k + l + m$ is the maximum value that can be obtained for this testbed, which is 15 in this case.

For the Total Order Multicast protocol, the total time for the delivery of ordered messages were measured with respect to the number of concurrent messages present in the whole network as shown in Fig. 4. Frame period was set to 170 ms, level count was three and the measurements were taken for 1, 5 and 10 concurrent messages in the system for varying total number of nodes. As seen from these values, message delivery times are similar, about 7*170 ms, as they depend only on the frame issue period. This would also mean that however the network is partitioned for a constant sum of k, l and m, whether as 3:3:9 or 5:5:5 nodes in the rings, we would still have the same message delivery time for RT-TOM. However, we would have 81 nodes in the first case, but 125 nodes in the second. The number of nodes, as long as their sum is smaller than the value required for that particular T_{lmin}, and the number of concurrent messages do not have any effect on the message delivery time for RT-TOM.

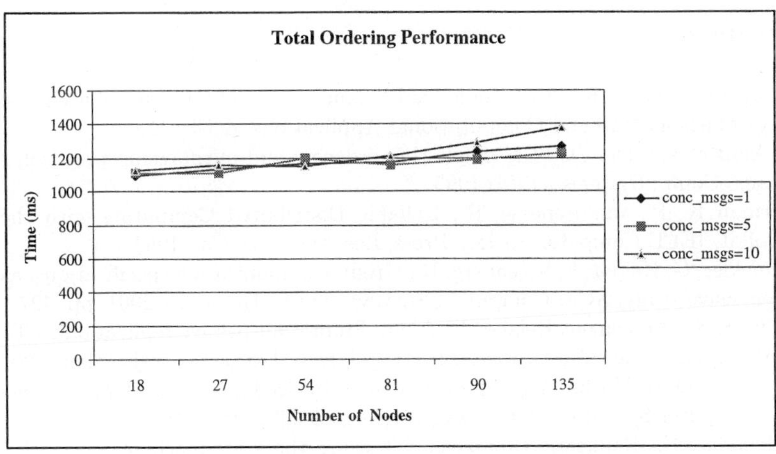

Fig. 4. Total Order Message Delivery Time

7 Conclusions

With this study, a synchronous, fault-tolerant, hierarchical and distributed real-time system model was designed and implemented. Using this model, a distributed clock synchronization module and a real-time total ordering protocol with the use of group management is realized. The main properties of the module are the total ordering of messages and atomicity where atomicity is performed by observing the group view. The main conclusions we draw are as follows. Due to its parallel operation in the rings, the normal ring operation and the RT-TOM protocol provides significant speedups with respect to a single level ring operation as stated in Theoerem 1. The performance for normal packet transfer obtained is scalable as shown by the test results. The RT-TOM protocol message delivery time is only dependent on the frame period, not on the number of concurrent messages in transit as stated by Corollary 1 and shown by the group message delivery tests. It is dependent on the number of nodes since token circulation period is dependent on the number of nodes, however, up to a pre-determined value for the period under consideration, the message delivery time is almost constant as shown in Fig. 4. Further enhancements of RT-TOM are possible by piggybacking the acknowledgement from the nodes to the *set* token during atomicity checking phase to reduce message delivery time. The privileged nodes may fail and the recovery procedures are as in [8] which is not discussed here. Based on the foregoing, we can conclude that the protocol designed can be applied for TOM in hard real-time systems with stringent deadlines, possibly by decreasing the number of levels to 2 if the system under consideration is small.

References

1. Akay, O., Erciyes, K., A dynamic load balancing model for a distributed system , J. of Mathematical and Computational Applications ,8(1-3), 2003.
2. Y.Amir et all, The TOTEM single ring ordering and membership protocol, ACM Trans. Comp. Systems., 13(4),1995.
3. Birman K. P., van Renesse, R., Reliable Distributed Computing with the Isis Toolkit, IEEE Computer Society Press, Los Alamitos, Ca., 1994.
4. Chockler, G, Keidar, I., Vitenberg, R., Group communication specifications: a comprehensive study, ACM Computing Surveys, 33 (4), December 2001, pp. 427 - 469.
5. Defago, X., Agreement Related Problem: From semi-passive replication to Totally Ordered Broadcast. Ph.D. thesis, Ecole Polytech. Lausanne, Switzerland, 2000.
6. Keidar, I. et al, Moshe: A group membership service for WANs, ACM Transactions on Computer Systems (TOCS) 20 (3) , August 2002, 191-238.
7. Schenider, F., Replication management using the state-machine approach, *Distributed Systems*, pages 169-198, Ed. Sape Mullender, 2nd ed., 1993.
8. Tunali, T, Erciyes,K., Soysert, Z., A hierarchical fault-tolerant ring protocol for a distributed real-time system, Special issue of Parallel and Distributed Computing Practices on Parallel and Distributed Real-Time Systems, 2(1), 2000.
9. Van Renesse R., Birman K. P.,Maffeis S., Horus : A Flexible Group communication System, CACM Special section on Group Communication, 39(4), April 1996

A Rule-Based Intrusion Alert Correlation System
for Integrated Security Management*

Seong-Ho Lee[1], Hyung-Hyo Lee[2], and Bong-Nam Noh[1]

[1] Department of Computer Science, Chonnam National University,
Gwangju, Korea 500-757
shlee2@lsrc.jnu.ac.kr, bongnam@jnu.ac.kr

[2] Division of Information and EC, Wonkwang University,
Iksan, Korea, 570-749
hlee@wonkwang.ac.kr

Abstract. As traditional host- and network-based IDSs are to detect a single intrusion based on log data or packet information respectively, they inherently generate a huge number of false alerts due to lack of information on interrelated alarms. In order to reduce the number of false alarms and then detect a real intrusion, a new alert analyzing system is needed. In this paper, we propose a rule-based alert correlation system to reduce the number of false alerts, correlate them, and decide which alerts are parts of the real attack. Our alert correlation system consists of an alert manager, an alert preprocessor, an alert correlator. An alert manager takes charge of storing filtered alerts into our alert database. An alert preprocessor reduces stored alerts to facilitate further correlation analysis. An alert correlator reports global attack plans.

1 Introduction

IDSs have evolved significantly over the past two decades since their inception in the early eighties. The simple IDSs of those early days were based on the use of simple rule-based logic to detect very specific patterns of intrusive behavior or relied on historical activity profiles to confirm legitimate behavior. In contrast, we now have IDSs which use data mining and machine learning techniques to automatically discover what constitutes intrusive behavior and quite sophisticated attack specification languages which allow for the identification of more generalized attack patterns[1].

Attackers usually try to intrude a system after collecting and analyzing the vulnerabilities of the victim. As traditional host- and network-based IDSs are to detect a single intrusion based on log data or packet information respectively, they inherently generate a huge number of false alerts due to lack of information on interrelated alarms. To address this problem, a research to correlate the alerts of several IDSs has emerged recently. These researches include CRIM[2,3] and Hyper-alert Correlation Graph[4,5]. Those correlation methods are based on attack specification with pre- and

---

* This research was supported by University IT Research Center Project.

M. Bubak et al. (Eds.): ICCS 2004, LNCS 3036, pp. 365–372, 2004.

post-condition of an attack. However, those correlation methods have some disadvantages. First, if the specifications are not correct, those correlation methods do not provide useful results. Second, it is difficult to cover all attack specifications in those correlation methods because new attacks are developed continuously.

So, we propose a rule-based alert correlation system to reduce the number of false alerts, correlate them, and decide which alerts are parts of the real attack. Our alert correlation system is composed of an alert manager, an alert preprocessor, and an alert correlator. An alert manager takes charge of storing filtered alerts into our alert database. An alert preprocessor reduces stored alerts to facilitate further correlation analysis. An alert correlator reports global attack plans.

2   Related Work

2.1   CRIM

CRIM is an IDS cooperation module developed within MIRADOR project[2,3]. This project is initiated by French Defense Agency to build a cooperative and adaptive IDS platforms. As figure 1 in the below, CRIM is composed of 5 functions.

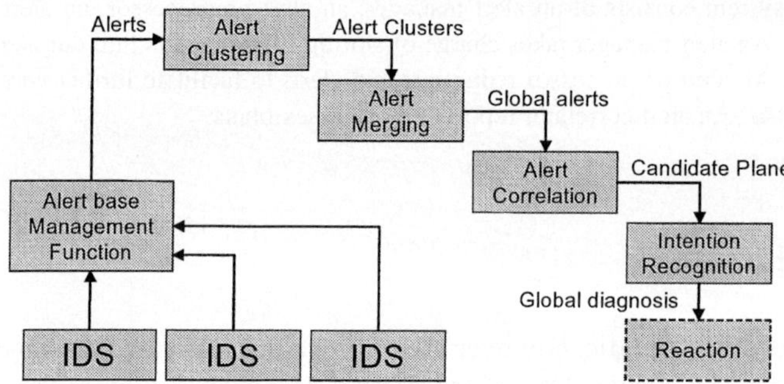

Fig. 1. CRIM architecture

The alert base management function receives the alerts generated by different IDSs and stores them for further analysis by cooperation module. When an attack occurs, the IDS connected to CRIM may generate several alerts for this attack. The clustering function attempts to recognize the alerts that actually correspond to the same occurrence of an attack. These alerts are brought into a cluster. Each cluster is then sent to the alert merging function. For each cluster, this function creates a new alert that is representative of the information contained in the various alerts belonging to this cluster.

Alert correlation function further analyzers the cluster alerts provided as outputs by the merging function. The result of the correlation function is a set of candidate plans that correspond to the intrusion under execution by the intruder. The purpose of the intention recognition function is to extrapolate these candidate plans in order to anticipate the intruder actions. The result of this function is to be used by the reaction function to help the system administrator to choose the best counter measure to be launched to prevent malicious actions performed by the intruder.

2.2 Hyper-alert Correlation Graph

Peng Ning in North Carolina State University analyzes alert correlation visually and constructs attack scenarios using a hyper-alert correlation graph[4,5]. Figure 2 depicts the architecture of an intrusion alert correlator.

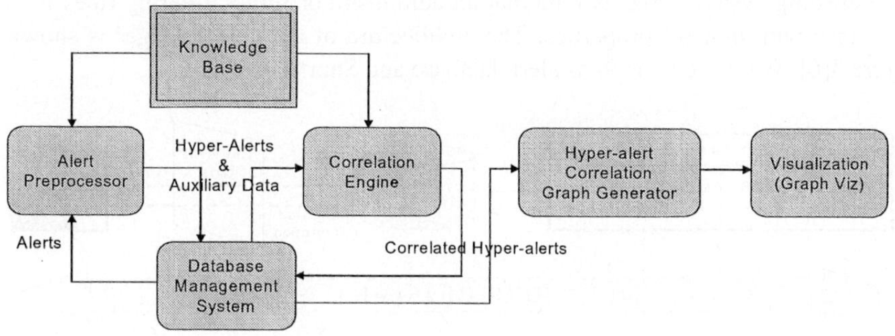

Fig. 2. An architecture of the intrusion alert correlator

It consists of a knowledge base, an alert preprocessor, a correlation engine, a hyper-alert correlation graph generator, and a visualization component. All these components except for the visualization component interact with a DBMS, which provides persistent storage for the intermediate data as well as the correlated alerts.

The knowledge base contains the necessary information about hyper-alert type as well as implication relationships between predicates. In their current implementation, the hyper-alert types and the relationship between predicates are specified in an XML file. When the alert correlator is initialized, it reads the XML file, and then converts and stores the information in the knowledge base.

Their current implementation assumes the alerts provided by IDSs are stored in the database. Using the information in the knowledge base, the alert preprocessor generates hyper-alerts as well as an auxiliary data from the original alerts. The correlation engine then performs the actual correlation task using the hyper-alerts and the auxiliary data. After alert correlation, the hyper-alert correlation graph generator extracts the correlated alerts from the database, and generated the graph files in the format accepted by GraphViz. As the final step of alert correlation, GraphViz is used to visualize the hyper-alert correlation graphs.

3 Rule-Based Intrusion Alert Correlation System

3.1 Alert Manager

The alert manager stores the alerts received from IDSs into alert database. Before sending an alert message, each IDS checks whether an issued alert satisfies filtering rules. In our work, filtering rules are based on the information of protection domain. That is, we select the alerts targeted at interesting systems and store them into alert database. We intend to facilitate further analysis by excluding the alerts targeted at uninteresting systems. We assume that an administrator setups filtering rules in conformance with domain properties. The architecture of an alert manager is shown in figure 3[6]. We use Oracle 9*i* as alert database and Snort 1.8 as IDS.

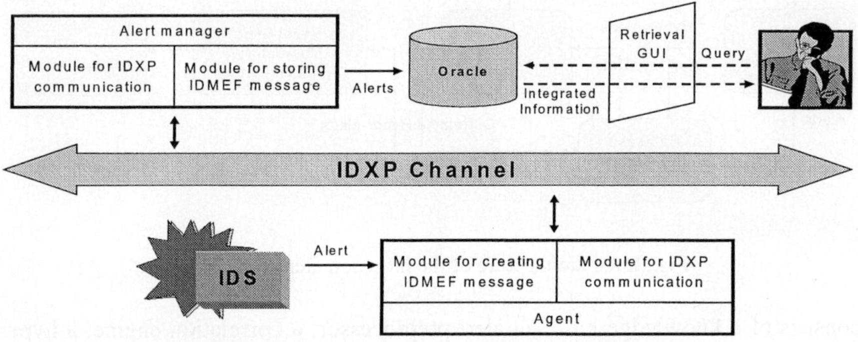

Fig. 3. The architecture of an alert manager

Each IDS sends filtered alerts in an IDMEF format. To facilitate further analysis, we store the information necessary for analysis, not all the information contained in an IDMEF message. This improves the efficiency of storing and retrieval during alert analysis. So, we use only one table for alert storage.

3.2 Alert Preprocessor

3.2.1 Deleting Duplicate Alerts

Duplicate alerts are generated because several identical type of IDSs issue them on seeing the suspicious log data or data packets. So, if the identical type of IDSs are installed on a network, the IDSs generate the alerts with the same source address and port number, the same target address and port number, and the same classification name. We identify these duplicate alerts and delete them except one to facilitate further analysis.

3.2.2 Alert Merging

Several similar alerts can be generated over an identical event. If these alerts are merged into one with minimizing information loss, the next step, correlation analysis will be performed more easily. An algorithm for merging similar alerts is figure 4.

```
mergeSimilarAlerts( ) {
   for each alert
      flag = 0;
      for each Queue in memory
         if (Creation time of this alert - time of Queue > 30 seconds) then
            Create new tuple;
            Store this new tuple into DB;
            Free the Queue;
         endIf
      endFor
      for each Queue in memory
         if (the attributes of this alert and this Queue are equal) then
            Put this alert into this Queue;
            Set Creation time of this alert to the time of this Queue;
            flag = 1;
            break;
         endIf
      endFor
      if flag == 0 then
         Create new Queue;
         Set the attributes of this alert to this Queue;
      endIf
   endFor
}
```

Fig. 4. An algorithm for merging alerts

3.3 Alert Correlator

In an alert correlator, the process for finding correlations among different kinds of alerts such as vulnerability gathering alerts and U2R attack alerts, is conducted. For this, we extract correlation data by the similarity of alert attributes and verify the usefulness of the extracted correlation data through inference rules.

3.3.1 Extraction of Correlation Data

There are two cases in correlation. One is a consecutive attack to the same target and the other is an attack by an identical intruder. To do such a correlation analysis, we extract correlation data by 'Target address' and 'Source address' respectively. These extracted data are chained in time order.

3.3.2 Verification of Extracted Correlation Data

In this section, we evaluate if the extracted correlation data is useful. Since all correlation data are not useful, we identify meaningful correlation data. To achieve this, we apply inference rules to correlation data. Inference rules are described like Prolog predicates. Figure 5 shows an example of inference rules.

```
sameTargetAddress(A,B).
classificationName(A,'WEB-MISC backup access').
classificationName(B,'WEB-CGI scriptalias access').
correlate(A,B) :- sameTargetAddress(A,B),
                  classificationName(A,'WEB-MISC backup access'),
                  classificationName(B,'WEB-CGI scriptalias access').
```

Fig. 5. An example of inference rules

4   Experiments and Discussion

We use '99 DARPA dataset as experiment data in our work to ensure certified experimental results. We consider the hosts labeled with "Victim" in Simulation Network '99 as protection domain. In our work, Snort reads tcpdump files in the dataset and issues alerts. Then, the alerts are stored in Oracle database. Table 1 shows an example of duplicate alerts. This example shows the alert of "WEB-IIS *.idc attempt" dated March 8th among the alerts generated from '99 DARPA dataset.

Table 1. An example of duplicate alerts

No.	Alert ID	Creation time	Source address	Source port	Target address	Target port	Analyzer
1	10	1999-03-08 13:01:59	206.48.44.18	1058	172.16.112.100	80	Inside IDS
2	176632	1999-03-08 13:01:13					Outside IDS

Table 2 shows an example of alerts to be merged. This example is an alert of "WEB-CGI scriptalias access" dated March 8th among the alerts generated from '99 DARPA dataset. These 241 alerts are caused by 'Back' attack, which is a DoS(Denial of Service) attack over an Apache web server. All the attributes except 'Creation time' and 'Source port' number are equal.

Table 2. An example of alerts to be merged

No.	Alert ID	Creation time	Source address	Source port	Target address	Target port	Analyzer
1	532	1999-03-08 14:39:12	199.174.194.16	1028	172.16.114.50	80	Inside IDS
			...				
241	772	1999-03-08 14:40:11	199.174.194.16	1379	172.16.114.50	80	Inside IDS

Table 3 shows an example of merging the aforementioned alerts into one. When alerts are merged into one, we should decide how we can merge several different 'Creation time' into one. We minimize the information loss of Creation time by introducing range from the first creation time to the last. 'Alert ID' is newly given during merging process. A new 'Alert ID' is the form of 'M + the first Alert ID'. Source port is processed as '#' if several Source port number exist. 'Analyzer' is dropped as it is not utilized at the next step.

Table 3. An example of merging alerts

Alert ID	First Creation time	Last Creation time	Source address	Source port	Target address	Target port
M532	1999-03-08 14:39:12	1999-03-08 14:40:11	199.174.194.16	#	172.16.114.50	80

Table 4 shows an example of correlation by the same target address and the same target port number. An alert of Alert ID 471 is 'WEB-MISC backup access'. This is regarded as an alert by an attack for information gathering about a web server in 172.16.114.50. So, we guess that an attacker investigates if 172.16.114.50 provides a web service and initiates 'Back' attack.

Table 4. An example of correlation by target address and port number

Alert ID	First Creation time	Last Creation time	Source address	Source port	Target address	Target port
471	1999-03-08 14:26:54	#	197.182.91.233	6266	172.16.114.50	80
M532	1999-03-08 14:39:12	1999-03-08 14:40:11	199.174.194.16	#	172.16.114.50	80

In our correlation experiments, we find out various global attack plans such as the scan of an entire network and a large-scale attack. However, our correlation technique should be refined elaborately. To achieve this, we will consider the logs of Secure OS and HIDS(Host-based IDS).

The advantages of our research are as follows. First, we facilitated correlation analysis by reducing alerts to analyze minimizing information loss. Second, we decreased the possibility of false correlation analysis by wrong attack specification. We use the classification name of an alert and if IDSs provide the correct classification name of a single attack, we can get more reliable correlation results. In the contrary, our research has some shortcomings. As our correlation technique is based on the classification name of an alert, it has low efficiency. In addition to them, our correlation technique should be generalized to be applied to IDSs other than Snort.

5   Conclusion and Future Work

In this paper, we presented a rule-based alert correlation system. Our alert correlation system consists of an alert manager, an alert preprocessor, an alert correlator. An alert manager takes charge of storing filtered alerts into our alert database. Before alert correlation analysis, stored alerts go through preprocessing. An alert preprocessor includes a module of deleting duplicate alerts and an alert merging module. This preprocessing reduces stored alerts to facilitate further correlation analysis. We found out that filtering and preprocessing resulted in reducing a number of alerts through experiments. After that, we did correlation analysis over preprocessed alerts. An alert correlator includes an module for extracting correlation data and an module for verifying the extracted data. In the result, the alert correlator reported global attack plans such as the scan of the entire network and a large-scale attack.

In the future, we will refine our correlation technique. And, we will correlate the logs of HIDS and secure OS and get more reliable experiment results. In addition, we would like to establish a reaction strategy by maintaining an attacker list and a main victim list.

References

1. N. Carey, A. Clark, G. Mohay, "IDS Interoperability and Correlation Using IDMEF and Commodity Systems," ICICS 2002, LNCS 2513, pp. 252-264, 2002
2. F. Cuppens, "Managing Alerts in a Multi-Intrusion Detection Environment," In Proc. Of Annual Computer Security Applications Conference (ACSAC 2001), Dec. 10-14, 2001, New Orleans, Louisiana
3. F. Cuppens, A. Miege, "Alert Correlation in a Cooperative Intrusion Detection Framework," In Proc. Of the 2002 IEEE Symposium on Security and Privacy, May 2002
4. P. Ning, Y. Cui, D. S. Reeves, "Constructing Attack Scenarios through Correlation of Intrusion Alerts," 9th ACM conference on computer and communications security, pp. 245-254, Nov 18-22, 2002
5. P. Ning, Y. Cui, D. S. Reeves, "Analyzing Intensive Intrusion Alerts via Correlation," In Proc. Of the 5th Int'l Symposium on Recent Advances in Intrusion Detection (RAID 2002), Oct 2002
6. S. H. Lee, Y. C. Park, H. H. Lee, B. N. Noh, "The Construction of the Testbed for the Integrated Intrusion Detection Management System," In Proc. Of 19th KIPS Spring Conference, Vol. 10, No. 1, pp. 1969-1972, May 16-17, 2003
7. H. Debar, M. Dacier, A. Wespi, "Research Report: A Revised Taxonomy for Intrusion Detection Systems," Annales des telecommunications, 55(7-8), pp. 361-378, Jul-Aug 1997
8. T. Buchheim, M. Erlinger, B. Feinsteing, G. Matthews, R. Pollock, J. Bester, A. Walther, "Implementing the Intrusion Detection Exchange Protocol," In Proc. Of 17th Annual Computer Security Applications Conference (ACSAC 2001), New Orleans, Louisiana
9. H. Debar, A. Wespi, "Aggregation and Correlation of Intrusion Detection Alerts," In Proc. Of the 4th Int'l Symposium on Recent Advances in Intrusion Detection (RAID 2001), LNCS 2212, pp. 85-103, 2001

Stable Neighbor Based Adaptive Replica Allocation in Mobile Ad Hoc Networks*

Zheng Jing, Su Jinshu, Yang Kan, and Wang Yijie

School of Computer Science, National University of Defense Technology,
Changsha 410073, Hunan,China
zhengjing621@hotmail.com

Abstract. In mobile ad hoc networks (MANET), nodes move freely and the replica allocation in such a dynamic environment is a significant challenge. In this paper, a dynamic adaptive replica allocation algorithm that can adapt to the nodes motion is proposed to minimize the communication cost of the object access. When changes occur in the access requests of the object or the network topology, each replica node collects access requests from its neighbors and makes decisions locally to expand the replica to neighbors or to relinquish the replica. This algorithm dynamically adjusts the replica allocation scheme towards a local optimal one. To reduce the oscillation of replica allocation, a statistical method based on history information is utilized to choose stable neighbors and to expand the replica to relatively stable nodes. Simulation results show that our algorithms efficiently reduce the communication cost of object access in MANET environment.

1 Introduction

MANET (Mobile Ad hoc Network) is a collection of wireless autonomous mobile nodes without any fixed backbone infrastructure, in which nodes are free to move. MANET can be used in many situations where temporary network connectivity is required, for example in battlefields and in the disaster recovery. Such a dynamic environment brings about significant challenges to the replica allocation mechanism, which is one of the key technologies to improve accessibility, reliability and performance of the system. The replica allocation algorithm proposed in this paper addresses the issue of the performance of the data access in the MANET environment.

Replica allocation for performance improvement in the field of fixed networks has been an extensive research topic. In many researches, the communication cost is used as cost function. However, because these researches are for fixed networks, they don't consider the effect on the data replication caused by the nodes mobility. In [1], a minimum- spanning- tree (MST) write policy is introduced. However, this cost model is not suitable for the MANET environment because the communication cost and the algorithm complexity of building a spanning tree are very high in MANET. In [2], nodes forward read requests to the nearest replica node and write requests to all

---

* This research was supported by the National Natural Science Foundation of China (No. 90104001).

M. Bubak et al. (Eds.): ICCS 2004, LNCS 3036, pp. 373–380, 2004.

replica nodes along the shortest path. However, this scheme requires that every node should maintain information of all replica nodes. When a replica node changes, every node must be notified. Thus it isn't suitable for the mobile environment as well.

Several strategies [3,4,5] for replicating or caching data have been proposed in traditional wireless mobile networks. These data replication strategies emphasize reducing the one hop wireless communication cost induced by keeping consistency between the data in a base station and their replicas in mobile nodes. However, these strategies are completely different from our approach which is designed for the multi-hop MANET network without base stations.

Only a few replica allocation algorithms have been proposed for the MANET environment recently. In [6], much information needs to be exchanged among nodes, especially when the topology of network changes rapidly. In [7], an algorithm is proposed to predict the network partitioning and to allocate replicas to ensure the service availability. It is know, all these algorithms [6,7] only focus on improving the data accessibility during the network partitioning.

In this paper, a distributed dynamic adaptive replica allocation algorithm is proposed for the MANET environment. The communication cost is used as the cost function in the algorithm because the communication cost becomes the most important factor which influences the performance of data access in this environment. Our algorithm can dynamically adjust the replica allocation scheme towards a local optimal one according to the access requests distribution and topology changes. The concept of " stable neighbor" is proposed in our algorithm and the access requests are collected only from stable neighbors while replica nodes expanding or relinquishing the replica. Thereby the replicas are stored on relatively stable nodes and the oscillation of replica allocation is reduced while nodes move rapidly.

The rest of the paper is organized as follows: in section 2 the cost model is defined; in section 3 a new distributed dynamic adaptive replica allocation algorithm is presented in detail; in section 4 the simulation results are given demonstrated; and finally in section 5, the summary and some future work are presented.

2   The Cost Model

In our research, hops is used as the metric of the communication cost of data access. In the MANET environment, the communication cost between two nodes includes the wireless bandwidth cost, energy consumption, the delay of the communication and so on. All these factors are related to hops, so we use the hops between two nodes to measure the communication costs between these two nodes.

We suppose that the access request for an object is sent to the closest replica in the network. The read request is served by the closest replica node, and the write request is propagated from the closest replica to all other replicas along the shortest path. Therefore the information of replication allocation just needs to be maintained on the replica nodes. The ROWA (READ-ONE-WRITE- ALL) policy is used to ensure the consistency of the replicas, and we assume that each individual access is independent.

Definition 1:The replica allocation scheme of an object O, denoted by F, is the set of nodes at which O is replicated.

The set of mobile nodes is denoted by V. For $i, j \in V$, $d(i, j)$ is the least hops between i and j. Thus the cost of a single read request by node i is $d(i,F) = \min_{j \in F} d(i,j)$. The cost of a single write request by node i is $d(i,F) + \sum_{k \in F} d(j,k)$. Where j is the node which satisfies $d(i,j) = d(i,F)$. Therefore during the interval t, the total communication cost of F, denoted by $cost(F)$, can be computed as below:

$$cost(F) = \sum_{i \in V} W(i)d(i,F) + \sum_{s \in F} \sum_{j \in F} Wre(s)d(s,j) + \sum_{i \in V} R(i)d(i,F)$$

$$= costW_{forward}(F) + costW_{up}(F) + costR(F) \tag{1}$$

In this equation, $W(i)$, $W_{re}(i)$, and $R(i)$ are statistical values acquired during the interval t. $R(i)$ and $W(i)$ are the number of the read and write requests to O issued by i, $W_{re}(i)$ is the total number of the write requests to O that i receives from itself or its non-replica neighbors. $costW_{forward}(F) = \sum_{i \in V} W(i)d(i,F)$, is the cost of forwarding write requests to replica nodes; $costW_{up}(F) = \sum_{s \in F} \sum_{j \in F} Wre(s)d(s,j)$, represents the cost of propagating write requests among replica nodes; $costR(F) = \sum_{i \in V} R(i)d(i,F)$, refers to the total access cost of read requests.

Definition 2: *The read-write pattern for an object O is the number of reads and writes to O issued by each node.*

For general static networks, the problem of finding an optimal replica allocation scheme has been proved to be NP-complete for different cost model [8, 2]. As for the cost model defined by (1), this problem is also proved to be NP-complete in [10]. As for the MANET environment, it is more difficult to find the optimal replica allocation. Thus a distributed adaptive replica allocation algorithm is proposed to find the near-optimal replica allocation scheme.

3   Adaptive Replica Allocation Algorithms

3.1   The Adaptive Replica Allocation Algorithm

In the fixed networks, the optimal replica allocation scheme of an object depends on the read-write pattern, but in the MANET environment it depends not only on the read-write pattern but also on the nodes motion. The *ARAM (the Adaptive Replica Allocation Algorithm In MANET)* algorithm has been proposed in [10]. In the *ARAM* algorithm, each replica node collects access requests from its neighbors and makes decisions locally to update the replica allocation scheme. Thus the *ARAM* algorithm adapts to the dynamic MANET environment.

The algorithm is executed at each replica node periodically and independently. The duration of the period t is a uniform system parameter. The period tends to be shorter with more frequent topological changes and read-write pattern changes.

The *ARAM* algorithm is executed on each replica node s at the end of each interval t and is shown in Table 1.

Table 1. The ARAM algorithm

```
execute Expansion_test for each non-replica neighbor u of s
if exist node u*, u* satisfies the condition of expansion
   then expand replica to u*, F=F∪{ u*}, return 1
 endif
execute Switch_test for each non-replica neighbor u of s
if exist node u*, u* satisfies the condition of switch
   then switch replica from s to u*, F= F-{s}+{ u*}, return 2
 endif
execute Relinquishment_test on s
if the condition of the relinquishment is satisfied then
relinquish s, F=F-{s},   return 3
 endif
return 0
```

The *Expansion_test, Relinquishment_ test*, and *Switch_test* operation in the ARAM algorithm will be discussed as below.

1. *Expansion_test.* For the neighbor u of the replica node s and $u \notin F$, if the replica is expanded to u, one hop will be decreased for some nodes to access the replica, but the cost will increase for propagating write requests to the new replica node u. If the condition of expansion is satisfied (which means that when a replica is expanded to u, the decrease of the access cost is greater than the increase of the update cost, thus the total communication cost declines.), a replica is expanded to $u*$ and $F'=F \cup \{u*\}$.

2. *Relinquishment_ test.* If the number of update request received by replica node s from other replicas is larger than that of the read and write requests received by s from itself and those non-replica nodes, then s requires to relinquish the replica.

3. *Switch_test.* The switch test will allocate the replica to the neighbor node which receives more read and write requests. Then $u*$ becomes a replica node while s isn't a replica node any more.

THEOREM 1. *For a static network, suppose that the read-write pattern doesn't change and the ARAM algorithm is executed at the ends of every interval. Then the communication cost of object access will decrease once any operation of Expansion_test, Relinquishment_ test, and Switch_test succeeds until the replica allocation scheme reaches a local optimal one.* The details of the proof are manifested in [10].

In the *ARAM* algorithm, information of all access paths is collected, which makes the algorithm complex. Meanwhile, in the *ARAM* the replica expansion and relinquishment operation are executed only when the strict conditions are satisfied. Such a policy can ensure the communication cost decreasing each time, but the chances to achieve a more optimal result are lost. Therefore, an improved algorithm based on *ARAM—EARAM* is proposed [10]. The *EARAM (The Enhanced ARAM Algorithm)* algorithm ignores the changes of the access path caused by the changes in replica allocation scheme. Thus the replica expansion condition can be simplified as below:

$$R_{from}(u) > W_{re}(s) - W_{from}(u) + \sum_{i \in F} W_{re}(i)d(i,s) \qquad (2)$$

Similarly, the relinquishment condition can be simplified as below:

$$\sum_{i \in F} W_{re}(i)d(i,s) > (R_{re}(s) + W_{re}(s))d(s, F - \{s\}) \qquad (3)$$

The switch condition is achieved as below:

$$R_{from}(u) + W_{from}(u) > 1/2(R_{re}(s) + W_{re}(s)) \qquad (4)$$

Where $R_{re}(s)$ is the read requests received by s from itself and other nodes; $R_{from}(u)$ and $W_{from}(u)$ are read and write requests received by s from u respectively.

The *EARAM* algorithm is the same as the *ARAM* algorithm except that the expansion, relinquishment, and switch conditions in the *ARAM* algorithm are replaced by the condition (2), (3) and (4) respectively. The information collected by the *EARAM* algorithm is not enough to ensure that the communication cost of data access decreases once the *EARAM* algorithm is executed, and replica may be mistakenly expanded or relinquished. But the total communication cost tends to decline.

3.2   The EARAM_SN Algorithm

In the MANET environment, the changes of network topology caused by nodes motion may cause the replica allocation scheme to be oscillated. The main idea of the *EARAM_SN(The EARAM Algorithm Based On The Stable Neighbor)* algorithm is to find the relatively *stable neighbors* of replica nodes in a distributed way and to expand replicas only to stable neighbors. Also in the *EARAM_SN* algorithm the access requests are collected only from stable neighbors while expanding or relinquishing the replica. The algorithm enables the replicas to be stored on relatively stable nodes, thus the oscillation of replica allocation is reduced while nodes move rapidly.

Now the details of this algorithm are discussed. The distance between two neighbors is used to measure the *neighborhood stability* of them (the distance between neighbors can be achieved by GPS). Suppose that the effective wireless communication area of the mobile node h is a circle with the center h and the radius r, and the area is divided into n sub-areas, i.e. n cirques H_1, H_2, H_3...H_n according to their distance from h (H_1 is the furthest from h and H_n is the nearest to h).

Definition 3: Neighbor g's vicinity on node h, denoted by $Rd(h,g)$. If node g is the neighbor of h and g is in area H_i of h, then $Rd(h,g)=i$; else if node g isn't the neighbor of h, then $Rd(h, g)=-n$.

For each neighbor g of node h, g's vicinity on h can be estimated by its history information. Denoting $r_k(h,g)$ as the estimated value of g's *vicinity* at the k interval, and $Rd_{k-1}(h,g)$ as the actual value of g's *vicinity* at the $k-1$ interval, we can get the value of $r_k(h,g)$ from the estimated value and the actual value at the $k-1$ interval, shown as below:

$$r_k(h,g) = \frac{r_{k-1}(h,g)\alpha + Rd_{k-1}(h,g)}{\alpha + 1} \ , \ r_1(h,g) = Rd_1(h,g);$$

In this equation, a is a smooth factor and $a>0$. If $r_k(h,g) > \tilde{C}$ ($\tilde{C}$ is a threshold), g can be regarded as the **stable neighbor** of node h. $S(h)$ is denoted as the **stable neighbor set** of h, i.e., $S(h)=\{g|\ r_n(h,g) > \tilde{C}\ \}$.

Definition 4. Stable Path, the path $Path(i, j)$ which is comprised of nodes $i,\ c_1.c_2...c_k$, $j\ (k\geq0)$ is called a stable path when $i\in S(c_1),\ c_1\in S(c_2)...c_k\in S(j)$.

Definition 5. Stable neighbor group, the set $T(h)$ is a stable neighbor group of node h if for each $i\in T$, there is at least a stable path between h and i.

The *EARAM_SN* algorithm improves the *EARAM* algorithm by replacing the neighbor set with the stable neighbor set while every node selects the stable path as its access path if there is a stable path to the replica node. In the *EARAM_SN* algorithm, all the replica nodes and their stable neighbor groups form a relatively stable topology. Only the access requests issued by nodes in the stable path can have impact on the replica allocation scheme and hence the oscillation of replica allocation caused by nodes motion can be reduced.

4   Simulation and Analysis

In this section, simulation results are shown to evaluate the performance of our algorithms. The program is written by C++ with event-driven method.

The simulation parameters are presented as follows: area of motion is 1000m$\cong$1000m; the number of mobile nodes is 100; the velocity of motion is 0 m/s - 10 m/s; the range of motion director is 0 - 2π; the communication radius of nodes is 200m; the number of object to be replicated is 1; the interval of algorithm executed is 0.1s; the initial number of replica is 10; the ratio between reads and writes is 5:1. The mobile nodes move in Random Waypoint Mobility Model [9]. Each experiment is performed 10 times to acquire the average values.

Firstly, we compare the performance of algorithms in static networks. In this experiment, our main concern is the effect of read-write pattern on algorithms. We compare four algorithms: the *ARAM*, the *EARAM*, the *ADR_G* [1], and the Static Replica Allocation algorithm (i.e. *SRA*, replicas are distributed on m nodes, and the replica allocation scheme doesn't change during the whole process of simulation). In the simulation, the read-write pattern doesn't change during each 10 intervals. The simulation result is presented in Fig.1.

Fig.1 shows that when the read-write pattern is fixed, the communication cost of object access keeps decreasing until reaches a stable value in the *ARAM* and the *ADR_G* algorithm. When the read-write pattern changes, because the current read-write patterns of nodes cannot be estimated by the statistical values of write and read requests in the last interval, the communication cost increases rapidly. However, in the following 10 intervals, the similar process is repeated. This result validates the conclusion of Theorem 1. From Fig.1, it is inferred that the mean communication costs in the *ADR_G* is the lowest. The reason is that the write requests are propagated among replica nodes along the MST in the *ADR_G* algorithm and the communication cost is $|F|-1$.

Secondly, we compare the performance of algorithms in the MANET environment. We compare the *ARAM*, the *EARAM*, the *EARAM_SN*, and the *ADR_G* algorithms with the *SRA* algorithm. Now our main concern is the effect of nodes mobility on algorithms. In the simulation, the read request issued by every node conforms to the random distribution and is fixed during the whole process. In the *EARAM_SN* algorithm, $n=10$, $a=0.5$, and $\tilde{C}=0.25*10$.

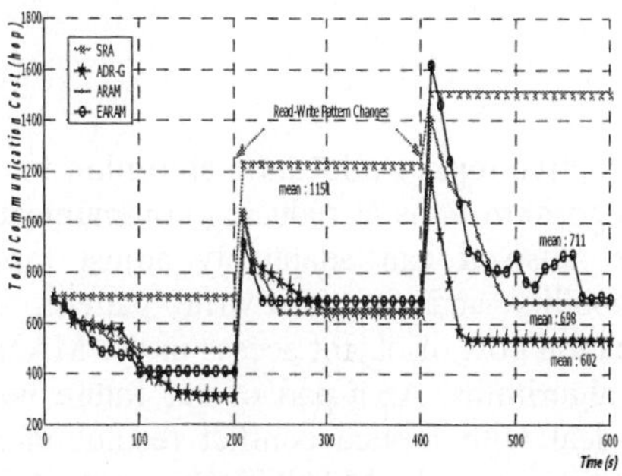

| Fig. 1. Performance in Static Networks | Fig. 2. Performance in MANET |

Fig.2 shows that compared with the *ADR_G* algorithm, the mean cost is 23% less in the *ARAM* algorithm, 25% less in the *EARAM* algorithm and is 31% less in the *EARAM_SN* algorithm. The reason is that in the *ADR_G* algorithm, the write requests are propagated among replica nodes along the MST of the replicas and the resulting optimal configuration tends to locate replicas in nodes adjacent to each other. While nodes move and the network topology changes, the replica nodes are no longer adjacent physically. Therefore the data access communication cost in the *ADR_G* algorithm increases rapidly when the network topology changes. From Fig.2 we know that the communication cost of object access in the MANET environment is greatly reduced in our algorithms. The simulation result also indicates that the communication cost is the most stable in the *EARAM_SN* algorithm among that in other algorithms.

Table 2. Changes of Replica Allocation Scheme

No.	EARAM		EARAM_SN	
	cost(F)	F（No.of node）	cost(F)	F（No. of node）
1	266	0 81	371	2 81 23 30
2	329	0 81	367	2 77 23 30
3	442	0 81	372	2 5 7 30
4	341	0 9 81	333	5 77 9 30
5	411	0 9 26	361	2 77 9 30
6	360	9 19 26	385	77 9 11 30
7	398	9 26 37	412	77 9 11 15 30
8	398	9 26 33	384	77 9 11 13 30

Table 2. presents 8 sequential replica allocation schemes generated by the *EARAM* algorithm and the *EARAM_SN* algorithm respectively in one experiment.

Table 2. indicates that the replica allocation scheme generated by the *EARAM_SN* algorithm is more stable than that generated by the *EARAM* algorithm. The number of nodes that are expanded to be replica node in one interval and relinquish replica in the next interval is reduced. Therefore the oscillation of replica allocation is cut down in the *EARAM_SN* algorithm.

5   Conclusion

In this paper, a new distributed dynamic adaptive replica allocation algorithm for the MANET environments is proposed. The algorithm aims at reducing communication cost and improving system performance. Also it can adaptively adjust replica allocation scheme according to node mobility and the read-write pattern. The simulation results show that the communication cost of object access in the MANET environment is reduced efficiently in our algorithms. As a part of our future work, these algorithms should be improved to deal with replica conflict resolution and reconciliation problem during network partitioning in the MANET environment. The replica consistency protocol should also be investigated.

References

1. O. Wolfson, S. Jajodia, and Y.Huang, "An Adaptive Data Replication Algorithm", ACM Transactions on Database System, vol. 22, no. 4, 1997, pp. 255-314.
2. S.A Cook, J.K Pachl, and I.S Pressman, "The Optimal Location of Replicas in A Network Using A Read-One-Write-All Policy", Distribute Computing, vol.15, no.1, 2002, pp. 7-17.
3. D Barbara, T.Imielinski, "Sleeper and Workholics: Caching Strategies in Mobile Environment", in: Proceedings of ACM SIGMOD'94,1994, pp. 1-12.
4. J Cay, K.L Tan, B.C.Ooi, " On Incremental Cache Coherency Schemes in Mobile Computing Environments", In Proceeding of IEEE ICDE'97, 1977, pp. 114-123.
5. Y Huang, S.Pistla, and O. Wolfson, "Data Replication for Mobile Computer", In Proceedings of ACM SIGMOD'94,1994, pp.13-24.
6. T.Hare, "Replica Allocation in Ad hoc Networks with Periodic Data Update", In Proceedings of Int'l Conference on Mobile Data Management(MDM 2002), 2002, pp.79-86.
7. K Wang, B.Li, "Efficient and Guaranteed Service Coverage in Partitionable Mobile Ad-hoc Networks", In IEEE Joint Conference of Computer and Communication Societies (INFOCOM'02) , New York City, New York, 2002, June 23-27, pp.1089-1098.
8. O.Wolfson, A.Milo, "The Mulicast Policy and Its Relationship to Replicated Data Placement", ACM Transaction on Database System, vol.16, no.1, 1991, pp.181-205.
9. T Camp, J Boleng, v.Davies, "A Survey of Mobility Models for Ad Hoc Network Research," Wireless Communication & Mobile Computing (WCMC): Special Issue On Mobile Ad Hoc Networking: Reach, Tends and Applications, vol.2, no.5, September 2002, pp. 483- 502.
10. Zhengjing, et al., "An Adaptive Replica Allocation Algorithm in MANET Environment," Tech. Rep. PDD-2003-9, School of computer, National University of Defense Technology, 2003.

Mobile-Based Synchronization Model for Presentation of Multimedia Objects

Keun-Wang Lee[1], Hyeon-Seob Cho[2], and Kwang-Hyung Lee[3]

[1] Dept. of Multimedia Science Chungwoon University, Korea
kwlee@cwunet.ac.kr
[2] Dept. of Electronics Engineering Chungwoon University, Korea
[3] School of Computing, Soong-sil University, Korea

Abstract. This paper presents a synchronization model that implements the presentation of multimedia objects as well as a synchronization scheme that improves media data latency and quality of service (QoS) in mobile environments. The proposed model meets synchronization requirements among multimedia data because it employs an approach to not only adjusting synchronization intervals using the maximum available delay variation or jitter at the base station (BS) but also flexibly dealing with variable latencies due to variations in delay time.

1   Introduction

Synchronization is a critical prerequisite to ensuring an adequate QoS in providing multimedia services. The reason behind that lies in the disruption of original timing relations resulting from the occurrence of random delays in mobile networks or mobile host(MH) system clock inconsistencies among inter-media, all of which is attributable to difference in the time of arrival of data being transferred from the server side to the MH over wireless communications networks. Therefore, for the multimedia data in which timing relations are disrupted, an artificially synchronized presentation of data streams is required to ensure similarity or identicalness to the original ones, through the use of the requirements for application services or the limitations of human perception toward loss and delays occurring in individual media [1][2][3].

This paper deals with live synchronization involving the real-time synchronization of multimedia information as well as synthetic synchronization for stored media. For this example, let us assume that we are taking 3D animation training for electrical safety from a MH. While a lecturer is providing verbal explanations, his/her voice, moving images, texts, and 3D animation should be presented simultaneously. As in wired environments, smooth audio/visual presentations should be delivered in mobile environments. In this regard, this paper proposes a synchronization model that allows for simultaneous presentations of multimedia objects such as voice, 3D animation, moving images and texts on a MH as well as presents an intra-media and inter-media synchronization scheme that enables mobile multimedia services.

2   Related Work

Previous studies have focused on describing synchronization models for multimedia applications [5][6]. Among them, Petri-net based specification models are effective in

M. Bubak et al. (Eds.): ICCS 2004, LNCS 3036, pp. 381–388, 2004.
© Springer-Verlag Berlin Heidelberg 2004

specifying temporal relationships among media objects. These models allow various media to be integrated and give easy descriptions of QoS requirements. However, previous extended forms of Petri-net modeling like Object Composition Petri-Net (OCPN) and Real Time Synchronization Model (RTSM) exhibit limitations as far as QoS parametric modeling is concerned.

Many existing studies pose significant problems in that delay in wireless networks causes a reduction in the maximum media playout latency. That's because conventional research places focus on determining the sequence and playout time of the media created by the server side and fixing the queuing time at the buffer on the MH side.

Delay jitter causes discontinuity. If any discontinuity is permissible at a MH, it should be smaller than the worst end-to-end playout delay. This case leaves you with two strategy options; one with I-strategy in which belatedly arrived frames from the BS are discarded, and the other with E-strategy in which belatedly arrived frames from the BS are played out.

This paper proposes a dynamic synchronization scheme that adjusts intervals between synchronization activity using the maximum available delay variation or jitter, reduces data loss caused by variations in latency, and synchronizes inter-media and intra-media.

3   Proposed Synchronization Model

The proposed synchronization model is Petri-net based standard model that enables the presentation of multimedia objects.

3.1   Definition of Synchronization Model

The proposed synchronization model for specifying Petri-net in any BS is defined as follows:

The model is specified by the tuple $[P, T, K, A, Re, M]$

   where

$P = p_1, p_2, \ldots, p_n$; *Regular places(single circles).*
$T = t_1, T_2, \ldots, t_m$; *Transitions.*
$K = k_1, k_2, \ldots, k_i$; *Key places.*
$X = P \cup K$; *All places.*
$A = (X \times T) \cup (T \times X) \to I, I, = 1, 2, 3, \ldots$; *Directed arcs.*
$Re : X \to r_1, r_2, \ldots, r_k$; *Type of media.*
$M : X \to I, I = 0, 1, 2$; *State of places*

The "Place" is used to represent a medium unit and its action. It may have tokens. A place without any token indicates that it is currently in an inactive state. A place with a token stands in an active state and may be in either a blocked or unblocked state.

The information determining the firing of a transition t_i is delivered by a control medium. When a key medium arrives within the time specified by an absolute time, the transition t_i is immediately fired at the corresponding transition's input place that has open tokens. However, in the event that the key medium reaches beyond the time specified by an absolute time, the absolute time initiates the firing.

3.2   Control Medium

As a control medium, CT has the information about the number of input places and key media while transmitting this information to a subsequent transition. The roles of a control medium is addressed by the following:

(1) Check the set of input places;

(2) Determine the number of objects selected as key media

(3) Transmit the number of input places and key media as well as an absolute time to a subsequent transition.

Figure 1 shows the state of a key medium and the active state stored in the control medium, which will be transmitted to subsequent transitions. 7 bits at the front and 7 bits at the back indicate the information of a key medium and the information of its active state, respectively.

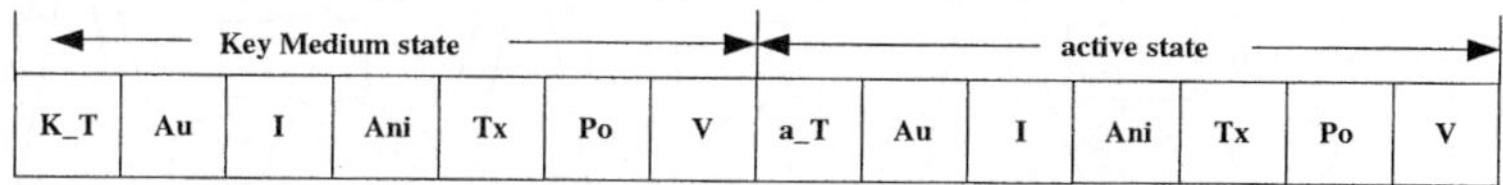

	Key Medium state						active state						
K_T	Au	I	Ani	Tx	Po	V	a_T	Au	I	Ani	Tx	Po	V

Fig. 1. Information of Control Medium

All media are transmitted over networks, so real-time constraints are likely to be exceeded. In the event of the delay of a key medium, a firing is available via the absolute time of the control medium.

3.3   Presentation of Synchronization Model

Figure 2 shows an overall view of Petri-nets. 1'(125,3,1,1,1,0,0,0,6,1,1,1,1,1,1) indicates the information of a control medium. The first parameter 125 refers to the absolute time, 3 the number of key media, and 6 the number of input places. As Control indicates type, Control_in is applicable to all conditions. When an event (i.e. number of key media and number of input places) occurs, the control medium transmits the information to subsequent transitions.

HS indicates a hierarchical submodule that goes to a lower level module for job processing before going to the next phase. HS goes to a relative duration routine - one of its lower level modules - to find the relative duration before executing After_duration. Therefore, HS has synchronization intervals three times, which results in effective synchronizations in a way that computes the relative duration before executing *sync1,* computes the compensation time for jitters before executing *sync2,* and computes the flexible playout time before executing *sync3.*

Verifications of the proposed multimedia synchronization model have been performed in order to make sure that it corresponds to the analysis methods of petri-nets including a reachability graph and a matrix-equation.

Verification of Reachability Graph. The reachability tree represents the reachability set of petri-nets. In the proposed model, the initial marking is (1, 1, 1, 1, 1,

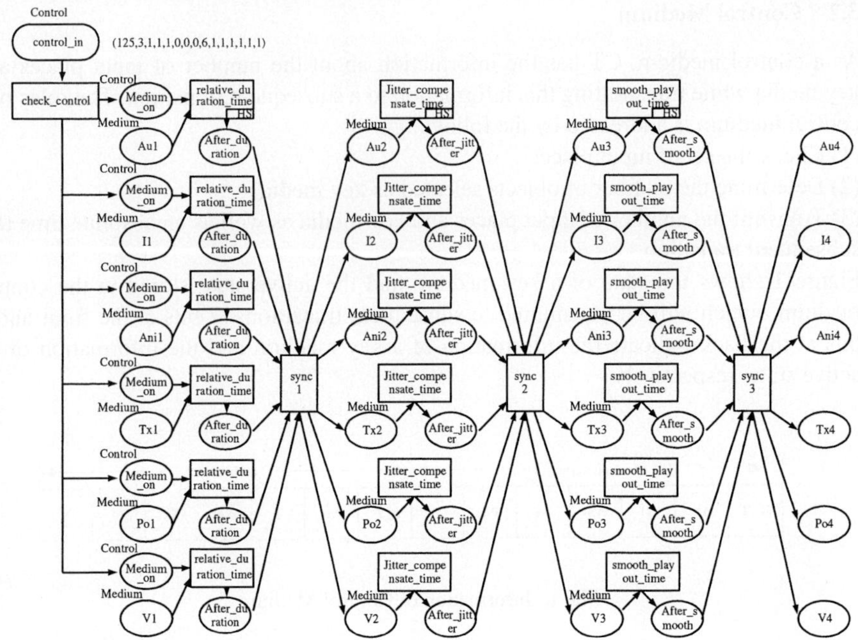

Fig. 2. Presentation of Synchronization Model

1, 0, 0, 0, 0, 0, 0, 0, 0, 0, 0, 0, 0). The transition t_1 is generated from this initial marking. If t_1 is fired, t_1[giving(0, 0, 0, 0, 0, 0, 1, 1, 1, 1, 1, 1, 0, 0, 0, 0, 0, 0)] is obtainable, and a transition to t_2 is possible. If t_2 is fired, t_2 [giving(0, 0, 0, 0, 0, 0, 0, 0, 0, 0, 0, 0, 1, 1, 1, 1, 1, 1)] is obtainable. The result of this tree is represented as follows:

$$(1, 1, 1, 1, 1, 1, 0, 0, 0, 0, 0, 0, 0, 0, 0, 0, 0, 0)$$
$$\downarrow t_1$$
$$(0, 0, 0, 0, 0, 0, 1, 1, 1, 1, 1, 1, 0, 0, 0, 0, 0, 0)$$
$$\downarrow t_2$$
$$(0, 0, 0, 0, 0, 0, 0, 0, 0, 0, 0, 0, 1, 1, 1, 1, 1, 1)$$

Verification of Matrix-Equation. Verifications of the proposed model were performed using matrix-equations as the second analysis method of the petri-nets. The following are two matrices D^- and D^+ indicating input function and output function, respectively. Matrices D^- and D^+ are expressed as follows:

$$D^- = \begin{bmatrix} 1, 1, 1, 1, 1, 1, 0, 0, 0, 0, 0, 0, 0, 0, 0, 0, 0, 0 \\ 0, 0, 0, 0, 0, 0, 1, 1, 1, 1, 1, 1, 0, 0, 0, 0, 0, 0 \end{bmatrix}$$

$$D^+ = \begin{bmatrix} 0, 0, 0, 0, 0, 0, 1, 1, 1, 1, 1, 1, 0, 0, 0, 0, 0, 0 \\ 0, 0, 0, 0, 0, 0, 0, 0, 0, 0, 0, 0, 1, 1, 1, 1, 1, 1 \end{bmatrix}$$

And the matrix D is expressed as follows:

$$D = D^+ - D^- =$$
$$\begin{bmatrix} -1, & -1, & -1, & -1, & -1, & -1, & 1, & 1, & 1, & 1, & 1, & 1, & 0, 0, 0, 0, 0, 0 \\ 0, & 0, & 0, & 0, & 0, & 0, & -1, & -1, & -1, & -1, & -1, & -1, & 1, 1, 1, 1, 1, 1 \end{bmatrix}$$

The result of applying the equation $\mu' = \mu + x \cdot D$
to the matrices is addressed by the following :

$(0,0,0,0,0,0,0,0,0,0,0,0,0,1,1,1,1,1,1) =$
$(1,1,1,1,1,1,0,0,0,0,0,0,0,0,0,0,0,0,0)$

$$+$$

$$x \begin{bmatrix} -1, & -1, & -1, & -1, & -1, & -1, & 1, & 1, & 1, & 1, & 1, & 1, & 0, 0, 0, 0, 0, 0 \\ 0, & 0, & 0, & 0, & 0, & 0, & -1, & -1, & -1, & -1, & -1, & -1, & 1, 1, 1, 1, 1, 1 \end{bmatrix}$$

$(-1,-1,-1,-1,-1,-1,0,0,0,0,0,0,1,1,1,1,1,1)=$

$$X \begin{bmatrix} -1, & -1, & -1, & -1, & -1, & -1, & 1, & 1, & 1, & 1, & 1, & 1, & 0, 0, 0, 0, 0, 0 \\ 0, & 0, & 0, & 0, & 0, & 0, & -1, & -1, & -1, & -1, & -1, & -1, & 1, 1, 1, 1, 1, 1 \end{bmatrix}$$

where x is (x_1, x_2)
$-1 = -1 \cdot x_1$
$0 = x_1 - x_2$
$1 = x_2$
$1 = x_2$
For (1), $x_1 = 1$;
For (2), $x_1 = 1$;
Therefore, for (3), $x_1 - x_2 = 0$

Through the verification described above, the proposed synchronization model
proved to be consistent with the reachability graph and matrix-equation.

4   Synchronization Scheme

This chapter describes a delay jitter scheme. Delay time or latency is an important
factor in evaluating the playout quality of a frame. This paper provides a description
of delay jitter strategies such as I-strategy and E-strategy. It also makes a comparison
between those two strategies and the proposed delay jitter strategy using the waiting
time in the queue as well as playout time.

[Theorem] A frame waits for as long as variable latency or jitter times until they are
presented.

[Proof] Where the maximum latency or jitter time is below 10ms, temporary discontinuity is tolerable for voice media without causing any impact to quality of service. If
the discontinuity tolerance is δ, an extended synchronization interval becomes $\Delta' = \Delta'$

+ δ. Therefore, if the instantaneous discontinuity tolerance permitted by the corresponding media data is δ, an extended synchronization interval becomes $\Delta' = \Delta + \delta$. If the $(j+1)$th packet arrival time at the buffer is $B_{i(j\cdot i)}$, the synchronization requirements are met when $B_{i(j\cdot 1)}$ is smaller than or equal to the frame playout time $M_{i(j\cdot i)}$. Formula 1 meets the synchronization requirements.

$$B_{ij} < M_{i(j\cdot 1)}$$
$$B_{ij} < M_{ij} + \Delta'$$
$$B_{ij} < M_{ij} + \Delta + \delta$$
$$B_{ij} < M_{ij} + 1/N + \delta \cdots (Formula 1)$$

The proposed strategy illustrates that, if the (j)th packet for media data stream i is presented at the time M_{ij}, the $(j+1)$th packet is presented at the time $M_{i(j+1)} = M_{ij} + \Delta'$. In other words, (j)th and $(j+1)$ th packets meet Formula 1 within synchronization intervals.

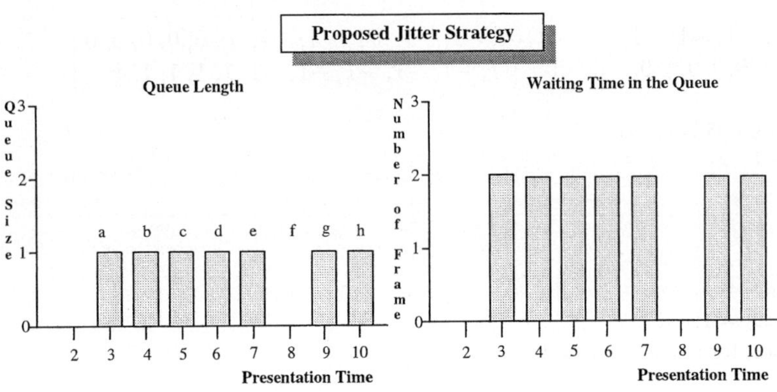

Fig. 3. Proposed Strategy Applied Under Reduced Network Traffic

Figure 3 shows the proposed jitter strategy that complements the shortcomings of both the I-strategy and E-strategy. Belatedly arrived frames wait for as long as latency or jitter times until they are presented, instead of being deleted unconditionally or waiting indefinitely until they are presented at the next playout time. As depicted in Figure 3, the frames b and c are played out in the units 4 and 6 by compensating for the maximum latency or jitter. The frame f indicates that it has not arrived within the variable delay jitter time. In this case, the frame f cannot be played out even if it waits for as long as the variable latency or jitter times.
Therefore, the unit 8 in the frame f indicates that it cannot be compensated for the maximum delay jitter time due to excessive delay.

Figure 4 shows that the unit 4 in the frame c was skipped due to excessive delay and the unit 5 in the frame d was compensated by applying the maximum delay jitter time.

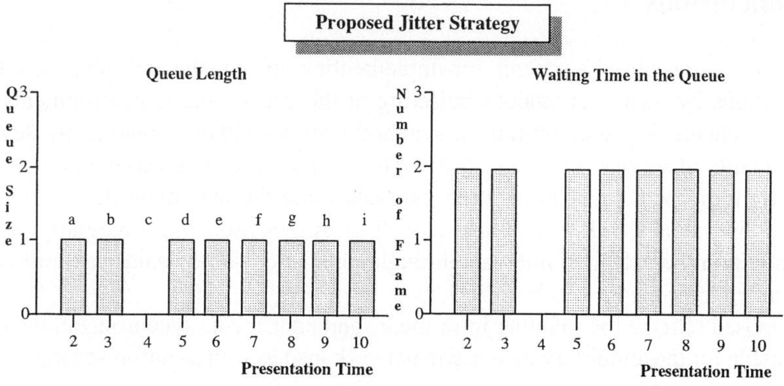

Fig. 4. Proposed Strategy Applied Under Heavy Network Congestion

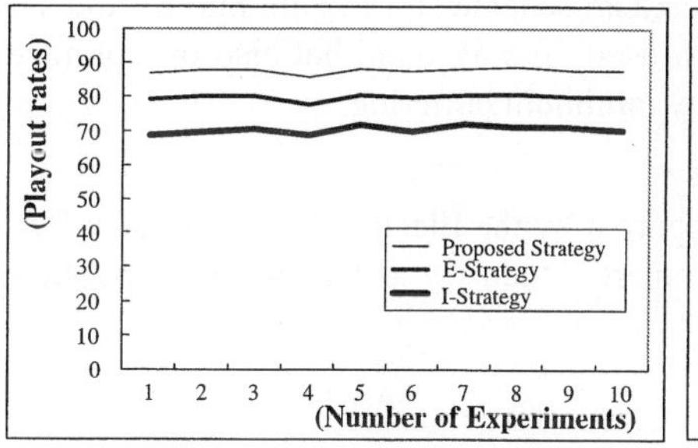

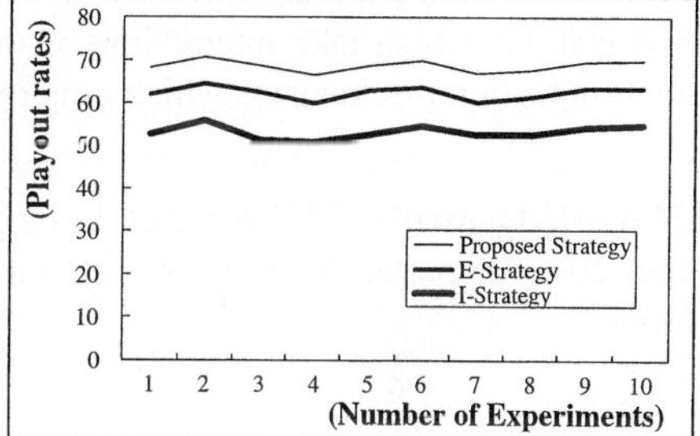

Fig. 5. Comparison of Playout Rates   **Fig. 6.** Comparison of Playout Rates

5   Performance Evaluation

With focus on the playout time and loss time, we performed a comparative analysis between the existing scheme and the delay jitter & playout scheme using the maximum delay jitter time. Comparisons among the I-strategy, E-strategy, and the proposed strategy were made in consideration of the following two cases; Circumstances of reduced network traffic and circumstances of heavy network congestion. As a result, the proposed strategy proved to be vastly superior to other two strategies. For the purpose of this paper, we have assumed that the average delay is 100ms and the variance is 20ms in the event of delay in the audio stream.

Figure 5 shows the result of comparisons of playout rates among the I-strategy, E-strategy, and the proposed strategy that were applied under reduced network traffic. The playout rates of the proposed strategy were obtained by conducting experiments ten times. The proposed strategy showed more improved playout rates than the I-strategy and E-strategy by 17.33 % and 7.8%, respectively.

As shown in Figure 6, under heavy network congestion, the proposed strategy showed more improved playout rates than the I-strategy and E-strategy by 15.48% and 6.3%, respectively.

6   Conclusions

This paper has proposed a scheme for implementing intra-media and inter-media synchronizations by means of smooth buffering at the BS in mobile environments. The proposed scheme delivered optimized synchronizations without causing any degradation of quality of service. The superiority of the proposed scheme was demonstrated by extending intra-media synchronization intervals using the maximum delay jitter time of the audio media as a key medium, as well as by synchronizing irregularly arriving packets within the extended intervals through applications of the said maximum delay jitter time to inter-media synchronizations.

The proposed scheme for enabling intra-media and inter-media synchronizations is ideally suitable for the temporary increase in network load as well as unforeseeable disconnections.

Furthermore, it allows us to take 3D animation training for electrical safety while on the road.

Future work needs to focus not only on standard schemes for mobile multimedia synchronizations which take interactions with users into account, but also on optimized synchronization mechanisms which employ minimum buffering.

Acknowledgements. This research is supported by the Electric Power Industry R&D Fund 2003 supported by Ministry of Commerce, Industry and Energy in republic of Korea.

References

1. D. H. Nam and S. K. Park, "Adaptive Multimedia Stream Presentation in Mobile Computing Environment," Proceedings of IEEE TENCON, 1999.
2. A. Boukerche, S. Hong and T. Jacob, "MoSync: A Synchronization Scheme for Cellular Wireless and Mobile Multimedia Systems," Proceedings of the Ninth International Symposium on Modeling, Analysis and Simulation of Computer and Telecommunication Systems IEEE, 2001.
3. M. Woo, N. Prabhu and A. Grafoor, "Dynamic Resource Allocation for Multimedia Services in Mobile Communication Environments,"IEEE J. selected Areas in Communications, Vol.13, No.5, June. 1995.
4. D. H. Nam, S. K. Park, "A Smooth Playback Mechanism of Media Streams in Mobile Computing Environment," ITC-CSCC'98, 1998.
5. P. W. Jardetzky, and C. J. Sreenan, and R. M. Needham, "Storage and synchronization for distributed continuous media," Multimedia Systems/Springer-Verlag, 1995.
6. C.-C. Yang and J.-H. Huang, "A Multimedia Synchronization Model and Its Implementation in Transport Protocols," IEEE J. selected Areas in Communications, Vol.14, No.1, Jan. 1996.

Synchronization Scheme of Multimedia Streams in Mobile Handoff Control

Gi-Sung Lee

Dept. of Computer Science, Howon Univ., Korea
ygslee@sunny.howon.ac.kr

Abstract. This paper presents a synchronization scheme for enabling a smooth presentation of multimedia streams during a handoff in mobile environments. As changes are made to base stations in wireless environments, mobile host-initiated handoffs result in multimedia data loss in a base station as well as low quality of service (QoS) for multimedia streams. As a result of evaluations, the proposed schemes, compared to the previons schemes, delivered a continuous playout of multimedia streams while achieving low packet loss.

1   Introduction

Driven by the drastically increasing use of the Internet, mobile computing-based services are emerging in rapid succession. Presentations of multimedia data stored in a mobile host are not readily implemented, largely due to the dynamic nature of mobile network's connectivity, for example, higher data loss rate, higher delays, and lower network bandwidth[6]. For this reason, buffers are used in a lot of distributed multimedia systems connected to wireless networks in order to address network-specific problems including latency and data loss. A Base Station(BS) transmits subframes from its several multimedia servers. If, due to unexpected delay and increased traffic, their projected playout time is faster than their actual arrival time, the playout of subframes cannot be implemented[4,7]. In order to address these problems, buffering at a BS is required to reduce packet delay and jitter between a multimedia server and a BS. One advantage of a mobile network is host mobility. Though, this advantage can become a disadvantage due to some cumbersome processes; not only resources must be secured for transmission of multimedia streams but also multimedia streams transmitted to a BS must be transmitted again. With the purpose of minimizing those drawbacks, this paper places focus on configuring two jitter buffers for a BS and one jitter buffer for a mobile host. Streams occurring and lost during a handoff are transmitted to an Old BS (BS_{old}), and substreams in odd numbered buffers are transmitted to a New Base Station (BS_{new}). During this process, a mobile host is adjusted to the maximum playout delay jitter. This scheme implements the playout of substreams in BSnew buffer that require retransmission, eliminating the need for retransmission. However, media loss adversely affects playout.

M. Bubak et al. (Eds.): ICCS 2004, LNCS 3036, pp. 389–396, 2004.

2   Related Work

Research has reached a level where mobile-based synchronization schemes and previous schemes are converging. M. Woo, N. U. Qazi, and A. Ghafoor defined the BS as wired/wireless interface. For wired networks, the interface is defined as a BS buffer used to reduce delay and jitter between packets. The disadvantage of this approach is its attempt to apply synchronization to buffers through assignment of existing wireless communication channels [5]. D. H. Nam and S. K. Park proposed a buffer management and feedback scheme that allows transmitted multimedia unit storage in the buffer of a MH(MH), and that adjusts playout time to the medium buffer length [2,3]. However, there is a disadvantage to this approach that the adjustment of palyout time lowers the quality of multimedia data even when a mobile network is in a normal state. That's because the stable state does not correspond to the buffer of a MH. Azzedine Boukerche proposed an algorithm that enables message transmission among multimedia servers, BS and mobile hosts while interfering with multimedia servers as well as startup time for transmission [1]. The merit of this approach is a well-defined role of the multimedia server, BS, and MH. Nonetheless, this scheme has some disadvantages in terms of its lack of a playout policy as well as the application of only startup time for communications transmission as means of solving the synchronization problem.

3   Mobile Network-Based Buffer Management Scheme

3.1   System Configuration

This system supports a k multimedia server node, m BS, and n MH. The BS communicates with an i mobile host in mth cell. The MH is necessarily supposed to access the server via the BS. This system allows a BS to manage variations of startup time for transmission as well as buffers, using variables such as delay jitter and arrival time of subframes transmitted from the multimedia server. Therefore, this system contributes to effectively dealing with mobile communications-specific small memory footprint and low bandwidth. Some of the advantages of the proposed scheme are reduced server startup time, lighter network traffic, and reduced buffer size, all of which are achieved by storing a movie in the k server. The multimedia server saves multimedia data in a distributed way that splits data streams according to logical time. Data streams are compressed data split by logical time. Accordingly, their configurations are made according to a synchronization group, not the same byte. Such split streams are called substreams, and this technique is referred to as subframe stripping. Figure 1 shows the configuration of the proposed system. The Message Manager in the multimedia server skips and transmits subframes using the offset control mode notified by the Feedback Manager at the BS. The BS saves the arrival time of the subframes received by individual servers, and it sends/receives a dummy packet to identify the startup time for transmission. The Communication Manager in the multimedia server distinguishes feedback packets from messages notified by the BS, and the Communication Manager in the BS arranges the sub-frames received from individual servers in an orderly manner.

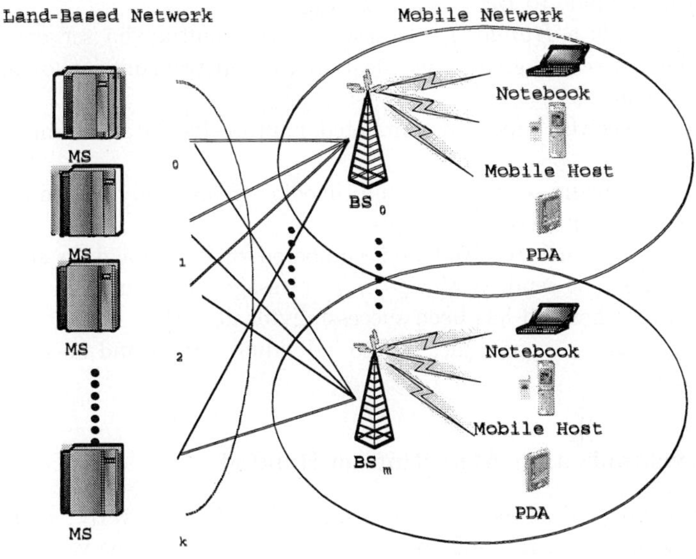

Fig. 1. System Configuration

3.2   Handoff Management

Due to its use of soft handoff, CDMA systems have some problems involving multiple BSs. Handoff management eventually leads to the increase in channel capacity. The information such as multimedia units, data, and voice is transmitted via multiple BSs. We also present an algorithm that enables a mobile host to implement the playout of media streams, without any multimedia data loss, or without any additional delay accompanying message transmission, or within the reasonable QoS limits during a handoff. In this scheme, the MH (HM_i) transmits multimedia services via the BS (BS_m) from the Server (S_k). The BS is classified into the primary and non-primary BSs. The former is responsible for sending multimedia streams to the MS, and the latter adjoins the primary BS. The BS buffer is of the two-jitter size, and the MH buffer is of the one-jitter size. Compared against buffer configurations made only for the BS, buffer configurations like this gives the advantages of no multimedia data stream loss during a handoff while complementing the MH with small memory footprint. Figure 2 shows the system configuration applied when a handoff is being processed. A handoff occurs when MH_i moves from $BS_{current}$ to BS_{new}. $BS_{current}$ sends a handoff message to the multimedia server, and starts to implement handoff processing in the following sequence:

1. MH_i implements playout as long as (1jitter $\times$ r) $\times$ λ time, regardless of handoff. /* r refers to the playout speed of a substream, and λ means the maximum jitter within the media */
2. Setting is made to $BS_{old} = BS_{current}$.
3. Setting is made to $BS_{current} = BS_{new}$.
4. BS_{old} sends a $Handoff_{on}$ message to the multimedia server. Then, the multimedia server gives a notification to stop transmission of multimedia data streams.
5. BS_{new} receives only odd numbered multimedia data streams present in BS_{old}.
6. The multimedia server sends multimedia streams only to the servers that contain odd numbered subscripts.
7. Once BS_{new} has received odd numbered multimedia data streams from BS_{old} buffer, it sends a $Handoff_{off}$ message to the multimedia server to inform that handoff has been successfully dealt with.
8. BS_{new} sends a request message to individual servers and normal transmission is made.

3.3   Synchronization Algorithm on Handoff

We now formally describe the algorithm for a BS_m, a MH_n, and a, S_k by means of pseudocode. Handoff control algorithm describes the variables used by Pseudocode. The handoff occurs when MH_m moves from the current BS_m to another BS_m.

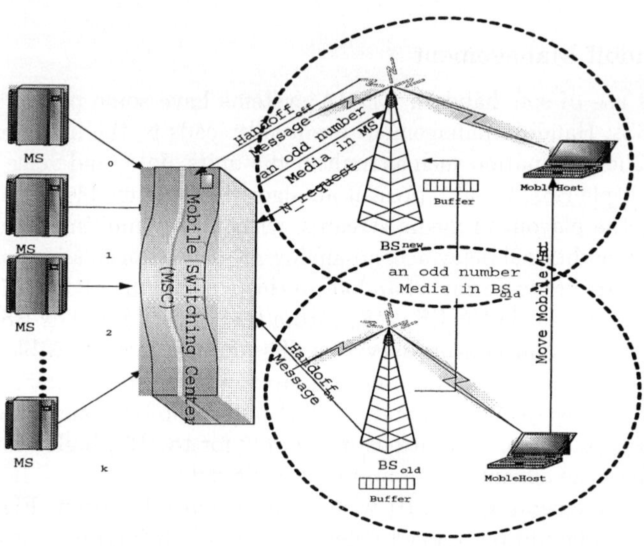

Fig. 2. Handoff Process Structure

During the handoff control process, multimedia data streams stored in are sent to a new BS_m for playout within the QoS limits. The algorithm is designed to solve the problems arising out of handoffs by the use of an approach to multimedia synchronization. This paper deals with two types of messages, which are addressed as follows:

1. Send (source, action; argument);
2. Receive (source, action; argument);

where action $\leftarrow$ Request| N-request| Reply | Update | INT | $Handoff_{off}$ | $Handoff_{on}$ | Done , and arguments $\subset \{MS_k^i, DT_k^i, RT_k^i, ST_k^i, S_k\}$.
The algorithm is described as follows:

Synchronization on Handoff

Mutimedia Server$_k$

Receive $(BS_m$, DummyPacket);{
    Send $(BS_m$, reply; CurrentMultiserverTime);}
Receive $(BS_m$, Request; MMU$_k^i$);{
   For i $= 0$, i $<$ k, i $++$ Do
     Send(BS_m, reply; MMU$_k^i$);}
Receive $(BS_m$, N $-$ Request; MMU$_k^i$);{/∗MMU$_k^i$ is i-th multimedia steame∗/
   Make schedule for (N $-$ 1) requests
   Save RT_k^i and D_{k}^i;   **/ ∗ RT_k^i is Round trip time∗/**
   Send $(BS_m$, reply; MMU$_k^i$);}/ ∗ D_k^i is Delay time∗/
Receive $(BS_m$, Handoff$_{on}$);{
Do
   Wait;
While (Receive(BS_{new}, Handoff$_{off}$))}

Mobile Host$_i$

   IF (Receive(BS_m,Handoff$_{on}$)){
   Play $-$ out is MMU$_k^i$ at τ_t+10ms;}
   Else{
   Send(BS_m,CurrentMultiserverTime, Requst : MMU);
   Receive(BS_m, MMU$_k^i$; τ_t);
   Play $-$ out is MMU$_k^i$ at τ_t;}

Base Station$_m$

   BufferControl(); {
  ρ_i =125ms;
    If (buffer_point $==$ NormalLevel)
      $\tau_i = \rho_i$; Send(MH_i, MMU_k^i, τ_t);
    Else If (buffer_point $==$ UpperLevel){

$BL_{allsize} = BL_{allsize} + 1;$ /*BLallsize is overall buff size*/
$BL_{psize} =$ CurrentBuffer;   /*BL psize is buffer Level size*/
$\omega = 0;$ /* Weight Value */ $\rho_i = 125;$
$\omega = BL_{allsize}$ $/ BL_{psize};$ $\tau_t = \rho_i - (\rho_i \times \omega);$
Send$(MH_i,\ MMU_k^i,\ \tau_t);\}$
    Else IF (buffer_point == LowerLevel){
      $\omega = BL_{allsize} - BL_{psize};$ $\omega = \omega / BL_{allsize};$ $\tau_t = \rho_i + \lambda \times \omega;$
      Send$(MH_i,\ MMU_k^i,\ \tau_t);\}\}$ /*λ is Maximum delay jitter*/
Receive$(BS_n, Handoff_{on})$
  $BS_{old} = BS_{current};$
  $BS_{current} = BS_{new};$
  Send$(MH_i, Handoff_{on});$
  Send$(S_k, Handoff_{on});$
  For k=1, k $\leq$ K, k+2 Do
    $BS_{new}(MMU_k^i) = BS_{old}(MMU_k^i);$
  Send$(S_k, Handoff_{on});$ /* from BS_{new} to S_k */
Send$(S_k,$ **Request;** $MMU_k^i);$ }

Main(){   /* Main Program */
Call Start – UpTime;   /*Δ is jitter*/
Start-Up=MAX $D_k^i - D_k^i;$
Set Start Up Time for Server $S_k;$
Send$(S_k,$ **Start Up Time);**
$\Delta = DT_k^{max} - DT_k^{max}$ /* Δ is jitter */
If (Buffer_point == NomalLevel)
Send$(S_k,\ MMU_k^i,$ **Request;**$MMU_k^i)$
Else If (Buffer_point==UpperLevel){
    If $(DT_k^i > \Delta)\{$
    Feedback_value = Call Feedback;
    Send$(S_k,$**Feedback_value,** $D_k^i);\}$
    Else{
    Continue;
    Call BufferControl(UpperLevel);}}
Else If (Buffer_point == LowerLevel){
    If $(DT_k^i > \Delta)\{$
    Feedback_value = Call Feedback;
    Send$(S_k,$**Feedback_value,** $D_k^i);\}$
    Else{
    Continue;
    Call BufferControl(UpperLevel);}
  }}**end of main**

4   Performance Evaluation

Experiments were carried out using IBM-compatible PCs with Pentium processor, and interfaces and algorithms were implemented using Java development

Kit JDK 1.3. Outputs obtained from the experiments were stored in Microsoft MDB as simulation.mdb files. 1Kbyte audio data were encoded using a PCM encoding technique, and the video frames used had a resolution of 120 X 120 pixels. The frames used in the experiment were those encoded with 24 frames per second. This paper assumes simulations have been performed in a mobile environment. In order to process individual packets properly, the information used in the actual simulations was applied equally to mobile networks using Poisson distribution. One thousand frames were used in performance evaluation experiments where the maximum delay jitter time of 600 ms was applied. This paper presents our work on the algorithm that disallows any loss of data in the BS buffer to smooth the playout of multimedia streams during a handoff. The comparative evaluation presented in this paper is focused not only on the feedback policy toward maintaining the buffer at a normal level but also on the play-out and loss time where the buffer level control-based playout policy is applied Figure 3 shows how much the level of the buffer varies. As illustrated in Figure 3, the data stored in BS_{old} buffer are sent to BS_{new} buffer in order to prevent streams against loss in the frames numbered from 800 through 960 during a handoff. Additionally, during the handoff, MH_i delays the playout time of MMU_k^i to the maximum delay jitter of 10ms in order to secure more time for moving substream data in BS_{old} buffer to BS_{new} buffer. As shown in Figure 3, existing conventional schemes appear to be less effective in dealing with handoffs, thus suffering from loss of substream data stored in BS_{old} buffer. Figure 3 shows that overflow occurs in the frames numbered from 150 through 200, and also shows that the buffer level of the proposed scheme is kept in a more stable state than existing schemes. As it were, existing conventional schemes could not avoid starvation occurring in the buffer due to handoff, as the results obtained in figure 3. However, the proposed scheme was able to prevent starvation from occurring in the frames numbered from 800 through 960. Furthermore, it could prevent overflow from occurring in the frames numbered between 150 and 200.

Figure 4 depicts what effect the proposed scheme has on the buffer level-based playout policy. The proposed scheme delays playout time even though a handoff has occurred in the frames numbered from 800 through 960. For existing conventional schemes, however, the occurrence of a handoff results in the loss of data in BS_m buffer, causing not only a failure of the playout of data streams but also skipped frames numbered between 150 and 200. For the proposed scheme, the playout time of the frames numbered from 800 through 960 varies according to the playout policy while the playout time of the frames numbered from 150 through 200 decreases gradually, ultimately leading to the reduction in overflow.

5   Conclusions

This paper presents a scheme that disallows lower playout rates within the QoS limits for multimedia presentations during a handoff in mobile networks. The proposed scheme enables a base station to manage buffers and playout policies, and further deals with handoffs in a fast and efficient way. Additionally, this scheme offers a suitable approach to effectively dealing with limiting factors for

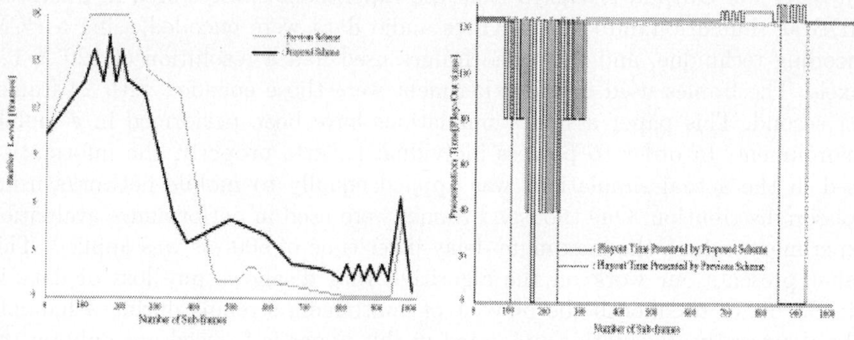

Fig. 3. Buffering Fluctuations at BS

Fig. 4. Playout Time Presented by Proposed Scheme

mobile communications such as small memory footprint and low bandwidth. The proposed scheme's adaptive playout time delivers soft and smooth handoff performance where the entire playout time is maintained within the playout time of the original's.

References

1. Azzedine Boukerche, Sungbum Hong and Tom Jacob, "MoSync : A Synchronization Scheme for Cellular Wireless and Mobile Multimedia System", Proceedings of the Ninth International Symposium on Modeling, Analysis and Simulation of Computer and Telecommunication Systems IEEE 2001
2. D. H. Nam and S. K. Park, "Adaptive Multimedia Stream Presentation in Mobile Computing Environment," Proceedings of IEEE TENCON, 1999.
3. Ernst Biersack, Werner Geyer, "Synchronization Delivery and Play-out of Distributed Stored Multimedia Streams", Multimedia Systems , V.7 N.1 , 70-90, 1999
4. M. Woo, N. U. Qazi, and A. Ghafoor, "A Synchronization Framework for Communication of Pre-orchestrated Multimedia Information," IEEE Network, Jan./Feb. 1994
5. T. D. C. Little, and Arif Ghafoor, "Multimedia Synchronization Protocols for Broadband Integrated Services," IEEE Journal on selected Areas in Comm., Vol. 9, No.9, Dec. 1991.
6. W. Geyer, "Stream Synchronization in a Scalable Video Server Array," Master's thesis, Institute Eurecom, Sophia Antipolis, France, Sept., 1995.
7. Gi-Sung Lee, Jeung-gyu Jee, Sok-Pal Cho, "Buffering Management Scheme for Multimedia Synchronization in Mobile Information System," Lecture Notes in Computer Science Vol. 2660, pp 545-554, June, 2003.

The Development of a Language for Specifying Structure of a Distributed and Parallel Application

Robert Dew, Peter Horan, and Andrzej Goscinski

School of Information Technology
Deakin University, Australia
{rad, peter, ang}@deakin.edu.au

Abstract. A common characteristic of distributed and parallel programming languages is that the one language is used to specify both the organisation of the application, and its functionality. Large distributed and parallel applications will benefit if connectivity and functionality of processes are specified separately. Research is being carried out to develop two new programming languages for the specification and development of distributed and parallel applications. Here, we present a language for specifying process connectivity.

1 Introduction

Existing languages and tools for developing distributed and parallel applications specify connectivity with functionality explicitly or implicitly or provide no support [8]. Furthermore, connectivity is specified in different ways.

Tightly Integrated Connectivity. A parallel Orca application starts with one process and forks to create additional processes [3]. Connectivity of an Orca application is implicit in the functionality of the application, and a specification is not captured formally during design. PVM, a library of low level tools, has a similar characteristic [7,10]. Connectivity relies on send/receive calls and programmer supplied identifiers.

Implied Connectivity. Linda [1] and SR [2] use shared memory and permit virtual memory to be in the scope of several distributed processes. A single chunk of memory is shared by all Linda processes and there is no explicit connectivity. SR uses shared memory but processes share several message queues. Connectivity in OO systems depends on objects accessing data or function members of other objects [9]. CORBA [11] delivers messages from clients to objects and returns results. Clients provide the ORB with object references, only available during run-time by object creation or as the result of a request to a directory service. Thus, connectivity is implied.

No Support for Specifying Connectivity. MPI is a popular message passing library, but not a system for specifying complete distributed applications [6]. This is because MPI lacks both process management and tools for application development.

Explicit Connectivity. HeNCE and CODE are environments specifying parallel applications as graphs which contain nodes to represent procedures and arcs, data and control flow. CODE is similar to HeNCE [4], but arcs represent data flow only.

However, large distributed and parallel applications will benefit if connectivity and functionality of processes are specified separately.

M. Bubak et al. (Eds.): ICCS 2004, LNCS 3036, pp. 397–400, 2004.

The aim of our project is to develop a method and tools for building distributed and parallel applications consisting of concurrently running processes distributed on a network, using asynchronous message passing. Messages are typed arrays of data sent by one process and automatically placed into a buffer in the receiving process [5].

The tasks to be solved to achieve this aim are: specify application connectivity and functionality independently; specify connectivity in a modular fashion to permit re-use of specifications; build applications by combining functional and connectivity specifications, and then compiling to produce object modules; distribute these modules on a cluster; and manage running distributed applications.

2   A New Approach to Programming Distributed Applications

A complete system (to show this work) is distributed over a network. Applications are managed by a run-time system implemented as a single process. Machines are loosely coupled, supported by a file system. Fig. 1 shows two applications of several processes connected by channels, and the run-time management process that builds, compiles, instantiates and manages concurrently executing applications. Connectivity is an essential characteristic of distributed applications and is independent of the internal workings of processes. So, the design of an application can be partitioned into designs of connectivity and functionality.

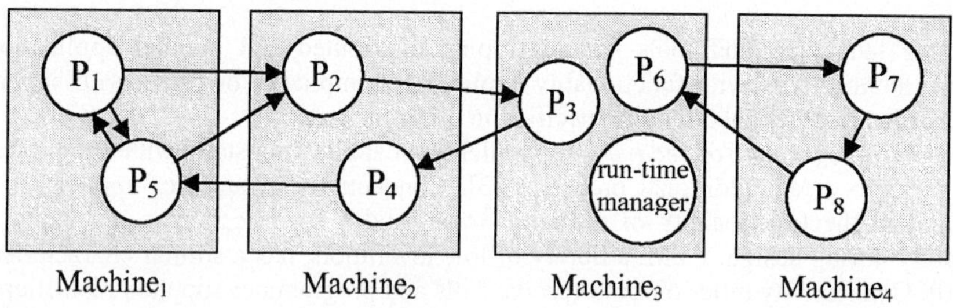

Fig. 1. Two applications and the run-time management process distributed on four machines

2.1   Two Languages for the Specification of Structure and Functionality

The specification of distributed applications has independent parts: structure and functionality, which can be specified by two languages: DAL (Distributed Application Language) and DAPL (DAPL is the subject of another paper), respectively. The programmer uses statements of the structure specification language to specify structure, namely, where processes execute on a network and how they are connected by communication channels. Also, this language supports modularity by allowing existing specifications to be re-used. The structure specification language does not specify functionality; it provides hooks to associate modules of functionality using a second language. These hooks allow the creation of applications that have the same structure but similar or different functionality.

2.2 Designing the DAL Language

The DAL language was designed to specify the structure of a distributed and parallel application, in particular, the connectivity of several processes and where they may execute. The DAL grammar is defined using BNF [REF5]. DAL supports:
- naming the application,
- listing machine names on which the application executes,
- naming processes and listing the order in which they commence execution,
- listing machine names on which a particular process may execute,
- declaring buffers and communication channels for asynchronous message passing,
- declaring the application's structure by associating one end of a communication channel with one process and the other with a receiving buffer of another process,
- re-using prewritten DAL specifications as components of this specification.

2.3 Implementing and Testing the DAL Language

The high level statements written in the DAL language and programs written in DAPL are translated to C code which is then used by a C compiler to generate object code. These generated object modules are used by the run-time management system to instantiate the applications processes on a network.

The DAL parser was tested to ensure that specifications written in DAL conform not only to the DAL grammar but also to the DAL language. Tests were performed by applying the DAL parser to known valid and invalid DAL specifications and comparing the parser's output with expected output. The DAL translator was tested to ensure that its generated C code is the correct C code. Tests were performed by comparing actual C code generated to expected C code that was manually developed.

3   Example

The following DAL specification specifies a simple application called `example` with processes restricted to machines `host_1`, `host_2` and `host_3`. The creation clause dictates that `proc_1` is created last.

```
application example
machines host_1, host_2, host_3
creation proc_1
processes
   process proc_1                      process proc_2
   machines host_1,host_2              machines host_3,host_2
   buffers int buf_1                   buffers int buf_2
   outputs int a:buf_2:proc_2          outputs int b:buf_1:proc_1
   end                                 end
end
```

The first process clause locates process proc_1 on either host_1 or host_2, but not host_3. It has one integer buffer, buf_1, and one integer output, a. Its output clause specifies connection to buffer buf_2 in proc_2.

The second process clause locates proc_2, which is very similar to proc_1, on either host_3 or host_2. It has an integer buffer, buf_2 and an integer output, b. This process is connected to proc_1 via buffer buf_1 of proc_1. This DAL specification implies that, given the proper functionality of both processes, either process can send data to the other process. The data here is simply arrays of integers.

4   Conclusion

A new programming language called DAL has been developed as part of this research. It is used to specify connectivity of distributed and parallel applications with respect to application processes and communication channels, independently of the application's functionality. The DAL language also locates processes on a set of machines, the order in which these processes begin execution, and the types of communications channels and buffers. Existing DAL specifications can be re-used when developing a new DAL specification to form a hierarchical DAL specification, enabling modular development of distributed applications.

References

1.   Ahuja, S., Carriero, N. and Gelernter, D.: Linda and Friends. IEEE Computer, Vol. 19, No. 8 (1986) 26-34
2.   Andrews, G. R. and Olsson, R. A.: The SR Programming Language: Concurrency in Practice. Benjamin-Cummings (1993)
3.   Bal, H.: Programming Distributed Systems. Prentice-Hall (1990)
4.   Browne, J. C., Dongarra, J., Hyder, S. I., Moore, K. and Newton, P.: Experiences with CODE and HeNCE in Visual Programming for Parallel Computing. IEEE Parallel and Distributed Technology, Vol. 3, No. 1 (1994) 75-83
5.   Dew, R. A.: The Development of DAL and DAPL Languages for Building Distributed Applications. Deakin University (2002)
6.   Dongarra, J., Otto, S. W., Suir, M. and Walker, D.: An Introduction to the MPI Standard. CS-95-274 (1995a)
7.   Geist, A., Beguelin, A., Dongarra, J., Jiang, W., Manchek, R. and Sunderam. V.: PVM: Parallel Virtual Machine - A Users' Guide and Tutorial for Networked Parallel Computing. The MIT Press (1994)
8.   Goscinski, A.: Finding, Expressing and Managing Parallelism in Programs Executed on Clusters of Workstations, Special Issue on Network-based Parallel and Distributed Computing, Computer Communications, Vol. 22, No. 11 (1999)
9.   Horan, P.: Eiffel Assertions and the External Structure of Classes and Objects. Journal of Object Technology, Vol. 1, No. 4 (2002) 105-118
      http://www.jot.fm/issues/issue_2002_09/article1
10.   Sunderam, V. S.: PVM: A Framework for Parallel Distributed Computing, Concurrency: Practice and Experience, Vol. 2, No. 4 (1990) 315-339
11.   Vinoski, S.: Distributed Object Computing with CORBA. C++ Report Magazine (1993)

Communication Primitives for Minimally Synchronous Parallel ML

Frédéric Loulergue

Laboratory of Algorithms, Complexity and Logic, Créteil, France
`loulergue@univ-paris12.fr`

Abstract. Minimally Synchronous Parallel ML is a functional parallel language whose execution time can then be estimated and dead-locks and indeterminism are avoided. Programs are written as usual ML programs but using a small set of additional primitives. It follows the cost model of the Message Passing Machine model (MPM). This paper explore two versions of an additional communication function: one uses this small set of primitives, the other one is considered as a primitive and implemented at a lower level.

1   Introduction

Bulk Synchronous Parallel ML (BSML) is a functional parallel language for the programming of Bulk Synchronous Parallel (BSP) [4] algorithms. It is an extension of the ML family of languages by a small set of parallel operations taken from a confluent extension of the λ-calculus. It is thus a deterministic and dead-lock free language.

We designed a new functional parallel language, without the synchronization barriers of BSML, called Minimally Synchronous Parallel ML (MSPML) [1]. As a first phase we aimed at having (almost) the same source language and high level semantics (programming view) than BSML (in particular to be able to use with MSPML work done on type system and proof of parallel BSML programs), but with a different (and more efficient for unbalanced programs) low-level semantics and implementation.

MSPML will also be our framework to investigate extensions which are not suitable for BSML, such as the nesting of parallel values or which are not intuitive enough in BSML, such as spatial parallel composition. We could also mix MSPML and BSML for meta-computing. Several BSML programs could run on several parallel machines and being coordinated by a MSPML-like program.

MSPML Programs are written as usual ML programs but using a small set of additional functions. Provided functions are used to access the parameters of the parallel machine and to create and operate on a parallel data structure. This paper explore the writing of an additional communication function using this small set of primitives. This function could also be considered as a primitive. These two versions are compared.

M. Bubak et al. (Eds.): ICCS 2004, LNCS 3036, pp. 401–404, 2004.

2   Minimally Synchronous Parallel ML

BSPWB, for *BSP Without Barrier*, is a model directly inspired by the BSP model [4]. It proposes to replace the notion of super-step by the notion of m-step defined as: at each m-step, each process performs a sequential computation phase then a communication phase. During this communication phase the processes exchange the data they need for the next m-step.

The parallel machine in this model is characterized by three parameters (expressed as multiples of the processors speed): the number of processes p, the latency L of the network, the time g which is taken to one word to be exchanged between two processes. This model could be applied to MSPML but it will be not accurate enough because the bounds used in the cost model are too coarse.

The Message Passing Machine model (MPM) [3] gives a better bound $\Phi_{s,i}$. The parameters of the Message Passing Machine are the same than those of the BSPWB model. The model uses the set $\Omega_{s,i}$ for a process i and a m-step s defined as:

$$\Omega_{s,i} \;=\; \{j/\text{process } j \text{ sends a message to process } i \text{ at m-step } s\} \cup \{i\}$$

Processes included in $\Omega_{s,i}$ are called "incoming partners" of process i at m-step s. $\Phi_{s,i}$ is inductively defined as:

$$\begin{cases} \Phi_{1,i} = & \max\{w_{1,j}/j \in \Omega_{1,i}\} + (g \times h_{1,i} + L) \\ \Phi_{s,i} = & \max\{\Phi_{s-1,j} + w_{s-1,j}/j \in \Omega_{s,i}\} + (g \times h_{s,i} + L) \end{cases}$$

where $h_{s,i} = \max\{h_{s,i}^{+}, h_{s,i}^{-}\}$ for $i \in \{0, \dots, p-1\}$ and $s \in \{2, \dots, R\}$. Execution time for a program is thus bounded by: $\Psi = \max\{\Phi_{R,j}/j \in \{0, 1, \dots, p-1\}\}$.

The MPM model takes into account that a process only synchronizes with each of its incoming partners and is therefore more accurate. The MPM model is used as the execution and cost model for our Minimally Synchronous Parallel ML language.

There is no implementation of a full Minimally Synchronous Parallel ML (MSPML) language but rather a partial implementation as a library for the functional programming language Objective Caml [2] (using TCP/IP for communications). The so-called MSPML library is based on the following elements:

```
bsp_p: unit->int              apply: ('a->'b) par->'a par->'b par
mkpar: (int->'a)->'a par    get: 'a par->int par->'a par
```

It gives access to the parameters of the underling architecture which is considered as a Message Passing Machine (MPM). In particular, **p()** is p, the static number of processes of the parallel machine. The value of this variable does not change during execution. There is also an abstract polymorphic type **'a par** which represents the type of p-wide parallel vectors of objects of type **'a**, one per process. The nesting of **par** types is prohibited. This can be ensured by a type system .

The parallel constructs of MSPML operate on parallel vectors. Those parallel vectors are created by: **mkpar** so that **(mkpar f)** stores **(f i)** on process i for

i between 0 and $(p-1)$. We usually write **fun pid-¿e** for **f** to show that the expression **e** may be different on each processor. This expression **e** is said to be *local*. The expression (**mkpar f**) is a parallel object and it is said to be *global*.

In the MPM model, an algorithm is expressed as a combination of asynchronous local computations and phases of communication. Asynchronous phases are programmed with **mkpar** and the point-wise parallel application **apply** which is such as **apply(mkpar f)(mkpar e)** stores (**f i**)(**e i**) on process i.

The communication phases are expressed by **get**. Its semantics is given by:

$$\text{get} \langle\, v_0 \,,\dots,\, v_{p-1} \,\rangle\langle\, i_0 \,,\dots,\, i_{p-1} \,\rangle = \langle\, v_{i_0 \% p} \,,\dots,\, v_{i_{(p-1)\%p}} \,\rangle$$

When **get** is called, each process i stores the value v_i in its communication environment. This value can then be requested by a process j which arrived at a later m-step.

3   A New Communication Function

The communication function described below uses functions from the MSPML standard library (http://mspml.free.fr) and some sequential functions. Their semantics follow:

$$\mathbf{ft}\ n_1\ n_2 = [n_1; n_1 + 1; \dots ; n_2] \qquad\qquad \mathbf{procs}() = [0, \dots, p()]$$

$$\mathbf{fill}\ e\ [x_1; \dots ; x_n]\ m = [x_1; \dots ; x_n; \underbrace{e; \dots ; e}_{m-n}]$$

$$\mathbf{nfirst}\ n\ [x_1; \dots\ x_n] = [x_1; \dots ; x_k]\ \text{where}\ k = \min\{m, n\}$$

The **get_list** function is similar to **get** but it takes as second argument a parallel vector of lists of integers rather than a parallel vector of integers. Its semantics is thus given by:

$$\mathbf{get\_list}\langle\, v_0 \,,\dots,\, v_{p-1} \,\rangle\langle[i_1^0; \dots ; i_{k_0}^0], \dots, [i_1^j; \dots ; i_{k_j}^j], \dots, [i_1^{p-1}; \dots ; i_{k_{p-1}}^j]\rangle$$
$$= \langle\, [v_{i_1^0}; \dots ; v_{i_{k_0}^0}] \,,\dots,\, [v_{i_1^j}; \dots ; v_{i_{k_j}^j}] \,,\dots,\, [v_{i_1^{p-1}}; \dots ; v_{i_{k_{p-1}}^j}] \,\rangle$$

The problem is that the lists may not have the same length. Thus we will proceed in two phases. First we define an auxiliary **get_list_sl** function which takes a third argument: the length of the lists which are assumed to have the same length. Then we use it to define the general **get_list** function:

```
let rec get_list_sl vv vl = function     0 -> replicate []
  | n -> let vh=parfun List.hd vl and vt=parfun List.tl vl in
    let vg = get vv vh in
      parfun2 (fun h t->h::t) vg (get_list_sl vv vt (n-1))
let get_list vv vl =
  let vlen=parfun List.length vl in let mlen=reduce max vlen in
  let vl2=apply2 (mkpar fill) vl mlen in
  parfun2 nfirst vlen (get_list_sl vv vl2 (unsafe_proj mlen))
```

The **reduce** function has the following semantics:

$$\mathbf{reduce}\ \oplus \langle\, v_0 \,,\dots,\, v_{p-1} \,\rangle = \langle\, \oplus_{0 \le k < p} v_k \,,\dots,\, \oplus_{0 \le k < p} v_k \,,\dots,\, \oplus_{0 \le k < p} v_k \,\rangle.$$

The **get_list** function implemented using the **get** primitive has MPM cost given by the following formula: $n + t_{\text{reduce}} + n \times (s \times g + L)$ where n is the length of the biggest list, s is the size of each element of the lists and t_{reduce} is the time required to compute the maximum length. For a direct **reduce** function $t_{\text{reduce}} = p + 2 \times (g \times (p - 1) + L)$.

The low level implementation does not need the computation of the maximum length. Furthermore it is possible to use threads for the requests: a process will send sequentially its requests to the processes given by the lists without waiting for the answers. Thus the cost formula is: $L + n_i \times s \times g$ plus an additional overhead introduced by the use of the threads. This formula is given for a process i where n_i is the length of the integer list at process i.

We performed tests to compare the two **get_list** using the following programs:

```
let mshift d l v=get_list v (mkpar(fun i->ft (i+d) (i+d+1)))
let pre n v=get_list v (mkpar(fun i->ft n (n+1+(nmod(i-n)(p())))))
```

We performed the tests on a cluster of 11 Pentium III processors with a fast Ethernet dedicated network. Let summarizes the results where the values are the average (10 tests from 2 to 11 processors were run) efficiency of the high-level implementation with respect to the low-level implementation. For the **mshift** function the ratio range from 20% to 60% for size between 1 and 1000. For sizes greater than 10K the efficiency is almost the same for the two versions. For the **pre** function the ratio is between 15% and 70%. The advantage of the primitive decreases with the size but the asynchronous nature of the function makes the advantage still interesting.

4   Conclusions and Future Work

We have explored how to write communication functions for the Minimally Synchronous Parallel ML language using only the unary **get** communication primitive: the **get_list** function allows to receive messages from several processes. It could also be considered as a communication primitive: the low-level parallel implementation described in this paper follows the execution model of the Message Passing Model. This implementation is more efficient but the proof of correctness is not done, while it is simple for the first version.

References

1. M. Arapinis, F. Loulergue, F. Gava, and F. Dabrowski. Semantics of Minimally Synchronous Parallel ML. In W. Dosch and R. Y. Lee, editors, *SNPD'03*, pages 260–267. ACIS, 2003.
2. Xavier Leroy. The Objective Caml System 3.07, 2003. web pages at www.ocaml.org.
3. J. L. Roda, C. Rodríguez, D. G. Morales, and F. Almeida. Predicting the execution time of message passing models. *Concurrency: Practice and Experience*, 11(9):461–477, 1999.
4. D. B. Skillicorn, J. M. D. Hill, and W. F. McColl. Questions and Answers about BSP. *Scientific Programming*, 6(3):249–274, 1997.

Dependence Analysis of Concurrent Programs Based on Reachability Graph and Its Applications

Xiaofang Qi and Baowen Xu

Department of Computer Science and Engineering, Southeast University,
210096 Nanjing, China
{xfqi, bwxu}@seu.edu.cn

Abstract. This paper presents task synchronization reachability graph(TSRG) for analyzing concurrent Ada programs. Based on TSRG, we can precisely determine synchronization activities in programs and construct a new type of program dependence graph, TSRG-based Program Dependence Graph(RPDG), which is more precise than previous program dependence graphs and solves the intransitivity problem of dependence relation in concurrent programs in some extent. Various applications of RPDG including program understanding, debugging, maintenance, optimization, measurement are discussed.

1 Introduction

As concurrent systems are intensively used day by day, approaches to analyze, comprehend, test and maintain concurrent programs are imperatively demanded. Since determining dependencies between statements is indispensable and crucial to such activities, dependence analysis gradually attracts many researchers to make efforts[1, 2]. Present studies on dependence analysis for concurrent programs are mostly based on concurrent program flow graph. With the model, Krinke and Nanda have computed dependence information of concurrent programs without synchronization[3, 4]. Zhao and Cheng have considered effects of synchronization. However, they analyzed synchronization activities merely by syntactical matching[1, 5]. This processing may produce spurious results leading to inaccurate dependence analysis in most case because some of these synchronization activities are possible to happen while some of them not. We have proposed an adapted MHP(May Happen in Parallel) algorithm to increase the precision of determining synchronization activities[6]. Unfortunately, this approach is still conservative because MHP algorithm only calculates a conservative approximation of MHP statement pairs. Reachability graph, recording all possible reachable states and describing executions of concurrent programs, includes various precise information related to dependence analysis[7, 8]. To improve the accuracy of dependence analysis, we employ reachability graph as the model for analysis and present a new method of dependence analysis for concurrent Ada programs.

M. Bubak et al. (Eds.): ICCS 2004, LNCS 3036, pp. 405–408, 2004.

2 Task Synchronization Reachability Graph

A concurrent Ada program consists of one or more tasks. Each task proceeds independently and concurrently between the points(called by synchronization points) where it interacts with other tasks by inter task synchronization activities during its lifecycle. Statements, like *new, entry call, accept, select, select-else*, indicate such synchronization activities. Each segment extracted between synchronization points is called a task region.

Definition 2.1. Task synchronization graph(TSG) is a labeled directed graph $G_T =$ $<N, E, n_s, F, L>$, where N is the set of nodes corresponding to task regions, $E \subseteq N \times N$, is the set of edges representing synchronization activities, L is the mapping function, n_s is the initial node in which the statement *begin* appears, and F is the final nodes in which the statement *end* appears.

For a given entry E, the starting and ending edges of the entry call(accept) are labeled with E.cs, E.ce(E.as, E.ae) or reduced as E.c, E.a for no accept body. If task s_1, s_2, ..., s_n are activated by parent task p in some activation, the edge is labeled with $(p>(s_1, s_2, ..., s_n))$. The edge labeled with $(m<(d_1, d_2, ..., d_n))$ specifies that master task m is to wait for the terminations of task $d_1, d_2, ..., d_n$ before its termination.

TSG emphatically describes synchronization and concisely represents the execution for single task. Task synchronization reachability graph gives the behavior of an entire concurrent program and is constructed from the TSGs of the tasks that compose the program. Suppose that a concurrent Ada program is composed of k tasks(the main program is processed as the first task) and the TSG of the *i*th task is denoted by TSG_i $= <N_i, E_i, n_s^i, F_i, L_i>$ (1<= i <= k), then a TSRG-node m is a k-tuple of TSG-nodes (m[1], m[2], ..., m[k]) where $m[i] \in N_i \cup \{\bot\}$, $\bot$ indicates the corresponding task is inactive.

Definition 2.2. Task synchronization reachability graph(TSRG) is a labeled directed graph $G_R = <M, E, L, m_s, F>$, where M is the set of TSRG-nodes, one for indicating an execution state of the program, $E \subseteq M \times M$, is the set of edges, each corresponding to one possible inter task synchronization activity, L is the mapping function, m_s is the initial node, $m_s = (n_s^1, \bot, ..., \bot)$, F is the final nodes representing final state. There is an edge from m to m′ iff any of the following conditions holds(i, j, $l =$ 1, 2, ..., k) where k is the number of tasks:

(1) $\exists i ((m[i], m'[i]) \in E_i \wedge L(m[i], m'[i]) = i>(s_1, s_2, ..., s_n) \wedge$
      $(\forall j(j = s_1, s_2, ..., s_n) m[j] = \bot \wedge m'[j]=n_s^j))$ $(l \neq i, j, m[l] = m'[l])$

(2) $\exists i ((m[i], m'[i]) \in E_i \wedge L(m[i], m'[i]) = m<(d_1, d_2, ..., d_n) \wedge$
      $(\forall j(j = d_1, d_2, ..., d_n) m[j] \in F_j \wedge m'[j]= \bot))$ $(l \neq i, j, m[l] = m'[l])$

(3) $\exists i \exists j ((m[i], m'[i]) \in E_i \wedge (m[j], m'[j]) \in E_j \wedge$
      $((L(m[i], m'[i]) = E.cs \wedge L(m[j], m'[j]) = E.as) \vee$
      $(L(m[i], m'[i]) = E.ce \wedge L(m[j], m'[j]) = E.ae) \vee$
      $(L(m[i], m'[i]) = E.c \wedge L(m[j], m'[j]) = E.a)))$ $(l \neq i, j, m[l] = m'[l]).$

In definition 2.2, labels are similar to those in definition2.1 except that the starting(ending) of rendezvous are labeled with E.s(E.e) or reduced as E. Three conditions correspond to task activation, waiting for termination and rendezvous respectively. By TSRG, we can precisely determine synchronization activities and get more accurate MHP statement pairs.

3 TSRG-Based Program Dependence Graph and Its Applications

Considering that TSRG provides all global reachable states and one statement may appears simultaneously in more than one TSRG-nodes which may reside in different control flow branches of TSRG, then we propose a new paradigm of dependency between one statement binding with its TSRG-node and another.

Definition 3.1. TSRG-based program dependence graph (RPDG) of a concurrent Ada program is a directed graph $G_D = <M, S, MS, E>$, where M is the set of TSRG-nodes, S is the set of statements, $MS = M \times S$, is the set of RPDG-nodes, $E \subseteq MS \times MS$, is the set of edges, $E = \{(<m_1, s_1>, <m_2, s_2>) \mid Dep(<m_1, s_1>, <m_2, s_2>), Dep \in \{DepDc, DepAc, DepRc, DepCc, DepVc, DepSd, DepCd\}\}$.

In definition 3.1, various dependencies can be primarily classified into control and data dependencies. DepDc, DepAc, DepRc, DepCc, DepVc represent direct, activation, rendezvous, competence, virtual control flow dependencies respectively. Direct control flow dependency exists in task regions, similar to control dependency appearing in sequential programs. Activation, rendezvous, competence control dependency exist between task regions and are induced respectively by task activation, rendezvous, competence for a same entry. Virtual control dependency contributes to keep the connectivity of intra task control dependency on the border between task regions. Since statements of definition and reference on variables may execute concurrently or sequentially in some execution, we classify data dependencies into concurrent and sequential data dependency, denoted by DepSd, DepCd.

Dependencies in RPDG possess special property in transitivity, which do not appear in traditional program dependence graphs(PDG) where dependencies are defined between statements. Below, we analyze two main cases in concurrent programs where intransitive dependency happens in PDG:

(1) When multiple tasks compete for one resource (e.g. *accept* statement), only one of them can occupy and consume it. This will lead to several exclusive program segments from those tasks taking part in the competence, i.e., only one of the segments can be executed in one execution of the program. Obviously, it's impossible that there exists dependency among these exclusive segments.

(2) From one statement s_1 in $task_i$, the dependency propagates back into another statement s_2 in $task_i$ by inter task dependency sequence. When s_1 and s_2 appear in different branches of control flow or s_1 always executes before s_2 in any execution of $task_i$, s_1 is impossible to indirectly depend on s_2.

However, for such two cases imprecise transitivity of dependency sequence may be hindered in RPDG in some extent. In (1), there must exist multiple TSRG-nodes representing each competence and those exclusive segments will reside in different branches in TSRG. If s_1, s_2 are such exclusive statements and $Dep(<m_2, s_2>, <m_2, s>)$, $Dep(<m_1, s>, <m_1, s_1>)$ holds, then there's no transitive dependency between s_1 and s_2 because s appears respectively in two different TSRG-nodes m_1 and m_2 residing in different branches. Similar situation may happen in the former case of (2). Although the dependency is intransitive in the latter case of (2), we can say dependencies in RPDG is transitive in most of cases.

In addition to having better transitivity than traditional PDG, dependence analysis in RPDG are more accurate because of precisely detecting synchronization activities and MHP pairs by TSRG. Thus, RPDG may be used in various software engineering activities including program understanding, slicing, debugging, optimization, complexity measurement, maintenance and etc. Given a concurrent Ada program consisting of k tasks, n statements including c *entry call* and *accept* statements, the cost of RPDG is $O(n(2c/k+3)^k)$ in worst case.

4 Conclusions

Based on task synchronization reachability graph, we have constructed a new type of program dependence graph – TSRG-based Program Dependence Graph(RPDG) for concurrent Ada programs. RPDG is more precise than previous program dependence graphs and solves the intransitivity problem of dependence relation in concurrent programs in some extent. Nevertheless, we have mainly consider several primary aspects of task mechanism. Some constructs, such as communications by shared variables, protected object, arrays of tasks and etc, have not been discussed in detail. In the future work, we will promote our research more systematically and extend to other concurrent languages to facilitate analysis for more concurrent programs.

References

1. Cheng, J.: Task dependence nets for concurrent systems with Ada 95 and its applications. In: ACM TRI-Ada International Conference, St. Louis, Missouri, USA: ACM Press (1997) 67–78
2. Tip, F.: A Survey of Program Slicing Techniques. Journal of Programming Languages, Vol. 3 (1995) 121–189
3. Krinke, J.: Static slicing of threaded program. ACM SIGPLAN Notices, Vol. 7 (1998) 35–42
4. Nanda, M.G., Ramesh, S.: Slicing concurrent programs. ACM SIGSOFT Software Engineering Notes, Vol. 25 (2000) 180–190
5. Zhao, J.: Multithreaded dependence graphs for concurrent Java programs. In: International Symposium on Software Engineering for Parallel and Distributed Systems, Los Angeles, California, USA: IEEE CS press (1999)13–23
6. Chen, Z.Q., Xu, B.W.: An Approach to Analyzing Dependence of Concurrent Programs. Journal of Computer Research and Development, Vol. 39 (2002) 159–164
7. Qi, X.F, Chen, Z.Q., Xu, B.W.: A Petri Net Representation of Concurrent Ada Program and Its Application for Communication Slice. Journal of Nanjing University, Vol. 38 (2002) 37–42
8. Dwyer, M. B., Clarke, L. A.: A Compact Petri Net Representation and Its Implication for Analysis. IEEE Transaction on Software Engineering, Vol. 22 (1996) 794–811

Applying Loop Tiling and Unrolling to a Sparse Kernel Code

E. Herruzo[1], G. Bandera[2], and O. Plata[2]

[1] Dept. Electronics, University of Córdoba, Spain
[2] Dept. of Computer Architecture, University of Málaga, Spain

Abstract. Code transformations to optimize the performance work well where a very precise data dependence analysis can be done at compile time. However, current compilers usually do not optimize irregular codes, because they contain input dependent and/or dynamic memory access patterns. This paper presents how we can adapt two representative loop transformations, tiling and unrolling, to codes with irregular computations, obtaining a significant performance improvement over the original non-transformed code. Experiments of our proposals are conducted on three different hardware platforms. A very known sparse kernel code is used as an example code to show performance improvements.

1 Introduction

Over the years, the ratio between the main memory latency and processor cycle time has been increasing. Computer architects have proposed several hardware mechanisms that reduce the impact of the memory latency problem: lockup-free caches, prefetching, out-of-order execution, etc... The efficiency of architectural improvements depends on the compiler ability to change the structure of programs for taking full advantage of them.

When optimizing a program an important performance improvement will come from optimizing the most time-consuming code regions, that is, repetitive sentence blocks. In this way, a number of compiler strategies have been developed to enhance the performance: (1) strategies that change the original data layout of the array variables [4,3]; (2) strategies based on loop restructuring transformations that reduce the number of executed instructions and/or change the order in which statements are executed [1]. Some works also attempt to integrate both strategies as a single algorithm [2].

Most of the compiler optimizations were designed for regular computations [5]. If access patterns are input or code conditional dependent, compiler optimizations are much more difficult to decide and apply. There are many special memory access patterns, that appear frequently in irregular applications, where no data dependence analysis is needed in order to apply some optimization transformations. The majority of compilers, however, do not take into account these special situations, loosing the opportunity of obtaining a better object code.

This paper analyzes one of the most important special irregular access patterns, resulting from the multiplication of a sparse matrix by a dense array

M. Bubak et al. (Eds.): ICCS 2004, LNCS 3036, pp. 409–412, 2004.

(spMxV). We will show that various commercial compilers, from Compaq, Silicon Graphics and Cray, do not optimize codes with this kind of computational structure. However, manually applying powerful optimization techniques, a significant performance improvement is obtained.

2 Optimizing Sparse Codes

Conventional data dependence techniques are not usually applicable to irregular codes, due to the variant nature of the reference patterns. This is one of the main reasons why compilers usually cannot decide to apply loop optimizations and then the obtained object code is frequently sub-optimal. In this section we will analyze two widely used loop transformation techniques, *loop tiling* and *loop unrolling*, in an irregular kernel code example (spMxV). We have selected these techniques because we have found that the powerful loop tiling method slightly improve the performance of the object code, while, a much simpler technique, loop unrolling, is able to significantly reduce the execution time of the code.

To simplify our analysis, from now on we will focus our attention on the sparse computation spMxV and compressed data storages. In this paper, we have considered the **CRS** (*Compressed Row Storage*) and **CCS** (*Compressed Column Storage*) formats, which do not restrict our range of application nor store any unnecessary element. Basically, CRS (CCS) permits to represent the sparse matrix using three arrays (DA, CO and RO). For CRS, the first array stores the non-zero values of the matrix as they are traversed in a row-wise fashion, the second array retains the column index of each non-zero element in A, and the latter array marks the beginning of the data for each matrix row. For CCS the elements are stored as traversed by columns.

2.1 Sparse Tiling

Loop tiling is a well known loop transformation that can be used automatically by the compiler to create block algorithms and to exploit locality. It alters the way in which individual iterations are executed so that iterations from outer loops are carried out before completing all the inner loop iterations. The use of tiling with sparse matrices is not as easy. Compressed representations make difficult both the selection of the tile size and the code transformation to divide the iteration space.

When using the CRS format, the problem is that while the column coordinate of a non-null $DA(j)$ is stored in the array entry $CO(j)$, the row number is not stored in $RO(j)$. As the RO array stores a list of indices pointing to some compressed array cells, the block tiling will require the modification of this array to visit the entries by blocks. This transformation can be done during the matrix reading. Fig. 1.a sketches the tiled code for the spMxV kernel. The benefit in performance on the spMxV kernel is expected to be not very high. While arrays Y, DA, RO and CO are traversed linearly, array X is accessed more randomly. Thus, tiling can only improve locality in that array X, but with the side effect of reducing locality access to array Y.

```
                                              rr1 = 1
                                                DO i = 1, n
                                                  rr2 = RO(i+1)
                                                  rr3 = rr2-rr1
  rr1 = 1                                         rr4 = 0
    DO j = 1, n/B                                 DO WHILE (rr4 .LT. rr3)
      DO i = 1, n                                   Y(i) = Y(i)+DA(rr1+rr4)*X(CO(rr1+rr4))
        rr2 = NRO(i+1+n*(j-1))                      rr4 = rr4+1
        rr3 = rr2-rr1                               DO WHILE (rr4 .LT. rr3-5)
          DO rr4 = 0,rr3-1                             Y(i) += DA(rr1+rr4)*X(CO(rr1+rr4))
            Y(i)+= NDA(rr1+rr4)*X(NCO(rr1+rr4))       Y(i) += DA(rr1+rr4+1)*X(CO(rr1+rr4+1))
          ENDDO                                       Y(i) += DA(rr1+rr4+2)*X(CO(rr1+rr4+2))
        rr1 = rr2                                     Y(i) += DA(rr1+rr4+3)*X(CO(rr1+rr4+3))
      ENDDO                                           Y(i) += DA(rr1+rr4+4)*X(CO(rr1+rr4+4))
    ENDDO                                             rr4 = rr4+5
                                                    ENDDO
                                                  ENDDO
                                                  rr1 = rr2
                                                ENDDO
                  (a)                                                  (b)
```

Fig. 1. (a) SpMxV after the sparse tiling using the modified CRS representation; (b) SpMxV after the sparse loop unrolling of size 5 ($\delta = 5$)

2.2   Sparse Unrolling

This technique cannot be directly applied in sparse compressed representations, because the number of non-null entries per dimension is not known at compile-time. In the spMxV code the inner loop uses the index array RO to traverse the non-nulls of a matrix row. As the content of this array is unknown during the compilation, the optimal unrolling step should be selected depending on matrix features, as matrix homogeneity. As this kind of information is not known by the compiler, we can do it manually, as no data dependence relation is violated by the transformation in any case.

Fig. 1.b shows the spMxV kernel code with the inner loop unrolled by an example factor of $\delta = 5$. Two **DO WHILE** appear instead of the inner loop j. The inner while loop iterates a block of 5 consecutive sentences of the original j loop, while the outer while loop is used to execute the residual number of sentences. An important parameter of this transformation is the selection of the unrolling factor δ. Its value depends on sparse matrix properties, as its size, the sparsity pattern and the amount of non-null entries by row. Other facts are related to properties of the machine processor, as the amount of internal CPU registers or its ILP (Instruction Level Parallelism) capacity.

3   Experimental Results and Conclusions

In this section we present some experimental results for the codes presented before, conducted on different hardware platforms and using different compilers. We only discuss here results for the unrolling transformation, because sparse loop tiling for the spMxV kernel is predicted to have a small effect on its performance.

We have evaluated the unrolled spMxV code on three platforms: a Digital AlphaServer 4100 with a 400 MHz Alpha 21164 processor; a SGI Origin2000 with a 195 MHz MIPS R10000 processor; and a Cray T3E with a 450 MHz Alpha 21164 processor. For the purposes of an experimental validation, we run

Matrix			Alpha 21164			R10000			Cray T3E		
Name	Size	Density	δ	spMxV	**Improv.**	δ	spMxV	**Improv.**	δ	spMxV	**Improv.**
Psmigr3.rua	3140x3140	5.51%	14	22.80	**35%**	15	91.64	**22%**	6	18.82	**62.7%**
Fidapm37.rua	9152x9152	0.91%	14	32.51	**32%**	14	116.01	**45%**	6	19.70	**73.7%**
Beaflw.rra	497x507	21.2%	11	1.66	**99%**	12	13.48	**34%**	6	1.27	**79.4%**
Af23560.rua	17281x17281	0.18%	9	20.52	**23%**	3	363.1	**11%**	6	13.03	**34.7%**
S3dkq4m2.dat	90449x90449	0.03%	12	114.9	**30%**	3	1371.2	**8%**	4	54.61	**29.7%**

Fig. 2. Improvement of the spMxV kernel on the 3 platforms with different sparse matrices and different δ values (time in milliseconds)

the sparse Conjugate Gradient (CG) algorithm, the oldest, best known, and most effective of the non-stationary iterative methods for the solution of symmetric positive definite systems. Since the features of the input matrix become paramount in the algorithm behavior, we have selected a set of very different matrices from the Harwell-Boeing Collection (HB).

Fig. 2 shows the execution times for the spMxV kernel code and the improvement of the sparse unrolling, for different HB sparse matrices and using different compilers. The unrolling factor δ in table corresponds to the value with the best performance improvement. This value has a large variation range because it depends strongly on input sparse data (sparsity pattern).

Our main conclusions from this work are that tiling requires a high cost preprocessing stage to modify the storage format, an extra memory cost (NRO vector is bigger than RO), and the locality exploited in the source vector X is missed in destination (vector Y). We can not exploit temporal locality because the sparse matrix and the dense vector are linearly stored, and then does not exist any data reusage in cache line. Another conclusion is that unrolling reduces the loop overhead and obtain in most real situations a significant improvement in performance. The selection of the unrolling factor (δ) depends on the matrix sparsity pattern and size and the processor internal characteristics.

References

1. S. Carr, K.S. Mckinley and C. Tseng, *Compiler Optimizations for Improving Data Locality*, 6th International Conference on Architectural Support for Programming Languages and Operating Systems, San Jose, CA, October 1994.
2. M. Kandemir and J. Ramanujam, *Data Relaton Vectors: A New Abstraction for Data Optimizations*, IEEE Transactions on Computers, Vol. 50, No. 8, August 2001.
3. M. O'Boyle and P. Knijnenburg, *Integrating Loop and Data Transformations for Global Optimizations*, IEEE International Conference on Parallel Architectures and Compilation Techniques, Paris, France, October 1998.
4. G. Rivera and C-W. Tseng, *Data Transformations for Eliminating Conflict Misses*, ACM SIGPLAN Conference on Programming Language Design and Implementation, Montreal, Canada, June 1998.
5. M. Wolfe, *High Performance Compilers for Parallel Computing*, Addison–Wesley Pub., Redwood City, CA, 1996.

A Combined Method for Texture Analysis and Its Application

Yongping Zhang[1] and Ruili Wang[2]

[1]Bioengineering Institute, The University of Auckland
Level 6, 70 Symonds St., Auckland, New Zealand
Zhangyp1963@yahoo.com
[2]Institute of Information Sciences and Technology, Massey University
Private Bag 11 222, Palmerston North, New Zealand
R.Wang@Massey.ac.nz

Abstract. In this paper, a rotational invariant feature set is introduced for texture classification, based on wavelet transformation in combination with co-occurrence probabilities. Using this combined method, through wavelet decomposition and reconstruction, an approximation image and a new details image are generated. Beside of using the statistic approximation and the new details respectively, the joint distribution of the original and the new detail image is computed, and seven novel digital features are derived from the joint probability. By combination with a MLP neural network, our method has successfully applied to pollen discrimination. In experiments with sixteen types of airborne pollen grains, more than 95 percent pollen images are correctly classified.

1 Introduction

Texture feature is one of the most widely used visual features in pattern recognition and computer vision. Texture contains important information about the surface structure of objects and their relationship to the surrounding environment [1].

Many studies have shown that use of wavelet transforms for texture description can achieve good classification performance [2-10]. Smith and Chang used the statistic feature of subbands as the texture representation [4]. Gross et al. characterised texture by using wavelet transform in combination with KL expansion [5].

A combined approach of wavelet transform with co-occurrence matrix was also carried out by Thyagarajan et al. in [6]. In this paper, a novel co-occurrence matrix is introduced for texture description based on wavelet transforms, such matrix is corresponding to the joint distribution of the original greyscale image and the details derived from wavelet transforms.

Distinguishing the pollen species through the analysis of their surface images has become a new application field of computer vision [11-16].

In the present work, we use our combined algorithm to extract texture features of surface images of pollen grains and use a multilayer perceptron (MLP) neural network to classify the extracted feature vectors.

M. Bubak et al. (Eds.): ICCS 2004, LNCS 3036, pp. 413–416, 2004.

2 Wavlet-Based Feature Set

In this section, we introduce the combined features. First we define the details reconstruction and then we introduce a novel method for feature extraction.

On the each level of orthogonal wavelet decomposition, one approximation image and three details images can be got; those details images are horizontal, vertical and diagonal details respectively. For image analysis, the four subband images are usually assumed to be independent. In the present study, we reconstruct a new details image using the three details subbands, and compute the corresponding statistic features of the approximation sunband and the new details image, and to compute the joint distribution of the original image and the new details image. The scheme for feature extraction as shown in Fig. 1.

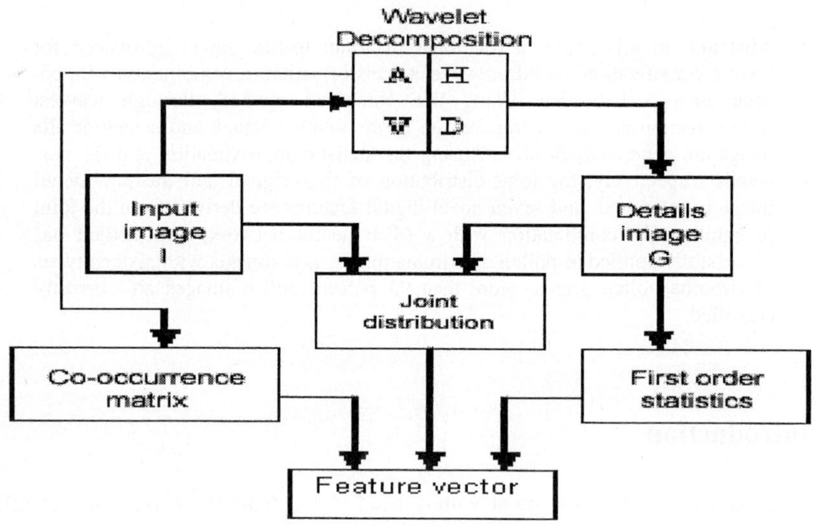

Fig. 1. The scheme of texture feature extraction based on wavelet decomposition

To form effective feature vector, we compute the joint distribution of the original image and its detail image. First of all, like the gray level co_occurrence matrix, a co-occurrence matrix Q is calculated as:

$$q(i, j) = \frac{\#\{p \in I \mid I(p) = i, G(p) = j\}}{\#I}$$

where G presents the details image reconstructed from the wavelet decomposition of the input image I, which is re-quantized (discretized) to certain grey levels.

Based on this co-occurrence matrix Q, we compute the following seven features: small detail emphasis (SDE), large detail emphasis (LGE), gray distribution non-uniformity (GDNU), details distribution non-uniformity (DDNU), energy (ENE), Entropy (ENT) and inverse difference moment (IDM). We can call the above seven features as grey-detail co-occurrence matrix (GDCM) features.

3 Pollen Image Classification

We have applied the new texture features to discriminate the airborne pollen grains. The classification results verify the robustness of our combined method by providing high percentage of correct classification for texture images obtained from sixteen types of pollen grains.

In this research, sixteen types of airborne pollen grains are considered and their typical surface images shown in Fig. 2.

We computed 15 features for each sample using our method. For feature selection, the 15 features were extracted from 15 samples of each type of pollen grains, and the ratio of interclass/intraclass distance was used. The resultant 7 features: GLCM features ENT and IDM, GDCM features GDNU, ENE and IDM, detail features mean and standard deviation are selected to represent pollen texture.

For classification of pollen images, the MLP neural network of 7×15×16 was employed, and back-propagation algorithm was used for network training. In this experiment, 95.4 percent of pollen images are correctly classified.

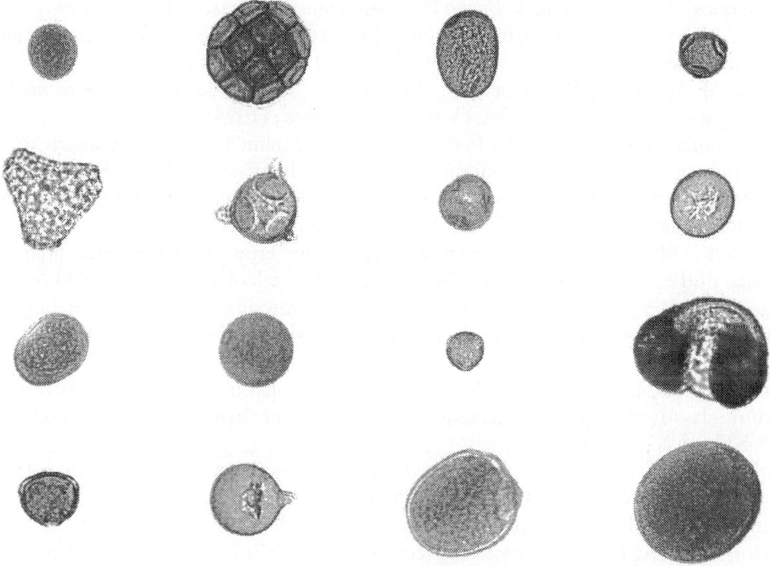

Fig. 2. The pollen types analysed in this research. From left to right and top to bottom: AC (*Agrostis capillaris*), AD(*Acacia dealbata*), AP(*Alopecuris pratensis*), BP(*Betula pendula*), CD(*Cyathea dealbata*), CL(*Corylus*), CR(*Coprosma robusta*), CT(*Cedar tree*), FA(*Festuca arundibaceae*), PM(*Phalaris minor*), PO(*Platanus orientalis*), PR(*Pinus radiata*), QR (*Quercus robur*), SG(*Sequiadendron gigantea*), TA(*Triticum aestivum*), ZM(*Zea mays*)

4   Conclusions

A novel combined feature set has been developed and evaluated for texture classification.

The proposed method also has been successfully applied to identification of pollen grains. By feature selection, seven features are used as texture descriptors to present pollen images, and the MLP neural network is used to discriminate features of sixteen type pollen grains. A classification rate of more than 95 percent is achieved.

References

1. Ojala, T., Pietikainen, M.: A comparative study of texture measures with classification based on feature distributions, Pattern Recognition, Vol. 29(1), (1996) 51-59.
2. Randen, T., HusØy, H.J.: Filtering for texture classification: A comparative study, IEEE Transactions on Pattern Analysis and Machine Intelligence, Vol. 21(4), (1999) 291-310.
3. Reed, T.R., Buf, J.M.H.: A review of recent texture segmentation and feature extraction techniques", Computer Vision, Image Processing and Graphics, Vol. 57(3) (1993) 359-372.
4. Smith, J.R., Chang, S.: Transform features for texture classification and discrimination in large image databases, In Proc. IEEE Int. Conf. on Image Proc. (1994).
5. Gross, M. H., Koch, R., Lippert, L., Dreger. A.: "Multiscale image texture analysis in wavelet spaces, In Proc. IEEE Int. Conf. on Image Proc. (1994).
6. Thyagarajan, K.S., Nguyen, T., Persons, C.: A maximum likelihood approach to texture classification using wavelet transform, In Proc. IEEE Int. Conf. on Image Proc.(1994).
7. Do, M.N., Vetterli, M.: Texture similarity measurement using Kullback-Leibler distance on wavelet subbands, In Proc. of IEEE Int. Conf. on Image Proc. (2000).
8. Ma, W.Y., Manjunath, B.S.: A texture thesaurus for browsing large aerial photographs, Journal of the American Society for Information Science Vol. 49(7), (1998) 633-648.
9. Manjunath, B.S., Ma, W.Y.: Texture features for browsing and retrieval of large image data, IEEE Transactions on Pattern Analysis and Machine Intelligence, Vol. 18, (1996) 837-842.
10. Charalampidis, D., Kasparis, T.: Wavelet-based rotational invariant roughness features for texture classification and segmentation, IEEE Trans. on Image Proc., Vol. 11(8), (2002) 825-837.
11. Stillman, E.C., Flenley, J.R.: The needs and prospects for automation in palynology, Quaternary Science Reviews Vol. 15 (1996) 15.
12. Fountain, D.W.: Pollen and inhalant allergy, Biologist Vol.49 (1), (2002) 5-9.
13. Trelor, W.J.: Digital image processing techniques and their application to the automation of palynology, Ph. D. Thesis (1992), University of Hull, Hull UK.
14. Li, P., Flenley J.R.: Pollen texture identification using neural networks, Grana Vol. 38 (1999) 59-64
15. Langford, M., Taylor, G.E., Flenley, J.R.: Computerised identification of pollen grains by texture analysis, Review of Palaeobotany and Palynology, Vol. 64, (1990) 197-203
16. Ronneberger, O.: Automated pollen recognition using grey scale invariants on 3D volume image data. 2nd European Symposium on Aerobiology. (2000) p.3. Vienna, Austria.
17. Mallat, S.: A Wavelet Tour of Signal Processing, Academic Press (1998), San Diego
18. Pandya, A.S., Macy, R.B.: Pattern Recognition with Neural Networks in C++. CRC and IEEE Press, (1996) Florida.

Reliability of Cluster System with a Lot of Software Instances

Magdalena Szymczyk and Piotr Szymczyk

Department of Control Systems, University of Mining and Metallurgy, al. Mickiewicza 30
30-059 Kraków, Poland
{mszm, pszm}@ia.agh.edu.pl

Abstract. This paper presents model of complex fault – tolerant system with multiple software instances and hardware clusters. We want to show influence of number software and hardware components on overall reliability of the system. Previously, other models have been developed only for software or hardware systems. Our model assumes that failure of each component is statistically independent.

1 Introduction

Today computers are much more sophisticated then their earlier version. They are made up of software and hardware components.

These days, in order to ensure effective performance of the computer, software and hardware need to function with considerable amount of reliability. There are number of methods to improve reliability of software products: reliable software design, fault-tolerance design, formal methods and testing [1] [2] [3].

Reliability is the probability that system works properly by a certain time. Reliability of serial systems where failures of components in system are mutually independent equals with the probability that all components are working [4] [5].

If the probability that a component A_i is operational equals $P(A_i)$, and its reliability equals $R_i=P(A_i)$, then reliability of the whole system (probability that system is functioning properly) with the assumption of independency of failures equals:

$$R_s = \prod_{i-1}^{n} R_i \tag{1}$$

For parallel system reliability R_p equals to:

$$R_p = 1 - \prod_{i=1}^{n} (1 - R_i) \tag{2}$$

under the assumption of independence of failure events.

M. Bubak et al. (Eds.): ICCS 2004, LNCS 3036, pp. 417–420, 2004.

2 Optimization of the Number of Software and Hardware

Let's make some assumption in our model. A repeated task and each repetition of task are independent. We assume that software task has a fixed probability of failure. Software failures of different version of code are independent, but still have the same of reliability. For this architecture we have such RBD diagram [5]:

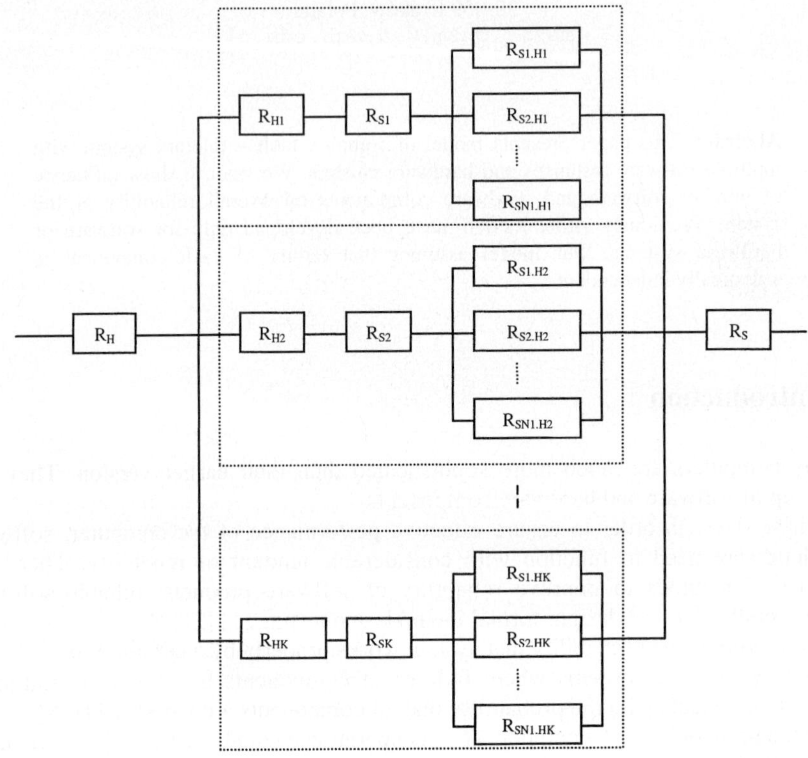

Fig. 1. RBD diagram for considered system

Notation:

K – number of hardware clusters in systems
N – number of software instances on j-th cluster node
H_j – j-th hardware node
S_i,H_j – i-th software instance on j-th cluster node
R_H – common reliability of hardware
R_{Hj} – reliability of each hardware node
R_{Sj} – reliability of common software on j-th hardware node
$R_{Si,Hj}$ – reliability of each software instance on j-th hardware node
R_S – reliability of common software

Reliability of the system is given by:

$$R = R_H * \left(1 - \prod_{j=1}^{K} \left(1 - R_{H_j} * R_{S_j} * \left(1 - \prod_{i=1}^{N_j} \left(1 - R_{S_i,H_j} \right) \right) \right) \right) * R_S \tag{3}$$

from this equation , if all R_{Hj} are equal, R_{Sj} and $R_{Sj,Hj}$ too, we get:

$$R = R_H * R_S * \left(1 - \left(1 - R_{H_K} * R_{S_N} * \left(1 - \left(1 - R_{S_N,H_K} \right)^N \right) \right)^K \right) \tag{4}$$

Results can be presented after some calculations in a form of charts. For simplicity and better visualization of influence R_{HK} and $R_{SN,HK}$ on R it is assumed that R_H, R_S, and R_{SN} are equal to 1. Values R_H, R_S and R_{SN} in all cases we have a direct influence on R, so it should have the highest value possible.

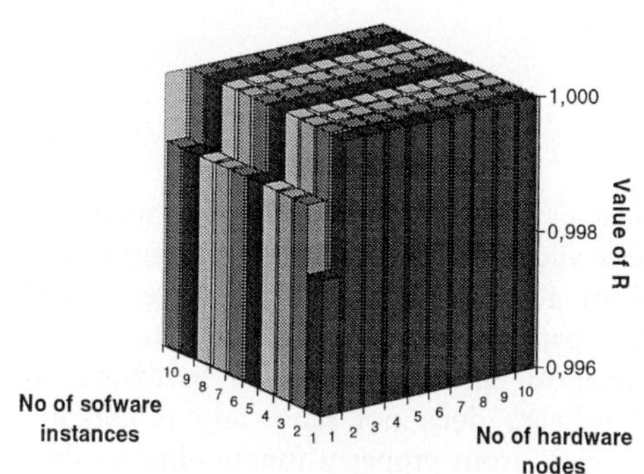

Fig. 2. Charts for R_H=1, R_S=1, R_{HK}=0.999, R_{SN}=1, $R_{SN,HK}$=0.999

From Fig. 2 it follows that for software instances greater than 4 and for the same numbers of nodes the value of R does not change at all. So if we give greater numbers of software and clusters, R doesn't change dramatically. When R_{HK} and $R_{SN,HK}$ are equal, increasing number of hardware nodes to 2 produce good result on R.

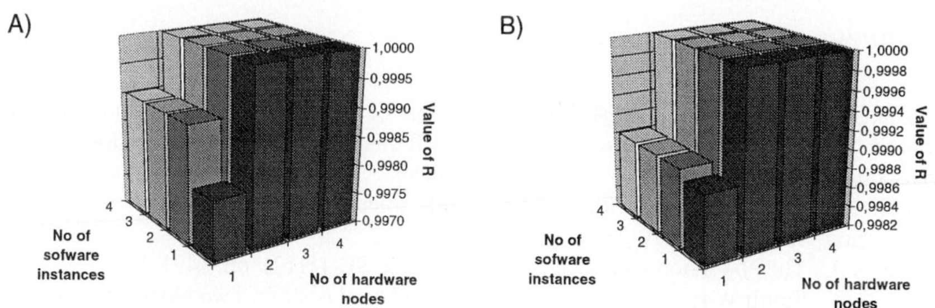

Fig. 3. Charts for R_H=1, R_S=1 R_{SN}=1, R_{HK}=0.999   A) $R_{SN,HK}$=0.999;   B) $R_{SN,HK}$=0.9999;

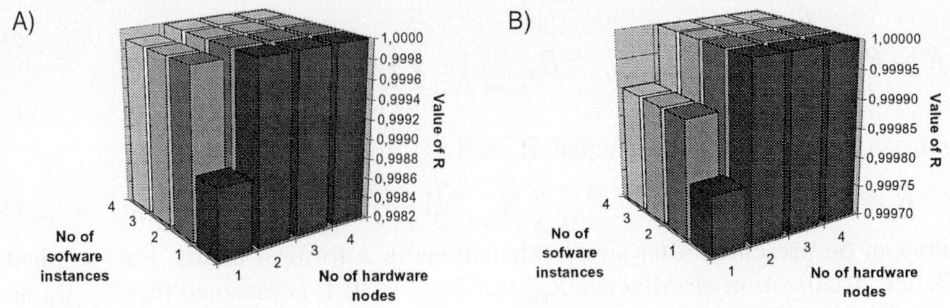

Fig. 4. Charts for $R_H=1$, $R_S=1$ $R_{SN}=1$, $R_{HK}=0.9999$ A) $R_{SN,HK}=0.999$;   B) $R_{SN,HK}=0.9999$;

3   Conclusions

We consider dependencies in the system reliability with respect to the number of hardware cluster nodes and software instances for fault-tolerant systems. Analyzed system is composed of K clusters and N software instances. For such system reliability R is most sensitive on changes from 1 to 2 in hardware nodes in all considered cases. The most optimal number of hardware nodes equals 2 because further increase in hardware does not have any remarkable influence on R. In principle, software depends from properly functioning hardware so if we have very high $R_{SN,HK}$ values with relation to R_{HK}, increase in software number dose not produce the same effects as increase in node number. It can raise reliability R if $R_{SN,HK} >> R_{HK}$ (Fig.3 B) and software instances changes from 1 to 2. Such solution can be economically justified in the case when the implementation of two different instances of software is significantly less than the cost of buying two hardware nodes and $R_{SN,HK} >> R_{HK}$. Increasing above two software instances and two hardware nodes don't have significant influence on the system reliability. It is necessary to focus our attention on obtaining high level of values $R_{SN,HK}$ and R_{HK}.

References

1. Kim K.H. and Welch H.O.: Distributed execution of recovery blocks: an approach for uniform treatment of hardware and software faults in real-time applications, IEEE Transaction on Computers, 38 (5): pp. 626-636, 1989.
2. Randell B.: System Structure for software fault tolerance. IEEE Transaction Software Engineering, SE-1(2): pp. 220-232, 1975.
3. Avizienis A.: The N-version approach to fault-tolerant, SE-11(12): pp. 1491-1501, 1985.
4. Hunter S.W., Smith W.E.: Availability Modeling and Analysis of a Two Node Cluster.
5. Bobbio A. and Trivedi K.: Reliability Theory and Methods

A Structural Complexity Measure for UML Class Diagrams*

Baowen Xu[1,2], Dazhou Kang[1], and Jianjiang Lu[1,2,3]

[1]Department of Computer Science and Engineering, Southeast University, Nanjing, 210096, China
[2]Jiangsu Institute of Software Quality, Nanjing, 210096, China
[3]PLA University of Science and Technology, Nanjing, 210007, China
bwxu@seu.edu.cn

Abstract. UML class diagrams constitute a key artifact in the conceptual modeling phase and their quality can have a significant impact on the quality of the system. The structural complexity measure is one of the most important measures to evaluate the quality of a UML class diagram. This paper uses weighted class dependence graphs to represent a given class diagrams, and then presents a structure complexity measure for the UML class diagrams based on entropy distance. It considers complexity of both classes and relationships between the classes, and presents rules for transforming complexity value of classes and different kinds of relations into a weighted class dependence graphs. This method of measure has many good properties; therefore it can measure the structure complexity of class diagrams objectively.

1 Introduction

One of the principal goals of software engineering is to assure the quality of object oriented software from the early phases of the life-cycle, such as conceptual modeling phase. UML [1] class diagrams constitute a key artifact in this phase. The structural complexity measure is one of the most important metrics to evaluate the quality of a UML class diagram [2].

Chidamber and Kemerer proposed a set of design metrics defined at class level [3]. Lorenz and Kidd proposed a group of metrics deal with the static characteristics of software design [4]. Brito, Abreu and Melo proposed a set of metrics at system level [5]. Marhchesi proposed a set of metrics to measure UML class diagrams at the analysis phase, but did not take into account some UML measurable elements [2]. Genero proposed new metrics to cover the necessity of measuring these relationships [6]. Manso and Genero used 8 metrics for measuring the structural complexity and the size of UML class diagrams and their maintainability [7]. But they did not give a single complexity measuring integrate all these metrics.

---

* This work was supported in part by the Young Scientist's Fund of NSFC (60373066, 60303024), National Grand Fundamental Research 973 Program of China (2002CB312000), National Re-search Foundation for the Doctoral Program of Higher Education of China.

M. Bubak et al. (Eds.): ICCS 2004, LNCS 3036, pp. 421–424, 2004.

2   Weighted Class Dependence Graphs

2.1 Complexity Measure for Classes and Relations

Classes and relations are the basic elements of class diagrams. We use suitable metrics proposed by others to measure the complexity of classes which satisfied that: only one metric for measuring each class; the value is above zero and denotes the complexity of the class; it take both the class structure and inheritance into account. The cohesion measure sometimes can be used here.

There are mainly three kinds of relations in UML class diagrams: associations, generalizations, and dependencies. Complexity of different kinds of relations can be compared. Different kinds of relations influence the dependency between classes in different degrees. We fractionize relations in UML class graph into totally 10 kinds and their different influences on the dependency between classes can be weighted. It forms Table 1.

Table 1. Dependency weight value of relations

No.	Relation	Weight
1	Common Dependency	H1
2	Common association	H2
3	Qualified association	H3
4	Association class	H4
5	Aggregation association	H5
6	Composition association	H6
7	Generalization (parent class is concrete)	H7
8	Binding	H8
9	Generalization (parent class is abstract)	H9
10	Realize	H10

Comparing the complexity between these kinds of relations, it has $H1 \leq H2 \leq H3 \leq H4 \leq H6$ and $H1 < H5 < H6 < H7 < H8 < H9 < H10$.

Each ends of a relation should link to a certain class. So relations of the same kind may have different complexity because they related to different classes. This can be calculated in the Weighted Class Dependence Graphs (WCDG) which denote the given class diagram abstractly.

2.2   Weighted Class Dependence Graphs

Definition 1. D denotes a given class diagram. WCDG is defined as $G(D) = (N, E)$, where $N = V(D)$, $E = R(D)$, i.e. the nodes and edges. $V(D) = \{c \mid c$ is a class in $D\}$, $R(D) = \{(n_1, n_2, W(n_1, n_2)) \mid n_1, n_2 \in V(D) \wedge ($ there are relations in T from n_1 to n_2 or $n_1 = n_2)\}$. When $(n_1 \neq n_2)$, $W(n_1, n_2) = \sum W_i$, W_i is the dependence weighted value of each relation from n_1 to n_2; when $(n_1 = n_2)$, $W(n_1, n_2)$ should be added $H1 * C(n_1)$, $C(n_1)$ is the complexity of the class n_1 denotes.

Every node in the WCDG is corresponding to a class in the class diagram; relations are transformed to edges between the nodes. Firstly we form the structure of the WCDG from the class diagram. Then calculate the dependence weight value of all the relations in the class diagram.

Let the complexity measure of class A and B are C(A) and C(B), the weight value of dependencies is H. The dependence weighted value W of a relation from B to A can be calculated as follows: when the relation has no destination multiplicity , such as dependency including Generalization, Binding and Realize, W = H*C(A); when the relation has a destination multiplicity of n, such as association including Aggregation and Composition, W = H*(2–1/n)*C(A), if n is *, let 1/n = 0; when it is a qualified association, W = H3*(2–1/n)*C(B)+b, b denotes the complexity of the qualifier; association class is transformed to a new node has relations with A and B.

Now we can calculate the weight value of the edges and nodes by the dependence weight value of relations. The WDCG can also be expressed in a matrix, in which W[i][j] = W(n$_i$, n$_j$). It will predigest the calculation of complexity.

3   A Structure Complexity Measure Based on Entropy Distance

X and Y are discrete stochastic variables: $A_x = \{ x_i \mid 1 \leq i \leq m \}, A_y = \{ y_j \mid 1 \leq j \leq n \}$.
The entropy of their joint distribution is:

$$H(X,Y) = - \sum_{x_i \in A_x \wedge y_j \subset A_y} p(x_i, y_j) \log p(x_i, y_j) \tag{1}$$

The entropy of X when Y happened:

$$H(X \mid Y) = \sum_{x_i \in A_x \wedge y_j \in A_y} P(x_i, y_j) \log \frac{1}{P(x_i \mid y_j)} \tag{2}$$

The mutual-information of X and Y:

$$I(X,Y) = H(X) - H(X \mid Y) = H(Y) - H(Y \mid X) \tag{3}$$

Let D be a given class diagram; G(D) is the WCDG corresponding to D; N(D) is the set of all nodes in the WCDG.

We can use the entropy distance to measure the complexity of G(D). Use stochastic variables X and Y to denote the output and input edges weight of each node. Let $A_x = A_y = N(D)$, for each $x_i \in A_x$, each $y_j \in A_y$, we have

$$p(x_i) = \frac{\sum_{n2 \in N(D)} W(x_i, n2)}{\sum_{n1 \in N(D)} \sum_{n2 \in N(D)} W(n1, n2)}, \quad p(y_j) = \frac{\sum_{n1 \in N(D)} W(n1, y_j)}{\sum_{n1 \in N(D)} \sum_{n2 \in N(D)} W(n1, n2)} \tag{4}$$

$$P_{X,Y}(x_i, y_j) = \frac{W(x_i, y_j)}{\sum_{n1 \in N(D)} \sum_{n2 \in N(D)} W(n1, n2)}, \quad P_{X,Y}(x_i \mid y_j) = \frac{W(x_i, y_j)}{\sum_{n1 \in N(D)} W(n1, y_j)} \tag{5}$$

Definition 2. The complexity of D is defined to be the entropy distance of X and Y:

$$\text{Complexity(D)} = DH(X,Y) = H(X,Y) - I(X,Y) \tag{6}$$

Specially, when D=ϕ, Complexity(D)=0. For any class diagram D, it has that:

$$0 \leq Complexity(\text{D}) \leq 2\log|V(\text{D})| \tag{7}$$

where V(D) is the set of classes in class diagram D.

We use Weyuker's properties [8] for complexity measures to evaluate our measure method. It contains 9 properties. We can prove that our measure satisfies 7 of the total 9 properties. We think our method can measure the structure complexity of class diagrams objectively.

4   Conclusions

We uses weighted class dependence graphs to represent a given class diagrams, and then present a structure complexity measure for UML class diagrams based on entropy distance. It considers complexity of both classes and relationships between the classes. This method of measure has many good properties; therefore it can measure the structure complexity of class diagrams objectively.

The UML class diagrams can only represent static model of the software. When dealing with dynamic knowledge, UML dynamic diagrams and state diagrams should be used. How to measure the quality of these diagrams is the future work.

References

1.   Rumbaugh, J., Jacobson, I., Booch, G.: The Unified Modeling Language Reference Maual. Addison-Wesley, Reading, MA, USA (1999)
2.   Marchesi, M.: OOA metrics for the United Modeling Languages. Proceedings of 2nd Euromicro Conference on Software Maintenance and Reengineering. Palazzo degli Affari, Italy (1998) 67-73
3.   Chidamber, S., Kemerer, C.: A Metrics Suite for Object Oriented Design. IEEE Transactions on Software Engineering, 20(6). (1994) 476-493
4.   Lorenz., M., Kidd, J.: Object-Oriented Software Metrics: A Practical Guide. Prentice Hall, Englewood Cliffs, New Jersey (1994)
5.   Brito, E., Abreu, F., Melo, W.: Evaluating the Impact of Object-Oriented Design on Software Quality. Proceedings of 3rd International Metric Symposium (1996) 90-99
6.   Genero, M., Piattini, M.: Empirical validation of measures for class diagram structural complexity through controlled experiments. Proceedings of 5th International ECOOP Workshop on Quantitative Approaches in Object-Oriented Software Engineering. Budapest, Hungary (2001) 87-95
7.   Manso1, M.E., Genero, M., Piattini, M.: No-Redundant Metrics for UML Class Diagram Structural Complexity. CAiSE 2003 The 15th Conference On Advanced Information Systems Engineering, LNCS 2681, 127-142
8.   Weyuker, E.J.: Evaluating Software Complexity Measures. IEEE Transaction on Software Engineering. (1988) 1357-1365

Parallelizing Flood Models with MPI: Approaches and Experiences[1]

Viet D. Tran and Ladislav Hluchy

Institute of Informatics, Slovak Academy of Sciences
Dubravska cesta 9, 842 37 Bratislava, Slovakia
viet.ui@savba.sk

Abstract. Parallelizing large sequential programs is known as a challenging problem. This paper focuses on problems encountered during parallelization process of different flood models and on the approaches used for solving them. The approaches are focused on reducing development time, which can help programmers make a parallel version of existing sequential programs within a short time.

1 Introduction

Over the past few years, floods have caused widespread damages throughout the world. Most of the continents werc heavily threatened. Therefore, modeling and simulation of floods in order to forecast and to make necessary prevention is very important. As Linux clusters are widely used as low-cost high performance platforms, it is important to make the parallel versions of the flood models running on Linux clusters. That limits the possibility of using OpenMP or parallel compilers for parallelization. Therefore, programmers have to rely on MPI or other message-passing libraries for developing the parallel version of the flood models.

This paper focuses on the problems encountered during parallelizing flood models using MPI and solutions for them. In Section 2, the flood models are introduced. The problems encountered during parallelization and their solutions are discussed in Section 3. Section 4 gives the results of the parallelization and Section 5 concludes the paper.

2 Numerical Flood Models

At the beginning of ANFAS project [4], many surface-water flow models were studied in order to find a suitable high-performance model for pilot sites at Vah river in Slovakia and Loire river in France. The result of the study showed that many models exist only in sequential forms. Two models were chosen for the pilot site; one is FESWMS [3] which is based on finite element approach and is distributed with

---

[1] This work is supported by EU 5FP CROSSGRID IST-2001-32243 RTD and the Slovak Scientific Grant Agency within Research Project No. 2/3132/23

M. Bubak et al. (Eds.): ICCS 2004, LNCS 3036, pp. 425–428, 2004.

commercial package SMS [9] by EMS-I. The second model is DaveF, a new model based on time-explicit, cell-centered, Godunov-type, finite volume scheme.

Although both models are used for modeling water flow, they are based on completely different numerical approaches. Detailed descriptions of the numerical approaches of the models can be found in [5]. This paper focuses on problem encountered during its parallelization and solutions for the problems. Therefore, the following descriptions of computational approaches are purely from the view of parallel programming.

3   Problems Encountered during Parallelization with MPI

Understanding Algorithms

Although the mathematical approaches (finite elements, finite volumes) of the models may be well-known, there are many hydraulic details in the algorithms such as different boundary conditions, wetting/drying, raining/infiltration and different tricks to stabilize the solutions. Such details complicate the programs considerably and are not easy to understand for the programming experts who parallelize the source.

However, in our parallelization approaches (see later in data and code duplications), the programmers do not have to understand all the details in the algorithms. From the view of parallelization, every finite-element model consists of three steps: generating the matrix, solving the matrix and updating solutions. The programmers do not have to learn what governing differential equations are used in the models or Galerkin method works, as the implementation already exists in the source code and they are not going to change them. Similarly, the finite-volume algorithm can be simplified as follows: in every time step, each cell updates its values from its current values and the values of its neighbors. That frees the programmers from learning all details in hydraulics and allows them to parallelize applications they are not familiar with.

Understanding the Source Codes

The source code of FESWMS has about 65 thousand lines, i.e. about one thousand pages; DaveF is nearly the same. Reading and understanding the source codes (especially when programmers do not understand all details in algorithms) for identifying the critical part may take a long time. Therefore, profiling tools (e.g. gprof in Linux) are extremely useful for parallelizing sequential programs. By using profiling tools, programmers can easily identify the computation-intensive parts in the source code (computation kernel), see the call graphs and analyze the performance of the program. Programmers then can concentrate on studying the computation kernel needed to parallelize/optimize, and consider the rests of the source code as "black boxes". Discussion with the original authors of the models is also useful for understanding the most important data structures (e.g. global arrays of nodes/cells/elements).

Writing the Parallel Code

After analyzing the algorithm and the source code, the programmers can start to write parallel versions of the models. Paralleling with MPI for Linux clusters adds some more problems. It may be argued whether writing a parallel program from scratch with MPI on distributed-memory architectures such as Linux clusters is easier or more difficult than with OpenMP on share-memory systems such as supercomputers [6]. However, to parallelize existing sequential programs it is much easier to use OpenMP, because OpenMP does not change program and data structure.

In our approach, data and codes, which are not interesting in parallelization, are duplicated to reduce the development time. The data and code duplication greatly reduce the amount of codes that need to be modified during parallelization. That also allows programmers to ignore the implementation details in the parts of the codes that are duplicated. The programmers have to understand only the very basic computation scheme of the algorithms used in the models, and study/modify only few routines in the computation kernel of the models during parallelization.

4   Experimental Results

Experiments have been carried out on Linux cluster at the Institute of Informatics (II-SAS) in Slovakia. The Linux cluster at II-SAS consists of 16 computational nodes, each with a Pentium IV 1800 MHz processor and 256 MB RAM. All of the nodes are connected by an Ethernet 100Mb/s switch. Input data for the experiments are taken from Vah river in Slovakia and Loire river in France.

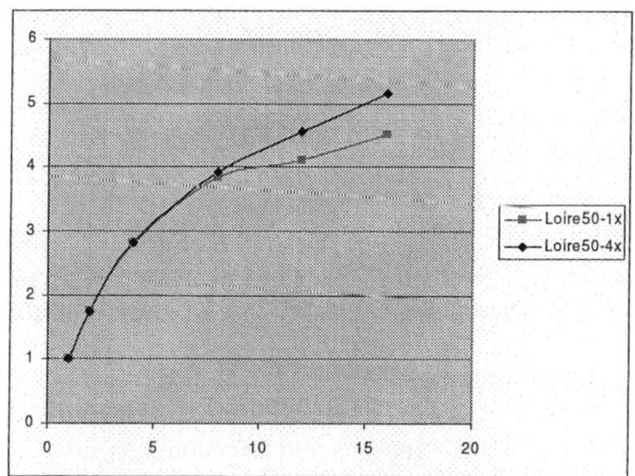

Fig. 1. Speedup of DaveF on II SAS cluster

In parallel version of FESWMS, only about 50 lines of code are modified from the 65000 lines in original sequential version, and 150 new lines of code are created for the new parallel iterative matrix solver from PETSC library, which replaces the frontal solver from the original sequential version. The speedup of FESWMS is difficult to describe in a table or graph. There are several iterative solvers and preconditioners, each of them has also several additional parameters. According to our experiments, the combination of BiCGStab method and ILU preconditioner is the quickest one (about 10x faster than the original frontal method), but GMRES/ILU is the most stable combination in the sequential version. Generally the speedup of FESWMS is about 5-7 on 16 nodes.

The parallel version of DaveF has less than modified 100 lines from its 45000 lines of code in the original sequential version and it is developed in 3 days. That clearly proves the advantages of our approach: to develop a parallel version in a very short time. Fig.1 shows the speedup of DaveF on II-SAS with two different input data from Loire river, the one being four times larger than the other one. It is easy to see that the speedup is increased with the size of input data, especially for larger number of processors. The reason is the fine granularity of DaveF; the more processors are used the larger is the granularity performance effect.

5   Conclusion and Future Work

This paper has presented an approach to parallelizing sequential flood models. The approach allows programmers to produce parallel versions within very short time. The approach is proved with two different flood models that are used in the ANFAS project.

At the moment, both models have been ported to Grid environment in CrossGrid project [7] and are running in CrossGrid testbed [8]. The details of Grid-aware Flood Virtual Orgranization, where DaveF is used, are described in a separate paper [2].

References

1.   L. Hluchy, V. D. Tran, J. Astalos, M. Dobrucky, G. T. Nguyen, D. Froehlich: Parallel Flood Modeling Systems. International Conference on Computational Science ICCS'2002, pp. 543-551.
2.   L. Hluchy, V. D. Tran, O. Habala, J. Astalos, B. Simo, D. Froehlich: Problem Solving Environment for Flood Forecasting. Recent Advances in Parallel Virtual Machine and Message Passing Interface, 9th European PVM/MPI Users' Group Meeting 2002, pp. 105-113.
3.   FESWMS - Finite Element Surface Water Modeling.
     `http://www.bossintl.com/html/feswms.html`
4.   ANFAS Data Fusion for Flood Analysis and Decision Support.
     `http://www.ercim.org/anfas/`
5.   D. Froehlich: IMPACT Project Field Tests 1 and 2: "Blind" Simulation by DaveF. 2002.
6.   OpenMP, MPI and HPF: Comparing The Three: Which Programming Model Is Best? The Portland Group, Inc. http://www.pgroup.com/SLC2000/omp_hpf_mpi_files/frame.htm.
7.   EU 5FP project CROSSGRID. `http://www.crossgrid.org/`
8.   Marco, R.: Detailed Planning for Testbed Setup. The CrossGrid Project, 2002.
     `http://grid.ifca.unican.es/crossgrid/wp4/deliverables/CG-4-D4.1-001-PLAN.pdf`
9.   Surface-water Modeling System. `http://www.ems-i.com/SMS/sms.html`

Using Parallelism in Experimenting and Fine Tuning of Parameters for Metaheuristics[*]

Maria Blesa and Fatos Xhafa

Universitat Politècnica de Catalunya
C6 Campus Nord, E-08034 Barcelona, Spain
{mjblesa,fatos}@lsi.upc.es

Abstract. We show that parallel implementations of metaheuristics are efficient tools for both experimenting and fine tuning of parameters.

1 Introduction

Metaheuristics were introduced in the last two decades as a new kind of approximate algorithms for solving combinatorial optimization problems which combine heuristic methods with higher level frameworks [5]. Their implementation is usually a complex task, since they involve three main concepts: (1) *Main method,* (2) *internal/external heuristics,* and (3) *setting of search parameters.*

Measuring the performance of a metaheuristic implementation requires testing on a large set of instances and on real world instances usually of big and very big size. Moreover, finding of right values for the search parameters of the metaheuristic is almost indispensable for the success of the metaheuristic implementation. Considerable efforts have been done by researchers and practitioners to provide, on the one hand, a methodology and rigorous basis for experimental evaluation of heuristics [11,2] and, on the other, to find efficient approaches for fine tuning of parameters such as developing specific software [1], use of experimental design [7] and self-adaptive procedures [10].

We address the issue of using parallel implementations as a mean for efficient experimenting and fine tuning of parameters for metaheuristics. Our proposal is based on two parallel models and, to illustrate our proposal, we have applied it in experimenting and fine tuning of parameters for the Tabu Search method applied to the 0-1 Multidimensional Knapsack problem. High quality solutions as compared with best known up-to-date results for the problem are obtained.

2 Parallel Models for Experimenting and Fine Tuning

Parallelism has been usually used to reduce computation times. For our purpose we describe here two simple parallel models: the *Independent Runs* (IR) and

* Partially supported by the CICYT Project TIC2002-04498-C05-03 (TRACER) and by the Catalan Research Council of the Generalitat de Catalunya (grant no. 2001FI-00659). *For a longer version of this work, see [4].*

the *Independent Runs with Autonomous Strategies* (IRAS). Although closely related, they are used here with different objectives: the IR model is intended for experimenting while IRAS model for the fine tuning of parameters.

In the **Independent Runs model (IR)** there is a coordinator processor sending the problem instance and parameters' setup and receiving the results, and each processor runs the same instance of the program. Observe that this model make sense as far as the program is non-deterministic. This is precisely the case of metaheuristic implementations which take randomized or probabilistic decisions. Running the same implementation in p different processors leads to exploring different areas of the search space and it is equivalent to performing p sequential executions, thus scaling down the experimentation time with a factor of up to p. For this reason, this model is then very suitable for experimental evaluation of metaheuristic implementations. The **Independent Runs with Autonomous Strategies (IRAS)** can be seen as a special case of the IR model in which the processors are given, additionally, a *strategy* to be used for its own search. A *strategy* is defined as an m-tuple $\langle parameter_1, \dots, parameter_m \rangle$, where each $parameter_i$ is a different parameter of the metaheuristic. For each processor $proc_i$, the *coordinator processor* $proc_0$ computes a strategy S_i, and then sends it together with the problem instance to the processor $proc_i$. Clearly, using this parallel model we can efficiently make the fine tuning of parameters.

Our implementation of both parallel models is fully generic and independent of the (sequential) metaheuristic implementation at hand. This is achieved through a careful class design and implementation in C++ using the MPI as a communication library. The class `Solver_LAN` will be be in charge of running the parallel program while the sub-classes `Solver_IR` and `Solver_IRAS` will implement specifically the task of coordinator and slave processors. `Solver_Seq` denotes the sequential implementation of the metaheuristic, through which we can declare an instance of such implementation and run the main method. There is a one-to-one relationship between `Solver_LAN` and `Solver_Seq` since the former will use instances of the latter as a black-box. The classes `Instance`, `Setup`, `Strategy` and `Solution` represent the problem instance data, parameters, strategy and a feasible solution to the problem, respectively. Those entities are problem-dependent and will be implemented according to the problem at hand. The parallel program provides their interfaces and hence can use them as black-boxes. Any of the models is run via the method `run()` of the corresponding class `Solver_IR` or `Solver_IRAS`. This generic way of designing and implementing the framework has important benefits, like genericity and reusability. Once the execution is finished, different information of the search process can be accessed, e.g., the best solution found or the time required to find it.

3   Tabu Search for the 0-1 Multidimensional Knapsack

To illustrate our proposal, we implement the Tabu Search (TS) for the 0-1 Multidimensional Knapsack problem both in the IR model (which is intended for experimenting) and the IRAS model (which is intended for the fine tuning.)

Table 1. Numerical values for the parameters.

nb_iterations	independent_runs	tabu_list_size	max_neighbors
$100n$ (small and middle) $1000n$ (big size)	2	$[3\ldots15]$	full exploration

nb_best_sols	nb_intensifications	history_rep	nb_diversifications
$[10\ldots15]$	$\simeq 10$	$80-95\%$	$\simeq 10$

Table 2. Results for the 0-1MKNP. Best and average costs obtained over 20 executions, respectively. The 7th column is the deviation of the sample wrt. the average. The last two columns indicate the number of iterations performed and time spent on it.

Instance	n	m	Optimum	Best cost	Avg. cost	deviation	Iters.	time (s)
KNAP15	15	10	4015	4015	4014.1	0	3000	3.6
KNAP20	20	10	6120	6120	6120.0	0	1600	2.5
KNAP50	50	5	16537	16520	16441.0	0.001	10000	21.3
SENTO1	60	30	7772	7772	7772.0	0	5000	53.5
SENTO2	60	30	8722	8722	8720.6	0	5000	55.5
OR10x100-00	100	10	23064	22478	22360.4	0.025	85629	600
OR10x250-00	250	10	59187	56213	55945.6	0.050	55132	900
OR10x500-00	500	10	117726	111773	111486.7	0.051	35487	1200
OR30x100-00	100	30	21946	21614	21520.7	0.015	31615	600
OR30x250-00	250	30	56693	54711	54534.6	0.035	15215	900
OR30x500-00	500	30	115868	111272	110942.4	0.040	10459	1200

Tabu Search [9] belongs to the family of local search algorithms but here the search is done in a *guided* way in order to overcome the local optima. The search process tries to avoid cycling by forbidding or penalizing moves which take the solution, in the next iteration, to solutions previously visited (called *tabu*). To this aim, TS keeps a *tabu list* which constitutes the tabu search memory. The role of the memory can change as the algorithm proceeds. At initialization the goal is to make a coarse examination of the solution space and further on the search is focused to produce local optima solutions in a process of *intensification* or make a *diversification* in order to explore new regions of the solution space.

The NP-hard 0-1 Multidimensional Knapsack problem (0-1MKNP) consists in selecting a subset of n given objects in such a way that the total profit of the selected objects is maximized while a set of knapsack constraints are satisfied. The 0-1MKNP problem can be stated as: maximize $c \cdot x$, subject to: $Ax \le b$, $x \in \{0,1\}^n$, where $c \in \mathbb{N}^n$, $A \in \mathbb{N}^{m \times n}$, and $b \in \mathbb{N}^m$. The binary components x_j of x are decision variables: $x_j = 1$ if the object j is selected, and $x_j = 0$ otherwise. The profit associated to j is denoted by c_j. Each $A_i x \le b_i$ is a capacity constraint.

Parameters involved, fine tuning and computational results. Five parameters define the 0-1MKNP: the number of objects n, the number of constraints m, the profits of the objects $c \in \mathbb{N}^n$, the matrix of constraints $A \in \mathbb{N}^{m \times n}$, and the capacities $b \in \mathbb{N}^m$. Every fixed set of values for these parameters defines an instance of the problem and, according to them, instances can be easier or harder to solve. This is an important feature to consider when studying the robustness and the performance of an algorithm. The basic parameters controlling TS are concerned with stopping conditions (nb_iterations, and independent_runs) and the influence of the historical search memory (tabu_list_size). Other parameters control the search process, specially the neighborhood exploration (max_neighbors), the diversification (history_rep and nb_diversifications), and the intensification (nb_best_sols and nb_intensifications). All those parameters are mutually and strongly dependent. For the success of the method, appropriate values for those parameters have to be find. We have tuned the Tabu Search parameters by using the IRAS model introduced above.

After tuning these parameters (see Table 1), we have run the 0-1MKNP implementation in a cluster of computers AMD K6-11 with 450 MHz processors and 256Mb of memory (see [4]). To obtain some statistical significance about the robustness of the algorithm, the same instance should be run several times with the same parameters setting and average results should be provided. We test small ($n \leq 50$), middle-sized ($50 < n \leq 100$) and big instances ($100 < n \leq 500$) taken from the literature [8,6,3]. Since our aim is to test how does our generic implementation and parallel fine tuning of parameters behave, we have chosen instances for which the optimum value (obtained through computationally expensive exact methods) is known (see Table 2). The low values on the deviation of the cost of our solutions from the optimum shows both that the values of the parameters that we found through our approach are appropriate, and also the robustness of our approach in the sense that the values we found for the parameters perform very well for a large set of different instances.

References

1. B. Adenso-Diaz and M. Laguna. (2002). Fine tuning of Algorithms Using Fractional Experimental Designs and Local Search. Submitted.
2. R.S. Barr, B.L. Golden, J. Kelly, W.R. Stewart, M.G.C. Resende. (2001) Designing and Reporting Computational Experiments with Heuristic Methods. *Journal of Heuristics*, 1(1):9–32.
3. J.E. Beasley. (1990). OR-Library: Distributing Test Problems by Electronic Mail. *Journal of the Operational Research Society*, 41:1069–1072.
4. M. Blesa and F. Xhafa. (2003). Using Parallelism in Experimenting and Fine Tuning of Parameters for Metaheuristics. Technical Report no. LSI-03-56-R, UPC.
5. C. Blum and A. Roli. (2003). Metaheuristics in combinatorial optimization: Overview and conceptual comparison. *ACM Computing Surveys*, 35(3):268–308.
6. C. Cotta and J.M. Troya. (1998). A Hybrid Genetic Algorithm for the 0-1 Multiple Knapsack Problem. In *Artificial Neural Nets and Genetic Algorithms*, chapter 3, pp. 251–255. Springer-Verlag.
7. S.P. Coy, B.L. Golden, G.C. Runer, E.A. Wasil. (2000). Using Experimental Design to Find Effective Parameter Settings for Heuristics *Journal of Heuristics*, 7:77–97.
8. A. Freville and G. Plateau. (1990). Hard 0-1 multiknapsack test problems for size reduction methods. *Investigation Operativa*, 1:251–270.
9. F. Glover and M. Laguna. (1997). *Tabu Search*. Kluwer Academic Publishers.
10. J. Kivijärvi, P. Fränti and O. Nevalainen. (2003). Self-Adaptive Genetic algorithm for Clustering. *Journal of Heuristics*, 9:113–129.
11. R.L. Rardin and R. Uzsoy. (2001). Experimental Evaluation of Heuristic Optimization Algorithms: A Tutorial. *Journal of Heuristics*, 7:261–304.

DEVMA: Developing Virtual Environments with Awareness Models

Pilar Herrero and Angélica de Antonio

Facultad de Informática. Universidad Politécnica de Madrid.
Campus de Montegancedo S/N.
28.660 Boadilla del Monte. Madrid. Spain
{pherrero,angelica}@fi.upm.es

Abstract. In this paper, we present an application, called DEVMA, developed at the Universidad Politécnica de Madrid with the aim of introducing into a 2D virtual environment some of the key concepts of the *Spatial Model of Interaction* (SMI) – an awareness model designed for Computer Supported Collaborative Work (CSCW). This application also takes into account how these concepts can be deformed by the presence of boundaries in the environment and how these deformations could have an influence on the awareness of interaction between them.

1 The Spatial Model of Interaction (SMI)

The aim of this research was to study how the key concepts of one of the most successful models of awareness in Computer Supported Cooperative Work (CSCW), called the Spatial Model of Interaction (SMI) [1,2], are deformed by the presence of objects that were acting as boundaries in the environment. It allows objects in a virtual world to govern their interaction through some key concepts: medium, aura, awareness, focus, nimbus, adapters and boundaries.

Aura is the sub-space which effectively bounds the presence of an object within a given medium and which acts as an enabler of potential interaction. In each particular medium, it is possible to delimit the observing object's interest. This area is called *focus* "The more an object is within your focus the more aware you are of it". The focus concept has been implemented in the SMI as an "triangle" cone limited by the object's aura.

In the same way, it is possible to represent the observed object's projection in a particular medium. This area is called *nimbus*: "The more an object is within your nimbus the more aware it is of you". The nimbus concept, as it was defined in the Spatial Model of Interaction, has always been implemented as an circumference in a visual medium. The radio of this circumference has an "ideal" infinite value, although in practice, it is limited by the object's aura.

The main concept involved in controlling interaction between objects is *"awareness"*. One object's awareness of another object quantifies the subjective importance or relevance of that object. The awareness relationship between every pair of objects is achieved on the basis of quantifiable *levels* of awareness between them

M. Bubak et al. (Eds.): ICCS 2004, LNCS 3036, pp. 433–436, 2004.
© Springer-Verlag Berlin Heidelberg 2004

and it is unidirectional and specific to each medium. Awareness between objects in a given medium is manipulated via *focus* and *nimbus*. Moreover, an object's aura, focus, nimbus, and hence awareness, can be modified through *boundaries* and some artefacts called *adapters*.

For a simple discrete model of focus and nimbus, there are tree possible classifications of awareness values when two objects are negotiating unidirectional awareness [3]:

- *Full awareness*: The awareness that object A has of object B in a medium M is "full" when object B is inside A's focus and object A is inside B's nimbus (Figure 1).
- *Peripheral awareness*: The awareness that object A has of object B in a medium M is "peripheral" when (Figure 2) object B is outside A's focus but object A is inside B's nimbus, or object B is inside A's focus but object A is outside B's nimbus.
- *No awareness*: An object A has no awareness of object B in a medium M when object B is outside A's focus and object A is outside B's nimbus.

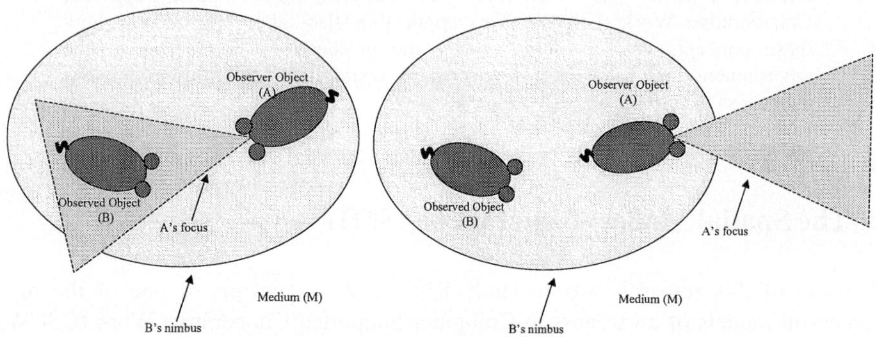

Fig. 1. Full Awareness   **Fig. 2.** Peripheral Awareness

In this application we have concentrated on four of these concepts: focus, nimbus, awareness and boundaries. More specifically, we have concentrated on how to introduce the geometrical modifications that boundaries can produce on the focus (and nimbus) shape and how these modifications can have an influence on the awareness of interaction between participants.

2 The DEVMA Application

This application shows the avatar's focus and nimbus deformation while it is moving around the environment and interfering with some boundaries. The user can introduce one or more avatars, having the chance of modifying its spatial properties – such as the avatar's position.

The application also allows to modify some of the focus properties such as the focus' length (see "longitud" in the Figure 3) or the focus' angle (see "Ángulo" in the Fig. 3). In the same way, the DEVMA application allows the user to modify the nimbus' ratio [4].

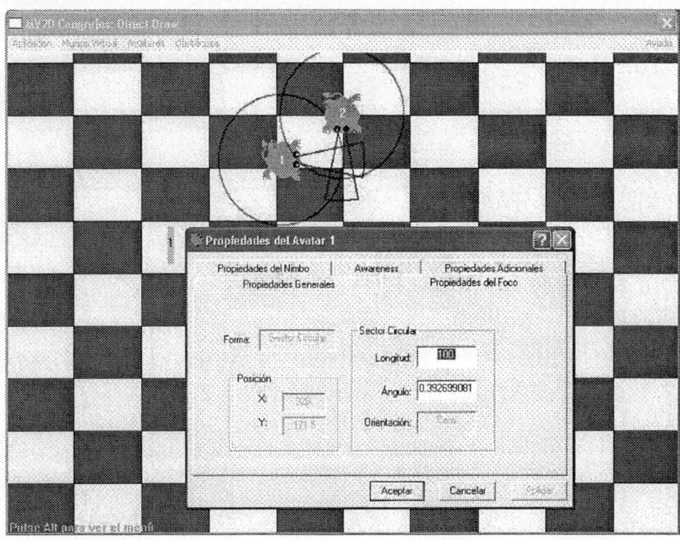

Fig. 3. Focus' Properties

The user also has the possibility of introducing one or more boundaries, interacting with each of them using the keyboard, mouse or application menu. When the avatar's focus (or nimbus) intersect with an obstacle, the focus's (or nimbus') shape will be deformed as it is showed in the figures 4.

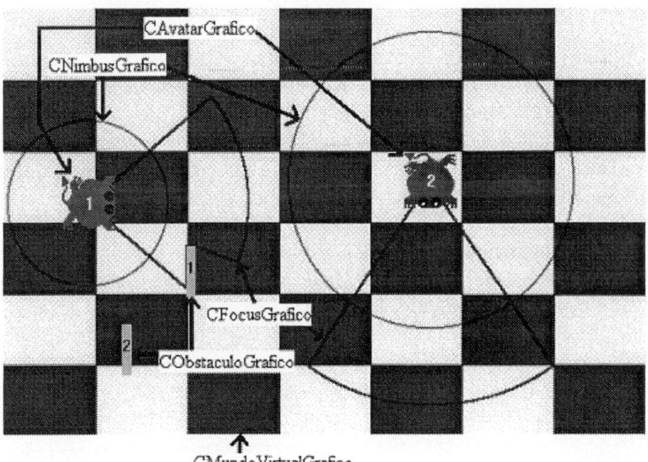

Fig. 4. Avatar's Focus and Nimbus Deformation in the Environment

The application also determine if there is any kind of awareness between avatars and, if later, it can classify it as *full awareness*, *peripheral awareness* or *no awareness*. In the figure 5 it is possible to appreciate how there is a "Full Awareness" between the avatars 1 and 2.

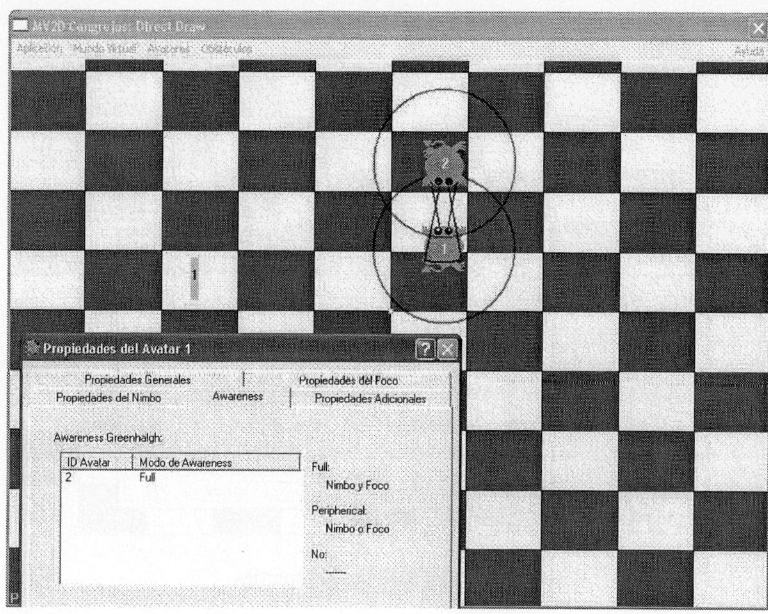

Fig. 5. Full Awareness between Avatars

References

1. Benford, S. Prinz, W. Mariani, J. Rodden, T. Navarro, L. Bignoli, E. Grant Brown C. and Naslund, T. *MOCCA - A Distributed Environment For Collaboratio*n, Available from the MOCCA Working Group of Co-Tech.
2. Benford, S., and Fahlén, L.E. *A spatial model of interaction in large virtual environments*, in Proc. Third European Conference on Computer Supported Cooperative Work (ECSCW'93), Milano, Italy. Kluwer Academic Publishers, pp. 109-124.
3. Greenhalgh, C., *Large Scale Collaborative Virtual Environments*, Doctoral Thesis. University of Nottingham. October 1997.
4. Herrero P. *A Human-Like Perceptual Model for Intelligent Virtual Agents* PhD Thesis. Universidad Politécnica de Madrid, June 2003.

A Two-Leveled Mobile Agent System for E-commerce with Constraint-Based Filtering

Ozgur Koray Sahingoz[1] and Nadia Erdogan[2]

[1] Air Force Academy, Computer Engineering Department, Yesilyurt, Istanbul, TURKEY,
sahingoz@hho.edu.tr

[2] Istanbul Technical University, Electrical-Electronics Faculty, Computer Engineering Department Ayazaga, 80626, Istanbul, TURKEY
erdogan@cs.itu.edu.tr

Abstract. This paper presents a two-leveled mobile agent system for electronic commerce. It is based on mobile agents as mediators and uses the publish/subscribe paradigm for registration and transaction processing. In the system we present, suppliers can connect, register or unregister to the system at any time, thus preserving the dynamic nature of the system. To reduce network load and the flow of unnecessary information in the product brokering part of the buying process, a rule based subscription methodology is used for necessary filtering operations.

1 Two-Leveled Mobile Agent System

Our work consists of a framework for a large-scale electronic commerce system that uses the publish/subscribe paradigm and exploits mobile agent technology extensively. It not only supports activities of buyers and suppliers, but also facilitates parallel computation by running mobile agents on suppliers concurrently. Our electronic commerce system involves three actors. Buyers look for purchase services from suppliers. Suppliers or sellers offer the services or products and a Dispatch Service facilitates communication between buyers and suppliers.

The system consists of mobile agents that belong to two different levels of execution and responsibilities as shown in Figure 1: Broker level Mobile Agent (BMA) and Supplier level Mobile Agents (SMA). A BMA is created by a Buyer Agent (BA) and is sent to the Broker. This BMA creates SMAs and sends them to suppliers in order to search their databases, to select among products, and to negotiate with the supplier (if necessary). The system does not rely on only a single mobile agent that visits every supplier one by one. Instead, we send a replica of the mobile agent to each of the suppliers concurrently, and thus benefit the advantages of parallel processing. This model of parallel computation is especially important as more suppliers can be searched in a shorter time.

In a dynamically changing electronic marketplace, a system should have the ability to adapt itself to this dynamic world. To meet this requirement, we have designed an architecture that utilizes the publish/subscribe paradigm for registration and dispatching operations, to increase efficiency and effectiveness of the procurement process in terms of costs, quality, performance, and time for both buyers and suppliers. The execution flow of the procurement process is as the following:

M. Bubak et al. (Eds.): ICCS 2004, LNCS 3036, pp. 437–440, 2004.
© Springer-Verlag Berlin Heidelberg 2004

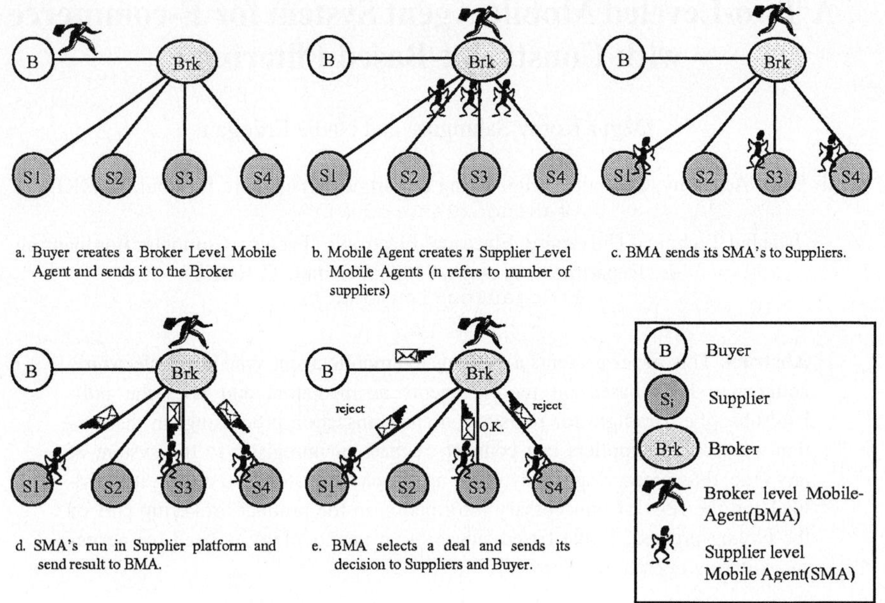

a. Buyer creates a Broker Level Mobile Agent and sends it to the Broker

b. Mobile Agent creates *n* Supplier Level Mobile Agents (n refers to number of suppliers)

c. BMA sends its SMA's to Suppliers.

d. SMA's run in Supplier platform and send result to BMA.

e. BMA selects a deal and sends its decision to Suppliers and Buyer.

Fig. 1. Buying a product from a Supplier in Two-Leveled Mobile Agent System

A number of buyers and suppliers have been subscribed to the system with their products and services. Their number in the system can vary randomly, increasing or decreasing at any time. When a user wants to buy a product, he has to make a request from the Buyer subsystem. A BA gets the request, creates a Mobile Agent, initializes its instance variables and sends it to the Broker as shown in Figure 1.a. When this Mobile Agent (we call it Broker level Mobile Agent (BMA)) arrives at the Broker, it checks the Knowledge Base of the Broker, selects the suppliers, which provide the requested products, and creates a new Supplier level Mobile Agent (SMA) for each of the selected suppliers. Thereafter, BMA sends each of these agents to a different supplier, as shown in Figure 1.b and Figure 1.c. Each SMA searches the product database of its supplier, negotiates with the supplier agent and sends results back to the BMA, as depicted in Figure 1.d. After all results have been collected from SMAs (or a specified timeout period has expired), BMA selects the best dealing one and sends an approval message to the SMA owning the deal, demanding it to buy the product and then to destroy itself. BMA sends rejection messages to all other SMA's and they destroy themselves. BMA also sends a message that reports the negotiation and its result to the Buyer Agent, as depicted in Figure 1.e.

We will now describe in more detail the main components of the system, the buyer and supplier subsystems and the dispatch system, explaining their functionality and discussing the major design decisions.

Buyer Subsystem: To request a purchase order from the system, a buyer has to initialize a Buyer Subsystem on its machine. Buyers have to know the address (URL) of the broker agent that they will connect to, just as the URLs of well known web sites. The function of a Buyer Subsystem/Buyer Agent(BA) is to search for product information and to perform goods or services acquisition. When a BA receives a purchase request from a user, it creates a Mobile Agent to search for product information and to perform goods or services acquisition in the system. BA specifies the criteria for the acquisition of the product and dispatches BMA to the broker. When BMA reaches the best deal, it sends a result report to BA and this information is added to its Database.

As several transaction scenarios are possible, allowing users to generate Broker level Mobile Agents with different behavioral characteristics increases the flexibility of the system. A human user interacts with the BA via a Buyer GUI module. In the beginning of a transaction, the user supplies the necessary information (i.e. name, maximum price, quantity required and required delivery date of the product, etc.). The buyer GUI allows users to control and monitor the progress of transactions, and to query past transactions from its Database.

Supplier Subsystem: A supplier has to initialize a Supplier Subsystem on its machine to join the system. When a supplier system is created for the first time, it subscribes to the system providing its address and the names of its products. If a supplier starts delivering a new product or ends the delivery of a former one, it again needs to subscribe or unsubscribe with product information, respectively. Every supplier agent has to know the address of the Broker so that it can make a connection. Supplier agent subscribes to the broker by sending its product definitions and waits for buyers to make requests for his products.

Each SMA on the supplier side can access the Product Database in read mode, according to its interests. It determines whether the required quantity is already in the inventory and thus available to offer, and then makes negotiations (if necessary). If so, supplier agent gives an immediate quotation to the BMA and sends a result message.

Dispatch Service: The Dispatch Service plays an important role in cyberspace. It is a logically (also physically in our system) centralized party which mediates between buyers and suppliers in a marketplace. The main component of the Dispatch Service is the "broker". Broker is useful when a marketplace has a number of buyers and suppliers, when the search cost is relatively high or when trust services are necessary. In the system that we present, the broker implements the publish/subscribe paradigm in which purchase events are published and made available to the supplier components of the system through notifications.

Filtering by Rules: To reduce unnecessary information in the product brokering, several filtering methods [1] have been widely used. In most of the E-commerce systems, only buyers make use of filters, but our system enables filtering for suppliers also. In other words, supplier selection process is carried out according to not only the buyer's criteria, but also the supplier's criteria. Buyers filtering criteria is loaded to BMA at the creation time by Buyer Agent, and BMA makes supplier selection process according to this. On the other hand, a supplier uploads its filtering criteria of a product as a rule in a subscription message at registration process of a product. This rule is store at the Subscription Table and BMA carries out it by examining the Subscription Table.

Our prototype system exploits a constraint-based filtering (as a rule), which makes use of textual annotations coded with content-based technique usually describing the products offered by e-commerce sites. Constraint-based filtering technique requires the definition of variables, domains and constraints. Therefore, many powerful algorithms such as backtracking, constraint propagation and variable ordering could be applied. PersonaLogic is an example of constraint-based filtering [2].

A rule is a set of attribute filters and it is sent in a subscription message at registration process of a product. Each attribute constraint is a tupple specifies a type, a name, a boolean binary operator, and a value for an attribute, as depicted in Figure 2. When a rule is used in a subscription, multiple constraints for the same attribute are interpreted as a conjunction; all such constraints must be matched one by one.

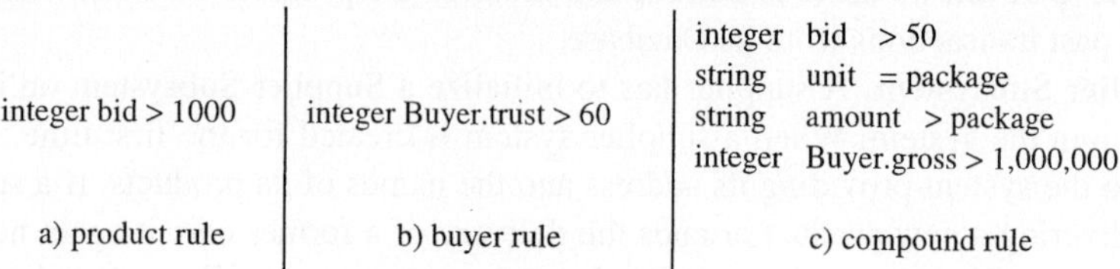

integer bid > 1000	integer Buyer.trust > 60	integer bid > 50 string unit = package string amount > package integer Buyer.gross > 1,000,000
a) product rule	b) buyer rule	c) compound rule

Fig. 2. Types of rules

When a subscriber registers its product to the system, he can also define constraint-based filtering criteria via rules. Three types of rules can be used in our system. A product rule defines constraints only the properties of the product (as in Figure 2.a). A buyer rule defines constraints according to buyer statistics stored at Broker (as in Figure 2.b). A compound rule is the conjunction of these two rules (as in Figure 2.c).

2 Conclusions and Directions for Future Work

In this paper, we have presented a general framework for a Two-Leveled Mobile Agent based E-Commerce systems including mechanism for a communication infrastructure based on publish/subscribe paradigm. The system relies on the utilization of mobile agents as mediators between buyers and suppliers. The publish/subscribe protocol allows participants to join and leave the system dynamically, extending the flexibility and adaptability of the system. By using a two-leveled agent model, we have also made use parallel computation to enhance performance. This is especially important as a larger number of suppliers can be searched concurrently in a shorter time to provide buyers with better choices in their decision-making.

References

1. Maes P., Guttman, R. H., Moukas, A. G. "Agents that Buy and Sell: Transforming Commerce as we Know It", Communications of the ACM, March 1999
2. PersonaLogic URL. http://www.personalogic.com/.

ABSDM: Agent Based Service Discovery Mechanism in Internet[1]

Shijian Li, Congfu Xu, Zhaohui Wu, Yunhe Pan, and Xuelan Li

Zhejiang University Artificial Intelligence Institute, Hangzhou, 310027
Jian1231@yahoo.com, {Xucong, wzh}@cs.zju.edu.cn,
panyh@sun.zju.edu.cn, gracelxl@sina.com

Abstract. To improve popular services discovery mechanism (UDDI mainly), we propose an agent-based services discovery mechanism. In this mechanism, services information is stored in distributed servers that are regarded as independent agents. These servers are joined into a tree structure; and then, a recursive algorithm was used to distribute searching request over the whole tree. Based on this condition, searching request could be rapidly parallel dealt with.

1   Introduction

Nowadays, UDDI (Universal Description, Discovery, and Integration) was used mostly for services discovery process [1]. But it has some disadvantages as follows [2]: (1) The condition that tremendous WSDL documents are stored on centralized UBR (UDDI Business Registry) servers potentially makes servers become the bottleneck of the system. (2) The information stored in UBR servers is static. (3) To keep the coherence and validity of the UBR servers' information is difficult. Therefore, distributed storage and independent management could be the basal ideology to improve UDDI. Agent is a conception developed in the field of the artificial intelligence these years [3]. Meanwhile, agent has been an important method to resolve the problems of distributed systems [4]. In this paper, we introduced an agent based service discovery mechanism. Its essential is that every node, which provides web services, is an agent and WSDL documents are distributed stored in nodes; A web-services searching tree composed of agents comes into being so that the service requestor can query information along it.

2   The Kernel Algorithms of ABSDM

In ABSDM, the web service management is achieved during the management of service agent (for short: SAgent). A new server connecting the network, a corresponding SAgent is created, where there are WSDL documents describing the web services in the new server. The SAgent structure is described in Backus-Naur Form as follows:

---

[1] The Project was supported by Zhejiang Provincial Natural Seience Foundation of China. No. 602045 and M603169

M. Bubak et al. (Eds.): ICCS 2004, LNCS 3036, pp. 441–444, 2004.

SAgent ::= <function module><information library>
function module ::= <Information library management><send/receive service re-
                quest ><receive result>
Information library ::= "local SAgent information" <Service description>
                "father-SAgent information" <Service description>
                {["son-SAgent information "<Service description>]}
Service description ::= <Web service interface><Web service implement>
Service interface   ::= <type><message><operation><port Type><binding><port>
Service implement ::= <service>

The service interface and implementation are described with a WSDL document, and the format is explained in W3C note [5].

According to the order of SAgent joining to the network, we could create a web-service-searching tree. The Searching-Tree Creating Algorithm (STCA) could be divided into several steps as follows: (1) SAgent creating. When new service was published to Internet, a new SAgent should be created according to the format mentioned. (2) SAgent registering. At beginning, the new SAgent starts searching a whole tree, finding the nearest SAgent as its father-Sagent, and registering in it. If there were no existing searching tree, the new SAgent would become the root of a new tree. During registering, services and requests that the father could provide would be add to the information library of the son. If there are several SAgents that are equidistant to the new SAgent, it will choose one as it father-SAgent randomly. (3) Leaving. At first, the SAgent tending to leave will ask its father to remove its information from the father's information library. After that, the son will choose one of his sons randomly as its replacer and send the replacer's information to the father. Then, the son will ask all but the replacer of its sons to register to the replacer. (Certainly, if the leaving SAgent has no father, it just needs to do the last step.)

Based on the tree structure, A Service Discovery Algorithm (SDA) was brought forward for service discovery. SDA could be divided into several orderly steps: ①The SAgent received a search request will search its information library at first. ② If there is no target document in local information library, the searching request will be referred to the whole son tree of current SAgent.③ If there were no target document in current SAgent and its sons, it would refer the request to its father. The father will repeat the same operation in ①. If current SAgent is the root of the searching tree, searching attempt will fail.

Compared with UDDI, the characteristics of ABSDM are presented as follows: i) Managing resource dynamically. In ABSDM, service managing and publishing was dynamically implemented during the process of SAgent creating and registering. (ii) Distributing deposited service information. There is no so much information in SAgent as in UBR. Further more, service providers in UDDI should login on a UBR server to maintain service information, while every SAgent could maintain information locally. (iii) Service searching synchronously. As described in SDA, searching request could be distributed by a SAgent to its sons and performed there synchronously. In this way, searching request could be synchronously performed in many SAgents.

3 An Example

We will illustrate how ABSDM was implemented with the following example from a project, the Application Platform of Virtual Research Center (VRC), main function of which is sharing files among VRCs. The structure of its network is showed in Fig.1 including root servers and 18 VRCs every of which is of 5 sub-VRCs. Because the files saved in nodes are just like WSDL documents, the file-sharing system could be a typical example of ABSDM.

Given that there are 10000 documents in every node, and every document is different from the others. The time used to search one in 10000 documents is denoted as T_{DB}, the time used to send searching request and result from node to node is denoted as T_{net}, T_{net} = (sending request (or result) data)/(network bandwidth). With the universality, we suppose that searching request were sent by 1A, the first sub-VRC of the first VRC, and the document that 1A requested were kept in sub-VRC 18E. The whole searching process is shown in table1. The time every step spending is also listed.

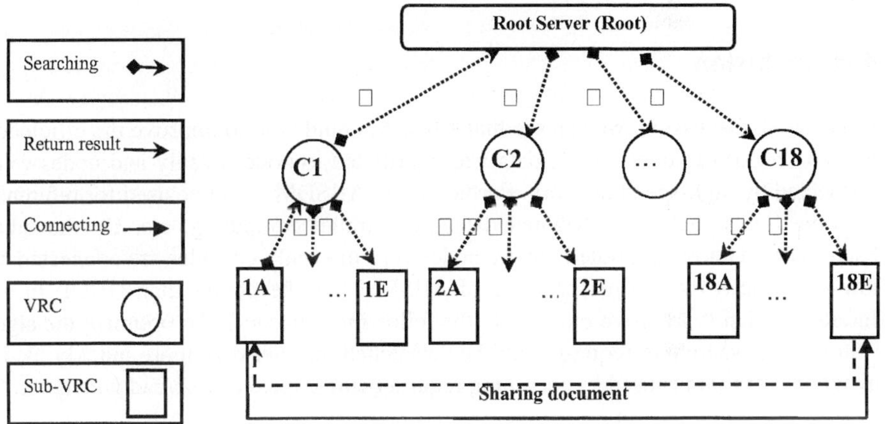

Fig. 1. An example

In this example, we could suppose that there is equal probability for every document to be requested. Because that the total number of document (denoted as totalnum) = (node number)*(document number in every node) = (1+18+18*5)*10000=1090000, then: (the occurred probability of every step) = (searched documents number in this step)/totalnum. The result is listed in Table 1.

Based on the result, we could compute the average time cost: $T_{ABSDM} = \sum_{cond=1}^{6} T_{all} = 9.27$

$T_{net} + 5.63\, T_{DB}$.

If the file-searching process is in the way of UDDI, the time cost is $T_{UDDI} = 2\,T_{net} + T_{DB}'$, the mark T_{DB}' denotes the time used to search one in 1090000 documents. From experiment, we get the value of T_{net}, T_{DB}, T_{DB}' as follows: $T_{net} = 20ms$, $T_{DB} = 28ms$, T_{DB}'

= 532ms. So in our deduction in theory and experiment, $T_{ABSDM}/T_{UDDI} \approx 0.33$, ABSDM could be more efficient than UDDI.

Table 1. The process of document searching

	Document location	Probability	Document number	Time wasted			Next Nodes
				Used	Current	Totally time to the step	
Step 1	1A	0.92%	10000	0	T_{DB}	T_{DB}	VRC1
Step 2	VRC 1	0.92%	10000	T_{DB}	$2T_{net}+T_{DB}$	$2T_{net}+2T_{DB}$	Son of VRC1
Step 3	Son of VRC1	3.67%	40000	$2T_{net}+2T_{DB}$	$2T_{net}+T_{DB}$	$4T_{net}+3T_{DB}$	Root
Step 4	Root	0.92%	10000	$4T_{net}+3T_{DB}$	$2T_{net}+T_{DB}$	$6T_{net}+4T_{DB}$	Other VRCs
Step 5	Other VRCs	15.60%	17*10000	$6T_{net}+4T_{DB}$	$2T_{not}+T_{DB}$	$8T_{net}+5T_{DB}$	Son of other VRCs
Step 6	Son of other VRCs	77.98%	17*5 *10000	$8T_{net}+5T_{DB}$	$2T_{net}+T_{DB}$	$10T_{net}+6T_{DB}$	No nodes left, process ended.

Explanation: *At one time, the searching action was performed concurrently in many searching tree nodes so all of these actions just needed one T_{DB}.*

4 Conclusion

From above discussion, we can see that ABSDM could help to improve the efficiency of web service discovery while data was distributed to nodes evenly and nodes were connected by high-speed network. Further more, ABSDM could be used for reference in some fields such as distributed database, parallel computing, etc. On the other hand, when data is distributed among nodes very unevenly or nodes are connected by low-speed network, the performance of ABSDM may be poorer than UDDI. In the future, we plan to improve our idea in the following aspects: 1. To optimize the algorithm so that searching request could be dispatched into network more quickly. 2. To avoid network jam caused by searching request being copied and spread widely.

References

1. F. Curbera et al., Unraveling the Web Services: An Introduction to SOAP, WSDL, and UDDI, IEEE Internet Computing, vol. 6, no. 2, Mar./Apr. 2002, pp. 86–93.
2. Wolfgang Hoschek. The Web Service Discovery Architecture. In Proc. of the Int'l. IEEE/ACM Supercomputing Conference (SC2002), Baltimore, USA, November 2002. IEEE Computer Society Press.
3. Wooldridge M., and Jennings N. R. Intelligent agents: theory and practice. Knowledge Engineering Review, 1995, 10(2):115-152
4. F. Zambonelli, N. R., Jennings,etc., Agent -Oriented Software Engineering for Internet Applications, 326-346. Springer Verlag. 2001.
5. Ariba Inc., IBM Corp., and Microsoft Corp.,Web Services Description Language (WSDL) 1.1, http://www.w3.org/TR/wsdl, W3C Note, 2001

Meta Scheduling Framework
for Workflow Service on the Grids

Seogchan Hwang[1], Jaeyoung Choi[1], and Hyeongwoo Park[2]

[1] School of Computing, Soongsil University,
1-1 Sangdo-5dong, Dongjak-gu, Seoul 156-743, Korea
`seogchan@ss.ssu.ac.kr, choi@comp.ssu.ac.kr`
[2] Supercomputing Center, Korea Institute of Science and Technology Information,
52 Eoeun-dong, Yusung-gu, Daejun 305-333, Korea
`hwpark@supercomputing.re.kr`

Abstract. The Globus becomes a standard to construct a Grid and provides core services such as resource management, security, data transfer, information services, and so on. Workflow management becomes a main service as one of the important grid services for grid applications. We propose a Meta Scheduling Framework (MSF) in this paper. The MSF provides an XML-based Job Control Markup Language for describing information and procedures of applications, and a workflow management service for scheduling job flow.

1 Introduction

Grid computing [1] is a new infrastructure to provide computing environments for grand challenge problems by sharing large-scale resources. The Globus toolkit is a standard to construct a Grid and provides essential grid services such as security, resource management, data transfer, information service, and so on. However, it still needs more works and researches to satisfy the requirements of various grid applications. Workflow management is emerging as one of the most important grid services. It is difficult to use the grid resources for general applications because the grid resources have various characteristics such as heterogeneity and dynamic organization. So many research groups have been working on a workflow related projects.

GridFlow [2] is a workflow management system using agent-based resource management and local resource scheduling system, Titan. It focuses on the scheduling of time-critical grid applications in a cross-domain and highly dynamic grid environment using a fuzzy timing technique and performance prediction of application. MyGrid [3] provides a service for integration such as resource discovery, workflow enactment and distributed query processing. It is a research project middleware to support biology environments on a Grid. And Condor [4] provides a workload management system of compute intensive jobs, and a scheduling of dependencies between jobs using DAGMan. This project provides similar functionalities but require their own specific infrastructures.

M. Bubak et al. (Eds.): ICCS 2004, LNCS 3036, pp. 445–448, 2004.

In this paper, we introduce a system, called Meta Scheduling Framework (MSF), for grid computational environments. MSF is developing high-level middleware to support process with complexity for general applications. MSF provides a Job Control Markup Language that is able to specify the job flow for general applications which are not developed for grid environments. It also provides a workflow management service, based on Globus, to execute the job, and a graphical user interface to facilitate the composition of grid workflow elements and the access to additional grid resources.

2 Meta Scheduling Framework

The goal of this research is to develop a framework to provide a workflow service to applications using the Globus. To accomplish this, we designed a workflow description language, called Job Control Markup Language (JCML), and a workflow management system. The JCML is designed to describe a process of tasks. And the workflow management system provides services to control the flows of tasks.

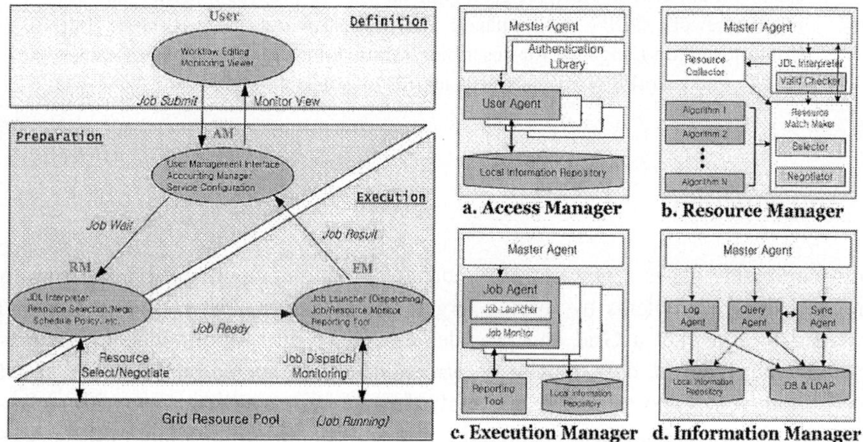

Fig. 1. Meta Scheduling Framework Architecture

Figure 1 shows the architecture of MSF and its major components. A user describes a job flow using the MSF console. The Access Manager (AM) provides services which include user authentication, environment setup, and job submit service. The Resource Manager (RM) provides resource discovery and matching. The Execution Manager (EM) provides job launching, monitoring, and reporting.

MSF consists of three phases; definition, preparation, and execution. During the definition phase jobs are defined to specify a Job Definition List (JDL), which describes a task flow using JCML. In this phase, users connect to the AM for authentication of the MSF and then the AM creates a user proxy for Globus. In the preparation phase, resources are searched and assigned to the matched tasks. The AM creates an agent to provide a proxy service to the user. The agent passes the JDL and traces the status of the job. The RM receives the JDL from the AM and analyzes it. After finding appropriate grid resources for the job, the RM assigns them to tasks and generates a worklist that includes information on activities and their executing order. Finally, during the execution phase the tasks on the worklist are executed, the status is monitored, and the results are reported.

A job description language to specify the flow of an application task has to provide a way to describe various grid environments and task information including execution environments, such as arguments, sequence, data dependency, prefetching, and so on. The JCML is a workflow description language based on the Graph eXchange Language (GXL) [5], which defines a graph-based XML representation for specifying the dependencies among components. The JCML consists of four major elements; Info, Resource, Component, and Dependency.

Info: This element lists the document name, scope, target namespaces, authoring date, and so on.

Resource: This element describes the resources of hardware and software required to execute a job. The hardware includes architecture, CPU, memory, disk, and network bandwidth. The software includes an operating system, installed application, and local scheduler. The Time represents the deadline to be executed.

Component: This element lists all of the task-related information. A Node is an executing program or computer in the workflow. A node is classified into a task and a resource. A task node is an executing program in workflow. And a resource node is an assistant which represents the physical computing resources, such as storage and database, to support the task node. A task node includes execution file, input, output, arguments, and resource configuration, and a resource node includes data location and the access method. A group is a node logically, if it is necessary to refer a series of nodes to a single node according to a job flow logic. And one group can include another group(s).

Dependency: This element describes the dependencies of a workflow. Each line is an edge which represents an execution order and a dependency between two objects (node or group). It has two types of links, PriorityOrder and Datalink, which have a direction that expresses a starting point and an end point of linked nodes. The PriorityOrder just represents an execution order between the two linked nodes. The Datalink displays a flow of data which is used for the input or output file of each task.

A workflow management system guarantees that a flow of task will be executed exactly. A job is processed by a workflow management system as follows: interpreting the job, mapping resources, generating a worklist, and scheduling tasks, as shown in Figure 1. A user specifies a JDL to execute a grid application. After interpreting the JDL, the JDL Interpreter searches resources and assigns resources to the job, selected by the Resource Match Maker, which uses an external information service such as NWS [6], and then generates the worklist. The worklist has the execution information of the task and its execution order. Each task consists of activities.

The Job Launcher executes activities according to the worklist. At the workflow system, it is required to transform activities in the worklist to RSL type in order to execute the real task in a local grid scheduler.

3   Implementation and Conclusion

We developed a prototype of MSF using pure Java (JDK 1.4.0) and Java CoG Kit (version 0.9.13) on Globus 2. We also implemented and executed a Virtual Screening on MSF. A Virtual Screening is one of the design methods, called *docking* in Bioinformatics. In order to execute docking, the format of material must be changed. We chose the AutoDock application for this experiment. Figure 2 displays snapshots of the progress windows for AutoDock processing

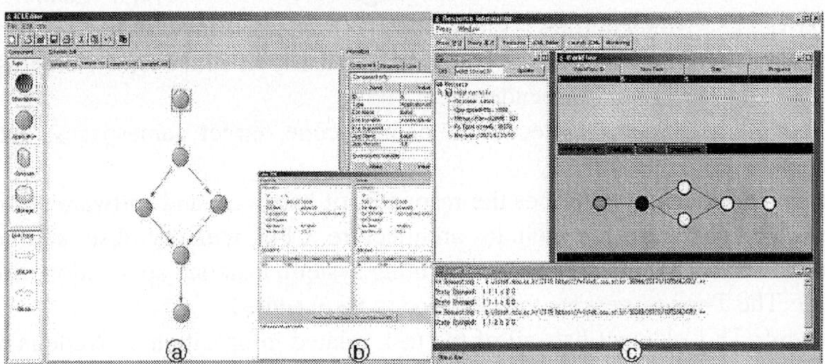

Fig. 2. (a) JCML Main Editing Window, (b) Edge Window, (c) Monitoring Console

MSF provides a workflow service for general applications in a Grid environment. So we designed and implemented a workflow description language, JCML, to describe the flow of application including complexity and dependencies, and a workflow management system to execute and monitor the flow. Currently we are working to extend the architecture in order to enhance efficiency and availability and to describe jobs in more detail with JCML. And MSF will support the Globus 3.0, which integrates scientific and enterprise environments based on web service.

References

1. I. Foster and C. Kesselman, ed., The Grid: Blueprint for a New Computing Infrastructure, Morgan Kaufmann, (1998).
2. J. Cao, S. A. Jarvis, S. Saini and G. R. Nudd, GridFlow: Workflow Management for Grid Computing, 3rd International Symposium on Cluster Computing and the Grid, (2003), 12-15.
3. R. Stevens, A. robinson and C. Goble, myGrid: personalized bioinformatics on the information grid, Bioinformatics, 19(1), (2003), 302-304.
4. M. Litzkow, M. Livny and M. Mutka, Condor - A Hunter of Idle Workstations, 8th International Conference of Distributed Computing Systems, (1998), 13-17.
5. A. Winter, B. Kullbach and V. Riediger, An Overview of the GXL Graph Exchange Language, Software Visualization. LNCS, Vol. 2269, (2002), 324-336.
6. R. Wolski, N. Spring and J. Hayes, The Network Weather Service: A Distributed Resource Performance Forecasting Service for Metacomputing, Future Generation Computing Systems, 15(5/6), (1999), 757-768.

Resources Virtualization in Fault-Tolerance and Migration Issues

G. Jankowski, R. Mikolajczak, R. Januszewski, N. Meyer, and M. Stroinski

Poznan Supercomputing and Networking Center
61-704 Poznan, ul. Noskowskiego 12/14, Poland
{gracjan,fujisan,radekj,meyer,stroins}@man.poznan.pl

Abstract. One of the low-level services Grids can benefit from is checkpointing. Unfortunately, the checkpointing technology imposes many common problems that are still open. The example of such often-encountered problems is recovering the identifiers of some checkpointed resources. This paper describes the idea of low-level virtualization of such identifiers. The mechanism allows overcoming the semantically imposed limitations.

1 Introduction

In this paper we focus on the coherency of restored process memory with reference to the surrounding environment and resources rather than recovering the processes themselves. Some variables in the process memory can hold the value of the identifiers that point to an external resource. After migration it can turn out that although the value has been restored properly, the external resource cannot be linked with this value any more. The solution to this problem can be the usage of transparent (from the program's viewpoint) virtualization of these resources identifiers.

2 Considered Resources' Description

Actually each resource type has its own semantics. Although the general idea of resources virtualization is invariable, the particular cases may require a little tune attention. In this paper we opt for tackling the System V IPC objects and their identifiers and keys. The System V IPC mechanisms include messages, shared memory and semaphores objects. Each object is unambiguously identified by its type and the key value. In order to gain access to a particular object, the process has to use one of the system calls *msgget()*, *shmget()* or *semget()* for messages, shared memory and semaphores, respectively. The key value, which determines the desired object, is passed to these functions and the return value is the object identifier through which further interaction with the object is performed. The operating system can forbid access to the object for any reasons (e.g. lack of rights to the object). If the process requires the object of IPC_PRIVATE key value then the operating system supplies a completely new IPC object, i.e. not

M. Bubak et al. (Eds.): ICCS 2004, LNCS 3036, pp. 449–452, 2004.

shared with anyone. The problem is that during the recovery phase some other processes can already occupy the IPC objects that were originally used. Furthermore, the physical identifier of re-gained object will most probably have a different value than it originally did.

3   Virtualization

The main idea of the presented solution is intercepting the crucial system calls and replacing the virtual identifiers (passed by the process) with the physical ones. The function that is executed as a result of interception is called the *intercepting system call*. It is actually a wrapper. The wrapped function is named the *original system call*. The *original system call* is finally invoked by the *intercepting system call*. In some cases, the values that are returned by the *original system calls* are replaced, too.

The virtual identifiers that are used by the user programs are mapped to the physical ones by means of the *mapping tables*. Each mapping table defines a separated *mapping domain* of identifiers of the resources of the same type. There can be multiple *mapping domains* of a given kind of resources in the system at the same time, but one resource instance can be associated with only one *mapping domain*. The format of the *mapping table* depends on the kind of resources that are virtualized by this table. If any program has access to a *mapping table*, we say that this program is associated with the *mapping domain* that is defined by this table. Moreover, if a process is associated with the *mapping domain*, all its children inherit this association. To increase the legibility of the presented algorithms, we make an assumption that a single *mapping domain* can be associated with only one application (single- or multi-process). All domains associated with the application must be the domains of different type of resources. It means that for a given application and a given resource type, only one domain (i.e. only one *mapping table*) can exist. The next assumption that simplifies further study is that programs associated with different mapping domains should not communicate with each other. The *mapping tables* must be held in the shared memory. Access to this memory must be synchronized. The values of virtual identifiers must be unique within the scope of the *mapping domain*. However, the virtual identifiers from different domains can have the same values, even if these are the domains of resources of the same type. The first process that is created as a part of the multi-process program as well as the single process (the single-process application) is called the *root process*. All processes that are descended from the *root process* are named *branch processes*.

The virtual key and virtual identifier of all types of the System V IPCs have the same semantics. For this reason, in this section we refer to IPC objects generally instead of considering each of them separately. In order to simplify the following description, a special case when the object key equals IPC_PRIVATE was omitted.

The *mapping table* used for managing the *mapping domain* of the System V IPC object consists of four columns. The VKEY column holds the virtual key of an IPC object. The RKEY column holds the physical value of the IPC object. The VID column is the virtual value of the identifier of the IPC object and the RID column is the physical identifier of the IPC object.

During the initialization of the process, if it is the *root process* that is initialized, the memory for a *mapping table* it allocated, or if it is the *branch process*, it is attached to the *mapping domain* of its parent.

When the *xxxget()*[1] *intercepting system call* is invoked, first by means of the *mapping table* the virtual key (the one which is passed to the *intercepting system call*) is translated to the physical one. If the *mapping table* does not contain mapping for the current virtual key, the physical one is assigned the same value as the virtual one. The physical key is passed to the *xxxget() original system call*. If this function returns the error code, it is forwarded to the user process and the execution of the *intercepting system call* is finished. If the *original system call* returns the correct IPC object identifier (physical identifier), the RID column of the *mapping table* is searched for it. If the searching succeeds, the row that contains that value is marked and the virtual identifier is given the value of the VID column of the marked row. Otherwise, if the current *mapping domain* does not contain the virtual identifier of the same value as the one returned by the *original system call*, the new virtual identifier is given the same value as the just obtained physical one. If the physical identifier is not in the RID but in the VID column of the *mapping table*, the new virtual identifier is given an arbitrary value that is different from all the other within the VID column. If the VID column of the *mapping table* does not contain the just established virtual identifier yet, the new row is added to this table. The VKEY, RKEY, VID and RID columns are given the values of virtual key, real key, virtual identifiers and real identifiers, respectively. Finally, the value of the virtual identifier is returned as a result of the *intercepting system call*.

The *intercepting system calls* that operate on the IPC objects take the virtual IPC object identifier as a parameter. Before the *original system call* is called, the virtual object identifier must be translated into the physical one. To achieve that, the VID column of the *mapping table* is searched for the virtual value that has been passed to the *intercepting system call*. The row that contains the searched value is marked. The physical identifier that is passed to the *original system call* is taken from the RID column of the marked row. The value returned by the *original system call* is forwarded to the user process.

The values of physical keys and objects identifiers are the same as the values of the related virtual ones until the execution is not interrupted by migration or failure. The presented algorithm can be applied by the user-level and kernel-level checkpointing solutions, but it better fits the former case. To simplify the description below, the algorithms made by the *root process* and the *branch processes* are described separately.

The *root process* is recovered as the first one. It reallocates and restores the *mapping table* and then, for each row in this table, tries to request from the system the IPC object of the same key value as it was before the recovery phase

---

[1] *xxxget()* stands for *msgget()* or *shmget()* or *semget()*.

(by means of the *xxxget() original system call*). If, unluckily, the originally used key is occupied by another process, the new physical key is given an arbitrary value that is different from all other keys that are currently occupied. The value returned by the *xxxget() original system call* is the new physical value of the IPC object identifier. The RKEY and RID columns of the current row of the *mapping table* are updated with the values of the new physical key and identifier, respectively. If the type of the recovered IPC object is the shared memory and if the *root process* had it attached to its own memory space, the memory is attached to the *root process'* address space. Finally, the state or content (depending on the case) of the just recovered object is restored.

When the *root process* finishes the recovery phase, each *branch process* is attached to the *mapping table*. Generally, from the point of view of System V IPC objects, access to a correctly filled *mapping table* is sufficient for the *branch processes* to be executed properly. However, if the recovered IPC object is the shared memory, the *branch process* for each row in the *mapping table* additionally makes one more following step. If the shared memory, which is associated with the current row in the *mapping table*, was attached to the current process, it is reattached.

When a system call which releases the IPC object ends with success, the row that is correlated with it must be removed from the *mapping table*. When the *root process* is finished, the shared memory containing the *mapping table* must be freed.

4   Conclusion

The idea presented above is a kind of hint how to look at the resources virtualization issue rather than any formal and stiff standard proposition. An attempt to define and classify the basic notions and terms associated with resources virtualization has been made. The authors of this paper have implemented the presented conception in *psncLibCkpt* package, a checkpointing library used on user level, which supports the System V IPC objects virtualization (PROGRESS project: `http://progress.psnc.pl`).

References

1. J.S. Plank, M.Beck, G. Kingsley, and K.LI. Libckpt: Transparent Checkpointing Under UNIX, Conference Proceedings, Usenix Winter 1995. Technical Conference, pages 213-223. January 1995.
2. Hua Zhong and Jason Nieh. CRACK: Linux Checkpointing / Restart As a Kernel Module. Technical Raport CUCS-014-01. Department of Computer Science. Columbia University, November 2002.
3. Eduardo Pinheiro. Truly-Transparent Checkpointing of Parallel Applications. Federal University of Rio de Janeiro UFRJ.
   `http://www.research.rutgers.edu/ edpin/epckpt/paper_html/`.

On the Availability of Information Dispersal Scheme for Distributed Storage Systems*

Sung Keun Song[1], Hee Yong Youn[1], Gyung-Leen Park[2], and Kang Soo Tae[3]

[1]School of Information and Communications Engineering, Sungkyunkwan University
Suwon, Korea
kskk103@skku.edu, youn@ece.skku.ac.kr
[2]Department of Computer Science and Statistics, Cheju National University,
Cheju, Korea
kstae@jeonju.ac.kr
[3]Dept. of Computer Engineering, Jeonju University, Korea
glpark@cheju.ac.kr

Abstract. For distributed storage systems the way how the data are replicated and distributed significantly affects availability, security, and performance. Here many factors such as the number of nodes and partitions, replications, node survivability, etc. are interrelated. This paper investigates the availability of Information Dispersal Scheme that can be used for distributed storage system. It will help construct a large distributed system allowing high availability.

1 Introduction

The survivable storage system [1-2] requires to encode and distribute data over multiple storage nodes to survive failures and malicious attacks. It also needs to replicate data to enhance availability. For distributed storage systems the way how the data are replicated and distributed significantly affects the availability, security, and performance.

There exist various data replication and distribution schemes such as replication, splitting, information dispersal, and secret sharing [3-8]. The schemes display different availability, security, and performance trade-off since many factors such as the number of nodes, storage space, operation speed, etc. affect each other. Therefore, finding an optimal scheme for a given condition is very difficult. In this paper we formally define a data replication and distribution scheme called information dispersal scheme (IDS), and investigate the availability of the IDS for deciding an optimal IDS with a given condition. It will help construct a large distributed system allowing

---

* This work was supported in part by 21C Frontier Ubiquitous Computing and Networking, Korea Research Foundation Grant (KRF - 2003 - 041 - D20421) and the Brain Korea 21 Project in 2003. Corresponding author: Hee Yong Youn

M. Bubak et al. (Eds.): ICCS 2004, LNCS 3036, pp. 453–457, 2004.

454   S.K. Song et al.

high availability. The rest of the paper is organized as follows. Section 2 investigates
the availability of the IDS. We conclude the paper in Section 3.

2 The Availability of IDS

The basic properties of IDS are reported in [9]. This section focuses on the availability of IDS. The notations are as follows. The (m, n)-IDS is a data distribution scheme
where m pieces of the original data are replicated into n pieces which are stored in n
nodes respectively.

$k(=n/m)$	Information Expansion Ratio (IER); $k \geq 1$
P	node survivability; $0 < P < 1$
$P(m, n)$	availability of the (m, n)-IDS
$P^*((i, j), (m, n))$	critical node survivability which allows $P(i, j)=P(m, n)$
$Class_i$	all IDS's whose k is i
$(m, n)_{(i, j)}$-IDS	boundary IDS of $Class_{n/m}$; for example, if $m > s$ and $n > t$, (i, j)-IDS

and (s, t)-IDS of $Class_{n/m}$ do not have a critical node survivability. However, if $m \leq s$ and $n \leq t$, there exists a critical node survivability.

Availability of (m, n)-IDS is as follows [9-10].

$$P(m, n) = \left(\sum_{i=1}^{k} \binom{k}{i} P^i (1-P)^{k-i} \right)^m, \quad k = \frac{n}{m} \text{ (In what follows } k \text{ is assumed to be an integer)} \tag{1}$$

Some important properties of IDS based on this availability formula are as follows.

Theorem 1: If m and n increase by the same ratio, the availability decreases. Say,

$$P(k_1 m, k_1 n) > P(k_2 m, k_2 n) \text{ if } k_1 < k_2 \tag{2}$$

Proof: $(k_1 m, k_1 n)$-IDS and $(k_2 m, k_2 n)$-IDS have the same IER and belong to the same
class. Then,

$$P(m, n) > P(m+i, n+j) \text{ if } k=n/m=(n+j)/(m+i), i, j \geq 1$$

Therefore, $P(k_1 m, k_1 n) > P(k_2 m, k_2 n)$. ∎

Theorem 2: For two IDS's, A and B, if the number of partitions of A is smaller than
that of B while the k value of A is larger, then the availability of A is larger than that
of B. Say,

$$P(i, j) > P(m, n) \text{ if } i < m \text{ and } j/i \geq n/m \tag{3}$$

Proof: If $j/i = n/m$, then this is the following case.

$$P(m, n) > P(m+i, n+j) \text{ if } k=n/m=(n+j)/(m+i), i, j \geq 1$$

If $j/i > n/m$, then $j > (ni)/m$. Since $P(m, n) < P(m, n+mi)$ for $i > 1$, $P(i, j) > P(i,$
$(ni)/m)$. $P(i, (ni)/m) > P(m, n)$ due to Theorem 1 since $((ni)/m)/i = n/m$. As a result,
$P(i, j) > P(m, n)$. ∎

Theorem 2 reveals that availability increases if the number of partitions decreases and
the number of replications increases.

Theorem 3: Given *Class$_a$* and *Class$_b$* of different IER values, if $a < b$, an $(m, n)_{(i, j)}$-IDS of *Class$_b$* always exists, where $m > i$ and $n > j$. If $a > b$, an $(m, n)_{(i, j)}$-IDS exists, where $m < i$ and $n < j$.

Proof: In the case of $a < b$, if m and n are smaller than i and j respectively, for all P ranges, (m, n)-IDS is more available than (i, j)-IDS by Theorem 2. That is, critical node survivability does not exists. Note that if $l \to \infty$, $P(l, bl) \to 0$ by Theorem 1. Therefore, a boundary IDS, $(m, n)_{(i, j)}$-IDS exists, where $m > i$ and $n > j$. Similarly, in the case of $a > b$, if m and n are larger than i and j respectively, for all P ranges, (i, j)-IDS is more available than (m, n)-IDS by Theorem 2. If $l \to 0$, $P(l, bl) \to 1$ by Theorem 1. Therefore, a boundary IDS, $(m, n)_{(i, j)}$-IDS exists, where $m < i$ and $n < j$. ∎

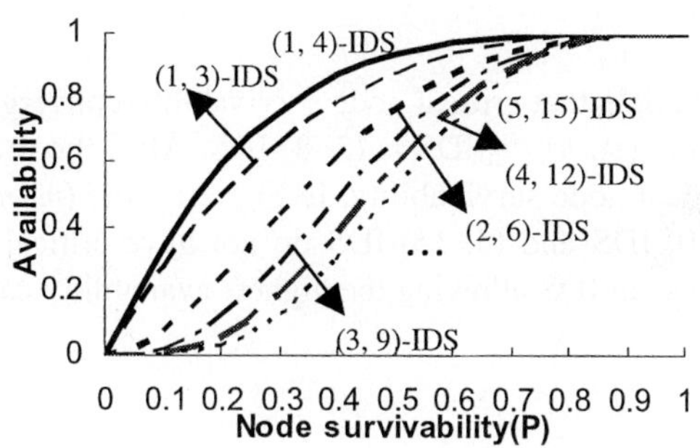

Fig. 1. The availabilities for (1, 4) IDS and IDS's of *Class$_3$*

We know that if (m, n)-IDS and *Class$_a$* do not have a boundary IDS and the IER of the (m, n)-IDS is larger than a, the (m, n)-IDS is more available than all IDS's of *Class$_a$* for entire P range. Fig. 1 shows an example of Theorem 3. Generally, the IDS of $(1, 1)$, $(1, 2)$, $(1, 3)$, •••, $(1, n)$ and the classes that have smaller IER's than (m, n)-IDS do not have a boundary IDS.

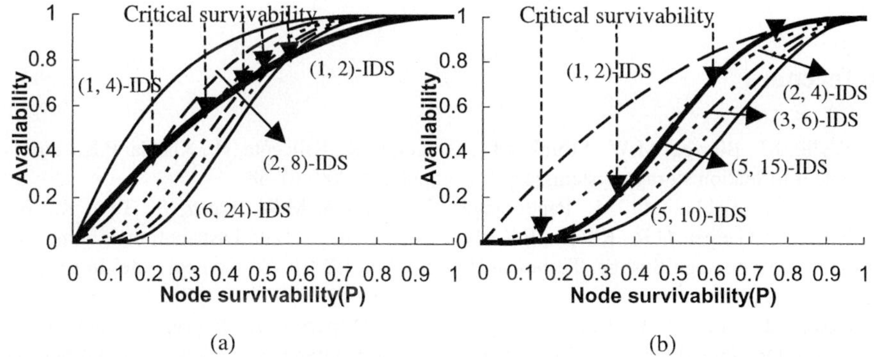

(a)   (b)

Fig. 2. The availabilities for (1, 2)-IDS and the IDS's of *Class$_4$* and the availabilities for (5, 15)-IDS and the IDS's of *Class$_2$*

Theorem 4: Suppose that an (i, j)-IDS of $Class_a$ and $Class_b$ have $(m, n)_{(i, j)}$-IDS of $Class_b$ and $a < b$. If m and n increase, critical node survivability converges to 1. On the contrary, suppose that an (i, j)-IDS of $Class_a$ and $Class_b$ have $(m, n)_{(i, j)}$-IDS and $a > b$. If m and n increase, critical node survivability converges to 0.

$$P^*((i, j), (m, n)) \rightarrow 1 \text{ if } j/i < n/m \text{ and } m,n \rightarrow \infty$$
$$P^*((i, j), (m, n)) \rightarrow 0 \text{ if } j/i > n/m \text{ and } m,n \rightarrow \infty \tag{4}$$

Proof: The IDS of the largest availability in the $Class_b$ is $(1, b)$-IDS by Theorem 1. Here, if $l \rightarrow \infty$, $P(l, bl) \rightarrow 0$. Therefore, when m and n increase, while $a < b$, critical node survivability of (i, j)-IDS of $Class_a$ and $(m, n)_{(i, j)}$-IDS converges to 1. If $a > b$, it converges to 0.   ∎

Fig. 2 shows an example that critical node survivability converges to 1 or 0, respectively. In Fig. 2(a), the $(m, n)_{(1, 2)}$-IDS is (2, 8)-IDS. Also, the (1, 4)-IDS and (1, 2)-IDS do not have critical node survivability. In Fig. 2(b), the $(m, n)_{(5, 15)}$-IDS is (4, 8)-IDS. Also the (5, 10)-IDS and (5, 15)-IDS do not have critical node survivability. Using these properties, an IDS allowing the highest availability can be determined for a given condition.

3   Conclusion

In this paper we have studied the availability of information dispersal schemes that can be used for survivable storage systems. It will help construct a large distributed system allowing high availability. In the study, we made some assumptions in deriving the models. We will develop a more vigorous model without such assumptions which allows the best IDS in real environment. We will also investigate the properties of IDS in terms of both security and availability with which secure and highly available IDS can be obtained.

References

1. Wylie, J.J., Bigrigg, M.W., Strunk, J.D., Ganger, G.R., Kiliccote, H., Khosla, P.K.: Survivable information storage systems. IEEE Computer. (2000) 61-68
2. Wylie, J.J, Bakkaloglu, M., Pandurangan, V., Bigrigg, M.W., Oguz, S., Tew, K., Williams, C., Ganger, G.R., Khosla, P.K.: Selecting the Right Data Distribution Scheme for a Survivable Storage System. Technical Report CMU-CS-01-120 Carnegie Mel-lon University. (2001)
3. Choi, S.J., Youn, H.Y., Choi, J.S.: An Efficient Dispersal and Encryp-tion Scheme for Secure Distributed formation Storage, International Conference on Computational Science, Springer-Verlag, (2003) 958-967
4. Rabin, M.O.: Efficient Dispersal of Information for Security. Load Balancing and Fault Tolerance. ACM (1989) 335-348

5. Shamir, A.: How to Share a Secret: Comm. ACM. (1979) 612-613
6. Blakley, G.R., Catherine Meadows: Security of ramp scheme: Advances in Cryptol-ogy, Springer-Verlag, (1985) 242-268
7. Karnin, E., Greene, J., Hellman, M.: On Secret Sharing Systems: IEEE Trans. Informa-tion Theory (1983) 35-41
8. A. De Santis and B. Masucci: Multiple Ramp Schemes: IEEE Trans. Information Theory (1999) 1720-1728
9. Song, S.K., Youn, H.Y., Park, J.K.: Deciding Optimal Information Dispersal for Parallel Computing with Failures: International Conference on Parallel Computing Technologies, Springer-Verlag, (2003) 332-335
10. Hung-Min Sun., Shiuh-Pyng Shieh.: Optimal Information Dispersal for Increasing the Reliability of a Distributed Service. IEEE Trans. Vol. 46. (1997) 462-472

Virtual Storage System
for the Grid Environment

Darin Nikolow[1], Renata Słota[1], Jacek Kitowski[1,2], and Łukasz Skitał[1]

[1] Institute of Computer Science, AGH-UST, al.Mickiewicza 30, Cracow, Poland
[2] Academic Computer Center CYFRONET AGH, ul.Nawojki 11, Cracow, Poland

Abstract. Experiments in a Grid-based virtual laboratory, as well as, simulation and visualization grid computing usually deal with large data kept in different locations often far away from each other. These data need to be archived. The goal of the Virtual Storage System (VSS) for grid-based accessing is to integrate the mass storage resources distributed geographically into a common storage service. In this paper the architecture of a virtual storage system is discussed and implementation details are presented.

1   Introduction

Grid computing provides computational, visualization and data storage services, by using geographically distributed resources. Some of the grid projects concerns high performance computing and visualization for virtual laboratory applications. Visualization applications running on the grid need to access large amounts of data possibly distributed among the participating sites. Grid data management is an important topic in many grid-related research projects [1,2, 3]. The data obtained during experiments in the Virtual Laboratory (VLAB) or the data being results of simulation or visualization often need to be archived. A Virtual Storage System (VSS) for grid-based accessing, providing the demanded archiving service, is under development for the SGIgrid project [4]. The main goal of VSS is to integrate the mass storage resources residing in the participating computer centers into a common storage service. High Performance Computing (HPC) sites use tertiary storage (like tape libraries and optical jukeboxes) to economically store vast amounts of data. Usually, in such cases the tertiary storage is managed by the Hierarchical Storage Management (HSM) type of software. In the participating sites the DiskXtender HSM software [5] by Legato Systems is used.

Different grid based data management systems for replicated data sets are being currently developed. Storage Resource Broker (SRB) has been developed at San Diego Supercomputing Center [6]. SRB is a client-server middleware providing unified interface for connecting different type mass storage facilities over network. The Reptor system is a prototype of the replica management service developed as a part of the EU DataGrid project [1]. Data Management System (DMS) has been developed as a part of the Progress project [7]. DMS

M. Bubak et al. (Eds.): ICCS 2004, LNCS 3036, pp. 458–461, 2004.
© Springer-Verlag Berlin Heidelberg 2004

is aimed at providing access to distributed mass storage through integrating the data in a virtual filesystem for the purpose of the computational portal.

The proposed VSS differs from other systems for managing distributed storage resources by its specific functionalities. Each of the mentioned above systems could be used as a base for developing the VSS, by extending the existing systems functionalities.

The rest of the paper is organized as follows: VSS new functionalities are described in Section 2. Some implementation details are given in Section 3. In the last section we conclude the paper and provide some insight into future work.

2   Functionality Details

The VSS provides the following add-on functionalities: data access time estimation, file ordering, replica management, automatic generation and selection of replica, file fragment access, API for the user application, described below.

Access Time Estimation. Access time for data kept in HSM systems can take values from wide range (few miliseconds to tens of minutes). Therefore, it is essential to know in advance the access time for such data, for example for the replica selection algorithm [8]. The HSM access time estimation subsystem attempts to estimate the latency and transfer times for the file, which is eventually going to be requested [9].

File Ordering. The user has ability to order a file, which means to inform the system, when he will need to access the file and how long it will be required. If the file is located on slow media (in the mean of access time), the system forces to transfer it to the fast disks cache and locks it for a given period of time. The next problem is optimization of staging operations for the ordered files, i.e., to select the right moment of issuing a file staging request to the HSM system. Results obtained with *Access time estimation* can be helpful in making the proper selection of the moment to start copying; in order to have some safety margin we compute the *scheduled time* in the following way: $T = order\_time - ETA*X + Y$, where ETA (Estimated Time of Arrival) is the latency time, returned by the HSM estimator, X, Y variables (or functions) describing our safety margin.

Replica management, automatic generation and selection. Replication has two purposes: to increase data safety in the case of destroying or damaging and to increase data availability. In the first case the user has ability to mark a file as "important", which forces the system to replicate this file. In the second case the replication is done automatically.

Selection of optimal replica is based on the local HSM access time estimation and the network transfer rate between the client and the site keeping the required data sets [8].

File Fragment Access. The user has ability to access a specified fragment of the file. This is very useful in a case, when the user needs access to some data in a large, well ordered file. Advantages are the lower latency and the shorter transfer time. Access to file fragments is done by file ordering.

Application Programming Interface. VSS API allows programmers to omit details of specification of communication protocol between client and VSS and focus onto usage of the system. API has been developed for Java programming language.

3   VSS Architecture

DMS [7] has been chosen as a data storage and management system for the task, which is responsible for developing the VLAB virtual laboratory system in the SGIgrid project. In order to keep the project consistent we decided to use DMS as a base for developing VSS.

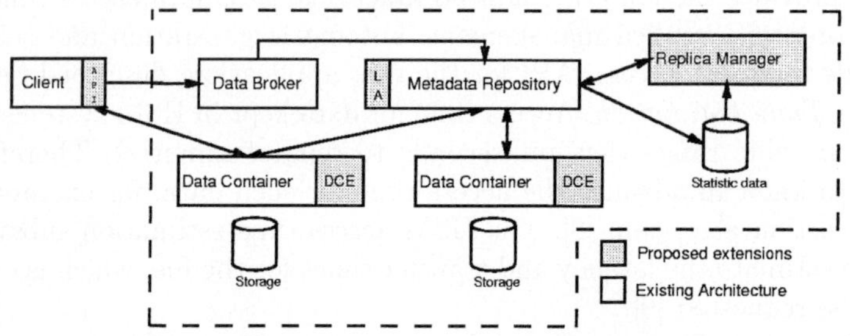

Fig. 1. VSS architecture based on DMS.

In Fig. 1 the DMS-based architecture of VSS is shown. The DMS consists of three main modules: Data Broker, Metadata Repository and Data Container. The Data Broker receives data access requests from the Client, checks permissions, updates the metadata via the Metadata repository module and sends back to the client a handle for accessing the physical data stored on the Data Containers. Metadata Repository keeps the meta data records in a general purpose database. DMS uses the web services technology and SOAP for communicating between components.

In order to develop VSS using DMS as a base, some modules need to be extended with new functionality or new modules need to be added to the architecture. The gray painted boxes (see Fig. 1) indicate the proposed extensions. The DCE (Data Container Extension) module provides access time estimation, file fragment access and file ordering capabilities. The following SOAP methods has been implemeted: estimateFile(), which estimates the acces time for a given physical file; addFileOrder(), removeFileOrder() and updateFileOrder(), which are responsible for managing the file ordering; tranferFile(), which realizes file transfers between data containers. The Replica Manager module is responsible for automatic data replication based on statistical data. The optimization algorithm for replication decides, based on these data, which files have to be replicated. At present, a basic algorithm, taking into account frequency of file usage and user category, has been implemented. The LA (Log Analyzer) obtains file access statistical information (like the number of references and file transfer performance chracteristics) from DMS logs. It is implemented in the Perl language.

4   Conclusions

In this paper the design and implementation details of the virtual storage system developed as part of the SGIgrid project has been presented. This storage system is aimed at integrating the mass storage equipment installed in the participating sites into a common data archivization service. By using replica automatic generation and selection the system is flexible and suitable for efficient and reliable usage of the distributed storage resources by the grid-enabled applications. The described system is different from the ones being developed in similar projects since it provides additional functionality for the HSM-based storage resources like data access time estimation, file ordering and efficient access to file fragments.

Acknowledgments. The work described in this paper was supported by the Polish Committee for Scientific Research (KBN) project "SGIgrid" 6 T11 0052 2002 C/05836 and in part by KBN project 4 T11C 028 24 and by AGH grant. ACC CYFRONET-AGH is acknowledged.

References

1. "DataGrid – Research and Technological Development for an International Data Grid", EU Project IST-2000-25182.
2. "CROSSGRID – Development of Grid Environment for Interactive Applications", EU Project IST-2001-32243.
3. Dutka, Ł., Słota, R., Nikolow, D., Kitowski, J., "Optimization of Data Access for Grid Environment", presented at 1st European Across Grids Conference, Universidad de Santiago de Compostela, Spain, February 13-14, 2003.
4. SGIgrid: Large-scale computing and visualization for virtual laboratory using SGI cluster (in Polish), KBN Project, http://www.wcss.wroc.pl/pb/sgigrid/
5. Legato Systems, Inc. - DiskXtender Unix/Linux,
   `http://www.legato.com/products/diskxtender/diskxtenderunix.cfm`.
6. Storage Resource Broker, `http://www.npaci.edu/DICE/SRB/`.
7. PROGRESS, `http://progress.man.poznan.pl/`.
8. Stockinger, K., Stokinger, H., Dutka, Ł., Słota, R., Nikolow, D., Kitowski, J., "Access Cost Estimation for Unified Grid Storage Systems", 4-th International Workshop on Grid Computing (Grid 2003), Phoenix, Arizona, November 17, 2003, IEEE Computer Society Press.
9. Nikolow, D., Słota, R., Kitowski, J. "Gray Box Based Data Access Time Estimation for Tertiary Storage in Grid Environment", 5-th Int. Conf. Parallel Processing and Applied Mathematics, Czestochowa, Poland, September 7-10, 2003, LNCS vol.3019.

Performance Measurement Model
in the G-PM Tool[*]

Roland Wismüller[1], Marian Bubak[2,3], Włodzimierz Funika[2,3],
Tomasz Arodź[2,3], and Marcin Kurdziel[2,3]

[1] LRR-TUM, Institut für Informatik, Technische Universität München, D-85747
Garching, Germany
`wismuell@in.tum.de`
[2] Institute of Computer Science, AGH, al. Mickiewicza 30, 30-059 Kraków, Poland
[3] Academic Computer Centre – CYFRONET, Nawojki 11, 30-950 Kraków, Poland
`{bubak,funika}@uci.agh.edu.pl`
phone: (+48 12) 617 39 64, *fax:* (+48 12) 633 80 54, *phone:* (+49 89) 289 17676

Abstract. This paper focuses on the model of the performance analysis
of distributed grid-enabled applications within the G-PM tool. The
major focus is at the issues that arises as a consequence of the on-line
application monitoring paradigm the G-PM follows. In particular, two
major issues are presented - the consequences of a discrete nature of
measuring the function based performance quantities, and the aggregation of performance values that were measured at distant locations, and
thus may be desynchronized.

Keywords: Grid, performance analysis, performance monitoring

1   Introduction

The introduction of grid computing has a great impact on the development of
parallel applications. In order to meet the new requirements of the users, the
application development tools also have to be adapted to the new computing
paradigm.

In this paper, we describe the evolution of one these tools, i.e., the performance analysis tool for grid applications – G-PM [3]. The G-PM is developed
as part of the CrossGrid project [4]. The tool uses the OCM-G [1] as a low level
monitoring system and was designed to allow for on-line performance analysis.
During its development, a number of issues concerning the implementation of
measurements have to be addressed in order to meet the constraints of the grid
computing. This resulted in a relaxation of the strict on-line scheme of the tool.
Furthermore, the initially assumed pure pull model of communicating with the
OCM-G was augmented to include reservation requests.

---

[*] This work was partly funded by the European Commission, project IST-2001-32243,
CrossGrid

M. Bubak et al. (Eds.): ICCS 2004, LNCS 3036, pp. 462–465, 2004.
© Springer-Verlag Berlin Heidelberg 2004

2   Issues with Measuring the Function-Based Metrics

The G-PM is mainly focused on measuring various quantities (metrics) related to function calls. This may be e.g. duration of the *MPI_Send()* calls since the application start-up or the amount of data received with *MPI_Recv()* function. The appropriate library functions are instrumented, so the values of these metrics are gathered at the start and the end of the individual function calls.

Consider the metric measuring the total time spent in *MPI_Send()* since some start time, denoted as a function $R(t)$ of time t. It is not possible to measure the value of R directly. What can be measured is the function $V(t)$, i.e. the total wall-clock time since the beginning of the measurement (in such a case, $V(t) = t$). The value of R can then be obtained in the following way: $R(t) = \sum_{i:t_i^b, t_i^e \leq t} (V(t_i^e) - V(t_i^b))$. The t_i^b and t_i^e are the time stamps of the ith begin and end of the function call. In the same approach, the $V(t)$ may, e.g., denote the total amount of data received by *MPI_Recv()* until the time t.

The G-PM tool inspects the value of R at fixed intervals, i.e., at moments t_j. The scheme outlined above works fine, as long as t_j falls into the range (t_i^e, t_{i+1}^b) and not in (t_i^b, t_i^e), i.e. falls between consecutive function calls. In the latter case, the measured value of R does not reflect the real value of the metric. This is because the value of R is not updated until the end of the function call. Only then $V(t_i^e)$ is known, and difference between the values of V at the end and beginning of the function call (i.e. difference between $V(t_i^e)$ $V(t_i^b)$) can be added to the previous value of R. For example, in case of the *MPI_Send()* delay, the value of R, if inspected during the call to *MPI_Send()*, does not take into account the time spent in this call.

We propose two solutions to this problem. In case the value of $V(t)$ can be accessed at any moment in time, as it is in the case of $V(t) = t$, we modify the function R if the time t_j of the query falls within some function call. The current value of V at time t_j of the data query (i.e. $V(t_j)$) is used instead of the yet unknown value $V(t_i^e)$ at yet unknown time t_i^e of the end of the call.

The scheme outlined above is not suitable if the value of $V(t)$ cannot be queried at any time. That is the case of e.g. *MPI_Recv()* total data volume transferred. The amount of the data received can be known only at the times of the end of the function call. Therefore, if R is inspected during the call to *MPI_Recv()*, the value $R(t_j)$ gives only the amount of data received before the beginning of the call. The data received between the beginning of the function call and the querying of the value of R is not taken into account. Thus, the values of the metrics can be misleading, especially when the duration of the function calls is long.

A solution to the problem is to make use of the asynchronous nature of the underlying OCM-G monitoring system. The replies to the data requests are sent to the G-PM asynchronously. Originally, the reply is sent immediately as the data is gathered from the sites of the grid where the monitored application is running. However, this reply can be postponed until the current function call ends, and the value of R reflects the real value. In this way, the G-PM tool will always display correct data. However, the visualization may be temporarily delayed, as the replies will no longer arrive immediately after the requests for measurement data to be displayed.

3   Performance Measurement under Large Network Latency Conditions

The G-PM makes it possible to narrow the measurement of performance properties of the application to any set of processes specified by the user. The partial results from this locations can be aggregated to produce the final performance measurement value. This is a strong advantage of the tool that facilitates to find performance bottlenecks in the application. On the other hand, the aggregated value is only meaningful, if the measurement was performed on all nodes at the same time interval. Consequently, the monitoring must be synchronized across grid nodes. Currently, the G-PM communicates with the OCM-G in a purely *pull* model. Each time when a new performance value is needed the requests from the G-PM are sent to OCM-G service manager and then broadcast to local monitors. This preserves the G-PM user's workstation from being overloaded by messages from the OCM-G. However, if the network latency is high and varies significantly across links, the pooling may make the required synchronization impossible.

For this reasons, a hybrid model that shares some properties of both *pull* and *push* approaches, is being considered to replace the current communication scheme between G-PM and the OCMG-G. In this hybrid approach, the G-PM asks the OCM-G to measure a performance property at given time intervals. This request is time stamped in OCM-G's service manager and broadcast to local monitors. The local monitors compute the results at requested time intervals and store them in a temporal buffer. Provided that the clocks on local nodes are synchronized (which can be assured within a reasonable accuracy), the time stamp of the initial G-PM request will allow to synchronize the performance measurement across the nodes. To preserve the user workstation from being overloaded, the results are sent from the OCM-G buffers only at an explicit request from the G-PM. An important issue that must be solved in this approach is handling the OCM-G buffers overflow. In this case, the OCM-G may simply perform a partial aggregation of the results, (e.g. via averaging) to reduce the number of data points.

The synchronization of performance measurement across nodes has an additional advantage. Suppose, that the user monitors two performance properties: a mean value of the communication send volume per one second and a total communication send volume from the application startup. Both this performance properties are based on the metric that measures the send volume in communication operations. The desynchronization across the nodes may result in the measurements being performed at different points in time. Consequently, if the synchronization is kept, the values of the metric can be shared.

4 Conclusions

The G-PM tool measures the performance of the distributed application in an on-line fashion. This is a strong advantage in the grid computing where long running applications are common. On the other hand, a number of issues arise which result from the requirements posed by the on-line analysis. Some of these, like e.g., the influence of a discrete nature of the performance monitoring of the function based quantities on the measurement results, are already addressed within the G-PM. Others, like e.g. meaningful aggregation of measurement results from different nodes, still require further study.

An illustration of the evolution of the performance measurement model the G-PM follows is the communication schema between the tool and underlying monitoring layer - OCM-G. Originally, the communication was designed in a strictly *pull* model. The rationale behind this was the minimization of the probability that the user workstation will be overloaded by performance measurement data. Currently, a new schema is being worked out that is based on a hybrid *pull* and *push* model. This schema is believed to still prevent the G-PM from being overloaded, while enabling performance measurements of the application in an on-line, grid-wide synchronized fashion.

References

1. Baliś, B., Bubak, M., Funika, W., Szepieniec, T., and Wismüller, R.: An Infrastructure for Grid Application Monitoring. In: Kranzlmüller, D. et al. (Eds.), Recent Advances in Parallel Virtual Machine and Message Passing Interface, 9th European PVM/MPI Users' Group Meeting, Sept. - Oct. 2002, Linz, Austria, Lecture Notes in Computer Science 2474, pp. 41-49, Springer-Verlag, 2002.
2. Baliś, B., Bubak, M., Funika, W., Szepieniec, T., and Wismüller, R.: Monitoring and Performance Analysis of Grid Application. In: P.M.A. Sloot et al. (Eds.), Computational Science - ICCS 2003, June 2003, St. Petersburg, Russia, Lecture Notes in Computer Science 2657, pp. 214-224, Springer-Verlag, 2003.
3. Bubak, M., Funika, W., Wismüller, R., Arodz, T., and Kurdziel, M.: The G-PM Tool for Grid-oriented Performance Analysis. In: 1st European Across Grids Conference, Santiago de Compostela, Spain, Feb. 2003.
   `http://wwwbode.in.tum.de/~wismuell/pub/santiago03a.ps.gz`
4. CrossGrid - Development of Grid Environment for interactive Applications, EU Project, IST-2001-32243, Technical Annex. `http://www.eu-crossgrid.org`
5. Wismüller, R., Bubak, M., Funika, W., Arodź, T., and Kurdziel, M.: Support for User-Defined Metrics in the Online Performance Analysis Tool G-PM. Accepted to 2nd European AcrossGrids Conference, Nicosia, Cyprus 2004.
6. Wismüller, R., Bubak, M., Funika, W., and Baliś, B.: A Performance Analysis Tool for Interactive Applications on the Grid. In: Performance Analysis and Grid Computing, Proc. Workshop on Clusters and Computational Grids for Scientific Computing, Sept. 2002, Le Chateau de Faberges de la Tour, France. Kluwer, 2003. In print.
7. Wismüller, R. , Oberhuber, M. , Krammer, J. , and Hansen,O. : Interactive debugging and performance analysis of massively parallel applications. In: Parallel Computing, 22(3):415-442, March 1996.

Paramedir: A Tool for Programmable Performance Analysis

Gabriele Jost[1]*, Jesus Labarta[2], and Judit Gimenez[2]

[1]NAS Division, NASA Ames Research Center, Moffett Field, CA 94035-1000 USA
`gjost@nas.nasa.gov`
[2]European Center for Parallelism of Barcelona-Technical University of Catalonia (CEPBA-UPC), cr. Jordi Girona 1-3, Modul D6,08034 – Barcelona, Spain
`{jesus,judit}@cepba.upc.es`

Abstract. Performance analysis of parallel scientific applications is time consuming and requires great expertise in areas such as programming paradigms, system software, and computer hardware architectures. In this paper we describe an extension to the Paraver performance analysis system that facilitates the programmability of performance metric calculations thereby allowing the automation of the analysis and reducing the application development time.

1 Introduction

Successful performance analysis is one of the great challenges when developing efficient parallel applications. Meaningful interpretation of a large amount of performance data requires significant time and effort. A variety of software tools have been developed to assist the programmer in this task. An example of a commercial product is Vampir [9] which allows tracing and trace visualization of message passing and OpenMP [6] applications. In order to analyze the performance the user will typically inspect timeline views of processes and threads, calculate performance statistics for parts of the code, and try to identify the problem. There are several research efforts on the way with the goal to automate this process. We can only name a few. The URSA MINOR project [8] at the Purdue University uses program analysis information as well as performance trace data in order to guide the user through the program optimization process. The Paradyn Performance Consultant [4] automatically searches for a set of performance bottlenecks. The SUIF Explorer [3] Parallelization Guru developed at Stanford University uses profiling data to bring the user's attention to the most time consuming sections of the code. KOJAK [2] is a collaborative project of the University of Tennessee and the Research Centre Juelich for the development of a generic automatic performance analysis environment for parallel programs aiming at the automatic detection of performance bottlenecks.

---

* The author is an employee of Computer Sciences Corporation.

M. Bubak et al. (Eds.): ICCS 2004, LNCS 3036, pp. 466–469, 2004.

One crucial issue that has not been sufficiently addressed in the previous work, according to our opinion, is the flexibility in the calculation of performance metrics. The reasons for poor performance will usually be due to the interaction of a plethora of factors stemming from hardware, system software, and choice of the programming paradigm. Every new computer architecture or programming model will give rise to new metrics that need to be checked and compared during the performance analysis. The tool presented in this paper is based on the Paraver [7] performance analysis system. It allows the automatic calculation of complex performance metrics that have been predefined by expert users based on visual inspection of the trace data. We describe the extension of Paraver that provides for the programmability of the performance analysis process in Section 2 and draw our conclusions in Section 3.

2 Paramedir

Paraver is a performance analysis system that is being developed and maintained at the European Center for Parallelism of Barcelona-Technical University of Catalonia (CEPBA-UPC). It supports a variety of programming paradigms and enables the user to obtain a qualitative global perception of the application's behavior as well as a detailed quantitative analysis of the performance. The Paraver distribution includes a tracing package, OMPItrace [5], with a simple but very flexible format. The trace file contains a wealth of information, which must be filtered and interpreted in order to obtain meaningful statistics. Paraver has filter and semantic modules that provide a high degree of flexibility for the specification of time line views to be displayed by the Paraver graphical user interface (GUI). An analysis module allows the calculation of meaningful statistics. The specification of filters, semantics, and metric calculations can be saved to re-usable configuration files. This way know-how can be transferred from the experienced to the novice user. Nevertheless, the displayed information still needs to be visually inspected in order to draw conclusions. At this point we should mention that *ver* in the name Para**ver** is Spanish for *to see*.

We have extended the Paraver system by a non-graphical user interface to the Paraver analysis module. The new module, Paramedir (**Para**llel **Medir**, where *medir* is Spanish for *to measure*) is a command line tool that takes a performance trace file and a Paraver analysis configuration file as input. It generates an ASCII table containing the requested performance metrics, which can be used for further processing. Paramedir accepts the same trace and configuration files as Paraver. This way the same information can be captured in both systems. The internal structure of Paraver and Paramedir and their relationship is shown in Figure 1.

Paramedir supports the programmability of performance analysis in the sense that complex performance metrics, determined by an expert user, can be automatically computed and processed. An example for the usage of Paramedir is to automate the detection of reasons for poor performance. Paramdir was used within the prototype implementation of an expert system for automatic performance analysis [1]. A whole set of configuration files is applied to the same trace file and the performance metrics

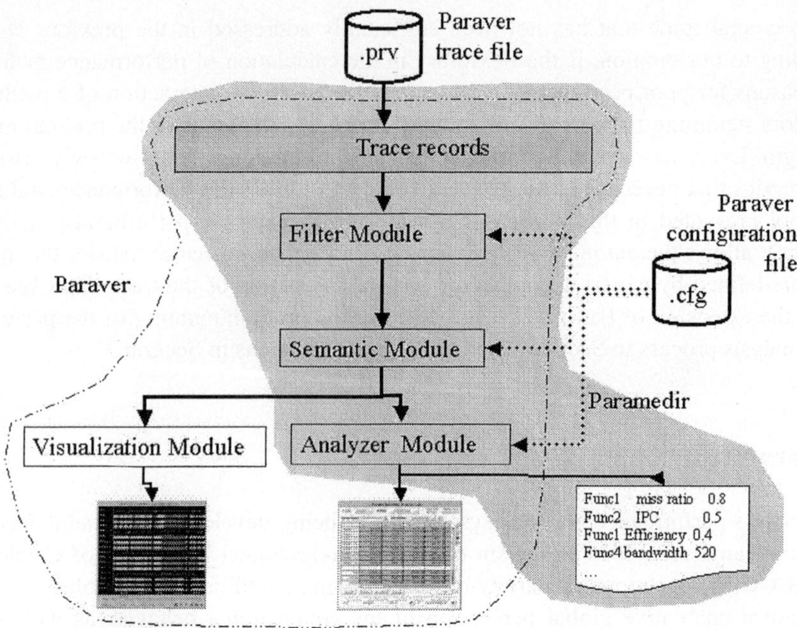

Fig. 1. Internal structure of Paraver and Paramedir. Paramedir uses the shaded components. Paramedir is a command line tool that takes a performance trace file and a Paraver analysis configuration file as input. It generates an ASCII table containing the requested performance metrics

are compared against empirically determined thresholds. Performance problems are determined by applying a set of rules to the outcome of the threshold tests. A high degree of flexibility in the calculation of performance metrics is essential in this scenario. For example, in order to determine why there is an imbalance in the computation time among the threads of a parallel program, many metrics need to be checked. Large sequential sections or an imbalance in the computational workload within the parallel sections are potential reasons. On NUMA architectures, the experienced user may also want to check the cost of L2 cache misses, in order to detect problems related to memory placement. If we denote the number of instructions by *Instr*, the number of L2 misses by *L2misses* and the ideal number of instructions per seconds for the system by *idealMPIS* , then we can estimate the cost for an L2 miss as:

$$Estimated\ L2cost = (Elapsed\ Time - Instr/(idealMIPS))/(L2misses)$$

The Paraver configuration files provide sufficient flexibility to specify this metric and Paramedir allows its calculation in batch mode. The automatic checking of performance metrics saves time for the experienced user and points the novice user, who may often not know what to look for, to potential performance problems.

3   Conclusions

We have extended the Paraver performance analysis system by a tool that facilitates the programmability of performance metric calculations and discussed a usage scenario within an expert system for performance analysis. The first conclusion we draw is, that the great challenge in automatic performance analysis is to de-mangle the factors that influence the performance of the program in the right way. The extensive tracing and analysis capabilities of Paraver are crucial to meet this challenge. Secondly, we found it to be very important that the GUI based Paraver system and the command line based Paramedir module share the same configuration files. This makes it possible to switch from one tool to the other at any point during the analysis process. The automated analysis using Paramedir rapidly guides the user to code segments that require further detailed analysis with Paraver. The detailed analysis will often lead to the design of new analysis configuration files which can then, in turn be included in the automated process.

Acknowledgements. This work was supported by NASA contract DTTS59-99-D-00437/A61812D with Computer Sciences Corporation/AMTI, by the Spanish Ministry of Science and Technology, by the European Union FEDER program under contract TIC2001-0995-C02-01, and by the European Center for Parallelism of Barcelona (CEPBA).

References

1.   G. Jost, R. Chun, H. Jin, J. Labarta, and J. Gimenez, *"An Expert System for the Development of Efficient Parallel Code"*, to be presented at PARA'04, Kopenhagen, Denmark, June 2004.
2.   KOJAK Kit for Objective Judgment and Knowledge based Detection of Performance Bottlenecks, http://www.fz-juelich.de/zam/kojak/.
3.   S. Liao, A. Diwan, R. P. Bosch, A. Ghuloum, M. Lam, *"SUIF Explorer: An interactive and Interprocedural Parallelizer"*, 7th ACM SIGPLAN Symposium on Principles & Practice of Parallel Programming, Atlanta, Georgia, (1999), 37-48.
4.   B.P. Miller, M.D. Callaghan, J. M. Cargille, J. K. Hollingsworth, R. B. Irvin, K.L. Karavanic, K. Kunchithhapdam and T. Newhall, *"The Paradyn Parallel Performance Measurement Tools"*, IEEE Computer 28, 11, pp.37-47 (1995).
5.   OMPItrace User's Guide, https://www.cepba.upc.es/paraver/manual_i.htm
6.   OpenMP Fortran/C Application Program Interface, http://www.openmp.org/.
7.   Paraver, http://www.cepba.upc.es/paraver/.
8.   I. Park, M. J. Voss, B. Armstrong, R. Eigenmann, *"Supporting Users' Reasoning in Performance Evaluation and Tuning of Parallel Applications"*, Proceedings of PDCS'2000, Las Vegas, NV, 2000.
9.   VAMPIR User's Guide, Pallas GmbH, http://www.pallas.de.

Semantic Browser: An Intelligent Client for Dart-Grid

Yuxin Mao, Zhaohui Wu, and Huajun Chen

Grid Computing Lab, College of Computer Science, Zhejiang University,
Hangzhou 310027, China
{maoyx, wzh, huajunsir}@zju.edu.cn

Abstract. In this paper, we propose a generic architecture of Semantic Browser for Dart-Grid, which is an intelligent Grid client and provides users with a series of Semantic functions. Extensible plug-in mechanism enables Semantic Browser to extend its functions dynamically; Semantic Browser converts various format of semantic information into uniform semantic graph with Semantic Graph Language (SGL); a semantic graph is composed of operational vectographic components. An application of Semantic Browser on Traditional Chinese Medicine (TCM) is also described.

1 Introduction

The evolution of the Web has resulted in a great deal of distributed database and knowledge base (KB) [1] resources. In such an environment, sharing and utilizing large-scale information resources has become a central issue to be addressed. Traditional architecture of Web browser is quite insufficient for these requirements. So some novel browser should be brought forth and developed. Semantic Browser is just such a new type of browser, aimed at sharing and managing information from distributed KBs and databases for Dart-Grid [2], which is an OGSA [3] -based system developed by Grid Computing Lab of Zhejiang University and intends to support information resource management in the open, dynamic wide-area environment. This paper discusses the architecture of Semantic Browser based on Grid [4] and Semantic Web [5] and introduces the key technologies of implementation, as well as an application on TCM.

2 Overview

We indicate the architecture of Semantic Browser for Dart-Grid in the frame with dotted line border in figure 1. Semantic Browser views distributed information at semantic layer and acts as an intelligent client to Dart-Grid. It provides richer interaction to end-users of Dart-Grid for querying and managing information.

M. Bubak et al. (Eds.): ICCS 2004, LNCS 3036, pp. 470–473, 2004.

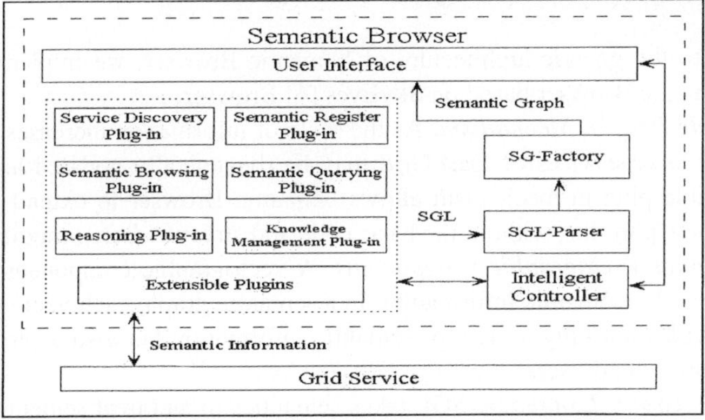

Fig. 1. The architecture of Semantic Browser

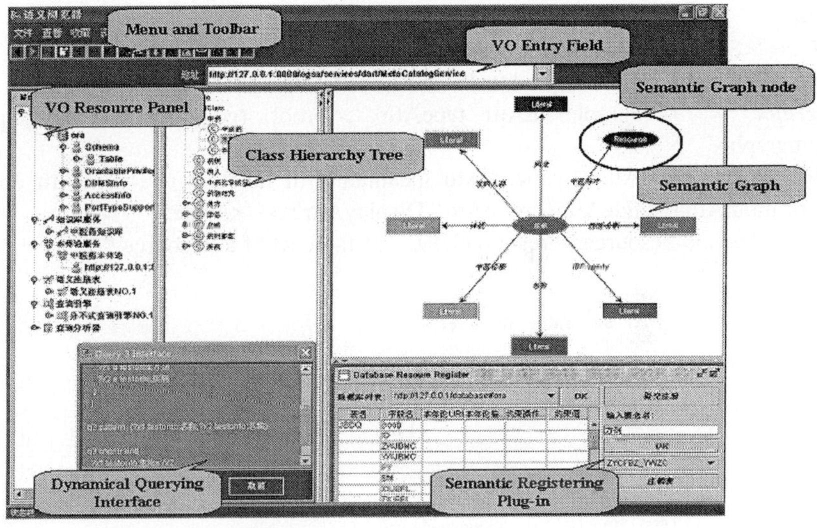

Fig. 2. The screen shot of Semantic Browser

A Customizable User Interface. The information view is intuitive and users can customize the interface to fit their specific needs as shown in figure 2.

Semantic Browser Plug-ins. Semantic Browser accesses Grid services through Grid interface by semantic plug-ins, which usually contain some specific stubs [3] and gain access to a Grid service instance through GSH and GSR.

An Intelligent Controller. Intelligent Controller coordinates and schedules plug-ins, enabling proper plug-ins to access proper services.

An SGL-parser and an SG-factory. Semantic Browser needs an SGL-parser to parse and process SGL data stream. An SG-Factory will produce uniform semantic graph based on SGL, despite formats of semantic information.

3   Implementation and Key Technologies

According to the generic architecture of Semantic Browser, we implement a proto-type of Semantic Browser based on SkyEyes [6] Browser.

Extensible Plug-in Mechanism. As the scale of information increases or users' re-quirements vary, services of Dart-Grid may be dynamically updated and delivered. The extensible plug-in mechanism allows Semantic Browser to extend its functions by adding new plug-ins, without the basic code and structure being modified.

Operational Vectographic Components. A vectographic component is used as a proxy or view for semantic information. In a semantic graph, each vectographic com-ponent provides not only a view of semantic information but also a series of intelli-gent functions to end-users.

Semantic Graph Language. SGL takes semantics in and treat semantics as part of graph elements. We can convert various Semantic Web languages like RDF(S) [7] into SGL and use SGL to describe both the semantics and the appearance of a seman-tic graph. SGL is an XML-based language and here is a small part of SGL BNF defi-nition.

SGL ::= '<SGL>' namespacelist,graph* '</SGL>'

graph ::= '<graph' idAttr depthAttr '>' subgraph* '</graph>'

subgraph ::= '<subgraph' idAttr typeAttr '>' root, (edge, node | subgraph)* '</subgraph>'

node ::= '<node' idAttr ((resourceAttr localnameAttr labelAttr) | (literalAttr opera-torAttr inputAttr)) angleAttr? spaceAttr? DisplayAttr?'>' '</node>'

resourceAttr ::= 'resource="'resourceURI'"' /* the URI of a Resource */

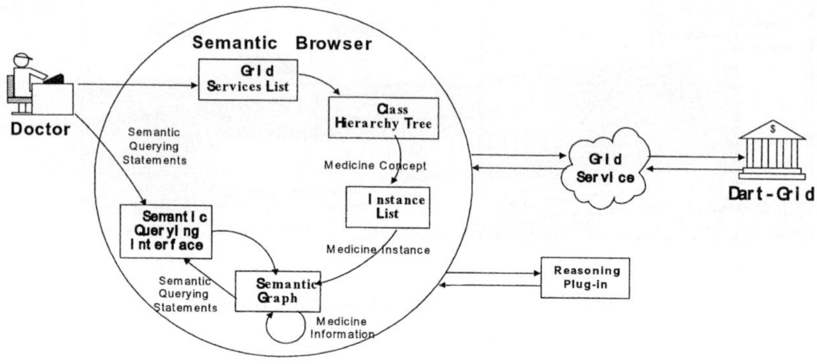

Fig. 3. A use case of Semantic Browser

4   Application on Traditional Chinese Medicine

In collaboration with the China Academy of Traditional Chinese Medicine, we have finished the development of TCM Information sharing platform based on Dart-Grid, which involves tens of large databases from 17 institutes. Users can use Semantic Browser to acquire useful TCM information. For example, if a doctor is not sure about the use of a new medicine, he can take Semantic Browsing, Semantic Querying and reasoning in Semantic Browser to get useful information about the new medicine quickly and effectively, as shown in figure 3.

5   Summary

Semantic Browser differs itself from traditional Web browsers as an intelligent client for Dart-Grid. It's based on dynamical and open Grid environment and accesses Grid services to solve more complex problems. Browsing and Querying with semantic links can get information more exactly and effectively. Vectographic components provide users with excellent semantic information view and interactive functions. Extensible plug-in mechanism enables extending functions dynamically according to the change of Grid services. Heterogeneous information is converted into uniform semantic graphs based on SGL. Our future work is to improve the functions of Semantic Browser further, especially on reasoning services and knowledge services.

Acknowledgement. This work is supported in part by the Data Grid for Traditional Chinese Medicine, subprogram of the Fundamental Technology and Research Program, China Ministry of Science and Technology, and the China 863 Research Program on Core Workflow Technologies supporting Components-library-based Coordinated Software Development under Contract 2001AA113142, and the China 211 Research Program on Network-based Intelligence and Graphics Technology.

References

1. Wu Zhaohui, Chen Huajun, Xu Jiefeng. Knowledge Base Grid: A Generic Grid Architecture for Semantic Web. JCST Vol.18, No.4, July, 2003.
2. Wu Zhaohui, Chen Huajun, Huang Lican et al. Dart-InfoGrid: Towards an Information Grid Supporting Knowledge-based Information Sharing and Scalable Process Coordination. CNCC, 2003.
3. I. Foster et al. The Physiology of the Grid: An Open Grid Services Architecture for Distributed Systems Integration. Tech. report, Glous Project.
4. I. Foster, C. Kesselman, S. Tuecke. The Anatomy of the Grid: Enabling Scalable Virtual Organizations. Int'l J. High-Performance Computing Applications, 2001.
5. Berners-Lee, T., Hendler, J., Lassila, O. The Semantic Web. Scientific American, 2001.
6. Resource Description Framework (RDF) Model and Syntax Specification. http://www.w3.org/TR/1999/REC-rdf-syntax-19990222/.
7. Mao Yuxin, Wu Zhaohui, Chen Huajun. SkyEyes: A Semantic Browser for the KB-Grid. GCC, 2003.

On Identity-Based Cryptography and GRID Computing[*]

H.W. Lim and M.J.B. Robshaw

Information Security Group
Royal Holloway, University of London
Egham, Surrey, TW20 0EX, UK
{h.lim, m.robshaw}@rhul.ac.uk

Abstract. In this exploratory paper we consider the use of Identity-Based Cryptography (IBC) in a GRID security architecture. IBC has properties that align well with the demands of GRID computing and we illustrate some trade-offs in deploying IBC within a GRID system.

1 Introduction

Continual improvements to computing power, storage capacity, and network bandwidth are permitting computing technologies of previously unheard sophistication. Nevertheless, there remain increasing demands for more computational power and resources and GRID computing has been proposed as a mechanism to provide such demands. In order to realise the GRID vision, a sound and effective security architecture is of the utmost importance, and there are many complications due to the interoperable, heterogenous, scalable, and dynamic qualities of a GRID deployment.

Independently of GRID computing, a variant of traditional public key technologies called Identity-based Cryptography (IBC) has recently received considerable attention. In [4] Shamir introduced identity-based cryptosystems in which the public key can be generated from a publicly identifiable information such as a person's e-mail address. The corresponding private key is generated and maintained by a Private Key Generator (PKG) (or Trusted Authority). More recently, work by Boneh and Franklin [1] on identity-based encryption has inspired much new research in the field. The potential of IBC to provide more immediate flexibility to entities in a security infrastructure may well match the qualities demanded by GRID computing. In particular, the properties of IBC that allow generation of keying information on the fly offers a good opportunity to consider IBC as an alternative approach to GRID security.

---

[*] The full paper is available at http://www.isg.rhul.ac.uk/~hwlim/.

M. Bubak et al. (Eds.): ICCS 2004, LNCS 3036, pp. 474–477, 2004.

2   Alternative Approach to GRID Security

The Globus Toolkit (GT)[1], in its latest 3.0 version, is currently the most popular open source software toolkit for building GRID systems. The security services provided by the GT rely upon a security architecture called Grid Security Infrastructure (GSI), which is based on PKI and Transport Layer Security (TLS) communication protocol. The focus is primarily on authentication (including cross-domain), message protection, and single sign-on and identity delegation through proxy credentials [3]. The major components are as follows.

Virtual Organisation (VO). A dynamic collection of users and resources that potentially span multiple administrative domains and governed by a set of defined sharing rules.

User. A subscriber to a GRID. The user can belong to a VO or multiple VOs and he may share part or all of his local resource to other users.

Resource. This comprises from any sharable resource including hardware and software. A user can be a resource to other users if he could offer part or all of his local resource.

Community Authorization Service (CAS). Each VO has its own CAS to maintain a set of policies and communicate those policies to the resources.

Community Policy. A local database that stores policies imposed on each user of a VO.

Grid Certificate Authority (CA). An independent, trusted and potentially shared CA for a VO. It certifies and signs certificates for the VO members.

PKI is an authentication enabling technology and it is widely used in the GT. Using a combination of secret key and public key cryptography, it enables a number of other security services including data confidentiality, data integrity, and key management. Within the GT, when a user (say Alice) wishes to send a job request to a resource (Resource X), she needs to authenticate herself to her CAS Server. The CAS Server establishes Alice's identity and rights using a local policy database maintained at Community Policy. It then issues Alice a signed policy assertion containing her identity and rights. Alice sends the policy assertion and her certificate to Resource X. Resource X authenticates Alice and verifies her VO membership. It also enforces VO's policies stated in the assertion and local policies in regard with VO and Alice herself. Once these are done, Alice is authorized to use Resource X.

The identity token used in GRID is provided through an X.509 public key certificate. It contains a public key, a subject name in the form of a distinguished name (DN), and a validity period that is signed by a Grid CA [6]. Each entity's public key can be transmitted in a X.509 certificate as part of a TLS connection handshake. Upon completion of a handshake protocol which includes key exchange messages from both parties, the parties can begin to transfer data securely over the established communication channel. In GRID, a private/public key pair is usually generated by each individual (user/resource). Should a user

---

[1] The Globus Toolkit, http://www-unix.globus.org/toolkit/

realise or suspect that his private key has been compromised, then the holder himself is held accountable for the notification of the exposed key to the Grid CA in order to have his certificate revoked as soon as possible.

Despite the importance of PKI, there is an increased research focus on IBC. The main stimulus for this trend is the problem of managing certificates and their associated keys using PKI (refer to the full paper for further description). In the full paper we explore whether IBC may be used to alleviate similar problem in a GRID environment and perhaps provide other advantages. To illustrate the properties of IBC, suppose Alice, through an IBC-based system, wants to send an encrypted message to Bob using an identity-based cryptosystem. Alice does not need to verify the authenticity of Bob's public key (by retrieving Bob's public key certificate). Instead Alice simply encrypts the message with an identifying public key, e.g. `bob@xyz.com`. Clearly, Alice needs to know the public parameters or system parameters of Bob's PKG. If Bob does not already possess the corresponding private key, he has to obtain it from his PKG. If the PKG is satisfied that Bob is the legitimate receiver of the message, the PKG uses a master key to generate the private key that matches Bob's public key string. The major technical difference between IBC and PKI is the binding between the public/private keys and the individual. This is achieved by using certificates in PKI. For IBC, the public key is bound to the transmitted data while the binding between the private key and the individual is managed by the PKG [2].

To see how IBC could be applied in a GRID environment, we presume that Alice and Bob both belong to the same VO. When Alice wishes to communicate securely with Bob, in principle she could simply encrypt the message with public key string:

`'Bob's DN ‖ timestamp'.`

She neither needs Bob's public key certificate nor verifies his identity as the authentication task has been indirectly transferred to the PKG. Note that Bob needs to authenticate himself to the PKG before he receives the appropriate corresponding private key. In addition, one can add more granularity to impose restrictions on the receiving party. For instance, if Alice wants to ensure that her job request can be read by Resource X only and no other resource, she can in principle encrypt her job descriptions and the associated policy assertion with public key string that includes Resource X's role:

`'Resource X's DN ‖ role ‖ timestamp'.`

The potential shown by IBC in generating public key instantly without performing certificate lookup and verification offers the flexibility that closely matches the dynamic qualities of the entities within the GRID environment as they join and leave the VO. However, IBC-based cryptography also has a drawback since each entity needs an authenticated and secure channel with the PKG when retrieving his private key. Thus IBC is a relatively new technology in comparison with PKI and the full implications have yet to be considered in its application to a GRID system. However, some first steps are taking place and we note that Stading [5] has recently developed an IBC-based key management mechanism for use within a distributed system.

3   Conclusions

The development of GRID computing is one of today's most important technical problems and the security issues within a GRID deployment are numerous and complicated. Current implementations rely heavily on traditional PKI as a way of supporting many security services. In the full paper we explore the potential benefits of IBC within a GRID infrastructure and we suggest that IBC might have the right properties to provide an alternative security solution for GRID systems. However, true possibilities of integrating identity-based mechanisms within a GRID infrastructure will become clearer with more research. Nevertheless, this interaction could be promising, and the inherent qualities of IBC appear to closely match the demands of a dynamic environment like GRID where the availability of new or current resources can change swiftly over time.

References

1. D. Boneh and M. Franklin. Identity-Based Encryption from the Weil Pairing. In J. Kilian, editor, Proceedings of Advances in Cryptology - CRYPTO 2001, pages 213-229. Springer-Verlag LNCS 2139, 2001.
2. K.C. Paterson and G. Price. A Comparison between Traditional Public Key Infrastructures and Identity-Based Cryptography. Information Security Technical Report, 8(3):57-72, 2003.
3. L. Pearlman, V. Welch, I. Foster, C. Kesselman, and S. Tuecke. A Community Authorization Service for Group Collaboration. In Proceedings of the 3rd IEEE International Workshop on Policies for Distributed Systems and Networks (POLICY'02), pages 50-59, June 2002.
4. A. Shamir. Identity-based cryptosystems and signature schemes. In G. R. Blakley and D. Chaum, editors, Proceedings of Advances in Cryptology - CRYPTO '84, pages 47-53. Springer-Verlag LNCS 196, 1984.
5. T. Stading. Secure Communication in a Distributed System Using Identity Based Encryption. In Proceedings of 3rd IEEE International Symposium on Cluster Computing and the Grid (CCGrid 2003), pages 414-420, May 2003.
6. M.R. Thompson, D. Olson, R. Cowles, S. Mullen, and M. Helm. CA-based Trust Model for Grid Authentication and Identity Delegation. Global Grid Forum (GGF) Grid Certificate Policy Working Group, June 2003. Available at http://www.gridforum.org/documents/GFD/GFD-I.17.pdf, last accessed in November 2003.

The Cambridge CFD Grid Portal for Large-Scale Distributed CFD Applications

Xiaobo Yang[1], Mark Hayes[1], Karl Jenkins[2], and Stewart Cant[2]

[1] Cambridge eScience Centre, University of Cambridge
Wilberforce Road, Cambridge CB3 0WA, United Kingdom
`xy216,mah1002@cam.ac.uk`
[2] Department of Engineering, University of Cambridge
Trumpington Street, Cambridge CB2 1PZ, United Kingdom
`kwj20,rsc10@eng.cam.ac.uk`

Abstract. The Cambridge CFD (computational fluid dynamics) Web Portal (CamCFDWP) has been set up in the Cambridge eScience Centre to provide transparent integration of CFD applications to non-computer scientist end users who have access to the Cambridge CFD Grid. Besides the basic services provided as other web portals such as authentication, job submission and file transfer through a web browser, the CamCFDWP makes use of the XML (extensible markup language) techniques which make it possible to easily share datasets between different groups of users.

1 Introduction

CFD is now widely used in aerodynamics, automotive industry, etc. In order to satisfy the increased demands of understanding complex flows, increased computing power becomes more and more important for large-scale CFD applications. With the emerging Grid technique [1], the integration of resources belonging to different organisations is now practical. The Cambridge CFD Grid, a distributed problem solving environment between the Cambridge eScience Centre and the CFD Lab at the Cambridge University Engineering Department has been set up as a testbed for such large-scale distributed CFD applications. At the same time, the Cambridge CFD Web Portal (CamCFDWP) [2] has been developed in the Cambridge eScience Centre to provide end users transparently access to the power of computing resources contributed to the Cambridge CFD Grid through a web browser.

In this paper, we first briefly describe the Cambridge CFD Grid. Then the CamCFDWP with application of the XML techniques is depicted in detail. Finally our conclusions are presented.

2 Cambridge CFD Grid

As mentioned above, the Cambridge CFD Grid is a distributed problem solving environment. The Globus group [3] defined the Grid as "an infrastructure

M. Bubak et al. (Eds.): ICCS 2004, LNCS 3036, pp. 478–481, 2004.

that enables the integrated, collaborative use of high-end computers, networks, databases, and scientific instruments owned and managed by multiple organisations." Detailed information on the Grid technique was given by Foster *et al.* [1,4,5] in their publications. An introduction to the Grid technique in CFD is reported by Yang *et al.* [6]

Currently, the Cambridge CFD Grid comprises two dedicated linux clusters, a web server, database and dedicated data storage machines. The network link between the two sites is currently investigated by considering a virtual private network (VPN) for security although this has not been fully tested yet. Once setup, the VPN will provide a route around the departmental firewalls. The clusters run the Globus Toolkit [7] and Condor [8] for remote job submission, file transfer and batch queue management.

SENGA, a parallel combustion DNS (direct numerical simulation) code developed by Jenkins *et al.* [9] at the Cambridge CFD Lab, has been tested in the Cambridge CFD Grid. The CFD code is used to study the effects of a turbulent flame kernel, in which there exists a strong coupling between turbulence, chemical kinetics and heat release.

3   Cambridge CFD Web Portal

The Globus Toolkit v2.4.3 used in the Cambridge CFD Grid provides a set of command line tools to manage remote computing resources. This means extra work for end users to get accustomed to these commands. In order to provide transparent access to remote resources including computing resources, large datasets, etc., many web portals such as the ASC Portal [10], the Telescience Project [11] and PACI HotPage [12] have been set up. Basically these portals enable end users to run large-scale simulations through web interfaces. The aim of the CamCFDWP is also to hide command line tools of the Globus Toolkit and resources behind a simple but user friendly interface, i.e., web interface. The CamCFDWP provides the ability to guide users through running the SENGA CFD code inside the Cambridge CFD Grid.

The current version of the CamCFDWP was developed based on the Grid Portal Toolkit (Gridport) 2.2 [13] with the following capabilities through a web browser: 1) login/logout through MyProxy [14] delegation, 2) remote job submission either interactively or in batch mode, 3) a batch job manager, 4) file transfer including upload, download and third party transfer, and 5) a database (Xindice [15]) manager. Fig. 1 shows the architecture of the CamCFDWP. The portal web server plays a key role. Whatever an end user wants to do on remote computing resources, he or she only needs to contact the portal web server through a web browser, from which he/she can execute his/her job.

As XML is fast becoming an industry standard because of its intrinsic merit for data exchange, we adopted XML techniques in order to store information about each job. Without too much modification of the legacy FORTRAN CFD code (SENGA, mainly to read in new parameters), a user inputs parameters through a web form in the CamCFDWP. These parameters will first be saved as an XML file, which will then be validated against a schema [16] designed for

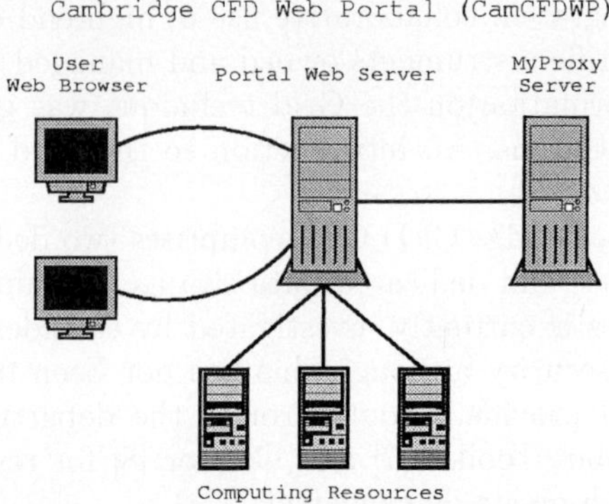

Fig. 1. Architecture of the Cambridge CFD Web Portal (CamCFDWP)

SENGA. Inside the schema, all the input parameters are described as precisely as possible so that they can be set up correctly for SENGA. Xerces-C++ [17] is used to validate the XML file against the schema. If the validation is successful, a plain text file with all input parameters will then be created and transferred to SENGA. Inside the XML file, extra information such as the creator and date are also saved.

When the numerical simulation has finished (on remote machines), all output data are transferred to a file server. During this stage, the location of these data will be recorded in the same XML file mentioned above. Thus for each calculation, the input parameters, location of output data, creator, date, etc. are all recorded in one XML file. Apache Xindice [15], a native XML database has been adopted to manage these small XML files (each job has one XML file accordingly). According to our tests, it has the ability to query an element in an XML database and return elements only or whole XML files. For example, a user may be interested in querying all data created by user "xyang", or all simulations done with the "*Reynolds number*" equals "30.0".

While developing the CamCFDWP, we have also developed a similar web portal for the Cambridge EM (electromagnetic scattering from aircraft) Grid. Basically, we simply modified a configuration file of the CamCFDWP. Although it is really easy to do such work, we realise that for centres like the Cambridge eScience Centre with many projects hosted it is not a good idea to develop one web portal for each project with similar interface. Thus, we are now developing some portlets. These portlets are divided into two classes. First, general portlets for authentication, file transfer, etc. These portlets should be available to all grid users. Second, particular portlets for particular projects. For instance, a RunSENGA portlet which should only be available to CFD people. With the help of Jetspeed [18], a portlet container, each user can customise his/her own web interface, he/she should have permission to run all general portlets and any special portlet. But he/she will not have permission to run portlets for other projects.

4   Conclusions

This paper describes the Cambridge CFD Web Portal for the Cambridge CFD Grid. Through a web browser, the CamCFDWP provides a user friendly interface, which makes the Grid transparent to end users. Besides the basic services of authentication, job submission and file transfer, XML techniques have been introduced to the project. At the current stage, XML brings us two benefits. First, an XML schema has been developed which makes it easy to validate user input parameters through the CamCFDWP. Second, Xindice, a native XML database has been set up which manages all the necessary information on each numerical simulation including all input parameters, user name, date and data location for possible future datasets sharing with other groups of users.

Acknowledgements. We thank the anonymous reviewers for their insightful comments helped to improve this paper. This work was undertaken at the Cambridge eScience Centre supported by EPSRC and the DTI under the UK eScience Programme.

References

1. Foster, I. and Kesselman, C., "The Grid: Blueprint for a New Computing Infrastructure", Morgan Kaufman, San Francisco, Calif, 1999.
2. https://www.escience.cam.ac.uk/portals/CamCFDWP/.
3. http://www.globus.org/.
4. Foster, I. and Kesselman, C., "Globus: A Metacomputing Infrastructure Toolkit", *Int. J. Supercomputer Applications*, 11(2):115-128, 1997.
5. Foster, I., Kesselman, C. and Tuecke, S., "The Anatomy of the Grid: Enabling Scalable Virtual Organizations", *Int. J. Supercomputer Applications*, 15(3), 2001.
6. Yang, X. and Hayes, M., "Application of Grid Technique in the CFD Field", *Integrating CFD and Experiments in Aerodynamics*, Glasgow, UK, 8-9 September 2003.
7. http://www-unix.globus.org/toolkit/.
8. http://www.cs.wisc.edu/condor/.
9. Jenkins, K. and Cant, R.S., "Direct Numerical Simulation of Turbulent Flame Kernels", *Recent Advances in DNS and LES, eds. Knight, D. and Sakell, L.*, pp. 191-202, Kluwer Academic Publishers, New York, 1999.
10. Russel, M., Allen, G., Foster, I., Seidel, E., Novotny, J., Shalf, J., von Laszewski, G. and Daues, G., "The Astrophysics Simulation Collboratroy: A Science Portal Enabling Community Software Development", *Proceedings of High-Performance Distributed Computing 10 (HPDC-10)*, pp. 207-215, San Francisco, CA, 7-9 August 2001.
11. https://telescience.ucsd.edu/.
12. https://hotpage.npaci.edu/.
13. https://gridport.npaci.edu/.
14. http://grid.ncsa.uiuc.edu/myproxy/.
15. http://xml.apache.org/xindice/.
16. http://www.escience.cam.ac.uk/projects/cfd/senga.xsd.
17. http://xml.apache.org/xerces-c/index.html.
18. http://jakarta.apache.org/jetspeed/.

Grid Computing Based Simulations of the Electrical Activity of the Heart

J.M. Alonso, V. Hernández, and G. Moltó

Departamento de Sistemas Informáticos y Computación.
Universidad Politécnica de Valencia. Camino de Vera s/n 46022 Valencia, Spain
{jmalonso,vhernand,gmolto}@dsic.upv.es
Tel. +34963877356, Fax +34963877359

Abstract. Simulation of the electrical activity of the heart by modelization of the action potential propagation is a computational and memory intensive process. In addition many studies, such as the investigation of the ischemia phenomenon, require the execution of lots of parametric simulations, what increases the computational problem by several orders of magnitude. This paper presents the integration of a parallel simulator for the action potential propagation on cardiac tissues into a Grid infrastructure under the Globus Toolkit and InnerGrid middlewares.

1   Introduction

Electrical activity is a key indicator of the state of the heart. Its modellization and simulation allows a better understanding of the electrical behaviour. Cardiac tissue simulations present high computational and memory requirements. Moreover, many cardiac research studies require the execution of a huge amount of parametric simulations. Studies of vulnerable window in ischemia require to vary the time interval between two consecutive stimulus in order to detect the range of values which provoke a reentry, a phenomenon that can derive into heart fibrillation. Besides, to study the effects of late ischemia it is necessary to vary the coupling resistances in all the dimensions of the tissue and observe the evolution of the electrical activity for different anisotropy ratios.

A MPI-based parallel simulation system has been already developed [1] in order to reduce the simulation time on beowulf architectures and allowing the study of larger tissues. Nevertheless, the integration of parallel concurrently executed simulations in a Grid infrastructure seems the key combination to offer a substantial increase in productivity.

2   Grid Computing System Developed

2.1   Portability and Interoperability

To enable portability, the simulation system has been statically linked, so that no external dependencies are required, creating one simulator for the 32-bit Intel

M. Bubak et al. (Eds.): ICCS 2004, LNCS 3036, pp. 482–485, 2004.

architectures and other one for the 64-bit Intel Itanium architectures. Even the MPI library has been introduced into the executable. Besides, all the platform-dependent optimizations have been switched off, such as optimized versions of BLAS and LAPACK, as they may result in potentially executing illegal instructions in the remote machine. This way, it is possible to achieve a self-contained parallel simulation system that can be executed on different Linux platforms.

2.2   Globus Toolkit Developments

We have designed a software layer [2], based upon the Globus Toolkit 2.4 [3] and results from the GridWay project [4]. Basic scheduling support has been added to allocate tasks to nodes on the grid. The total number of available processors is guessed querying the MDS (Monitoring and Discovery Service) server of the execution hosts, assigning a number of simulations to each host proportional to its available computational resources. No attempt is performed to investigate the workload of remote workstations, as they do not offer a local queue system to be queried for free nodes.

Once the scheduler has decided the best computational resource, the stage in phase takes place, compressing the executable and the input data and transferring them to the execution node. A temporary folder is created on the execution machine which will act as a container for the job execution. The data transfer is internally achieved via the Globus GASS (Global Access to Secondary Storage) service, launching a GASS server on the job submission machine. A decompression of the files is performed. Next, the simulation system is executed in parallel integrating, if configured, with the queue manager of the execution node (PBS, LoadLeveler, etc), thus respecting the execution policies of the remote organization. The binary file results obtained are compressed, transferred back to the submission node and saved on the appropriate local folder created for this simulation. Finally, all the temporary created files in the execution node are deleted.

2.3   InnerGrid Developments

InnerGrid [5] is a multi-platform commercial product, that comprises a set of tools that allow to manage an heterogeneous platform of computers. It consists of a server that distributes the pending tasks among the agents, which control each execution. This middleware offers a web-based single point of entry to the Grid, allowing the definition of new tasks, controlling the execution, managing the state of the Grid and accessing the file results of the simulations in a centralized manner. This software implements a fault-tolerance scheme that guarantees the finish of the tasks as long as there are living nodes in the Grid.

InnerGrid does not provide mechanisms for parallel execution of MPI-based applications and so it is restricted to the execution of sequential parametric simulations. In our case, a new module has been created, what represents a definition pattern of all the possible parametric tasks. This module defines the memory and storage space required for the parametric simulations, and allows to specify a different executable file for each architecture supported by InnerGrid. In addition, the module specifies the command-line arguments of the simulation system that are going to be parametric.

A task is the instantiation of a module, where the user specifies the range of values for the varying parameters indicated in the module, as well as the priority level under which the simulations will be run. InnerGrid built-in scheduler is in charge of allocating the pending tasks to the idle nodes of the Grid.

3   Experimental Results

3.1   Case Study

To analyze the vulnerability to reentry of a cardiac tissue that has been locally affected by ischemia, it is necessary to perform different parametric simulations where the interval between two consecutive stimulus is changed. The vulnerable window for reentry represents the time interval between the two stimulus, in which reentrant activity on the tissue is detected. For a three-dimensional 60x60x60 cell cardiac tissue, a vulnerable window of up to 40 ms has been studied, varying the injection delay of the second stimulus from 1 to 40 ms with respect to the application of the first stimulus and analyzing whether it has resulted in reentry or not. 250 ms of time will be simulated in this example. This results in 40 independent and different parametric simulations of action potential propagation that can be performed simultaneously on a Grid infrastructure.

3.2   Execution Results

The available testbed is composed of two clusters and a workstation. Cluster A has 20 Pentium Xeon 2.0 Ghz biprocessors, with 1 GByte of RAM. Cluster B consists of 12 Pentium III 866 Mhz biprocessors, with 512 MBytes of RAM. An Intel Itanium 2 900 Mhz biprocessor, with 4 GBytes of RAM, has been introduced in this heterogeneous testbed.

InnerGrid does not have an Itanium version yet and therefore only both clusters have this middleware installed. The InnerGrid server was setup in a separate machine, while the agents were run on every node of both clusters. On the other hand, the Globus Toolkit 2.4 has been installed in all the machines.

For each machine, Table 1 shows the execution time and the number of simulations performed, for the case study presented. Parallel simulations are executed with a quarter of the total available processors, a polite policy that allows the execution of several simultaneous simulations. Global execution time corresponds to the slowest machine, 39.16 hours for Globus based executions. On the other hand, simulations ran through InnerGrid needed an 8.5% extra time (42.5 hours) to conclude, executing on nodes of each cluster of the testbed. Sequential execution of the case study in one node of Cluster A required over 563.3 hours, while 9-processor parallel executions in the same cluster, which allows two concurrent simulations, lasted for 45.6 hours.

Table 1. Execution times (in hours) for the simulations of the case study. Numbers in parentheses indicate the number of processors involved in each parallel simulation

		Cluster A	Cluster B	Itanium
Globus	Simulations	24 (9 proc.)	13 (5 proc.)	3 (1 proc.)
	Execution time	38.13	39.16	25.6
InnerGrid	Simulations	28 (1 proc.)	12 (1 proc.)	-
	Execution time	34.3	42.5	-

Large tissues enforce a serious memory requirement and thus, they may not be succesfully executed on sequential platforms, an important handicap for InnerGrid simulations. While InnerGrid seems appropiate to take advantage of idle computers in single-organizational Grids, the Globus Toolkit is focused on running on dedicated resources in different organizational Grids. Therefore, a Globus-based solution is much more appropiate for the cardiac electrical simulation problem, as it offers transparent access to distant computational resources.

4   Conclusions

This paper has presented the integration of a parallel system for the simulation of electrical activity on cardiac tissues into a Globus-based Grid infrastructure. The application features state-of-the-art capabilities such as data compression, self-contained executable and dependencies migration, cross-linux portability, and parallel execution of simulations on the multiprocessor machines. In addition, InnerGrid commercial product has been tested as an easy-to-use alternative middleware to create single-organizational Grids. A new module has been developed, allowing the user to vary several parameters and managing the execution of the different parametric tasks. Having available a parallel simulation system that can be integrated in a Grid infrastructure enables to focus both on speedup, running on a cluster, and productivity, taking advantage of the power of a Grid.

References

1. Alonso, J-M., Ferrero, J-M., Hernández, V., Moltó, G., Monserrat, M., Saiz, J.: High Performance Cardiac Tissue Electrical Activity Simulation on a Parallel Environment. Proc. of the First European HealthGrid Conf., January 16-17. 2003, 84-91
2. Alonso, J-M., Hernández, V., Moltó, G.: Grid Computing Based Simulations of Action Potential Propagation on Cardiac Tissues. Technical Report DSIC-II/05/04. Universidad Politécnica de Valencia (2004)
3. Foster, I., Kesselman, C.: Globus: A Metacomputing Infrastructure Toolkit. The International Journal of Supercomputer Applications and High Performance Computing. 11(2), 115-128
4. Huedo, E., Montero, R-S., Llorente, I-M.: A Framework for Adaptive Execution on Grids. Software Practice and Experience. 2004 (to appear)
5. InnerGrid Nitya Technical Specifications. GridSystems S.A., 2003.

Artificial Neural Networks and the Grid

Erich Schikuta and Thomas Weishäupl

Department of Computer Science and Business Informatics
University of Vienna
Rathausstraße 19/9, A-1010 Vienna, Austria
{erich.schikuta,thomas.weishaeupl}@univie.ac.at

Abstract. We introduce a novel system for the usage of neural network resources on a world-wide basis. Our approach employs the upcoming infrastructure of the Grid as a transparent environment to allow users the exchange of information (neural network objects, neural network paradigms) and exploit the available computing resources for neural network specific tasks leading to a Grid based, world-wide distributed, neural network simulation system, which we call *N2Grid*. Our system uses only standard protocols and services in a service oriented architecture, aiming for a wide dissemination of this Grid application.

1   Introduction

A Grid based computational infrastructure couples a wide variety of geographically distributed resources and presents them as a unified integrated resource which can be shared transparently by communities (virtual organizations).

The Grid started out as a means for sharing resources and was mainly focusing high performance computing. By the integration of Web Services as inherent part of the Grid infrastructure the focus evolved to the sharing of knowledge to enable collaborations between different virtual organizations or subjects.

The focus of the presented paper is the development of N2Grid, a neural network environment based on the Grid. It implements a highly sophisticated connectionist problem solution environment within a Knowledge Grid [1].

2   N2Grid Architecture

The N2Grid system is a neural network simulator using the Grid infrastructure as deploying and running environment. It is an evolution of the existing NeuroWeb [2] system. The idea of this system was to see all components of an artificial neural network as data objects in a database. Now we extend this approach and see them as objects of the arising world wide Grid infrastructure.

Figure 1 shows the overall application model of the N2Grid system in a final implementation phase. We assume a sophisticated Grid infrastructure, including independent resource brokers and replica manager. A N2Grid client does not care for the execution hosts or data sources. It can control an artificial neural network

M. Bubak et al. (Eds.): ICCS 2004, LNCS 3036, pp. 486–489, 2004.

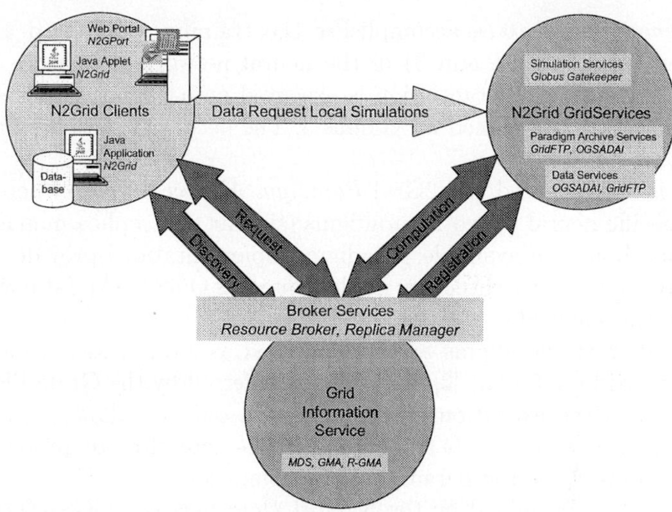

Fig. 1. N2Grid Application Model

simulation locally or remotely in the Grid infrastructure in a transparent way. Parts of the remote execution are the authentication to the Grid resources, the processing of the job descriptions and the usage of Grid data sources.

The components of the N2Grid are derived from the client-server model of NeuroWeb aiming for a novel, service oriented, tripartite Grid application. *N2Grid Services* realize Simulation Services, Paradigm Archive Services and Data Services. The end user can choose between three different *N2Grid clients*. These are a Java application with the possibility of direct data base connections, a Java Applet running in a web-browser, and a Web portal respectively. N2Grid services and clients are interconnected by standard Grid broker services, using standard Grid information services.

2.1 N2Grid Services

N2Grid Services are Grid Services hosted by the Grid infrastructure. They execute neural network simulation tasks (as creation, training and simulation), which are submitted to the Grid and do not consume local client resources. By the submission the work load is minimized on a local machine. Nevertheless the N2Grid system provides also the possibility for local artificial neural network simulations. We can run the following tasks remotely in the Grid:

1. Training of neural networks
2. Evaluation of neural networks
3. Processing of data by a neural network
4. Archiving of paradigms
5. Archiving of network objects (nodes, structure, weights, etc.)
6. Using data sources and archives

The *Simulation Service* accomplishes the training, evaluation and propagation function (task 1, 2,and 3) of the neural network simulator. A selected paradigm and network instantiation is executed on a Globus Gatekeeper (Version 2). Migrations are planed for Globus 3. The necessary data are provided by other N2Grid services.

Task 4 is implemented as N2Grid *Paradigm Archive Service*, where the users can find specific neural network paradigms (similar to a replica manager). More information about the available paradigm implementations provides the paradigm archive service by a directory implemented as OGSADAI database (XML), residing on the same site.

Neural network paradigms are implemented as Java classes using the Java Commodity Grid (CoG) Kit [3]. They are transferred by the GridFTP protocol. In special cases the execution of the class will only be allowed on a specific N2Grid Simulation Service (e.g. one specific remote site) to protect securely intellectual property of the paradigm implementation.

Task 5 and 6 are unified by the N2Grid *Data Services*. OGSADAI provides the access to a database storing all training-, evaluation-, propagation-data and network objects (nodes, structure, weights, etc.). To provide more flexibility, references (GridFTP URLs) to flat files can be registered in the database which can be accessed directly by the neural network simulation system.

2.2   N2Grid Clients

We propose three different clients as shown in Figure 1.

First, there exists a N2Grid *Java Application* client for an advanced user, who can run also a local database storing his own data. The user can extend the functionality of his client by including own paradigm Java classes without influencing the rest of the N2Grid system.

Second, we provide a N2Grid *Java Applet* executing within a standard Web browser having a similar user interface as the Java application but with limited functionality. Because of the sandbox principle local database and file accesses are not allowed.

Third, for the purpose of thin clients a simple web browser can be used as a front end of the N2Grid system by accessing a web-portal called *N2GPort*. It provides control over running simulation jobs on the N2Grid services and presents their results.

3   Use Cases – Scenarios

For the N2Grid system we propose several use cases, depending on the state of the dynamic and changing Grid infrastructure. Table 1 shows the categorization of the scenarios according to the Grid layers. The prototype is under development, three scenarios are already fully implemented, which are the scenarios *Data Pull (GET)*, *Data Push (PUT)*, and *Stand-alone Local Execution*. The other scenarios are implemented partly until now.

Table 1. Grid Layers Mapping

Layer	N2Grid Architecture	Use Case
Knowledge Grid 2-dimensional	N2Grid Paradim Serv. N2Grid Java clients N2GPort web portal client	Search Paradigm Search Net Object Create Neuroal System
Information Grid 1-dimensional	N2Grid Simulation Service Resource Broker Replica Manager	Directed Remote Execution Data-driven Remote Execution Computation-dirven Remote Ex. Paradigm-driven Remote Exec. Stand-alone Local Execution
Data Grid 0-dimensional	N2Grid Data Service N2Grid Paradigm Archive Service	Data Pull (GET) Data Push (PUT) Paradigm Pull (GET) Net object Pull (GET) Net object Push (PUT)

4   Conclusion and Future Research

We presented the N2Grid project, as a next step in the program evolutions
for neural network simulation. It is a framework for the usage of neural network
resources on a world-wide basis by the upcoming Grid infrastructure. Our system
uses only standard protocols and services to allow a wide dissemination and
acceptance. To reach the full capability of our system and to develop more
sophisticated systems further research has to be done in the following two areas,

- The description of the paradigm has to be enhanced, to establish easier sharing between paradigm providers and customers. These semantic description
  is a key concept specified only rudimentary until now by a directory implementation. We will define a semantic paradigm description language by a
  pattern and/or scheme approach using XML.
- The actual N2Grid client controls single simulation runs. To allow the building of large connectionist systems consisting of several neural network in-
  stantiations (possibly of different paradigms) an extension of the N2Grid
  system is on the way, which allows to control a flow of simulations by a
  specific neural network workflow language.

References

1. Cannataro, M., Talia, D.: The Knowledge Grid. Communications of the ACM **46**
   (2003) 89–93
2. Schikuta, E.: NeuroWeb: an Internet-based neural network simulator. In: 14th
   IEEE International Conference on Tools with Artificial Intelligence, Washington
   D.C., IEEE (2002) 407–412
3. von Laszewski, G., Foster, I., Gawor, J., Lane, P.: A Java Commodity Grid Kit.
   Concurrency and Computation: Practice and Experience **13** (2001) 643–662

Towards a Grid-Aware Computer Algebra System

Dana Petcu[1,2], Diana Dubu[1,2], and Marcin Paprzycki[3]

[1] Computer Science Department, Western University
[2] Institute e-Austria, Timişoara
[3] Computer Science Department, Oklahoma State University
{petcu,ddubu}@info.uvt.ro, marcin@cs.okstate.edu

Abstract. One of the developments that can lead to a wider usage of grid technologies is grid-enabling of application software, among them computer algebra systems. A case study described here involves Maple. The proposed maple2g package allows the connection between the current version of Maple and the computational grid based on Globus.

1   Introduction

Computer algebra systems (CASs) are frequently used by mathematicians or engineers to perform complicated calculations and are rightfully seen as one of major sources of user's productivity. In practice it is often desirable to be able to augment the CAS with functionality from an external software artifact (e.g. pakage, application etc.). Nowadays, in this process one can rely on already available solutions, such as the grid technology.

Several projects aim at providing APIs to execute scientific libraries or programs over the grid. NetSolve [1] is a grid based server that supports Matlab and Mathematica as native clients for grid computing. MathLink [8] enables Mathematica to interface with external programs via an API interface. Math-GridLink [6] permits the access to the grid service within Mathematica, and the deployment of new grid services entirely from within Mathematica. The Geodise toolkit [3] is a suite of tools for grid-services which are presented to the user as Matlab functions, calling Java classes which in turn access the Java CoG API. MapleNet [4] offers a software platform to effective large-scale deployment of comprehensive content involving live math computations.

To be able to facilitate development of parallel grid distributed CAS applications, a CAS interface to a message-passing library is needed. There exist more then 30 parallel Matlab projects [2]. gridMathematica [8] allows the distribution of Mathematica tasks among different kernels in a distributed environment.

Maple is utilized as the CAS of choice in our attempt to couple a CAS and a computational grid. The main reason for this choice is that, despite its robustness and ease of use, we were not able to locate efforts to link Maple with grids. Second, it is well known that Maple excels other CAS in solving selected classess of problems like systems of nonlinear equations or inequalities [7].

M. Bubak et al. (Eds.): ICCS 2004, LNCS 3036, pp. 490–494, 2004.
© Springer-Verlag Berlin Heidelberg 2004

Furthermore, Maple has already a sockets library for communicating over the Internet, and a library for parsing XML (a data-exchange standard widely utilized in the grid community). Finally, distributed versions of Maple have been recently reported in [5].

To obtain our goal we proceeded by developing Maple2g: the grid-wrapper for Maple. It consists of two parts: one which is CAS-dependent and the other, which is grid-dependent and thus any change in the CAS or the grid needs to be reflected only in one part of the proposed system. The CAS-dependent part is relatively simple and can esily be ported to support another CAS or a legacy code.

2   Developing a Grid-Aware Maple Extension

Our analysis of the grid aware CAS systems indicates that any such a system must have at least the following facilities:

Ability to accept inputs from the grid: the CAS must be opened to augment its facilities with external modules, in particular it should be able to explore grid facilities, to connect to a specific grid service, to use the grid service, and to translate its results for the CAS interface.

Being a source of an input for the grid: the CAS or some of its facilities must be seen as grid services and activated by remote users in appropriate security and licensing conditions; furthermore, deployment of grid services must be done in an easy way from the inside of the CAS.

Ability to communicate and cooperate over the grid: similar or different kernels of CASs must be able to cooperate within a grid in solving general problems; in order to have the same CAS on different computational nodes a grid-version of the CAS must be available; in the case of different CASs, appropriate interfaces between them must be developed and implemented or a common languages for inter-communication must be adopted.

Rewriting a CAS kernel in order to improve its functionality towards grids can be a complicated and high-cost solution. Wrapping the existing CAS kernel in code acting as the interface between the grid, the user and the CAS can be done relatively easily as an added-functionality to the CAS. In addition, it can also be adapted on-the-fly when new versions of the CAS in question become available.

Maple2g is a prototype of a grid/cluster-enabling wrapper for Maple. As described below it consists of two components, MGProxy, a Java interface between Maple and the grid/cluster environment, and m2g, a Maple library of functions allowing the Maple user to interact with the grid/cluster middleware.

MGProxy has three operating modes:

1. *User mode*: activated from inside of the Maple environment (by the m2g_MGProxy_start command), receives the user command from the user's Maple interface via a socket interface, contacts the grid/cluster services (including also other MGProxy processes), queries the user requests to the contacted services, and sends the results of the queries to the main Maple interface.

Table 1. Functions available in m2g library

Function	Description
m2g_connect()	Connection via Java COG to the grid
m2g_getservice(s, l)	Search for a service s and retrieve its location l
m2g_jobsubmit(t, c)	Allows a job submission in the grid environment labeled with the number t: the command c is a string in the RSL format
m2g_results(t)	Retrieve the results of the submitted job labeled t
m2g_maple(p)	Starts p processs MGProxy in parallel modes
m2g_send(d, t, c)	Send to the destination kernel d a message labeled t containing the command c; d – 'all' or a number, t – number, c – string
m2g_recv(s, t)	Receive from the kernel labeled s results from the command labeled t; s – 'all' or a number, t – number
m2g_rank	MGProxy rank in the MPI World, can be used in a command
m2g_size	Number of MGProxy processes, can be used in a command

2. *Server mode*: activates a Maple twin process (which enters in a infinite cycle of interpreting commands incoming via the socket interface from MGProxy), acts as a server waiting for external calls, interprets the requests, sends the authentications requests to the Maple twin process, receives the Maple results, and sends them back to the user.

3. *Parallel mode*: is activated from user's interface with several other MGProxy copies; the copy with the rank 0 enters in user mode and runs in the user environment, while the others enter in server mode; the communication between different kernels is established through a standard message passing interface.

The current version of Maple2g has a minimal set of functions (described in Table 1) allowing access to the grid services. These functions are implemented in the Maple language, and they call MGProxy which accesses the Java CoG API.

For example, accessing a grid-service can be done in the steps described in Fig.1.

```
> with(m2g): m2g_MGProxy_start(); m2g_connect();
  [m2g_connect, m2g_getservice, m2g_jobstop, m2g_jobsubmit, m2g_maple,
   m2g_MGProxy_end, m2g_MGProxy_start, m2g_rank, m2g_recv, m2g_results,
   m2g_send,m2g_size]
  Grid connection established
> m2g_getservice("gauss",'service_location');
   ["&(executable=/home/Diana/m2g/Gauss.sh)","&(executable=/tmp/gauss)"]
> m2g_jobsubmit(3,service_location[1]);
   job submitted
> m2g_results(3); m2g_MGProxy_end();
  Solving linear syst. with Gauss method: Input in.txt, Output out.txt
  Grid connection closed
```

Fig. 1. Accessing in Maple an external linear solver, available as a grid service

```
>with(m2g): m2g_MGProxy_start(); m2g_maple(4): d:='all':
>m2g_send(d,1,"f:=(x,y)->(x^2-y^2+0.32, 2*x*y+0.043):"):
>m2g_send(d,2,"g:=(x,y)->x^2+y^2:"):
>m2g_send(d,3,"h:=(x,y)->if g((f@@130)(x,y))<1 then 0 else 1 fi:"):
>m2g_send(d,4,"plot3d('h(x,y)',grid=[400/mg_size,400],y=-1.15..1.15,
>   x=-1+2*mg_rank/mg_size..-1+2*(mg_rank+1)/mg_size,style=point,
>   view=[-1..1,-1.15..1.15,0..0.1],orientation=[90,0]);"):
>plots[display3d](m2g_recv('all',4)); m2g_MGProxy_end();
```

Fig. 2. A Julia fractal: the plotting time of order $O(10^3)$ s in the sequential case can be reduced by a speedup factor of 3.5 using 4 Maple kernels treating equal vertical slices

The component responsible for accessing Maple as a grid-service is similar to that of the MapleNet [4]. In the current version of the Maple2g prototype, the access to the fully functional Maple kernel is allowed from the grid: MGProxy acting as CAS-grid interface implements only an account check procedure in order to verify the user rights to access the licensed version of Maple residing on the grid.

Parallel codes using MPICH as their message-passing interface can be easily ported to grid environments due to the existence of a MPICH-G version which runs on top of the Globus Toolkit. On other hand, the latest Globus Toolkit is build on Java, and the Java clients are easier to write. This being the case, we selected the mpiJava as the message-passing interface between Maple kernels.

In Maple2g a small number of commands is available to the user, for sending commands to other Maple kernels and for receiving their results (Table 1). These facilities are similar to those introduced in PVMaple [5]. The user's Maple interface is seen as the master process, while the other Maple kernels are working in a slave mode. Command sending is possible not only from the user's Maple interface, but also from one kernel to another (i.e. a user command can contain inside a send/receive command between slaves).

To test the feasibility of this approach to developing distributed Maple applications, tests have been performed on a small PC cluster (8 Intel P4 1500 MHz processors, connected by a Myrinet switch at 2Gb/s). When splitting the time-consuming computations we have observed an almost linear speedup. While a detailed report on parallel Maple2g is outside of the scope of this note, in Fig.2 we give an example of a parallel Maple2g code.

At this stage Maple2g exists as a demonstrator system; however it already shows its potential. In the near future it will be further developed to include facilities existing in other systems, in order for it to become comparably robust as NetSolve or Geodise. Tests on grid on a large domain of problems will help guide further development of the system. Deployment of grid services from Maple in other languages than Maple using the code generation tools will be also taken into consideration. Finally, the next version of MGProxy will allow the cooperation between different CAS kernels residing on the grid.

References

1. Casanova H. and Dongarra J.: NetSolve: a network server for solving computational science problems. Inter.J. Supercomputer Appls. & HPC, 11-3 (1997) 212–223
2. Choy R., Edelman A.: Matlab*P 2.0: a unified parallel MATLAB, In Procs. 2nd Singapore-MIT Alliance Symp. (2003), in print.
3. Eres M. H. et al: Implementation of a grid-enabled problem solving environment in Matlab. In Procs. WCPSE03 (2003), in print, www.geodise.org
4. MapleNet. www.maplesoft.com/maplenet/
5. Petcu D., PVMaple: A distributed approach to cooperative work of Maple processes. LNCS 1908, eds. J.Dongarra et al., Springer (2000) 216–224
6. Tepeneu D. and Ida T.: MathGridLink - A bridge between Mathematica and the Grid. In Procs. JSSST03 (2003), in print.
7. Wester M.: A critique of the mathematical abilities of CA systems. In CASs: A Practical Guide, ed. M.Wester, J.Wiley (1999), math.unm.edu/~wester/cas_review
8. Wolfram Research: MathLink & gridMathematica, www.wolfram.com.

Grid Computing and Component-Based Software Engineering in Computer Supported Collaborative Learning*

Miguel L. Bote-Lorenzo, Juan I. Asensio-Pérez, Guillermo Vega-Gorgojo,
Luis M. Vaquero-González, Eduardo Gómez-Sánchez, and Yannis A. Dimitriadis

School of Telecommunications Engineering, University of Valladolid
Camino Viejo del Cementerio s/n, 47011 Valladolid, Spain
`{migbot,juaase,guiveg,lvaqgon,edugom,yannis}@tel.uva.es`

Abstract. This paper presents our research efforts towards enabling the use of
grid infrastructures for supporting Computer Supported Collaborative Learning
(CSCL) applications developed according to the principles of Component-
Based Software Engineering (CBSE). An illustrative example of a grid-
supported component-based collaborative learning application is presented and
discussed. This discussion leads to the study of application scheduling and
component hosting problems for CSCL applications within a grid context based
on the Open Grid Services Architecture (OGSA).

1 Introduction

CSCL [1] is a discipline devoted to research in educational technologies that focuses
on the use of Information and Communications Technology (ICT) as mediational
tools within collaborative methods (e.g. peer learning and tutoring, reciprocal
teaching, project or problem-based learning, games) of learning [2]. The effort of
developing CSCL applications is only justified if they can be used in a large number
of learning situations and if they can survive the evolution of functional requirements
and technology changes [3]. In this sense, CBSE appeared as an enabling technology
for the development of reusable, customizable, and integrated CSCL software tools.

In addition, there is a remarkable synergy between CBSE and grid computing:
several ongoing research efforts, such as ICENI [4], suggest the suitability of grid
computing for supporting the distributed execution of component-based applications.
In this same direction, OGSA [5], which has emerged as the *de facto* standard for the
construction of grid systems, recognizes the suitability of software component
containers for implementing the functionality of *Grid Services*.

Besides these two synergies, CBSE with CSCL and CBSE with grid computing, a
third relationship can be established: grid computing and CSCL. Education is
considered to be a "very natural and important application of grid technologies" [6],
and CSCL is one of the major research fields in technology-enabled education. The
analysis of main grid characteristics [7] also supports the idea that the use of a grid
infrastructure can provide major benefits for CSCL applications: large scale of grid
infrastructures, wide distribution of resources, inter-organization relationship support

---

* This work is supported by Spanish projects TIC2002-04258-C03-02, TIC2000-1054 and VA
117/01.

M. Bubak et al. (Eds.): ICCS 2004, LNCS 3036, pp. 495–498, 2004.

and heterogeneous nature of shared resources are some of the most relevant characteristics of grid computing for the CSCL domain.

This paper presents our work towards merging CSCL, CBSE, and grid technologies. With this aim, a scenario combining both CBSE and grid principles within a CSCL context is defined and discussed in section 2. This study identifies two research issues that must be tackled so as to allow CSCL applications to profit from CBSE and grid computing. First, the CSCL application scheduling is dealt in section 3. Second, the component-hosting problem is studied in section 4. Preliminary research results are also described for both issues. Finally, conclusions and future work may be found in section 5.

2   Grid-Supported Component-Based CSCL Application Scenario

The joint use of grid support and CBSE principles can be very valuable for CSCL applications such as the following: an electronic magazine published by children from different schools by collaboratively interacting both synchronously and asynchronously. Learning objectives of this scenario include the acquisition of writing abilities as well as the understanding of concepts related with the articles they write.

The CSCL application supporting this scenario should provide children with a synchronous collaborative editor (for writing articles) and with a conceptualization tool in order to collaboratively organize the ideas that they intend to include in their articles. The latter tool, eventually aided by an intelligent peer that may be computationally intensive, would generate the so-called "cognitive maps". The tool should also provide support for conflict resolution, so that children can propose new concepts and relationships, then discuss them and finally produce a cognitive map that includes the contributions they agree with. Furthermore, this tool would enable the children to access information sources (e.g. previous articles, web pages, etc) and link them to the concepts and relationships they propose. If CBSE development principles are used, the CSCL application supporting the above scenario could be the result of assembling different software components. The functionality of components could be replicated and executed in multiple grid nodes (potentially from different schools) taking advantage of the aforementioned large scale of the grid and its wide geographical distribution. These would enable allow a large number of participants reading and/or writing articles while keeping low response and notification times.

Achieving the benefits identified in this scenario implies the availability of certain mechanisms in grid infrastructures supporting component-based CSCL application: (1) A component-based CSCL application scheduler that decides what software components are migrated/replicated and over what grid nodes (according to both the availability of resources and, in the example, the distribution of schools and children). (2) A component hosting service offered by third-party organizations that allows the dynamic deployment and execution of CSCL software components within grid nodes chosen by the scheduler. These research issues are further studied in the next sections.

3   Component-Based CSCL Application Scheduling

Application scheduling is a research problem widely studied in grid literature. This may suggest that schedulers already available from the grid community could be employed for CSCL application scheduling. However, schedulers are highly dependent on the domain of the application to be scheduled [8], and CSCL applications are significantly different from typical grid applications so far (e.g. supercomputing or high-performance applications). Therefore, existing schedulers cannot be reused and new schedulers must be developed for CSCL applications.

CSCL applications promote learning by enabling and enhancing collaboration between students. However, for this collaboration to be fruitful from the educational point of view, CSCL applications must yield good performance, e.g., a collaborative editor application is not feasible if it does not perform as good as to quickly distribute to all users every change that is made to the document being edited. Scheduling can improve CSCL application performance and, consequently, collaboration.

In the case of component-based CSCL applications, an acceptable performance level can be met (if possible) by properly distributing (i.e. deploying) application components within the available resources. Hence, a CSCL application scheduler should be able to dynamically *select the resources* where components are to be deployed, *allocate each component replica* to one of the selected resources and *configure the communication* between component instances. The CSCL scheduling problem can thus be regarded as the exploration of a solution space defined by all possible combinations of selection, allocation and configuration for a given application. Valid solutions can be found within this space if the following elements are provided: (1) An *application model* describing the decomposition of the CSCL application in components as well as the communication relationships between component instances, (2) *Selection criteria* defining the variables that quantify application performance as well as the conditions on these variables that must be met by solution points. (3) *Resource characteristics* describing the state of grid resources at the time a scheduling decision must be made. (4) A *performance model*, so as to estimate the value of performance variables according to the solution point to be evaluated. (5) An *exploration method*, defining the way the solution space is searched.

As a proof of concept, a specific scheduler has been developed for a simple synchronous collaborative editor. Simulation results show that editor performance, measured in terms of notification time (i.e. time elapsed since a user makes a change in the text being edited until this change is delivered to all users), is improved as much as 60% if components are distributed by the proposed scheduler when compared to traditional non-scheduled component distributions.

4   Component Hosting Service

A very important idea underlying Grid Services, as promoted by OGSA, is that they hide the way organizations implement the service they offer and the resources they use for their provision. The typical grid computing problem of *resource selection* is thus somehow moved towards the problem of *grid service selection*. This implies that a potential component-based CSCL application scheduler should select a suitable grid service allowing the dynamic deployment of the components that make up a CSCL application.

In this sense, an open problem stems from the fact that, although OGSA considers the possibility of using software components for implementing the functionality offered by Grid Services, OGSA has not defined any standard means for the dynamic deployment of software components over grid nodes offered by organizations. Therefore, part of the ongoing research described in this paper is devoted to the definition of a *component hosting service*: a Grid Service offered by organizations capable of hosting the execution of software components of CSCL applications and used by component-based CSCL application schedulers.

We have already developed a prototype of such a component hosting service for Globus Toolkit 3 (GT3). This service allows automatic deployment and hosting of Enterprise Java Bean (EJB) components in a JBoss component application server. EJB technology was chosen in order to allow the deployment of CSCL applications already developed by our research group according to J2EE standards. JBoss is preferred to other component applications servers supported by GT3 because it is freely available. This prototype is limited to the deployment of only one component.

5   Conclusions and Future Work

This paper has presented arguments supporting the feasibility of merging CSCL, CBSE and grid technologies. An illustrative example of a grid-supported component-based collaborative learning scenario has been presented and discussed leading to the study of scheduling and component hosting problems within a CSCL context. Future work includes development of schedulers for representative CSCL applications and their integration with a fully implemented component hosting service.

References

1. Dillenbourg, P.: Collaborative Learning: Cognitive and Computational Approaches. Elsevier Science, Oxford, UK (1999)
2. Wasson, B. Computer Supported Collaborative Learning: an Overview. Lecture Notes from IVP 482, University of Bergen, Norway (1998)
3. Roschelle, J., DiGiano, C., Koutlis, M., Repenning, A., Phillips, J., Jackiw, N., Suthers, D.: Developing Educational Software Components. Computer. 32 (9) (1999) 50-58
4. Furmento, N., Mayer, A., McGough, S., Newhouse, S., Field, T., Dalington, J.: ICENI: Optimisation of Component Applications Within a Grid Environment. Parallel Computing. 28 (2002) 1753-1772
5. Foster, I., Kesselman, C., Nick, J. M., Tuecke, S.: The Physiology of the Grid. In: Berman, F., Fox, G. , Hey, A. (eds.): Grid Computing: Making the Global Infrastructure a Reality. John Wiley & Sons, Chichester, UK (2003) 217-249
6. Fox, G.: Education and the Enterprise With the Grid. In: Berman, F., Fox, G., Hey, A. (eds.): Grid Computing: Making the Global Infrastructure a Reality. John Wiley & Sons, Chichester, UK (2003) 963-976
7. Bote-Lorenzo, M.L., Dimitriadis, Y.A., Gómez-Sánchez, E.: Grid Characteristics and Uses: a Grid Definition. Proc. of the 1st European Across Grids Conference, Santiago, Spain (2003)
8. Berman, F.: High-Performance Schedulers. In: Foster, I., Kesselman, C. (eds.): The Grid: Blueprint for a Future Computing Infrastructure. Morgan Kaufmann Publishers, San Francisco, CA, USA (1998) 279-309

An NAT-Based Communication Relay Scheme for Private-IP-Enabled MPI over Grid Environments

Siyoul Choi[1], Kumrye Park[1], Saeyoung Han[1], Sungyong Park[1],
Ohyoung Kwon[2], Yoonhee Kim[3], and Hyoungwoo Park[4]

[1] Dept. of Computer Science, Sogang University, Seoul, Korea
`{adore, namul, syhan, parksy}@sogang.ac.kr`
[2] Korea University of Technology and Education, Chonan, Korea
[3] Sookmyung Women's University, Seoul, Korea
[4] Korea Institute of Science and Technology Information, Daejeon, Korea

Abstract. In this paper we propose a communication relay scheme combining the NAT and a user-level proxy to support private IP clusters in Grid environments. Compared with the user-level two-proxy scheme used in PACX-MPI and Firewall-enabled MPICH-G, the proposed scheme shows performance improvement in terms of latency and bandwidth between the nodes located in two private IP clusters. Since the proposed scheme is portable and provides high performance, it can be easily applied to any private IP enabled solutions including the private IP enabled MPICH solution for Globus toolkit.

1 Introduction

As cluster systems become more widely available, it becomes feasible to run parallel applications across multiple private clusters at different geographic locations as a Grid environment. However, in the MPICH-G2 library [1], an implementation of the Message Passing Interface standard over Grid environment, it is impossible for any two nodes located in different private clusters to communicate with each other directly across the public network until additional functions are added to the library.

In PACX-MPI [2], another implementation of MPI aiming to support the coupling of high performance computing systems distributed in a Grid, the communications among multiple private IP clusters are handled by two user-level daemons that allow the library to bundle communications and avoid having thousands of open connections between systems. However, since these daemons are implemented as proxies running in user space, the total bandwidth is only about half of the bandwidth obtained from kernel-level solutions [3]. It also suffers from higher latency due to the additional overhead of TCP/IP stack traversal and switching between kernel and user mode.

This paper proposes an NAT-based communication relay scheme, combining the NAT service with a user level proxy, for private IP enabled MPI solution over Grid environments. In our approach, only incoming messages are handled by a user-level proxy to relay them into proper nodes inside the cluster, while the outgoing messages are handled by the NAT service at the front-end node of the cluster. This brings

M. Bubak et al. (Eds.): ICCS 2004, LNCS 3036, pp. 499–502, 2004.

performance improvement since we use the user-level proxy only once. By using the NAT service, which is generally provided by traditional operating systems, we can also easily apply our proposed scheme to any private IP enabled solutions without modifying operating system kernel. We have benchmarked our scheme and compared it with the user-level two-proxy scheme used in PACX-MPI [2] and Firewall-enabled MPICH-G [4]. The experimental results show that our NAT-based scheme outperforms the user-level two-proxy scheme.

The rest of the paper is organized as follows. Section 2 explains three communication relay schemes used for private IP enabled MPI, and provides the detailed mechanism of the NAT-proxy relay scheme. The experimental results are presented in section 3. Section 4 concludes the paper.

2   Communication Relay Schemes

In order to support the communication between private IP clusters in a Grid environment, we consider three communication relay schemes such as kernel-level two-proxy scheme, user-level two-proxy scheme, and NAT-proxy scheme.

In the kernel-level two-proxy scheme, we can implement a kernel-level proxy process in each of the front-end node within the cluster. Although this scheme is expected to have the best performance among the others described here, it is not used in general due to its poor portability.

In the user-level two-proxy scheme, we can implement a user-level proxy process in each of the front-end node within the cluster. A user-level proxy is easy to implement but has performance overheads such as those incurred by TCP/IP stack traversal and context switching between kernel and user mode. In this scheme, all the packets sent from one node to the other nodes located in other cluster have to go through the user-level proxy twice, which decreases the performance further. Despite its poor performance, this scheme has been widely used due to its highly portable nature. The PACX-MPI [2] and Firewall-enabled MPICH-G [4] use this scheme.

The NAT-proxy scheme is a combination of previous two solutions. The proxy implemented as a user-level program is responsible for forwarding only the incoming streams into the appropriate nodes within the cluster, while the outgoing streams go through the NAT service. Using a user-level proxy, no kernel modification is necessary. Moreover, since only incoming packets go through the proxy, the performance problems introduced by proxy can be minimized. Furthermore, using the NAT service for outgoing streams, multiple connections can be efficiently managed between front-end nodes of the clusters, which improves the communication performance further.

Fig. 1 depicts the NAT-proxy communication relay scheme proposed in this paper. In order to implement this scheme, each cluster should activate the NAT service in the front-end node. A user-level proxy, called stream relay daemon (SRD), is implemented in each front-end node. The SRD forwards incoming streams from the nodes in other clusters into their computation nodes. The outgoing streams from the computation nodes of one cluster go through the NAT service in the front-end node to reach the destination.

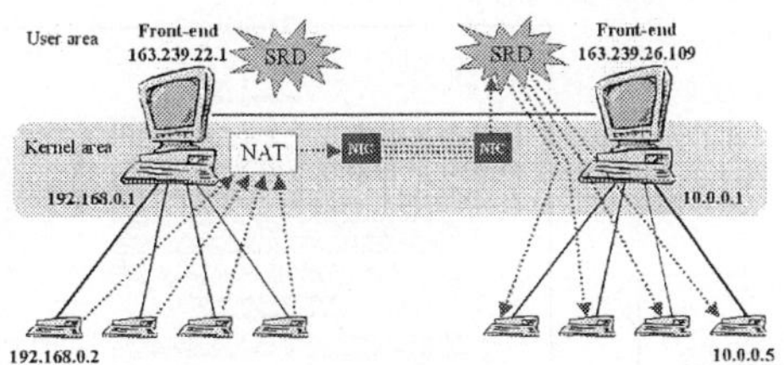

Fig. 1. The NAT-proxy communication relay scheme

3 Experimental Results

We have conducted our experiments over two private IP clusters, each of which has four computation nodes and one front-end node, respectively. The two clusters and all the nodes within the clusters are connected via 100Mbps Fast Ethernet cards. The two front-end nodes are configured to have both public and private IP addresses and each computation node is configured to have only private IP address.

In this benchmark, we compare the performance of our NAT-proxy scheme with that of the user-level two-proxy scheme. For the comparison, we measure the latency and the bandwidth between two private IP clusters using various traffic patterns.

Fig. 2 shows the latency between two private IP clusters. The latency was measured via *ping-pong* program using small sized messages (i.e., 128 bytes). As we can see from Fig. 2, our NAT-proxy scheme shows large performance improvement over two-proxy scheme by about 144%. For example, the measured latency using NAT and proxy was 1923 *usec*, while the latency using two user-level proxies was 2756 *usec*. It is clear from the result that the overhead incurred by using NAT was much lower than that of using two user-level proxies.

Fig. 3 compares the performance of our scheme with that of user-level two-proxy scheme by varying traffic patterns (one to one (1:1), many to one (2:1 and 4:1), and many to many (4:4) patterns) and varying message size from 1 Kbytes to 1024 Kbytes. As we can see from Fig. 3, the overall bandwidth obtained by using our scheme was much larger than that of using two user-level proxies. Furthermore, as we increase the message size, the performance gap is widening.

This can be explained by the following observations. In the user-level two-proxy scheme, the context-switching overhead (including message copy overhead between user space and kernel space) is bigger than that of our scheme, and the overhead becomes bigger as we increase the message size. If we apply the proposed relaying scheme to wide-area clusters, the performance improvement can be amortized to some extent, especially in small sized messages, due to the long delay (propagation delay) incurred between two front-end nodes. However, for the clusters transferring large messages and located in relatively near distance can benefit from the proposed scheme.

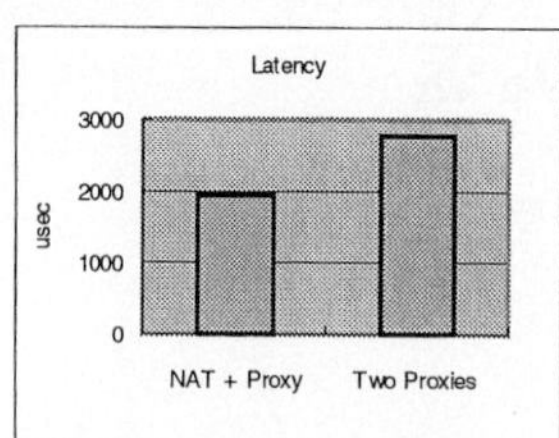

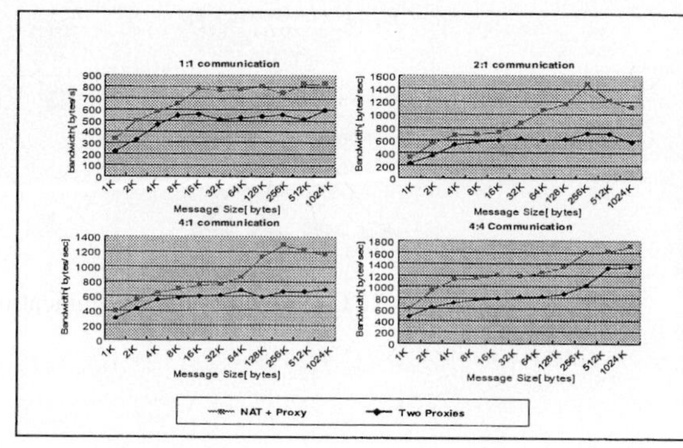

Fig. 2. Latency between two clusters   **Fig. 3.** Bandwidth between two clusters

4   Conclusion

In this paper, we have proposed a communication relay scheme based on the NAT and a user-level proxy, and compared our scheme with that of user-level two-proxy scheme that is implemented in PACX-MPI and Firewall-enabled MPICH-G. From the experiments, we showed that the performance of our scheme was better than that of user-level two-proxy scheme and the performance improvement became larger as we increase the message size. Considering that our scheme provides better performance and also does not require modifying kernel code to improve the performance, we can easily incorporate our scheme into any private IP enabled solutions.

Currently, we are working on developing a private IP enabled MPICH solution for Globus toolkit (i.e., MPICH-G2) using the scheme proposed in this paper.

References

1. Karonis, N.T., Toonen, B., Foster, I.: MPICH-G2: A Grid-Enabled Implementation of the Message Passing Interface (2002), http://www3.niu.edu/mpi/
2. Gabriel, E., Resch, M, Beisel, T., Keller, R.: Distributed computing in a heterogeneous computing environment, in Alexandor V., Dongarra, J. (eds.): Recent advances in Parallel Virtual Machine and Message Passing Interface, Vol. 1497 of Lecture notes of Computer Science, 180-188. Springer (1998). 5th European PVN/MPI User's Group Meeting.
3. Müller, M., Hess, M., Gabriel, E.: Grid enabled MPI solutions for Clusters, in Proceedings of the 3rd IEEE/ACM International Symposium on Cluster Computing and the Grid (CCGRID'03) (2003)
4. Tanaka, Y., Sata, M., Hirano, M., Nakata, H., Sekiguchi, S.: Performance Evaluation of a Firewall-compliant Globus-based Wide-area Cluster System, in Proceedings of the Ninth IEEE International Symposium on High Performance Distributed Computing, 121-128. IEEE Computing Society (2000)

A Knowledge Fusion Framework in the Grid Environment

Jin Gou[1], Jiangang Yang[1], and Hengnian Qi[2]

[1] College of Computer Science of Zhejiang University,
310027 Hangzhou, China
{goujin, yangjg}@zju.edu.cn
[2] School of Information Engineering of Zhejiang Forestry College,
311300 Hangzhou, China
qhn@zjfc.edu.cn

Abstract. The paper presents a knowledge fusion architecture based on the grid platform. The proposed framework suggests a semi-structural paradigm that emphasizes connotation of distributed knowledge resources in a grid environment. Our approach involves an extractive process of meta-knowledge sets which predigests the diversion among multi-source knowledge, the Genetic Fusion Algorithm which can generate a new knowledge space, a resources allocation method with meta-knowledge directory service. Experimental results of a case study show the feasibility of design rationale underling knowledge grid.

1 Introduction

Knowledge fusion is an important component of the knowledge science and engineering, which can transform and integrate diversiform knowledge resources to generate new information [1]. So information and knowledge from distributed nodes can be shared and cooperating. Multi-agent technology and grid computation method are used to integrate specifically knowledge in a way, which require plenty of data conversion operations and mapping procedures [2][3][4]. The paper proposes the architecture based on meta-knowledge and ontology bases to replace the complex interchanging process among diverse knowledge bases with extraction of meta-information. In order to fuse knowledge according to not formats but connotation, the paper contributes the Genetic Fusion Algorithm (GFA).

Knowledge fusion will result in an enormous amount of knowledge resources on the web. In such settings, we encounter resources management and other challenges. Since grid technologies carry the promise to enable widespread sharing and coordinated use of networked resources with effective scheme, we adopt a kind of semi-structure data model to encapsulate data resources on the knowledge grid platform [5].

The purpose of this paper is to present a knowledge fusion architecture with an autonomous resource allocation method used in the grid and the GFA for generating a new knowledge space.

M. Bubak et al. (Eds.): ICCS 2004, LNCS 3036, pp. 503–506, 2004.

2 Architecture

The knowledge fusion is constructed on the Globus grid services. Figure 1 shows the overall architecture.

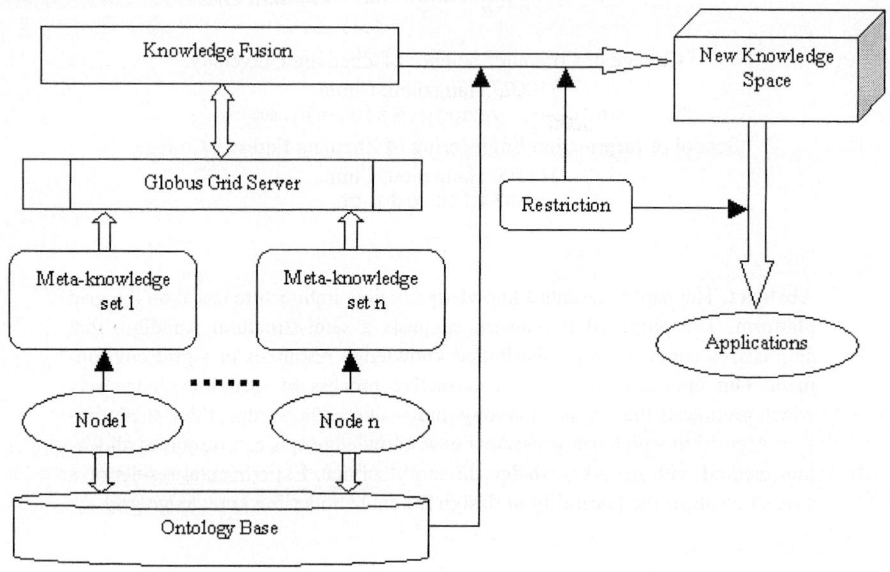

Fig. 1. Framework of the grid based knowledge fusion system

The Globus project is the infrastructure of Grid computation. Globus Toolkit[6] can run on several operating systems. When we extract meta-knowledge from any node, the ontology should be described underlying certain rules and added into the ontology base. In the following, we elaborate on this issue.

Ontology base is a complex description sets. It is much more difficult to model inference in such a distributed knowledge grid environment. Let O and O_i denote the ontology base and the ith object instance in it. The O_i will be used in the paper, which is defined by

$$O_i = \{(P^i_j,\ T^i_j,\ D^i_j)\} . \tag{1}$$

where P^i_j denotes the jth attribute of the object O_i, T^i_j denotes the type of P^i_j, D^i_j denotes its value, expression or behavior. The domain for variant j is decided by idiographic scope knowledge ontology.

Let S_k denotes meta-knowledge sets of the kth knowledge base node which can be defined by

$$S_k = \{(C^k_1, E^k_1),\ (C^k_2, E^k_2),\ \dots,\ (C^k_n, E^k_n)\} . \tag{2}$$

where C^k_i denotes the ith character object of S_k, E^k_i denotes the description content of C^k_i.

Relationships among character objects are not defined here because those will be described in the ontology base. In order to implement the interchanged process between meta-knowledge sets and knowledge space, the definiendum $(C_{,i}^k, E_{,i}^k)$ stands for not embodied characters but denotative objects of knowledge ontology. That operation must be synchronous with the initializtion of meta-knowledge sets. Meta-knowledge sets and ontology base are formalized to be fuse and generate new knowledge elements.

3 Fusion Algorithm

With a resource allocation method mentioned above, we contribute the GFA to generate a new knowledge space according to the embodiment of knowledge ontology.

Describe the supposition as constructing meta-knowledge sets, function as fusing and generating new knowledge space.

$$GFA\ (Fitness,\ Fitness\_threshold, p, r, m) \tag{3}$$

where *Fitness* denotes assessing function for fusion matching grade, which can endow the given supposition sets with a matching grade. *Fitness_threshold* denotes the threshold value beyond which fusion process can not persist. p is a supposed number which should be adjusted according to the result of unifying among diversiform meta-knowledge sets. If dimension of some sets are too small, NULL can be used to instead of it. r and m are intercross and aberrance percent.

Some description of the algorithm can be found in [7], the major steps are summarized as follows:

Select: Choose any S_i and append it to H_s, let the counter $c=1$

   if $c< (1-r)p$, do the following operation circularly:

   choose any S_j not belonged to H_s from $\{S_k\}$ - S_i

   if $\exists\ (P_m^i, T_m^i, D_m^i) \in O_i$, and $P_m^i = $ "$R(C_*^i, C_*^j)$" and $D_m^i \neq$ NULL

   or $\exists (P_m^i, T_m^i, D_m^i) \in O_i$, and $P_m^i = $ "$R(C_*^i, C_*^j)$" and $D_m^i \neq$ NULL

   append S_j to H_s, c++ .

Intercross: Selected result $<S_1, S_2>$ must make the following expression right:

$$E_1^l \cap E_1^2 \neq \phi \tag{4}$$

this means the intersection of ontology relating to suppositions can not be temp. Any element except the first one meets the requirement above can be intercrossed.

Aberrance: For each (C_j^i, E_j^i) , reverse its value as follows:

   for every member $S^k \neq S^i$ of H_s ,

   if $O_l = \{(P_m^l, T_m^l, D_m^l)\}$ which can meet the requirement: $\quad j'\exists P_m^l = R(C_j^i, C_{j'}^k)$

and $E_{j'}^k \neq E_j^k$

   $E_{j'}^k$ is a result of reversing operation on E_j^k.

Solution knowledge can be generated as follows:

Create ontology corresponding with question (O_p) and meta-knowledge (restriction) sets (S_p). Search all knowledge states in K for S_a whose relationship grade to question state is the max. For each knowledge state related to question ontology, seek out its relationship grade. It is also the percent of related knowledge points in knowledge state and those in question state.

Knowledge state S_k relates to question ontology O_p must meet requirement as follows: $\exists\, (P^p_j, T^p_j, D^p_j) \in O_p$, $(P^k_j, T^k_j, D^k_j) \in O_k$.

Table 1. Result of a case study

	Transform times	Losing connotation	Generate new knowledge	Find solution	Can be reused	Uniform
Based on interchange	600	Yes	No	No	Can	Yes
Method in the paper	50	No	Yes	Yes	Can	Yes

4 A Referential Application and Summary

Let us shift focus beyond the abstract architecture and point to a case study.

As shown above, the framework in the paper can minimize cost in a knowledge grid especially when a knowledge fusion procedure runs on it. And it can also improve the reuse performance of knowledge elements.

We present a new knowledge fusion framework in a grid environment. Compared with traditional resource management system, method given in the paper gives more flexibility on task requirements and resource utilization. We have also presented the GFA which can fuse diversiform knowledge and generate new knowledge space according to the connotation of ontology. In the future, we will apply the framework to more finely granularity knowledge grid and optimize the matching process.

References

1. LU, R.Q.: Knowledge Science and Computation Science. Tsinghua University Press, Beijing China (2003)
2. James, M.: Structured Knowledge Source Integration and Its Applications to Information Fusion. Proceedings of The 5th International Conference on Information Fusion. Maryland: IEEE, (2002) 1340-1346
3. Mario, C., Domenico, T.: The Knowledge Grid. Communications of The ACM. 1 (2003) 89-93
4. Tomas, M., Zsolt, B., Ferenc, B., *et al*: Building an Information and Knowledge Fusion System. Proceedings of The 14th International Conference on Industrial and Engineering Applications of AI and Expert System. Budapest: ACM, (2001) 82-91
5. Ian Foster, Carl Kesselman, Steven Tuecke: The Anatomy of The Grid: Enabling Scalable Virtual Organizations. Lecture Notes in Computer Science, Vol. 2150, (2001) 1-26
6. The Globus Project http://www.globus.org
7. Mitchell, T.M.: Machine Learning. China machine press, Beijing China (2003)

A Research of Grid Manufacturing and Its Application in Custom Artificial Joint

Li Chen [1], Hong Deng [1], Qianni Deng [2], and Zhenyu Wu [1]

[1] Shanghai Jiao Tong University, School of Mechanical and Power Engineering,
Shanghai 20 00 30,
62932905 Shanghai, China
{chen_li, denghong76, wzy}@sjtu.edu.cn
[2] Shanghai Jiao Tong University, Department of Computer and Science, Shanghai 20 00 30,
62932632 Shanghai, China
deng-qn@cs.sjtu.edu.cn

Abstract. This paper presents the framework of Grid Manufacturing, which neatly combines Grid technology with the infrastructure of advanced manufacturing technology. It studies the Grid-oriented knowledge description and acquisition, and constructs the distributed Knowledge Grid model. It also deals with the protocol of node description in collaborative design, and builds up the distributed collaborative design model. And the research on the protocol and technology of node constructing leads to the collaborative production model of Grid Manufacturing. Finally, the framework of Grid Manufacturing is applied in the design and manufacturing of custom artificial joint and the joint product is produced more efficiently.

1 Introduction

With the rapid technological innovations of the networked manufacturing, much more is learned about the inherent limitations of the network technology. Grid is regarded as the next generation Internet as well as Grid Manufacturing is then presented as an advanced solution for the bottleneck of networked manufacturing. The research on Grid will build up solid theoretical and technological fundaments to realize a great stride in manufacturing [1-4].

2 Data and Knowledge Management of Grid Manufacturing

Data and knowledge management of Grid Manufacturing includes the following steps:
Firstly analyzing the storage mode and structure of heterogeneous data on the Grid nodes; then building up the general and open knowledge description, internal encapsulation protocol and exchange standard, and various information and knowledge

M. Bubak et al. (Eds.): ICCS 2004, LNCS 3036, pp. 507–510, 2004.

required in Grid Manufacturing encapsulation (including the heterogeneous database, design and operating know-how, thinking process on various nodes, etc.); finally displaying these knowledge by the uniform external interactive protocols and inter-faces[5].

3 Collaborative Design and Production Model of Grid Manufacturing

The distributed design mechanism based on Grid Manufacturing technology is seeking to enhance the interactions of collaborative design between the dynamic union of enterprises to the level of high efficiency, high speed, large scale, and massive data traffic [6].

The construction protocol and technology of manufacturing nodes is to realize the high performance scheduling and dynamic collaboration of Grid Manufacturing resources.

The kernel of the collaborative production model, the resources scheduling and manufacturing collaboration will be realized by the five-level Grid structure and related functional modules. With the functions and services provided by the five-level Grid structure, the manufacturing collaboration can be finally achieved when these sub-models are realized by the technical support of the corresponding level. The production collaboration model is shown in Fig. 1.

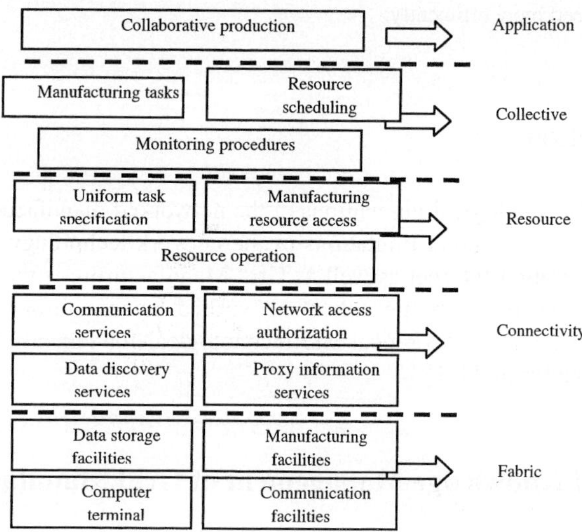

Fig. 1. Collaborative production model

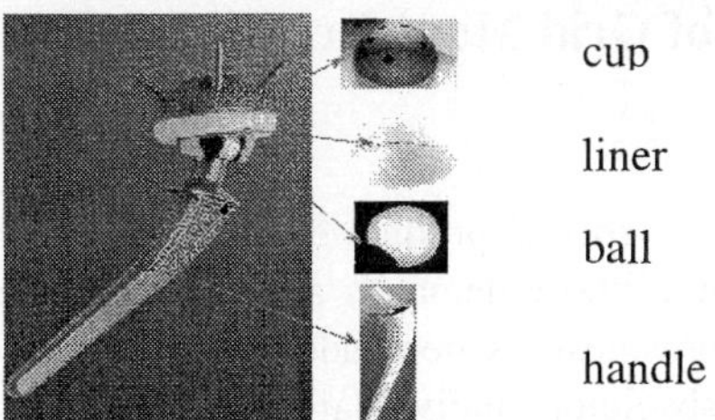

Fig. 2. A product of custom artificial hip joint

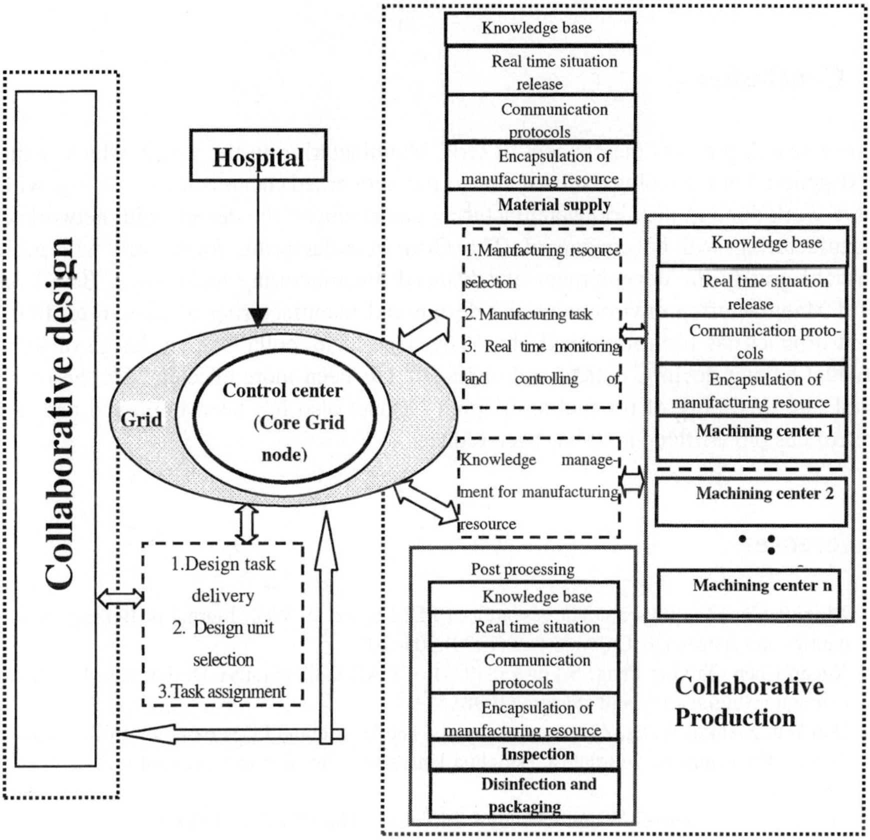

Fig. 3. The workflow of custom artificial joint in Grid Manufacturing environment

4 An Application of Grid Manufacturing in the Custom Artificial Joint

Custom artificial joint is a special product designed for perfectly matching an individual's medullary cavity, whose demands are unforeseeable and urgent[7]. Once a custom artificial joint prosthesis is demanded, it should be supplied as quickly as possible. Everyone is absolutely individualized.A product of custom artificial hip joint is illustrated in Fig.2.

The workflow of custom artificial joint in Grid Manufacturing environment is illustrated in Fig.3.

5 Conclusion

This research presents the concept of Grid Manufacturing in the world, which is the next-generation technology subsequent to the networked manufacturing. Along with more work done in the Grid Manufacturing environment, the defects with networked manufacturing will be conquered. The Grid Manufacturing framework will push forward greatly the development of advanced manufacturing technology. Based on Grid Manufacturing environment, the design and manufacturing of custom artificial joint product has been improved obviously. One hand, collaborative design of joint product among doctors, patient and engineers has been more efficient. On the other hand, the flexibility of production of joint product also has been improved and the cost of custom artificial joint has been reduced.

References

1. Xingjun Chu, Yuqing Fan: The research of PDM based on Web. Journal of Beijing Aeronautics and Astronauts University 2 (1999) 205–207
2. Xucan Chen, Yuxing Peng, Sikun Li: PDM of CAD Collaborative Design based on C-S. Computer Engineering and Design (1998) 53
3. Dan Wu, Xiankui Wang, Zhiqiang Wei et al.: The Distributed PDM based on Collaborative Service Environment. Journal of Tsinghua University (Science and Technology) 6 (2002) 791–781
4. Zhiqiang Wei, Xiankui Wang, Chengyin Liu et al.: The Distributed PDM under the Environment of Agile Manufacturing. Journal of Tsinghua University(Science and Technology) 8 (2000) 45
5. H.Zhuge: A Knowledge Grid Model and Platform for Global Knowledge Sharing. Expert System with Applications, Vol. 22. no.4 (2002)
6. G. von Laszewski, I. Foster, J. Gawor et al.: Designing Grid-based problem solving environments and portals. Proceedings of the 34th Annual Hawaii International Conference on System Sciences, IEEE Press (2001)
7. Shang Peng: The multi-agent design system of custom artificial hip joint. Journal of
8. Chinese Biomedical Engineering 2 (2001)

Toward a Virtual Grid Service of High Availability

Xiaoli Zhi and Weiqin Tong

School of Computer Engineering and Science, Shanghai University,
Shanghai 200072,P.R. China
`{xlzhi, wqtong}@mail.shu.edu.cn`

Abstract. A new regulation approach is proposed to obtain a virtual resource service of high availability and service capacity on the basis of resources of low availability and small capacity. Some regulation algorithms with distinct characteristics are introduced.

1 Introduction

With the widespread proliferation of Grid services, quality of service (QoS) will become a significant factor in distinguishing the success of service providers. In OGSA, anything providing some functions to the public can be treated as a virtual service. QoS of a service refers to its non-functional properties such as performance, reliability, availability, etc [1]. E.g., QoS of a processor can be measured by availability, computation capacity (in MIPS). QoS of a storage can be measured by availability, storage capacity (as in Terabytes) and so on. This paper will pay more interest in availability and service capacity (computation capacity or storage capacity).

Grid computing borrowed its term from the "Electric Power Grid" [2]. We got the idea of regulation from comparison of computational grids with the electrical power industry. Regulation of grid services is to achieve more stable, high available service from unstable source services just as the rectifier in the power grid get direct current from alternating current.

2 Regulation Algorithms

In a regulation system, several services, termed as source services, are organized into the 'backend' of a regulated service, as shown in Fig.1. The main task for a regulated service is to delegate service request to some appropriate source services according to its regulation algorithm. A regulated service appears nothing special externally just as a normal resource service on which a grid resource scheduler or broker can act.

The regulated service and its source service form a service aggregation. Different from the classic purpose of service aggregation, this aggregation is to serve as a buffer or stabilizer to passivate the sensitivity of service failure or variation.

M. Bubak et al. (Eds.): ICCS 2004, LNCS 3036, pp. 511–514, 2004.
© Springer-Verlag Berlin Heidelberg 2004

The most important thing in service regulation is the regulation algorithm. There can be various regulation algorithms under different conditions to meet different requirements. Here presents only a small yet typical and important directory of regulation paradigms. The directory is open to evolve.

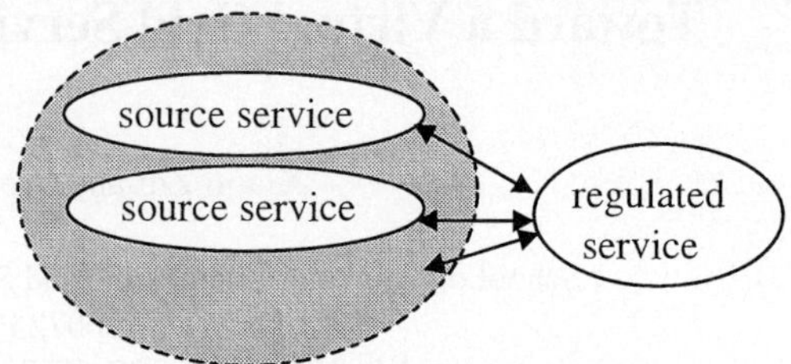

Fig. 1. Diagrammatic view of a regulated service

2.1 Paradigm 1: Multiple-to-One

Paradigm 1 is used to integrate service capacity or increase service availability.

Paradigm11: Heaping. The regulated service treats source services as a heap of service power and uses up the source services one by one. It can add up source services' capacity to make a service of bigger capacity and higher availability than a single component service.

Assume there have M source services. Source services can provide normal capacity $C_1, C_2,\ldots$ with the probabilities $p_1, p_2, p_3,\ldots$, that is, availabilities for source services with normal capacity. Unavailabilities or downtime probabilities are $1\text{-}p_1$, $1\text{-}p_2$, $1\text{-}p_3\ldots$, respectively. For simplicity, we presume source services only have two states: normal work or downtime/repair state. Then availability for the regulated service with service capacity c_r:

$$\text{Pro}(c_r = \sum_{i \in \text{subset of } \{1,2,\ldots,M\}} C_i) = \prod_{i \in \text{subset of } \{1,2,\ldots,M\}} p_i \tag{1}$$

The unavailability of the regulated service:

$$\prod_i (1 - p_i) \tag{2}$$

Paradigm12: Stripping. In this paradigm, workload is down into parts and each part is assigned to a separate source service. The service capacity and some performance parameters such as I/O speed for data transfer is greatly improved. However, it does offer a few disadvantages. It is more vulnerable than its source service. Availability for the regulated service with service capacity c_r:

$$\text{Pro}(c_r = \sum_i C_i) = \prod_i p_i \tag{3}$$

The unavailability of the regulated service:

$$1 - \prod_i p_i \tag{4}$$

Paradigm13: Fault tolerant configuration. The source services in this paradigm are configured as an active-active fault tolerant system. And workload is mirrored identically in every source service. This paradigm has the highest availability among paradigms introduced in this paper. Availability of the regulated service:

$$1 - \prod_i (1 - p_i) \tag{5}$$

2.2 Paradigm 2: Multiple-to-Multiple

Paradigm 1 normally demotes the utilization coefficient of source services although it can provide an integrated service of bigger capacity or higher availability. This paradigm will promote the utility factors as well as availability by grasping idle source service to serve in the place of a failed component service. A regulated service in Paradigm 2 has a designated main source service (Note every source service is the main service of a regulated service). The regulated service is just a transparent broker when its main source service runs normally. It acts what its main source service acts. But it will draft an idle other source service for the incoming task using some scheduling algorithm when its main source service is down.

Assume idle coefficients for source services (nothing to do while the service is ready to do something) are $q_1, q_2, q_3, \ldots$. Then Availability for the regulated service $_k$ (assume the number of its main source service is also k):

$$p'_k = p_k + (1 - p_k)(1 - \prod_{j \neq k}(p_j \times (1 - q_j) + (1 - p_j)) \tag{6}$$

Under the assumption of random choosing idle source services, idle coefficient for the source service k after it participates into a regulation system:

$$q''_k = q_k - \sum_j ((p'_j - p_j) \times (1 - q_j)) \times (\frac{p_k \times q_k}{1 - q_k}) / \sum_j (\frac{p_j \times q_j}{1 - q_j}) \tag{7}$$

The above formula is an approximation due to the highly intricacies of the accurate computation of q''_k. Actually, the idle coefficient for the source service after regulation is too difficult to be formulated mathematically when the scheduling algorithm is not a random one.

2.3 Paradigm 3: One-to-One

In the power industry, a rectifier has element of energy storage and transformation (e.g., capacitance or inductance) to regulate a fluctuating current into a smoother one. Paradigm 3 adopts a task buffer to serve as something like 'energy storage and transformation'. The buffer revokes the source service when available to process buffered tasks and return tasks' results at proper time. This paradigm is suitalbe for batch processing, asynchronous applications but not areas with high time demand.

Assume availability of the source service is p; the service capacity for the source service is C, B for the regulated service ($B<C$), then availability for the regulated service:

$$p' = \frac{Cp}{B} > p \tag{8}$$

3 Discussions

A regulation service is composed of several source services as a service aggregation is. But they implement different purposes with completely disparate techniques. A service aggregation is to finish a task through cooperation of its components while regulation is targeted for improving services' QoS. Component services of a service aggregation are usually playing different roles with different functions, while those in regulation have similar service capability.

In some sense, regulation seems like a resource broker. But a regulated service actually distinguishes itself by different status in the grid. A regulated service is just a resource service. A resource broker is a part of or a base service in the grid middleware. And the scheduling (if any), interaction of a regulated service with its source services are hidden completely from the grid user while a resource broker doesn't. In addition, the scheduling algorithm and communication protocol, implemented in a regulated service, tend to be more likely proprietary, while they may not be adopted in a resource broker for out of standardization.

In summary, advantages of regulation are the following:

- Partly the resource management function will be spread around to virtual regulated services. This will alleviate the burden of resource management and enhance the grid middleware's reliability with higher available low-level resource services.
- Various proprietary resource scheduling algorithm or composition technique can be utilized in regulation within a relatively local scope.
- No additional complexity to the grid system middleware will be induced.

References

1. A. Mani, A. Nagarajan: Understanding quality of service for Web services---Improving the performance of your Web services. January 2002,
    `http://www-900.ibm.com/developerWorks/cn/webservices/ws-quality/ index_eng.shtml`
2. I. Foster and C. Kesselman: The Grid: Blueprint for a New Computing Infrastructure, Morgan Kaufmann, San Fransisco, CA, 1999.
3. G. Mateescu: Quality of service on the grid via metascheduling with resource co-scheduling and co-reservation. International Journal of High Performance Computing Applications, vol.17, no.3, 2003, p 209-218

The Measurement Architecture of the Virtual Traffic Laboratory

A Design of a Study in a Data Driven Traffic Analysis

Arnoud Visser, Joost Zoetebier, Hakan Yakali, and Bob Hertzberger

Informatics Institute, University of Amsterdam

Abstract. In this paper we introduce the measurement architecuture of an application for the Virtual Traffic Laboratory. We have seamlessly integrated the analyses of aggregated information from simulation and measurements in a Matlab environment in which one can concentrate on finding the dependencies of the different parameters, select subsets in the measurements, and extrapolate the measurements via simulation. Available aggregated information is directly displayed and new aggregate information, produced in the background, is displayed as soon as it is available.

1   Introduction

Our ability to regulate and manage the traffic on our road-infrastructure, essential for the economic welfare of a country, relies on an accurate understanding of the dynamics of such system. Recent studies have shown very complex structures in the traffic flow [1]. This state is called the synchronized state, which has to be distinguished from a free flowing and congested state. The difficulty to understand the dynamics originates from the difficulty to relate the observed dynamics in speed and density to the underlying dynamics of the drivers behaviors, and the changes therein as function of the circumstances and driver motivation.

Simulations play an essential role in evaluating different aspects of the dynamics of traffic systems. As in most application areas, the available computing power is the determining factor with respect to the level of detail that can be simulated [2] and, consequently, lack of it leads to more abstract models [3]. To be able to afford more detailed situations, we looked how we could use the resources provided by for instance Condor or the Grid.

Simulation and real world experimentation both generate huge amount of data. Much of the effort in the computer sciences groups is directed into giving scientists smooth access to storage and visualization resources; the so called middle-ware on top of the grid-technology. Yet, for a scientist seamless integration of the ineformation from simulated data and measurements is the most important issue, the so called data-driven approach.

In this article we show our experience with building our Virtual Traffic Laboratory as a data driven experimentation environment. This experience can be used as input for the future development of the Virtual Laboratory on other application domains.

M. Bubak et al. (Eds.): ICCS 2004, LNCS 3036, pp. 515–518, 2004.

2 Measurement Analysis Architecture

Traffic flow on the Dutch highway A12 is investigated for a wide variety of circumstances in the years 1999-2001. This location has the unique characteristic that, although the flow of traffic is high, traffic jams are very sporadically seen. In this sense it is a unique measurement point to gather experimental facts to understand the microscopic structures in synchronized traffic states [1], which was not reported outside of Germany yet.

For the understanding the microscopic structures in synchronized traffic states the relations between several aggregates of single vehicle measurements have to be made. Important aggregate measurements are for instance average speed, average flow, headway distribution and speed difference distribution. The dynamics of these one-minute aggregates over 5-10 minutes periods are important for a correct identification of the state.

To facilitate the analysis of aggregate measurements over time we designed the following architecture:

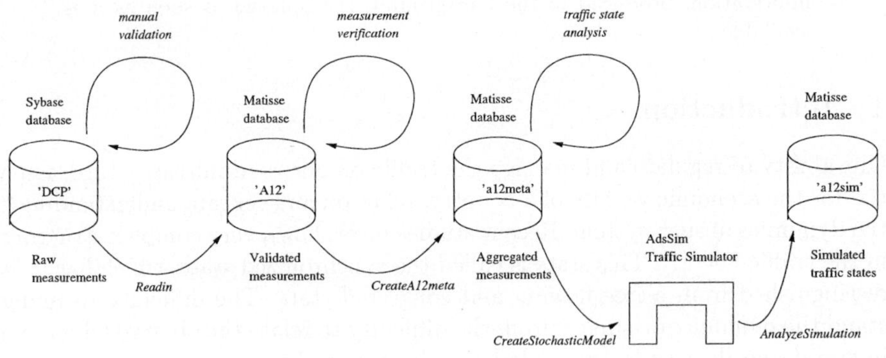

Fig. 1. The measurement analysis architecture

Along the A12 there was a relational database from Sybase that collected the measurements from two independent measurement systems. One system was based on inductive loops in the road, the other on an optical system on a gantry above the road. Although both were quality systems, some discrepancies occur between measurements due to different physical principles. Video recordings were used to manually decide the ground truth when the measurements were not clear.

After this validation process, the measurements were converted to an object oriented database from Matisse. This database was used to verify the quality of the measurement systems themselves. While the manual validation process was used to get the overall statistics of errors in the measurements, the object oriented database was used to analyze the circumstances of the measurement errors.

The validated measurements were used to generate the statistics that characterize the traffic flow. Different measurements-periods could be combined based on different criteria. The right combination of criteria results in candidate traffic flow states. The statistics that are important to characterize the microscopic structure of the traffic flow require fits of complex (non-Gaussian) probability density functions. The statistics were stored as meta-data in a separate database.

An example of such analysis is given in figure 2, where the average speed is given as a function of the flow (as percentage of the maximum flow) and the fraction of lorries (as percentage of the number of passages).

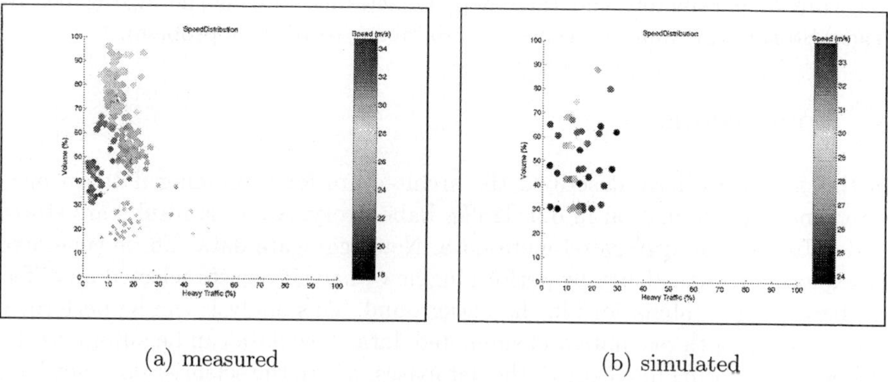

(a) measured   (b) simulated

Fig. 2. The average speed as function of the flow and the fraction heavy traffic

The average speed is indicated with a colorcode, red (top of the bar) indicates high speeds, blue (bottom of the bar) indicates low speeds. Each point indicates an aggregate over longer period (30-60 minutes), which are typically equivalent with a few thousand passages.

Combinations of measurement-periods that showed the same patterns on their aggregated traffic flow measurements over time were candidate traffic flow states. These aggregated measurements could be translated into the parameters of a microscopic traffic simulator, AdsSim [4], which is based on the microscopic Mixic model [5].

The characteristics of the simulated data were aggregated in the same way as the real data, and the resulting dynamics were compared to the original dynamics, to see if the model was complete (see figure 2). As one can see, the simulated points are more homogeneous spread over the spectrum because one can generate a dependency grid. Yet, the results are less to be trusted when one has to extrapolate far from actual measured parameter-combinations space.

3   Discussion

We have chosen this application, because of the complexity of both the measurement analysis and the traffic flow model. For instance, the Mixic model has

68 parameters in its traffic flow model [5], and most parameters are described as functions of single vehicle data such as lane, speed and headway. For AdsSim this resulted in 585 variables that can be adjusted to a specific traffic condition. Compare this with the 150 keywords in the standard application in molecular dynamics in the UniCore environment [6].

To be able to calibrate such a model for a certain traffic state, one needs to be able to select characteristic subsets in the bulk of measurements, and visualize the dynamics of the aggregates in different ways. It is no problem that it takes some time to generate aggregates, as long as one is able to switch fast between diagrams of parameters and their dependencies as soon as the aggregates are ready. Storing the analysis results in a database solves this problem.

4   Conclusions

In this article we have described the architecture for combining data on measurements and simulation in our Traffic Laboratory. Analysis results are stored in databases with aggregated meta-data. New aggregate data can be generated by exploring a dependency by performing new analysis on sets selected on different parameter-combinations in the background. This analysis can be performed seamlessly on both real data and simulated data. New data can be automatically displayed by adding monitors to the databases, where the scientist does not have to worry that too rigorous filtering will force him to do the aggregation again.

References

1. L. Neubert, et al., "Single-vehicle data of highway traffic: A statistical analysis", Physical Review E, Vol. 60, No. 6, December 1999.
2. K. Nagel, M.Rickert, "Dynamic traffic assignment on parallel computers in TRANSIMS", in: Future Generation Computer Systems, vol. 17, 2001, pp.637-648.
3. A. Visser et al. "An hierarchical view on modelling the reliability of a DSRC-link for ETC applications", IEEE Transactions on ITS, Vol. 3: No. 2, June 2002.
4. A. Visser et al. "Calibration of a traffic generator for high-density traffic, using the data collected during a road pricing project", paper 4052 to the 9th World congress on Intelligent Transport Systems, Chicago, Illinois, October 2002
5. C. Tampére, C. "A Random Traffic Generator for Microscopic Simulation", Proceedings 78th TRB Annual Meeting, Jan. 1999, Washington DC, USA.
6. D.W. Erwin et al. "UNICORE: A Grid Computing Environment", in LNCS 2150, p. 825-839, Springer-Verlag, 2001.

Adaptive QoS Framework
for Multiview 3D Streaming[*]

Jin Ryong Kim[1], Youjip Won[2], and Yuichi Iwadate[3]

[1] Digital Contents Research Division, Electronics and Telecommunications Research
Institute, Daejeon, Korea
`jessekim@etri.re.kr`
[2] Div. of Electrical and Computer Engineering, Hanyang University, Seoul, Korea
`yjwon@ece.hanyang.ac.kr`
[3] NHK Science & Technical Research Laboratories, Japan Broadcasting Corporation,
Tokyo, Japan
`iwadate.y-ja@nhk.or.jp`

Abstract. We present the adaptive QoS framework for multi-view 3D
streaming to deliver the media in time and at the same time, we provide
an optimal solution to minimize the quality variation. We dynamically
adjust the number of polygons in 3D model so that it can support
constant frame rate. We also propose to minimize the frequencies in QoS
transition to provide better user perceptive streaming. As a result, the
stable frame transmission rate is guaranteed and the quality fluctuation
becomes smoother.

Keywords: QoS, Virtual Studio, Interactive TV, Multimedia Streaming, 3D Streaming

1   Introduction

When 3D media move downstream to viewers, transmission rate in the network
will be varied depending on the network traffic condition. We propose an adaptive
QoS management scheme to efficiently stream time-critical media and optimal
quality adaptation scheduling algorithm for multiview 3D streaming.

NHK[3] developed a 3D model generation system using multiple cameras with
multi-baseline stereo algorithm and the volume intersection method. It is designed to generate 3D model media contents and support any viewpoint. Salehi[5]
proposed an optimal rate smoothing algorithm based on the traffic smoothing
technique to achieve minimum variability of the transmission rate. Cuetos et
al.[2] proposed to find a shortest path to minimize variability. Nelakuditi et
al.[4] accomplished the maximum reduction of quality variability for layered
CBR video using bidirectional layer selection. This paper extends NHK's multiple camera system by developing the QoS architecture to develop 3D streaming
system for immersive environment. It is designed for delivering visual contents
from the studio to the consumer platform via the Internet.

---

[*] This work is in part funded by KOSEF through Statistical Research Center for
Complex System at Seoul National University.

M. Bubak et al. (Eds.): ICCS 2004, LNCS 3036, pp. 519–522, 2004.

2   Rate Adaptive Transmission

The idea of our scheme is to monitor the network bandwidth availability and send the 3D video having appropriate bit rate. Each time slot represents the time unit for playing a video. Let k be the time slot at t_k. C is the number of frames in a time slot and it is set to $C=20$ for every time slot to have a fixed frame rate for every time slot. We define W_k as a quantitative amount of the available network bandwidth at each time slot k. We assume that the current W_k is known. Given a maximum network bandwidth, W_k is fluctuating over a wide range. The available network bandwidth curve is divided into time slots and quantified under $\delta \geq \phi$ for $k = 1...N$ where N is the total number of time slots, ϕ_k is inner portion of network bandwidth curve, and δ_k is outer portion of network bandwidth curve. W_k is an optimal value at each time slot k from the network available bandwidth curve. QoS level at time slot k is determined as follow: if available bandwidth is greater than r_E, then QoS_k is assigned as 'E', where QoS_k is QoS level at k and r_{lv} is bit rate for QoS level lv. If available bandwidth is between r_E and r_G, then QoS_k is assigned as 'G'. If available bandwidth is between r_G and r_F, QoS_k is assigned as 'F'. If available bandwidth is between r_F and r_P, then QoS_k is assigned as 'P'. If available bandwidth is lower than r_P, then QoS_k is assigned as 'B'. Qos_k of 'E', 'G', 'F', 'P', and 'B' can be mapped into quality scale Q_k of 5, 4, 3, 2, and 1 for simplicity.

3   Optimizing Quality Variation

Fig. 1(a) illustrates the unstable video transmission if the server transmits only one quality of the video sequence and Fig. 1(b) illustrates how the video sequence is adaptively transmitted under variable network bandwidth using our existing scheme. It also illustrates the quality fluctuation in Fig. 1 (b). Each colored time slot represents the distinct quality. The basic idea of our extended scheme is that instead of changing the quality of the time slot at each point of time, we keep the same quality and raise the quality level at some point. We accomplish this scheme by prefetching some portions of the next time slot as shown in Fig. 1(c). This enables us to maximize the usage of available network bandwidth and minimize the quality variation.

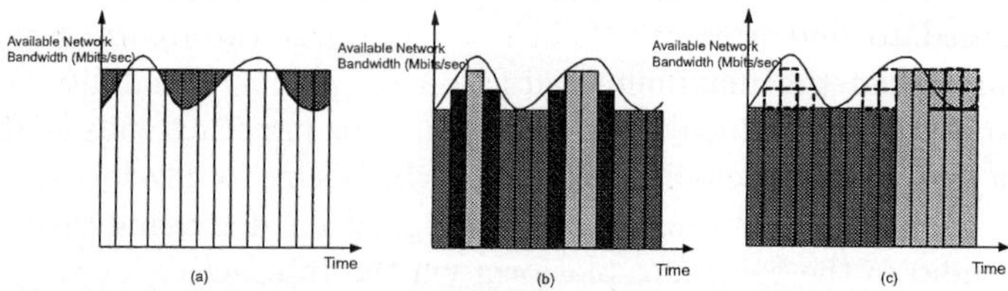

Fig. 1. Example of quality optimization

We forecast the future network bandwidth availability using double exponential smoothing based predictor (DESP)[1]. Using $DESP$, the future available network bandwidth can be forecasted as $P(k) = \alpha W(k) + (1-\alpha)(W(k-1) + b(k-1))$ and $b(k) = \gamma(P(k) - P(k-1)) + (1-\gamma)b(k-1)$ where $P(k)$ is smoothed value at k and $b(k)$ is a trend equation. α and γ are smoothing and trend constants, respectively, and $\alpha \in [0,1]$ and $\gamma \in [0,1]$. Forecast equation $Z(k)$ is defined as $Z(k+u) = P(k) + ub(k)$ where u denotes $u - period$ ahead forecast.

We model the quality adaptation for CBR video by replacing re-scheduled time slots to maintain a uniform quality scale. In formulating optimal quality adaptation, we consider a discrete-time model. We assume that there are 5 seconds of buffers in the client. The time slots are scheduled in the server based on $DESP$. We set u as 4 to have 4 time slots in the server. We consider the future available network bandwidth trend as a reference quality scale for every point of time. Let i be the index and N be the number of time slots in the server. Then, the mean quality scale of the time slots in the server is $Q_{avg} = \lfloor \frac{\sum_{i=k+1}^{N+k} Q_i}{N} \rfloor$. We define a majority quality scale in the system for selecting optimal quality scale. Let $f_R(r,k)$ denotes a set of frequencies of quality scales where $r \in \{E, G, F, P, B\}$. Then, the majority quality scale is $Q_{majority} = max\{f_R(r,k)\}$. We now introduce our prefetching algorithm. The time slots are divided into layers with the same size and it is represented as granules. When the server transmits the time slot with rescheduled quality scale, some vacant granules become available. Then, we take advantage of using these empty spots to prefetch some granules from the next time slot. We keep track of residual bandwidth $RB(k)$ and it is defined as $RB(k) = RB(k-1) + W_k - r_{lv}$. $L(k)$ denotes the depth of layer at k and it is defined as $L(k) = \frac{r_{lv}}{l}$ where l denotes the size of granule. $e(k)$ denotes the number of empty granules and it is $e(k) = \frac{RB(k)}{l}$. Using $L(k)$ and $e(k)$, we prefetch granules in $k+1$ as $Prefetch = \sum_{j=L(k+1)-e(k)+1}^{L(k+1)} Q_{k+1}^j$ where j denotes the layer index.

4 Performance Experiment

We examine the effectiveness of rate adaptive transmission. To measure the smoothness of the quality scale, we use average run length(ARL) metric proposed by [4]. ARL is the metric to measure a sequence of consecutive frames in a layer. This metric attempts to measure the smoothness in the perceived quality of a layered video. It is defined by $ARL = \frac{1}{L} \sum_{i=1}^{L} \frac{\sum_{j=1}^{k_i} n_j}{k_i}$ where k_i is the number of runs in the ith layer, and n_j is the length of the jth run. Fig. 2 illustrates the performance of rate adaptation transmission scheme between normal and optimal transmission. Fig. 2(a) illustrates the quality fluctuation with the traffic interferences at 5 Mbps in normal rate adaptive transmission. In Fig. 2(a), the majority quality scale in this figure is 'F' and some transitions are occurred between 'G' and 'F'. ARL is 1.263, which is very small run length. As a result, the quality fluctuation and degradation of perceptual quality is highly occurred.

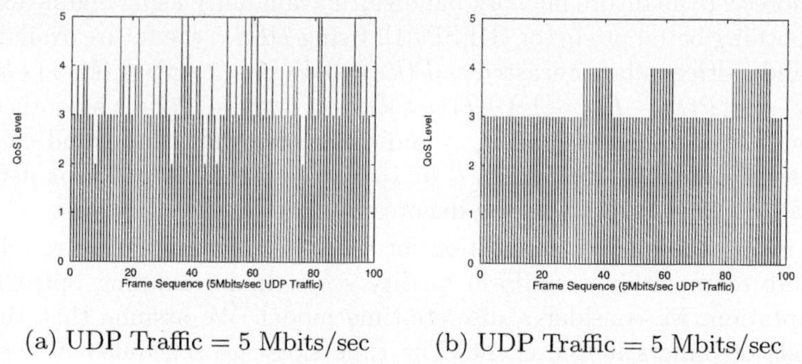

(a) UDP Traffic = 5 Mbits/sec   (b) UDP Traffic = 5 Mbits/sec

Fig. 2. Performance of Normal and Optimal Rate Adaptive Transmission

Fig. 2(b) is the performance of optimal rate transmission with the traffic interferences at 5 Mbps. In Fig. 2(b), *ARL* is 14.286. Note that *ARL* is longer in optimal rate transmission than normal rate adaptation scheme. It is also noticed that only two quality scales are used and an extremely smoothed result is accomplished. Overall, the results show that the optimal rate transmission algorithm exhibits smoother quality fluctuation compared to the normal rate adaptive transmission scheme.

5   Conclusion

In this paper, we proposed adaptive QoS management to reduce time delay and guarantee the constant playback rate in delivering high-quality, 3D contents. We also provided optimal quality adaptation scheme to minimize the quality fluctuation. Our results show that proposed QoS architecture can effectively utilize the available network bandwidth and minimize the quality variation.

References

1. C. Chatfield. Time-series forecasting. *Chapman and Hall/CRC*, 2001.
2. P. de Cuetos and K. Ross. Adaptive rate control for streaming stored fine grained scalable video. In *Proceedings of NOSSDAV 2002*. ACM, May 2002.
3. Y. Iwadate, M. Katayama, K. Tomiyama, and H. Imaizumi. Vrml animation from multi-view images. In *ICME2002 IEEE International Conference on Multimedia and Expo*, pages 881–884. IEEE, August 2002.
4. S. Nelakuditi, R. Harinath, E. Kusmierek, and Z.-L. Zhang. Providing smoother quality layered video stream. In *Proceedings of NOSSDAV 2000*. ACM, June 2000.
5. J. Salehi, Z.-L. Zhang, J. Kurose, and D. Towsley. Supporting stored video: Reducing rate variability and end-to-end resource requirements through optimal smoothing. *IEEE/ACM Trans. Networking*, 6(4):397–410, August 1998.

CORBA-Based Open Platform for Processes Monitoring. An Application to a Complex Electromechanical Process

Karina Cantillo[1], Rodolfo E. Haber[1,2], Jose E. Jiménez[1], Ángel Alique[1], and Ramón Galán[3]

[1]Instituto de Automática Industrial – CSIC, Campo Real km 0.200,
Arganda del Rey, Madrid 28500
{cantillo, rhaber, jejc, a.alique}@iai.csic.es
[2] Escuela Politécnica Superior. Ciudad Universitaria de Cantoblanco
Ctra. de Colmenar Viejo, km. 15. 28049 - Spain
Rodolfo.Haber@ii.uam.es
[3]E.T.S. de Ingenieros Industriales, Universidad Politécnica de Madrid,
c/ José Gutiérrez Abascal Nº2, Madrid 28006.
rgalan@etsii.upm.es

Abstract. The goal of this work is to develop an open software platform called SYNERGY, supported by portable, low cost and worldwide-accepted technologies (i.e., Real Time CORBA), focused on networked control systems. Preliminary results of SYNERGY corroborate the viability for networked control, supervision and monitoring of complex electromechanical processes like high speed machining (HSM), on the basis of current communications and computation technologies upon open architectures.

1 Introduction

During the last decade, successful applications of distributed real time systems have grown considerably due mainly to the availability of new standards and open architectures based on distributed objects (e.g., middleware Common Object Request Broker Architecture CORBA). Indeed, the combination of these technologies with the current control and supervision techniques based on classical and Artificial Intelligence paradigms are the foundation for the development of the new generation of networked control systems (NCS) [1,2,3]. In order to deal with communication constraints in NCS, has increased the use of CORBA, which is structured in layers, ORB core, services and the application layer, easing the development of distributed applications [4].

This paper is organized as follows. The main characteristics of TAO (CORBA implementation) are shown in section 2. A brief description of HSM process is presented in section 3. The design and implementation of SYNERGY software platform and results concerning networked real-time monitoring of HSM process are presented in Section 4. Finally, some conclusions and remarks are provided.

M. Bubak et al. (Eds.): ICCS 2004, LNCS 3036, pp. 523–526, 2004.

2 Background

TAO, The ACE (Adaptive Communication Environment) ORB, unlike the most of CORBA implementations in market (MT-Orbix, CORBAplus, Visiobroker, miniCOOL, Orbacus), provides a predictable behaviour. TAO real time ORB core shares a minimum part of ORB resources, reducing substantially the synchronization costs and the priority inversion between the process threads. These characteristics are responsible of a better performance of CORBA applications. Besides, TAO implement the specification Real-Time CORBA (RT CORBA), to support real-time distributed requirements, defining mechanisms and policies to control processor, communication and memory resources [5,6,7].

3 High Speed Machining Process

In High speed machining (HSM), cutting force is considered to be the variable that best describes the cutting process. This can be used to evaluate the quality and geometric profile of the cutting surface, the tool wear and the tool breakage [8].

Relevant variables involved in HSM process are: cutting tool position (x_p, y_p, z_p) $mm]$, spindle speed $(s)[rpm]$, feed speed $(f)[mm/min]$, cutting power $(P_c)[kW]$, cutting force $(F)[N]$, radial cutting depth $(a)[mm]$ and cutting-tool diameter $(d)[mm]$.

The laboratory at the CSIC is equipped with a KONDIA HS-1000 HSM centre and Siemens SINUMERIK840D open computerized numerical control (CNC). The communication between the CNC and the applications is done by a multiport interface (MPI). External signals of the sensors and acquisition cards are acquired and processed using a LABVIEW-based program called *SignalAcquisition*. The application *NCDDE Server*, supplied by Siemens [9], allows to access real time data in a machining centre.

4 Results

SYNERGY software platform consists in two main parts. A server application called *Monitoring-Server* that comprises data acquisition, identification and communication modules, and a client application called *Remote-Monitoring*, which includes communication and control modules. This work is focused on the data acquisition and the communication modules of the server and client application.

The communication modules were developed based in TAO (i.e., ACE5.3 TAO1.3 version). The TAO services and policies of RT CORBA used are: *naming service, RT ORB, RT POA, RT Current, Priority Mappings, Server_Declared Priority Model, Server Protocol Policy, Explicit Binding, Private Connections*, POA *Threadpools*. The communication interface defined is depicted below.

```
module rtcontrol{
        typedef sequence<string> arraydata;

    interface monitor{
        string request(in short item);
```

```
      long inidataloop();
      void dataloop(in long indini,out long indend,
            in short item, out arraydata vcdts);};
interface controller{
      typedef sequence<arraydata,6> sample;
      void iniobtsample(out long ind);
      long obtsample(in long ind,out sample mact);
};};
```

The acquisition module of *Monitoring-Server* application enables communications with the *NCDDE Server* and *SignalAcquisition* applications (Figure 1). Data are stored in a temporal matrix. Remote CORBA Object accesses the data matrix and recover data.

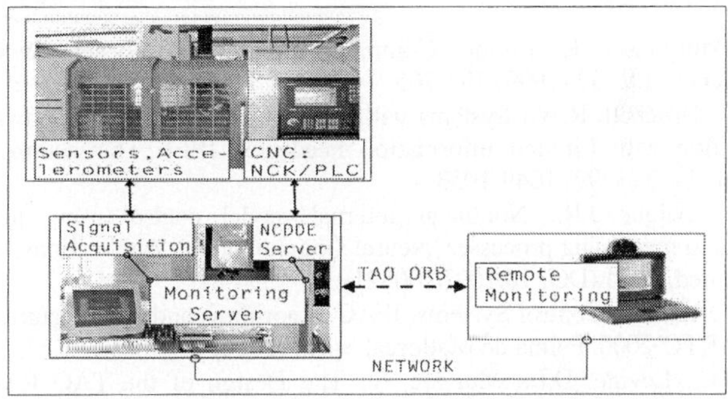

Fig. 1. Platform for networked control, supervision and monitoring of high speed machining

The developed platform was evaluated under real HSM operations. The condition of the cutting tool was also considered. The experiments were conducted considering high and low traffic in network for assessing the effect of the network congestion in the application performance. For the sake of space, only two cases are shown in Figure 2. For a case study with high network traffic the mean delay was 10.89e-3 seconds and the variance of 6.897e-5 sec. In presence of low network traffic the mean delay was 8.8826e-3 sec. and the variance was 2.312e-4 sec.

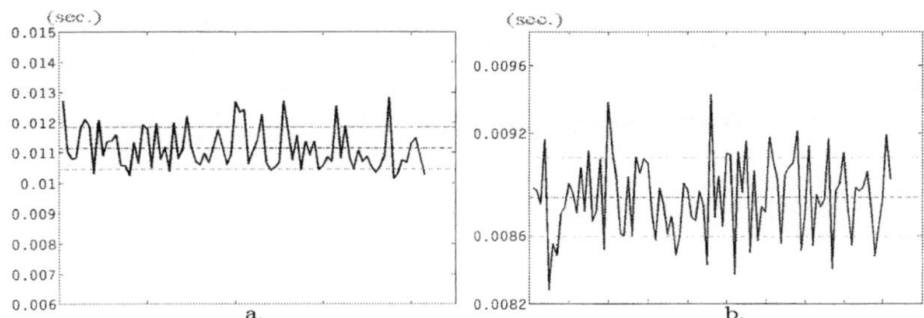

Fig. 2. Figure 2a show the behaviour of the communication delay (measure in seconds) for high traffic and 2b for low traffic

5 Conclusions

An open architecture for measuring variables in a HSM process has been developed. Additionally, SYNERGY software platform has been designed for networked control, supervision and monitoring for HSM with the following characteristics:
- Portability, low cost design and easy implementation of the developed software, including the easiness to incorporate new technologies and control methodologies.
- Superior performance of real time TAO ORB for real time applications, providing a deterministic and predictable behaviour.

References

1. Hristu, D., Morgansen, K.: Limited Communication Control. Systems & Control Letters, Elsevier Science B.V. 37 (1999) 193-205
2. Wong, W.S., Brockett, R.W.: Systems with Finite Communication Bandwidth Constraints II. Stabilization with Limited Information Feedback. IEEE Transactions on Automatic Control, Vol. 44, 5 (1999) 1049-1053
3. Haber R.E., Alique J.R.: Nonlinear internal model control using neural networks: Applications to machining processes, Neural Computing and Applications, Springer-Verlag London Limited, 2004 (DOI 10.1007/s00521-003-0394-8).
4. Sanz, R.: CORBA for Control Systems. IFAC Algorithms and Architectures for Real-Time Control, AARTC'2000. Palma de Mallorca, Spain (2000)
5. Schmidt, D.C., Levine, D.L., Mungee, S.: The Design of the TAO Real Time Object Request Broker. Computer Communications, Vol. 21, 4 (1998) 294-324
6. TAO Developer's Guide, Building a Standard in Performance. Object Computing, Inc. TAO version 1.2 a, Vol. 1, 2. St. Louis (2002)
7. Schmidt, D.C., Mungee, S., Gaitan, S.F, Gokhale, A.: Software Architectures for Reducing Priority Inversion and Non-determinism in Real-time Object Request Brokers. Journal of Real-time Systems, Special issue on Real-time Computing in the Age of the Web and the Internet Vol. 21, 1-2, (2001) 77-125
8. Haber R.E., Alique A., Alique J.R., Haber-Haber R., Ros S., Current trend and future developments of new control systems based on fuzzy logic and its application to high speed machining, Revista Metalurgia Madrid Vol. 38 (2002) 124-133
9. SINUMERIK 840D/810D/FM-NC. User Manual. Edition 09 2000, Siemens

An Approach to Web-Oriented Discrete Event Simulation Modeling

Ewa Ochmańska

Warsaw University of Technology, Faculty of Transport
00-662 Warsaw, Poland
och@it.pw.edu.pl

Abstract. The paper describes a methodology for creating simulation models of discrete event systems and for executing them on the Web platform. Models defined as extended Petri nets are built following schemas describing their available elements and admissible structures. Simulation portal provides access to Java class libraries of model elements, to XML documents of process schemas and defined models, as well as to several functions concerning model definition, execution of simulation jobs, analysis and visualization of results.

1 Introduction

Some of recent efforts concentrated on sharing computational resources concern Grid environments, providing middleware platform to organize transparent controlled use of advanced computing resources and cooperative Web-based technologies [1, 2].

This paper presents an approach to construct a frame for cooperative Web-based DES environment, founded on a particular method for defining, building and executing simulation models based on extended Petri nets. The method, implemented in simulation modeling of transport processes [3,4], comprises data-driven construction of object-oriented models of semantic classes of net elements, following predefined schemas. Such modeling approach can be implemented in the collaborative simulation portal based on Java / XML and Grid technologies, giving common access to the program and data resources and permitting to exploit and develop them collectively.

2 Modeling Principles and Implementation of Simulation Models

Model of a process is represented by Petri net: a bi-graph with two disjoint subsets of nodes, transitions and places; places are passive containers for tokens; dynamic transitions change net states by consuming tokens from input places and producing them in output places, as defined in [5], according to so-called enabling rule.

Simulation of a process represented by Petri net consists in changing its states by dynamic behavior of transitions. Various extensions, in particular concerning timing and control rules [6], were proposed to increase semantic expressiveness of the Petri net formalism. Some of them have been adopted in the presented modeling approach:

- Data structures assigned to tokens describe processing or processed entities.
- Timestamps assigned to tokens record a time of entity appearance or creation.

M. Bubak et al. (Eds.): ICCS 2004, LNCS 3036, pp. 527–531, 2004.
© Springer-Verlag Berlin Heidelberg 2004

- Contents of all tokens, including timestamps, describe current state of a process.
- Predicates, defined on values of input tokens, extend enabling rules of transitions.
- Transitions perform actions transforming values of tokens consumed on input to values of output tokens, timestamps comprised, resulting in new process states.
- The model has the dual structure shown on Fig. 1, which comprises a bi-graph of Petri net along with a list of transitions forming a queue of planned events.

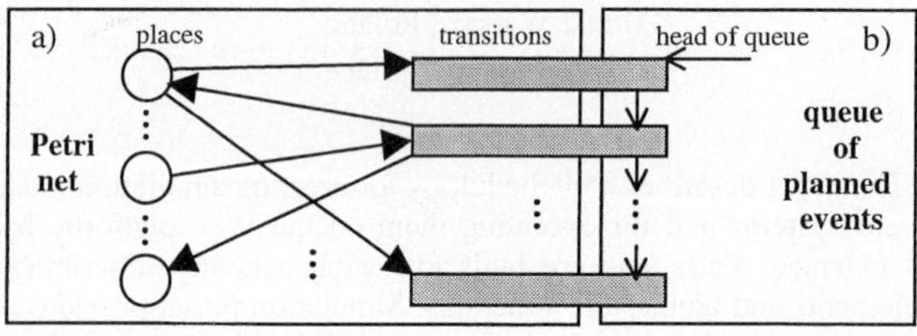

Fig. 1. Dual structure of a model: a) bi-graph of places and transitions, b) list of transitions

Process schemas define families of models for various categories of processes, with particular structure and semantics, by specifying following construction rules:
- a set of semantic place classes of with proper classes of tokens (data structures)
- a set of semantic transition classes with proper of input and output place classes
- partial ordering of transition subclasses according to the processing flow.

Java classes of model components are implemented in category-specific libraries as sub-classes of four base classes: token, place, transition and process equipped with following attributes and methods, resumed in Table 1. Program items marked by *(o)* in the table are overridden (redefined) in semantic subclasses of a process category.

Table 1. Attributes and methods of base object classes of an extended Petri net model

Java class	Attributes	Methods
token	time TimeStamp object DataStructure *(o)* token NextToken	void RemoveFromPlace () void InsertTo (Place P)
place	token TokenList *//semantic place subclasses differ just by the classes of contained tokens* transition[] OnInput, OnOutput	
transition	token[][] CandidateTuples place[] Input, Output transition NextTransition	token[][] EnablingPredicate () *(o)* void Action () *(o)* time TimeFunction ()
process	transition FirstTransition time CurrentTime, TimeLimit	transition[][] OperativePredicate () transition[] DecisivePredicate (transition[][] Sets) *(o)* void ExecuteSimulation (), void SimulationStep ()

Simulation program is in fact an instance of a category-specific subclass of process, say myProcess. It calls ExecuteSimulation method to execute a loop of simulation steps:

```
while (myProcess.FirstTransition.TimeFunction < TimeLimit) {
   myProcess.CurrentTime = myProcess.FirstTransition.TimeFunction;
   SimulationStep ();
}
```

During a simulation step, EnablingPredicate computes CandidateTuples of tokens for each transition enabled at CurrentTime. OperativePredicate returns a decision space for current state of simulation i.e. alternate non-conflict subsets of enabled transitions with proper CandidateTuples. Category-specific DecisivePredicate chooses one of these subsets. Action transforms the chosen candidate tuple of tokens from Input places to a new tuple of tokens in Output places for each of the chosen ActivatedTransitions:

```
void SimulationStep () {
   transition T = process.FirstTransition;
   while (T.TimeFunction == CurrentTime) {
      T.CandidateTuples = T.EnablingPredicate;
      T = T.NextTransition; }
   int j;
   transition[] ActivatedTransitions = new(DecisivePredicate (OperativePredicate ()));
   for (j=0; j < ActivatedTransitions.Length; j++) {
      ActivatedTransitions[j].Action (); }
}
```

Executing multi-thread model of concurrent processes demands synchronization of their local times, involving current communication between processes executed as separate program threads. Running all threads on single machine, it can be implemented by a meta-process class using some of the common synchronization strategies [7] to control cooperation of processes. Distributed meta-process simulation requires autonomous mechanism for suspending/unrolling built into process instances.

XML technologies provide means to define simulation models as XML documents. Actual simulation program is synthesized by parsing such document and building specified net structure of proper subclasses of components. Formal and semantic correctness of model definition is controlled by an XML Schema for process category describing model structure and data types of semantic token subclasses. XSLT/XPath techniques permit to automatically generate context-dependent user interface for defining simulation models and tasks from XML process Schemas.

3   Web Based Simulation Environment

Functionality of simulation environment for processing input data, including user activity, to produce results specific for different phases of simulation experiment, is resumed in Tab. 2 in the context of previously described concepts.

Web-based simulation environment can be constructed as a virtual grid application in OGSA architecture [8], accessed via specialized Web simulation portal. Client-side activities are localized in the frame of a Web browser providing user with dynamic, context-dependent GUI for interacting with particular functionalities of simulation environment. Several tools suitable for such purpose are available. XML/XSL standards can be used in connection with Java based scripting technologies such as JSP, in order to transform XML process schemas in adaptable user interfaces for

Table 2. Input and output data in the phases of simulation experiment

Phase	User activity	Input	Output
Defining simulation model	Choosing process categories. Building model configuration	XML Shemas for process categories	XML definition of simulation model
Specifying simulation task	Defining model parameters. Specifying initial states	XML model definition & process XML Shemas	XML specification of simulation task
Synthetizing simulation model	Demanding execution of specified simulation job	XML model definition	Java program with model instantiation
Running simulation job	"	XML job specification parameters, initial states	Row result recorded by simulation passes
Elaboration of simulation results	Queries on simulation output; choosing presentation forms	Recorded simulation results	Analytical/synthetic views of results, visualization

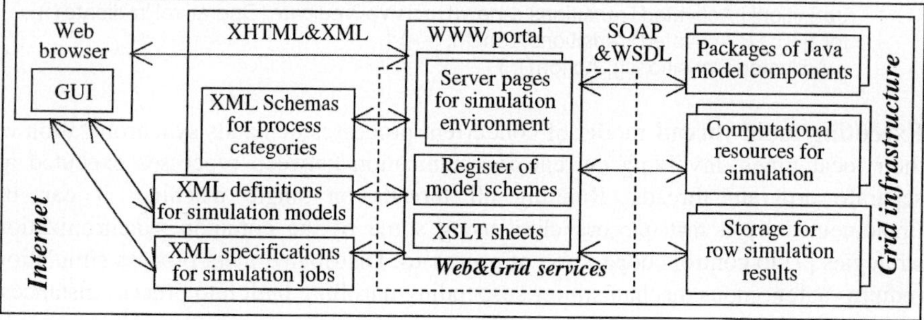

Fig. 2. The functional structure of a Web-based simulation environment

defining valid models and specifying well formulated tasks to perform simulation research. A middleware layer, with GT3 implementing OGSI on top of the Web service SOAP and WSDL protocols, can organize transparent and secure access to virtual simulation machine composed of distributed software, computation and storage resources. A functional structure of such an environment is outlined on Fig. 2.

Model definitions can be stored locally by users or archived by simulation portal to be shared among cooperating groups of users. Besides, all kinds of resources may be situated anywhere in the Web. XML process schemas defined with namespaces are related to providers of component implementations for various process categories.

References

1. http://www.computingportals.org
2. Nemeth Z., Sunderam V.: A Comparison of Conventional Distributed Computing Environments and Computational Grids. In: Computational Science - ICCS 2002. Part II, Vol. 2330 of LNCS, Springer-Verlag (2002)
3. Ochmańska E.: System Simulating Technological Processes. ESM'97, Proceedings of the 11th European Simulation Multiconference, Istambul (1997)
4. Ochmańska E., Wawrzynski W.: Simulation Model of Control System at Railway Station. Archives of Transport. Polish Academy of Science, Committee of Transport. Warsaw 2002-

5. Desel J., Reisig W., Place/Transition Petri Nets. In: Lectures on Petri Nets I; Basic Models, Vol. 1491 of LNCS, Springer-Verlag (1998)
6. Ghezzi C., Mandrioli D., Morasca S., Pezzè M.: A General Way to Put Time in Petri Nets. Proceedings of the 5th International Workshop on Software Specification and Design, IEEE-CS Press, Pittsburg (1989)
7. Yi-Bing Lin, Fishwick P.A.: Asynchronous Parallel Discrete Event Simulation. `http://www.cis.ufl.edu/~fishwick/tr/tr95-005.html`
8. Foster I. at al.: The Physiology of the Grid. `http://www.globus.org/research/papers.html`

Query Execution Algorithm in Web Environment with Limited Availability of Statistics

Juliusz Jezierski and Tadeusz Morzy

Poznan University of Technology
Piotrowo 3a, 60-965 Poznan, Poland
{jjezierski, tmorzy}@cs.put.poznan.pl

Abstract. Traditional static cost-based query optimization approach uses data statistics to evaluate costs of potential query execution plans for a given query. Unfortunately, this approach cannot be directly applied to Web environment due to limited availability of statistics and unpredictable delays in access to data sources. To cope with lack or limited availability of statistics we propose a novel competitive query execution strategy. The basic idea is to initiate simultaneously several equivalent query execution plans and measure dynamically their progress. Processing of the most promising plan is continued, whereas processing of remaining plans is stopped. We also present in the paper results of performance evaluation of the proposed strategy.

1   Introduction

There is increasing interest in query optimization and execution strategies for Web environment that can cope with two specific properties of this environment: lack or limited availability of data statistics and unpredictable delays in access to data sources. Typically, in Web environment query processing parameters may change significantly over time or they may be simply not available to query engines. Web sites that disseminate data in Web environment in the form of files, dynamically generated documents and data streams usually do not allow access to internal data statistics. The second specific property of Web environment is unexpected delay phenomenon in access to external data sources. Such delays may cause significant increase of system response time. They appear due to variable load of network devices resulting from a varying activity of users, and also, due to breakdowns. As a result, traditional static optimization and execution techniques cannot be directly applied to Web environment.

In the paper, we present the novel competition strategy of query execution in Web environment that solves or reduces limitations of previous solutions (e.g. [1,2,3,4]). Our approach consists in simultaneous execution of a set of alternative query execution plans for a given query. The system monitors execution of these plans, and the most attractive plans are promoted, while execution of the most expensive plans is canceled. Final query result is delivered to the user by the

M. Bubak et al. (Eds.): ICCS 2004, LNCS 3036, pp. 532–536, 2004.

plan that has won the competition according to rules defined by the strategy implementation.

2   Competition Strategy of Query Execution

In traditional database systems query specified by user is transferred to the query optimizer, which chooses optimal query execution plan (QEP) during query compilation. The query optimization process depends on: (1) a cost function used to evaluate the cost of a query, (2) the search space of all possible QEPs for a given query, and (3) a search strategy used to penetrate the search space of QEPs. Final QEP generated by the query optimizer is static and it does not change during its execution.

Our query execution strategy is based on the idea that the query optimization process should be continuous and interactive, which means that the search space of QEPs should be also analyzed during query execution. Query optimizer improves the initial QEP by taking into account data statistics gathered during query execution. Formally, the competition strategy of query execution can be defined as a triple: CSQE = {PGR, CC FR}, where: PGR - rules of plan generation, CC - competition criteria, FR - feedback rules.

PGR denotes a set of rules used to generate QEPs participating in the competition. The important issue is the proper selection of initial QEPs. On the one hand, in order to reduce the overhead related to simultaneous processing of many QEPs, it is necessary to restrict the number of initiated plans. On the other hand, if the number of initiated plans is too small, then, the adaptation to changing conditions of runtime environment is automatically restricted. CC denotes competition criterion used to evaluate attractiveness of different QEPs (i.e. response time, evaluation cost). Proper definition of CC allows to limit overhead related to simultaneous processing of many plans by pruning ineffective QEPs. FR denotes a set of rules that control the competition process (e.g., start new plans). FR allows to adapt query execution process accordingly to changes of runtime environment parameters, e.g., delays in access to data sources.

The competition strategy has an "open" character and it can be implemented in many different ways. To illustrate the approach, we implemented greedy algorithm (abbreviated as GA), which implements our strategy. We assume no availability of statistics. All necessary data are dynamically established or estimated during query execution.

3   Experimental Evaluation

To demonstrate the practical relevance and universality of our strategy, we compare it with simple "brute force" strategy (abbreviated as BFS), which generate all possible query execution plans for a given query, and to comparison we have taken into account the average value of their results. We considered two basic performance evaluation criteria: system response time and utilized CPU time. Two main goals of experiments were: analysis of the impact of transfer rates and initial delays on performance evaluation criteria. Arbitrary, three transfer rates were assumed for experiments: 20Kbytes/s (slow), 200Kbytes/s (normal) and 2Mbytes/s (fast). The algorithm was implemented in Java 1.4 and experiments were computed on PC Intel 1000Mhz, 512MB RAM under control of MS Windows 2000. We analyzed the cycle SQL query Q1 given below:

```
select * from A, B, C where A.b=B.a and B.c=C.b (Q1)
```

The aim of the first series of experiments was comparison of our strategy with BFS in case where costs of potential query execution plans significantly differ from each other. Thus, we generated data with large range of values of join selectivity coefficients: $sel(A \bowtie B)=2*10^{-3}$, $sel(A \bowtie C)=2*10^{-4}$ and $sel(B \bowtie C)=2*10^{-5}$. The volumes of the sources we assumed as follows: $A-800KB(5*10^3$ tuples), $B-1500KB(10^4$ tuples), $C-2000KB(2*10^4$ tuples).

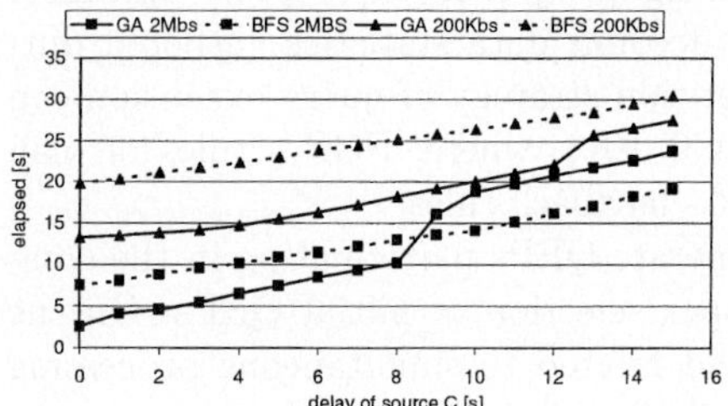

Fig. 1. Elapsed time of Q1 execution versus initial delay of source C

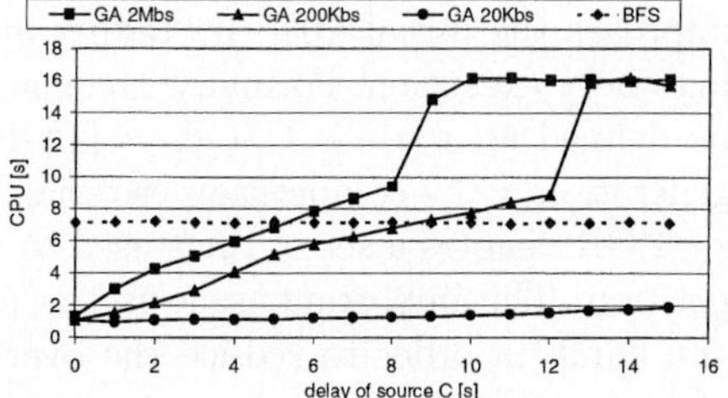

Fig. 2. Utilized CPU time of Q1 execution versus initial delay of source C

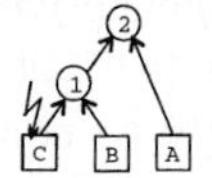

Fig. 3. QEP1 of Q1

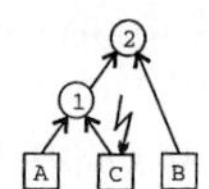

Fig. 4. QEP2 of Q1

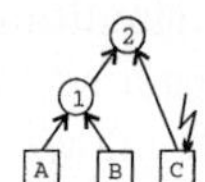

Fig. 5. QEP3 of Q1

Figure 1 presents the system response time for the query Q1 versus initial delay of the source C and different values of the transfer rate. The cost ranking of QEP for Q1 is the following (from cheapest to the most expensive): 1. QEP1, 2. QEP2, 3. QEP3 (Fig. 3, 4, and 5). Most attractive plans require access to source C in the first step. For 2Mbytes/s transfer rate, we observe that the algorithm switches from QEP1 to QEP2 for 8 seconds delay. This switch appears when delay in access to source C is so long that execution of the subplan 1 of QEP3 is finished before the algorithm collects statistically reliable samples from executions of subplan 1 of QEP2 and subplan 1 of QEP3. From the figure follows, that the GA outperforms BFS strategy before switch from QEP1 to QEP3 occurred, and longer response time after the switch. For 200Kbytes/s transfer rate, response time provided by the algorithm is always better than response time provided by BFS. For 20Kbytes/s transfer rate, we do not observe any switch. QEP1 always wins the competition. Moreover, for the whole range of delays for the source C, the GA outperforms BFS. However, for readability of Fig.1, we omit in the figure the results for 20Kbytes/s transfer rate.

Figure 2 presents utilized CPU time (i.e. overhead) versus initial delay of source C and different values of the transfer rate. The overhead depends on delay in access to most attractive sources. Increasing delay in data transfer from the source C delays a moment of competition termination, and, thus, extends also execution time of QEPs, belonging to a competition group, and CPU consumption. The largest overhead is observed for 2Mbytes/s transfer rate, while the smallest one is observed for 20Kbytes/s transfer rate. This phenomenon can be explained as follows: for a given delay of source C, in case of higher transfer rate, a large part of unattractive subplans (i.e. subplan 1 of QEP3) will be executed until the competition process stops their processing. In case of lower transfer rate, unattractive subplans consume less CPU since the algorithm cancels their processing "earlier". Notice that for 2Mbytes/s and 200Kbytes/s transfer rates, if delay of the source C is rather small, i.e. does not exceed several seconds, GA is cheaper than the BFS. For 20kbytes/s transfer rate, GA is several times cheaper than the BFS.

We also performed a series of experiments, which tested different transfer rates for other data sources. If attractive data sources transfer data with higher rate than other sources, then the competition overhead decreases. We also tested the performance of the GA with the 5-way join query (Q2). In this case, the GA produced query result few times faster than the BFS. It can be explained as follows: the query Q2 is more complex than Q1, and, therefore, a set of QEPs for Q2 is much larger than that of Q1. Therefore, an average cost of these plans taken into account in the comparison, is relatively large, whereas the GA generated nearly optimal QEP.

4   Summary

In this paper, we proposed novel strategy of dynamic query optimization and execution in Web environment, which cope with limited availability of data statistics and unexpected delay in access to data sources. We evaluated our strategy by a set of experiments for different transfer rates and different delay scenario, and proved its feasibility. As the experiments show, our strategy is especially appropriate for small and medium transfer rates (20Kbytes/s and 200Kbytes/s). The strategy is efficient also for large transfer rate (2Mbytes/s) and relatively small delays (several seconds) in access to attractive sources. The algorithm prefers bushy QEPs, which, when compared to linear QEPs produced by traditional static cost-based optimization algorithms provide usually better response times.

References

1. Urhan, T., Franklin, M.J., Amsaleg, L.: Cost based query scrambling for initial delays. In: Proc. ACM SIGMOD Conf., June 2-4, 1998, Seattle, USA, 130–141
2. Kabra, N., DeWitt, D.J.: Efficient mid-query re-optimization of sub-optimal query execution plans. In: Proc. ACM SIGMOD Conf., June 2-4, 1998, Seattle, USA, ACM Press (1998) 106–117
3. Avnur, R., Hellerstein, J.M.: Eddies: Continuously adaptive query processing. In: Proc. ACM SIGMOD Conf., May 16-18, 2000, Dallas, USA, 261–272
4. Viglas, S., Naughton, J.F., Burger, J.: Maximizing the output rate of multi-way join queries over streaming information sources. In: Proc. VLDB Conf., September 9-12, 2003, Berlin, Germany, Morgan Kaufmann (2003) 285–296

Using Adaptive Priority Controls for Service Differentiation in QoS-Enabled Web Servers

Mário Meireles Teixeira[2,1], Marcos José Santana[1], and
Regina H. Carlucci Santana[1]*

[1] University of São Paulo
Institute of Mathematics and Computer Science
São Carlos, SP, Brazil 13560-000
{mjs, rcs}@icmc.usp.br
[2] Federal University of Maranhão
Department of Informatics
São Luís, MA, Brazil 65085-580
mario@icmc.usp.br

Abstract. We propose an architecture for the provision of differentiated
services at the web server level. The architecture is validated by means of
a simulation model and real web server traces are used as workload. We
implement an adaptive algorithm which allows the tuning of the priority
level provided and determines how strict the use of priorities will be. The
server can then adapt itself to various workloads, an essential feature in
a highly dynamic environment such as the Web.

1   Introduction

The service currently provided on the Internet is based on a best-effort model,
which treats all traffic uniformly, without any type of service differentiation or
prioritization, a characteristic we find even in the design of critical Internet
services, such as the Web. However, not all types of traffic are equivalent or have
the same priority to their users [1]. Therefore, it is essential to provide service
differentiation with different levels of quality of service (QoS) to different request
types [2].

In this paper, we propose a novel architecture for a web server capable of
providing differentiated services to its users and applications. We consider two
classes of users and analyze the implementation of an *adaptive priority mechanism*, an innovative solution for service differentiation at the application domain.

There are a few studies in the literature which use priorities for service differentiation [3] [4] [5]. However, all of them are based on some sort of strict priority
scheme and do not provide mechanisms for its adaptation, as is the case of our
algorithm.

* The authors would like to thank Brazilian funding agencies CAPES, CNPq and
  FAPESP for their support to the research projects at LaSDPC-ICMC-USP

M. Bubak et al. (Eds.): ICCS 2004, LNCS 3036, pp. 537–540, 2004.

2 Service Differentiating Web Server Model

In this section, we propose a generic model for a *Service Differentiating Web Server* (SWDS, in Portuguese) which should be able to provide different levels of service to its clients with quality of service guarantees. Figure 1 describes the proposed architecture, composed of the following modules: a Classifier, an Admission Control module and a cluster of web server processes.

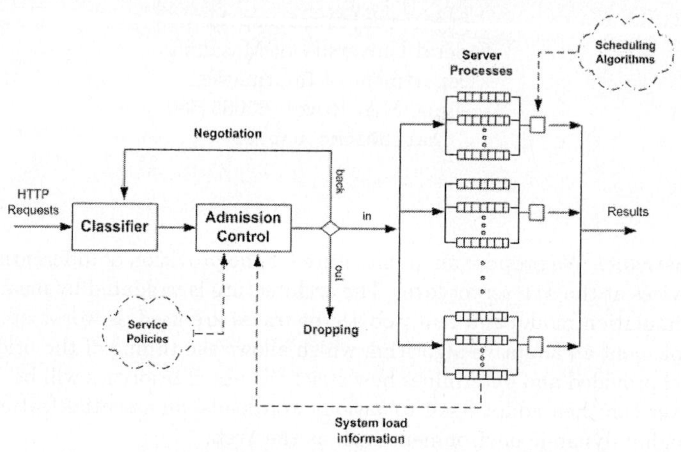

Fig. 1. Service Differentiating Web Server (SWDS)

The *Classifier* is the element responsible for receiving requests upon arrival at the server and for dividing them into classes following some previously defined criteria. The *Admission Control* module manages the acceptance of new requests by the server taking into account current service policies and system workload information. In case of system overload, a request may be either rejected (Dropping) or have its QoS requirements downgraded (Negotiation), so that it can be accepted in a lower priority class. After being admitted to the system, the request is assigned to one of the nodes of the web server cluster and is serviced according to the scheduling or service differentiating algorithm currently in operation. After processing, the results are sent back to the clients.

In this work, each cluster node is viewed as a plain web server with a CPU, a disk, a network interface and other resources. The nodes could have also been abstracted as processes, tasks or even CPU's in a parallel computer, since the model does not necessarily imply that the cluster is composed by computers in a distributed system.

3   Adaptive Priority Mechanism

To implement the adaptive algorithm, each server process is defined with a single waiting queue where requests are inserted in strict arrival order. The algorithm uses a look-ahead parameter (k) that specifies the maximum number of positions that will be searched from the head of the queue looking for requests of a given priority (class). If no request of the desired priority is found, the algorithm is repeated for the next lower level and so on. In the worst case, the first request of the queue will be chosen for processing. The higher the value of k, the better the treatment given to higher priority requests. For $k = 1$, requests will be serviced in strict arrival order, i.e., without any service differentiation.

The model is validated by means of a discrete-event simulation using the *SimPack* simulation package. We used log files collected from the 1998 World Cup web site [6] for workload generation. We assume four homogeneous web servers in the cluster. Arriving requests are divided into two service classes (high and low priority) with 50% of the requests in each class. The admission control module is disabled so as not to interfere with the performance evaluation of the algorithm. Therefore, the Classifier works as a dispatcher for the requests and server queues are unlimited.

Initially, we analyzed the behavior of request mean response time for different values of the look-ahead, as shown in Fig. 2. For $k = 1$, the curves overlap, since the same treatment is given to both service classes. However, for $k = 3000$, the service differentiation becomes evident and the service provided to high priority requests is noticeably better, as initially intended.

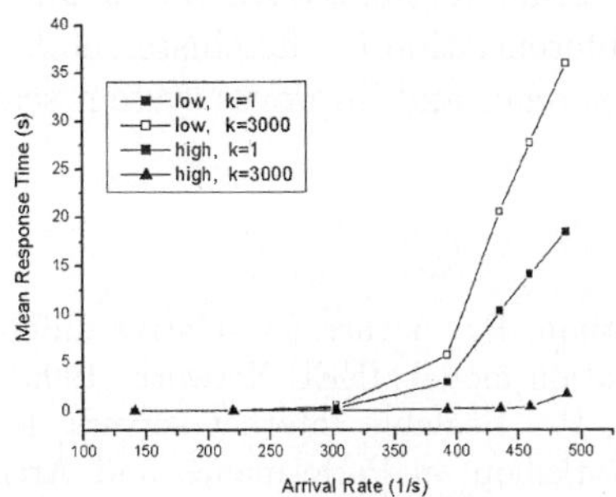

Fig. 2. Request response time using adaptive priority scheduling

The following experiments analyzed the behavior of the ratio of completed high priority requests with respect to the arrival rate. Look-ahead values range from 1 to 4,500. For $k = 1$, the service received by both classes of requests is virtually the same. However, higher values of k gradually increase the ratio of high priority requests that reach a successful completion (Fig. 3), to the point where strict priority scheduling is enforced. In this case, the treatment provided to low priority requests becomes much worse.

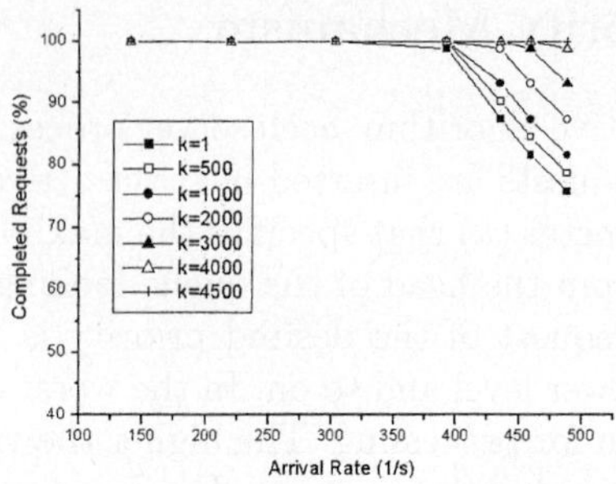

Fig. 3. Ratio of completed requests for different values of the look-ahead

4   Conclusions

We proposed an architecture for a service differentiating web server, the SWDS server, which can provide different levels of service to different classes of users. Our model is an evolution from conventional web server architectures, which service clients using an FCFS discipline, without considering the demands of any particular group of users or applications.

We proposed and implemented an adaptive priority mechanism in the SWDS server model, an innovative solution at the application domain. It employs a look-ahead parameter in the cluster's waiting queues in order to fine-tune the prioritization level used by the system. Thus, the server can support varying degrees of QoS-awareness according to the real time system load. The adaptive algorithm brings adaptability to the SWDS server and shifts the workload associated with service differentiation to the cluster nodes, which in turn reduces the workload of the dispatcher and improves system scalability.

References

1. Dovrolis, C., Ramanathan, P.: A case for relative differentiated services and the proportional differentiation model. IEEE Network (1999)
2. Kant, K., Mohapatra, P.: Scalable Internet servers: Issues and challenges. In: Proceedings of the Workshop on Performance and Architecture of Web Servers (PAWS), ACM SIGMETRICS (2000)
3. Chen, X., Mohapatra, P.: Providing differentiated services from an Internet server. In: Proceedings of the IEEE International Conference on Computer Communications and Networks. (1999) 214–217
4. Eggert, L., Heidemann, J.: Application-level differentiated services for web servers. World Wide Web Journal **3** (1999) 133–42
5. Rao, G., Ramamurthy, B.: DiffServer: Application level differentiated services for web servers. In: Proceedings of the IEEE International Conference on Communications. (2001)
6. Arlitt, M., Jin, T.: Workload characterization of the 1998 World Cup web site. Technical Report HPL-1999-35, HP Laboratories (1999)

On the Evaluation of x86 Web Servers Using Simics: Limitations and Trade-Offs

Francisco J. Villa, Manuel E. Acacio, and José M. García

Universidad de Murcia, Departamento de Ingeniería y Tecnología de Computadores
30071 Murcia (Spain)
{fj.villa,meacacio,jmgarcia}@ditec.um.es

Abstract. In this paper, we present our first experiences using *Simics*, a simulator which allows full-system simulation of multiprocessor architectures. We carry out a detailed performance study of a static web content server, showing how changes in some architectural parameters affect final performance. The results we have obtained corroborate the intuition of increasing performance of a dual-processor web server opposite to a single-processor one, and at the same time, allow us to check out *Simics* limitations. Finally, we compare these results with those that are obtained on real machines.

1 Introduction

Multiprocessor systems are increasingly being used for executing commercial applications, among which we can find web servers or On-Line Transaction Processing (OLTP) applications. As a consequence of the use of multiprocessors in these fields, simulating multiprocessor architectures running commercial applications accurately becomes important. Opposite to scientific applications, there are some characteristics of commercial workloads that make their simulation challenging. In particular, the activity of the operating system is very important, as well as the interaction with memory hierarchy, storage system and communication network. *Simics* [1] is a full-system simulator which allows us to simulate all these aspects and obtain accurate simulation results.

In this paper, we use *Simics* to evaluate three different architectures executing a static web content server, being *Apache* the web server and *httperf* the utility which places the workload at the server.

2 Related Work

Up to not long ago, the methodology used for evaluating commercial workloads in multiprocessors consisted in firstly generating memory references of applications, and then, using these references to feed a user-level simulator. For example, in [2] Ranganathan *et al.* study the performance of OLTP and decision support systems based on this methodology.

The appearance of full-system simulators, like *SimOS* [3] or *Simics* [1], has significantly simplified the evaluation of commercial workloads, as these simulators allow modelling elements such as the operating system, the I/O subsystem and so on. Recently,

M. Bubak et al. (Eds.): ICCS 2004, LNCS 3036, pp. 541–544, 2004.
© Springer-Verlag Berlin Heidelberg 2004

several studies have appeared in which *Simics* is used as the simulation tool employed for the evaluation. In [4,5], it is presented an exhaustive study of several commercial applications, including a static content web server and the TPC-C benchmark. The authors also identify one of the problems concerned with simulation of commercial applications: the variability they show.

3   Simulation Results and Limitations

In this Section, we present the results that we have obtained using *Simics* and compare them to the results obtained using real machines. In our evaluations, we have considered three different server architectures: two single-processor architectures with L2 cache sizes of 512 KB and 1024 KB respectively, and a dual-processor architecture in which each processor has a L2 cache of 512 KB. In the case of real machines, the single-processor architecture with a L2 cache of 1024 KB has not been analysed.

We measure the response time of *Apache* in each case as a function of the number of requests that are received. For this, we have executed 1000 requests referred to 10 web pages with an average page size of 537 bytes. This page size has been selected in order to avoid the influence of the interconnection network on the results.

We have carried out eight tests for each sever architecture, in which the total number of requests that *Apache* must process has been set to 25, 50, 75, 100, 125, 150, 175 and 200 respectively. Starting with the results of the simulations, Figure 1(a) shows the average response time that has been obtained in each case. This metric is provided by *httperf*. As we can see, the dual-processor server has greater performance than those that employ a single-processor, with an average response time of approximately half the response time of the single-processor severs (which show almost the same response time).

On the other hand, Figure 2(a) shows the evolution of the number of requests that are dispatched as a function of the total number of requests. This metric is provided by the *Apache* server. Although the dual-processor server is able to dispatch more requests than the single-processor architectures, the performance difference is lower than the observed for the response time.

Once we have seen how *Simics* can help us to analyse the behavior of a commercial web server, we want to check how accurate are the results the simulator provides. For this, we have repeated the experiments, but this time we have employed real computers. Figures 1(b) and 2(b) show the results we have observed for these tests.

Comparing these results to the obtained with *Simics*, we find that there are notable differences between them. In the case of the response time, it is scaled down by a factor of almost 100. In fact, the performance difference between dual and single-processor real servers is negligible. Something similar occurs with the number of requests that are dispatched. Although simulation results showed that the dual-processor server could sustain a larger request per second rate than the single-processor one, in the real environment we find that for the experiments we have carried out, single and dual-processor servers provide almost the same results in terms of the number of requests that are dispatched.

Therefore, we can conclude that the low detail level when modeling x86-like processors prevents *Simics* from be able to reproduce the results that would be reached in

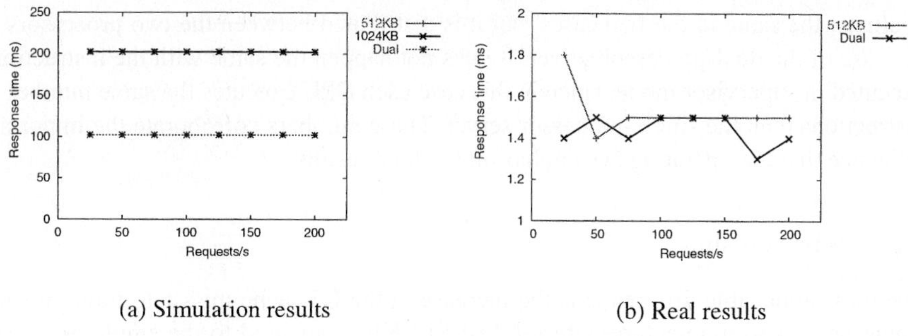

(a) Simulation results   (b) Real results

Fig. 1. Average response time as a function of the requests received per second.

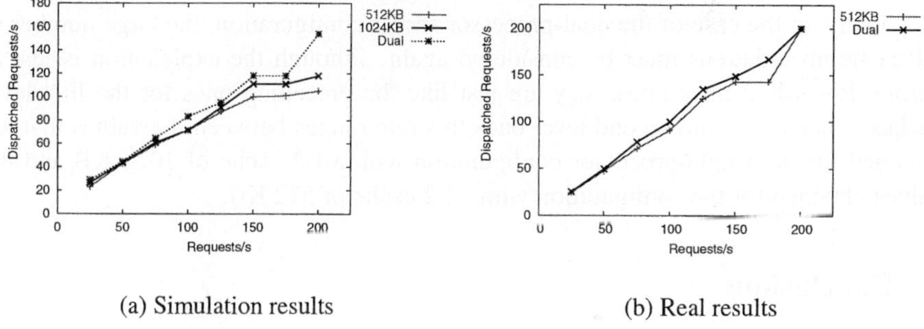

(a) Simulation results   (b) Real results

Fig. 2. Dispatched requests per second as a function of the requests received per second

the real world. Specifically, *Simics* doesn't implement out-of-order execution for these processors. In this way, we think that the *x86-Simics* machine is appropiate as functional simulator but not as timing simulator.

4   Additional Information Obtained with *Simics*

Using *Simics* we can easily obtain statistics of the processor and memory hierarchy, one of the main advantages of the simulator compared to real machines, for which collecting these measures is harder. In this Section, we analyse CPU and cache statistics, exploring their influence in the performance of the architectures that are evaluated.

4.1   CPU Statistics

The first important fact is that the number of instructions executed in user mode is 50 times lower than the number of instructions executed in supervisor mode. Comparing the statistics obtained for the single-processor server with a L2 cache of 1 MB to the dual-processor server, we notice that the number of instructions executed in user mode is almost the same in the two cases, but it is distributed between the two processors in the case of the dual-processor server. It does not happen the same with the instructions executed in supervisor mode, since in this case each CPU executes the same number of instructions than the single-processor server. These numbers corroborate the important influence that the operating system has on the final results.

4.2   Cache Statistics

The most noticeable difference is the increase in the L2 cache miss rate found for the single-processor architecture with a L2 of 512 KB, compared to the single-processor architecture with a L2 of 1024 KB. The increasing in the number of L1 cache invalidations is also a remarkable result. This fact is a consequence of the increased number of replacements (what is caused by the larger number of misses), which leads to invalidate more L1 blocks in order to maintain the inclusion property.

Finally, in the case of the dual-processor server configuration, the large number of L1 cache invalidations must be considered again, although the explanation is just as before. Regarding miss rates, they are just like the preceding ones for the first level caches, whereas for the second level ones this rate ranges between the values that are obtained for the single-processor configuration with a L2 cache of 1024 KB and the values obtained for the configuration with a L2 cache of 512 KB.

5   Conclusions

In this paper, we have introduced the evaluation of a functional simulator which allows us to simulate all the aspects that are critical in the execution of commercial workloads, such as the I/O subsystem and the operating system. However, we have found that the simulator does not provide an accurate model for the x86 family of processors, which leads to obtain different results than those that would be obtained using real computers. We think that the impossibility of using an out-of-order execution model for this family has a negative influence in the results that we have obtained.

References

1. Magnusson, P. S. *et al.*: Simics: A Full System Simulation Platform. IEEE Computer **35** (2002) 50–58
2. Ranganathan, P. *et al.*: Performance of Database Workloads on Shared-Memory Systems with Out-of-Order Processors. In: ASPLOS-VIII. (1998) 307–318
3. Rosemblum, M. *et al.*: Complete Computer System Simulation: The SimOS Approach. IEEE Parallel and Distributed Technology: Systems and Applications (1995) 34–43
4. Alameldeen, A.R. *et al.*: Simulating a $2M Commercial Server on a $2K PC. IEEE Computer **36** (2003) 50–57
5. Alameldeen, A.R. *et al.*: Evaluating Non-deterministic Multi-threaded Commercial Workloads. In: CAECW-02 (2002) 30–38

MADEW: Modelling a Constraint Awareness Model to Web-Based Learning Environments

Pilar Herrero and Angélica de Antonio

Facultad de Informática. Universidad Politécnica de Madrid.
Campus de Montegancedo S/N.
28.660 Boadilla del Monte. Madrid. Spain
{pherrero,angelica}@fi.upm.es

Abstract. In this paper, we present a web application developed at the Universidad Politécnica de Madrid with an special peculiarity: this web application is based on the extension and reinterpretation of one of the most successful models of awareness in Computer Supported Cooperative Work (CSCW), called the Spatial Model of Interaction (SMI), which manage awareness in Collaborative Virtual Environments (CVEs) through a set of key concepts. MADEW implements the key concepts of the SMI, introducing some extensions –associated to human-like factors such as Sense Acuity and Internal Filters- and providing some reinterpretations of these key concepts for the context of Web applications.

1 Introduction

The concept of awareness of other users assumes very different meanings depending on the situation. In 3D web-based collaborative environments, awareness of other participants may have a physical interpretation, while awareness in non-graphical environments must be interpreted in a more abstract way.

The aim of this research line started up at the Universidad Politécnica de Madrid is to make a new formal awareness model based on the reinterpretation and extension of one of the most successful models of awareness in Computer Supported Co-operative Work (CSCW), called the Spatial Model of Interaction (SMI).

Our model not only extends and reinterprets the key concepts of the SMI, but also takes into account some human-like factors – like, for example, Sense Acuity and Internal Filters. The new abstract reinterpretation that we are going to develop will be applied to the context of asynchronous WEB applications, 3D Web-based Collaborative Environments and web based learning environments.

2 The Spatial Model of Interaction (SMI)

As we mentioned in previous sections, these key concepts are based on the main concepts of a CSCW awareness model known as *The Spatial Model of Interaction* (SMI) [1].

M. Bubak et al. (Eds.): ICCS 2004, LNCS 3036, pp. 545–548, 2004.
© Springer-Verlag Berlin Heidelberg 2004

The spatial model, as its name suggests, uses the properties of space as the basis for mediating interaction. It was proposed as a way to control the flow of information of the environment in CVEs (Collaborative Virtual Environments). It allows objects in a virtual world to govern their interaction through some key concepts: medium, aura, awareness, focus, nimbus, adapters and boundaries.

Aura is the sub-space which effectively bounds the presence of an object within a given medium and which acts as an enabler of potential interaction. In each particular medium, it is possible to delimit the observing object's interest. This area is called *focus* "The more an object is within your focus the more aware you are of it". The focus concept has been implemented in the SMI as an "ideal" cone limited by the object's aura.

In the same way, it is possible to represent the observed object's projection in a particular medium. This area is called *nimbus*: "The more an object is within your nimbus the more aware it is of you". The nimbus concept, as it was defined in the Spatial Model of Interaction, has always been implemented as an sphere in a visual medium. The radio of this sphere has an "ideal" infinite value, although in practice, it is limited by the object's aura.

The implementations of these concepts –focus and nimbus- in the SMI didn't have in mind human aspects, thus reducing the level of coherence between the real and the virtual agent behaviour.

The main concept involved in controlling interaction between objects is *"awareness"*. One object's awareness of another object quantifies the subjective importance or relevance of that object. The awareness relationship between every pair of objects is achieved on the basis of quantifiable *levels* of awareness between them and it is unidirectional and specific to each medium. Awareness between objects in a given medium is manipulated via *focus* and *nimbus*. Moreover, an object's aura, focus, nimbus, and hence awareness, can be modified through *boundaries* and some artefacts called *adapters*.

3   Introducing Some Human-Like Factors

The SMI was integrated with different versions of the MASSIVE (Model, Architecture and System for Spatial Interaction in Virtual Environments) platform with some controlling parameters [2]. However, any of these implementations reflected properly real life for two reasons. The first one is that any of this implementations has considered all the key concepts of the SMI at the same time. The second and very important reason is that the SMI didn't consider human-like factors such as the "Sense Acuity" - the sense's specific ability to resolve fine details - or the "Internal Filters" – the selection of those objects that we are interested in.

4 An Asynchronous Interpretation of our Key Awareness Concepts

Some research has already been carried out by our research group to make this extension possible. An example of this is MADEW [3,4]. We also have some publications as the paper published at the Workshop on Awareness and the www in the ACM Conference on Computer Supported Cooperative Work 2000 (CSCW'00) [4].

The outcome of this research has been an abstract and preliminary interpretation in the context of an asynchronous collaboration of both the key SMI concepts and some of the human-like factors introduced in this dissertation. In this interpretation, all these key concepts have been defined as:

- *Awareness*: This concept will quantify the degree, nature or quality of asynchronous interaction between a user and the WEB-based environment.

- *Focus*: It can be interpreted as the subset of the web space on which the user has focused his attention. It can relate both to content and to other users. Regarding content, it can be computed by collecting information about the set of places that the user has visited while navigating through the Web and the set of resources that have been used. Regarding other users, it can be computed by collecting information about areas of common interest and effective past interactions.

- *Nimbus*: It is the user's projection over the WWW space. It can be defined as the set of owned resources that the user is interested in sharing with others and the kind of other users that could or should be informed about the user's activities.

- *Aura*: As in CVEs, this concept will be used to determine the potential for user interactions.

- *Boundaries*: They are used to divide the web space into different areas and regions and provide mechanisms for marking territory, controlling movement and for influencing the interaction properties of the web space.

- *Sense Acuity*: This concept will be used to limit the depth of search for interesting contents or users and the kind of information that the user can receive from the web site. The maximum number of links to be crossed and the format of the information can be established. The concept of *Visual Acuity*, which has been used in CVEs, can be interpreted as the extent of restrictions on the visual information that the user can receive from the web. A maximum acuity value will authorise the user to get all kinds of visual information (images and videos) from the web, while a minimum value will forbid him to acquire visual information. Similarly, *Sound Acuity* can be interpreted as the level of permission to receive sound effects from the information that is displayed at the web site. Just as in UNIX with its files and directories, it could be interesting to define a series of permissions to control the reception of information from the web: T (General Acuity): Permit access to just text information; V xxx (Visual Acuity): Permit xxx types and amount of visual information; S xxx (Sound Acuity): Permit xxx types and amount of sound effects.

- *Internal Filters:* Focus and nimbus could be restricted by the user's internal state and desires. For instance, focus could be restricted through potential collaborator's profiles and through content filters. We will only be aware of the users that are within our focus and fall into our defined profiles. The history of previous interactions and their effects on our mood or internal state can also restrict our focus or nimbus. Thus, a successful interaction will increase our level of attention to users or contents that fall into a similar profile.

5 An Implementation of This Interpretation

This asynchronous interpretation of these awareness concepts has already been implemented in a prototype system, called MADEW (Awareness Models developed in Web Environments) to be used for training and educational purposes. MADEW was carried out at the Universidad Politécnica de Madrid and it was tested with quite successful results [3].

MADEW was implemented as an electronic trademark course that an enterprise offered to its employees. Besides the typical set operations associates to a web course and to the management of users in a software application –such as introduce new users, remove users or modify user's details -, this course controlled employee access to some specific web areas, the format in which employees could access this information (visual or auditory) and the kind of information they could pick up from the course. The hierarchy of permissions was established by the enterprise depending on the position of the employee in the enterprise.

References

1.   Benford, S., and Fahlén, L.E. *A spatial model of interaction in large virtual environments*, in Proc. Third European Conference on Computer Supported Cooperative Work (ECSCW'93), Milano, Italy. Kluwer Academic Publishers, pp. 109-124.
2.   Greenhalgh, C., *Large Scale Collaborative Virtual Environments*, Doctoral Thesis. University of Nottingham. October 1997.
3.   Fernández E. *MADEW: Modelos de Awareness Desarrollados en Entornos Web*. End of Career Works supervised by P. Herrero. School of Computer Science. Universidad Politécnica de Madrid, 2002.
4.   Herrero P., De Antonio A., *A Formal Awareness Model for 3D Web-Based Collaborative Environments*. Published in Proceedings of the Workshop on Awareness and the www. ACM 2000 Conference on Computer Supported Cooperative Work (CSCW 2000). Philadelphia, Pennsylvania, USA, 2000.

An EC Services System Using Evolutionary Algorithm

Whe Dar Lin

The Overseas Chinese Institute of Technology
Dept of Information Management,
No. 100, Chiao Kwang Road, Taichung 40721, Taiwan

Abstract. Our new evolutionary method allows electronic commerce (EC) services on distinct distribution channels. Launching EC services on the Internet require careful on mobile agents. It supports EC transition flows written in XML. Our algorithm resolves the concurrent data-accessing problem among EC services databases. To create a better algorithm, we have analyzed a variety of transaction schemes compatible with standards and developed a modeling framework on which maintaining good consistency. With our EC transaction method, we can make use of different techniques and organize an EC framework with clients, agents, and EC application servers all included form an integrated EC system management mechanism. Our proposed system can improve the relationship between EC service systems and transaction agents for supply-chain management.

1 Introduction

To keep mobile agents on track toward making a purchase, EC systems must provide an effective function of local applications with applications running on remote servers. M-services do pose challenges to database management and transactions on EC services platforms in order to support greater workgroups and achieve better organizational productivity. The ultimate goal is to provide a richer and more user-friendly environment of information by integrating the user's desktop facilities with information exchange and collaboration infrastructures including groupware platforms and shared database servers. In a business setting, these information services are typically part of an EC service system [1], [2], [6], [8], [9].

One of the advantages of mobile agents can control their own shared resources. A commercial deal usually involves several transactions including the transfer of contract documents, billing, and settlement of payment. Sometimes several transactions need to be integrated, as when billing and settlement are to be processed at the same time. In addition, the definition of priority transactions is required for defining the entire commercial deal, so that should any individual transaction fail, the entire deal can be discarded [4], [5].

Evolutionary methods have been applied to a variety of different Web-based problems. In this paper, an algorithm for EC services system based on an evolutionary model is proposed.

M. Bubak et al. (Eds.): ICCS 2004, LNCS 3036, pp. 549–552, 2004.

2 Our Evolutionary Algorithm

In our evolutionary model, the reinforcements can be either positive or negative, depending on whether the realized channel cost is greater or less than what the EC services need. Given the evolutionary approach method set E_i of EC agent i, where E_i = {$e_{i,1}$, $e_{i,2}$, $e_{i,3}$, $\cdots$, $e_{i,Mi}$} respectively, there are alternative pure evolutionary approach to be performed by EC agent i, (i = 1, ..., M). EC agent i at each period uses an evolutionary approach method, and the state of the system in period t is denoted by $C_{t,i}$. Note that here in this place $C_{t,i}$ =($C_{t,i}(e_{i,1})$, $C_{t,i}(e_{i,2})$, $C_{t,i}(e_{i,3})$, ..., $C_{t,i}(e_{i,Mi})$) is the probability distribution of the evolutionary approach method set E_i in period t by EC agent i. If EC agent i plays evolutionary method $C_{t,i}$ in period t, then the resultant loading balance value is $C_{loading}(t, i, e_{t,i})$. The EC agent's communication channel cost is denoted by $C_{linking}(t, i, e_{t,i})$, and we set the loading balance value as $C_{balance}(t, i, e_{t,i})$ =$C_{loading}(t, i, e_{t,i})$ - $C_{linking}(t, i, e_{t,i})$. The C_{index} value is iif($e_{i,k}$ =$e_{i,t}$, 1, 0)=C_{index}. Then, for i = 1,...,N and k =1, 2,...,Mi, the system state evolves in the following way:

$$C_{t+1,i}(e_{k,i}) = (1+\left|C_{balance}(t,i,e_{t,i})\right|) * C_{t,i}(e_{k,i}) + C_{index} * C_{balance}(t,i,e_{t,i}) \tag{1}$$

Thus, it can be seen that if $C_{balance}(t, i, e_{t,i})$ is positive, that means the EC agent is pleased with the outcome, and then the probability associated with the strategy will increase. In our proposed algorithm, EC transactions can be calculated in terms of link capacity, buffer size, queue length, etc. In addition, we can even update the switching function on the arrival of every transaction. The key idea behind our proposed algorithm is to update the switching probability according to the loading strategy rather than the instantaneous or average loading weight, maintaining a single probability $C_{probablity}(t, i, e_{t,i})$ to transfer enqueued transactions.
I: Computing switching function

$$iif\left(L_{load}(t,i,e_{t,i}) \geq L_{threshhold}, 1, (L-(L_{threshhold} - L_{load}(t,i,e_{t,i})))/L\right) \tag{2}$$

$$= L_{switch}(t,i,e_{t,i})$$

The system loading value in period t is denoted by $C_{load}(t, i, e_{t,i})$. We set a loading weight threshold, $C_{threshhold}$.
II: Computing moving probability

$$iif\left(L_{switch}(t,i,e_{t,i}) \geq L_{switch}(t-1,i,e_{t-1,i}),\right. \tag{3}$$

$$L_{switch}(t,i,e_{t,i}) + (1-C_{probablity}(t,i,e_{t,i})) * C_{balance}(t,i,e_{t,i}),$$

$$L_{switch}(t,i,e_{t,i}) + C_{probablity}(t,i,e_{t,i}) * C_{balance}(t,i,e_{t,i}))$$

$$= L_{switch}(t+1,i,e_{t+1,i})$$

This result can be derived from equations listed in Sect. 3. Thus, when the outcome satisfies the EC transaction services, the loading probability is increased. However, the switching probability is increased when the EC services are dissatisfied.
In the next section, we shall present our simulation results on our proposed algorithm and see how it compares with other algorithms in the same network environment. We will show the validity and features of our proposed EC services algorithm.

3 Performance with Our Evolutionary Algorithm

Simulation results show that our system outperforms such existing EC services schedulers as earliest deadline, highest value and hierarchical earliest deadline when an application requires an EC transaction model.

We examined EC transactions under various conditions. According to the metrics of Commit times and throughput, our method has the best performance for distributed EC Web services using EC transaction models.

The setting for these basic parameters is based on our experiment, we varied the arrival rate from 1 transactions/second to 5 trans/sec.

Table 1. Commit time simulation results of different method

Channel availability	Commit time (millisec)				
	0.4	0.6	0.8	1	1.2
Earliest deadline	123	99	92	89	89
Highest value	115	90	91	87	86
Hierarchical earliest deadline	111	86	80	69	66
Our Evolutionary Algorithm	100	81	72	60	54

Table 2. Throughput simulation results of different method

Channel availability	Throughput (transaction/sec)				
	0.4	0.6	0.8	1	1.2
Earliest deadline	0.10	0.15	0.15	0.20	0.20
Highest value	0.15	0.15	0.15	0.20	0.20
Hierarchical earliest deadline	0.15	0.20	0.20	0.20	0.25
Our Evolutionary Algorithm	0.20	0.25	0.30	0.35	0.40

Tables 1 and 2 show the commit time and throughput results for real time EC services transactions. The results for EC services transactions. The performance orders are Our Evolutionary Algorithm > Hierarchical earliest deadline > Highest value> Earliest deadline.

Simulation results show that our system outperforms the others on throughput and commit time. An EC transaction based on evolutionary algorithm, to each transaction resides in the ready queue with the highest will be executed. The appropriate setting for the communication delay of the real time transactions can meet their loading balance value on time under the simulation results.

To begin with, the consideration of the loading characteristic in Web services gives a higher weight in the formula in the evolutionary model at the arrival of a transaction, since such a transaction requires an expensive cost for accessing data objects in the database. However, the loading policy also depends on the reward ratio and loading balance value as well as the slack time of the system. In addition, the communication delays in our evolutionary algorithm will result in a slightly higher weight for a remote transaction; hence, a local transaction will have a better chance to be executed completely under the adjustment of a transaction's reward ratio.

4 Conclusion

In this paper, we have presented a new algorithm to handle electronic commerce (EC) transactions on Web-based systems. Evolutionary methods have been used to solve a wide variety of Web-based systems problems. We have demonstrated that it is capable of offering smooth transaction services at an extremely low loss rate with little delay in supply chain management. We can enable the EC Web server to adapt to various network conditions and traffic characteristics intelligently. Simulation results show that our system outperforms others on throughput and commit time. It prevents the queue from turning into overflow and decreases the loss rate due to buffer overflow. All the parameters used in the algorithm can be derived and adjusted by using measured and estimated information. Indeed, the complexity of our new algorithm is lower than those of many other algorithms. Our evolutionary method responses rapidly to the changes of the network load by adjusting the switching probability quickly. The concepts presented in this paper can be further developed into a set of networks that will help identify the best design alternative for high balance loading management based on the characteristics and parameters of given transactions on EC service applications in supply chain management. The performance of our method in complex network topologies is not yet clear. We will work on that in the future. In addition, we shall also focus on the development of new service algorithms and differentiated service support in supply chain management.

References

1. C. M. Weng and P. W. Huang, "More Efficient Location Tracking in PCS Systems Using a Novel Distributed Database System," IEEE transactions on vehicular technology, Vol. 51, No.4, pp277-289, 2002.
2. D. Fudenberg and D.K. Levine, The Theory of Learning in Games, The MIT Press, 1998.
3. El-Sayed, A.A., Hassanein, H.S., and El-Sharkawi, M.E. "Effect of shaping characteristics on the performance of transactions." Information and Software Technology 43(10): 579-590,2001.
4. Haritsa, J.R., Ramamritham, K., and Gupta, R. "The PROMPT real-time commit protocol." IEEE Trans. Parallel and Distributed Systems 11(2):160-181, 2000.
5. Jain, R. "The Art of Computer Systems Performance Analysis: Techniques for Experimental Design, Measurement, Simulation, and Modeling." WILEY, 1991.
6. J. W. Weibull, Evolutionary Game Theory, The MIT Press, 1995.
7. K. K. Leung, Y. Levy, "Global Mobility Management by Replicated Databases in Personal Communication Networks," IEEE Journal on selected areas in communications, Vol. 15, No. 8, pp1582-1596, 1997.
8. R. Somegawa, K. Cho, Y. Sekiya, and S. Yamaguchi, "The Effect of Server Placement and Server Selection for Internet Services," IEICE Trans. on Communications, Vol.E86-B, No.2, PP.542-552, 2003.
9. V. Kanitkar and A. Delis, "Real-Time Processing in Client-Server Databases," IEEE transactions on computers, Vol. 51, No.3, pp269-288, 2002.

A Fast and Efficient Method for Processing Web Documents

Dániel Szegő

Budapest University of Technology and Economics
Department of Measurement and Information Systems
H-1521, pf. 91, Budapest, Hungary
`szegod@mit.bme.hu`

Abstract. This paper investigates the possibility of realizing some Web document processing tasks in the context of modal, especially description logics, providing a precise theoretical framework with well-analyzable computational properties. A fragment of SHIQ description logic which can primarily be used in document processing is introduced. The paper also presents a linear time algorithm for model checking Web documents proving that the logical approach can compete even in efficiency with other industrial solutions.

1 Introduction

During the last ten years, the success of World Wide Web was increasing and it has become part of our daily life. Due to this enormous success, several techniques for processing, transforming or searching Web documents, like XML or HTML, have been developed. Unfortunately, these techniques are usually based on different theoretical approaches, no uniform representation is known. Primary consequence of different theoretical frameworks is that several parts of them are reinvented and re-implemented at each of the techniques. Hence, some of these frameworks are lack of simple formal semantics or efficient algorithms.

Description logics are simple logical formalisms which primarily focus on describing terminologies and graph style knowledge [1,2]. Therefore, they seem to be an adequate basis for developing a common computational environment for several Web document processing tasks [3].

The origin of this work was motivated by a Web filter project. Several elements of the project and logic presented in this paper were previously published in [4,5]. However, non of the algorithmic aspects were considered yet.

The reminder of this paper is organized as follows. The fragment of SHIQ, and some of its application areas are introduced in section 2. Section 3 presents the basic idea behind the model checking algorithm. Last but not least section 4 draws some conclusions.

M. Bubak et al. (Eds.): ICCS 2004, LNCS 3036, pp. 553–556, 2004.

2 A Logical Approach for Processing Web Documents

This section briefly introduces a fragment of SHIQ description logic, which fragment has primary importance in Web document processing. First of all, the model of the logic has to be specified exactly, which is practically a formalized view of a web document. The model of a document is basically an ordered tree which nodes are associated with atomic predicates.

The *document model* is a six tuple $<V, AP, top, c, ap, n>$.

1. V is a set of nodes of the graph, AP is a set of atomic predicate, $top \in V$ is the top node.

2. c, ap and n binary relations describe the structure of an ordered tree which nodes are labeled by atomic predicates.

This definition seems natural for an XML document. For example, tags can be translated to nodes and embedding of tags represents the children relation. The definition is less trivial for an HTML document, consequently pre-transformations and pre-filters need to be applied.

Syntax and semantics of the logic are based on roles and concepts (Table 1.). In order to define a formal semantics of the syntax, an I interpretation function is considered, which assigns to every concept a set of nodes of a given 'd' document model and to every role a binary relation over $V \times V$.

Table 1. Syntax and semantics of the logical framework.

Constructor	Syntax	Semantics
Concept Constructors		
atomic concept	a	$a^I = \{\, v \in V \mid a \in ap(v)\}$
disjunction	or	$(C_1 \text{ or } C_2)^I = C_1^I \cup C_2^I$
conjunction	and	$(C_1 \text{ and } C_2)^I = C_1^I \cap C_2^I$
complement	not	$(\text{not } C)^I = V \setminus C^I$
universal quant.	all	$(\text{all } R.C)^I = \{v \in V \mid \forall w.\ <v,w> \in R^I \text{ implies } w \in C^I\}$
existential quant.	some	$(\text{some } R.C)^I = \{v \in V \mid \exists w.\ <v,w> \in R^I \text{ and } w \in C^I\}$
top concept	every	$\text{every}^I = V$
bottom concept	none	$\text{none}^I = \varnothing$
Role Constructors		
next role	next	$\text{next}^I = n$
children role	child	$\text{child}^I = c$
inverse role	inverse	$(\text{inverse } R)^I = \{<w,v> \in V \times V \mid <v,w> \in R^I\}$
transitive closure	infinite	$(\text{infinite } R)^I = \cup_{j>=1}(R^I)^j$

Using a logic in real life applications requires the existence of several basic reasoning services and efficient algorithms for computing these services. One of the most important and most efficient basic reasoning service is model checking but others like equivalence, querying or subsumption could also be used widely.

Basic reasoning services can be used in a wide variation of document processing tasks. Simple model checking is the basic reasoning mechanism of a searching process (e.g. searching in an XML database or searching the WWW). A logical expression could be the searching statement and documents, for which the evaluation of the statement is not an empty set, form the result of the search. Beside search, model checking can be used in several other areas, e.g. document categorization. In document transformation (e.g. XSLT, XQuery) or information extraction, the principal problem is to select some tags of the document which match with a predefined template. It is the most natural application area of querying because logical expressions can easily be regarded as templates. Last but not least, document checking (e.g. DTD) can be efficiently supported by subsumption or equivalence of model checking. For example, the following statement would be true for only those XML documents in which every slideshow tag contains only slide or title tags: 'slide $\Rightarrow$ all child.(title or slide)'.

3   Model Checking Algorithm

The model checking algorithm is based on the algebraic approach of the semantics. Expressions are interpreted as sets so concept constructors can be interpreted as operations between sets. For example, a conjunction can be regarded as a binary operation which transforms two input sets to an output one ('and': $2^V \times 2^V \rightarrow 2^V$). Similarly, universal or existential quantifications can be interpreted as unary operations associating input sets with output ones ('all child': $2^V \rightarrow 2^V$). Role constructors are manifested as variations in the unary operations. For instance, 'all child' represents a different unary operation as 'all infinite child' does.

The only question which highly effects efficiency is how to represent sets and relations of the document model. In our approach, nodes of the document model are labeled by integers in the $[0....|V|-1]$ domain, where $|V|$ denotes the cardinality of the node set. Each node has exactly one integer label. Primary consequence of this labeling is that most part of the algorithm can be built on hash tables and simplified hash joins.

The structure of the document is stored in five tables. For example, 'parenttable' is an array of integers associating each integer label of a node with the integer label of its parent node (according to the inverse of 'c' binary relation of the document model). The algorithm implements a realization for each operation. As an example, we can consider 'some infinite child' operator which requires the identification of nodes that can be reached from a given set of nodes. It can be implemented by a depth-first search of the graph described by 'parenttable'. Since the number edges of the graph are linear in the size of nodes, depth first search runs linear time in the size of nodes of the document model.

This approach has the following important property:

Proposition.
If the number of possible atomic predicates of each node is bound, the model checking algorithm has $O(l*|V|)$ time and space complexity (where l is the length of the logical expression, and $|V|$ is the number of nodes of the document model).

Beside theoretical investigation, an experimental architecture has also been implemented in C# to test the concepts and algorithms between real circumstances. The architecture realizes XML and HTML parsers which load the administration tables directly and an algorithm for evaluating logical expressions over document models.

4   Conclusion

This paper analyzed the possibilities of using description logics in web document processing. It has identified a fragment of SHIQ which has primary importance in document processing and briefly introduced how specific document processing problems can be solved by this fragment. It has several benefits comparing to other industrial solutions of document processing. It provides a uniform knowledge representation with well defined syntax, semantics and algorithms, which representation is sometimes more expressive than industrial ones. Hence, description logic integrates several previously unrelated document processing problems like categorization or document checking into one common framework. Besides, the article introduced an efficient algorithm for evaluating logical expressions over Web documents. Since the algorithm is linear, it can compete even in efficiency with other industrial solutions.

References

1. Baader, F., Nutt, W.: Basic Description Logics, In the Description Logic Handbook, edited by F. Baader, D. Calvanese, D.L. McGuinness, D. Nardi, P.F. Patel-Schneider, Cambridge University Press (2002) 47-100
2. Borgida, A., Brachman, R. J.: Conceptual Modeling with Description Logics In the Description Logic Handbook, edited by F. Baader, D. Calvanese, D.L. McGuinness, D. Nardi, P.F. Patel-Schneider, Cambridge University Press (2002) 359-381
3. Calvanese, D., Giacomo, G., Lenzerini, M.: Representing and reasoning on XML documents: A description logic approach Journal of Logic and Computation, 9(3) (1999) 295-318
4. Szegő, D.: Using Description Logics in Web Document Processing, SOFSEM vol. II. (2004) 256-263
5. Szegő, D.: A Logical Framework for Analyzing Properties of Multimedia Web Documents, Workshop on Multimedia Discovery and Mining, ECML/PKDD-2003, (2003) 19-30.

Online Internet Monitoring System
of Sea Regions

Michał Piotrowski and Henryk Krawczyk

Department of Computer Architecture, Gdańsk University of Technology
bastian@eti.pg.gda.pl, hkrawk@pg.gda.pl

Abstract. The paper describes design and implementation problems of on-line Web monitoring and visualization systems. The three-layer architecture is proposed and example of oceanographic map documents (XML) server is presented. A graphical user interface representing user functionality is also given.

1   Introduction

Digital maps are becoming an integral part of many monitoring systems. The representative example is a measurement system, which determines different parameters such as: water temperature, drift speed and direction referring to a given geographic region. The system also can simulate different emergency events, such as oil spill at the sea and its dissipation in time. The general architecture is given in Figure 1 a). It is an on-line Web oriented application consisted of three basic components: communication media (Internet), monitoring servers which gather data from measurement systems and clients which display measurement data in a way convenient to end users [1]. In case of high time consuming simulation parallel processing can be used.

Our main goal was to create visualization part of our system as elastic as possible. Classical applications use bitmaps [2], but their big disadvantage is that for every user action like panning, zooming map and toggling layers, a server needs to generate the new map image. Besides, a bitmap could not be effectively utilized in other user applications. Using a vector format more processing is oriented on client, because it stores a whole map and it can zoom, pan and toggle layers without extra connection to server. However there is still one problem: logical structure of the map document is not available, so we cannot effectively process them after generating map images. Using a subset of XML called Scalable Vector Graphics (SVG) we could preserve logical structure of a map document. Eventually we decided to create and use XML Map Documents.

In the paper we present architecture of the designed system and describe its main layers. Next we concentrate on XML Map Documents and its server implementation.

M. Bubak et al. (Eds.): ICCS 2004, LNCS 3036, pp. 557–560, 2004.

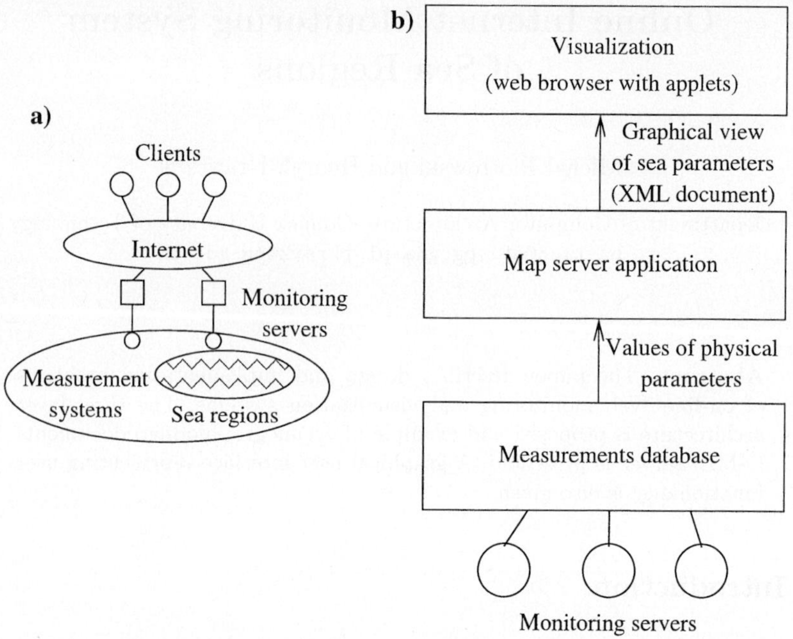

Fig. 1. Idea of Monitoring System (a), Layered System Architecture (b)

2   Monitoring System Architecture

Figure 1 b) shows more details of the proposed architecture suitable for on-line monitoring system of a sea area. The system consists of three standard tiers: business logic (measurements database with map server application), presentation logic (map server application) and visualization tier.

Measurement data are originally inserted into the database as points. Each point represents a geographic place (it has defined latitude and longitude) and suitable sea or atmospheric parameter like water or salty levels. This data can be obtained from oceanographic models like HIROMB or ICM. Points of measurement data are converted into various map objects, what is done by special scripts or corresponding developed code. It needs many calculations and resources so it is often impossible to make calculations in real time, while serving map document. Therefore we preprocess measurement data and cache them in a database.

In the presentation logic tier, the map and preprocessed measurement data are converted in SVG (XML) document. For implementing such functions we decided to use PHP. The map server takes a map data from the database and convert them to the SVG document.

The last tier – visualization is located on users computer. It is a client application which displays SVG map, allows user to zoom, pan map and select displayed map's layers etc. Apart from displaying maps it manages map documents and it has JavaScript interface for dynamic visualization of simulations, like oil diffusion in the sea. To create client application we use Java applet technology and Batik library which is a part of The Apache XML Project [3].

3   XML Map Documents

We use SVG format to create map documents. Graphical objects in SVG can be grouped into layers. There is possibility to define a user coordinate system and use scripting languages for processing events (e.g. mouse click on a graphical object). SVG allows to insert private application data into file and to create new tags and attributes which defines special shape types. Besides, SVG images can be imported into popular vector image editing applications developed by Adobe, Corel and many more. These applications will ignore a private application data and they will display a graphical content of the file.

One of the challenges was to use SVG in such way, that graphical applications could display as much of the map as possible and we maintain logical structure of the map. Bearing this in mind we designed coding of semantic data referring to layers, map's legend etc. The part of data which corresponds to logical structure of map is ignored by graphical applications but it is used by our applet. Additionally each graphical object corresponds to one map object and is labeled with measured value. This label will be interpreted by our applet and by graphical applications.

4   Map Server Implementation and Testing

Map documents generator we have implemented in PHP. Server code performs calculations connected to changing of coordinate system into screen coordinates, generates appropriate map layers and adds XML (SVG) headers.

For developing the client application we have used Batik library which allows to create extensions. We used the extension mechanism to implement a class used to display special map objects (special XML tag added to SVG maps). This possibility is used because, we have encountered some problems with implementation of displaying map symbols, which have special behavior: they must not change their size while zooming and they must not intersect with other symbols on the same layer. Creating our own extension allowed us to optimize symbols rendering speed by more than 5 times.

Figure 2 illustrates clients user interface. On the left side there is a list of available layers, and we can choose layers to display. On the top, there is a toolbar for choosing date of measures, zooming and panning. At the bottom of this applet there is a status bar with cursor's current coordinates and messages. The main part of applet's window displays the map. We can see wind measurements (speed and direction) showed as colored regions and symbols. There is also a small window with some information about selected symbol (list of some other measurements).

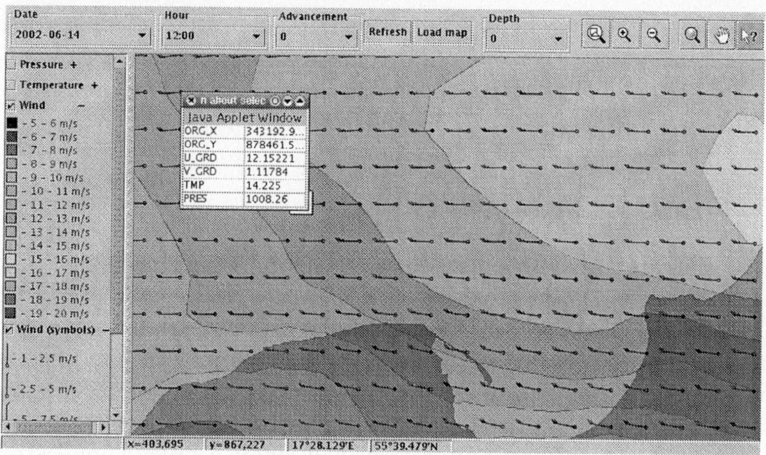

Fig. 2. Applet's graphical user interface with map of wind's speed and direction

5 Remarks

The client-server application creates measurement maps without loosing any significant data and delivers always present values of measured parameters. Big advantage of our solution is that it can be used as a base for visualizing different simulation. Map displaying applet has a suitable JavaScript interface, which allows viewing of dynamic changes of environment and simulated phenomenon.

The basic features of the proposed solution are as follows:

- flexibility – it can be used in various environments, because a map is created as a XML document, its logical structure can be interpreted. Besides, it can be used as visualization of sea simulations results;
- interchangeability – XML map documents can be used not only with clients (map viewer) but also with many other graphical manipulation applications;
- usability – all maps are on-line updated, so users have always access to latest measurement data;

The obtained Web application is a good example of utilization of Web technologies for creating moder scientfic applications.

References

1. Michał Piotrowski, *MSc. Thesis: Graphical Web Browser Interface To Oceanographic Database*, Gdańsk University of Technology, Department of Computer Architecture, 2002
2. Plewe Brandon, *GIS Online: Information Retrieval, Mapping, and the Internet*, Santa Fe, 1997
3. Apache Software Foundation, *The Apache XML Project*, http://xml.apache.org, 2001

Modeling a 3G Power Control Algorithm in the MAC Layer for Multimedia Support

Ulises Pineda[1], César Vargas[2], Jesús Acosta-Elías[1], J.M. Luna[1],
Gustavo Pérez[1], and Enrique Stevens[1]

[1] Facultad de Ciencias, Universidad Autonoma de
San Luis Potosí, Av. Salvador Nava s/n, Zona Universitaria,
San Luis Potosí, S.L.P., 78290, México.
Tel: +52 (444) 826 2316, Fax: +52(444) 826 2321
{u_pineda, estevens, jacosta, mlr}@fc.uaslp.mx,
http://www.fc.uaslp.mx
[2] ITESM-CET, Monterrey, N.L., 64849, Mexico,
cvargas@itesm.mx

Abstract. Modern Third Generation Wireless Networks demand more
and more resources in order to satisfy customers' needs. And these resources can only be provided by a good Power Control. However, power
control needs an algorithm to work at the margin of the Quality of Service (QoS) requirements. This work proposes a power control algorithm
modeled under probabilistic criteria. By means of applying a Markovian
model to a MAC Protocol (power control algorithm), to optimize the
power assignment to each user in the system. This protocol is highly
interrelated to the power control functionality to extract the maximum
capacity and flexibility out of the WCDMA scheme.

1 Introduction

Recently, extensive investigations have been carried out into the application of
a Code Division Multiple Access (CDMA) as an air interface multiple access
scheme for IMT-2000 (International Mobile Telecommunications System) - 2000
/ UMTS (Universal Mobile Telecommunication System). CDMA is the technology for the third generation wireless personal communication systems.[1]

Power control is the single most important system requirement for CDMA
based wireless networks systems. In the absence of power control the effect of
near/far phenomena is dominant, and the capacity of the CDMA mobile system
is very low. Power control allows users to share system resources equally between
themselves. Besides furthermore, with a proper power control it is possible to
lower transmitting power of the mobiles and prolong the battery life.

With this in mind, we will improve a MAC power control algorithm previously proposed in [2] and enhance its capabilities for an specific application: two
services (voice and multimedia) with different rates in a 3G wireless network.

M. Bubak et al. (Eds.): ICCS 2004, LNCS 3036, pp. 561–564, 2004.

2   Model Description

Based on a Markovian process, the *On-Off model* or *Eng-Set distribution* offers an accurate information about transmission activity or inactivity of users. Using this and adding it to a practical MAC protocol algorithm for power control, we can determine with accuracy the activity of users transmitting, and to establish the power vector P_R. The power vector P_R describes the total power transmitted.

2.1   Proposed Model

Taking the multimedia MAC protocol proposed in [2], and thinking about to enhance their capabilities, we extend the protocol to consider more than one service. The next step is to establish a power control vector determined by an *On-Off* model with the purpose of knowing accurately how many users are in activity, and give them sufficient power without causing interference to the rest and keeping the *QoS* requirements.

Determination of the Number of Users. The number of packet transmission that could be supported in the next frame is calculated on a frame by frame basis to ensure the different Bit Error Rate (BER) requirements of all type of users. The general case for the power vector P_R is defined as

$$min\left\{ P_R = \sum_{n_1=0}^{N_1} \sum_{n_2=0}^{N_2} \cdots \sum_{n_k=0}^{N_k} P_{n_1,n_2,\ldots,n_k} \cdot \left(\sum_{j_1=1}^{n_1} P_{1,j_1} + \sum_{j_2=1}^{n_2} P_{2,j_2} + \cdots + \sum_{j_k=1}^{n_k} P_{k,j_k} \right) \right\}, \tag{1}$$

where the services are subject to the *QoS* constraints established in [2] and $P_{n,m}$ represents the states of a bidimensional Markov chain of N_k elements ($k = 1, 2$), $P_{1,j}$ and $P_{2,k}$ are the power assigned to users j and k of services 1 and 2, respectively, n_1 and n_2 determine how many users of the N_1 and N_2 are active transmitting. Therefore the optimal power vector can be obtained by solving the linear *QoS* equations in the powers. Since we are considering a single cell scenario, we will drop the subindices with respect to cells and redefine them to consider the single cell scenario with two-classes of traffic. In this way the Energy-bit to Noise ratio (E_b/N_0) is established as

$$\gamma_{i,j} = \frac{W}{R_{i,j}} \frac{P_{i,j}G_{i,j}}{\eta_0 W + \sum_{\hat{i}=0} \sum_{\hat{j}=0} P_{\hat{i},\hat{j}}G_{\hat{i},\hat{j}}}, \tag{2}$$

where $\gamma_{i,j}$ is the $(Eb/N_0)_{Target}$ of service i for user j, $P_{i,j}$ the power transmitted by user j of service i, $G_{i,j}$, the channel gain of user j of service i, $R_{i,j}$ the bit rate of user j and service i and $\hat{i}, \hat{j}$, represents the rest of users who are transmitting and in consequence, interfering with i, j. In the case of $G_{\hat{i},\hat{j}}$ and $P_{\hat{i},\hat{j}}$, these are channel gain and power of the users interfering in the system to the uplink transmission of user j of service i.

2.2   Algorithm Description

Hence, in order to show how the algorithm works, or how we get results from the equations, the next procedure can help us to have a better understanding of the functioning of this algorithm. Figure 1.a shows a block diagram of the procedure to follow during the execution of the algorithm for evaluation purposes.

First, we have to define the activity parameters of the Markov chain (α, β, δ and λ, see Figure 1.b), number of users, N_1 and N_2, in the system, also the power received in the uplink in the Base Station (BS) -this is in order to obtain channel gains and the necessary power from it-, bandwidth W, rates R_1 and R_2 for each service, and the noise density η_0.

Once defining these variables, we obtain by generating random numbers, the position of each of the N_1+N_2 users. This provides a random distance from the BS, and then we obtain $P_{i,j}$ and $G_{i,j}$, for each user in the system, independently of the service they require. This is done since users are located within the cell.

So, with this parameter we evaluate Equation (2) in order to obtain γ_{Target} for both services.

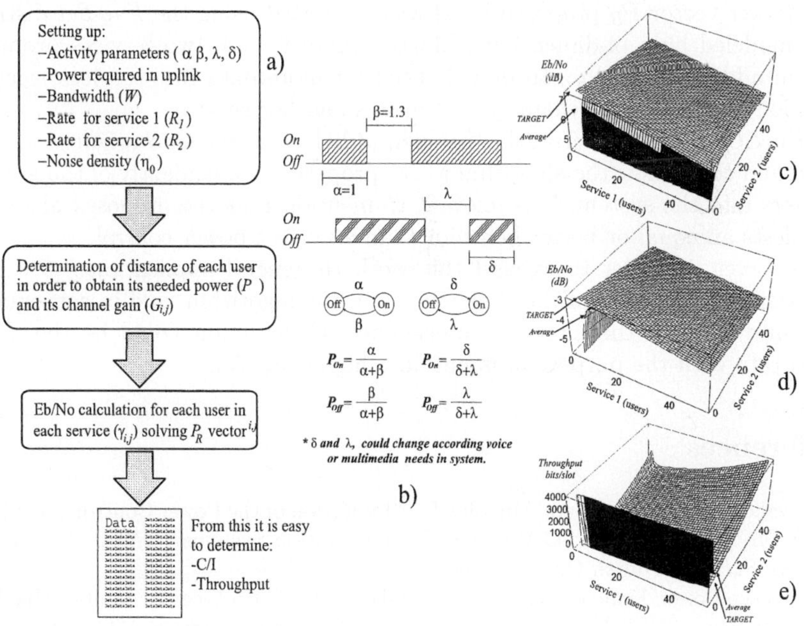

Fig. 1. a) Block diagram of the proposed algorithm, b) activity parameters of the Markov chain, c) and d) are the behavior of the system seen from service 1 and 2 respectively in response to the arriving of users, and e) is the throughput average and target of the system according to the arriving of users

3   Numerical Results

Two services in the system are established, voice and multimedia data with a rate of 12.2 kbps and 128 kbps respectively. The activity parameters α and β (that rule the Markov chain states, see Figure 1.b) were set in 1 second, δ and λ were fixed with 0.5 and 0.65 seconds respectively. The diameter of the cell was established as 100 m. Evaluation parameters accomplish with the WCDMA standard [3,4]. In figures 1.c and 1.d are shown the system behavior according to the arriving of users in transmission process in comparison with perfect power control. Notice that independently of the fading due the demand of the users, all of them manifest the same data rate asked or assigned since the beginning. In addition, because the E_b/N_0 does not manifest abrupt changes, the Carrier to Interference ratio (C/I) obtained from it let us to get a constant throughput through the arriving of transmitting users, see Figure 1.e. This last figure is also compared with perfect power control.

4   Conclusions and Future Work

The power vector P_R proposed in [2] was optimized using the *Eng-Set* distribution modeled by a bi-dimensional Markov chain for the two services proposed. But in addition, the P_R can be enhanced for more data rate services, each dimension of the Markov chain system means another service.

Since C/I does not affect the throughput behavior, the QoS required was satisfied and shows how the algorithm works properly independently of the arriving of users into the system. It is relevant to mention that the proposed algorithm manifests an equal or better behavior than a perfect power control.

However, in order to expand this work to general applications, it will be necessary add-on the multi cell capability in the algorithm, and in consequence take into account the inter-cell interference. Other point could be the use of multi rate with the purpose of maintain a higher E_b/N_0.

References

1. Novakovic Dejan M., Dukic Miroslav L., "Evolution of the Power Control Techniques for DS-CDMA Toward 3G Wireless Communication Systems", *IEEE Communications Surveys*, Fourth Quarter 2000.
2. Carrasco Loren, Femenias Guillem, "W-CDMA MAC Protocol for Multimedia Traffic Support", *IEEE Vehicular Technology Conference Proceedings*, VTC 2000-Spring Tokyo, Vol. 3, pp. 2193-2197, 2000.
3. Rappaport, Theodore S., *Wireless Communications: Principles & Practice*, Prentice Hall Inc., 2002.
4. Laiho Jaana, Wacker Achim, Novosad Tomás, *Radio Network Planning and Optimisation for UMTS*, John Wiley & sons, LTD., 2002.

Network Probabilistic Connectivity: Exact Calculation with Use of Chains*

Olga K. Rodionova[1], Alexey S. Rodionov[1], and Hyunseung Choo[2]

[1] Institute of Computational Mathematics and Mathematical Geophysics
Siberian Division of the Russian Academy of Science
Novosibirsk, RUSSIA +383-2-396211
alrod@rav.sscc.ru
[2] School of Information and Communication Engineering
Sungkyunkwan University
440-746, Suwon, KOREA +82-31-290-7145
choo@ece.skku.ac.kr

Abstract. The algorithmic techniques which allow high efficiency in the exact calculation of reliability of an undirected graph with absolutely reliable nodes and unreliable edges are considered in this paper. The new variant of the branching algorithm that allow branching by chains is presented along with improvement of series-parallel reduction method that permits the reduction of a long chain by one step.

1 Introduction

The task of calculating or estimating the probability of whether the network is connected (often referred to as its *reliability*, is the subject of much research due to its significance in a lot of applications, communication networks included. The problem is known to be NP-hard irrelative of whether the unreliable edges or nodes or both are considered. Most explored is the case of absolutely reliable nodes and unreliable edges that corresponds to real networks in which the reliability of nodes is much higher than that of edges. The transport and radio networks are good examples. We show that, by taking into consideration some special features of real network structures and using modern high-speed computers, we can conduct the exact calculation of reliability for networks with dimension of a practical interest.

The well-known branching algorithm [1] uses branching on the alternative states of an arbitrary edge. Our first approach is to branch by the whole chain if it exists. Another well-known approach that uses series-parallel reduction owes to its spreading mostly to A.M. Shooman [2,3]. In the reduction of series this method uses consequent reduction of pairs of edges. We propose to reduce the entire chain at once thereby increasing in calculation speed.

---

* This work was supported in parts by BK21, University ITRC and RFBR. Dr. H.Choo is the corresponding author.

M. Bubak et al. (Eds.): ICCS 2004, LNCS 3036, pp. 565–568, 2004.

The programming of the proposed algorithms is non-trivial. In this paper we are trying give a proper attention to this task. Special notice is given to the problem of computer storage economy.

2   Using Chains in the Calculation of Network Reliability

As the treating of dangling nodes, articulation nodes and bridges in the reliability calculation is well-known we consider the initial network structures that are free of them.

Our extended branching method (branching by chain) is based on the following theorem.

Theorem 1. *Let a graph G have a simple chain $Ch = e_1, e_2, \ldots, e_k$ with edge reliabilities $p_1, p_2, \ldots, p_k$, respectively, connecting nodes s and t. Then the reliability of G is equal to*

$$R(G) = \prod_{j=1}^{k} p_j \cdot R(G^*(Ch)) + \sum_{i=1}^{k}(1 - p_i) \prod_{j \neq i} p_j \cdot R(G \backslash Ch), \tag{1}$$

if e_{st} does not exist and

$$R(G) = \left[(p_1 + p_{st} - p_1 p_{st}) \prod_{j=2}^{k} p_j + p_{st} \sum_{i=2}^{k}(1 - p_i) \prod_{j \neq i} p_j \right] \times R(G^*(Ch)) + \tag{2}$$

$$\left[(1 - p_1)(1 - p_{st}) \prod_{j=2}^{k} p_j + (1 - p_{st}) \sum_{i=2}^{k}(1 - p_i) \prod_{j \neq i} p_j \right] \times R(G \backslash Ch \backslash e_{st}),$$

otherwise, where $G^*(Ch)$ is a graph obtained from G by contracting by a chain, $G \backslash Ch$ is a graph obtained from G by deletion of this chain with nodes (except for terminal ones), and p_{st} is the reliability of an edge directly connecting the terminal nodes of the chain.

A.M. Shooman [2,3] has proposed substituting the parallel or subsequent pair of edges to one to speed up the reliability calculation. Thus the graph G is transformed to some graph G^* with lesser number of edge and, possibly, nodes. Reducing k parallel edges to one with reliability p is obvious and simple while the reducing of an consequent pair of edges leads to a graphs with a different reliability:

$$R(G) = r R(G^*), \tag{3}$$

$$p = \frac{p_1 p_2}{1 - (1 - p_1)(1 - p_2)} = \frac{p_1 p_2}{p_1 + p_2 - p_1 p_2}, \quad r = p_1 + p_2 - p_1 p_2.$$

Based on this result and the consequent reduction on pairs of edges for the chain with length $k > 2$ we derived the following

Theorem 2. *Let a graph $G(n, m)$ have a simple chain $Ch = e_1, e_2, \ldots, e_k$ with edge reliabilities $p_1, p_2, \ldots, p_k$, respectively, connecting nodes s and t. Then*

$$R(G(n, m)) = \prod_{i=1}^{k} p_i \left(\sum_{i=1}^{k} p_i^{-1} - k + 1 \right) R(G_2(n - k + 1, m - k + 1)), \quad (4)$$

where a graph $G_2(n - k + 1, m - k + 1)$ is derived from $G_1(n, m)$ by substituting the chain by a single edge with the probability of the edge existence

$$p = 1 / \left(\sum_{i=1}^{k} p_i^{-1} - k + 1 \right). \quad (5)$$

After substituting all chains by edges the reduced graph is calculated by the simple branching method. If during the process a new chain appears, then it is also substituted by an edge. Reducing all chains with consequent branching is faster than branching by chains as it leads to small-dimension graphs on earlier recursions.

3   Program Realization of the Algorithms and Case Studies

The problem of programming the proposed algorithms is not trivial by virtue of the high request to the memory, and of numerous recursions also. We discuss the following aspects in this section: (1) re-usage of memory in recursions; (2) finding chains for branching and reduction; (3) renumbering nodes; and (4) the final graphs that allow direct calculation.

The **re-usage of memory** is provided by considering the upper-left block of the same probability matrix on each recursion. To provide this we need **renumbering** of nodes: the chain should be contracted to a node with node number $n - k$ (dimension of the reduced graph), thus this number is assigned to one of its terminal nodes. The number $n - k + 1$ is assigned to the other one. Thus the numbers of nodes of the resolving chain (including terminal) should be $n - d, n - d + 1, \ldots, n$ after renumbering, where d is the number of edges for the chain, and n is the number of nodes for the graph under reduction.

On execution of branching it is necessary to take into account all **possible variants of the resulting graphs**. While performing the classical branching method there are only 3 possible results: the derivation of a disconnected graph at deletion of an edge, a graph of small dimension simple for calculation at contracting and a graph that is connected but not possible for direct calculation yet, to which the operation of branching is applied again. At usage of the branching by chain or chain reduction it is necessary to take additional variants into account. They are: (1) the resulting graph is a cycle; (2) the resolving chain is a cycle; (3) the dangling node appears. We specially treat the case (4) "the resulting graph is disconnected". The last means that any edge in the deleted chain is a bridge. Accordingly, by contracting we obtain a articulation point and the reliability of the graph is considered as the product of the reliabilities of two graphs G_1 and G_2 and probability of the existence of a resolving chain (or edge).

We conducted several experiments on the computer with the processor AMD Athlon 800MHz inside. We have made the comparisons among the algorithm with branching by chains (BC), basic branching algorithm (BB), branching algorithm with chain reduction (BR) and algorithm from [6] (RT).

In the example of the lattice (4×4) graph, that was used in [6], the number of basic recursion for RT is 2579141, time spent for calculation was about 47 seconds. Algorithm BC takes 0.17 seconds and only 407 recursions on this example. Note, that 200 chains were found during the calculation with average length 2.385. So on this example our algorithm is more than 200 times faster. The basic BB algorithm takes on this example 8.35 seconds, which is about 50 times slower than BC and takes 80619 recursions. However best results were shown by the BR algorithm which takes only 0.06 seconds on 93 recursions. When the dimension of a lattice was increased up to (5×5) the algorithm RT did not finished in 2 hours, and BB, BC and BR algorithms took 21 minutes, 15.05 and 2.47 seconds on 13817311, 51652 and 14581 recourses respectively.

Last we calculate the reliability of the graph with the structure of well-known ARPA network. This graph has 58 nodes and 71 edges. The algorithm BC takes approximately 20 minutes and BR – about one minute for calculation. With this the last algorithm takes only 31933 recursions.

Thus we can state that our modifications of branching method and method of parallel-subsequent reduction are faster than previous methods and allow the calculation of reliability of networks with tens of elements in reasonable time.

References

1. Moore, E.F., Shannon, C.E., "Reliable Circuits Using Less Reliable Relays," *J. Franclin Inst., 262, n. 4b*, pp. 191-208, 1956.
2. Shooman, A.M., Kershenbaum, A., "Exact Graph-Reduction Algorithms for Network Reliability Analysis," *Proc. GLOBECOM' 91. Vol. 2*, pp. 1412-1420, 1991.
3. Shooman, A.M., "Algorithms for Network Reliability and Connection Availability Analysis," *Electro/95 Int. Professional Program Proc.*, pp. 309-333, 1995.
4. Rodionov, A.S., Rodionova, O.K., "On a Problem of Practical Usage of the Moore-Shennon Formula for Calculating the Reliability of Local Networks," *Proc. 2nd Int. Workshop INFORADIO-2000*, Omsk, pp. 67-69, 2000.
5. Rodionova, O.K., "Some Methods for Speed up the Calculation of Information Networks Reliability," *Proc. XXX International Conf. "IT in Science, Education, Telecommunications and Business,"* Ukraine, Gurzuf, pp. 215-217, 2003.
6. Chen, Y., Li, J. Chen, J., "A new Algorithm for Network Probabilistic Connectivity," *Proc. MILCOM'99. IEEE, Vol. 2*, pp. 920-923, 1999.
7. Rodionova, O.K. "Application Package GRAPH-ES/3. Connectivity of the Multigraphs with Unreliable Edges (Atlas, procedures)," Preprint No. 356, Computing Center of the SB AS of the USSR, Novosibirsk, 1982. (in Russian)
8. T. Koide, S. Shinmori and H. Ishii, "Topological optimization with a network reliability constraint," *Discrete Appl. Math., vol. 115, Issues 1-3*, pp. 135-149, November 2001.

A Study of Anycast Application for Efficiency Improvement of Multicast Trees

Kwang-Jae Lee[1], Won-Hyuck Choi[2*], and Jung-Sun Kim[2]

[1] School of Electronics, Electronics and Multimedia, Seonam University, 702, Kwangchi-dong, Namwon-city, Jeollabuk-do, 590-711, Korea
`kjlee@seonam.ac.kr`
[2] School of Electronics, Telecommunication and Computer Engineering, Hankuk Aviation University, 200-1, Hwajeon-dong, Deokyang-gu, Koyang-city, Kyonggi-do, 412-791, Korea
`rbooo@korea.com, jskim@mail.hangkong.ac.kr`

Abstract. In this paper, we considered previously existing multicast routing algorism and protocol and especially put more attention on CBT routing protocol to analyze its strength and weakness. As a result, traffic was converged upon core router because of structural problem of CBT protocol and according to the converged traffic, there was congestion at the core link, thus it caused efficiency degrades of the whole routing. Therefore, we proposed a way of conversion to Anycast Routing method from the method of CBT multicast tree routing that was suitable for traffic decentralization even though there was high bandwidth depended on increment of traffic load. In order to support multimedia service that requires from small to large bandwidth and to consider multicast routing protocol to improve characteristic of multicast packet's delay, CBT/Anycast routing method can be proposed as an alternative plan for freedom of bandwidth in traffic.

Keywords: Internet applications, multicast routing, anycast routing, core base tree

1 Introduction

Multicast protocol classifies network users into specific groups and provides not only various but characterized services with communicating protocol to individuals, enterprises, and the government. It becomes a matter of concern and interest for internet communication.

The CBT (Core Base Tree) method, the representative protocol of the covalent tree, is one of methods to improve high-speed transmission of multicast packet and efficiency of communication by decreasing overhead from tree constitution's overlap. However, CBT (Core Base Tree) has several problems in structure and they work as its vulnerability (Core Base Tree) [3], [4].

The first problem of CBT is the phenomenon of transmitter's traffic concentration around Core Router. For instance, traffic density and surplus sign around Core router that are often seen in services like video, Telnet, Ftp, etc. Fig. 1 shows concentration problem in traffic and Fig. 2 is Poor Core phenomenon.

---

* The corresponding author will reply to any question and problem from this paper

M. Bubak et al. (Eds.): ICCS 2004, LNCS 3036, pp. 569–572, 2004.

The core's ideal position in traffic reception is right in the middle that correspondent with the size of distance from group members.

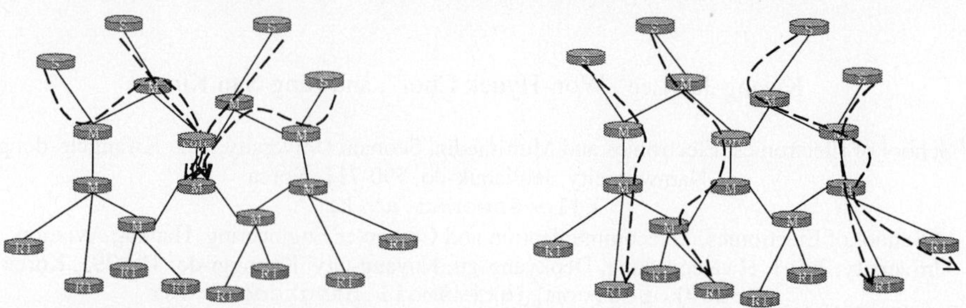

Fig. 1. Traffic concentration   **Fig. 2.** Poor core placement

However, if the core is positioned in an isolated area from transmitter-recipient of packet and used independently, then it becomes impossible to have right choice and practice even though it does not require much the high bandwidth and the maintenance space of routing information. Therefore, ABT (Anycast Based Tree) is proposed in the paper.

ABT does not limit core in specific position within network but let it actively be located so that the previously mentioned problems of CBT can be solved. The specific resolution is to use AIMD (Addictive Increase Multiple Decrease) algorism. The controlled transmission rate of traffic enables traffic that is concentrated in core router, to maintain average transmission rate and leads traffic to poor core so it helps to improve excess use in whole system and performs multicast service in high speed [6], [7].

2 ABT

The main characteristic of the suggested ABT is its treatment of multicast packet in a formation like CBT without having core router. In this process, however, it requires control mechanism that moves traffic to core in the other side when traffic gets concentrated in core more than threshold.

In multicast routing, the time for traffic to pass the core is called Core round trip time (crtt), and crtt becomes reset time for a table in a transmitter and a control parameter.

For increase factor of packet, the transmission time increment of core can be shown as $a/crtt$, and if there is increase in transmitter, the formula is like below:

$$R_{in} = R_{now} + \frac{a}{crtt}. \tag{1}$$

where, R_{in} is transmit packet and R_{now} is amount of packet in the present core. If there is decrease of packet to core, the formula becomes like below:

$$R_{in} = \frac{R_{now}}{b}. \tag{2}$$

where, b is factor for decrease. The increase of transmitted packet and the average transmitted rate based on decrease can be calculated at the core from a and b. Also, the transmission rate is calculated according to the size of packet from recipient and the minimum and maximum rate for transmission can be calculated with increase of recipient as follows:

$$R_{max} = R_{in} \cdot \frac{a}{crtt} \cdot n \cdot \qquad (3)$$

$$R_{min} = \frac{R_{max}}{b} \cdot \qquad (4)$$

R_{min} is minimum rate for transmission of core, R_{max} is maximum rate for transmission of core, and transmission time increase is n. The below formula is for the average rate for transmission by using minimum and maximum rates for transmission of core:

$$R_{ave} = \frac{R_{max} + R_{min}}{2} = \frac{a}{2} \cdot \frac{b+1}{b-1} \cdot \frac{n}{crtt} \cdot \qquad (5)$$

According to Eqn 5, Poor Core phenomenon that becomes the minimum rate for transmission of core and congestion around core that occurs it becomes the maximum rate of transmission can be controlled with the average rate of transmission.

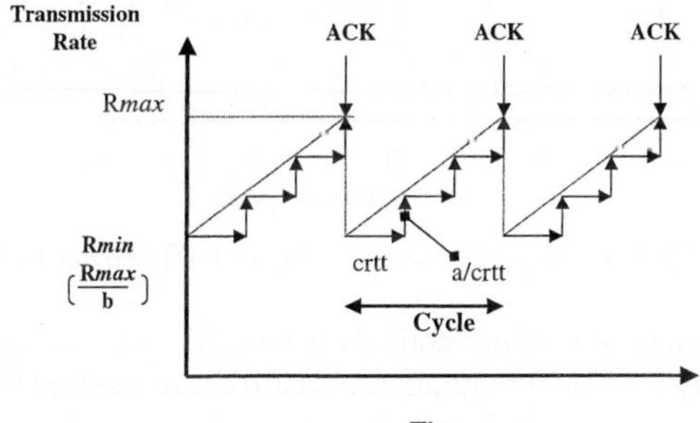

Fig. 3. The rate of transmission based on increase in transmission

Fig. 3 shows retransmission of ACK at the end of each cycle after the multicast packet (n, crtt, Rave) completed transmission. The formula for the average rate for transmission in core router is like below. At this time, loss of packet through retransmission of ACK is p.

$$Trans = \frac{1}{crtt} \sqrt{\frac{a}{2} \cdot \frac{b+1}{b-1} \cdot \frac{1}{p}} \cdot \qquad (6)$$

3 Simulation and Discussion

To the simulation model, each CBT routing protocol is applied and the numbers of multicast groups and transmitters are varied. Then the packet process condition of Core router is measured based on the sized of multicast data packet. Fig. 4 and Fig. 5 show the result of the simulation. The reason for this is that there is formation of initialization of multicast tree and is frequent *Join* and *Leave* of group, thus interval for packet's arrival becomes shorter and relatively increase in load of packet occurs as a result. Fig. 4 shows packet transmission delay of CBT and Fig. 5 shows cuing delay of core as the system is executed as Anycast routing protocol in CBT routing protocol.

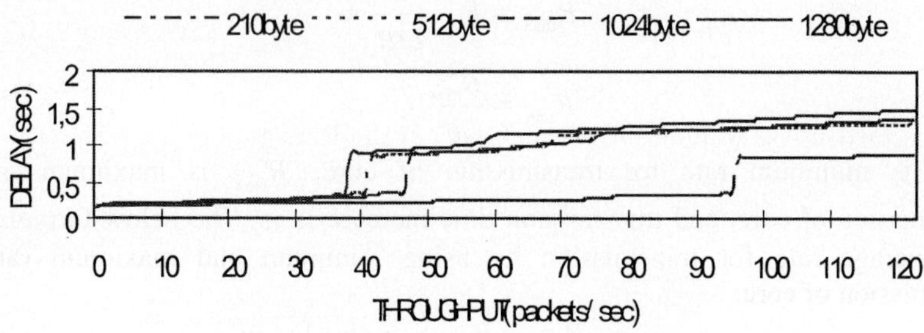

Fig. 4. Packet transmission delay of CBT

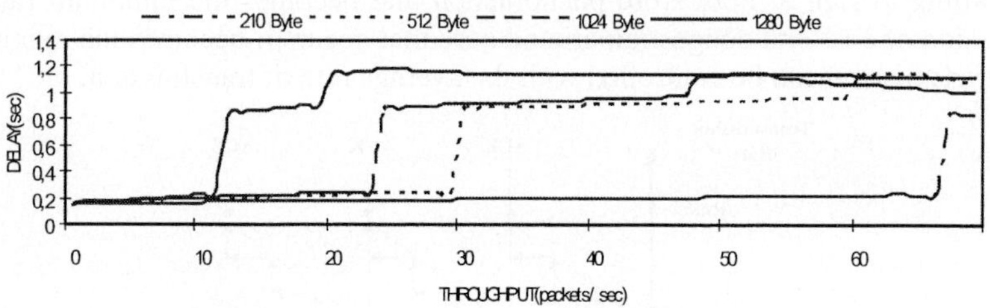

Fig. 5. Packet transmission delay of CBT/Anycast core

In this paper, change of routing methods from CBT shared tree routing method that is stable in relatively low bandwidth to Anycast routing method that is appropriate for traffic congestion even in high bandwidth depends on load of traffic. In the study, characteristics of delay according to the size change of multicast data packet when the system is changed from CBT to Anycast, were observed and evaluated.

References

1. Parsa, M., Garcia-Luna-Aceves, J. J.: A protocol for scalable loop-tree multicast routing. IEEE J. Select. Areas Commun. **15** (1997) 316-331
2. Jia, X., Wang, L.: A Group Multicast Routing Algorithm by using Multiple Minimum Steiner Trees. Computer Communications (1997) 750 -758
3. Ballardie, A.: Core Based Trees (CBT) Multicast Routing Architecture. RFC2201 (1997)
4. Ballardie, A.: Core Based Trees (CBT Version 2) Multicast Routing Protocol Specification. RFC2189 (1997)
5. Moy, J.: Multicast Extensions to OSPF. IETF RFC1584 (1994)
6. Ettikan, K.: An Analysis of Anycast Architecture And Transport Layer Problems. Asia Pacific Regional Internet Conference on Operational Technologies (2001)
7. Lin, J., Paul, S.: RMTP: A Reliable Multicast Transport Protocol. IEEE INFOCOM96 (1996)

Performance Analysis of IP-Based Multimedia Communication Networks to Support Video Traffic[*]

Alexander F. Yaroslavtsev[1], Tae-Jin Lee[2], Min Young Chung[2], and Hyunseung Choo[2]

[1] Institute of Mining, Siberian Branch of the Russian Academy of Science
Novosibirsk, Russia +7-3832-170930
`yar@misd.nsc.ru`
[2] School of Information and Communication Engineering, Sungkyunkwan University
440-776, Suwon, Korea +82-31-290-7145
`{tjlee,mychung,choo}@ece.skku.ac.kr`

Abstract. With the rapid growth of the communication equipment performance, it is possible that communication networks, which use TCP/IP, will be able to provide real-time applications such as the broadcasting of a multi-media traffic. These applications are sensitive to transmission delay and its variance, thus estimating QoS is very important. In this paper, we evaluate the performance of an IP-based multimedia network in terms of the end-to-end mean delivery time and the utilization of communication equipments.

Keywords: Performance evaluation, video traffic, MPEG encoding.

1   Introduction

In order to support multi-media services in IP-based networks, it is important to assure service qualities, e.g., delay, since IP networks inherently provide best effort service. In general, modeling of IP-based networks supporting multi-media services is complex and thus QoS estimation is challenging.

There have been previous works, which investigate transmission of multi-media traffic over IP–networks [1], [2], [3], [4], [5]. Usually simulation methods are used to research such communication networks. But this approach has a series of limitations. There are great difficulties in adequate representation of transmission of multi-media traffic by analytical models. In [6], Yaroslavtsev et al. proposed an analytic method based on queuing network theory to model IP-based high speed communication networks, which is more adequate than conventional analytical models, and has less computational load than simulation models.

---

[*] This paper was partially supported by BK21 program. Dr. Choo is the corresponding author.

M. Bubak et al. (Eds.): ICCS 2004, LNCS 3036, pp. 573–576, 2004.

In this paper, we simply evaluate probabilistic temporal characteristics of an IP-based multimedia network in terms of average delay of data flows and utilization of equipment. This paper is organized as follows. In Section 2, we discuss related work about performance evaluation of communication networks and modeling of multi-media traffic. And we present the model of the investigated network and the multi-media services as well. Section 3 shows some performance results on the model of the system.

2   Description of Modeled Communication Network

In this paper, as a representation of communication networks, we consider the video transmission in an IP network shown in Fig. 1. We have N_w clients connected to a video server via a switch. Each client is connected to the switch through Ethernet port and an information server is connected to the switch through fast Ethernet port. To describe the considered network, we use the set of traffic, hardware and software parameters.

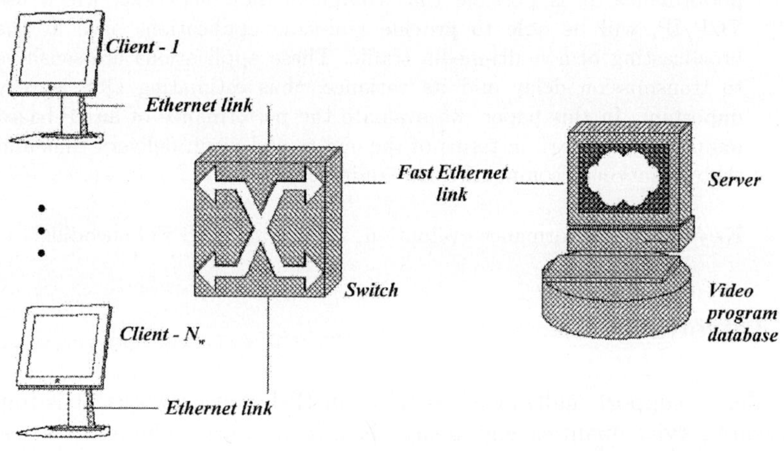

Fig. 1. A structure of IP network, for which QOS was estimated

Traffic parameters describe the characteristics of transferred traffic to the clients. Video stream must be encoded into an MPEG–4 format [7]. According to this standard, a digital video stream consists of a sequence of groups of video pictures (GoP). Each GoP consists of the M/Z–sequence of video frames (VOP – Video Object Plane). Each GoP or M/Z–sequence has fixed length and structure. In GoP three types of video frame are defined: I–frame (Intra), P–frame (Predictive) and B–frame (Bidirectional). In the sequel, we will designate the set of all frame types as $\Omega = \{I,P,B\}$. Each GoP in video stream has one I–frame, the first frame in GoP. The parameter M defines the number of frames in groups. The parameter Z is the distance between P–frames in GOP. For example, the structure of group of a typical digital video stream with parameters 15/3 has the following frames: *IBBPBBPBBPBBPBB*. Values of video traffic parameters are summarized in [6].

Hardware parameters describe the hardware characteristic associated with communication lines (distance, rate, and Bit Error Rate (BER)); performance of computers (server and nodes); rate of the switch (forwarding rate and size of its shared memory). Values of hardware parameters are given in [6].

Software parameters define the characteristics of the network software, which realize the protocol of video data transmission to clients. An end-to-end TCP connection between a client and the server is established. Video frames are encapsulated in TCP segments, and are transmitted along the network to the clients by IP packets. When IP packets are transmitted over the network, they undergo random delays, and they can be received with errors or lost. All these events cause retransmission of appropriate TCP segments and, hence, incur additional delays. Software parameters are in detail illustrated in [6].

3   Performance Evaluation and Summary

The proposed model allows to estimate a wide set of performance metrics for video traffic transmission over communication networks, e.g., mean number of IP packets associated with each client in each communication equipment, utilization of each communication equipment, size of required buffer in switch or server, and traffic rate and delay in transmitting each frame type. For illustration, we compute several performance metrics as functions of the number of clients. We have scaled up the traffic parameters so that the traffic rate is equal to 2Mbps for each client.

To evaluate the performance of the considered network, we define the mean delivery time of I/P/B frames from the server to a client node as the time interval between a new I, P, or B frame generated by the server and a correctly received by the client node. The mean delivery time of each type of frames is shown in Fig. 2. The mean delivery time of I-frame is considerably greater than that of B(P)-frame because the size of I-frame is greatly higher than that of other types. Most of these delivery times is caused by delays in the Ethernet link for less than 20 clients. The figure shows that there is considerable amount of remaining time for frames to be delivered to clients. Let λ^{VOP} denote the generating rate of VOP stream for each client. This remaining time is equal to $1/\lambda^{VOP} = 0.04$ sec. And, quality of broadcasting video stream is guaranteed in the communication network for less number of clients than 20. For more than 25 clients, these delivery times are mostly incurred from delays in fast Ethernet links and Server, and the metric shows exponential growth.

Fig. 3 shows the utilization of equipments such as fast ethernet link, switching fabric, and server in Fig. 1. The server and the fast Ethernet link have more considerable load than the other communication equipments. We see that these coefficients of utilization exhibit unreasonable level for the client number larger than 25. These equipments are bottlenecks and they can reduce QoS level considerably.

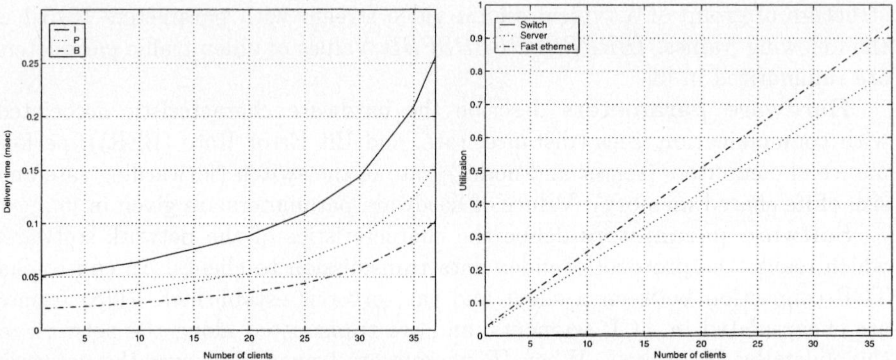

Fig. 2. Mean delivery time of ω-frames from the server to the first client node

Fig. 3. The utilization of various communication equipment

The results show that the modeled communication network, represented in Fig. 1, sufficiently transmits video traffic and provides simultaneous broadcasting of a digital video to 20-25 clients without significant loss of QoS. Its characteristics is almost linear until the number of clients increases to more than 25. For 35-40 client nodes, the modeled communication network is shown to be overloaded. Fast Ethernet link and server are bottlenecks in the modeled communication network. Thus the proposed modeling method can be used to investigate performance of communication networks with more complex topologies.

References

1. Zheng, L., and Zhang, L.: Modeling and Performance Analysis for IP Traffic with Multi-Class QoS in VPN. IEEE MILCOM Vol. 1 (2000) 330-334
2. Tian, T., Li, A.H., Wen, J., and Villasenor, J.D.: Priority Dropping in Network Transmission of Scalable Video. Int. Conf. on Image Processing Vol. 3 (2000) 400-403
3. Wu, D., Hou, Y.T., Zhang, Y.-Q., and Chao, H.J.: Optimal Mode Selection in Internet Video Communication: an End-to-End Approach. IEEEE ICC Vol. 1 (2000) 264-271.
4. F. Beritelli, G. Ruggeri and G. Schembra: TCP-friendly Transmission of Voice over IP. IEEE ICC Vol. 2 (2002) 1204-1208
5. de Carvalho Klingelfus, A.L. and Godoy Jr., W.: Mathematical Modeling, Performance Analysis and Simulation of Current Ethernet Computer Networks. 5th IEEE Int. Conf. on High Speed Networks and Multimedia Commun. (2002) 380-382
6. Yaroslavtsev, A.F., Lee, T.-J., Chung, M.Y., and Choo, H.: Performance Analysis of IP-Based Multimedia Communication Networks to Support Video Traffic. TR-ECE03-001 Sungkyunkwan University (2003)
7. Overview of the MPEG-4 standard. ISO/IEC JTC1/SC29/WG11 2459 (1998)

Limited Deflection Routing with QoS-Support

HyunSook Kim[1], SuKyoung Lee[2], and JooSeok Song[1]

[1] Dept. of Computer Science, Yonsei University, Seoul, Korea,
[2] Graduate School of Information&Communications, Sejong University, Seoul, Korea,

Abstract. In this paper, we propose a Limited Deflection Routing with Wavelength Conversion (LDR with WC) technique which decides alternative paths adaptively according to data burst's priority. Furthermore, this technique is extended to support multiple QoS. Performance is evaluated in terms of burst blocking probability as a performance metric. The proposed LDR with WC enables us to support various QoS in optical burst switching networks while still maintaining good performance.

1   Introduction

In Optical Burst Switching (OBS) networks, most of the research on contention resolution has focused on one particular method (e.g. wavelength conversion, deflection routing, Fiber Delay Line). Recently, however, several different methods have been integrated and hybrid approaches have emerged. Our scheme considers whether to use wavelength conversion on alternative future paths according to the QoS priority of each burst, when the alternative path for limited deflection routing is selected. Unlike traditional deflection routing, our proposed scheme determines a node with a wavelength converter to be over an alternative path on the basis of a certain performance metric, as well as includes limited deflection routing. Considering that generally, only some nodes are capable of converting a wavelength into different one or deflection routing in OBS networks, our scheme has a benefit of reducing the cost of wavelength conversion and the overhead due to unnecessary deflection routing in the overall network. In practice, as we use more highly efficient functions in switches such as deflection routing, wavelength conversion, and optical buffering, the overall network performance will gradually improve, but at a very high cost. To reduce this cost, we assume that only some nodes have the ability to convert wavelengths, otherwise try to find a cost-efficient alternative path that includes the nodes with wavelength converters selectively according to the QoS level of the contending burst.

2   Limited Deflection Routing with Wavelength Conversion

In OBS, important approaches for resolving contentions include deflection routing and wavelength conversion[1]. If all core nodes have these functions, the system cost becomes very high. Therefore, we consider the case in which just

M. Bubak et al. (Eds.): ICCS 2004, LNCS 3036, pp. 577–581, 2004.
© Springer-Verlag Berlin Heidelberg 2004

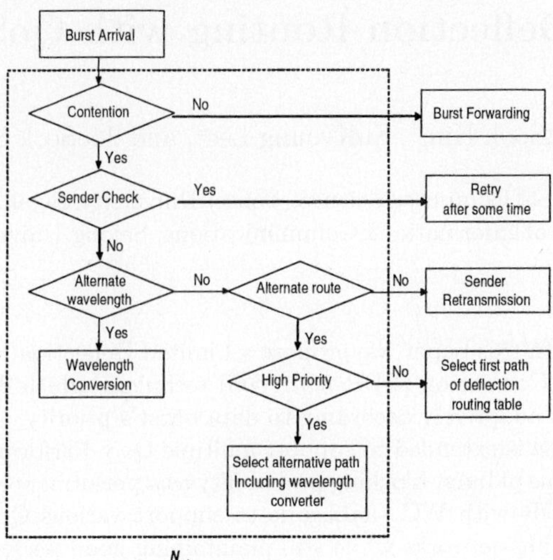

Fig. 1. Node with wavelength converter and deflection routing function

some nodes have the capability of resolving contentions using the function of wavelength conversion or deflection routing. There are different kinds of nodes within a network as follows : a node which has a wavelength converter, a node which can perform deflection routing in the event of a contention, a node which can both convert wavelengths and do deflection routing, and a node which is a normal node without wavelength conversion or deflection routing capability. Each core node has a statistical database which has the information of the network state including the network topology, the mean length of a burst, the average blocking rate over a path, and so on. With the burst blocking rate as a performance metric, each node tries to resolve contentions according to its node type. To support QoS, we modify the limited deflection routing scheme proposed in [2] as can be seen in Figure 1. This procedure checks whether the burst is generated in the congested node by sender check function, and then it tries wavelength conversion before starting limited deflection routing. To apply our integrated scheme to OBS, it is assumed that the fields of control packet include burst priority, offset time, burst size, and so on. When bursts and control packets are generated at an ingress node, the values of these fields in each packet are determined.

3   QoS-Support Algorithm of LDR with WC

In this section, we describe how LDR with WC supports various QoS. Since LDR with WC can choose an optimal alternative path flexibly, it satisfies the QoS while providing several ways of selecting an alternative path. It is very simple to select the shortest path among the various alternative paths. However, we aim to choose an cost-efficient path depending on the QoS class of the burst, which includes nodes with or without a wavelength converter. We assume that c_i^{wc} is zero for any node without a physical wavelength converter. Conversely, a node with a wavelength converter has a value that is proportional to the number of available wavelengths.

- Step 1: We collect the paths for which the value of C^{wc}, the sum of the wavelength conversion cost at each node, c_i^{wc}, is greater than zero. This means that we select some paths from the deflection routing database, on which the deflected burst can pass by one or more wavelength converters.
- Step 2 : For each candidate path gathered in Step 1, we compute the standard deviation of the number of available wavelengths per path. If the standard deviation of any path is smaller than a given threshold, the number of available wavelengths at each node on the path does not differ significantly. Although one path has several nodes with wavelength converters, if the number of available wavelengths is concentrated at one node, there is a high possibility of further blocking. Therefore, we do not consider the average number of available wavelengths, but the standard deviation of the number of available wavelengths.
- Step 3 : We examine the mean blocking probability on each candidate path gathered in Step 1. Since one of the important and challenging issues in optical networks is to reduce burst loss, we use the burst blocking probability as a QoS parameter.
- Step 4 : According to the priority of the burst, we must find a path by minimizing the difference between the respective value of the wavelength conversion cost, the standard deviation of the path, and the mean blocking probability, and threshold value.
- Step 5 : The alternative paths obtained by this process are stored in a deflection routing table at each node. Information necessary to compute the paths is also provided from the OAM control packet and the control packet for each burst. This process of updating the deflection routing table is also periodic, but the frequency is low. If it is high, it will deteriorate the performance of the whole network. That is, frequent updates of information about network status and the overload of exchanging control packets are always a trade-off.

4   Performance Evaluation

Simulations are carried out on NSFNET topology with 14 nodes where the number of wavelengths per link is 6, the average burst length is 1Mb and link bandwidth is 1Gbps. The type of node is determined randomly and the pairs of source and destination nodes are also selected randomly. At every ingress node, we assume a Poisson burst arrival with a mean rate of λ. When two or more bursts collide at the same node, one of the following policies is applied : No Conversion and No Deflection routing (NCND), Only Wavelength Conversion (WC), Only Deflection Routing (DR), LDR with Wavelength Conversion (LDRWC). Next, we define some QoS classes similar to [3] and the requested blocking probability

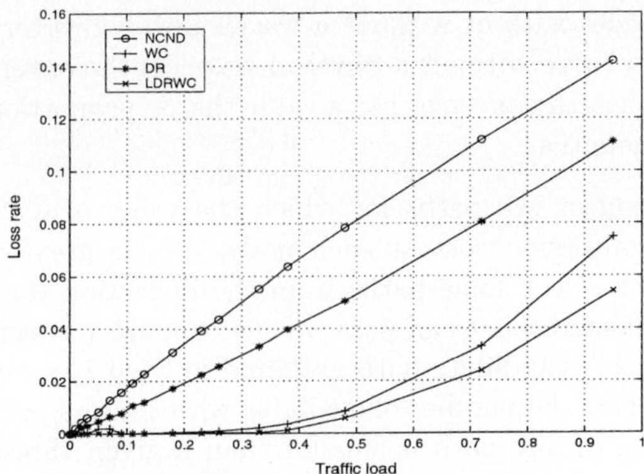

Fig. 2. Loss rate vs. Traffic load

as a QoS parameter is respectively 10^{-1}, 10^{-2}, 10^{-3}, and 10^{-4}. It is assumed that a burst is randomly generated with one of the four QoS classes at the edge router. Performance is evaluated in terms of loss rate and throughput.

Figure 2 shows that the mean loss rate is smaller in LDR with WC than in WC or DR. In addition, we observe that LDR with WC is more efficient at high loads because the difference between the performances at low and high loads is less than that in the other schemes. That is, the other schemes have a deteriorated performance at high load compared to that at low load. Accordingly, LDR with WC can be considered to provide a uniform performance consistently in the overall network, irrespective of traffic fluctuations. Our algorithm is also simulated with different QoS classes. Mean burst loss rate obtained from our simulation is 0.98×10^{-1}, 1.01×10^{-2}, 1.13×10^{-3} and 1.29×10^{-4} compared with the required mean burst loss rate for each class: 10^{-1}, 10^{-2}, 10^{-3} and 10^{-4}, while respectively keeping throughput of 0.90%, 0.98%, 0.99% and 0.99%. This result reveals that if we use the LDR with WC algorithm, we can isolate a specific QoS class and adequately support the QoS in terms of blocking probability.

5   Conclusions

Simulation results indicated that the proposed LDR with WC scheme achieves a lower loss rate than existing contention resolution schemes. Moreover, it was shown that the LDR with WC satisfies the burst loss rate required for each class while still keeping good throughput. Finally, the LDR with WC is more efficient in a large network environment with high traffic load.

References

1. S. Yao, B. Mukherjee, S.J.B. Yoo and S. Dixit: A Unified Study of Contention-Resolution Schemes in Optical Packet-Switched Networks. IEEE Journal of Lightwave Technology, Vol. 21, No. 3, March 2003, p. 672-683.

2. H.S. Kim, S.K. Lee and J.S. Song: Optical Burst Switching with Limited Deflection Routing Rules. IEICE Trans. on Commun., Vol. E86-B, No. 5, May 2003, p. 1550-1554.
3. W.H. So, H.C. Lee and Y.C. Kim: QoS Supporting Algorithms for Optical Internet Based on Optical Burst Switching. Photonic Network Communications, Vol. 5, No. 2, March 2003, p. 147-162.

Advanced Multicasting for DVBMT Solution[*]

Moonseong Kim[1], Young-Cheol Bang[2], and Hyunseung Choo[1]

[1] School of Information and Communication Engineering
Sungkyunkwan University
440-746, Suwon, Korea +82-31-290-7145
{moonseong,choo}@ece.skku.ac.kr
[2] Department of Computer Engineering
Korea Polytechnic University
429-793, Gyeonggi-Do, Korea +82-31-496-8292
ybang@kpu.ac.kr

Abstract. Our research subject in the present paper is concerned with the minimization of multicast delay variation under the multicast end-to-end delay constraint. The delay- and delay variation-bounded multicast tree (DVBMT) problem is NP-complete for high-bandwidth delay-sensitive applications in a point-to-point communication network. The problem is first defined and discussed in [3]. In this paper, comprehensive empirical study shows that our proposed algorithm performs very well in terms of average delay variation of the solution that it generates as compared to the existing algorithm.

1   Introduction

In real-time communications, messages must be transmitted to their destination nodes within a limited amount of time, otherwise the messages will be nullified. Computer networks have to guarantee an upper bound on the end-to-end delay from the source to each destination. This is known as the multicast end-to-end delay problem [1,5]. In addition, the multicast tree must also guarantee a bound on the variation among the delays along the individual source-destination paths [3]. In this paper, we propose a new algorithm for DVBMT problem. The time complexity of our algorithm is $O(mn^2)$.

The rest of the paper is organized as follows. In Section 2, we give a formal definition of the problem. Our proposed algorithm is presented in section 3 and simulation results are presented in section 4. Section 5 concludes this paper.

2   Problem Definition

We consider a computer network represented by a directed graph $G = (V, E)$, where V is a set of nodes and E is a set of links. Each link $(i, j) \in E$ is associated with delay $d_{(i,j)}$. Given a network G, we define a path as sequence

---

[*] This paper was supported in part by Brain Korea 21 and University ITRC project. Dr. H. Choo is the corresponding author.

M. Bubak et al. (Eds.): ICCS 2004, LNCS 3036, pp. 582–585, 2004.

of nodes u, i, j, ..., k, v, such that (u,i), (i,j), ..., and (k,v) belong to E. Let $P(u,v) = \{(u,i),\ (i,j),\ \ldots,\ (k,v)\}$ denote the path from node u to node v. If all elements of the path are distinct, then we say that it is a simple path. We define the length of the path $P(u,v)$, denoted by $n(P(u,v))$, as a number of links in $P(u,v)$. Let $\preceq$ be a binary relation on $P(u,v)$ defined by $(a,b) \preceq (c,d) \leftrightarrow n(P(u,b)) \leq n(P(u,d))$, $^{\forall}(a,b),\ (c,d) \in P(u,v)$. $(P(u,v), \preceq)$ is a totally ordered set. For given a source node $s \in V$ and a destination node $d \in V$, $(2^{s \Rightarrow d}, \infty)$ is the set of all possible paths from s to d. $(2^{s \Rightarrow d}, \infty) = \{\ P_k(s,d) \mid$ all possible paths from s to d, $^{\forall}s,\ d \in V,\ ^{\forall}k \in \Lambda\ \}$, where Λ is a index set. Both cost and delay of an arbitrary path P_k are assumed to be a function from $(2^{s \Rightarrow d}, \infty)$ to a nonnegative real number. Since $(P_k, \preceq)$ is a totally ordered set, if there exists a bijective function f_k then P_k is isomorphic to $\mathcal{N}_{n(P_k)}$. $f_k\ :\ P_k \longrightarrow \mathcal{N}_{n(P_k)}$. We define a function of delay along the path $\phi_D(P_k) = \sum_{r=1}^{n(P_k)} d_{f_k^{-1}(r)}$, $^{\forall}P_k \in (2^{s \Rightarrow d}, \infty)$. $(2^{s \Rightarrow d}, supD)$ is the set of paths from s to d for which the end-to-end delay is bounded by $supD$. Therefore $(2^{s \Rightarrow d}, supD) \subseteq (2^{s \Rightarrow d}, \infty)$. For multicast communications, messages need to be delivered to all receivers in the set $M \subseteq V \setminus \{s\}$ which is called multicast group, where $|M| = m$. The path traversed by messages from the source s to a multicast receiver, m_i, is given by $P(s, m_i)$. Thus multicast routing tree can be defined as $T(s,M) = \bigcup_{m_i \in M} P(s, m_i)$, and messages is sent from s to destination of M using $T(s,M)$. The multicast delay variation, δ, is the maximum difference between the end-to-end delays along the paths from the source to any two destination nodes. $\delta = max\{|\phi_D(P(s,m_i)) - \phi_D(P(s,m_j))|,\ ^{\forall}m_i, m_j \in M, i \neq j\}$. The DVBMT problem is to find the tree that satisfies $min\{\delta_\alpha \mid\ ^{\forall}m_i \in M,\ ^{\forall}P(s,m_i) \in (2^{s \Rightarrow m_i}, supD),\ ^{\forall}P(s,m_i) \subseteq T_\alpha,\ ^{\forall}\alpha \in \Lambda\}$, where T_α denotes any multicast tree spanning $M \cup \{s\}$, and is known to be NP-complete [3].

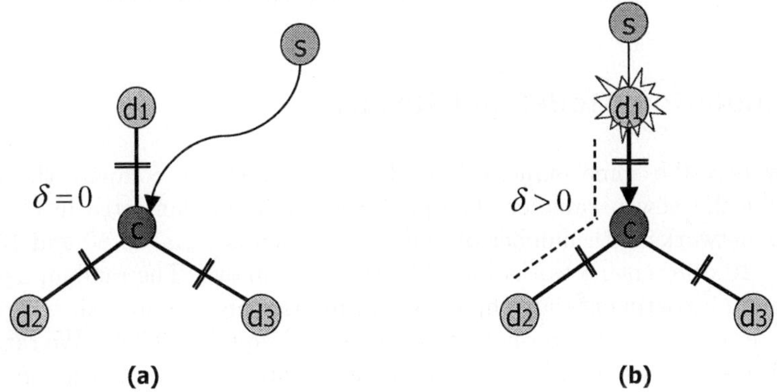

Fig. 1. The shortcoming of the DDVCA

3   An Illustration on New Heuristic

The proposed algorithm consists of a core node selection part and the multicast tree construction part. When candidate of core node is several nodes, the DDVCA [4] randomly choose a core node among candidates but our proposed algorithm is going to overcome a shortcoming of the DDVCA. See the Fig. 1. In selecting such a core node, we use the minimum delay path algorithm. The proposed algorithm calculates the minimum delay from each destination node and source node to each other node in the network. For each node, our method calculates the associated delay variation between the node and each destination node. We check whether any destination node is visited in the path from source node to each other node. If any destination node is visited, then the proposed algorithm records in '$pass_{v_i}$' data structure. And we conform $supD$ and select nodes with the minimum delay variation as the candidates of core node. As you shown in Fig. 2, our algorithm chooses the core node with $min\{\phi_D(P(s, v_i)) - min\{pass_{v_i}\}\}$. The time complexity of the proposed algorithm is $O(mn^2)$, which is the same as that of the DDVCA.

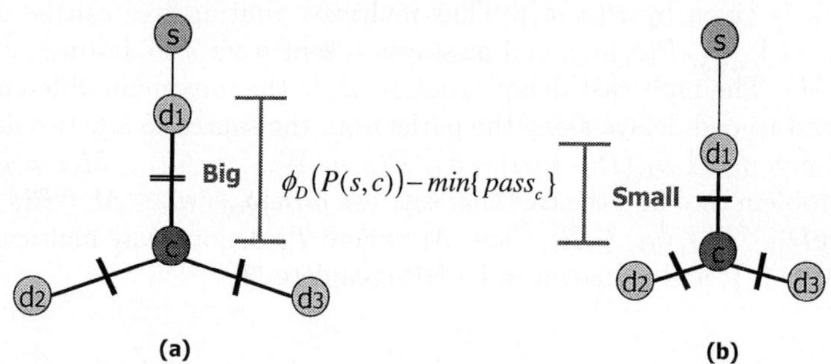

Fig. 2. The basic concept of the proposed algorithm

4   Simulation Model and Result

We now describe some numerical results with which we compare the performance for the new parameter. The proposed one is implemented in C++. We consider networks with number of nodes (n) which is equal to 50 and 100. We generate 10 different networks for each size given above. The random networks used in our experiments are directed, symmetric, and connected, where each node in networks has the probability of links (P_e) equal to 0.3 [2]. We randomly selected a source node. The destination nodes are picked uniformly from the set of nodes in the network topology. Moreover, the destination nodes in the multicast group will occupy 10, 20, 30, 40, 50, and 60% of the overall nodes on the network, respectively. We randomly choose $supD$. We simulate 1000 times ($10 \times 100 = 1000$) for each n and $P_e = 0.3$. For the performance comparison, we

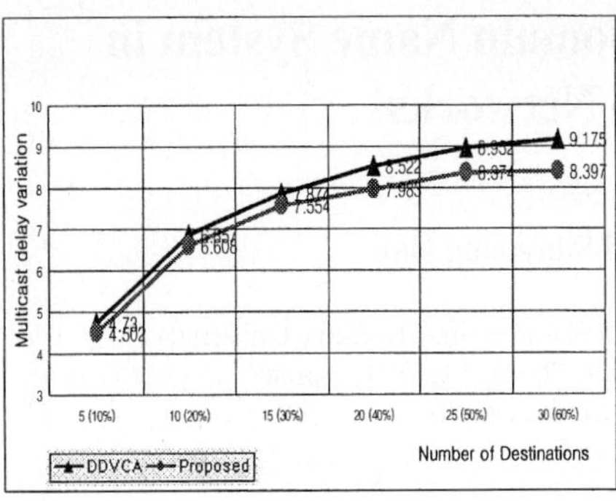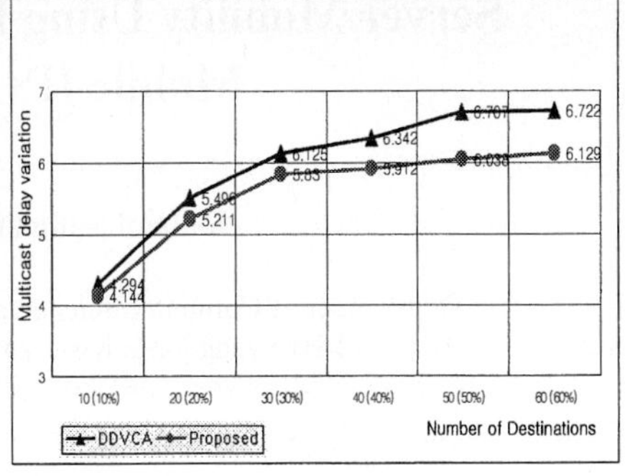

(a) P_e=0.3, $|V|$=50   (b) P_e=0.3, $|V|$=100

Fig. 3. The multicast delay variations of the three different networks and Normalized Surcharges versus number of nodes in networks

implement the DDVCA in the same simulation environment. Fig. 3 shows the simulation results of multicast delay variations. We easily notice that the proposed algorithm is always better than the DDVCA. The enhancement is up to about $100(9.18\text{-}8.39)/9.18\doteq9\%$ and $100(6.71\text{-}6.04)/6.71\doteq10\%$ for $|V| = 50$ and $|V| = 100$, respectively.

5   Conclusion

In this paper, we consider the transmission of a message that guarantees certain bounds on the end-to-end delays from a source to a set of destinations as well as on the multicast delay variations among these delays over a computer network. It has been shown that the DDVCA [4] outperforms the DVMA [3] slightly in terms of the multicast delay variation for the constructed tree. The comprehensive computer simulation results show that the proposed scheme obtains the better minimum multicast delay variation than the DDVCA.

References

1. V. P. Kompella, J. C. Pasquale, and G. C. Polyzos, "Multicast routing for multimedia communication," IEEE/ACM Trans. Networking, vol. 1, no. 3, pp. 286-292, June 1993.
2. A.S. Rodionov and H. Choo, "On Generating Random Network Structures: Trees," LNCS, vol. 2658, pp. 879-887, June 2003.
3. G. N. Rouskas and I. Baldine, "Multicast routing with end-to-end delay and delay variation constraints," IEEE JSAC, vol. 15, no. 3, pp. 346-356, April 1997.
4. P.-R. Sheu and S.-T. Chen, "A fast and efficient heuristic algorithm for the delay- and delay variation bound multicast tree problem," Information Networking, Proc. ICOIN-15 pp. 611-618, January 2001.
5. Q. Zhu, M. Parsa, and J. J. Garcia-Luna-Aceves, "A source-based algorithm for near-optimum delay-constrained multicasting," Proc. IEEE INFOCOM'95, pp. 377-385, March 1995.

Server Mobility Using Domain Name System in Mobile IPv6 Networks[1]

Hocheol Sung and Sunyoung Han

Department of Computer Science and Engineering, Konkuk University
1 Hwayangdong, Kwangin-gu, Seoul, 143-701, Korea
{bullyboy,syhan}@cclab.konkuk.ac.kr

Abstract. A mechanism using DNS to support server mobility in Mobile IPv6 networks is proposed in this paper. The name server in the mobile server's home domain maintains the mobile server's home address and care-of address. When the mobile server changes its link and gets a new care-of address, it sends a dynamic DNS update request to the name server to update its care-of address. Clients perform the DNS lookup to find the mobile server's home address and care-of address and set the connection directly to the mobile server.

1 Introduction

When a correspondent node begins communication with a mobile node in Mobile IPv6 networks, the correspondent node has no binding for the mobile node. Thus the data sent from the correspondent node should be routed to the mobile node via its home agent [1]. If a server is mobile, called a mobile server, the total number of clients that send requests to connect to mobile servers at the same time would increase in proportion to the number of mobile servers that are registered to a home agent. As a result, requests to mobile servers are concentrated in the home agent so that the load of the home agent increases. Moreover, as a mobile server becomes more distant from its home link, transmission delay over the tunnel between the home agent and the mobile server also increases. In the worst case, if a client cannot reach a mobile server's home agent because of the home agent crash or link failure, the client cannot even connect to the mobile server.

To connect to the mobile server as it changes its link, we take advantage of DNS and its ability to support dynamic updates. Most Internet users may use the hostname rather than complex IPv6 address at the beginning of a connection. The client application performs DNS lookup and receives the mobile server's IPv6 address as a DNS response [2]. When a mobile server moves to another link, it sends a dynamic DNS update to a name server in its home domain updating its current location [3]. As a result, the name server has the name-to-care-of address mapping for the mobile server and clients can get the mobile server's care-of address by means of DNS lookup.

---

[1] This work is supported by the Korean Science and Engineering Foundation under grant number R01-2001-000-00349-0(2003)

M. Bubak et al. (Eds.): ICCS 2004, LNCS 3036, pp. 586–589, 2004.

2 Operations

The following cases are described in this paper:
- When a client starts communication with a mobile server.
- When the mobile server responds to the client.
- When the mobile server moves to another link from the current link.

2.1 When a Client Starts Communication with a Mobile Server

A client that wants to connect to a mobile server performs a DNS lookup to find the mobile server's home address and the care-of address. According to the transport layer protocol or the application program, it is necessary that the transport layer connection such as TCP session be established before sending data and maintained during communication. In order to maintain the TCP connection when a mobile server moves to another link, clients have to establish the connection to the mobile server with its home address at the beginning of the connection. In this case, the client has to send the initial packet such as TCP SYN segment with the Routing header containing the mobile server' home address. Before the packet is sent, the mobile server's care-of address becomes the destination address of the packet and the mobile server's home address is moved to Address 1 in the Routing header. If the client is also mobile, the source address in the IPv6 header is set to the client's care-of address and the packet includes a Destination Options Header with the Home Address option containing the client's home address.

2.2 When the Mobile Server Responds to the Client

Upon receiving the packet from the client, the mobile server adds an entry for the client to its binding update list and sends a responding packet with a Destination Options header that contains the Home Address option and the Binding Update option. The Home Address option indicates the mobile server's home address. If the client is also mobile, the destination address in the IPv6 header of the packet is set to the client's care-of address and the packet includes a Routing header with the client's home address.

2.3 When the Mobile Server Moves to Another Link from the Current Link

As the mobile server moves to another link, it acquires a new care-of address through the conventional IPv6 mechanisms. Whenever the mobile server changes its care-of address, it sends a dynamic DNS update request to the name server in its home domain updating its current care-of address. In addition, the mobile server should also send Binding Updates to its home agent and clients as described in Mobile IPv6 [1].

3 Experiments

3.1 CAAAA Resource Record

To store the mobile server's care-of address, a new resource record, CAAAA, is defined in this paper. The type field is set to CAAAA and the data section of the record simply contains the mobile server's care-of address. A CAAAA resource record type is a record specific to the Internet class that stores a single mobile server's care-of address. The value of TTL field must be set to zero since the record applies only to the current transaction and should not be cached. A CAAAA query for a mobile server's host name in the Internet class returns the mobile server's care-of address in the additional section of the response.

3.2 Simulated Results

The following cases are simulated using OMNeT++, a network simulation tool, for estimating the approach proposed in this paper [4].

-   When the number of clients increases
-   When the mobile server becomes more distant from its home link

We simulated the RTT (round trip time) in each case for measuring the response time between a client and a mobile server. In all following figures, graph (a) shows the simulation result when the connecting request from the client is sent directly to the mobile server and the graph (b) shows the simulation result when the connecting request from the client is routed to the mobile server via the home agent.

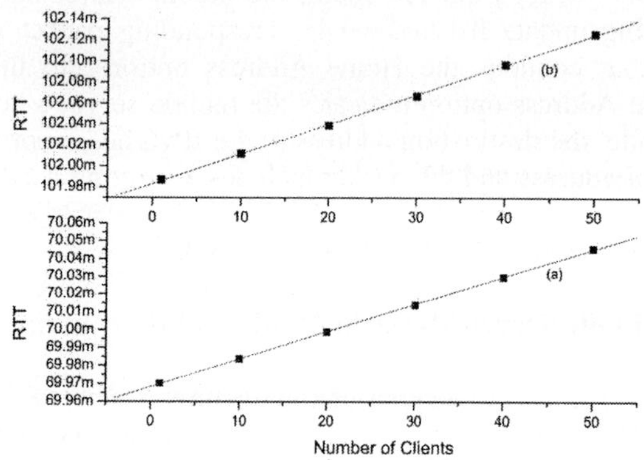

Fig. 1. RTT variation as the number of clients increase

Fig. 1 shows RTT variation caused by the increment of the number of the client. Transmission delay on the network, Processing time on the home agent and the mobile server are considered as parameters in this simulation. It is note that the RTT increases in case (b) much more than in case (a) according to the increment of the number of clients.

The following figures show RTT variation when the mobile server becomes more distant from its home link. Hop count variation indicates distance variation between the mobile server and its home agent or between the mobile server and the client in our simulation. Transmission delay on the network and per hop, processing time on the home agent and the mobile server are considered as parameters in this simulation.

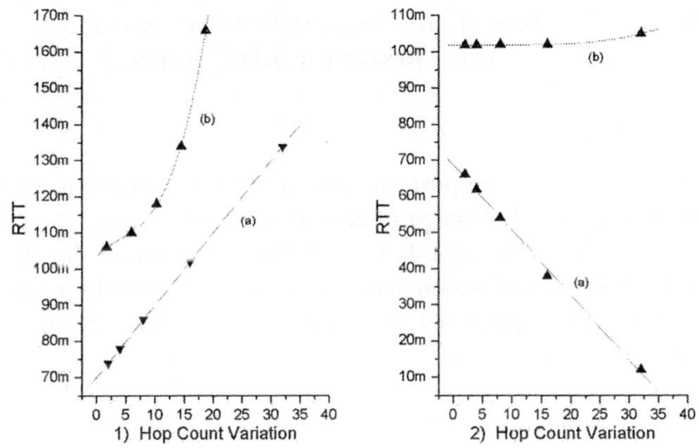

Fig. 2. As the mobile server becomes more distant from its home link, 1) the mobile server also becomes more distant from its client. 2) the mobile server becomes nearer to its client.

4   Conclusion

This paper presents a mechanism using DNS for server mobility in Mobile IPv6 networks. Although the extra processing time is needed for processing IPv6 extension headers, it is far smaller than the extra transmission delay caused by being routed via the home agent. The simulated results show that it can always shorten the round trip time when clients connect to the mobile server directly.

References

1. Perkins, C. E., and Johnson, D. B.: Mobility support in IPv6, Internet Draft, IETF, June 30, 2003 `draft-ietf-mobileip-ipv6-24.txt`.
2. Mockapetris, P.: Domain Names – Implementation and Specification, RFC 1035, IETF, November 1987.
3. Vixied (Ed.), P., Thomson, S., Rekhter, Y. and J. Bound: Dynamic Updates in the Domain Name System, RFC 2136, IETF, April 1997.
4. OMNeT++ Community Site, `http://www.omnetpp.org`.

Resource Reservation and Allocation Method for Next Generation Mobile Communication Systems

Jongchan Lee[1], Sok-Pal Cho[2], and Chiwon Kang[3]

[1]Senior Researcher, Mobile Access Research Team, ETRI, Korea
chan2000@etri.re.kr
[2] Dept. of C&C. Eng. Sungkyul Univ., Korea
[3] Senior Researcher, KDM, Korea

Abstract. This paper proposes a handoff scheme to transmit multimedia traffic based on the resource reservation procedure using direction estimation. The handoff requests for real-time sessions are handled based on the direction prediction and the resource reservation scheme. In simulation results, proposed method provides a better performance than the previous method.

1 Introduction

As mobile users move around, the network must continuously track them down and discover their new locations in order to be able to deliver data to them. Especially wireless resources availability varies frequently as users move from one access point to another [1, 2]. In order to deterministically guarantee QoS support for a mobile, the network must have prior exact knowledge of the mobile's path.

Majority of the previous schemes to support mobility make a reservation for resources in adjacent cells [3, 4]. The reserved resource approach offers a generic means of improving the probability of successful handoffs by simply reserving the corresponding resources exclusively for handoff sessions in each cell. The penalty is the reduction in the total carried traffic load due to the fact that fewer resources are granted to new sessions.

2 Proposed Structure

The base station reserves only the resources corresponding to the minimum transmission rate to the mobile. Based on the location and the direction of the mobile within a cell, the resource reservation is performed with the following order: unnecessary state, not necessary state, necessary state, and positively necessary state. If the reservation variable for the mobile is changed, the reservation is canceled and the resources have to be released with the reverse order and returned to the fool of available resource. The set of the reserved resources have its priorities depending on whether it can be

M. Bubak et al. (Eds.): ICCS 2004, LNCS 3036, pp. 590–593, 2004.

allocated to new sessions or not: a real-time handoff session (priority 1), a non-real-time handoff session (priority 2) and a non-real-time new session (priority 3). This strategy is explained in the following thing.

Resource Reservation ( )
while
If (*Unnecessary State*) **then**
    The resource reservation needs not be performed;
else if (*Not Necessary State*) **then**
    if (there are available resources in each of the estimated cells) **then**
        Reserve the resources;
    end if
    If (enough resources are not available for a new session in the estimated cells) **then**
        The reserved resources is occupied by the new sessions;
    end if
else if (*Necessary State*) **then**
    if (no resources are available for the reservation in the estimate cell) **then**
        Allocate and reserve the shared resources for a real-time session;
    end if
    If (there is no enough resource available to accommodate a new session in the estimated cells) **then**
        The reserved resources for real-time handoff sessions can be occupied by non-real-time new sessions;
    end if
else if (*Positively Necessary State*) **then**
    if (no resources are available for the reservation in the estimate cell) **then**
        Allocate and reserve the shared resources for a real-time sessions and non-real-time sessions;
    end if
    If (there is no enough resource available to accommodate a new session in the estimated cells) **then**
        New sessions cannot occupy the reserved resources;
    end if
end if

Resource Allocation ()
while
If (*handoff session*) **then**
    if (*Real-time class*) **then**
        if (there is reserved resource) **then**
            Admit the handoff session;
            Allocate the reserved resource;
        else if (there is available resource) **then**
            Admit the handoff session

```
                    Allocate the resource;
             else  Drop the session request;
             end if
       else     // Non-real-time class
             if (there is available resource) then
                    Admit the handoff session
                    Allocate the reserved resource;
             else  Buffer the session in a non-real-time queue;
             end if
       end if
    else          // New session
       if (Real-time class) then
          if (there is available resource) then
                    Admit the new session
                       Allocate the resource;
          else Block the new session;
          end if
       else     // Non-real-time class
          if (there is available resource) then
                    Admit the new session
                    Allocate the resource;
          else if (there is available resource) then
                    Admit the new session
                    Allocate the reserved resource;
          else Block the new session;
          end if
       end if
   end while
```

3 Performance Analysis

The simulation model is based on a B3G system proposed from ETRI, which is implemented using MOBILESimulatorV5. The simulation model composed of a single cell, which will keep contact with its six neighboring cells. Each cell contains a base station, which is responsible for the session setup and tear-down of new applications and to serve handoff applications. The moving path and the mobile velocity are affected by the road topology. The moving pattern is described by the changes in moving direction and velocity. Fig.1 and Fig. 2 shows the variation in the dropping rates in the different strategies when arrival rate of new session requests is increased. Results demonstrate that the dropping rate of the direction-based has decreased to about 20% and 15% for real-time and non-real-time sessions, as compared to the Fixed-based and dynamic-based, respectively. Handoff dropping rate for the dynamic scheme is much better than that for Fixed.

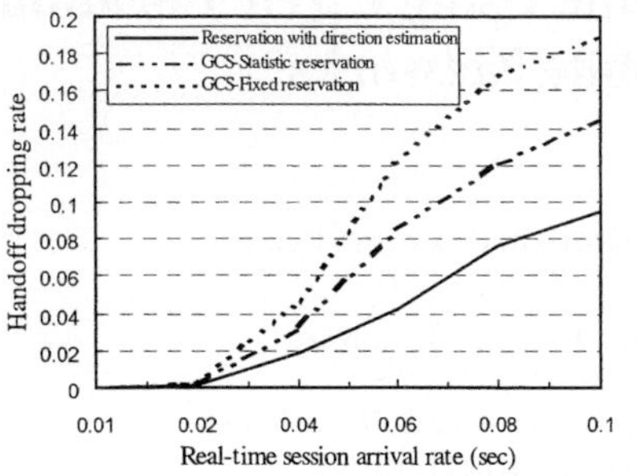

Fig. 1. Real-time sessions

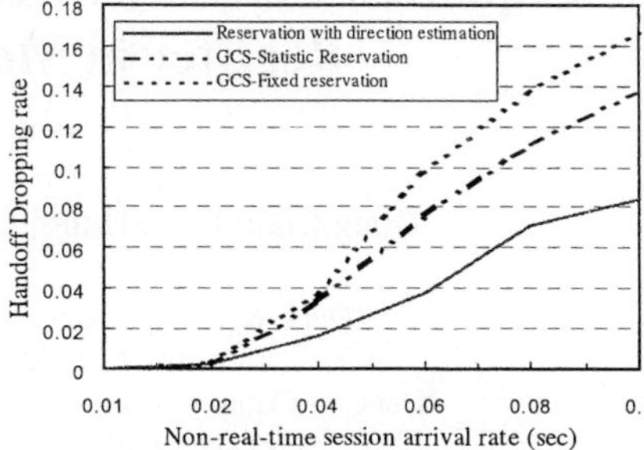

Fig. 2. Non-real-time sessions

4 Conclusions

This paper goal is to address the problem of guaranteeing an acceptable level of QoS requirements for mobile users as they move from one location to another. This is achieved through reservation variables such as the current location and the moving direction that is presented with a set of attributes that describes the user mobility. Based on reservation variables a scheme that provides predictive QoS guarantees in mobile multimedia networks is proposed. We have focused in improving the overall system performance.

References

1. Ab. Aljadhai, Taieb F. Znati, "Predictive Mobility Support for QoS Provisioning in Mobile Wireless Environments," *IEEE JSAC.*, Vol. 19, No. 10, Oct., 2001.
2. W. C. Y. Lee, "Smaller cells for greater performance," IEEE Com. Mag., 1999.
3. O. T. W. Yu and V. C. M. Leung, "Adaptive Resource Allocation for Prioritized Call Admission over an ATM-based Wireless PCN," *IEEE JSAC.*, Vol. 15, pp. 1208–1225, Sept. 1997.
4. L. Ortigoza-Guerrero and A. H. Aghvami, "A Prioritized Handoff Dynamic Channel Allocation Strategy for PCS," IEEE Trans. Vehic.Tech., vol. 48, No. 4, pp. 1203–1215, Jul. 1999.

Improved Location Scheme Using Circle Location Register in Mobile Networks

Dong Chun Lee[1], Hongjin Kim[2], and Il-Sun Hwang[3]

[1]Dept. of Computer Science, Howon Univ., Korea
`ldch@sunny.howon.ac.kr`
[2]Dept. of Computer Information KyungWon College, Korea
[3]R&D Network Management KISTI, Korea

Abstract. We propose Circle Location Register (CLR) scheme to solve Home Location Register (HLR) bottleneck problem and terminal's Ping-pong effect in Mobile Networks (MN). Each Visiting Location Register (VLR) has a given fixed circle Registration Area (RA) around itself and has IDs of other VLRs in this circle area. Whenever a terminal moves to another RA, system computes whether the terminal is located in the current CLR area, and sends the recent location information of terminal to the old or new CLR according to computing results. The proposed scheme reduces to location traffic cost compared with IS-41scheme.

1 Introduction

The Interim Standard-41(IS-41) and Global System for Mobile Communication(GSM) [1-3] based mobility management scheme which records all the movements of terminals in a centralized DB, HLR, is questionable considering that keeping track of lots of users in real time is not a simple task. This scheme has been the bottleneck problem on HLR which occurred in due lots of signal transfer between one HLR and many VLRs and Ping-pong effect which arise frequently in the boundary of RA because of the terminal's Ping-pong movement. For this case, frequent DB queries and call updates will degrade the system performance.

2 Proposed Structure

In this scheme, each VLR acts as a CLR and has a given fixed circle area (k-circle) around itself and IDs of VLRs which are included in its circle area. When a terminal powered on, the VLR which includes terminal becomes the CLR of terminal and the terminal's latest location information is sent to the CLR when terminal changes its RA. This state is maintained as long as the terminal is located in the current k-circle area. When the terminal moves to new VLR from the current CLR's k-circle, the new

M. Bubak et al. (Eds.): ICCS 2004, LNCS 3036, pp. 594–597, 2004.

VLR becomes the CLR of the terminal. By this manner, the k-circle of the terminal can be changed dynamically. This mechanism can be performed easily by comparing VLR_id which current CLR has with VLR_id where terminal moved. For the example in Fig. 1, suppose the 1-circle which consists of seven VLRs where current CLR is VLR_1 and the others are VLRs which are included in 1-circle area. The terminal is located in VLR_5 now. If the terminal moves to the new RA, VLR_6, the CLR isn't changed, and VLR_6 sends the terminal's new location information to VLR_1, current CLR. If the terminal moves to VLR_10, the current CLR of VLR_1, has no id of VLR_10. Thus the CLR is changed, and VLR_10 belongs to the new CLR of terminal.

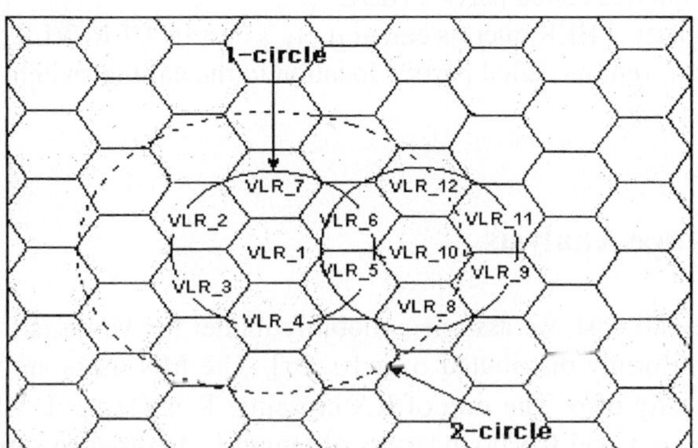

Fig. 1. CLR Structure

In mobility management algorithm, this following shows the Pseudo-code for location registration and call tracking algorithm.

Algorithm Location Registration
{
Terminal's current CLR id, VLR_xxx, received from old VLR;
Compare VLR_xxx with My_CLR_entry;
  If VLR_xxx exist in My_CLR_entry, then Send terminal_CURR_LOC to CLR;
      else { Write TID to MY_CLR_Area; //*belongs to a new CLR of the terminal
          Send terminal_CURR_LOC to HLR;
          Send REGCANC to VLR_xxx; //*REGCANC is registration cancel message
          }
  If call location update, then the terminal which moved to a new RA requests registration to the VLR of the new RA;.
      The new VLR inquires the id of the terminal's current CLR to the old VLR and the old VLR replies to the new VLR with ACK message including this information;
  The new VLR calculate and determines whether the id of current CLR exist in its VLR list of not;
  end if hit, then After sending the location information of the terminal,
      the new VLR send a registration cancel message to the old VLR;

else miss, then after transmitting location information of the terminal to HLR,
the new VLR transmits registration cancel message to old VLR and old CLR;
}

Algorithm Call tracking

```
CLR FIND( )
{
    Call to MN user is detected at local switch;
    If called party is in the same RA, then return;
    else {
        Switch queries called party's HLR;
        Called party's HLR queries called party's current CLR, VLR_xxx;
        VLR_xxx returns called party's location to the calling switch;
    } }
```

3 Performance Analysis

To estimate the call cost, we assume a mobility model for MN users. The direction of movement is uniformly distributed over $[0, 2\pi]$. The MN users are uniformly populated with a density of ρ. The rate of RA crossing, R is $(1/\pi)\rho vL$ where the average velocity of users is v and RA boundary is of length L. In order to simply performance analysis, call cost parameter and cost set are defined using formulas [4].

In Fig. 2, we can see that proposed method has lower cost than IS-41 scheme, even though it is a worst case of proposed method provides mostly same cost as IS-41 scheme. The worst case takes places when ratios of six cases are same. In other word, it occurs when LSTP connected with very few RAs (i.e., less then three VLR/MSC). But we know that a LSTP's coverage is more than that of three RAs generally. The worst case of proposed scheme seldom occurs in actual networks. We know that the next generation wireless system will adopt smaller RA. It means that a LSTP will cover more wide registration area. We can see that the proposed scheme is more efficient than IS-41scheme.

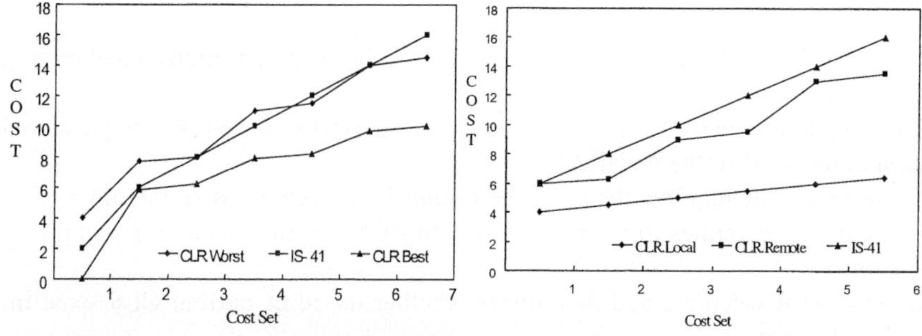

Fig. 2. Location registration cost   **Fig. 3.** Call tracking cost

In Fig. 3, we can see that proposed method has lower cost than IS-41 scheme, even though it is a worst case of proposed method provides mostly same cost as IS-41 scheme. The worst case takes places when ratios of six cases are same. In other word, it occurs when LSTP connected with very few RA. But we know that a LSTP's coverage is more than that of three RAs generally. The worst case of proposed scheme seldom occurs in actual networks. We know that the next generation wireless system will adopt smaller RA. It means that a LSTP will cover more wide registration area.

4 Conclusions

In this paper we proposed CLR method, which is effective for smaller cell and more frequent terminal moving pattern. Each VLR has a given fixed circle registration area around itself and has IDs of other VLRs which belong to the circle in proposed method. VLR only computes whether the terminal is located in the current CLR area or not by comparing the old VLR id with its ids. Then it sends recent location information of the terminal to the old or new CLR according to computing results

References

1. A. Bar-Noy and I. Kessler, "Tracking Mobile Users in Wireless Networks," Proc. of INFOCOM'98, 1998.
2. Y.B, Lin, "Determining the User Locations for Personal Communications Networks," IEEE Trans. Veh. Tech., 1994.
3. Y.B. Lin, "A Caching Strategy to Reduce Network Impacts of PCS," IEEEE J. SAC. Vol. 12, no.8, pp.1434 – 1444, Oct. 1994.
4. R Jain and Y.B.Lin, "An Auxiliary User Location Strategy Employing Forwarding Pointers to Reduce Network Impacts of PCS", ACM-Baltzer Journal of Wireless Network, Jul. 1995.
5. R. Jain, Y. B. Lin and S. Mohan, " A Caching strategy to Reduce Network Impacts of PCS," IEEE Journal in Comm., Vol. 12, No. 8, Oct., 1994.
6. S.J.PARK, Dong Chun Lee and J.S Song, "Locality Based Location Tracking using Virtually Hierarchical Link in Personal Communication Services," IEICE Trans. Com., Vol. Z81-B, No. 9,1998.

An Energy Efficient Broadcasting for Mobile Devices Using a Cache Scheme*

Kook-Hee Han[1], Jai-Hoon Kim[1], Young-Bae Ko[2], and Won-Sik Yoon[2]

[1]Graduate School of Information and Communication,
[2]College of Information Technology,
Ajou University
{justuniq, jaikim, youngko, wsyoon}@ajou.ac.kr

Abstract. Broadcasting mechanisms have been widely used to transfer information to a large number of clients. Most of the broadcast schemes try to minimize the average "access time". In this paper, we present a broadcasting mechanism which uses a cache to reduce not only access time but also energy consumption. There is a trade-off between energy saving by accessing data in cache and energy consumption by receiving broadcast data to update cache. Therefore, we determine the optimal size of cache to minimize energy consumption according to information access patterns and update characteristics.

1 Introduction

Transferring the information of common interest to mobile users is an important issue in personal wireless communications such as stock trading systems, weather information systems, and parking information systems. In these systems, broadcast mechanisms can be efficiently used, in which a broadcast server (e.g., satellite and base station) transfers the information of common interest to a large number of mobile users. An asymmetric environment is common in such systems. The downstream communication capacity (from server to clients) is much greater than the upstream capacity (from clients to server) [1]. Many schemes [1,2,3] have been proposed to broadcast information efficiently to a large number of users, and their main purpose is to minimize the average "access time" for the information needed. The access time is the time amount required for a client to wait for information that the client needs.

Mobile devices being used in personal wireless communications such as PDA, Palmtops, etc. are powered by small batteries without directly connecting to fixed power sources. Many hardware and software schemes are proposed and implemented to overcome such a power constraint. As one of the solutions, the index-based organization of data transmitted over wireless channels is proposed to reduce power

* This work was supported by grant no. R05-2003-000-10607-0 and R01-2003-00-0794-0 from Korea Science and Engineering Foundation, by ITA Professorship for Visiting Faculty Positions in Korea (International Joint Research Project) from Ministry of Information and Communication in Korea, by Korea Research Foundation Grant (KRF-2003-003-D00375), and by University IT Research Center Project.

M. Bubak et al. (Eds.): ICCS 2004, LNCS 3036, pp. 598–601, 2004.

consumption. Clients are interested in obtaining the individual data items from the broadcast[4,5]. If a directory index has information when a specific data item is transferred in the broadcast, then each client needs listening the channel in active mode selectively to obtain the required data, while in doze mode during the rest time to reduce energy consumption.

In this paper, we present a new index-based broadcasting mechanism using cache to reduce energy consumption. There is a trade-off between energy saving by accessing data in cache and energy consumption caused by cache memory itself and frequent cache update through broadcast. We propose an algorithm to decide the optimal size of cache to minimize energy consumption of mobile device in broadcast networks.

2 Index-Based Broadcast Using Cache Scheme

In general, a mobile node consumes much of its energy during data communication (e.g., broadcast). Table 1 shows the difference of energy dissipation between data access via wireless link and data access from cache memory [6,7]. In cache mechanisms, a mobile node holds data units which are most likely to be used in the future. It can certainly reduce energy consumption as well as access time.

Table 1. Per bit of energy cost

Operation	Energy Dissipate
Broadcast	50nJ/bit
Caching	1.65nJ/bit

When data in server is changed, data broadcast is needed to update cache data. In some situations, data update occurs very frequently, and it can cause increased energy consumption in mobile node to receive update data. Thus, there is a trade-off between energy saving by using cache instead of receiving data via wireless link and increased energy consumption caused by frequent broadcast for data update. We need to determine the optimal size of cache that minimizes energy consumption. We assume that popularity of data usage (locality of data access) follows zipfian distribution [8,9].

Equation (1) belong represents an expected power consumption ($P_{broadcast}$) to receive broadcast data not in the cache of mobile device. K_1 represents energy consumption to access one data unit not in the cache using broadcast, where c is the cache size.

In Equation (2), K_2 represents energy consumption to access one unit of data in cache. Remind that, as shown in Table 1, when the mobile node holds needed data in the cache, it consumes much energy. Of course, cache update requires additional energy consumption as shown in Equation (3). In Equations (2) and (3), T denotes an average update interval of cache. Now in Equation (4), P_{total} represents overall expected energy consumption when mobile device uses c units of cache. We can measure total energy consumption P_{total} as the sum of $P_{broadcast}$, P_{cache} and P_{update}.

$$P_{broadcast} = K_1 \times \sum_{x=c+1}^{n} \frac{1}{x^{\theta}} \tag{1}$$

$$P_{cache} = K_2 \times \sum_{x=1}^{c} \frac{1}{x^{\theta}} \tag{2}$$

$$P_{update} = \frac{K_1}{T} \times c \tag{3}$$

$$P_{total} = P_{broadcast} + P_{cache} + P_{update} = K_1 \left(\sum_{x=c+1}^{n} \frac{1}{x^{\theta}} + \frac{c}{T} \right) + K_2 \sum_{x=1}^{c} \frac{1}{x^{\theta}} \tag{4}$$

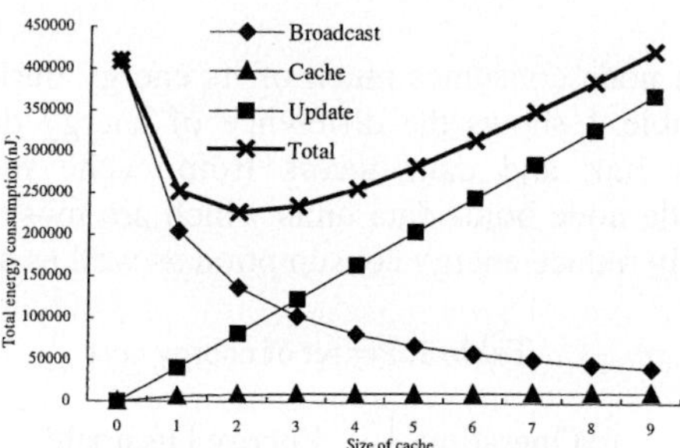

Fig. 1. Energy consumption in mobile node (when, T=10, θ=2)

Consequently, we find the optimal size of cache that minimizes total energy consumption for the broadcast scheme with cache.

Fig. 1 shows the overall energy consumption from broadcast, cache, and cache update, respectively, to access total 1024 bytes per period T. As shown in Fig. 1, mobile node consumes the least energy when 2 units of cache are used by mobile device in the system parameters ($T=10$, $\theta=2$).

The result can be changed by differentiating update rate and the distribution of data access popularity θ. Therefore, it is needed to reflect data update rate and data popularity to select the optimal size of cache. As shown in Fig. 2 and 3, the amount of energy consumption increases as the update rate ($1/T$) and data popularity (θ) decreases.

In Fig. 2, we observe energy consumption as the size of cache increases for different cache update rates. Cache update rate ($1/T$) represents how often cache update occurs on average. As shown in Fig. 2, energy consumption increases as the cache is frequently updated. Since each cache update needs additional energy consumption to receive broadcast data.

Fig. 3 shows energy consumption for different popularity (θ) of data access. As a result, energy consumption decreases as the data request popularity increases by accessing most of data in the cache.

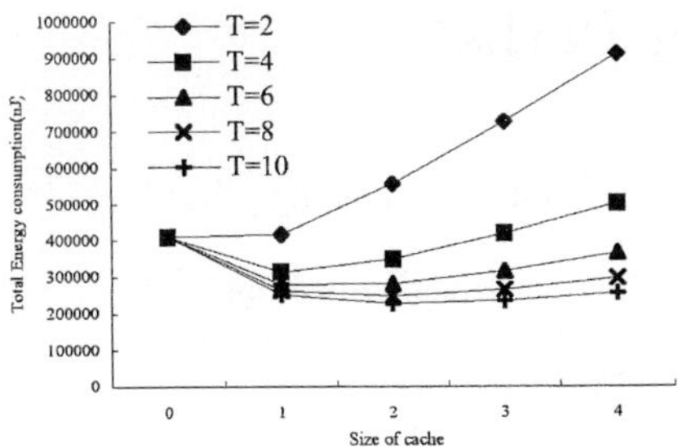

Fig. 2. Energy consumption in mobile node (when θ=2)

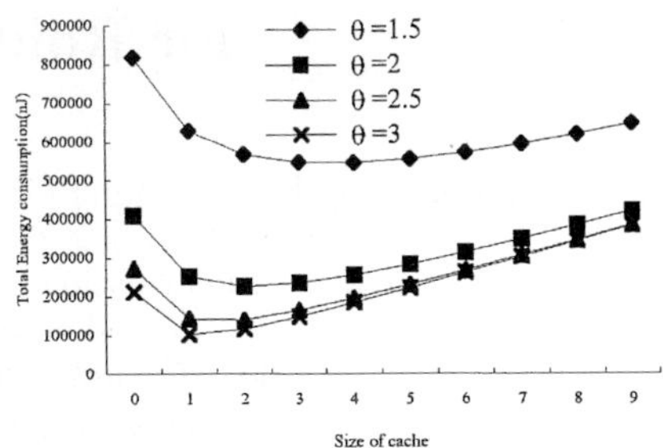

Fig. 3. Comparison of energy consumption (when T=10)

3 Conclusion

This paper presents energy saving broadcast using cache. We found that the optimal size of cache exists to minimize energy consumption. Concentration on specific data access, and data updating rate are important factor to decide the size of cache. We can decide the size of cache to minimize energy consumption for each system parameters.

References

1. S. Hameed and N. H. Vaidya, "Efficient Algorithms for Scheduling Data Broadcast," *ACM/Baltzer Wireless Networks (WINET)*, May 1999.
2. T. Imielinski, S. Viswanathan, and B. R. Badrinath, "Data on Air: Organization and Access," *IEEE Transactions on Knowledge and Data Engineering,* Vol. 9, No. 3, May/June 1997.
3. S. Acharya, M. Franklin, and S. Zdonik, "Dissemination-based data delivery using broadcast disks," *IEEE Personal Communication,* pp. 50-60, December 1995.
4. J. Dukes-Schlossberg, Y. Lee, N. Lehrer, "IIDS: Intellegent Information Dissemination Server," *Proc. of IEEE MILCOM '97*, Vol. 2, pp. 635-9.
5. R. Jain and J. Werth, "Airdisks and airRAID : Modeling and scheduling periodic wireless data broadcast (extended abstract)," *Tech. Rep. DIMACS Tech. Report 95-11*, Rutgers University, May 1995.
6. W. R. Heinzelman, A. Chandrakasan, H. Balakrishnan, "Energy-efficient communication protocol for wireless microsensor networks," *Hawaii International Conference on Systems Sciences*, 2000.
7. Mobile DRAM memory specification, http://www.sec.co.kr
8. John L. Casti, "Five More Golden Rules: Knots, Codes, Chaos, and Other Great Theories of 20th-Century Mathematics," *John Wiley & Sons Inc*, 2000.
9. Zipf Curves and Website Popularity, http://www.useit.com

On Balancing Delay and Cost
for Routing Paths[*]

Moonseong Kim[1], Young-Cheol Bang[2], and Hyunseung Choo[1]

[1] School of Information and Communication Engineering
Sungkyunkwan University
440-746, Suwon, Korea +82-31-290-7145
{moonseong,choo}@ece.skku.ac.kr
[2] Department of Computer Engineering
Korea Polytechnic University
429-793, Gyeonggi-Do, Korea +82-31-496-8292
ybang@kpu.ac.kr

Abstract. The distributed adaptive routing is the typical routing algorithm that is used in the current Internet. The path cost of the least delay (LD) path is relatively more expensive than that of the least cost (LC) path, and the path delay of the LC path is relatively higher than that of the LD path. In this paper, we propose an effective parameter that is the probabilistic combination of cost and delay. It significantly contributes to identify the low cost and low delay unicasting path, and improves the path cost with the acceptable delay.

1   Introduction

For distributed real-time applications, the path delay should be acceptable and also its cost should be as low as possible. We call it as the delay constrained least cost (DCLC) path problem [3,5]. It has been shown to be NP-hard [2]. As you see, the DCLC is desirable to find a path that considers the cost and the delay together. Even though there is a loss for the cost, two parameters should be carefully negotiated to reduce the delay. This is because the adjustment between the cost and the delay for the balance is important. Hence, we introduce the new parameter that takes in account both the cost and the delay at the same time.

The rest of paper is organized as follows. In section 2, we describe the network model, section 3 presents details of the new parameter. Then we analyze and evaluate the performance of the proposed parameter by simulation in section 4. Section 5 concludes this paper.

2   Network Model

We consider a computer network represented by a directed graph $G = (V, E)$, where V is a set of nodes and E is a set of links. Each link $(i, j) \in E$ is associ-

[*] This paper was supported in part by Brain Korea 21 and University ITRC project. Dr. H. Choo is the corresponding author.

ated with two parameters, namely available cost $c_{(i,j)}$ and delay $d_{(i,j)}$. Given a network G, we define a path as sequence of nodes u, i, j, $\ldots$, k, v, such that (u,i), (i,j), $\ldots$, and (k,v) belong to E. Let $P(u,v) = \{(u,i), (i,j), \ldots, (k,v)\}$ denote the path from node u to node v. If all elements of the path are distinct, then we say that it is a simple path. We define the length of the path $P(u,v)$, denoted by $n(P(u,v))$, as a number of links in $P(u,v)$. Let $\preceq$ be a binary relation on $P(u,v)$ defined by $(a,b) \preceq (c,d) \leftrightarrow n(P(u,b)) \leq n(P(u,d))$, $^\forall(a,b)$, $(c,d) \in P(u,v)$. $(P(u,v), \preceq)$ is a totally ordered set. For given a source node $s \in V$ and a destination node $d \in V$, $(2^{s \Rightarrow d}, \infty)$ is the set of all possible paths from s to d. $(2^{s \Rightarrow d}, \infty) = \{ P_k(s,d) \mid$ all possible paths from s to d, $^\forall s$, $d \in V$, $^\forall k \in \Lambda \}$, where Λ is a index set. Both cost and delay of an arbitrary path P_k are assumed to be a function from $(2^{s \Rightarrow d}, \infty)$ to a nonnegative real number. Since $(P_k, \preceq)$ is a totally ordered set, if there exists a bijective function f_k then P_k is isomorphic to $\mathcal{N}_{n(P_k)}$. $f_k : P_k \longrightarrow \mathcal{N}_{n(P_k)}$. We define a function of path cost $\phi_C(P_k) = \sum_{r=1}^{n(P_k)} c_{f_k^{-1}(r)}$ and a function of delay along the path $\phi_D(P_k) = \sum_{r=1}^{n(P_k)} d_{f_k^{-1}(r)}$, $^\forall P_k \in (2^{s \Rightarrow d}, \infty)$. $(2^{s \Rightarrow d}, supD)$ is the set of paths from s to d for which the end-to-end delay is bounded by $supD$. Therefore $(2^{s \Rightarrow d}, supD) \subseteq (2^{s \Rightarrow d}, \infty)$. The DCLC problem is to find the path that satisfies $min\{ \phi_C(P_k) \mid P_k \in (2^{s \Rightarrow d}, supD), ^\forall k \in \Lambda \}$.

3   Proposed Parameter for Low Cost and Low Delay

Since only link-delays are considered to compute P_{LD}, $\phi_C(P_{LD})$ is always greater than or equal to $\phi_C(P_{LC})$ [1]. If the cost of the path, $\phi_C(P_{LD})$, is decreased by $100(1 - \frac{\phi_C(P_{LC})}{\phi_C(P_{LD})})\%$, $\phi_C(P_{LD})$ is obviously equal to $\phi_C(P_{LC})$. Meanwhile, P_{LC} is computed by taking into account link-cost only. Because only link-costs are considered to compute P_{LC}, $\phi_D(P_{LC})$ is always greater than or equal to $\phi_D(P_{LD})$. If $\phi_D(P_{LC})$ is decreased by $100(1 - \frac{\phi_D(P_{LD})}{\phi_D(P_{LC})})\%$, then $\phi_D(P_{LC}) = \phi_D(P_{LD})$. The following steps explain a process for obtaining new parameter.

Steps to calculate the New Parameter

1. Compute two paths P_{LD} and P_{LC}
2. Compute $\bar{C} = \frac{\phi_C(P_{LD})}{n(P_{LD})}$ and $\bar{D} = \frac{\phi_D(P_{LC})}{n(P_{LC})}$
3. Compute $F^{-1}\left(\frac{3}{2} - \frac{\phi_C(P_{LC})}{\phi_C(P_{LD})}\right)$ and $F^{-1}\left(\frac{3}{2} - \frac{\phi_D(P_{LD})}{\phi_D(P_{LC})}\right)$ i.e., $z_{\alpha/2}^d$ and $z_{\alpha/2}^c$
4. Compute $post_{LD} = \bar{C} - z_{\alpha/2}^d \frac{S_{LD}}{\sqrt{n(P_{LD})}}$ and $post_{LC} = \bar{D} - z_{\alpha/2}^c \frac{S_{LC}}{\sqrt{n(P_{LC})}}$
   where S_{LD} and S_{LC} are the sample standard deviation
5. Compute $Cfct_{(i,j)}(c_{(i,j)}) = max\{ 1, 1 + (c_{(i,j)} - post_{LD}) \}$ and $Dfct_{(i,j)}(d_{(i,j)}) = max\{ 1, 1 + (d_{(i,j)} - post_{LC}) \}$
6. We obtain the new parameter $Cfct_{(i,j)}(c_{(i,j)}) \times Dfct_{(i,j)}(d_{(i,j)})$.

In order to obtain the percentile($z_{\alpha/2}^\beta$, $\beta = d, c$), we can use the cumulative distribution function (CDF). Ideally, the CDF is a discrete function but we assume that the CDF is a continuous function in convenience through out this paper. Let the CDF be $F(x) = \int_{-\infty}^{x} \frac{1}{\sqrt{2\pi}} e^{-\frac{y^2}{2}} \, dy$. Then, the percentile is a

604   M. Kim, Y.-C. Bang, and H. Choo

solution of the following equation. $F(z_{\alpha/2}^d) - \frac{1}{2} = 1 - \frac{\phi_C(P_{LC})}{\phi_C(P_{LD})}$ which means $z_{\alpha/2}^d = F^{-1}(\frac{3}{2} - \frac{\phi_C(P_{LC})}{\phi_C(P_{LD})})$ if $100(1 - \frac{\phi_C(P_{LC})}{\phi_C(P_{LD})})\% < 50\%$. Table 1 shows the percentile we have calculated.

Table 1. The percentile

$\eta = [\, 100\,(\, 1 - \frac{\phi_C(P_{LC})}{\phi_C(P_{LD})}\,)\,]\,\%$ where $[x]$ gives the integer closest to x.													
$z_{\alpha/2} = 3.29$ if $\eta \geq 50$ and $z_{\alpha/2} = 0.00$ if $\eta = 0$ ($z_{\alpha/2}$ is either $z_{\alpha/2}^d$ or $z_{\alpha/2}^c$)													
η	$z_{\alpha/2}$	η	$z_{\alpha/2}$	η	$z_{\alpha/2}$	η	$z_{\alpha/2}$	η	$z_{\alpha/2}$	η	$z_{\alpha/2}$	η	$z_{\alpha/2}$
49	2.33	48	2.05	47	1.88	46	1.75	45	1.65	44	1.56	43	1.48
42	1.41	41	1.34	40	1.28	39	1.23	38	1.18	37	1.13	36	1.08
35	1.04	34	0.99	33	0.95	32	0.92	31	0.88	30	0.84	29	0.81
28	0.77	27	0.74	26	0.71	25	0.67	24	0.64	23	0.61	22	0.58
21	0.55	20	0.52	19	0.50	18	0.47	17	0.44	16	0.41	15	0.39
14	0.36	13	0.33	12	0.31	11	0.28	10	0.25	9	0.23	8	0.20
7	0.18	6	0.15	5	0.13	4	0.10	3	0.08	2	0.05	1	0.03

Once the $Cfct_{(i,j)}(c_{(i,j)})$ and the $Dfct_{(i,j)}(d_{(i,j)})$ are found, we compute the value $Cfct_{(i,j)}(c_{(i,j)}) \times Dfct_{(i,j)}(d_{(i,j)})$ for each link of P. The best feasible selection is the link with the lowest cost per delay on initial P. In other words, the link with the highest 1/cost per delay could be selected. So then,
$$(\tfrac{1}{Cfct_{(i,j)}(c_{(i,j)})})/Dfct_{(i,j)}(d_{(i,j)}) = 1/(\, Cfct_{(i,j)}(c_{(i,j)}) \times Dfct_{(i,j)}(d_{(i,j)})\,)\,.$$
If the value of the formular is low, the performance should be poor. Therefore, we use Dijkstra's technique [1] with $Cfct_{(i,j)}(c_{(i,j)}) \times Dfct_{(i,j)}(d_{(i,j)})$.

4   Performance Evaluation

We compare our new parameter to only link-delays and only link-costs. Two performance measures - $\phi_C(P)$ and $\phi_D(P)$ - are combined our concern and investigated here. We now describe some numerical results with which we compare the performance for the new parameter. The proposed one is implemented in C++. We consider networks with number of nodes which is equal to 25, 50, 100, and 200. We generate 10 different networks for each size given above. The random networks used in our experiments are directed, symmetric, and connected, where each node in networks has the probability of links (P_e) equal to 0.3 [4]. Randomly selected source and destination nodes are picked uniformly. We simulate 1000 times $(10 \times 100 = 1000)$ for each n and P_e. Fig. 1 shows the average $\phi_C(P)$ and $\phi_D(P)$, where each path P is P_{LC}, P_{LD}, and P_{New}. As a result, the proposed new parameter ascertains that $\phi_C(P_{LC}) \leq \phi_C(\boldsymbol{P_{New}}) \leq \phi_C(P_{LD})$ and $\phi_D(P_{LD}) \leq \phi_D(\boldsymbol{P_{New}}) \leq \phi_D(P_{LC})$. For details on analyzing performance for the new parameter, refer to Fig. 1 (d). The path cost $\phi_C(P_{LC}) = 3.04$ is far superior, and $\phi_C(P_{LD}) = 13.51$ is the worst. Likewise the path delay $\phi_D(P_{LD}) = 3.03$

is far better, and $\phi_D(P_{LC}) = 13.53$ is the highest. Let us consider path P_{New} which is measured by the probabilistic combination of cost and delay at the same time. Because the $\phi_C(P_{New})$ occupies $\frac{5.92-3.04}{13.51-3.04} \times 100 = 27.5\%$ between $\phi_C(P_{LC})$ and $\phi_C(P_{LD})$, $\phi_C(P_{New})$ is somewhat expensive than $\phi_C(P_{LC})$ but becomes more superior than $\phi_C(P_{LD})$. In the same manner, the $\phi_D(P_{New})$ occupies $\frac{6.21-3.03}{13.53-3.03} \times 100 = 30.3\%$ between $\phi_D(P_{LD})$ and $\phi_D(P_{LC})$. In other words, the new parameter takes into account both cost and delay at the same time.

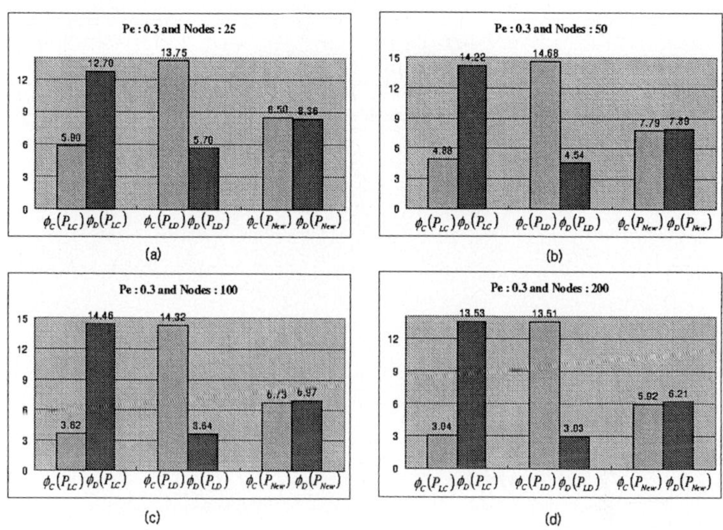

Fig. 1. Performance comparison for each P_e and n

5   Conclusion

In this paper, we have formulated the new parameter for DCLC path problem, which is known to be NP-hard [2]. Because the DCLC must consider together cost and delay at the same time, P_{LC} and P_{LD} are unsuitable to the DCLC problem. Hence the new parameter takes into consideration both cost and delay at the same time. We would like to extend the new parameter to the weighted parameter that can regulate as desirable $\phi_C(P)$ and $\phi_D(P)$.

References

1. D. Bertsekas and R. Gallager, Data Networks, 2nd ed. Englewood Cliffs, NJ: Prentice-Hall, 1992.
2. M. Garey and D. Johnson, Computers and intractability: A Guide to the Theory of NP-Completeness, New York: Freeman, 1979.

3. D.S. Reeves and H.F. Salama, "A distributed algorithm for delay-constrained unicast routing," IEEE/ACM Transac., vol. 8, pp. 239-250, April 2000.
4. A.S. Rodionov and H. Choo, "On Generating Random Network Structures: Trees," Springer-Verlag LNCS, vol. 2658, pp. 879-887, June 2003.
5. R. Widyono, "The Design and Evaluation of Routing Algorithms for Real-Time Channels," International Computer Science Institute, Univ. of California at Berkeley, Tech. Rep. ICSI TR-94-024, June 1994.

Performance of Optical Burst Switching in Time Division Multiplexed Wavelength-Routing Networks*

Tai-Won Um, YoungHwan Kwon, and Jun Kyun Choi

Information and Communications University,
P.O.Box 77, Yusong, Daejeon 305-348 Korea
{twum, yhkwon, jkchoi}@icu.ac.kr

Abstract. In this paper, we propose an optical burst switching architecture in time division multiplexed wavelength-routing networks, in which an edge OBS node requests time slots necessary to optical bursts to the time division multiplexed wavelength-routing network. Our scheme is attempt to improve the burst contention resolution and optical channel utilization.

1 Introduction

Researches on the optical internet set a goal of simplified and efficient transmission of IP traffic directly through the WDM layer. The conventional wavelength-routed optical network causes limited scalability and low channel utilization by assigning entire wavelength to a given session. This inefficiency can be reduced by adapting time-slot concepts to the wavelength-routed optical network, in which each individual wavelength is sliced in the time-domain into fixed-length time-slots. Multiple sessions are multiplexed on each wavelength by assigning a sub-set of the time slots to each session.

On the other hand, Optical Burst Switching (OBS) has been proposed as an efficient optical switching method to improve wavelength utilization, in which assembled packets called optical bursts follow a corresponding control packet after an offset time which is latency time required processing the control packets in intermediate OBS routers. Basically, this OBS architecture involves a critical collision problem, which occurs when burst packets contend for the same outgoing interface.

In this paper, we propose a new optical burst switching architecture over time division multiplexed wavelength-routing (TDM-WR) networks, in which an edge node requests time slots necessary to the burst, instead of a whole wavelength. Our scheme can improve the optical channel utilization by sharing the channel as well as it can provide a scalable optical network architecture with a guaranteed QoS. Following this introduction, the proposed network architecture and the control structure are described in section 2, and the results obtained from simulations will be discussed in section 3. Finally, we draw our conclusions in section 4.

---

* This work was supported in part by the Korea Science and Engineering Foundation (KOSEF) through the Ministry of Science and Technology (MOST) and Institute of Information Technology Assessment (IITA) through the Ministry of Information and Communication (MIC), Korea.

M. Bubak et al. (Eds.): ICCS 2004, LNCS 3036, pp. 607–610, 2004.

2 Network Architecture

On the basis of the OBS architecture over TDM-WR networks, which consists of TDM-WR intermediate nodes and OBS edge nodes [Fig. 1], we will investigate the network architecture and control structure in this section.

In the TDM-WR network described in [1], [2] and [3], every network node, which is made up of all-optical switching components, operates with a synchronized timing basis without O/E/O conversion. The wavelengths on their optical links are split into time-slotted wavelengths of a fixed-size interval divided by a time. One or more time-slotted wavelengths within a slotted wavelengths frame can be assigned for a request from an OBS edge node, and a time-slotted wavelength routed path to send the time-slot should be established between the edge OBS nodes by using the control node. On the time-slotted wavelength routed path, each time-slotted optical cross connect (OXC) has a role in switching each slotted wavelength to a destination node. To do this, every time-slotted OXC needs to maintain switching tables containing informa-tion entries to forward slotted wavelengths. Fig. 1 shows a time-slot reservation to dynamically control and assign time-slotted wavelength on demand of OBS control packet.

When a data packet arrives at an edge OBS node of the TDM-WR network, the edge node looks up the packet's destination node and pushes it to the corresponding queue to the destination. While the packets are aggregating in the queue, if the queue size reaches given threshold or timeout signal for delay-sensitive data, it has to send a control packet to the control node to request to assign time-slot for the burst. As de-scribed in [4], the control node estimates the traffic arrival time from the packet ac-cumulated when the control packet is sent and then establishes the fixed burst size by the time the acknowledgement arrives back at the sending edge router.

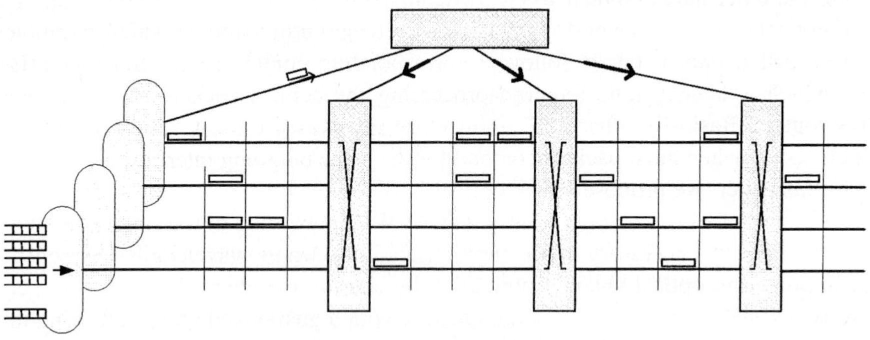

Fig. 1. Control packet delivery to request time-slot

When the control node receives the control packet, it decides whether it can accept this request or not. If, in the requested time-slot, all slotted wavelengths have been used for another request, then the control packet will be rejected. If there are available slots for that request, it will reserve the time slot and reply the edge node with an acknowledgement packet. When the edge OBS node receives the acknowledgement packet, it can send the optical burst to the outgoing time-slotted wavelength at the assigned time-slot.

The control plane of the TDM-WR OBS networks encompasses signaling, routing, scheduling, admission control and so on for each optical layer and also requires traffic engineering algorithms to efficiently utilize network resources and to maximize the number of time-slot assigned. To achieve performance objectives, plan of network capacity, selection of the explicit routes, wavelength assignment and time-slot arrangment should be considered.

3 Simulation Results

To analyze the proposed OBS scheme, we have developed an OBS simulator by extending NS2 simulator. The network topology and parameters for the simulation are given in Fig. 2. We assume that the average arrival rate is the same for all ingress nodes. Packets arrive at each ingress node according to a Poisson process with 1Gbps input bit rate. Packets are aggregated into a burst of 1.25Mbyte size at the ingress node and it is sent to the egress OBS node. The simulation results are obtained for the OBS and our scheme. The performance metrics are link utilization, edge queuing delay as a function of offered input traffic load.

Fig. 3 shows link utilization of the OBS and our OBS scheme using time-slotted wavelength assignment as a function of the offered traffic load per ingress node. Comparing the link utilization of the OBS with the proposed OBS shows that our scheme improves the utilization markedly. In the previous OBS, if an optical burst is collided with another burst, it should be dropped. However, in our scheme, by sending the control packet to the centralized control node, it can prove the available time-slots, if there is not any available time-slot, the edge node will try to reserve next slot again until it succeeds, instead of dropping optical burst. Fig. 4 shows the edge queueing delay versus the offered load. As described in the previous section 2, upon reaching the given threshold value, the edge node sends a control packet and an optical burst in the OBS. So, as the offered load increases, the burst aggregation time decreases and edge queueing delay decreases. However, in our scheme, if there is not available time-slot, the buffered data should wait until it reserves a time-slot, so it requires more edge delay then the previous OBS. In the previous OBS network, when a burst is blocked, the only way to recover lost packets is TCP retransmission. The TCP makes use of retransmission on timeouts and positive acknowledgments upon receipt of information. However, it could not provide fast recovery due to its host-to-host behavior and time-

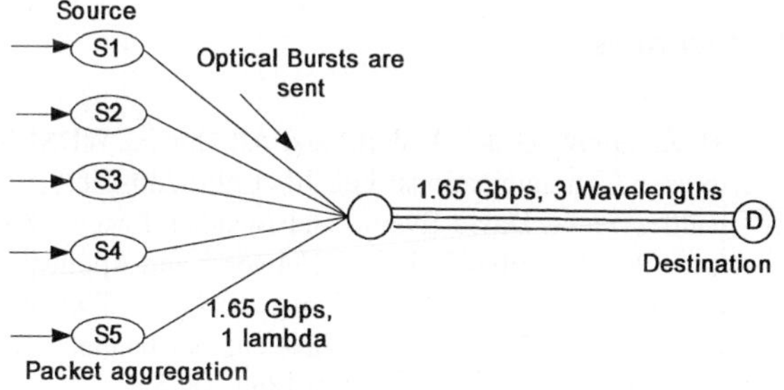

Fig. 2. Simulation topology

out mechanism. On the other hand, UDP is connectionless, which means that it can not provide error control and flow control.

Therefore, if we consider the TCP/UDP layer's retransmission of the lost packets in the previous OBS, the buffering at the OBS layer in our scheme may support better performance for upper layers. In this paper, we do not analyze the performance of the TCP layer, it will be remained for further studies.

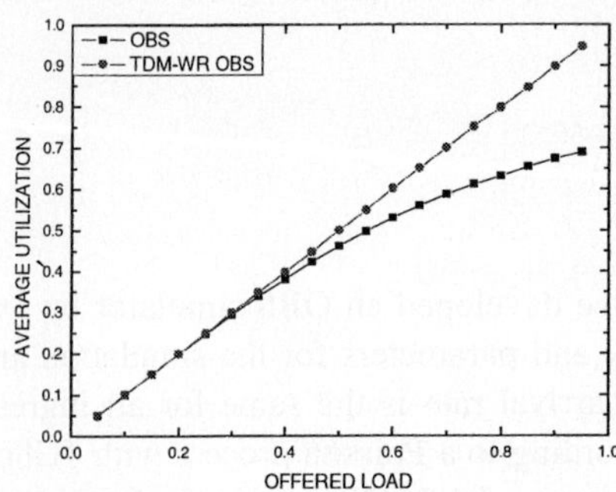

Fig. 3. Offered load vs. Link utilization

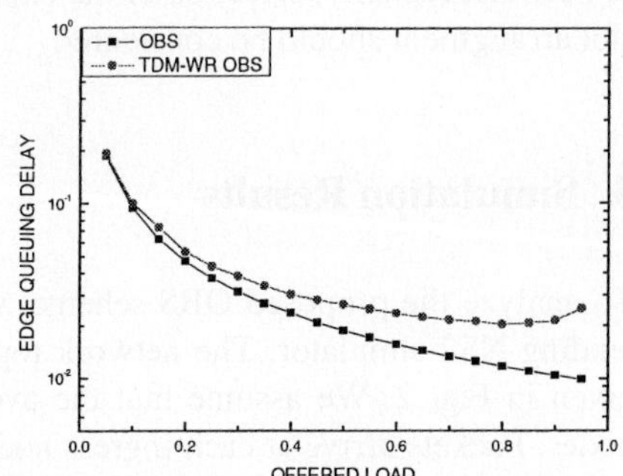

Fig. 4. Offered load vs. Edge queueing delay

4 Conclusion

This paper has introduced an optical burst switching over time division multiplexed wavelength-routing network architecture. Our scheme can improve the optical channel utilization by probing the reservation of time-slots as well as it can provide a scalable optical network architecture. Simulation results show that the link utilization of the OBS network is improved markedly at the expense of signaling and edge queueing delay.

References

1. I.P. Kaminow, et al.: A Wideband All-Optical WDM Network, IEEE Journal on Selected Areas in Communications. Vol. 14 (1996) 780–799
2. Jennifer Yates, David Everitt, and Jonathan Lacey: Blocking in Shared-Wavelength TDM Networks, Australian Telecom. Networks and Applications Conference (1995) 705–710
3. Nen-Fu Huang, Guan-Hsiung Liaw, and Chuan-Pwu Wang: A Novel All-Optical Transport Network with Time-Shared Wavelength Channels, IEEE Journal on Selected Areas in Communications, Vol. 18 (2000) 1863–1875
4. Michael Duser, Polina Bayvel: Anaysis of a Dynamically Wavelength-Routed Optical Burst Switched Network Architecture, Journal of lightwave technologies, Vol. 20, no 4 (2002) 573–585

On Algorithm for All-Pairs Most Reliable Quickest Paths*

Young-Cheol Bang[1], Inki Hong[1], and Hyunseung Choo[2,**]

[1] Department of Computer Engineering, Korea Polytechnic University
Kyunggi-Do, Korea

{ybang, isslhong}@kpu.ac.kr

[2] School of Information and Communications Engineering, Sungkyunkwan University
Suwon, Kyunggi-Do, Korea
choo@ece.skku.ac.kr

Abstract. The reliability problem of the quickest path deals with the transmission of a message of size σ from a source to a destination with both the minimum end-to-end delay and the reliability of the path over a network with bandwidth, delay, and probability of fault free on the links. For any value of message size σ, we present algorithm to compute all-pairs most-reliable quickest paths each with time complexity $O(n^2m)$, where n and m are the number of nodes and the number of arcs or links in the network, respectively.

1 Introduction

We consider point-to-point computer networks represented by a graph $G = (V, A)$ with n nodes and m arcs or links. Any node can be reached from any other node in this network, and two nodes are connected by at most single directed link in each direction. Each directed link $l = (i, j) \in A$ has a bandwidth $B(l) \geq 0$, delay $D(l) \geq 0$, and probability of fault free $0 \leq \pi(l) \leq 1$. A message of σ units can be sent along the link l in $T(l)$ $\sigma / B(l) + D(l)$ time with reliability $\pi(l)$ as in [4].

Consider a simple path P from i_0 to i_k given by $(i_0, i_1), (i_1, i_2), ..., (i_k - 1, i_k)$, where $(i_j, i_j + 1) \in A$, for $j = 0, 1, 2, ..., (k - 1)$, and all $i_0, i_1, ..., i_k$ are distinct. Subsequently, a simple path is referred to simply as a path. The delay of this path P, denoted by $D[P]$, is given by $\sum_{j=0}^{k-1} D(l_j)$, where $l_j = (i_j, i_j + 1)$. The bandwidth of this path is defined as

$B(P) = \min_{j=0}^{k-1} B(l_j)$. The reliability of P, denoted by $R(P)$, is $\prod_{j=0}^{k-1} \pi(i_j, i_{j-1})$. The end-to-end delay of the path P in transmitting a message of size σ is $T(P) = \sigma / B(P) + D[P]$ with reliability $R(P)$. Let $b_1 < b_2 < ... < b_r$ and G_b denote the distinct values of

---

* This paper was supported in part by Brain Korea 21 and University ITRC project.
** Dr. Choo is the corresponding author.

M. Bubak et al. (Eds.): ICCS 2004, LNCS 3036, pp. 611–614, 2004.

$B(l)$, $l \in E$ and the sub-network with all edges of G whose bandwidth is greater than or equal to b, respectively.

The path P from s to d is the most-reliable (MR) if $R(P)$ is the maximum among all paths from s to d. The path P is the quickest for message size σ if $T(P)$ is the minimum among all paths from s to d. The path P is the quickest most-reliable (QMR) if it is the quickest for σ among all MR paths from s to d. The P is the most-reliable quickest (MRQ) if it has highest reliability among all quickest paths from s to d for σ.

In this paper, we consider the all-pairs versions of computing MRQ paths with respect to any size of σ. The all-pairs version of the classical quickest path problem was solved in [2, 3] with time complexity of $O(n^2 m)$. By applying the algorithms of Xue [4] for each $s \in V$, we can compute MRQ paths between all pairs s and d with time complexity $O(nrm + rn^2 \log n)$; since $r \le m$, we have the complexity $O(nm^2 + n^2 m \log n)$. In this paper, we present $O(n^2 m)$ time algorithm to compute all-pairs MRQ paths, which match the best-known complexity for the all-pairs classical quickest path problem.

The rest of paper is organized as follows. In section 2, we present details of algorithm to compute the MRQ paths for all pairs of nodes in a given network. Section 3 summarizes our researches.

2 All-Pairs Most-Reliable Quickest Paths for Any Size of Message

To compute an MRQ path from s to d, we have to "account" for all quickest paths from s to d. Note that all-pairs quickest path algorithm (AQP) [3] returns a quickest path from s to d, which may not be a MRQ path, and hence a simple condition similar to line 8 of all-pairs QMR algorithm (AQMR) [1] does not work. In particular, it is not sufficient to check if an edge l is on a quickest path P_1 with bandwidth $B(l)$; in fact, l can be on a quickest path with any $b = B(P_1) \le B(l)$. In our algorithm, we compute the largest of such b and place l at an appropriate step in the computation, which is an iterative process similar to AQMR. Let $t[u, v]$ represent the end-to-end delay of quickest path from u to v for σ.

To compute MRQ paths, we first compute all-pairs quickest paths in G using AQP with the following enhancement. For each bandwidth value b_k and pair u, $v \in V$, we store a matrix $[d_{b_k}[u,v]]$ where $d_{b_k}[u,v]$ is the delay of the shortest path from u to v in G_{b_k}. These matrices can be computed for $b = b_r, b_{r-1}, ..., b_1$ during the execution of AQP. For each (σ_i, σ_{i+1}), $1 \le i \le r - 1$, we define

$$\Theta(\sigma_i, \sigma_{i+1}, u, v) = \{ b_k \mid B(P_k) \text{ such that } P_k \text{ is the quickest path for any } \sigma \in (\sigma_i, \sigma_{i+1}) \}$$

if such b_k exists, and $\varnothing$ otherwise. In this case, if $\sigma_i \ne \sigma_{i+1}$ then $\sigma \in (\sigma_i, \sigma_{i+1})$ represents $\sigma_i < \sigma < \sigma_{i+1}$, otherwise (σ_i, σ_{i+1}) equals to the intersection point $\sigma_{i,i+1}$.

Lemma 1. (i) $\Theta(\sigma_i, \sigma_{i+1}, u, v) \neq \varnothing$ if and only if there is a shortest path from u to v in G_b for some $b \in \Theta(\sigma_i, \sigma_{i+1}, u, v)$.

(ii) $\exists$ the quickest path from u to v for $b \in \Theta(\sigma_i, \sigma_{i+1}, u, v)$ if and only if $\exists$ b_k such that $b_k \in \Theta(\sigma_i, \sigma_{i+1}, u, v)$

Lemma 2. All $\Theta(\sigma_i, \sigma_{i+1}, u, v)$ can be computed with the time complexity of $O(n^2 m)$ for all $u, v \in E$.

In AMRQ, we organize the sets $\Theta(\sigma_i, \sigma_{i+1}, u, v)$'s as stacks with bandwidths decreasing top to bottom. Let $\Omega [u, v]$ denote the queue to store Θ s in order of which each Θ is computed. We use AQP [3] to compute $[d_b[u, v]]$ in line 1. In line 2-4, we compute all $\Theta(\sigma_i, \sigma_{i+1}, u, v)$'s for all pairs $u, v \in V$ with time complexity $O(n^2 m)$. There are $O(n^2 m)$ iterations in the rest of the algorithm, where links are considered in non-increasing order of bandwidth with which they participate in quickest paths (if at all). In each iteration, we consider the current link bandwidth $B(l)$, and pair $u, v \in V$. Lines 12-17 compute the maximum bandwidth with which the link l is used in a quickest path from u to v. The reliability of new path via l from u to v is then computed and the existing value and MRQ path are replaced appropriately in lines 18-20. Consider that as a result of while loop in lines 12-17, the retrieved bandwidth $b_{[u, v]}$ is strictly smaller than $B(l)$ if $b_{[u, v]}$ corresponds to link l_1, no more pop operations on $\Theta(\sigma_i, \sigma_{i+1}, u, v)$ will performed until all links with bandwidths in the range $[B(l_1),$ $B(l)]$ have been retrieved from the heap and processed. For each pair $u, v \in V$, this algorithm can be viewed in terms of alternating subsequences of top operations on arc_heap, dequeue operations on queue $\Omega [u, v]$, and pop operations on stack $\Theta(\sigma_i, \sigma_{i+1}, u, v)$ with no backtracking involved. In actual execution, however, all these subsequences corresponding to various $u - v$ pairs are intermingled among themselves as well as subsequences of top operations.

Algorithm AMRQ (G, D, B, π)
/* MRQP$(u, v, \sigma_i, \sigma_j)$ maintains MRQ path from u to v for (σ_i, σ_j) */
/* $P[u, v]$ maintains a currently selected MRQ path from u to v */

    1.   compute $[d_b[u, v]]$ using AQP (G, B, D)
    2.   **for** each pair $u, v \in V$ **do**
    3.   **for** each interval of σ **do**
    4.   compute stack $\Theta(u, v, \sigma_i, \sigma_j)$ and store to $\Omega [u, v]$ with $i \leq j$
    5.   **for** each pair $u, v \in V$ **do**
    6. $\Theta(u, v, \sigma_i, \sigma_j) = $ dequeue$(\Omega [u, v])$
    7. $b_{[u, v]} = $ pop$(\Theta(u, v, \sigma_i, \sigma_j))$;
    8.   arc_heap $= $ top_heavy heap of all edges of G according to the bandwidth

9.   **while** not arc_heap $\neq \varnothing$ **do**
10. $(i, j) = \text{top(arc\_heap)}$; let $l = (i, j)$;
11.   **for** each pair $u, v \in V$ **do**
12.   **while** $(B(l) < b_{[u, v]})$ **do**
13.   **if** $(\Theta(u, v, \sigma_i, \sigma_j) \neq \varnothing)$ **then**
14. $b_{[u, v]} = \text{pop}(\Theta(u, v, \sigma_i, \sigma_j))$;
15.   **else**
16. $\Theta(u, v, \sigma_i, \sigma_j) = \text{dequeue}(\Omega [u, v])$ if $\Omega [u, v] \neq \varnothing$
17. $b_{[u, v]} = \text{pop}(\Theta(u, v, \sigma_i, \sigma_j))$;
18.   **if** $(B(l) \geq b_{[u, v]})$ and
$(d_{b_{[u,v]}} [u, v] = d_{b_{[u,v]}} [u, i] + D(i, j) + d_{b_{[u,v]}} [j, v])$ **then**
19. $\Phi[u, v] \leftarrow min\{ \Phi[u, v] , \Phi[u, i] + \pi' (i, j) + \Phi[j, v] \}$;
/* update routing table for MRQ path from u to v for $[\sigma_i , \sigma_j]$ */
20. $\text{MRQ\_RT}(u, v, \sigma_i , \sigma_j) = P[u, i] + l(i, j) + P[j, v]$ if any

Theorem 1. The all-pairs most reliable quickest paths for any size of message can be computed by algorithm AMRQ with time complexity $O(n^2 m)$ and space complexity $O(n^2 m)$.

3 Conclusion

We presented algorithms to compute most-reliable quickest and quickest most-reliable paths between all pairs of nodes in a network. These algorithms match the best known computational complexity for the classical all-pairs quickest path problem, namely without the reliability considerations.

References

1. Y. C. Bang, H. Choo, and Y. Mun, Reliability Problem on All Pairs Quickest Paths, ICCS2003, LNCS 2660, pp. 518-523, 2003
2. G. H. Chen and Y. C. Hung, On the quickest path problem, *Information Processing Letters*, vol 46, pp. 125-128, 1993
3. D. T. Lee and E. Papadopoulou, The all-pairs quickest path problem, *Information Processing Letters*, vol. 45, pp. 261-267, 1993
4. G. Xue, End-to-end data paths: Quickest or most reliable?, *IEEE Communications Letters*, vol. 2, no. 6, pp. 156-158, 1998

Performance Evaluation of the Fast Consistency Algorithms in Large Decentralized Systems

Jesús Acosta-Elias[1] and Leandro Navarro-Moldes[2]

[1] Universidad Autónoma de San Luis Potosí, Av. Salvador Nava s/n, Zona Universitaria, San Luis Potosí, SLP 78000, México.
jacosta@fc.uaslp.mx
[2] Universitat Politecnica de Catalunya, J. Girona 1-3, C. Nord, Barcelona, Spain.
leandro@ac.upc.es

Abstract. Weak consistency algorithms allow us to propagate changes in a large, arbitrary changing storage network in a self-organizing way. These algorithms generate very little traffic overhead. In this paper we evaluate our own weak consistency algorithm, which is called the "Fast Consistency Algorithm", and whose main aim is optimizing the propagation of changes introducing a preference for nodes and zones of the network which have greatest demand. We conclude that considering application parameters such as demand in the event or change propagation mechanism to: 1) prioritize probabilistic interactions with neighbors with higher demand, and 2) including little changes on the logical topology, gives a surprising improvement in the speed of change propagation perceived by most users.

1   Introduction

A growing number of Internet applications need to run on a changing and unreliable network environment with a very large number of clients. Selective replication is one way to provide service to clients with low delay response, high degree of availability and autonomy (independent of unexpected backbone delays or link failures), and good scalability[3].

This paper presents a study, by means of simulation, of our "fast consistency" algorithm over several topologies and distributions of demand. Given that the worst case demand has a combination of high and low demand zones, the value of demand could be viewed as a landscape consisting of mountains and valleys of demand. For this purpose, we have developed a random demand generator with self-similar characteristics, in the form of mountains and valleys, using the diamond-square algorithm [1] from computer graphics. To evaluate the performance of the algorithm presented in this paper, a fast and weak consistency algorithm simulator has been constructed, over Network Simulator 2 [8].

The rest of the paper is organized as follows: Section 2 describes our system model. In section 3 we explain the methodology of simulation of our algorithms in terms of demand workload and performance metrics. In section 4 we discuss the simulation results for several cases. The paper concludes in section 5.

M. Bubak et al. (Eds.): ICCS 2004, LNCS 3036, pp. 615–618, 2004.
© Springer-Verlag Berlin Heidelberg 2004

2   System Model

The model of our distributed system consists of a number of N nodes that communicate via message passing. We assume a fully replicated system, i.e., all nodes must have exactly the same content. Every node is a server that gives services to a number of local clients. Clients make requests to a server, and every request is a "read" operation, a "write" operation, or both. When a client invokes a "write" operation in a server, this operation (change) must be propagated to all servers (replicas) in order to guarantee the consistency of the replicas. An update is a message that carries a "write" operation to the replica in other neighboring nodes. In this model, the demand of a server is measured as the number of service requests by their clients per time unit.

3   Simulation Methodology

To evaluate the performance of the fast consistency algorithm compared to Golding's algorithm[7], we simulate the behavior of the algorithms on a grid network with synthetic demand. In this section, we discuss the demand workloads that we use in our simulations and the performance metrics that we use as a basis for comparing the algorithms.

3.1   Demand Workload

In recent works of Yook et al. [9], and in [2] Anukool et al. demonstrated a similar fractal dimension (≈ 1.5) of routers, ASes, and population density. The demand is generated by the Internet users. If the geographic location of Internet users have fractal properties, we can infer that the demand have the same fractal properties. Other important characteristic is the existence of high demand regions and large regions of low demand [4].

3.2   Performance Metric

Every simulation calculates the pair (d_i, c_i) for all nodes, where d_i is the demand at node i, and c_i is the time when node i has received all changes. This pair can be expressed by the $c(n_i, t)$ function (an impulse function of value d_i):

$$c(n_i, t) = \begin{cases} d_i : t = c_i \\ 0 \end{cases} \qquad C(t) = \sum_{i=0}^{N} c(n_i, t) \qquad (1)$$

$C(t)$ is the sum of demand for all nodes that have reached a consistent state at a certain time t. In economic terms, we can define a utility function for each node $u(n_i, t)$. It represents the value of demand satisfied with up-to-date information at time t (a step function of value d_i).

$$u(n_i, t) = \begin{cases} d_i : t \geq c_i \\ 0 \end{cases} \qquad U(t) = \sum_{i=0}^{N} u(n_i, t) \qquad (2)$$

$U(t)$ is the sum of utility for all nodes that are consistent in time t. $U(t)$ expresses the satisfaction or benefit perceived by the community of users of our system. $U(t)$ roughly corresponds in economic terms with the Social Welfare function (SWF) defined in terms of global values as Benefit - Cost, given that the cost (total number of messages exchanged) does not change significantly. In time $t = 0$, all the nodes are in a non-consistent state, and as time passes more and more nodes will reach a consistent state and thus they will contribute to the SWF with their local demand d_i.

4   Simulation Results

In this section, we evaluate the performance of the various parts of the algorithm on a mesh topology using various demand workloads.

4.1   Mesh Topology with Fractal Demand

A fractal random demand is assigned to each node. This is done with the diamond-square algorithm in order to generate the demand that each node possesses. In other words, each node no longer possesses the same demand as the rest of the nodes on the network(Fig. 1). With this scenario, "fast consistency" (FC) shows a better performance than the weak consistency algorithms (WC). The FC algorithm in all nodes on the network reach a consistent state in a shorter period of time(Fig. 2.a). This occurs without any increase in use of resources for carrying out this task. Thus social welfare (SWF)(Fig. 2.b) grows much faster with FC.

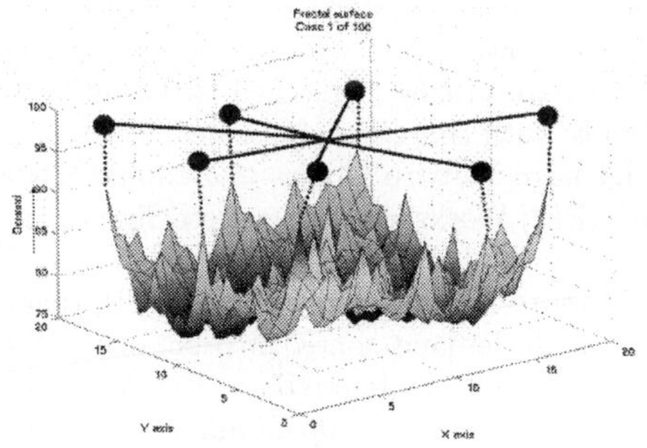

Fig. 1. Fractal demand of a grid. Z-axis corresponds to the demand. The hills are high demand zones. The black dots represent the nodes with high demand in logical star topology interconnection

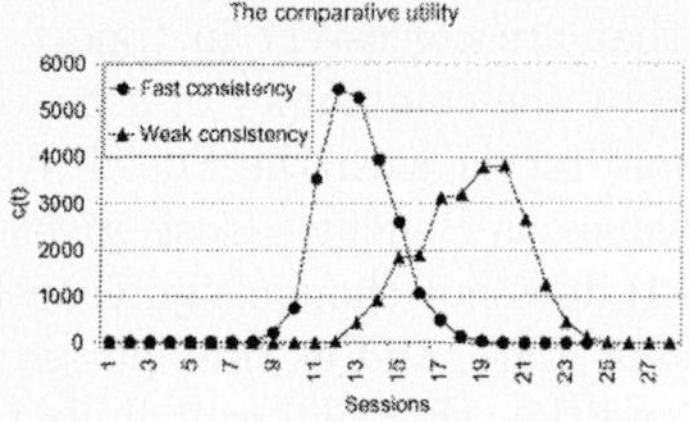

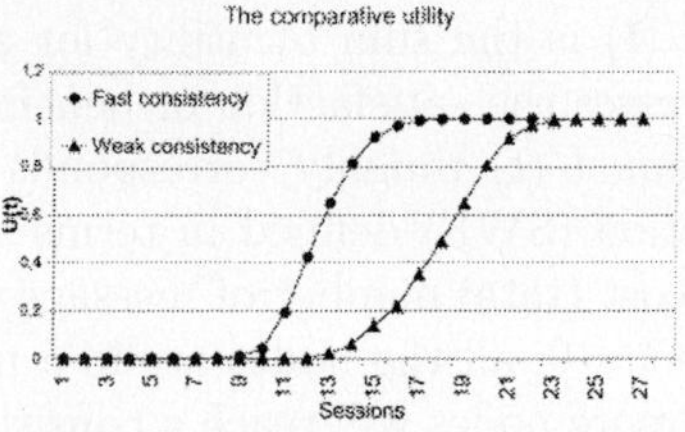

Fig. 2. In (a) We can observe that $C(t)$ for FC has a peak earlier than WC and in (b) the accumulated utility of FC grows faster, in less sessions (time), than WC

5   Conclusions

In this paper, we study the problem of propagating changes of replicated data on a Decentralized System in a system of any scale, with only little knowledge of a few neighbour nodes, using our "Fast consistency algorithm" and whose main aim is the propagation of changes with preference for nodes and zones of the network which have greatest demand. Employing, among other economic concepts, those such as utility and social welfare, we conclude that our "fast consistency" algorithm, optimizes the distribution of changes by prioritizing the nodes with greatest demand, independently of demand distribution. In other words, it satisfies the greatest demand in the shortest amount of time.

References

1. Alain Fournier, Don Fussell, and Loren Carpenter: Computer Rendering of Stochastic Models, Comm. of the ACM, Vol. 6, No. 6, June 1982, pages 371-384.
2. Anukool Lakhina, John Byers, Mark Crovella, Ibrahim Matta: On the Geographic Location of Internet Resources. Internet Measurement Workshop 2002 Marseille, France, Nov. 6-8, 2002
3. C.Neuman, "Scale in Distributed Systems. In Readings in Dist. Comp. Syst.", IEEE Computer Society Press, 1994
4. Jean Laherrere, D Sornette (1998): Stretched exponential distributions in Nature and Economy: 'Fat tails' with characteristic scales, Europ. Phys. Jour., B2:525-539.
5. Jesús Acosta Elias, Leandro Navarro Moldes. A Demand Based Algorithm for Rapid Updating of Replicas, IEEE Workshop on Resource Sharing in Massively Distributed Systems (RESH'02), July 2002.
6. Jesús Acosta Elias, Leandro Navarro Moldes: Generalization of the fast consistency algorithm to multiple high demand zones, in proc. of the Int. Conf. on Computational Science 2003 (ICCS2003). St.Petersburg, Russia, June. 2-4, 2003.
7. R. A. Golding, "Weak-Consistency Group Communication and Membership", PhD thesis, University of California, Santa Cruz, Computer and Information Sciences Technical Report UCSC-CRL-92-52, December 1992.
8. The Network Simulator: http://www.isi.edu/nsnam/ns/
9. Soon.-Hyung. Yook, H. Jeong, and A.-L. Barabási. Modeling the internet's large-scale topology. Tech. Report cond-mat/0107417, Cond. Matter Archive, xxx.lanl.gov, July 2001.

Building a Formal Framework for Mobile Ad Hoc Computing

Lu Yan and Jincheng Ni

Turku Centre for Computer Science (TUCS) and
Department of Computer Science, Åbo Akademi University,
FIN-20520 Turku, Finland.
{Lu.Yan, Jincheng.Ni}@abo.fi

Abstract. We present a formal framework towards a systematic design for MANET applications. In this paper, we define a layered architecture for mobile ad hoc computing and specify the system components with the B method and UML diagrams.

1   Introduction

We define a layered architecture in Fig. 1 for mobile ad hoc computing and propose a middleware layer with three key components between software application layer and ad hoc networking layer. We specify the system components with the B method [1], and model the interactions and message communications between components with UML diagrams.

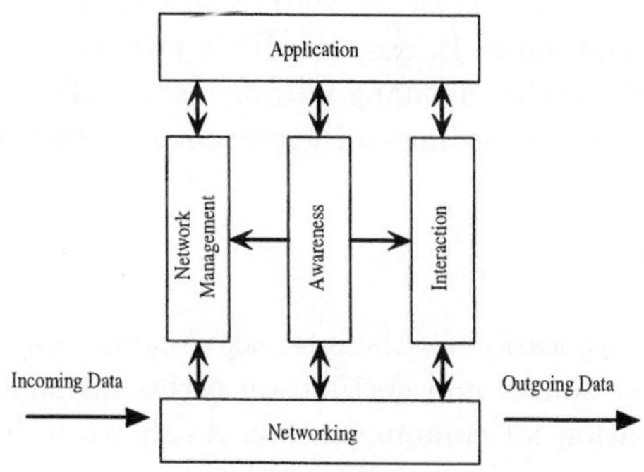

Fig. 1. MANET Architecture

2   Network Management

There is no constant topology or centralized manager in MANET. In order to form a self-organizing network, and support multi-hop routing by forwarding packets, it is necessary to have the network management in every node in MANET.

M. Bubak et al. (Eds.): ICCS 2004, LNCS 3036, pp. 619–622, 2004.

3   Awareness

As shown in Fig. 2, a node processes incoming messages according to the format of data packets. If the received message is a communication message, the system

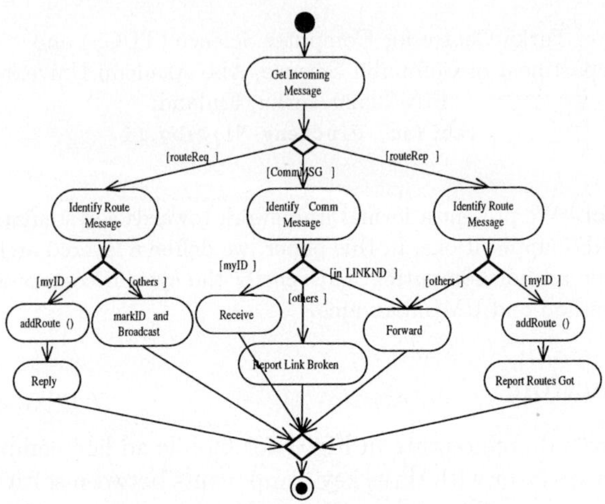

Fig. 2. Incoming Message Processing

checks the packet head, and then receives or forwards the packet according to the next hop ID of the route. In case the ID is unrecognizable, the system will report a broken route. If the incoming message is a routing message, the system will process the message according to the current routing protocols in MANET.

4   Interaction

We consider an opening session for the interactive communication between nodes. In such a session, the source and destination nodes exchange messages and update routing information for communication. As shown in Fig. 3, when the system opens such a session and starts interactive communication, the source node will select a route from the routing table or detect a new route to reach the destination node. If there is no available route or the destination node is not detected in the network, the opening session fails and a failure message is sent back to the source node. In a successful case, once a route is available, a communication session between the source node and destination node is created and the interactive communication starts.

During the interactive communication, the network topology might be changed and it might lead to a broken route. Thus route maintenance and recovery are needed for interactive communication. Figure 4 shows how a route is recovered when the system knows that the route is broken. In our design, it is

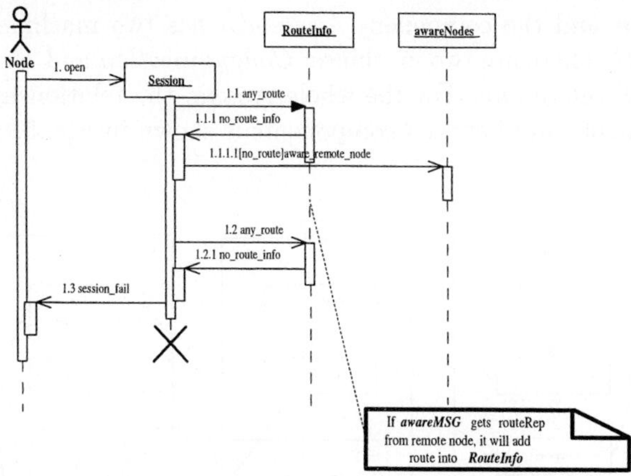

Fig. 3. Opening Session for Interactive Communication

assumed that multiple routes discovery protocols are used. For example, when source node S is communicating with destination node D, S sends data packets to D along with the selected route. During their communication, if S gets to know that the communication route is broken, S doesn't need to rediscover a new route immediately because S might have detected several routes in the previous discovery. It can then choose another available route and replace the broken one. Until all the routes are not reachable to the destination, the system will start route discovery again [2].

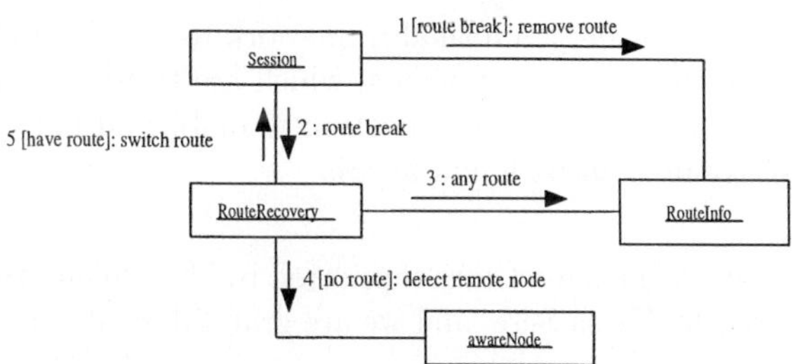

Fig. 4. Route Maintenance and Recovery

5   Relationship of Components

There are three components in the system specification, which are built up with nine B machines. Two pre-defined machines *AdHocNet* and *RouteInfo* are used to specify the context and environment of mobile ad hoc computing. The component *Network Management* is composed of three machines: *netManager*, *modeSet* and *Connector*, and the component *Awareness* has two machines: *awareNodes* and *awareMSG*. There are two machines: *Communication* and *RouteRecovery* in the component *Interaction*. For the whole system, the relationship of machines within components and between components is shown in Fig. 5.

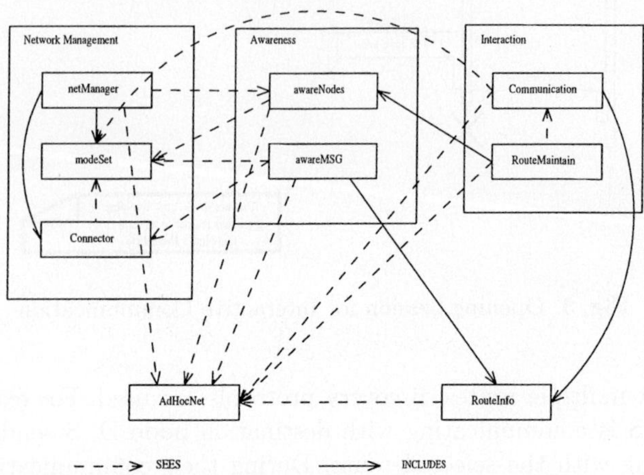

Fig. 5. Relationship of Components

6   Concluding Remarks

The goal of the specification is a formal framework to enable applications to be developed based on the three components, which are to be executed arbitrarily in MANET. A complete detailed specification of mobile ad hoc computing and some experiment results can be found at [3].

Acknowledgements. This work was supported by the *Mobile Ad Hoc Computing* project directed by Kaisa Sere, and we are grateful for Kaisa's contribution.

References

1. E. Sekerinski and K. Sere (Eds), *Program Development by Refinement: Case Studies Using the B Method*, Springer-Verlag, 1999.
2. Z. Ye, S. V. Krishnamurthy and S. K. Tripathi, *A Framework for Reliable Routing in Mobile Ad Hoc Networks*, Proceedings of the IEEE INFOCOM 2003, San Francisco, USA, 2003.
3. L. Yan, J. Ni and K. Sere, *Towards a Systematic Design for Ad hoc Network Applications*, Proceedings of the 15th Nordic Workshop on Programming Theory (NWPT'03), Turku, Finland, Oct. 2003.

Efficient Immunization Algorithm for Peer-to-Peer Networks*

Hao Chen, Hai Jin, Jianhua Sun, and Zongfen Han

Cluster and Grid Computing Lab
Huazhong University of Science and Technology, Wuhan, 430074, China
{haochen,hjin,jhsun,zfhan}@hust.edu.cn

Abstract. In this paper, we present a detail study about the immunization of viruses in Peer-to-Peer networks with power-law degree distributions. By comparing two different immunization strategies, we conclude that it is efficient to immunize the highly connected nodes in order to eradicate viruses from the network. Furthermore, we propose an efficient updating algorithm of global virus database according to the degree-based immunization strategy.

1 Introduction

Recently, a large proportion of research effort has been devoted to the study and modeling of a wide range of natural systems that can be regarded as networks, focusing on large scale statistical properties of networks other than single small networks. Some reviews on complex networks can be found [6]. From biology to social science to computer science, systems such as the Internet [5], the World-Wide-Web [2], social communities and biological networks can be represented as graphs, where nodes represent individuals and links represent interactions among them. Like these complex networks, one important characteristic of P2P networks is that they often show high degree of tolerance against random failures, while they are vulnerable under intentional attacks [3]. Such property has motivated us to carry out a study about the virus spreading phenomenon and some hacker's behaviors in P2P networks from a topological point of view. In our study, we choose Gnutella as our testbed. The main contributions of this paper are: first, an optimal immunization strategy is given; second, we propose an efficient information updating algorithm for P2P networks based on the immunization strategy.

The rest of this paper is organized as follows. Section 2 describes the immunization model of P2P networks. In Section 3, we propose an information updating algorithm for P2P networks. In Section 4, we give our conclusions and point some directions for future work.

2 Immunization Model of P2P Networks

2.1 Modeling Immunization of P2P Network

One widely used model of virus spreading is called SIS (*susceptible-infective-susceptible*) model [4]. This model assumes that the nodes in the network can be in

* This paper is supported by National Science Foundation of China under grant 60273076.

M. Bubak et al. (Eds.): ICCS 2004, LNCS 3036, pp. 623–626, 2004.

two states: *susceptible* (one node is healthy but could be infected by others), *infective* (one node has the virus, and can spread it to others). Each susceptible node is infected with rate ν if it is connected to one or more infected nodes. At the same time, an infected node is cured with rate δ, defining an effective spreading rate $\lambda = \nu/\delta$ for the virus.

A widely used theoretical model for power-law networks [3] is the Barabasi and Albert (BA) model [1]. In the following, we will use the BA model to deduce a theoretical framework of the prevalence of virus, and then compare with the real data obtained from Gnutella network [3].

In order to take into account the different connectivity of all the nodes, we denote the density of infected nodes with degree k by $\rho_k(t)$, where the parameter t indicates the time, and the average density of all infected nodes in the network by $\rho = \Sigma_k p(k)\rho_k$. According to the results in [7], we have the following equation:

$$\rho \simeq \frac{2e^{-1/m\lambda}}{1 - e^{-1/m\lambda}}. \tag{1}$$

The ρ is the stationary density of all infected nodes after time evolution of the stochastic cycle of SIS model.

2.2 Immunization Strategies of P2P Networks

The power-law networks exhibit different behaviors under random failures and intentional attacks [3], from which two intuitive immunization strategies are randomized and degree-based immunizations. In the randomized immunization strategy, a proportion of nodes randomly chosen in the network are immunized. Accordingly, in the degree-based strategy, nodes are chosen for immunization if their degrees are greater than a predefined value.

In the randomized case, for a fixed spreading rate λ, defining the fraction of immunized nodes in the network as f, we can get the effective spreading rate $\lambda(1-f)$, and substituting it into equation (1) we obtain

$$\rho_f = \frac{2e^{-1/m\lambda(1-f)}}{1 - e^{-1/m\lambda(1-f)}}. \tag{2}$$

Evidently, in the case of degree-based immunization, we can not use equation (2) to deduce an explicit formula as in the randomized case, but we will use simulations to compare the difference between the theoretical BA model and the real data of Gnutella network.

Our simulations are performed with a fixed spreading rate $\lambda = 0.15$, the smallest node degree $m = 3$ and the number of nodes $N = 34206$ the same as the real data of the topology collected from Gnutella network [3]. Initially we infect a proportion of healthy nodes in the network, and iterate the rules of SIS model. In Fig.1 (a), we plot the simulation results of degree-based immunization for BA network (line) and Gnutella network (square-line). With the increasing of f, ρ_f decays much faster in Gnutella network than in BA model, and the linear regression from the largest values of f yields the estimated thresholds $f_c \simeq 0.03$ in Gnutella network, $f_c \simeq 0.2$ in BA network. The value of f_c in Gnutella network indicates that the Gnutella network is

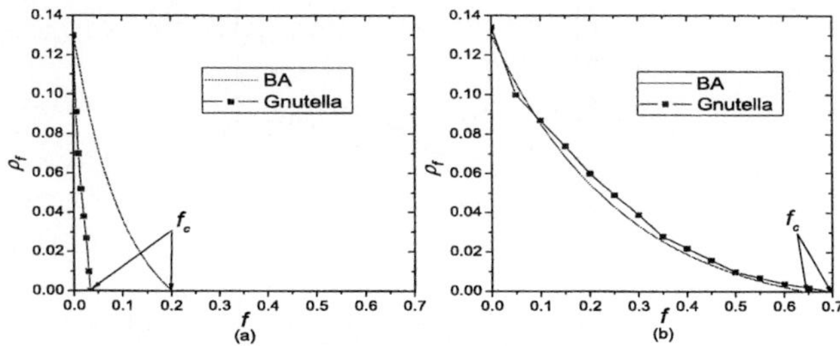

Fig. 1. Results for randomized and degree-based immunization measured by the density of infected nodes ρ_f as a function of the fraction of immunized nodes f.

very sensitive to the degree-based immunization, and the immunization of just a very small fraction (3%) of nodes will eradicate the spreading of virus. On the other hand, in Fig.1 (b), the simulation results of randomized immunization are plotted for Gnutella Network (square-line), which is in good agreement with the theoretical prediction (line) by equation (2), except for a larger value of $f_c \simeq 0.7$ compared with the value $f_c \simeq 0.64$ of BA network.

3 Efficient Immunization Algorithm for P2P Networks

Based on the analysis of immunization strategies, we use high degree nodes to transfer immunization information (when an intrusion or a virus is detected) to other nodes. First, we formulate the highest degree k_{max} in the network as a function of the network size. Given a specific degree distribution p_k, as stated in [6], we have $\frac{dp_k}{dk} \simeq -np_k^2$ [1]. For BA model, the probability distribution of degree is $p_k = 2m^2k^{-3}$. Substituting it into above equation, we have $k_{max} \simeq \sqrt{2m^2n/3}$. For simplicity, suppose that the degrees of the nodes in the transferring sequence, through which we update immunization information, are all approximate to k_{max}, then the number of steps needed to transfer the information in the network of size n is $s = n/k_{max} \simeq \sqrt{3n/2m^2}$.

We perform simulations of the real data of Gnutella network with a power-law exponent $\gamma = 2.0$ [3], and compare the simulation results with the theoretical prediction of BA network . The number of nodes range from $N = 10^3$ to $N = 10^4$. Fig.2 shows that the algorithm of transferring update information based on high degrees in Gnutella network is as efficient as the prediction of the theoretical BA model. We need only $s = 11$ steps to update all high degree nodes in Gnutella network with $N = 1000$ nodes, and $s = 36$ steps in Gnutella network with same number of nodes.

4 Conclusions

In this paper, based on the simple SIS model, we analyze the influence of virus spreading on P2P networks with two different immunization strategies, namely randomized and

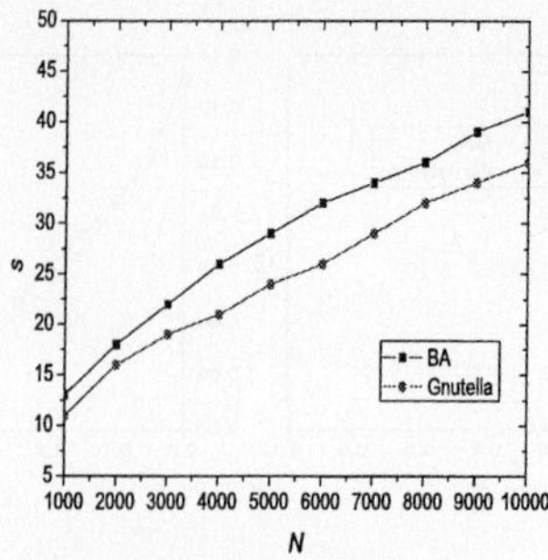

Fig. 2. The number of steps s needed to transfer information through high degree nodes as a function of the network size N.

degree-based immunization, and performe theoretical modeling and real data simulations. The results show that the degree-based strategy is more efficient than the randomized strategy, which also motivate us to design an effective immunization information transferring algorithm.

References

1. A. L. Barabasi and R. Albert, "Emergence of scaling in random networks", *Science*, Vol.286, pp.509, 1999.
2. A. Broder, R. Kumar, F. Maghoul, P. Raghavan, and R. Stata, "Graph structure in the web", *Computer Networks*, Vol.33, pp.309-320, 2000.
3. H. Chen, H. Jin, and J. H. Sun, "Analysis of Large-Scale Topological Properties for Peer-to-Peer Networks", *Proceedings of International Symposium on Cluster Computing and the Grid*, 2004.
4. O. Diekmann and J. A. P. Heesterbeek, *Mathematical epidemiology of infectious diseases: model building, analysis and interpretation*, JohnWiley & Sons, New York, 2000.
5. M. Faloutsos, P. Faloutos, and C. Faloutsos, "On Power-law Relationships of the Internet Topology", *Computer Communications Review*, Vol.29, pp.251-262, 1999.
6. M. E. J. Newman, "The structure and function of complex networks", *SIAM Review*, Vol.45, pp.167-256, 2003.
7. R. P. Satorras and A. Vespignani, "Epidemic Spreading in Scale-Free Networks", *Phys. Rev. Lett*, Vol.86, pp.3200-3203, 2001.

A Secure Process-Service Model

Shuiguang Deng, Zhaohui Wu, Zhen Yu, and Lican Huang

College of Computer Science, Zhejiang University,
Hangzhou 310027, PRC
{dengsg, wzh, yz,lchuang}@zju.edu.cn

Abstract. Encapsulating processes into process-services is a hot topic nowadays. Time management is an important issue for service providers to ensure the successful execution of process-services, and time information is also concerned by process-service consumers. Due to the security and secrecy factors in businesses, service providers are not willing to publish all information in process-services out. Thus process-services present as black boxes with only interfaces to consumers. As a result it is hard for consumers to engage in time management. We propose a secure process-service model, in which a process-service is divided into a public part and a private part.

1 Introduction

E-services have been announced as the next wave of internet-based business application that will dramatically change the use of the Internet [1]. The emergence of technologies and standards supporting the development of web services has unleashed a wave of opportunities for enterprises to form alliance by encapsulating processes into services and composing different services [2]. We give those services, which focus on processes, a name "process-services". Due to the security and secrecy factors in businesses, service providers tend to hide the details of the process from service consumers. Thus process-services present as black boxes with only interfaces to consumers. But on the contrary, process-service consumers want to know that information in order to use process-services well. How to deal with the contradiction between process-service providers and consumers is the focus of this paper. In our opinion, a process-service can be divided into two parts: a process-service body and a process-service declaration. The former is private to providers and contains all the details of the process information including its structure and time constraints. The latter, published to consumers, is abstracted from the former and contains some necessary information about the process in the process-service for consumers.

2 A Process-Service Model

Process is the center focus of a process-service, which achieves a special target through accomplishing serials of activities, between which there are structure and time constraints. In this section, we first introduce some basic elements and time

M. Bubak et al. (Eds.): ICCS 2004, LNCS 3036, pp. 627–630, 2004.
© Springer-Verlag Berlin Heidelberg 2004

constraints in process-service, and then present a process-service model with two parts: process-service body and process-service declaration.

2.1 Basic Elements and Time Constraints in Process-Service

Definition 1 (Activity). An activity is defined as a 2-tuple *<id, duration>* where *id* is the identity of the activity, *duration* is the execution time of the activity.

Definition 2 (Dependency). A dependency is defined as a 2-tuple, *<prev, succ>* which means activity *succ* must be executed after activity *prev*.

Definition 3 (Lower Time Constraint). A lower time constraint *LConstraint* is defined as a 5-tuple, *<src, P1, des, P2, limitation>*, where *src* and *des* are activities, *P1* and *P2* are from the set *{b, e}*, *b* represents the beginning time of the activity and *e* represents the end time of the activity, *limitation* represents a period of time.

A *LConstraint* means that the distance between the beginning time (or end time) of the activity *src* and the beginning time (or end time) of the activity *des* is greater than *limitation* time units.

Definition 4 (Upper Time Constraint). A upper time constraint *UConstraint* is defined as a 5-tuple, *<src, P1, des, P2, limitation>* which means that the *distance* between the beginning time (or end time) of the activity *src* and the beginning time (or end time) of the activity *des* is smaller than *limitation* time units.

2.2 Process-Service Body and Process-Service Declaration

In order to make process-service not only satisfy the security and secrecy requirements, and also provide enough structure and time information for consumers, we deem a process-service should have two parts. One is a process-service body, and the other is a process-service declaration. The former, containing the detail information about the realization of the process-service, is private to the service provider; and the latter, abstracted from the former according to a time equivalence principle, is public to service consumers.

Definition 5 (Process-Service Body or PSB). A process-service body is defined as a 5-tuple, *<ActSet, DepSet, In, Out, Constraints>*, where *ActSet* is a set of activities, *DepSet* is a set of dependencies, *In/Out* is a set of identities of the input/output activities which are all called interface activities. *Constraints* is a set of time constraints.

Definition 6 (Activity Time Assignment). For a *PSB* and *time:Identity × {b,e} → R* , if the following conditions are satisfied, the map time is called an activity time assignment of the process-service body.

i. $\forall activity \in Service.ActSet$, $time(activity.id.b) + activity.duration = time(activity.id,e)$

ii. $\forall dependency \in Service.DepSet$,
        $time(dependency.prev.id,e) \leq time(dependency.succ.id,b)$

iii. $\forall LConstraint \in Service.Constraints$
        $time(LConstraint.des.id,LConstraint.P2)$
        $- time(LConstraint.src.id,LConstraint.P1) \geq LConstraint.distance$

iv. $\forall UConstraint \in Service.Constraints$
        $time(UConstraint.des.id,UConstraint.P2)$
        $- time(UConstraint.src.id,UConstraint.P1) \leq UConstraint.distance$

Definition 7 (Interface Time Equivalent Principle) If two PSBs service1 and service2 fulfill the following conditions, they are interface time equivalent.

i. $Service1.In = Service2.In$; $Service1.Out = Service2.Out$
ii.   For any *activity time assignment* of *Service1*, *time1*, there exists an *activity time assignment* of *Service2*, *time2*, and they fulfill the conditions:
        $(\forall id \in Service1.In)(time1(id,b) = time2(id,b))$, $(\forall id \in Service1.Out)(time1(id,e) = time2(id,e))$
iii.   For any *activity time assignment* of *Service2*, *time2*, there exists an *activity time assignment* of *Service1*, *time1*. They fulfill the conditions:
        $(\forall id \in Service2.In)(time1(id,b) = time2(id,b))$, $(\forall id \in Service2.Out)(time1(id,e) = time2(id,e))$

Definition 8 (Process-Service Declaration or PSD) For a PSB, its PSD is a simple PSB that is interface time equivalent with the original PSB. This is expressed as *PSD=Declare (PSB)*. PSD is abstracted from its relative PSB and is open to process-service consumers and have some simple but essential structure and time information.

3   Algorithms to Automatically Generate PSD from a PSB

This section illustrates an example to automatically generate a corresponding PSD from a PSB shown in the left of fig. 1 using the algorithm introduced in [3]. The algorithm is $O(n^3)$ time complexity.

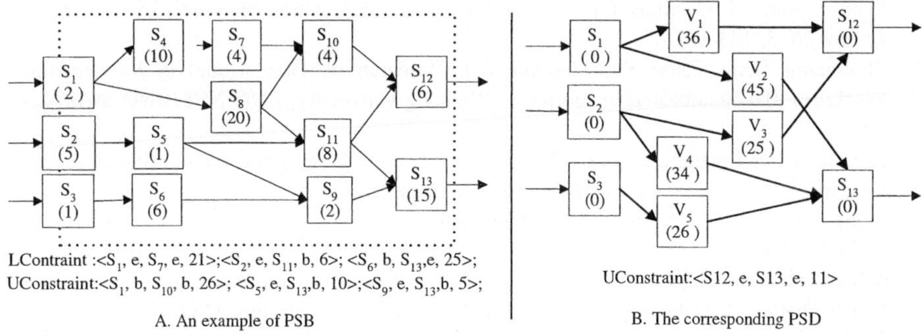

LContraint :<S_1, e, S_7, e, 21>;<S_2, e, S_{11}, b, 6>; <S_6, b, S_{13},e, 25>;
UConstraint:<S_1, b, S_{10}, b, 26>; <S_5, e, S_{13},b, 10>;<S_9, e, S_{13},b, 5>;

A. An example of PSB

UConstraint:<S12, e, S13, e, 11>

B. The corresponding PSD

Fig. 1. An example of process-service body

A comparison between the PSD and the corresponding PSB in the above fig. 1 shows that a PSD is deeply abstracted from its relative PSB and hides much detail information from consumers. If open the PSD to consumers instead of a black box with only interfaces, consumers know time constraints information between interface activities.

4 Conclusion

More and more enterprises pay attention to adopt web services to encapsulate processes. But web service model only emphasize on the publication of input/output interfaces and ignore processes in services. As a result, a process-service presents like a black box with only interfaces to consumers. It prevents consumers from learning more information about services. DAML-S [4] is another alternative model to describe process-service. But time information of processes in services is not included. Moreover it exposes the whole processes to public. However business enterprises are not willing to do that due to the consideration on security and secrecy factors. There are the same problems in WSFL [5], XLANG [6] and BPEL4WS [7]. We divide a process-service into two parts: a process-service body and a process-service declaration. This model not only satisfies the security and secrecy requirements from providers, but also provides enough information for consumers to engage in time management.

Acknowledgement. This work is supported by the National High Technology Development 863 Program of China under Grant No.2001AA414320 and No.2001AA113142; the Key Research Program of Zhejiang province under Grant No. 2003C21013.

References

1. Fabio Casati, M. C. Shan, et al. E-Service –Guest editorial. The VLDB Journal 10(1):1
2. Fabio Casati, M. C. Shan. Dynamic and adaptive composition of e-services. Information system 26, 3, 143-162.
3. Shuiguang Deng, Zhen Yu, Zhaohui Wu. Research of Time Model in Service-Based Workflow. To be appeared in Journal of Zhejiang University (ENGINEERING SCIENCE) 2004.
4. Ankolekar, M.Burstein, et al. DAML-S: Web Service Description for the Semantic Web. First International Semantic Web Conference, June, 2002
5. Frank Leymann. Web Services Flow Language.
    `http://www-4.ibm.com/software/solutions/ webservice/pdf/`
    `WSFL.pdf`, May 2001
6. Satish Thatte. XLANG: Web Services for Business Process Design. Microsoft Corporation 2001
7. BEA Systems, IBM, Microsoft, SAP AG and Siebel Systems, Business Process Execution Language for Web Services, May 2003

Multi-level Protection Building for Virus Protection Infrastructure

Si-Choon Noh[1], Dong Chun Lee[2], and Kuinam J. Kim[1]

[1] Dept. of Information Security, Kyonggi Univ., Korea
`nsc1@kt.co.kr`
[2] Dept. Computer Science, Howon Univ., Korea

Abstract. This paper proposes an improved multi-level virus protection infrastructure as a measure for correcting these weaknesses. Improved virus protection infrastructure filters unnecessary mail at the gateway stage to reduce the load on Server. As a numerical result, number of transmission accumulation decreases due to the reduction in the CPU load on the virus wall and increase in virus treatment rate.

1 Introduction

Computer virus is becoming increasingly sophisticated on a technical level. The base of the recommended virus protection strategy dictates that changed protection infrastructure is required to effectively address changed attack pattern [5]. Virus infection is infiltrated to local drive through floppy drive, email, Internet downloads, and various types of macro-enabled application. Moreover, HTTP based Web traffic, FTP based file transmission, and synchronized PDA data are received to this place [1]. Application of scanning in letters, stored in mail box, is difficult, and virus infection takes place in any case whether forwarding, opening, replying or using file [4]. Malignant code that infiltrated the target system through all types of routes begins to act as the same time when the user activates the operation. Proliferated worm virus continues to infect the inside while increasing the outbound traffic dramatically, which in turn increases session on the gateway level[2], [3].
The dualized protection method that is divided into server and PC, which does not have the function to block the virus that circulates in the network since it restricts the protection zone to the Server and PC. Characteristics of network traffic differs by the type of TCP/IP service, and single method based treatment by using one vaccine does not effectively block diverse infiltrations.

2 Multi-level Virus Protection Infrastructure

Multi-level virus infrastructure is applied on the network infrastructure, traffic route, protection zone, gateway area protection method, server protection method, anti-virus software configuration.

M. Bubak et al. (Eds.): ICCS 2004, LNCS 3036, pp. 631–634, 2004.

Protection Infrastructure Re-configuration: Gateway area and internal network virus wall area for the Web and email traffic filtering are added on to the structure, layered into three stages, from the exit point to the firewall, server and client areas to configure the defense layer consisting of five stages. The reason for SMPT gateway installation is that email protection gateway is re-set as the representative route for the influx of many emails in order to block virus before entering the intranet. Virus wall at the internal network is the function that blocks the internal circulation of the virus that is already infiltrated, and sets the new protection layer.

Traffic Route Re-setting: Network Traffic is divided into two channels by the type of channel, and the protection infra structure is made into stages accordingly. The traffic routes that are classified are internal intranet area and DMZ area. The internal intranet area ranges from the external contact point to the end user, and it is Exterior Router − > Web switch − > Firewall − > Web switch − > Interior Router − > Servers − > Client. DMZ area ranges from the external contact point to all types of server areas that are accommodated within the DMZ, and it is Exterior Router − > Web switch − > Firewall − > Web switch − > DMZ− > Servers. Re-set traffic route is normal traffic circulation route and it is also virus infiltration route.

Execution of Gateway Level Protection: There are mail gate, and web gate / internet proxy at gateway level. In order to protect data from virus infection, it is necessary to execute before virus reaches to the core information on the network, and it should be executed, targeting Web traffic and SMTP traffic. Virus filtering checks whether infected with virus or not at the packet unit level, and deletes if found whereas contents filtering is the function that blocks when a specific keyword is found in the email title and main body. Email filtering is the function that restricts the permitted size of email. And file filtering is the function that blocks by checking specifically attached file name or extension in advance. Spam filtering is the function that blocks the mail that is dispatched continuously [5]. SMTP Scanner executes scanning at a point where incoming, outgoing, email and attachment pass through email gateway. It protects file by mobilizing server based solution at the SMTP email server unit.

Execution of Virus Protection Against Internal Network Circulation: Vaccine for client blocks virus in PC, but it cannot inspect the mail's attachment file that is latent in server. When server's DB is infected, immensely adverse results face the overall network. Compared to many individual client protections that are placed widely, server unit protection is more effective and powerful. Blocking virus that circulates in the intranet requires separate installation of virus wall system at the foregoing part of the local file server or scanning with software embedded method on each server.

Virus Protection Zone Expansion: Shared folder is a form of database that is provided by the groupware exchange server to enable network user to share information and data. Groupware protection should enable real-time inspection of shared folder. Groupware (MS Exchange) provides NNTP to ensure relatively easy use of newsgroup, and real-time virus protection should be made possible for virus infection file when it comes to newsgroup.

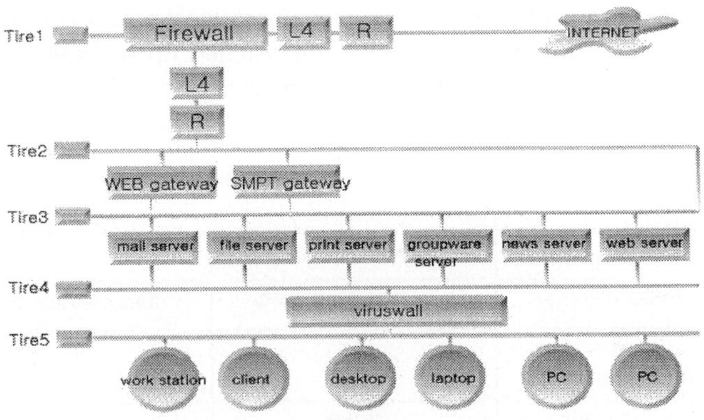

Fig. 1. Improved Protection Infrastructure

Fig.1 shows improved protection infrastructure. Infrastructure is transformed into five-stage blocking from two or three-stage blocking. And filtering and protection are executed for the entire server at the gateway level, and internet traffic route ensures differentiated protection by dividing into three, that is, SMTP, Web, and other traffic. Protection against virus that circulates in the network is conducted by collocating virus wall at the entry/exit point of local file server, and real-time protection network is configured for the client level, the infrastructure of gateway.

3   Performance Analysis

Gateway protection and virus blocking performance on the mail system are selected for measurement. Virus wall system for testing, used in this measurement, is E3500 system, and the software used is the virus wall of the Secureworks. Total throughput time increased to $2 \sim 3\%$ level after the installation of the Gateway virus wall. All types of powerful filtering functions filtered $30 \sim 40\%$ unnecessary mail among the total number of mails. When Sobig virus spills out 3,000 mails per hours, gateway Virus wall filters them first. Influx of malignant code is blocked by executing virus inspection by transmitted influx mail at the gateway first, and the mail, filtered once, is searched again by attachment file name, title, main body, and form of filtering to block abnormal mail from the network transmission process. Instead, it is stored for a specific period of time to enable re-dispatch of false positive mail that is normal mail classified as spam is enabled. Mail virus wall CPU load increased momentarily up to a maximum of 100% due to virus, which causes process processing delay and email transmission delay, but virus wall load stabilized at fewer than 60% on the improved structure due to virus blocking. When virus increased drastically, 10% of contaminated mail that could not be treated was transmitted to mail server, but this figure decreased to 3% after structure improvement. During the aggravation of virus wall load, the number of mails on standby for transmission reached up to 56,000 per day, but this number decreased considerably during the reduction of virus wall load.

Table 1. Comparison of Performance Analysis

Category	General structure	Improved Structure
Load of Virus wall CUP	Instant Maximum 10%	Instant Maximum 6%
Number of emails after Failure of cure of virus	10%	3%
Traffic/standby on the sending(daily)	56,000	500
Maximum number of Process on the system	3,000	10,000

4 Conclusion

This paper conducted research into improved effective model from the infrastructure configuration aspect. In order to ensure effective virus blocking, this paper emphasizes that a comprehensive approach through infrastructure improvement and combination of scanning tool is the only measure for preparing against today's environment of virus infiltration. The proposed method is a measure developed at a time when a permanent technological solution to virus is yet to be developed.

References

1. J. Hruska, "Computer Virus and Anti-virus Warfare" Ellis Horwood, 19962
2. P.Denning, "Computer under Attack Intruders, Worms, and Virus", Addison-Wesley, 1998
3. F.Cohen, "A short Course on Computer Viruses". ASP Press, 1990
4. F.Cohen." Computer Viruses. PhD thesis", University of Southern California, 1996.
5. Rainer Link, "Server-based Virus-protection on Unix/Linux", University of applied science Frut wangen, 2003.

Parallelization of the IDEA Algorithm

Vladimir Beletskyy and Dariusz Burak

Faculty of Computer Science & Information Systems, Technical University of Szczecin,
49 Żołnierska St, 71-210 Szczecin, Poland
{vbeletskyy, dburak}@wi.ps.pl

Abstract. In this paper, we present results of parallelizing the International Data Encryption Algorithm (IDEA). The data dependence analysis of loops was applied in order to parallelize this algorithm. The OpenMP standard is chosen for presenting the parallelism of the algorithm. The efficiency measurement for a parallel program is presented.

1 Introduction

Considering the fact that a relatively large part of the sequential C source code implementing the IDEA algorithm is filled in with "for" or "do-while" loops and the most of computation is comprised in these loops, there is an opportunity to parallelize this algorithm. A parallel IDEA algorithm permits us to reduce the time of running cryptographic tasks on multiprocessor computers. This problem is also connected with the current world tendency to hardware implementations of cryptographic algorithms (just because we also need parallel algorithms in this case).

The International Data Encryption Algorithm (IDEA), developed at Swiss Federal Institute of Technology in Zurich by James L. Massey and Xuejia Lai, published in 1990 (the algorithm was called IPES (Improved Proposed Encryption Standard) until 1991), and popularized by commercial versions of the PGP protocol, is used worldwide in various banking and industry applications.

The purpose of this paper is to present the IDEA algorithm parallelization.

2 Algorithm Parallelization

A C source code of the sequential IDEA algorithm in the ECB mode contains eight "for" or "do-while" loops (including no I/O function) [1].

We have used Petit to find dependences in source loops and the OpenMP standard to present parallelized loops. Developed at the University of Maryland under the Omega Project and freely available for both DOS and UNIX systems, Petit is a research tool for analyzing data dependences in sequential programs [2].

The OpenMP Application Program Interface (API) supports multi-platform shared memory parallel programming in C/C++ and Fortran on all architectures including Unix and Windows NT platforms. OpenMP is a collection of compiler directives,

M. Bubak et al. (Eds.): ICCS 2004, LNCS 3036, pp. 635–638, 2004.

library routines and environment variables that can be used to specify shared memory parallelism [3].

To build the valid parallel program, it is necessary to preserve all the dependences of the program [4].

The process of the IDEA algorithm parallelization can be divided into the following stages:

- carrying out the dependence analysis of a sequential source code in order to detect parallelizable loops,
- selecting parallelization and transformation methods,
- constructing sources of parallel loops in accordance with the OpenMP API requirements.

The most time-consuming are the idea_enc() and the idea_dec() functions presented below [1]:

2.1

```
void idea_enc (idea_ctx *c, unsigned char *data, int
blocks) {
int i;
unsigned char *d = data;
for (i=0; i<blocks; i++) {
        ideaCipher (d, d , c->ek);
        d += 8;
        }
}
```

2.2

```
void idea_dec (idea_ctx *c, unsigned char *data, int
blocks) {
int i;
unsigned char *d = data;
for (i=0; i<blocks; i++) {
        ideaCipher (d, d , c->dk);
        d += 8;
        }
}
```

Taking into account the strong similarity of these loops (there is the only difference between them — the first loop operates on variable "ek", the second does on "dk"; variables "ek" and "dk" are of the same type), we examine only the 2.1 "for" loop. However, this analysis is also valid in the case of the 2.2 "for" loop.

The parallelization process of the 2.1 loop consists of the five following steps:

- filling in the 2.1 "for" loop by the body of the function ideaCipher(d,d,c->ek) (otherwise, we cannot apply the data dependence analysis),
- conversion of the nested "do-while" loop [1] to an equivalent nested "for" loop,
- replacement of pointer operations with suitable array indexing for "in" and "out" variables,

- removal of the expression "d += 8;" located in the end of the original loop body and the insertion of the statements assigning values to the variables inbuf and outbuf, "inbuf = &d[8*i];" and "outbuf = &d[8*i];", respectively, in the beginning of the transformed loop body,
- appropriate variables privatization using OpenMP standard directives and clauses.

The skeleton of the parallel 2.1 "for" loop is the following:

```
#pragma omp parallel private
(i,ii,t16,t32,x1,x2,x3,x4,inbuf,outbuf,key,s2,s3,
in,out)
#pragma omp for
for (i=0;i<blocks;i++) {
        inbuf = &d[8*i];
        outbuf = &d[8*i];
        key = c->ek;
        in = (word16 *)inbuf;
        x1 = in[0];
        . . .
        for (ii=0;ii<8;ii++) {
        . . .
        }
        . . .
        out[0] = x1;
        . . .
}
```

The innermost "for" loop, included in the parallel 2.1 loop, is unparallelizable without applying advanced techniques of parallelization due to existing anti and output dependences and pointer operations.

The 2.2 "for" loop was parallelized in the same way as the 2.1 loop.

In the similar way, we have parallelized two "for" loops included in the ideaExpandKey() function and one loop included in the ideaInvertKey() function [1].

The remaining three sequentially iterated loops are unparallelizable. There are two reasons of this:
- occurrence of both data dependences and pointer operations in the loop body,
- occurrence of the instruction "return" in the loop body.

3   Experiments

In order to study the efficiency of the parallelization proposed, the Omni OpenMP Compiler has been used to run the IDEA sequential and parallel algorithms. The results received for a 15 megabytes input file using a PC computer with four processors Xeon, 2 GHz is shown in Table 1.

The total running time of the IDEA algorithm consists of the following operations: data receiving from an input file, data encryption and data decryption, and data writing to an output file (both encrypted and decrypted text).

The speed-up of the IDEA parallel algorithm depends considerably on the two factors: the parallelism degree of the idea_enc() and idea_dec() functions and choosing functions responsible for reading data from an input file and writing data to an output file.

The results confirm that the idea_enc() and idea_dec() functions are parallelizable with high speed-up (see Table 1).

The block method of reading data from an input file and writing data to an output file was used. The following C language functions and block sizes was applied: the fread() function and the 10-bytes block for data reading and the fwrite() function and the 512-bytes block for data writing.

The parallelization of the loops included in the ideaExpandKey() and the ideaInvertKey() functions has only a minimal influence on the speed-up value in the case of the software implementation but can be useful for hardware implementations of the parallel IDEA algorithm.

Table 1. Speed-ups of the sequential and the parallel IDEA algorithms

The number of processors	Total time of the IDEA sequential algorithm (sec)	Total time of the IDEA parallel algorithm 1 (sec)	Total time of the IDEA parallel algorithm 2 (sec)	**Total speed-up of** algorithm 1	Total speed up of algorithm 2	The idea_enc() time of the sequential algorithm (sec)	The idea_enc() time of the parallel 1 algorithm (sec)	**The des_enc() speed-up of** algorithm 1	The idea_dec() time of the sequential algorithm (sec)	The idea_dec() time of the parallel algorithm 1 (sec)	**The des_dec() speed-up** of algorithm 1
1	38,70	38,70	38,70	1	1	17,40	17,40	1	19,35	19,35	1
2	-	20,95	21,00	**1,85**	1,84	-	8,90	**1,96**	-	9,80	**1,97**
3	-	14,65	14,75	**2,64**	2,62	-	6,00	**2,90**	-	6,60	**2,93**
4	-	10,85	10,95	**3,57**	3,53	-	4,35	**4,00**	-	4,85	**3,99**

The idea parallel algorithm 1 contains parallel 2.1 and 2.2 "for" loops.
The idea parallel algorithm 2 contains five parallel "for" loops.

References

1. Bruce Schneier: Applied Cryptography: Protocols, Algorithms, and Source Code in C, Second Edition, John Wiley & Sons; 2 edition, 1995.
2. W. Kelly, V. Maslov, W. Pugh, E. Rosser, T. Shpeisman, D. Wonnacott: New User Interface for Petit and Other Extensions. User Guide. 1996.
3. OpenMP C and C++ Application Program Interface. Ver.2.0. 2002.
4. R. Allen, K. Kennedy: Optimizing compilers for modern architectures: A Dependence-based Approach, Morgan Kaufmann Publishers, Inc., 2001.

A New Authorization Model for Workflow Management System Using the RPI-RBAC Model *

SeungYong Lee[1], YongMin Kim[2], BongNam Noh[2], and HyungHyo Lee[3]**

[1] Dept. of Information Security, Chonnam National University, Gwangju, Korea 500-757
`birch@athena.chonnam.ac.kr`
[2] Dept. of Computer Science, Chonnam National University, Gwangju, Korea 500-757
`{ymkim, bongnam}@chonnam.ac.kr`
[3] Div. of Information and EC, Wonkwang University, Iksan, Korea 570–749
`hlee@wonkwang.ac.kr`

Abstract. The traditional Role Based Access Control (RBAC) model can be applied to WorkFlow Management System (WFMS) well, but there are some issues. Since the senior roles inherit all the permissions of the junior roles and all the permissions are accumulated for the top senior role, applying the traditional RBAC to WFMS does not meet the access control requirements: least privilege principle, Separation of Duty (SoD). To tackle these, we propose applying Restricted Permission Inheritance RBAC to WFMS authorization and evaluate the advantages and benefits of them in design time and runtime.

1 Introduction

WorkFlow Management System (WFMS) requires various policies involving access control since it defines the business processes and supports the enforcement of process control over processes. In this paper, when traditional Role Based Access Control (RBAC) is applied to WFMS to enforce an authorization [1], we investigate the problems that occur from inheritance of all permissions to senior roles. Then we evaluate the usefulness and benefits of employing the RPI-RBAC for the enforcement of access control polices in WFMS.

2 Related Work

YongHoon et al. introduced the concept of the sub-role to restrict the complete inheritance of junior permissions to senior roles [2]. In RBAC models, a senior role inherits the permissions of all its junior roles. One of the main security principles of RBAC models is to enable compliance with the least privilege principle. This unconditional permission inheritance can cause the violation of this principle. In order

---

* This paper was supported by University IT Research Center Project and partially by Wonkwang University in 2001
** Correspondent author

M. Bubak et al. (Eds.): ICCS 2004, LNCS 3036, pp. 639–643, 2004.

to address this drawback, they divide a role into a number of sub-roles based upon the characteristics of job functions and the degree of inheritance.

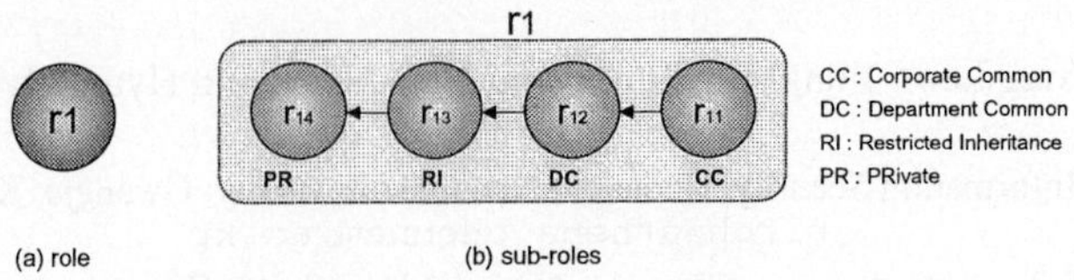

Fig. 1. Sub-role concept for corporate environment

Fig. 1. describes their proposed sub-role concept compared with the original role one in the traditional RBAC model.

3   Applying the RPI-RBAC Model to WFMS

In this section we mention a workflow scenario in a bank as an example the need for permission restrictions. We will use this scenario to illustrate how the RPI-RBAC model can be applied to this business process well.

Assume that a customer applies for a large loan from a bank. The deposit department uses the customer's credit information which it received from the credit review department to determine the amount of the money to loan to the customer, and a deposit clerk gives the money to the customer while a senior employee checks that the transfer was made successfully.

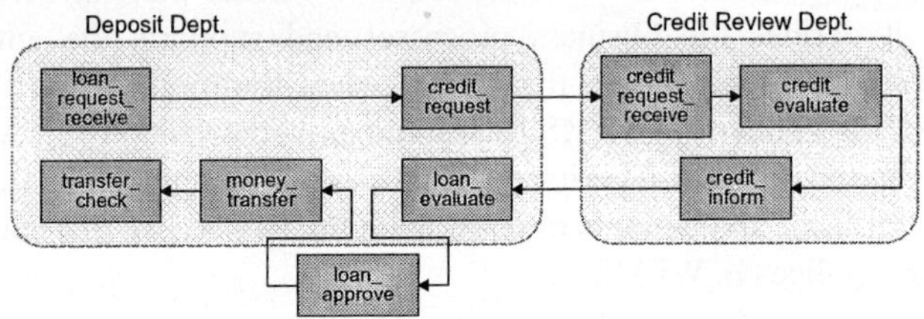

Fig. 2. Business process of loan in a bank

Fig. 2. shows the flow of tasks in the business process related to our scenario. There are two departments: the deposit and credit review departments in the bank scenario. Some tasks like *loan_request_receive, credit_request* and *transfer_money* are processed by the members of the deposit department whereas some tasks like *credit_request_receive* and *credit_evaluae* are executed by the members of the credit department. These business rules ensure that the Separation of Duty (SoD) requirement is met and that semantic integrity is maintained. In the rules, there are several constraints like who should do something, who can do something, who should not, etc.

In the case of employing traditional RBAC to WFMS, in which the senior roles inherit all the permissions of their junior roles, since the general manager has all the permissions of a manager, a supervisor and each departmental clerk, he or she has the permissions needed to execute all the tasks of all junior roles. This runs counter to the principle of least privilege and can cause many problems like abuse of rights and fraud. So, there is a need to restrict permission inheritance to satisfy the least privilege principle and prohibit an overuse of rights. Since the RPI-RBAC model limits the inheritance of permissions from a junior's rights, we can forbid the individual from committing frauds and improperly using his or her rights by applying RPI-RBAC to WFMS authorization.

We can consider the task's functionality, responsibility and interoperability to categorize them. These tasks can be categorized to four categories: *Private User Task*, *Department Task*, *Restricted Task* and *User Task*. We can assign each task to the proper sub-roles related to the jobs or tasks in design time. Because the permissions assigned to private user tasks should not be inherited by senior roles, if private user tasks are assigned to the private sub-role, we can prevent senior roles from executing the junior's own tasks. Fig. 3. shows the assignment of tasks to sub-roles.

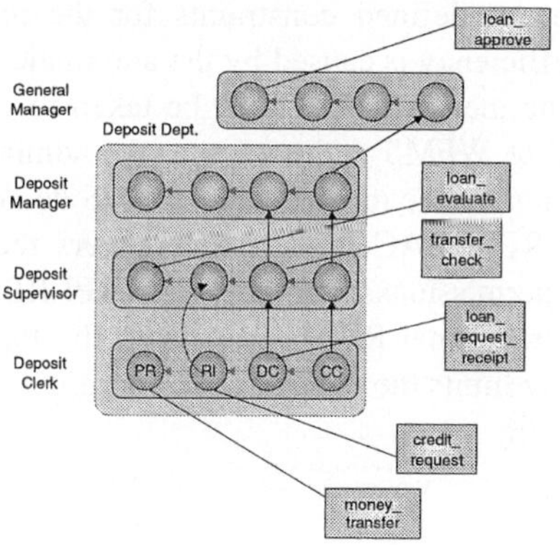

Fig. 3. Detailed assignment of tasks to sub-role

In this part, we examine the application of RPI-RBAC to WFMS authorization from the perspective of runtime and SoD. If it is possible that SoD will be violated by a task in a business process, WFMS prevents the task from being performed or voids the workflow instance by checking the associated constraints.

Assume that there is a business rules that no single individual should be allowed to process two tasks: *money_transfer* and *transfer_check*. If someone who executed a *money_transfer* can also execute a *transfer_check*, it is possible that he or she could commit fraud. This does not satisfy the requirement of SoD in the business processes.

In the case of applying traditional RBAC to WFMS authorization, the supervisor and other senior users have permission to execute both of the conflicting tasks. To satisfy the business rules described for the integrity, it is necessary to prevent a single user from executing both tasks. To cope with this requirement in traditional RBAC, some type of constraints must be defined and enforced during runtime. Bertino, Ferrari and Atluri (BFA) have introduced authorization constraints into WFMS authorization to deter fraud [3]. Sandhu et al. also presented a Transaction Control

Expression (TCE) for dynamic SoD [4]. In Table 1, we can use both BFA and TCE to resolve a SoD requirement. These authorization constraints require that any single user should not execute both *money_transfer* and *transfer_check* tasks.

Table 1. Constraints to tasks for SoD with BFA and TCE

Constraint by BFA
cannot_do$_u$(transfer_check, U$_i$) ← execute$_u$(U$_i$, money_transfer, k)
Constraint by TCE
money_transfer • deposit_clerk transfer_check • deposit_supervisor

From the perspective of WFMS authorization, applying the traditional RBAC model causes extra overhead on WFMS, since at runtime the WFMS must monitor the tasks and check the predefined constraints for the integrity during workflow execution time. This inefficiency is caused by the automatic permission inheritance in traditional RBAC and the measures that must be taken to circumvent it. If the RPI-RBAC model is applied to WFMS authorization, the administrator does not need to set or configure constraints for the integrity and WFMS do not need to check them for authorization since the RPI-RBAC model itself covers the authorization problems associated with these permissions. So, applying RPI-RBAC to WFMS is more efficient than applying traditional RBAC. However, the former is not as flexible as the later because it strictly limits the inheritance of permissions.

4 Conclusions

We have shown that RPI-RBAC satisfies the least privilege principle and separation of duty which are requirement needed for access control in WFMS, and that it works well when applied to WFMS authorization as an access control policy.

If traditional RBAC is applied to WFMS, some integrity constraints to WFMS must be defined and checked during workflow runtime to guarantee the integrity of business processes. But if RPI-RBAC is applied to WFMS, each role has the least amount of permissions and it does not have permission to other private ones. So, it can cut down on the possibility of rights abuse and fraud in advance and thus reduce the overhead of performing the constraints checks in runtime and satisfy SoD.

References

1. David F. Ferraiolo, D. Richard Kuhn and Ramaswamy Chandramouli, "Role-Based Access Control," Artech House Publishers, ISBN 1-58053-370-1

2. YongHoon Yi, Myongjae Kim, etc., "Applying RBAC Providing Restricted Permission Inheritance to a Corporate Web Environment," APWeb Conference, Lecture Notes in Computer Science(LNCS) 2642, Sep. 2003, pp. 287-292.
3. Elisa Bertino, Elena Ferrari and Vijay Atluri, "The Specification and Enforcement of Authorization Constraints in Workflow Management Systems," ACM Transactions on Information and System Security, Vol. 2, No. 1, Feb. 1999, pp. 65-104.
4. Savith Kandala, Ravi Sandhu, "Extending the BFA Workflow Authorization Model to Express Weighted Voting," Database Security XIII: Status and Prospects, Kluwer 2000.

Reducing the State Space of RC4 Stream Cipher*

Violeta Tomašević[1] and Slobodan Bojanić[2]

[1] Institute "Mihajlo Pupin", Volgina 15, 11050 Belgrade, Serbia and Montenegro
violeta@impcs.com, http://www.imp.bg.ac.yu
[2] Technical University of Madrid, Ciudad Universitaria s/n, 28040 Madrid, Spain
slobodan@die.upm.es

Abstract. The paper introduces an abstraction in form of general conditions for cryptanalytic managing of the information about the current state of the RC4 stream cipher. The general conditions based strategy is used to favor more promising values that should be assigned to unknown entries in the RC4 table. The estimated complexity of the cryptanalytic attack is lower than the best published result although the RC4 remains a quite secure cipher in practice.

1 Introduction

Based on the table-shuffling principle, the alleged RC4 stream cipher is designed for the fast software implementation and widely used in many commercial products and standards [1]. The RC4 cryptanalysis [2] has been mainly devoted to the statistical analysis of the output sequence [3], [4], or to the initialization weaknesses [5]-[7]. The most important results [8, 9] exploit the combinatorial nature of RC4. Although without a practical threat these attacks could be used in completing the internal state of the cipher, given some additional information [10]. They track the RC4 steps and assign the values to unknown table entries. The values are assigned one after one, thus some of them are favored without a reason. Since the number of the assignments until reaching the solution determines the complexity of attack, some improved strategy of selecting the values to be assigned could be very useful.

We propose the tree representation of the RC4 algorithm with the set of trees corresponding to each output symbol. The nodes and the branches encompass all information at given time. However, since the trees progressively increase, we could not practically exploit all available information. This problem is imminent for all other attacks, too. Therefore, we introduce an analytical abstraction, named *the general conditions* to represent all information from a subtree. We defined the examination strategy that favors the choice of the values by reordering the set of unassigned values. In addition to that, for each general condition, the probability that leads to the solution has been found, thus the most probable values are favored.

---

* This work has been partially supported by the Ministries of Science and Technology of Serbia (# IT.1.24.0041) and Spain (# TIC2003-09061-C03-02 and the "Ramon y Cajal" program).

M. Bubak et al. (Eds.): ICCS 2004, LNCS 3036, pp. 644–647, 2004.

2 General Conditions

The RC4 is a family of algorithms parameterized by a positive integer n (usually $n = 8$). The RC4 internal state at time t consists of a permutation table S_t of 2^n different n-bit values and of two n-bit pointers i_t and j_t [1]. The output n-bit symbol Z_t is:

$$Z_t = S_t(L_t), \qquad L_t = S_t(t) + S_t(j_t) . \tag{1}$$

The content of the S_t table can be given by:

$$S_t(t) = S_{t-1}(j_t) . \tag{2}$$

$$S_t(k{\neq}t) = S_{t-1}(k) , \; k \neq j_t \tag{3}$$

$$S_t(k{\neq}t) = S_{t-1}(t) , \; k = j_t. \tag{4}$$

Applying equations (2)-(4), there follows that equation (1) can be represented by the tree structure shown in Fig. 1.

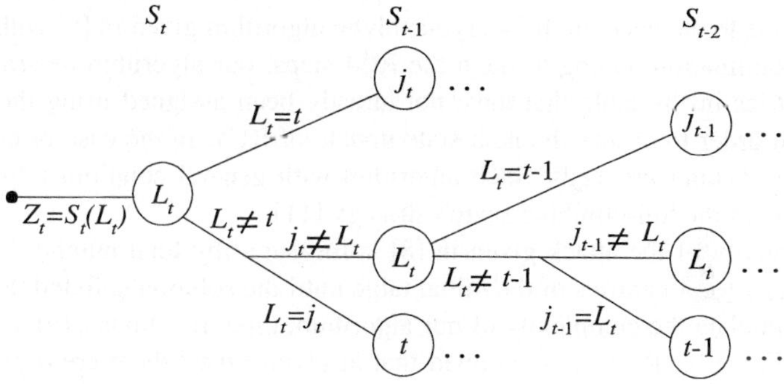

Fig. 1. Tree representation of the RC4 algorithm

Combining algorithm's equations, the whole sub-trees whose roots are given by the expressions like $S_{k-1}(k)$ and $S_{k-1}(j_k)$, can be described by more general conditions $C1...C5$ (Table 1) that can be checked instead of all conditions from the sub-tree. The $C3$ and $C4$ conditions both encompass t-1 conditions. The expression $e(Z_t)$ denotes the position of Z_t in the initial table. All calculations are modulo 2^n. By introducing the general conditions, we neglect some information, but significantly decrease the number of conditions that should be checked.

The probabilities that general conditions lead to the solution can be determined by multiplying the known probabilities of conditions on the path to the given node (Table 2). This fact inspired our further proposal intended to achieve an additional reduction of the search space by observing the probabilities of the general conditions. Consequently, these conditions should be checked in decreasing order of their probabilities while the existing approaches are based on arbitrary guessing.

Table 1. General conditions

$C1$	$C2$	$C3(p\in[2,t])$	$C4(p\in[2,t])$	$C5$
$j_t=t-Z_t+j_{t-1}$	$j_t=Z_t+j_{t-1}$	$Z_t=j_{t-p+1}-j_{t-p}$	$Z_t=L_{t-p+1}-j_{t-p+1}+j_{t-p}$	$L_t\neq t,..,1,j_t,..j_1$
$L_t=t$	$L_t=j_t\neq t$	$L_t=j_{t-p+1}\neq t,..,t-p+1,j_t,..j_{t-p+2}$	$L_t=t-p+1\neq j_t,..j_{t-p+2}$	$L_t=e(Z_t)$
$S_{t-1}(t)=t-Z_t$	$S_{t-1}(t)=Z_t$	$S_{t-p}(t-p+1)=Z_t$	$S_{t-p+1}(t-p+1)=Z_t$	$S_{t-1}(j_t)=e(Z_t)-j_t+j_{t-1}$
$S_{t-1}(j_t)=Z_t$	$S_{t-1}(j_t)=j_{t-1}$	$S_{t-1}(j_t)=j_{t-p+1}-j_t+j_{t-1}$ $S_{t-1}(t)=j_t-j_{t-1}$	$S_{t-1}(t)=j_t-j_{t-1}$ $S_{t-1}(j_t)=t-p+1-j_t+j_{t-1}$	$S_{t-1}(t)=j_t-j_{t-1}$

Table 2. Probabilities of the general conditions

$C1$	$C2$	$C3(p\in[2,t])$	$C4(p\in[2,t])$	$C5$
$1/2^n$	$(2^n-1)/2^{2n}$	$(2^n-1)^{p-1}(2^n-p)/2^{n\,(p+1)}$	$(2^n-1)^{p-1}/2^{np}$	$(2^n-1)^t(2^n-t)/2^{n\,(t+1)}$

3 Efficiency of the Attack

Our approach is to enhance the RC4 cryptanalytic algorithm given in [8] with general conditions examination. Going through the RC4 steps, our algorithm determines the values of entries in the table that have not already been assigned using the general conditions, in order to ensure the next state update of RC4. In the case of contradiction, the backtracking proceeds. Such algorithm with general conditions applies the basic principle of the hill-climbing search strategy [11].

The complexity of the attack given in [8] is measured by total number of the assignments made for all entries of the initial table until the solution is found. Similarly, in order to calculate the complexity of our algorithm, three functions $c_i(a)$, $i = 1, 2, 3$ are defined. For n-bit RC4, it is assumed that at given time t there are a previously assigned values in the table.

$$c_1(a)=(a/2^n)c_2(a)+(1-a/2^n)[(v(C1)+v(C2))c_1(a+1)+$$
$$+(v(C5)+\sum(v(C3(p))+v(C4(p)))))(2^n-a-(1-a/2^n))c_2(a+1)]$$
$$c_2(a)=(1-a/2^n)c_3(a)+(a/2^n)[(a/2^n)(1/2^n)c_1(a)+(1-a/2^n)(a/2^n+(1-a/2^n)c_1(a+1))]$$
$$c_3(a)=(1-a/2^n)[(1-a/2^n)(2^n-a-(1-a/2^n))+$$
$$+(a/2^n)(v(C1)+v(C5)(2^n-a-(1-a/2^n)))]((a+1)/2^n+(1-(a+1)/2^n)c_1(a+2))$$

The calculation of complexity of our attack starts with the known expressions $c_i(2^n)=0$. Then, going backwards, we calculate total number of the assignments given by $c_1(0)$. The results of the calculation for the RC4 versions with different word size n are presented in Table 3. Compared to the estimated complexity of the original algorithm, an improvement has been made definitely, yet not enough to practically menace the security of the alleged RC4 stream cipher.

Table 3. Complexity of the cryptanalytic attacks on n-bit RC4

Word size n	3	4	5	6	7	8
Knudsen et al. [8]	2^8	2^{21}	2^{53}	2^{132}	2^{324}	2^{779}
Our attack	2^5	2^{17}	2^{46}	2^{120}	2^{300}	2^{731}

4 Conclusions

We proposed a new technique to improve the cryptanalytic attack on the RC4 cipher. It is based on new information from the tree representation of the RC4 algorithm. To make a better choice for the assignment to unknown entries of the cipher's table, we represented analytically similar nodes and corresponding sub-trees by means of general conditions. Then, we defined a search strategy which uses the information derived from the general conditions, determined the probabilities that they lead to the solution, and incorporated it into the algorithm [8]. The complexity was estimated by an analytical calculation for different values of n. The results show that this complexity is lower than the best known result [8], although the RC4 remains a quite secure cipher for the practical applications. Our research is an additional argument which advocates for the security of the RC4 cipher.

References

1. Schneier, B.: Applied Cryptography. Willey, New York, (1996).
2. Tomašević, V., Bojanić, S., O. Nieto-Taladriz: On the Cryptanalysis of Alleged RC4 Stream Cipher. In C. Anias et al. (eds.): Telematics. Edit. Univ. F. Varela, Havana, (2002) 227-232.
3. Golić, J.: Linear Statistical Weakness of Alleged RC4 Keystream Generator. In: Advances in Cryptology - EUROCRYPT '97. LNCS, Vol. 1233, Springer-Verlag, (1997) 226–238.
4. Fluhrer, S., McGrew, D.: Statistical Analysis of the Alleged RC4 Keystream Generator. In: Fast Software Encryption - FSE 2000. LNCS, Vol. 1978, Springer-Verlag, (2000) 19–30.
5. Roos, A.: A Class of Weak Keys in the RC4 Stream Cipher. Sci.crypt. September 1995.
6. Grosul, A., Wallach, D.: A Related-Key Cryptanalysis of RC4. TR00-358, Rice University, October 2000.
7. Fluhrer, S., Mantin, I., Shamir, A.: Weakness in the Key Scheduling Algorithm of RC4, Selected Areas in Cryptography – SAC 2001. Vol. 2259, Springer-Verlag, (2001) 1–24.
8. Knudsen, L., Meier, W., Preneel, B., Rijmen, V., Verdoolaege, S.: Analysis Methods for (Alleged) RC4. In: ASIACRYPT '98. LNCS Vol. 1514, Springer-Verlag, (1998).
9. Mister, S., Tavares, S.: Cryptanalysis of RC4-like Ciphers, Selected Areas in Cryptography - SAC '98. Springer-Verlag, (1998) 136-148.
10. Mantin, I., Shamir, A.: A Practical Attack on Broadcast RC4. Fast Software Encryption FSE, 2001, LNCS Vol. 2355, Springer-Verlag (2002) 152-164.
11. Pearl, J.: Heuristics, Addison Wesley Publishing Company, 1984.

A Pair-Wise Key Agreement Scheme in Ad Hoc Networks

Woosuck Cha, Gicheol Wang, and Gihwan Cho

Department of Computer Science,
Chonbuk National University, Chonju, Korea
{wscha, gcwang, ghcho}@dcs.chonbuk.ac.kr

Abstract. Mobile Ad-hoc networks are exposed to various security threats because all traffics are carried in air and there is no central management authority. For the sake of secure communication in an Ad-hoc network, a scheme is inevitable to securely distribute security keys in the network. Based on the cluster structure and the verifiable secret sharing scheme[1], this paper proposes a pairwise key agreement scheme which is secure and induces a lower overhead. The proposed scheme is safe against a man-in-the-middle attack while not all private keys within the destination's cluster are exposed. In addition, our simulation result shows that the proposed scheme is more efficient and scalable than CABM(Clusterhead Authentication Based Method)[2].

1 Introduction

Ad-hoc networks are inherently vulnerable to security attacks because all transmissions are carried out using the opened air medium[2]. Therefore, a secure key agreement in Ad-hoc networks is more important than that of fixed networks. Since there is no TTP(Trusted Third Party), such as CA(Certification Authority) or KDC(Key Distribution Center) in Ad-hoc networks[3], all nodes are responsible of cooperative key distribution and management duty.

In this paper, a scheme is proposed to make a pair-wise key agreement between two Ad-hoc nodes securely with a lower overhead. The proposed scheme is based on the Diffie-Hellman protocol. Then each node hides its Diffie-Hellman value by encrypting with its counterpart's public key. As a result, only the counterpart can decrypt the Diffie-Hellman value. Also, the proposed scheme hides the source and destination of the key agreement so that an adversary cannot launch a man-in-the-middle attack easily while he cannot get all private keys within the destination's cluster.

2 Related Work

In literature [2], a key agreement scheme based on clueterhead authentication is proposed. In this scheme, two clusterheads representing their clusters respectively perform a mutual authentication using the counterpart's public key. Therefore, each clusterhead must distribute its public key to all other clusterheads. After the authenti-

M. Bubak et al. (Eds.): ICCS 2004, LNCS 3036, pp. 648–651, 2004.

cation between two clusterheads, a pair-wise key is exchanged between them, and eventually the key is distributed to the communicating members. Because this scheme forces each clusterhead to distribute its own public key to all other clusterheads, it causes high communication overhead. Also, since a communicating node must know the cluster to which its counterpart node belongs, it also requires additional communication overhead. Moreover, since the distribution of a pair-wise key is made through clusterheads, clusterheads can impersonate communicating nodes.

3 Pair-Wise Key Agreement Scheme Based on the Cluster Structure and Secret Sharing

To begin with, it is assumed that each node exchanges its own public key and clusterhead information with other nodes, by attaching them to the routing message. In addition, each node attaches a clue ($g^{f(id_i)}$) of its secret share into the routing message to prove its ownership. The security issues and solutions for routing messages have been addressed in a lot of literatures including [3]. However, these issues are beyond the scope of this paper.

The proposed scheme is based on the following assumptions.

- Dealer generates a polynomial $f(x) = a_{k-1}x^{k-1} + a_{k-2}x^{k-2} + ... + a_0 (\mathrm{mod}\ p)$.

- Each node knows a generator g of Z_p^*.

- Each node receives $g^{a_0}, g^{a_1}, ..., g^{a_{k-1}}$ that are witness of the coefficients of the sharing polynomial from the dealer.

- Each node receives its secret share($S_i = f(id_i)(\mathrm{mod}\ p)$) through an out-of-band method before entering Ad-hoc network. Each node employs its secret share to prove the ownership of its public key.

- Each node generates its public/private key pair and stores them after entering Ad-hoc network.

The proposed scheme for a pair-wise key agreement is as follows. For the sake of convenience, let's assume that node A wants to make a key agreement with node B belonging to other cluster.

1 Node A generates its random number ($R_A \in Z_p^*$) and computes a Diffie-Hellman value(g^{R_A}) for the key agreement. It encrypts only the Diffie-Hellman value, the source, and destination field in the key agreement packet with the public key of node B and adds an encapsulation header to the encrypted packet. The source of the encapsulation header is the clusterhead of node A and the destination of the encapsulation header is the clusterhead of node B. Node A sends the capsulated packet to its clusterhead.

2 The clusterhead of node A forwards the capsulated packet to the clusterhead of node B according to the routing path.

3 The intermediate nodes between the clusterheads of node A and B forward this capsulated packet along the established routing path.

4   The clusterhead of node B strips off the encapsulation header and broadcasts the encrypted inner packet.

5   Each member node decrypts the encrypted packet with its private key. Since the packet was encrypted with the public key of node B, only node B can decrypt the packet and achieve the Diffie-Hellman value of node A(g^{R_A}).

6   Node B repeats from step 1 to step 4 after setting node B to the source and node A to the destination.

4   Security and Performance Analyses

Our scheme is based on the Diffie-Hellman key exchange protocol. The Diffie-Hellman protocol is inherently vulnerable to a man-in-the-middle attack. However, since the proposed scheme forces a node to encrypt its Diffie-Hellman value with the counterpart's public key, it is very difficult for intermediate nodes to accomplish man-in-the-middle attacks without the corresponding private key. Furthermore, the proposed scheme hides the entities of key agreement by encrypting the source and destination field of the key agreement packet. Only the clusterheads of the source and destination of the key agreement are known through the encapsulation header. As a result, an adversary should know all private keys within the destination's cluster to launch a man-in-the-middle attack successfully.

Let's assume that n is the number of nodes, there are c compromised nodes holding k private keys of other nodes, and l is the number of cluster members belonging to a destination's cluster. If c compromised nodes exchange their key sets with one another, the probability p_l that all private keys of cluster members belonging to the destination's cluster are exposed to the c nodes is as follow.

$$p_l = \left(1 - \left(\frac{{}_{n-1-l}C_k}{{}_{n-1}C_k} \right)^{c} \right)^{l} \tag{1}$$

To evaluate the efficiency of the proposed scheme, an event-driven simulator for Ad-hoc networks was developed. The simulation has been done with 25 and 45 nodes on a plane 500m×500m during 600 seconds. Initially, 25 nodes were assumed to spread randomly on the plane, and they were assumed to move to any direction at any time randomly. The period of key agreement was set to 3 seconds and the entities of key agreement were selected randomly. Next, our simulation has been extended with 45 nodes while other parameters have been the same.

The proposed scheme was compared with the CABM[2]. The average number of sent messages was measured as the radio transmission range increased. The simulation result in Figure 1 shows that the proposed scheme reduces the number of sent messages significantly as compared with the CABM. It is also shown that the proposed scheme doesn't increase the number of sent messages so much though the number of nodes increases.

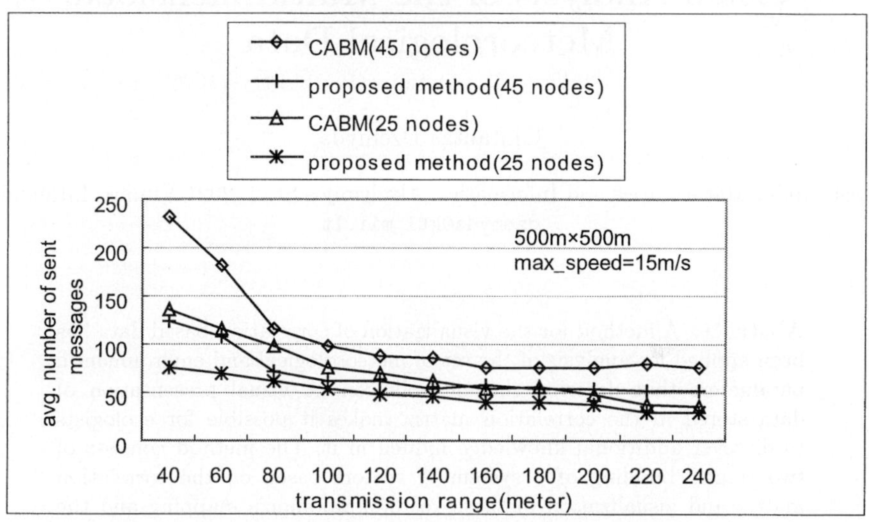

Fig. 1. Average number of sent messages against radio transmission range

5 Conclusion

In this paper, a pair-wise key agreement scheme based on the cluster structure and the verifiable secret sharing scheme has been proposed. Since the scheme forces each source to encrypt the source, the destination, and the Diffie-Hellman value in its key agreement packet with the destination's public key, it is very difficult for intermediate nodes to accomplish man-in-the-middle attacks. Also, the proposed scheme hides the entities of a key agreement using encapsulation header to prevent an adversary holding a few private keys from launching a man-in-the-middle attack easily. The simulation result shows that the proposed scheme offers a significant improvement over the CABM.

References

1. Schoenmakers, B.: A Simple Publicly Verifiable Secret Sharing Scheme and its Application to Electronic Voting. CRYPTO '99. Lecture Notes in Computer Science. Vol. 1666. Springer-Verlag, Berlin Heidelberg New York (1999) 148-164
2. Venkatraman, L., Agrawa, D.P.: A Novel Authentication scheme for Ad-hoc Networks. Proc. of IEEE WCNC 2000. Vol. 3. (2000) 1268-1273
3. Sanzgiri, K., Dahill, B., Levine, B., Shields, C., Belding-Royer, E.: A Secure Routing Protocol for Ad-hoc Networks. Proc. of ICNP'02. (2002) 78-87

Visual Analysis of the Multidimensional Meteorological Data

Gintautas Dzemyda

Institute of Mathematics and Informatics, Akademijos St. 4, 2021 Vilnius, Lithuania
dzemyda@ktl.mii.lt

Abstract. A method for the visualization of correlation-based data has been applied for analysis of the set of meteorological and environmental parameters that describe the air pollution. A visual presentation of data stored in the correlation matrix makes it possible for ecologists to discover additional knowledge hidden in it. The method consists of two stages: building of a system of vectors based on the correlation matrix and visualization of these vectors. Sammon's mapping and the self-organizing map were applied for visualization of the vectors.

1 Introduction

Any set of environmental objects (cases, vectors) may often be characterized by common parameters (variables, features). A combination of values of all the parameters characterizes a concrete object from the whole set. The values obtained by any parameter depend on the values of other parameters, i.e., the parameters are correlated. A problem of the analysis of correlations arises here. This problem as well as a lot of real correlation matrices became classical (see [1]). However, recent research and technology development applications produce correlation matrices and discover knowledge via their analysis, too. Correlations of meteorological and environmental parameters and their analysis appear in various studies. The references cover air pollution, vegetation of coastal dunes, groundwater chemistry, minimum temperature trends, zoobenthic species-environmental relationships, analysis of large environmental and taxonomic databases.

The goal of this paper is the illustration of a possibility to apply the visualization method to the analysis of correlation matrices of parameters that are of an environmental and ecological nature. The analysis is based on the correlation matrix of parameters that describe the air pollution. A visual presentation of data stored in the correlation matrix makes it possible for ecologists to discover additional knowledge hidden in the matrix and to make proper decisions about the interlocation of parameters and about their groups (clusters).

2 The Method of Visual Analysis of Correlation Matrices

One of the most popular methods of analysing correlations is the principal component analysis [2]. However, it does not show an interlocation of variables –

M. Bubak et al. (Eds.): ICCS 2004, LNCS 3036, pp. 652–656, 2004.

only their location around the zero-correlation. It means that we need more sophisticated means for the analysis of correlations.

A method for visualizing a set of parameters $x_1, ..., x_n$ characterized by their correlation matrix has been proposed in [1]. The method consists of two stages: building and visualization of a system of multidimensional vectors $Y_1, ..., Y_n \in S^n$ corresponding to the parameters $x_1, ..., x_n$. S^n is a subset of the n-dimensional Euclidean space R^n containing vectors of unit length.

There exist lots of methods that can be used for reducing the dimensionality of data (see, e.g., [3]). We apply here two popular and effective methods: Sammon's mapping [4] that is a nonlinear projection method closely related to the metric multidimensional scaling, and the self-organizing map (SOM) [3] that is a neural network.

Using Sammon's mapping we reduce the dimensionality of vectors $Y_1, ..., Y_n \in S^n$ by computing a correspondent system of two-dimensional vectors $Z_1, ..., Z_n \in R^2$. The self-organizing map (SOM) is a class of neural networks that are trained in an unsupervised manner. It is a well-known method for mapping a high dimensional space onto a low dimensional one. We consider here a mapping onto a two-dimensional grid of neurons. Using the SOM-based approach above, we can draw a table with cells corresponding to the neurons. The cells corresponding to the neurons-winners are filled with the order numbers of vectors $Y_1, ..., Y_n$. Some cells may remain empty (see Fig. 2a). One can make a decision visually on the distribution of the vectors $Y_1, ..., Y_n$ in the n-dimensional space in accordance with their distribution among the cells of the table. However, the table does not answer the question, how much the vectors of the neighboring cells are close in the n-dimensional space. This question may be answered by a combined mapping, i.e., by applying Sammon's mapping to visualize the n-dimensional vectors that are the numerical characteristics of neuron-winners.

3   Meteorological and Environmental Data Set

The experiment was carried out using the correlation matrix of 10 meteorological and environmental parameters that describe the air pollution in Vilnius city [5]: x_1, x_2, x_3 are the concentrations of carbon monoxide CO, nitrogen oxides NO_x, and ozone O_3; x_4 is the vertical temperature gradient measured at a 2–8 m height; x_5 is the intensity of solar radiation; x_6 is the boundary layer depth; x_7 is the amount of precipitation; x_8 is the temperature; x_9 is the wind speed; x_{10} is the stability class of atmosphere. The correlation matrix is presented in Table 1.

4   Results of the Analysis

In Fig. 1a, we present Sammon's mapping results of the vectors $Y_1, ..., Y_n$ calculated on the basis of the correlation matrix of parameters $x_1, ..., x_n (n = 10)$. Fig. 1a shows the distribution of vectors $Z_s \in R^2$, $s = \overline{1, n}$, obtained after the application of Sammon's algorithm to the vectors $Y_1, ..., Y_n$. In fact, we observe the distribution of parameters $x_1, ..., x_n$ on a plane. We do not present legends

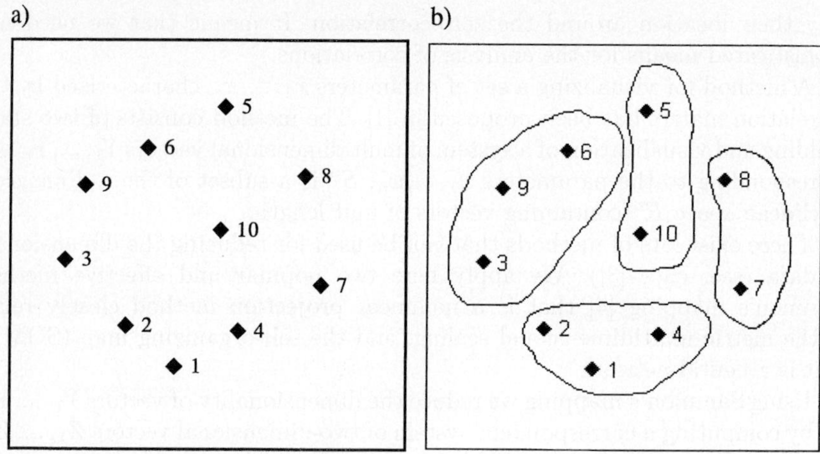

Fig. 1. Sammon's mapping results: a) mapping of 10 parameters; b) final conclusions on the clusters of parameters.

and units for both axes in the figure, because we are interested in observing the interlocation of points corresponding to the parameters only. The parameters are almost uniformly distributed after a direct compression of 10-dimensional points to a plane by Sammon's mapping, but more similar parameters are located nearer.

The SOM of size 4×4 was used in the experiments. The mapping results are presented in Fig. 2a. They indicate that there are at least three clusters of parameters. This estimate may be considered as the lower bounds for the number of clusters. What is the upper bound? The combined mapping should be used in search for the answer.

The results of combined mapping are presented in Fig. 2b. The figure shows the distribution of two-dimensional vectors obtained after an application of

Table 1. Correlation matrix of meteorological and environmental parameters

$i \backslash j$	1	2	3	4	5	6	7	8	9	10
1	1.00	0.78	−0.28	0.66	0.07	−0.33	−0.05	−0.09	−0.35	0.38
2	0.78	1.00	−0.37	0.63	−0.01	−0.31	−0.05	0.24	−0.38	0.37
3	−0.28	−0.37	1.00	−0.10	0.24	0.28	−0.11	0.18	0.64	0.04
4	0.66	0.63	−0.10	1.00	0.06	−0.45	−0.14	−0.06	−0.33	0.58
5	0.07	−0.01	0.24	0.06	1.00	−0.08	−0.05	0.09	−0.07	0.17
6	−0.33	−0.31	0.28	−0.45	−0.08	1.00	0.07	−0.10	0.60	−0.52
7	−0.05	−0.05	−0.11	−0.14	−0.05	0.07	1.00	−0.01	0.04	−0.11
8	−0.09	0.24	0.18	−0.06	0.09	−0.10	−0.01	1.00	0.01	0.23
9	−0.35	−0.38	0.64	−0.33	−0.07	0.60	0.04	0.01	1.00	−0.27
10	0.38	0.37	0.04	0.58	0.17	−0.52	−0.11	0.23	−0.27	1.00

Sammon's algorithm to the neurons-winners in the SOM from Fig. 2a. We can visually observe four clusters in Fig. 2b.

5   Conclusions

Comparing the results in Figures 1 and 2 on the parameters of air pollution, we can conclude that the meteorological and environmental parameters that describe the air pollution in Vilnius city form four clusters: $\{x_1, x_2, x_4\}$, $\{x_3, x_6, x_9\}$, $\{x_5, x_{10}\}$, $\{x_7, x_8\}$. These clusters are separated by curves in Fig. 1b. Data points in this figure repeat these of Fig. 1a.

One can see that the interlocation of parameters is similar in Figures 1a and 2a. The only difference is that clusters of parameters are more explicit in Fig. 2a. The clusters are more explicit in Fig. 2b as compared with Fig. 2a.

The analysis allows us to conclude that visualization is a powerful tool in data analysis. Its extension to the analysis of correlation matrices widened the range of applications. Most of problems with the correlation-based data sets may be solved in this way. In particular, the environmental problem was analysed successfully. The conclusions on the similarity of the measured parameters as well as on the possible number of clusters of similar parameters are drawn analysing the visual data presentation. This becomes possible due to the given method.

a)

5,10		7	
			8
6			4
3	9		1,2

b)

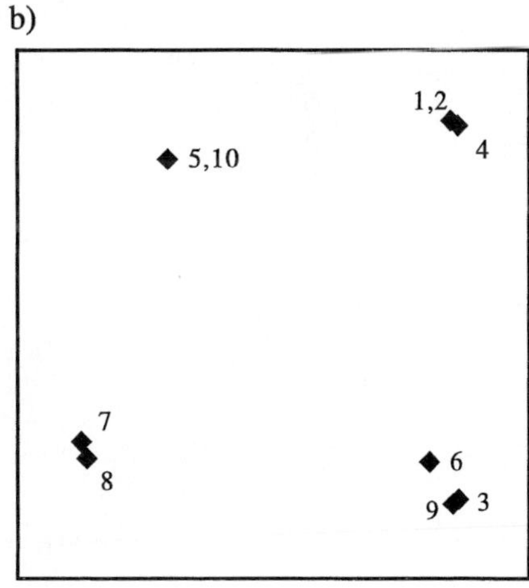

Fig. 2. Distribution of parameters characterizing the air pollution: a) 4×4 SOM; b) combined mapping (4×4 SOM + Sammon's mapping)

References

1. Dzemyda, G.: Visualization of a set of parameters characterized by their correlation matrix. Computational Statistics and Data Analysis **36(10)** (2001) 15–30

2. Jolliffe, I.T.: Principal Component Analysis. Springer (1986)
3. Kohonen, T.: Self-Organizing Maps. 3nd ed. Springer Series in Information Sciences, Springer–Verlag, Vol. 30 (2001)
4. Sammon, J.W.: A nonlinear mapping for data structure analysis. IEEE Transactions on Computers **18** (1969) 401–409
5. Zickus, M.: Influence of Meteorological Parameters on the Urban Air Pollution and Its Forecast. Thesis Presented for the Degree of Doctor in Physical Sciences (1998). http://vilnair.gamta.lt/thesis/content.html

Using Branch-Grafted R-trees for Spatial Data Mining

Priyanka Dubey[1], Zhengxin Chen[1], and Yong Shi[1, 2]

[1] College of Information Science and Technology
University of Nebraska at Omaha
Omaha, NE 68182, U. S. A.
{zchen,yshi}@mail.unomaha.edu
[2] Graduate School of Chinese Academy of Sciences, Beijing 100039, China

Abstract. Spatial data mining is a process of extraction of implicit information, such as weather patterns around latitudes, spatial features in a region, etc., with a goal of knowledge discovery. The work reported here is based on our earlier work on branch-grafted R trees. We have taken a bottom-up approach in our research: from efficient spatial data structure (i.e., branch-grafted R tree implementation), to efficient data access methods, and finally, to effective spatial data mining. Since previous experiments have shown that there are significant advantages of using branch-grafted implementation, this bottom-up approach exploits the performance advantages of the branch-grafted R-trees.

1 Introduction

Spatial data mining [4] refers to the extraction of implicit knowledge, spatial relationships, or other patterns, not explicitly stored in databases involving spatial data. *Spatial data* differs from data stored in relational databases. It carries topological information usually organized by multidimensional indexing structures, accessed by spatial data access methods. The focus of this paper is on using branch-grafted R-tree for data mining. Nevertheless, in order to make this paper self-contained, we also provide relevant background related to R trees and their branch-grafted implementation. Therefore, space limitation will only allow us to provide a general description of our approach, leaving much of the details to future papers.

We start our discussion with a brief review of necessary background. R-trees [2] are specialized forms of B+ trees adapted for the efficient representation, access and management of spatial data. An R-tree is a height-balanced tree that contains all leaves at the same level. The root node has at least two children. The non-leaf nodes contain $\lceil m/2 \rceil$ entries, where m is the maximum number of entries in a node. The spatial data is represented as rectangles called minimum bounding rectangle. A non-leaf node contains information of the bounding rectangle, in the form of the coordinates, of its child node. The leaf node of an R-tree stores the record of spatial data stored, in the form of the coordinates of a rectangle, in relational database. Although the original R-tree structure has many merits, it also leaves rooms for improvement. Variations of R-trees exist (such as R* tree). As an improvement proposed from our research group, the branch grafted algorithm uses the R-tree with

M. Bubak et al. (Eds.): ICCS 2004, LNCS 3036, pp. 657–660, 2004.
© Springer-Verlag Berlin Heidelberg 2004

better reorganization of records. The algorithm uses the grafting of a leaf or a branch node in the tree to reduce the number of nodes in the R-tree. The more accurate data structure improves the performance of query. The split in the original R-tree does not consider rest of the nodes in the tree while the branch-grafted tree scans the other nodes in order to find an opening. This opening can accommodate a rectangle. This could avoid the need of a split and reduces the number of node in the tree. The improvement in search time is a result of lesser number of nodes in the tree.

The branch-grafting algorithm improves the accuracy and speed of the node overflow operations. In case of a node overflow, the Branch Grafting algorithm first looks for records in the parent node that overlap the bounding directory of the full node. Individual records in the full node are then evaluated for placement under the overlapping parent nodes. Records are moved under a parent node if the resulting area of coverage for all nodes involved is smaller. If no records are moved to make room for the initially inserted record, then a split is performed using the original R-tree split algorithm.

Branch-grafted R-trees have been used to deal with location-based regional-queries and point-queries. Branch-grafted R-trees have an advantage when a huge amount of spatial data needs to be stored. Grafting reduces the height of the tree, which helps the data access from the tree. A regional query finds all data objects that intersect a given spatial rectangle formed around a spatial object that overlap or fall within the query window. A point-query is a special case of the region query in which an aligned rectangle is a single point. The data access is faster as the search algorithm traverses fewer search paths due to smaller number of nodes reduced by grafting.

Once the spatial data is organized using an efficient data structure such as branch grafted R-tree, data access mechanisms should be developed for further processing. Basic spatial data access methods developed in earlier work of our institution [1, 3, 6] include window search algorithm (also called overlap algorithm, which finds all the objects that overlap a specified search window and can be used for location-based queries) and its variation window search algorithm (also referred to as contains algorithm), search by distance algorithm and its extension nearest neighbor algorithm, etc.

2 Extending Spatial Queries for Spatial Data Mining

The representation and analysis of the information obtained as result of spatial queries, is important for the better understanding, usage, and meaningful insight into the spatial data. Recent studies of mining spatial rules have led to a set of interesting techniques to represent the patterns and features in a spatial data set. Strong spatial rules indicate the patterns and implication relationships in the large spatial data set. We can distinguish several kinds of spatial data mining rules:

- **Spatial Association Rules:** Spatial association is a rule that describes the implication of one or a set of features by another set of features in spatial database.

- **Spatial Classification Rules:** The process of spatial classification is to find a rule to partition a set of classified objects into a number of classes using spatial relations of the classified objects to other objects in the database.
- **Spatial Aggregation Rules:** Aggregate values for areas close to spatial objects plays a very important role in the analysis of many business objects like stores, gas stations, etc.
- **Spatial Discriminant Rules:** A spatial discriminant rule is a general description of the contrasting or discriminating features of a class of spatial-related data from other classes.

3 Spatial Data Mining Program Construction and Experiments

The spatial data mining program consists of three modules.
- Module 1 builds a data structure using input data files and stores the data in the branch grafted R-Tree.
- Module 2 is designed to retrieve relevant data stored in the branch grafted R-Tree with the help of queries and stores the relevant output in a text file.
- Module 3 describes the mining of the spatial rules viz. spatial association rule, aggregation rule, and discriminant rules from the pool of matching coordinates.

The main spatial operators used for the query design in the research can be mainly divided in two categories; topological and directional. The spatial operators such as 'Containment', 'intersection', 'adjacency', 'inside', and 'within' fall under the category of topological operators, while 'Northwest', 'distance_compared_to', 'near', 'close_to' are directional operators.

An example of spatial query that uses 'adjacent_to' function is shown below:

```
SELECT CITY_NAME, CITY_LOCATION
FROM   CITY, HYDRO
WHERE CITY_LOCATION ADJACENT_TO (WATER_BODY)
        AND     CLOSE_TO(US_BUNDARY,30)     AND     WATER_BODY     =
'ATLANTIC_OCEAN'
```

The spatial data for the work has been down loaded from TIGER/ LINE. The TIGER database contains spatial information about hydrology, transportation, and other objects like national and state parks, churches, universities etc. The data used in this work deals water bodies in the North America, cities of North America, and US highways.   More information about cities by counties was obtained from LANDVIEW3 site. This data contains the spatial and detailed non-spatial information as, census, industries, income level, etc. The raw data from different counties and states was rearranged and merged to make a master data file. The data format is as follows:

minimum x coordinate, minimum y coordinate, positive delta x, positive delta y, name

Several data sets have been used for experiments. For example, the data set hydro_tiger95 contains hydrographic spatial data, with number of records 360,330.

Below are some spatial rules discovered by the implemented program:

Spatial Association Rules

is_a (small_town) ^ adjacent_to(big_water_body) → adjacent_to (us_boundary): 78%

Spatial Aggregation Rules

Total number of small cities close_to(big_water_bodies) in North America: 134

Spatial Discriminant Rules

Large number of industries are located within(big_cities) while small number of cities located within(small_cities)

4   Conclusion

As indicated in our discussion, we have taken a bottom-up approach in our research: from efficient spatial data structure (i.e., branch-grafted R tree implementation) [1,6], to efficient data access methods [3], and finally, to effective spatial data mining [5]. Since previous experiments have shown that there are significant advantages of using branch-grafted implementation, this bottom-up approach exploits the performance advantages of the branch-grafted R-trees. However, due to space limitation, detailed description of algorithms, experiments and their analysis are omitted here. As for the future work, more experiments on larger data sets and more detailed analysis are still needed.

References

1.   B. Asato, Branch-Grafting Heuristic for R-tree Implementation, Working Paper, Department of Computer Science, University of Nebraska at Omaha , 1994.
2.   A. Guttman, R-trees: A Dynamic Index Structure for Spatial Searching, *Proc. ACM IGMOD*, 47-57, 1984
3.   M. Khanijo, Z. Chen and Q. Zhu, Spatial data access methods using branch-grafted R-trees, *Advances in Database and Knowledge-Based Systems: Data Mining and Data Warehousing* (G. E. Lasker and Z. Chen eds.), 31-36, 2000.
4.   K. Koperski, J. Han, and J. Adhikary, Mining Knowledge in Geographical Data, *IEEE Computer*, 1998.
5.   P. Dubey, Spatial Data Mining Using Branch-Grafted R-tree, MS thesis, University of Nebraska at Omaha, Omaha, 2003.
6.   T. Schreck and Z. Chen, Branch grafting method for R-tree implementation, *Journal of Systems and Software*, 53(1), 83-93, 2000.

Using Runtime Measurements and Historical Traces for Acquiring Knowledge in Parallel Applications

Luciano José Senger[1], Marcos José Santana[2], and Regina Helena Carlucci Santana[2]

[1] Universidade Estadual de Ponta Grossa, Departamento de Informática
Av. Carlos Cavalcanti, 4748 Zip Code 84030-900 Ponta Grossa, PR, Brazil
[2] Universidade de São Paulo, Instituto de Ciências Matemáticas e de Computação
Av. Trabalhador Saocarlense, 400 PO Box 668 Zip Code 13560-970 São Carlos, SP, Brazil
{ljsenger, mjs, rcs}@icmc.usp.br

Abstract. A new approach for acquiring knowledge of parallel applications regarding resource usage and for searching similarity on workload traces is presented. The main goal is to improve decision making in distributed system software scheduling, towards a better usage of system resources. Resource usage patterns are defined through runtime measurements and a self-organizing neural network architecture, yielding an useful model for classifying parallel applications. By means of an instance-based algorithm, it is produced another model which searches for similarity in workload traces aiming at making predictions about some attribute of a new submitted parallel application, such as run time or memory usage. These models allow effortless knowledge updating at the occurrence of new information. The paper describes these models as well as the results obtained applying these models to acquiring knowledge in both synthetic and real applications traces.

1   Introduction

One of the challenges in parallel and distributed is to develop scheduling algorithms that assigns the tasks of parallel applications to the heterogeneous machines. Many researchers have demonstrated that using parallel application knowledge may improve the scheduling decisions on multiprogrammed multiprocessor systems [1][2]. Nevertheless, most of the work has assumed that such knowledge is available *a priori* and does not provide effective indications to obtain it. There are commonly three main sources to obtain knowledge: the description of applications requirements provided by the user (or programmer) who submits the parallel application to the system; historical traces of all applications executed in a specific system over a time period, and runtime measurements from parallel applications. Among these knowledge sources, historical traces and runtime measurements have demonstrated a great potential to provide information aiming at classifying parallel applications and obtaining knowledge [3][4].

This paper presents two models for knowledge acquisition in parallel applications aiming at improving software scheduling decisions. The first model aims for parallel applications classifying, regarding behavior on resource usage. Compared to previous work, these models have two novel aspects. First, they allow updating of acquired knowledge at the occurrence of new information. Second, they can be used to improve different scheduling policies, not aimed at particular scheduling strategies.

M. Bubak et al. (Eds.): ICCS 2004, LNCS 3036, pp. 661–665, 2004.

2   Local Knowledge Acquisition

The main goal of the model described in this section is the knowledge acquisition from the execution of tasks which compose parallel applications, classifying parallel applications regarding its behavior on resource utilization by classes such as CPU Bound, I/O Bound and Communication Intensive [5]. This model allows the utilization of knowledge acquired in previous runs of the applications and the knowledge updating in presence of new information. The model aiming at identifying the resource usage phases during the parallel application execution using runtime measurements. Collected observations are separated by intervals of time T_o. A sample is created for each N_o observations, characterizing an Euclidean vector with dimension d. These samples samples are grouped in clusters. An important aspect of the model is how to carry out the clustering of the vectors. Since an on-line strategy is desired, it is adopted the utilization of the ART 2A self-organizing neural architecture [6] as the solution. The ART 2A neural network allows incremental clustering and classification of a pattern set composed of continuous values [7]. A significance matrix SM [8], which is composed by a set of significance values SV_{ij}, is employed to assign a label for each cluster. The significance values are obtained directly from the ART 2A long term memory.

The model was evaluated using an input feature vector with dimension $d = 7$: CPU user time; memory usage; read and write file operations and send and receive network operations. The model was tested on the PSTSWM (*Parallel Spectral Transform Shallow Water Model*) [9] application. The PSTSWM is a representative compact application that solves the nonlinear shallow water equations on a rotating sphere using the spectral transform method. The model parameters were $T_o = 1000ms$ and $N_o = 1$ and the results obtained are illustrated in Table 1. The majority of application phases (about 56%) were classified as communication intensive (cluster $k = 1$). The other phases were classified as communication intensive ($k = 2$ and $k = 4$) and CPU-Bound (cluster $k = 3$).

3   Global Knowledge Acquisition

The local knowledge acquisition model presented in the previous section aiming at knowledge acquiring of a particular application resource usage. Scheduling algorithms often need to know some application attribute in advance of its execution (e.g. *run time*, *memory usage* and *cpu time*). The workload traces can be treated as a database of previous experiences (*experience base*) about parallel applications and depending on the scheduling algorithm requirements, some attribute, like *run time* or *memory usage*, can be considered as the attribute to be predicted. The global knowledge acquisition model described in this section is constructed using instance-based learning. Instance-based learning (IBL) is an approach which find similar instances in an experience base aiming at approximating real-valued or discrete-valued target functions [10]. IBL algorithms compute the similarity between a new query instance and the experience base instances, returning a set of related instances as output.

The output attribute considered to predict is the parallel application run time, although any attribute can be used as output in our IBL algorithm implementation. Many authors have demonstrated that parallel applications run time knowledge can be very useful to space-sharing scheduling algorithms [11]. Parallel application traces recorded in two computing centers, namely SDSC (*San Diego Supercomputer Center*) and CTC (*Cornell Theory Center*), are used to evaluate the IBL model. The most relevant input attributes, selected according to [4], were used. The workload traces were organized through a number of disjoint sets. Each of these sets was partitioned considering 2/3 as experience base and 1/3 for testing, and a partial mean prediction error was computed for each set. The main IBL algorithm parameters are the neighborhood size K, which defines the number of relevant instances used for computing the estimate, and the σ^2, which defines the Gaussian function slope and the weighting values. The values experimented for K were $5, 10, 25$ and 50. For σ^2, the values were $0.125, 0.250, 0.500, 1.000$ and 2.000. The IBL algorithm achieved mean prediction errors that are between 50 and 58 percent of mean application run times (Table 2). For all workload traces, the best K value founded was 5 and there was not statistical difference among the prediction errors obtained by the experimented σ values. Nevertheless, it was observed a tendency for more higher prediction errors as the σ^2 value increases. The Table 3 shows the best IBL model prediction errors obtained compared to five previous approaches [12] [3] [4](the mean absolute error is used in order to compare the results with previous work).

Table 1. PSTSWM parallel application results

	ART-2A LTM centroid traces Cluster(k)				Significance Matrix			
Attribute	1	2	3	4	1	2	3	4
CPU time (user)	0.003	0.12	1.00	0.09	0.30%	10.94%	100.00%	6.26%
I/O (Bytes read)	0.00	0.00	0.00	0.00	0.00%	0.00%	0.00%	0.00%
I/O (Bytes write)	0.00	0.00	0.00	0.00	0.00%	0.00%	0.00%	0.00%
Network (TCP bytes read)	1.00	0.00	0.00	0.70	99.70%	0.00%	0.00%	46.87%
Network (TCP bytes write)	0.00	0.99	0.00	0.70	0.00%	89.06%	0.00%	46.87%
Network (UDP bytes read)	0.00	0.00	0.00	0.00	0.00%	0.00%	0.00%	0.00%
Network (UDP bytes write)	0.00	0.00	0.00	0.00	0.00%	0.00%	0.00%	0.00%
Frequency	56.07%	16.58%	14.79%	12.56%				

Table 2. Experiments details

Workload	Experience base size	Test size	Samples	Percentage of Mean Runtime Error
SDSC95	716	369	70	57.46±1.06
SDSC96	421	210	60	55.40±1.18
SDSC2000	621	320	60	50.05±1.59
CTC	688	355	60	50.45±1.29

4 Conclusions

This paper presented two models for acquiring knowledge in parallel applications. The aim of these models is to improve scheduling decisions, supporting the scheduler

Table 3. IBL algorithm results compared to previous work

| | Downey | | Gibbons | Smith | | |
Workload	Median Lifetime	Average Lifetime	Fixed Templates	Greedy Search	Genetic Search	**IBL**
SDSC95	82.44	171.00	74.05	67.63	59.65	**33.98**
SDSC96	102.04	168.24	122.55	76.20	74.56	**58.32**
CTC	179.46	201.34	124.06	118.05	106.73	**109.06**

with knowledge about the resource usage patterns of parallel applications. Through the experiments, the local knowledge acquisition model presented a good classification performance. Another aspect observed is the robustness of this model at the different computational loads and processing elements configurations [5]. The global knowledge acquisition model presented a great potential to define similarity among parallel applications, weighting the more relevant instances in an experience base to generate an output attribute estimate. Our IBL approach achieved a good prediction performance when compared to both static and adaptive templates prediction approaches.

Acknowledgments. This project is supported by CAPES/PICDT program. The authors would like to thank Reagan Moore, Allen Downey, Victor Hazelwood (San Diego Supercomputer Center) and the Cornell Theory Center for graciously providing the workload traces used in this work. Particular thanks to Warren Smith, for providing his prediction software.

References

1. Silva, F.A.B.D., Scherson, I.D.: Improving Parallel Job Scheduling Using Runtime Measurements. In Feitelson, D.G., Rudolph, L., eds.: Job Scheduling Strategies for Parallel Processing. Springer Verlag (2000) 18–38 Lect. Notes Comput. Sci. vol. 1911.
2. Naik, V.K., Setia, S.K., Squillante, M.S.: Processor Allocation in Multiprogrammed Distributed-memory Parallel Computer Systems. Journal of Parallel and Distributed Computing **47** (1997) 28–47
3. Gibbons, R.: A Historical Application Profiler for Use by Parallel Schedulers. In: Job Scheduling Strategies for Parallel Processing. Springer Verlag (1997) 58–77
4. Smith, W., Foster, I.T., Taylor, V.E.: Predicting Application Run Times Using Historical Information. In: JSSPP. (1998) 122–142
5. Senger, L.J., Santana, M.J., Santana, R.H.C.: A new approach fo acquiring knowledge of resource usage in parallel applications. In: Proceedings of International Symposium on Performance Evaluation of Computer and Telecommunication Systems (SPECTS'2003). (2003) 607–614
6. Carpenter, G.A., Grossberg, S., Rosen, D.B.: ART 2-A: An Adaptive Resonance Algorithm for Rapid Category Learning and Recognition. Neural Networks **4** (1991) 494–504
7. Whiteley, J.R., Davis, J.F.: Observations and problems applying ART2 for dynamic sensor pattern interpretation. IEEE Transactions on Systems, Man and Cybernetics-Part A: Systems and Humans **26** (1996) 423–437
8. Ultsch, A.: Self-organising neural networks for monitoring and knowledge acquisition of a chemical process. In: Proceedings of ICANN-93. (1993) 864–867

9. Foster, I.T., Worley, P.H.: Parallel algorithms for the spectral transform method. SIAM J. Sci. Stat. Comput. **3** (1997) 806–837

10. Aha, D.W., Kibler, D., Albert, M.K.: Instance-based learning algorithms. Machine Learning (1991) 37–66

11. Feitelson, D.G., Rudolph, L., Schwiegelshohn, U., Sevcik, K.C., Wong, P.: Theory and Practice in Parallel Job Scheduling. In: Job Scheduling Strategies for Parallel Processing. Volume 1291. Springer Verlag (1997) 1–34 Lect. Notes Comput. Sci. vol. 1291.

12. Downey, A.: Predicting queue times on space-sharing parallel computers. In: Proceedings of International Parallel Processing Symposium. (1997)

Words as Rules: Feature Selection in Text Categorization[*]

E. Montañés, E.F. Combarro[**], I. Díaz, J. Ranilla, and J.R. Quevedo

Artificial Intelligence Center, University of Oviedo, Spain
ir@aic.uniovi.es

Abstract. In Text Categorization problems usually there is a lot of noisy and irrelevant information present. In this paper we propose to apply some measures taken from the Machine Learning environment for Feature Selection. The classifier used is Support Vector Machines. The experiments over two different corpora show that some of the new measures perform better than the traditional Information Theory measures.

1 Introduction

Text Categorization (TC) [1] consists of assigning a set of documents to a set of categories. The removal of irrelevant or noisy features [2] improves the performance of the classifiers and reduces the computational cost.

The *bag of words* [1] is the most common document representation, using the absolute frequency (tf) to measure the relevance of the words over the documents [3]. *Stemming* and removing of *stop words* are usually performed. In this paper, words occurring in each category are used isolated from the rest [1] (local sets). The classification is tackled using the *one-against-the-rest* [4] approach and Support Vector Machines (SVM), since they perform fast and well [3] in TC.

This paper proposes some well-known impurity measures taken from the Machine Learning (ML) environment to perform Feature Selection (FS).

The rest of the paper introduces these measures, describes the corpora and the experiments and presents some conclusions and ideas for further research.

2 Feature Selection

FS is commonly performed in TC by keeping the words with highest score according to a measure of word relevance, like Information Theory (IT) measures. They consider the distribution of the words over the different categories. Some of the most adopted are *information gain* (IG) [5], *expected cross entropy for text* (CET) [6] and $S - \chi^2$ [7], a modification of the χ^2 statistic [2].

In the measures proposed here, fixed a category c, a word w is identified with the rule $w \rightarrow c$ which says: *If w is in a document, then the document belongs*

──────────

[*] This work has been supported under MCyT and Feder grant TIC2001-3579.
[**] The author acknowledges the support of research project FICYT PR-01-GE-15.

M. Bubak et al. (Eds.): ICCS 2004, LNCS 3036, pp. 666–669, 2004.

to c. Then, the relevance of w for c is identified with the quality of the rule $w \rightarrow c$ [8].

Many popular rule quality measures are based on the percentage of successes and failures of the application of the rule. Two examples are the Laplace measure (L) and the *difference* (D) [9]. The former is a slight modification of the precision. The latter establishes a balance between the documents containing w and penalizes the words from documents not belonging to c. They are defined by

$$L(w \rightarrow c) = \frac{a_{w,c} + 1}{a_{w,c} + b_{w,c} + s} \qquad D(w \rightarrow c) = a_{w,c} - b_{w,c}$$

where $a_{w,c}$ is the number of documents of c in which w appears, and $b_{w,c}$ is the number of documents contaning w but not belonging to c.

We also propose variants that consider the absence of the word in c, penalizing more aggressively those words which appear in few documents of c. Hence, we define

$$L_{ir}(w \rightarrow c) = \frac{a_{w,c} + 1}{a_{w,c} + b_{w,c} + c_{w,c} + s} \qquad D_{ir}(w \rightarrow c) = a_{w,c} - b_{w,c} - c_{w,c}$$

where $c_{w,c}$ is the number of documents from c not containing w.

3   Experiments

Before presenting the experiments, we describe the corpora. Reuters-21578 contains short economic news. The distribution of the documents over the categories is quite unbalanced and the words are little scattered. Considering the Apté split [4] we obtain 7063 training documents and 2742 test documents, assigned to 90 different categories.

Ohsumed is a MEDLINE subset from 270 medical journals. Here, we consider the first $20,000$ MEDLINE documents from 1991 with abstract (the first $10,000$ for train and the rest for test) and the 23 subcategories of diseases (C of MeSH[1]). The words here are quite more scattered than in Reuters and the distribution of documents over the categories is much more balanced.

In the experiments, the filtering levels (fl) range from 20% to 98%. One-tail paired t-tests at significance level of 95% are conducted between $F_1's$ for each pair of measures. Table 1 presents the *macroaverage* and the *microaverage* of F_1 [1]. Tables 2 and 3 show the *t-test* results[2].

For Reuters, L_{ir}, D_{ir} and IG produce, in general, the best *macroaverage* and *microaverage* among their variants. This may be because the words are little scattered, that is, each category has a high percentage of specific words. Hence, the best measures are either those which tend to select words frequent in

---

[1] Medical Subject Headings `http://www.nlm.nih.gov/mesh/2002/index.html`

[2] In them, "+" means that the first measure is significantly better than the seconnd, "-" means that the seconnd measure is better than the first, "=" means that there exists no significative difference.

Table 1. *Macroaverage* and *Microaverage* of F_1 for different variants

	Reuters													
	Macroaverage							*Microaverage*						
$fl(\%)$	L	L_{ir}	D	D_{ir}	CET	S-χ^2	IG	L	L_{ir}	D	D_{ir}	CET	S-χ^2	IG
20	47.15	47.53	46.41	46.71	46.92	46.35	46.07	82.71	85.03	80.99	81.54	83.39	83.35	84.78
40	48.31	48.90	45.17	45.75	46.90	47.27	48.03	81.53	85.16	79.42	80.49	83.03	83.19	84.96
60	48.40	49.01	44.06	44.61	46.89	46.67	47.84	80.16	85.26	78.91	79.79	83.03	83.28	85.02
80	43.73	48.66	42.41	43.77	48.55	48.87	48.57	77.69	85.21	78.68	79.99	83.33	83.40	85.42
85	43.74	48.40	41.61	44.06	48.84	47.69	48.44	76.80	85.05	78.63	80.25	83.40	83.17	85.53
90	41.64	48.09	40.30	42.48	48.12	47.87	48.69	75.26	84.81	77.77	80.26	82.99	83.30	85.48
95	33.45	47.42	38.87	41.42	47.61	47.82	49.19	74.15	84.47	75.63	79.26	83.22	83.57	85.40
98	30.43	45.73	37.85	40.68	47.55	46.30	48.41	74.76	83.43	78.40	80.37	83.02	83.01	84.76
	Ohsumed													
	Macroaverage							*Microaverage*						
$fl(\%)$	L	L_{ir}	D	D_{ir}	CET	$S-\chi^2$	IG	L	L_{ir}	D	D_{ir}	CET	$S-\chi^2$	IG
20	46.27	42.49	49.54	50.00	45.92	45.86	42.71	53.53	51.51	56.19	56.77	53.46	53.47	51.93
40	51.11	42.91	50.57	51.52	46.73	46.71	43.61	56.92	51.54	56.20	57.25	53.91	53.94	52.31
60	51.32	43.66	50.01	51.11	47.39	47.28	45.21	56.63	51.99	55.52	56.89	54.49	54.35	53.13
80	48.42	44.31	48.77	50.74	48.14	47.40	46.94	53.41	51.90	55.11	57.18	54.80	54.49	53.98
85	46.40	44.14	48.93	51.17	48.36	47.64	47.11	51.32	51.24	53.45	57.54	55.10	54.75	54.19
90	47.46	44.57	50.26	52.19	48.79	47.85	47.54	52.22	51.15	54.36	57.73	55.42	54.93	54.59
95	48.33	44.17	51.58	52.37	48.91	46.94	49.40	53.75	49.94	55.72	58.04	55.69	54.48	55.47
98	47.82	41.02	52.27	51.44	47.47	44.16	49.66	53.08	45.85	56.24	57.00	54.04	51.99	54.81

Table 2. t-tests among different variants

	Reuters					Ohsumed				
$fl(\%)$	L_{ir}-L	D_{ir}-D	IG-CET	IG-S-χ^2	CET-S-χ^2	L_{ir}-L	D_{ir}-D	IG-CET	IG-S-χ^2	CET-S-χ^2
20	=	+	=	=	=	-	+	-	-	=
40	=	+	+	=	=	-	+	-	-	=
60	=	=	=	=	=	-	+	-	-	=
80	+	+	=	=	=	-	+	-	=	+
85	+	+	=	=	=	=	+	-	=	+
90	+	+	=	=	=	-	+	-	=	+
95	+	+	+	+	=	-	=	=	+	+
98	+	+	=	+	+	-	=	+	+	+

the category (like L_{ir} and D_{ir}) or those that consider the absence of the word in the category (like IG). Among them, L_{ir} and IG are statistically better, with no significative differences between them.

Regarding Ohsumed, L, D_{ir} and CET are, in general, the best measures among their variants, in *macroaverage* and in *microaverage*. Here, there are not so many specific words in each category, since the words are slightly scattered. Hence, the best measures are those that select words frequent in the category, although they appear in the rest, like L, D_{ir} and CET. Among them, D_{ir} is statistically better than the rest.

4   Conclusions and Future Work

This paper proposes some measures taken from Machine Learning for Feature Selection in TC, comparing them with other traditional Information Theory measures.

The performance of the measures depends on the corpus. For Reuters, which has an unbalanced distribution of documents and has the words little scattered, it

Table 3. *t-tests* among the best variants

Filtering level	20	40	60	80	85	90	95	98		20	40	60	80	85	90	95	98
	Reuters									Ohsumed							
$L_{ir}-D_{ir}$	=	+	+	+	+	+	+	+	$L-D_{ir}$	-	=	=	-	-	-	-	-
$L_{ir}-IG$	+	=	=	=	=	=	-	=	$L-CET$	+	+	+	=	=	=	=	=
$D_{ir}-IG$	=	=	-	-	-	-	-	-	$D_{ir}-CET$	+	+	+	+	+	+	+	+

is better to penalize more those words which are not frequent in the category or to consider the absence of words in each category. For Ohsumed, whose distribution of documents is more uniform and which has the words highly scattered, it is better to reinforce the words of each category and not to penalize so much the words of the rest.

In our future work, we plan to use other classifiers and to find the optimal filtering levels for each measure, which may depend on properties of the category.

References

1. Sebastiani, F.: Machine learning in automated text categorisation. ACM Computing Survey **34** (2002)
2. Yang, T., Pedersen, J.P.: A comparative study on feature selection in text categorisation. In: Proceedings of ICML'97, 14th International Conference on Machine Learning. (1997) 412–420
3. Joachims, T.: Text categorization with support vector machines: learning with many relevant features. In Nédellec, C., Rouveirol, C., eds.: Proceedings of ECML-98, 10th European Conference on Machine Learning. Number 1398, Chemnitz, DE, Springer Verlag, Heidelberg, DE (1998) 137–142
4. Apte, C., Damerau, F., Weiss, S.: Automated learning of decision rules for text categorization. Information Systems **12** (1994) 233–251
5. Díaz, I., Ranilla, J., Montañés, E., Fernández, J., Combarro, E.: Improving performance of text categorization by combining filtering and support vector machines. (Journal of the American Society for Information Science and Technology (JASIST)) Accepted for publication.
6. Mladenic, D., Grobelnik, M.: Feature selection for unbalanced class distribution and naive bayes. In: Proceedings of 16th International Conference on Machine Learning ICML99, Bled, SL (1999) 258–267
7. Galavotti, L., Sebastiani, F., Simi, M.: Experiments on the use of feature selection and negative evidence in automated text categorization. In: Proceedings of ECDL-00, 4th European Conference on Research and Advanced Technology for Digital Libraries, Lisbon, Portugal, Morgan Kaufmann (2000) 59–68
8. Combarro, E.F., Montañés, E., Ranilla, J., Fernández, J.: A comparison of the performance of svm and arni on text categorization whit new filtering measures on an unbalanced collection. In: International Work-conference on Artificial and Natural Neural Network, IWANN2003, Lecture Notes of Springer-Verlag. (2003)
9. Muggleton, S.: Inverse entailment and prolog. New Generation Computing, Special issue on Inductive Logic Programming **13** (1995) 245–286

Proper Noun Learning from Unannotated Corpora for Information Extraction[1]

Seung-Shik Kang

School of Computer Science, Kookmin University & AITrc, Seoul 136-702, Korea
sskang@kookmin.ac.kr, http://nlp.kookmin.ac.kr/~sskang

Abstract. Named entity (NE) tagged corpus is an important resource for the learning of extraction patterns in information extraction system. We constructed Korean NE tagged corpus for economy, accident, and travel domain. As a semi-automatic approach to construct NE tagged corpus, a pattern learning method has been explored to extract personal names automatically from the raw corpus. Our NE tagging system has been trained for unannotated corpora to collect NE extraction patterns. Pattern extraction starts from a small set of proper names and NE extraction patterns are generated semi-automatically. Extracted patterns are used to automatically identify proper nouns from the text.

1 Introduction

Information retrieval system tends to extract too many search results for a given query. Manual selection of extracting useful information from search engines is a tedious work. Information extraction (IE) system automatically extracts predefined types of information from the extremely large set of information sources[1,2,3]. The target of discovering the knowledge in IE system is to find out meaningful named entities. Named entities are proper nouns that are extracted by IE system and they are the subjects of 5W1H. Typical named entities are person names, location names, organization names, product names, and numeric expressions like time, date, money, percent, and so on. So, the core function of the IE system is to identify the named entities and extract the predefined subject from the context of the text.

Information extraction started from the named entity contest of MUC and IREX workshop on Japanese text[4,5]. The major topics of the conferences are named entity extraction and coreference resolution on the specific domain. The basic language resources in IE system development are annotated corpora, language analyzer, and cue word dictionary. We constructed Korean NE tagged corpus on economy, accident, and travel domain. While constructing the corpus, we tried to automatically identify named entities from the text without using the lexicon or cue word dictionary. We trained our system on unannotated corpus to generate NE extraction templates, starting from the small set of proper names.

---

[1] This work was supported by the Korea Science and Engineering Foundation (KOSEF) through the Advanced Information Technology Research Center (AITrc).

M. Bubak et al. (Eds.): ICCS 2004, LNCS 3036, pp. 670–674, 2004.

2   Corpus Construction and Annotation

We collected a raw corpus and defined a named entity tagset for the construction of named entity tagged corpus. Raw corpus of Korean articles has been collected from newspaper articles and webpages. They are 3,000 articles of news on economy, travel, social events, and seminars. Manual NE tagging is performed through (1) initially by automatic tagging system, (2) manual tagging of untagged entities using tagging tool, and (3) manual cross checking. NE tagset is based on the tagsets of MUC and IREX. Tagset classes of named entities are proper nouns and numeric expressions that are valuable for information extraction. Proper nouns are divided into specific classes and artifacts are divided into title and description. Numeric expressions are date, time, money, percent, and quantity. In addition, phone and address are added to numeric expressions. Named entity tags are (1) proper nouns(PERSON, LOCATION, ORGANIZATION, TITLE, DESCRIPTION, URL), (2) numeric expressions(DATE, TIME, MONEY, PERCENT, QUANTITY, PHONE, ADDRESS), (3) ambiguity, and (4) reference(referent and its antecedent relationship).

Constructing tagged corpus is a laborious work with high cost. For tagging efficiency, we developed NE/CO(Named Entity & Coreference) tagging tool. It provides a good tagging environment by marking a named entity block and selecting a tag in a graphic user interface. Fig. 1 shows an example of NE/CO tagging. The tagging tool provides a pre-tagging function that pre-tagged named entities are automatically tagged as a named entity. It minimizes the tagging errors and increases the accuracy. In addition, it has a statistical report generation function that generates a named entity list with its frequency counts. It also helps to find mis-tagged entities merely by checking the entities with low frequency.

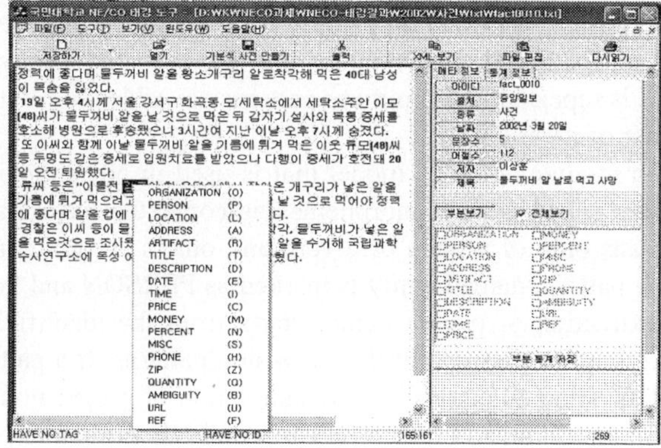

Fig. 1. NE/CO tagging workbench

3 Learning Extraction Patterns from Unannotated Corpora

There are two approaches in IE systems: knowledge engineering and automatically trainable approaches. Knowledge engineering technique is characterized as manually encoding extraction patterns. Knowledge engineers examine the corpus and write extract patterns. It requires lots of labors and the skill of a knowledge engineer affects on the performance of IE system. Automatically trainable approach needs an annotated corpus. Annotations are targeted to particular function like named entity recognition or coreference resolution. Once a suitable training corpus is constructed, the learning system automatically acquires extraction patterns from the corpus. In supervised learning system, the supervisor interacts with the learning system by indicating that the automatically acquired patterns are correct or not.

NE tagging system needs extraction patterns to identify named entities in the text. There are two tagging methods of manual construction and automatic construction. Manually constructed pattern is accurate and the performance of IE system is better than that of IE systems that are using automatically constructed patterns. Manual construction needs a continuous tuning with high cost. So, automatic construction is applied to get candidate patterns. Our NE tagging system has been trained for unannotated corpora to get NE extraction patterns that are included in the corpora. Extracting patterns start from a small set of proper names and NE extraction patterns are generated semi-automatically. Those extracted patterns are applied to automatically identify proper nouns when we construct NE tagged corpus.

The first step of the learning process of pattern templates is an identification of proper nouns that are given as an initial set of seeds. For each named entity that is given in the initial NE set, NE extraction patterns are automatically generated from the raw corpus. NE patterns are a sequence of part-of-speech or cue word. New patterns are mined from the corpus for the seed nouns in initial NE set, and they are added to the pattern set. So, extraction pattern set is expanded by repeating the learning process, and newly found proper nouns are added to the seed set. The pattern extraction process is repeated until no more patterns are added, or it stops after a predetermined number of iterations.

Pattern creation module adopts a model that is used in NE chunking/tagging and considers precedent or subsequent morphemes as contextual information. The template pattern consists only of a noun or a verb and others are not considered. When we generate a new pattern, named entity is marked as PERSON and the cue word is a morpheme. NE extraction of person names starts from the identification of named entities in the text that matches one of the extraction patterns. If a pattern is matched, then it is marked as a named entity. For the annotation of proper nouns, NE tagging system annotates proper nouns in the corpus by using NE extraction patterns. Named entity is described as a single word in the extraction pattern and multi-word proper nouns are combined into a noun phrase by NP chunking.

... Today	President	Kim ...	→	<<u>null : President : PERSON</u>>
adverb	noun	PERSON		extraction pattern

4 The Experiment

We performed an experiment for the automatic recognition of Korean person names. Test documents in this experiment are randomly selected from the raw corpus. Some person names that are found frequently in test corpus are given as an initial set of proper names. Experimentation has been performed for 2, 10, 20 person names as a seed name set, respectively. For the seed names, we run NE pattern extractor iteratively and the seed name set is expanded. The system runs until no more new names are added to the seed set. We automatically removed a pattern that has no information on both sides of the seed name and the patterns with common nouns. Table 1 shows the precision rates by the number of iterations. The precision ratio is 93%~94% for each experiment and the recall ratio is different according to the instances of the initial names. As a result, we found that NE pattern types depend on the document categories, and person names that are identified by NE patterns are also domain-dependent. When we set the initial seed set as politicians like 'Kim Dae-Jung' and 'Clinton', the system found most of the politically related person names, but it does not found the names of artists or novelist. Therefore, seed names should be given carefully that cover various categories of the documents.

Table 1. Precision of named entity extraction

no. seeds iteration	5	10	20
1	0.935	0.939	0.938
2	0.965	0.938	0.947
3	0.967	0.926	0.946
4	0.967	N/A	N/A

5 Conclusion

We have constructed an NE tagged corpus for 3,000 articles of economy, accident, and travel domain. As a semi-automatic construction of the corpus, we applied automatic annotation of NE tags by using a proper noun learning technique of extracting person names from the raw corpus. NE extraction patterns are automatically collected from unannotated corpus, starting from a small set of seed names and expanding extraction patterns by iteration method. Our automatic tagging system will be extended to extracting common proper nouns regardless of the area.

References

1. Appelt, D. E. and David J. Israel, "Introduction to Information Extraction Technology", A Tutorial Prepared for IJCAI-99, 1999.
2. Cardie, C., "Empirical Methods in Information Extraction", AAAI-97, pp.65-79, 1997.

3. Riloff, E, "Information Extraction as a Stepping Stone toward Story Understanding", In Computational Models of Reading and Understanding, Ashwin Ram and Kenneth Moorman, eds., MIT Press, 1999.
4. MUC, Proc. of 7th Message Understanding Conference(MUC-7), MUC, 1998.
5. Sekine, S., and Y. Eriguchi, "Japanese Named Entity Extraction Evaluation - Analysis of Results", the 18th International Conference on Computational Linguistics (COLING'2000), pp.1106-1110, 2000.

Proposition of Boosting Algorithm for Probabilistic Decision Support System

Michal Wozniak

Chair of Systems and Computer Networks, Wroclaw University of Technology,
Wybrzeze Wyspianskiego 27, 50-370 Wroclaw, Poland
Michal.Wozniak@ pwr.wroc.pl

Abstract. Different experts formulate the rules with different qualities. Additional we may get some information about problem from databases and qualities of information stored in the databases are different. We will propose the quality measure of knowledge we got. We will show how use it for decision process based on Bayes formulae and boosting concept.

1 Introduction

During designing decision support systems we get the rules from different sources (experts, databases) and their qualities are different. The following paper concerns on the decision making on the base on the different classifiers through voting procedure. This concept called *boosting* [4] will be used to the probabilistic decision making. The organization of voting system is based on the qualities of information sources.

The content of the work is as follow: Next section presents proposition of statistical knowledge quality measure and it shows how use proposed quality measure for boosting decision making. In section 3 the results of experimental investigation of proposed decision method are presented. The last section concluded the paper.

2 Boosting Concept for Probabilistic Reasoning

For the knowledge given by experts we can not assume that expert tell us true or the rule set is generated (by the machine learning algorithms) on the noise-free learning set. We postulate that we believe on it only with the γ factor ($P(\text{rule})=\gamma\leq1$), proposed as the confidence (quality) measure[6]. For the practical cases the value of proposed measure is constant for each rule obtained from the same expert or generated on the base on the same learning set. Therefore let $\gamma^{(K)}$ denotes confidence measure of K-th source of knowledge.

The Bayes decision theory consists of assumption [1] that the feature vector x and number of class j are the realization of the pair of the random variables X, J. The formalisation of the recognition in the case under consideration implies the setting of an optimal Bayes decision algorithm $\Psi(x)$, which minimizes probability of misclassification for 0-1 loss function:

M. Bubak et al. (Eds.): ICCS 2004, LNCS 3036, pp. 675–678, 2004.

$$\Psi(x) = i \ \text{if} \ p(i|x) = \max_{k \in \{1, \, ..., \, M\}} p(k|x).\tag{1}$$

In the real situation the *posterior* probabilities for each classes are usually unknown. Instead of them we can used the rules and/or the learning set for the constructing decision algorithms[5].

The analysis of different practical examples leads to the following form of rule $r_i^{(k)}$:

IF $x \in D_i^{(k)}$ **THEN** state of object is i **WITH** posterior probability $\beta_j^{(k)} = \int_{D_i^{(k)}} p(i|x)dx$

greater than $\underline{\beta}_i^{(k)}$ and less than $\overline{\beta}_i^{(k)}$.

Lets note the rule estimator will be more precise if rule decision region and differences between upper and lower bound of the probability given by expert will be smaller. For the logical knowledge representation the rule with the small decision area can be overfitting the training data [2]. For our proposition we respect this danger for the rule set obtained from learning data. For the estimation of the *posterior* probability from rule we assume the constant value of for the rule decision area. Therefore lets propose the relation "more specific" between the probabilistic rules pointed at the same class.

Definition. Rule $r_i^{(k)}$ is "more specific" than rule $r_i^{(l)}$ if

$$\left(\overline{\beta}_i^{(k)} - \underline{\beta}_i^{(k)}\right)\left(\int_{D_i^{(k)}} dx \bigg/ \int_X dx\right) < \left(\overline{\beta}_i^{(l)} - \underline{\beta}_i^{(l)}\right)\left(\int_{D_i^{(l)}} dx \bigg/ \int_X dx\right)\tag{2}$$

Hence the proposition of the *posterior* probability estimator $\hat{p}(i|x)$ is as follow: from subset of rules $R_i(x) = \left\{r_i^{(k)} : x \in D_i^{(k)}\right\}$ choose the "most specific" rule $r_i^{(m)}$

$$\hat{p}(i|k) = \left(\overline{\beta}_i^{(m)} - \underline{\beta}_i^{(m)}\right)\bigg/ \int_{D_i^{(m)}} dx\tag{3}$$

When only the set S is given, the obvious and conceptually simple method is to estimate *posterior* probabilities $\hat{p}(i|x)$ for each classes via estimation of unknown conditional probability density functions (CPDFs) and *prior* probabilities.

For the considered case, i.e. when some of rule sets and learning set are given, we propose the boosting probabilistic algorithm $\psi^{(B)}(x)$:

$$\psi^{(B)}(x) = i \ \text{if} \ p^{(B)}(i|x) = \max_{k \in M} p^{(B)}(k|x),\tag{4}$$

$$p^{(B)}(i|x) = \sum_{K=1}^{N} \gamma^{(K)} \hat{p}(i|x) \bigg/ \sum_{K=1}^{N} \gamma^{(K)}.\tag{5}$$

N denotes number of knowledge source (experts and learning sets), $\hat{p}(i|x)$ denotes estimator of *posterior* probability obtained on base on the learning set or rule one.

3 Experimental Investigations

In order to appreciate the proposed concept several experiments were made on the computer-generated data. We have restricted our considerations to the case of rules for whose the upper and lower bounds of the *posterior* probabilities are the same, rule defined region for each $i \in \{1, ..., M\}$ cover the whole feature space X and we have only one learning set. In experiments our choice of the CPDFsand the *prior* probabilities was deliberate. In experiments we considered a two-class recognition task with set of 6 and 10 rules, the Gaussian CPDFs of scalar feature x and with following parameters $p_1 = 0.333$, $p_2 = 0.667$, $f_1(x) = N(0,1), f_2(x) = N(2,1)$.

The value of quality measure of learning set has been counted using following heuristic formulae $\gamma^{(s)}$ = size of learning set * 0,001. The value of quality measure of rule has been counted using heuristic formulae $\gamma^{(R)}$ = number of rules * 0,1. $D^{(k)}$ denotes decision area of the rule, $\beta_1^{(k)}$ denotes *posterior* probability estimator of rule pointed at class 1, $\beta_2^{(k)}$ denotes *posterior* probability estimator of rule pointed at class 2.

Table 1. Rule sets for experiments

Experiment 1				Experiment 2			
k	$D^{(k)}$	$\beta_1^{(k)}$	$\beta_2^{(k)}$	k	$D^{(k)}$	$\beta_1^{(k)}$	$\beta_2^{(k)}$
1	[-3,0, -1,0)	0,984	0,016	1	[-3,0, -2,0)	0,997	0,003
2	[-1,0, 0,0)	0,887	0,113	2	[-2,0, -1,0)	0,982	0,018
3	[0,0, 1,0)	0,556	0,444	3	[-1,0, 0,0)	0,887	0,113
4	[1,0, 2,0)	0,166	0,834	4	[0,0, 0,5)	0,684	0,316
5	[2,0, 3,0)	0,031	0,969	5	[0,5, 1,0)	0,449	0,551
6	(3,0, 5,0]	0,004	0,996	6	(1,0, 1,5]	0,234	0,766
				7	(1,5, 2,0]	0,103	0,897
				8	(2,0, 3,0]	0,031	0,969
				9	(3,0, 4,0]	0,005	0,995
				10	(4,0, 5,0]	0,001	0,999

The results of experiments are shown on the Fig. 1.

The following conclusion may be drawn from the experiments:

- The frequency of correct classification depends on the value of the confidential measure of rules. The algorithms with bigger value always give the better results.
- Boosting algorithms lead to better or similar results compared to algorithm k-NN, especially for the big learning set.

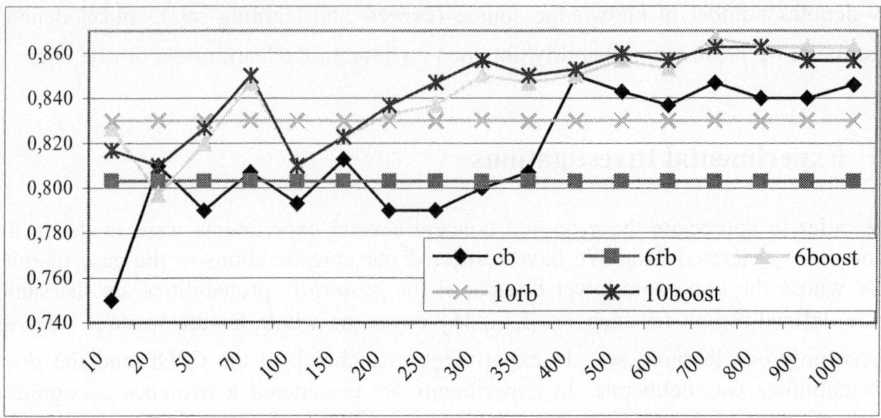

Fig. 1. Frequency of correct classification for the experiments. Cb denotes case-based algorithm, 6rb rule-based one which used 6 rules, 10rb rule based one with 10 rules, 6boost boosting algorithm for 6rb and cb, 10boost boosting one for 10rb and cd.

Drawing a general conclusion from such a limited scope of experiments as described above is of course risky. However results of experimental investigations encourage applying proposed algorithms in practise.

4 Conclusion

The paper concerned probabilistic reasoning and the proposition of the quality measure for that formulated decision problems. We presented how use the proposed measure in decision algorithm based on boosting concept.
Presented ideas need the analytical and simulation researches but the preliminary results of the experimental investigations are very promising.

References

1. Duda R.O., Hart P.E., Stork D.G., *Pattern Classification*, Wiley-Interscience, 2000.
2. Mitchell T., *Machine Learning*, McGraw Hill, 1997.
3. Puchala E., A Bayes Algorithm for the Multitask Pattern Recognition Problem – Direct Approach, *LNCS no 2659,* 2003.
4. Schapire R. E., The boosting approach to machine learning: An overview. *Proc. Of MSRI Workshop on Nonlinear Estimation and Classification*, Berkeley, CA, 2001.
5. Walkowiak K., A Branch and Bound Algorithm for Primary Routes Assignment in Survivable Connection Oriented Networks, *Computational Optimization and Applications*, Kluwer Academic Publishers, February 2004, Vol. 27.
6. Wozniak M., Concept of the Knowledge Quality Management for Rule-Based Decision System, W: Klopotek M.A. et al. [eds] *Intelligent Information Processing and Web Mining*, Springer Verlag 2003.

Efficient Algorithm for Linear Pattern Separation

Claude Tadonki[1] and Jean-Philippe Vial[2]

[1] University of Geneva, Centre Universitaire d'Informatique
24, rue Général Dufour, 1211 Genève 4 - Switzerland
`claude.tadonki@cui.unige.ch`
[2] HEC/LOGILAB, 40 Bd du Pont d'Arve, CH-1211 Geneva – Switzerland
`jean-philippe.vial@hec.unige.ch`

Abstract. We propose cutting plane algorithm for solving the linear pattern separation problem, which is a particular case of the more general topic of data mining. The solution we provided, based on convex programming, can also be applied to any other pattern separation scheme based on a convex discriminant like *linear piecewise* or *quadratic* models. Some experimentations are reported with different large databases together with a comparison with a direct implemention with two commercial specialized codes.

1 Introduction

Linear separation [1,5] is an important concept in data mining [3,6]. It is widely used and has been applied in many fields, e.g., *cancer diagnosis, human genome , game strategies, pattern recognition , decision/selection making*, and others. Many other separation rules can be found in the litterature, and our method can handle those of them that are based on a functional rule expressed by a convex form. In some cases the size of the data set [2] is so large that solving the mathematical programming problem becomes a challenge even with the state-of-the-art optimization software. In this paper we propose to resort a so-called cutting plane method to solve the problem efficiently, and we discuss ways to improve performance on the linear separation problem.

2 Problem Formulation

Given a set of points $\mathcal{A} = \{a_i \in \mathcal{R}^n, i = 1, 2, \cdots, N\}$, and a partition $\mathcal{S}_1 \cup \mathcal{S}_2$ of the set of indices $\mathcal{S} = \{1, 2, \cdots, N\}$, we wish to find $w \in \mathcal{R}^n$ and $\gamma \in \mathcal{R}$ such that the hyperplane $\{x \mid w^T x = \gamma\}$ separates the two subsets $\mathcal{A}(\mathcal{S}_1)$ and $\mathcal{A}(\mathcal{S}_2)$, where

$$\mathcal{A}(\mathcal{S}_1) = \{a_i \in \mathcal{A} \mid i \in \mathcal{S}_1\}, \tag{1}$$

$$\mathcal{A}(\mathcal{S}_2) = \{a_i \in \mathcal{A} \mid i \in \mathcal{S}_2\}. \tag{2}$$

M. Bubak et al. (Eds.): ICCS 2004, LNCS 3036, pp. 679–682, 2004.

For typographical convenience, we will write (w, γ) instead of (w^T, γ).

Actually, one looks for a strong separation. Thus, given a *separation margin* $\nu > 0$, we hope to achieve the separation properties (3-4) displayed bellow

$$\forall a_i \in \mathcal{A}(\mathcal{S}_1) \quad w^T a_i \geq \gamma + \nu, \tag{3}$$

$$\forall a_i \in \mathcal{A}(\mathcal{S}_2) \quad w^T a_i \leq \gamma - \nu. \tag{4}$$

In general, there is no guarantee that the two sets can be strongly separated. Therefore, for any choice of w and γ, we might observe *misclassification errors*, which we define as follows

$$e_i^1 = \frac{\max(-w^T a_i + \gamma + \nu, 0)}{||(w, \gamma, \nu)||}, \quad i \in \mathcal{S}_1, \tag{5}$$

$$e_i^2 = \frac{\max(w^T a_i - \gamma + \nu, 0)}{||(w, \gamma, \nu)||}, \quad i \in \mathcal{S}_2. \tag{6}$$

Our goal is then to build a separation hyperplane $\{x \mid w^T x = \gamma\}$ (i.e., compute w and γ) for which the total sum of *misclassification errors* is minimal. In other words, we want to find a vector w and a scalar γ such that the average sum of misclassifications errors is minimized [5].

The separation margin ν helps avoiding the useless trivial solution $(w, \gamma) = (0, 0)$. Its value is usually set to 1. In some cases the separation margin may lead to large values for w and γ. It may be necessary [2] to bound w to avoid this undesirable feature; so, we add the constraint $||w||_2 \leq k$.

Formally, we have to solve the following optimization problem

$$\min_{(w,\gamma) \in \mathcal{R}^n \times \mathcal{R}} \left[\frac{1}{|\mathcal{S}_1|} \sum_{i \in \mathcal{S}_1} \max(-w^T a_i + \gamma + \nu, 0) + \right.$$

$$\left. \frac{1}{|\mathcal{S}_2|} \sum_{i \in \mathcal{S}_2} \max(w^T a_i - \gamma + \nu, 0) \right] \tag{7}$$

$$\text{subject to:} \quad ||w||_2 \leq k. \tag{8}$$

In ACCPM the square normed in the constraint $||w||^2 \leq k^2$ is also treated as black box. If $\bar{w}$ is not feasible ($||\bar{w}||^2 > k^2$), the constraint

$$w + 2\langle \bar{,} w - \bar{w} \rangle \leq k^2 \tag{9}$$

holds for any feasible point.

Finally, let us give two bounds on f. Since $f(0, 0) = 2\nu$, then 2ν is an upper bound of the optimal value of the objective. A straightforward lower bound is 0, but this can be only attained if perfect classification is achieved.

Let us discuss now the formulation of problem (7)–(8) as a standard mathematical programming problem. Let z_i, $i \in \mathcal{S}$. be an auxiliary variable. The original problem becomes

$$\min_{\substack{(w,\gamma) \in \mathcal{R}^n \times \mathcal{R} \\ z \geq 0}} \frac{1}{|\mathcal{S}_1|} \sum_{i \in \mathcal{S}_1} z_i + \frac{1}{|\mathcal{S}_2|} \sum_{i \in \mathcal{S}_2} z_i \tag{10}$$

$$\text{subject to:} \quad z_i \geq \left(-w^T a_i + \gamma + \nu\right), \ i \in \mathcal{S}_1 \tag{11}$$

$$z_i \geq \left(w^T a_i - \gamma + \nu\right), \ i \in \mathcal{S}_2 \tag{12}$$

$$\|w\|_2 \leq k. \tag{13}$$

Note that the constraints (11)–(12) are numerous but linear. The problem is thus a large linear programming problem with one quadratic constraint (13). Some authors [2] prefer to replace the quadratic constraint by a quadratic penalty term in the objective. Another possibility consists in replacing the Euclidean norm in (13) by the ℓ_∞ norm [4]. The problem becomes then fully linear.

Table 1. Comparison with direct methods

n	m	MOSEK(1)	CPLEX(1)	MOSEK(2)	CPLEX(2)	ACCPM
10	10000	2.54	1.95	6.31	1.81	1.95
20	10000	2.21	2.47	11.37	2.35	2.63
30	10000	4.45	4.45	18.70	4.12	3.40
40	10000	6.34	6.11	23.78	5.99	4.50
50	10000	9.23	8.18	26.11	8.20	4.95
60	10000	11.84	11.10	30.78	11.10	6.30
70	10000	15.13	13.07	40.87	12.82	8.86
80	10000	19.87	15.00	50.21	14.21	10.16
90	10000	26.04	19.97	69 20	19.46	15.03
100	10000	30.22	22.19	62.63	21.29	16.81
10	100000	143.86	81.08	113.40	78.12	5.11
20	100000	120.47	109.53	132.25	108.29	8.22
30	100000	172.21	143.98	179.24	141.31	10.59
40	100000	253.38	194.34	215.89	190.40	16.31
50	100000	311.35	223.71	280.87	219.60	16.95
60	100000	576.77	273.46	303.09	285.38	18.97
70	100000	742.84	408.01	411.66	396.50	28.88
80	100000	850.15	427.15	478.61	406.19	31.25
90	100000	906.57	496.57	590.95	439.29	34.06
100	100000	1443.25	543.25	680.81	493.78	40.30

(1) simplex   (2) interior point

3   Implementation

In our experiments we have considered various sample data sets to test our algorithm, all generated radonmly using a normal distribution.

The problem were solved using the public release of ACCPM. All our experiments were conducted on a 500 Mhz SUN Ultrasparc with 256 MB of RAM. For problems with a very large data set, we performed an out-of-core computation. With a buffer of $200{,}000$ elements, five accesses to disk per iteration were required.

We compare our method with direct methods based on a standard linear programming (LP) formulation of the problem. Recall that the linear programming formulation was given in (10)–(12). (Note that we drooped the norm constraint (13).) The (dual) LP formulation can be solved using standard techniques of linear programming such as *simplex* or *interior point methods*. We have compared ACCPM with two linear programming codes: MOSEK and CPLEX. Both offer the options between a simplex and a primal-dual log barrier algorithm. Table 1 displays our experimental results.

4   Conclusion

We have proposed a cutting plane algorithm for the linear pattern separation problem. The main idea was to modeling the problem through the purpose of minimizing the total misclassification gap as a convex function. Experimentations and comparative study show that our method is quiet efficient even when consider very large databases.

Acknowledgement. The authors thank Olivier Péton and Cesar Beltran for their useful comments.

References

1. R.A. Bosh and J.A. Smith, *Separating Hyperplanes and the Authorship of the Disputed Federalist Papers*, American Mathematical Monthly, Volume 105, No 7, pp. 601-608, 1995.
2. M.C. Ferris and T.S. Munson, *Interior Point Methods for Massive Support Vector Machines*, Cours/Sèminaire du 3e cycle romand de recherche opèrationnelle, Zinal, Switzerland, march 2001.
3. J. Han and M. Kamber, *Data Mining: Concept and Techniques*, Morgan Kaufmann Publishers, 2000.
4. O.L. Mangasarian, *Linear and Non-linear Separation of Patterns by linear programming*, Operations Research, 13, pp. 444-452.
5. O.L. Mangasarian, R. Setino, and W. Wolberg, *Pattern Recognition via linear programming: Theory and Applications to Medical Diagnosis*, 1990.
6. M.S. Viveros, J.P. Nearhos, M.J. Rothman, *Applying Data Mining Techniques to a Health Insurance Information System*, 22nd VLDB Conference, Mumbai (Bombay), India, 1996, pp. 286-294.

Improved Face Detection Algorithm in Mobile Environment

Sang-Burm Rhee and Yong-Hwan Lee

Dept. of Elect. & Com. Eng., Dankook Univ., Korea
sang107@dku.edu

Abstract. In this paper we propose a new algorithm to be able to implement fast and accurate search for the characteristic points in a face by scanning with Mobile Camera Phone (MCP). The algorithm transforms the RGB color space to the YUV (i.e.,use a matrixed combination of Red, Green and Blue to reduce the amount of information in the signal.), and detects the face color reducing the influence of brightness by Min-max normalization and histogram equalization. Experimental results show that this algorithm has more accurately than previous method.

1   Introduction

Many researches to apply the face recognition to mobile systems such as mobile phones, and PDA have been proceeding in accordance with the mobile computing environment is rapidly advanced. The combination of the face recognition and the mobile system could be applicable to security checking for criminals and visual communication. We transmit an image from the mobile camera to a server using a PDA or a MCP, and then recognize the face by processing of the image transmitted [1][2]. The face recognition has been generally approached with the pattern recognition to use image halftoning or edge in a static image as its features. Kanade [3][4] provided a method that automatically recognizes faces using static features including face components such as the contour, the eye, the nose and the mouth. This method resulted in reliable face recognition at pictures containing face. But it is useful exclusively for the curves to have few parameters, and it requires long computation time and a lot of calculations and spaces, and the accuracy of the detection depends on the size to quantize the parameter spaces.

2   Authentication of the Face Entered with a MCP

Skin Color Detection using YUV: RGB values of pixels in the image are transformed to YUV color spaces by equation(1) in order to detect a face from the image entered with a MCP, the illumination calibration with Min-Max Normalization is prerequisite during the pre-processing for the accurate face detection. Histogram equalization enhances the performance of the image which

M. Bubak et al. (Eds.): ICCS 2004, LNCS 3036, pp. 683–686, 2004.

brightness is secund into the one direction can be used for the intensity equalization. After reducing the illumination impact by applying min-max normalization and histogram equalization to the brightness component in Y component, we transform YUV values to RGB form.

$$\begin{bmatrix} R \\ G \\ B \end{bmatrix} = \begin{bmatrix} 1.164 & 1.596 & 0 \\ 1.164 & -0.391 & 0.813 \\ 1.164 & 0 & 2.018 \end{bmatrix} \begin{bmatrix} Y - 16 \\ U - 128 \\ V - 128 \end{bmatrix} \tag{1}$$

The skin color is extracted by using the skin color model. It is the skin color extraction image at the image processed with the Min-max normalization. We defined light brightness as the optimized light of the skin color extraction in the skin color model. Defining both quite dark and bright images to the suitable light one by histogram equalization allows to extract the skin color regardless of light brightness.

Detection of Characteristic Points: For preventing errors from occurring when dark illumination is exposed or colors similar to pupil color are distributed, characteristic points in a face should be clearly expressed unlikely the skin color in order to find characteristic points such as the nose and the mouth in a face. we let the distance between both eyes L, and define L' for a distance to lower as much as L from the middle position between two eyes, and then would see a lot of features for the nose at the L'. The lip range is searched by using colors the lib has, by starting from L' with these feature.

Face Authentication using Support Vector Machine(SVM): SVM mainly distinguishes objects as two categories. Learning samples consist of N objects, and let's express the vector x_i, comprising of p variables for the ith object, and y_i for the category classified already, corresponding to the x_i. We assume that there are two categories for y_i, either $+1$ or -1. We consider a separable hyperplane for the positioning N objects, consisting of two categories, into a p-dimensional space. In this case, it is useful for the case that a hyperplane is not deterministic. let's consider two parallel hyperplanes as below.

$$H_1 : y = w'x + b = 1, H_2 : y = w'x + b = -1 \tag{2}$$

At this time, the hyperplanes H_1 and H_2 in equation(2) pass through the object nearest to the category $+1$ and category -1 respectively at the separating hyperplane. Therefore, the margin between H_1 and H_2 is $2/\|w\|$. So, the optimization problem for this case can be expressed as follow.

$$Max \frac{2}{w'w} (or Min \frac{w'w}{2}) \quad \text{subject to} \quad y_i(w'x_i + b) \geq 1 \tag{3}$$

Inducing Lagrange by introduction of non-negative Lagrange coefficient may cause the optimization problem, called Primal problem. Applying KKT condition into this problem with any object i which $\alpha_i > 0$, we can obtain the equation (4).In the optimal solution, object with $\alpha_i > 0$ is support vector, placed on H_1 or H_2, otherwise, $\alpha_i = 0$.

$$w = \sum_{i=1}^{N} \alpha_i y_i x_i, \quad b = \frac{1 - y_i w' x_i}{y_i} \tag{4}$$

Face Authentication: In order to authenticate a face using SVM, data learning process is needed. we assigned +1 and −1 as reference values for the case of being matched between faces and not being matched respectively, by analyzing the face group in the database and the face in images entered. We normalized SVM learning data to 80×80 size, and entered the images created by quantization by 3 bits from the normalized images into the input images for SVM. In the SVM learning process, we assigned +1 and −1 as the reference values for the case of being matched between faces and not being matched respectively, letting SVM to be learned.

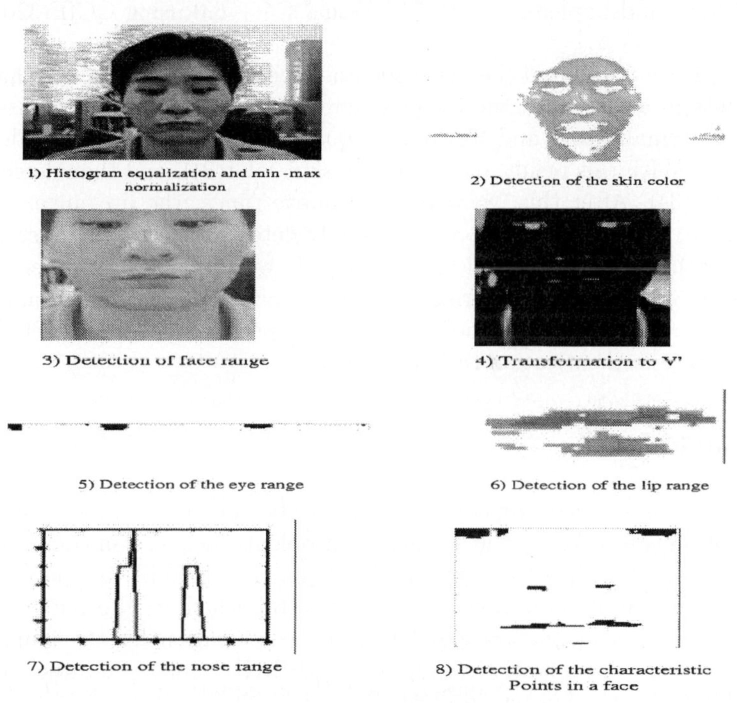

Fig. 1. An experimental process

Table 1. Comparison of searching time in a face

	Time
PCA	1.8 sec
PCA+LDA	2.3 sec
Proposed Method	1.4 sec

Table 2. Comparison of Authentication ratio

	The number of total data	The number of Authenticated data	%
PCA	200	158	79%
PCA+LDA	200	176	88%
Proposed method	200	185	92%

3   Experimental Environment and Result

To estimate the proposed method, we assume a system model for the face detection. In experiment environment, we use machines with running Windows XP Professional operation system, SCH-V300 Handset and BizCardReader 600c as input device, and implement with MS Visual C++ 6.0 using QCIF(320×240) image size.

In Fig.1, we illustrated the experimental process for detecting the characteristic points in a face from the image entered with a MCP. Pre-processed with Min-max normalization and histogram equalization for the accurate detection given in 1). With this result, extracted the skin color(2)), and then detected the face range (3)). After this, we detected the eye range, the lip range, the nose range(5), 6),and 7), respectively) and finally detected the characteristic points in a face(8)). As shown in table 1 and table 2, we compared proposed method with the previous method in terms of detection rate and authentication ratio for faces. The proposed method shows the fastest speed at the face search and the most accurate authentication ratio.

4   Conclusions

In this paper we utilized skin color detection to be able to extract the face region from color images of MCP, and settled the problems such as skin color definition and illumination, pointed out to the disadvantages included in the skin color detection. For the illumination problem, we established an alternative method using min-max normalization and histogram equalization for the illumination brightness, and used a statistical method with RGB values produced from two hundred data for the former problem.

References

1. Benjamin Miller, "Vital signs of identity", IEEE Spectrum pp,20-30, 1998.
2. D. Sanger, Y. Mlyake, "Algorithm for Face Extraction Based on Lip Detection," J. of Imaging Science and Technology, Vol.41, No.1, 1997.
3. B. Moghaddam, Was. Wahid, "Beyond Eigen Faces: Probabilistic Matching for Face Recognition," Prof. of IEEE ICAFGR'98, 1998.
4. T.Kandade, "Computer Recognition of Human faces," Birkhauser Verlag, 1997

Real-Time Face Recognition by the PCA (Principal Component Analysis) with Color Images

Jin Ok Kim[1], Sung Jin Seo[2], and Chin Hyun Chung[2]

[1] Faculty of Multimedia, Daegu Haany University,
290, Yugok-dong,Gyeongsan-si, Gyeongsangbuk-do, 712-715, KOREA
bit@dhu.ac.kr
[2] Department of Information and Control Engineering, Kwangwoon University,
447-1, Wolgye-dong, Nowon-gu, Seoul, 139-701, KOREA
chung@kw.ac.kr

Abstract. The face recognition by a CCD camera has the merit of being linked with other recognition systems such as an iris recognition to implement a multimodal recognition system. This paper is concerned with a new approach to face recognition that is automatically distinguished from moving pictures. Based on the research about recognition of color image by a CCD camera, we first find the proper value of color images in order to distinguish the tone of skin from other parts of face. Then, we look for the skin color among the regions of skin color converting RGB into Y, Cb, Cr to find skin parts of face. This new method can be applied to real-time biometric systems. We have developed the approach to face recognition with eigenface, focusing on the effects of eigenface to represent human face under several environment conditions.

1 Introduction

Face recognition is used to identify one or more persons from still images or a video image sequence of a scene by comparing input images with faces stored in a database. The application of face recognition technology can be categorized into two main parts: law enforcement application and commercial application. Face recognition technology is primarily used in law enforcement applications, especially mug shot albums and video surveillance. The commercial applications range over static matching of photographs on credit cards, ATM cards, passports, driver's licenses, and photo IDs to real-time matching with still images or video image sequences for access control. Each presents different constraints in terms of processing requirement.

This paper presents new technology that recognizes a face and automatically distinguishes it from moving pictures. Since difficult problems in recognizing color image by a CCD camera were handled in the prior research to make face region less sensitive [1]. We find out the proper value of color images in order to distinguish the tone of skin from other parts of face [2]. We look for the skin color among the regions of skin color converting RGB into Y, Cb, Cr and define the skin color by using Y, Cb, Cr to find skin parts [3] [4].

M. Bubak et al. (Eds.): ICCS 2004, LNCS 3036, pp. 687–690, 2004.

2   Face Detection and Face Recognition

In this section, we present a process of implementing face recognition and posing important factors. Lighting and angle of a pose have an effect on face detection. Particularly, it is hard to find the face color and face shape by an effect of lighting [5] [6]. Since most significant elements in facial detection can be changed by the angle of face, eyes and mouth, we should define the face's angle to obtain some necessary data within the significant elements.

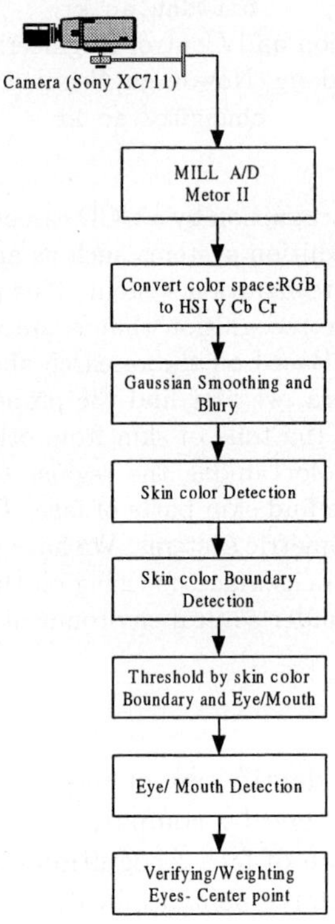

Fig. 1. Face detection algorithm

The proper value of color images is needed to distinguish the tone of skin from other parts of face [2] [7]. Moreover, we make sure that detection of the necessary facial part is done by using one of masks among 3×3, 5×5 or dynamic size. The detection of the characteristic color value is done with Gaussian distribution. After the face is detected, the rectangle area is extracted.

Color space taken by a CCD camera should be converted from RGB to HSI and Y, Cb, Cr to get specific ranges of H, S, Cb and Cr [8].

Figure 2 is about procedure of skin color detection.

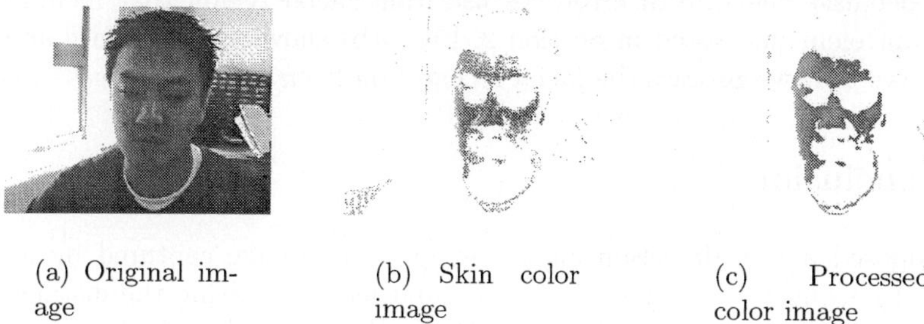

(a) Original im-
age

(b) Skin color
image

(c) Processed
color image

Fig. 2. Detection of skin regions

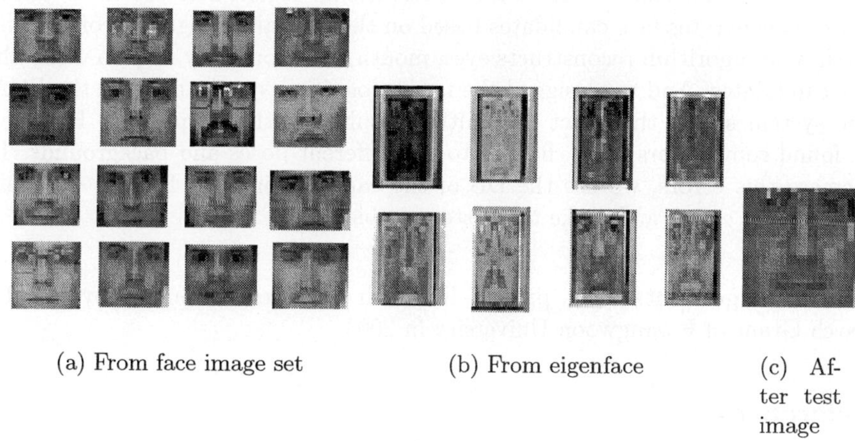

(a) From face image set

(b) From eigenface

(c) Af-
ter test
image

Fig. 3. Face detection

3   Experimental Results

The region of skin color on face can be detected by experiments. Specially, we
tried to search out the face, eyes and mouth by using division of Cb, Y, Cr
and RGB of skin color. Through the binarization, we found out the significant
elements from the face region.

Since the face could be calculated and displayed numerically, we could distinguish the face. With definition of the region of width or center value of the
eyes, we could distinguish the data when a face shows front pose.

Based on the result that error rate of detection decreases and recognition
rates increases, we could process the real-time recognition with the data.

We carried out the experiment for the facial recognition with 16 images. We
defined the eigenface as eight eigenvectors from each corresponding eigenvalue.

To decrease the rate of error, we use front facial region that includes the significant elements tested in Section 2. Fig. 3(b) shows the eigenface and Fig. 3(c) shows that we process the facial recognition from the test images

4   Conclusion

We proposed a face detection algorithm for color image captured by a CCD camera by using a skin-tone color model and features. Using the distance and width of significant elements like eyes and mouth, we could easily detect whether an object on monitor is an human or a picture. If face information captured several times is compared to the images in DB, we could recognize who he/she is. So, we could decrease the recognition error rate. Our method first corrects the color that automatically estimates the white reference pixel. We search Y, Cb, Cr skin color space. Our method detects skin region over the entire image and then generates face candidates based on the spatial arrangement of the skin patch. Our algorithm reconstructs eyes, mouth, and boundary map to verify the face candidates. And we designed the face recognition system through the PCA. The system shows the exact recognition results for the 16 persons. However, we found some errors after changes to the different poses and backgrounds. To decrease this errors, we use the DB of the captured images. Finally, we could decrease the errors and make the system stable.

Acknowledgements. The present Research has been conducted by the Research Grant of Kwangwoon University in 2004.

References

1. Hjelmas, E., Low, B.K.: Face detection: A survey. Computer Vision and Image Understanding **83** (2001) 236–274
2. Yang, J., Waibel, A.: A real-time face tracker. Proc. Third Workshop Application of Computer Vision (1996) 142–147
3. Donato, G., Bartlett, M.S., Hager, J.C., Ekman, P., Sejnowski, T.J.: Classfiying facial actions. IEEE Transactions on Pattern Analysis and Machine Intelligence **21** (2000) 974–989
4. Kjeldsen, R., Kender, J.: Finding skin and guesture recognition. In: Proc. 2nd Int'l Conf. on Automatic Face and Gesture Recognition. (1996) 312–317
5. Yang, M., Kriegman, D.J., Ahuja, N.: Detecting faces in images: A survey. IEEE Transactions on Pattern Analysis and Machine Intelligence **24** (2002) 34–58
6. Crow, I., Tock, Bennett, A.: Finding face features. In: Proc.Second European Conf., Computer Vision (1992) 92–96
7. Mckenna, S., Gong, S., Raja, Y.: Modelling facial colour and indentily with gaussian mixtures. Pattern Recognition **31** (1998) 1883–1892
8. Wang, Y., Osterman, J., Zhang, Y.Q.: Video processing and communications. Probability and Random Processes with applications to Signal Processing (2002) 24–25

Consistency of Global Checkpoints Based on Characteristics of Communication Events in Multimedia Applications

Masakazu Ono and Hiroaki Higaki

Department of Computers and Systems Engineering, Tokyo Denki University
Hatoyama-cho Ishizaka, Hiki, Saitama, 350–0394, Japan
{masa, hig}@higlab.k.dendai.ac.jp

Abstract. In order achieving fault-tolerant network systems, checkpoint-recovery has been researched and many protocols have been designed. A global checkpoint taken by the protocols have to be consistent. For conventionalnetworks, a global checkpoint is defined to be consistent if there is no inconsistent message in any communication channel [1]. For multimedia communication networks, there are additional requirements for time-constrained failure-free execution and large-size message transmissions where lost of a part of the message is acceptable. In addition, based on not characteristics of data of a message but characteristics of communication events for the message, restrictions of a consistent global checkpoint are determined. This reflects deterministic and non-deterministic property of the usage of communication buffers in layered system model. This paper proposes novel criteria, for global checkpoints in multimedia communication networks.

1 Multimedia Networks

In a multimedia network, it takes longer time to transmit and receive a message. Here, the following four *pseudo events* are defined for a multimedia message $\overline{m}$ transmitted through a communication channel $\langle p_i, p_j \rangle$; $sb(\overline{m})$, $se(\overline{m})$, $rb(\overline{m})$ and $re(\overline{m})$ for start and end of sending $\overline{m}$ in p_i and start and end of receiving $\overline{m}$ in p_j, respectively. In addition, $\overline{m}$ is decomposed into a sequence $\langle pa_1, \ldots, pa_l \rangle$ of multiple *packets* for transmission. Here, $s(pa_k)$ is a *packet sending event* and $r(pa_k)$ is a *packet receipt event* for a packet pa_k. A communication event in a multimedia network is characterized by when data of a transmitted packet is determined and when data of a received packet is accepted by an application.

Packet sending events are classified into *bulky* and *stream* ones. Data of a packet sent at a bulky packet sending event is determined at beginning of sending event of a multimedia message. Here, the following properties are held:

- If $s(pa)$ for $\overline{m}$ is bulky, $e \rightarrow s(pa)$ iff $e \Rightarrow sb(\overline{m})$ where $\rightarrow$ and $\Rightarrow$ represent causal precedence and temporal precedence between two events, respectively.
- If $s(pa)$ for $\exists pa \in \overline{m}$ is bulky, $s(pa')$ for $\forall pa' \in \overline{m}$ is bulky.

M. Bubak et al. (Eds.): ICCS 2004, LNCS 3036, pp. 691–694, 2004.
© Springer-Verlag Berlin Heidelberg 2004

On the other hand, data of a packet pa sent at a stream packet sending event is determined just at this event. Here, the following properties are held:

- If $s(pa)$ for $\overline{m}$ is stream, $e \rightarrow s(pa)$ iff $e \Rightarrow s(pa)$.
- If $s(pa)$ for $\exists pa \in \overline{m}$ is stream, $s(pa')$ for $\forall pa' \in \overline{m}$ is stream.

Same as a packet sending event, packet receipt events, are also classified into *bulky* and *stream* ones. Data of a packet received at a bulky packet receipt event is accepted at ending of receipt event of a multimedia message. Here, the following properties are held:

- If $r(pa)$ for $\overline{m}$ is bulky, $r(pa) \rightarrow e$ iff $re(\overline{m}) \Rightarrow e$.
- If $r(pa)$ for $\exists pa \in \overline{m}$ is bulky, $r(pa')$ for $\forall pa' \in \overline{m}$ is bulky.

On the other hand, data of a packet received at a stream packet sending event is accepted just at this event. Here, the following properties are held:

- If $r(pa)$ for $\overline{m}$ is stream, $r(pa) \rightarrow e$ iff $r(pa) \Rightarrow e$.
- If $r(pa)$ for $\exists pa \in \overline{m}$ is stream, $r(pa')$ for $\forall pa' \in \overline{m}$ is stream.

For a message m with conventional data, $s(m)$ and $r(m)$ are atomic. Here, each c_i in p_i is taken only when no other event occurs in p_i. However, a multimedia message is so larger that it takes longer time to transmit and receive the message. Thus, if p_i is required to take c_i during a communication event, it has to wait until an end of the event. Hence, timeliness requirement in a checkpoint protocol is not satisfied and communication overhead in recovery is increased. Therefore, c_i should be taken immediately when p_i is required to take it even during a communication event. That is, p_i sending $\overline{m} = \langle pa_1, \ldots, pa_l \rangle$ takes c_i between $s(pa_s)$ and $s(pa_{s+1})$ and p_j receiving $\overline{m}$ takes c_j between $r(pa_r)$ and $r(pa_{r+1})$. In addition, part of a multimedia message may be lost in a communication channel for an application. Such an application requires not to retransmit lost packets in recovery but to transmit packets with shorter transmission delay. Hence, an overhead for taking a checkpoint during failure-free execution is required to be reduced.

2   Consistency in Multimedia Networks

Global consistency Gc denotes degree of consistency for a global checkpoint $C_V = \{c_1, \ldots, c_n\}$. In a conventional network, Gc is defined as follows:

$$
Gc = \begin{cases} 1 & \text{no inconsistent message.} \\ 0 & \text{otherwise.} \end{cases} \tag{1}
$$

In a multimedia network, a local checkpoint is taken even during a communication event and it is acceptable to lose part of a multimedia message. Hence, a domain of Gc is a closed interval $[0, 1]$ instead of a discrete set $\{0, 1\}$.

2.1   Message Consistency

Message consistency. Mc_{ij}^u is degree of consistency for a set $\{c_i, c_j\}$ of local checkpoints and a multimedia message $\overline{m_u}$ through a communication channel $\langle p_i, p_j \rangle$. Here, we define an *inconsistent multimedia message*.

[**Inconsistent multimedia message.**] m_u is inconsistent iff m_u is a *lost* or an *orphan multimedia message*. m_u is a lost multimedia message iff $se(\overline{m_u})$ occurs before c_i in p_i and $rb(\overline{m_u})$ occurs after c_j in p_j. m_u is an orphan multimedia message iff $sb(\overline{m_u})$ occurs after c_i in p_i and $rb(\overline{m_u})$ occurs before c_j in p_j. $\square$
[**Consistency for inconsistent multimedia message.**] $Mc_{ij}^u = 0$ for an inconsistent multimedia message $\overline{m_u}$. $\square$

Next, inconsistent packets are introduced due to checkpoints during a communication event.

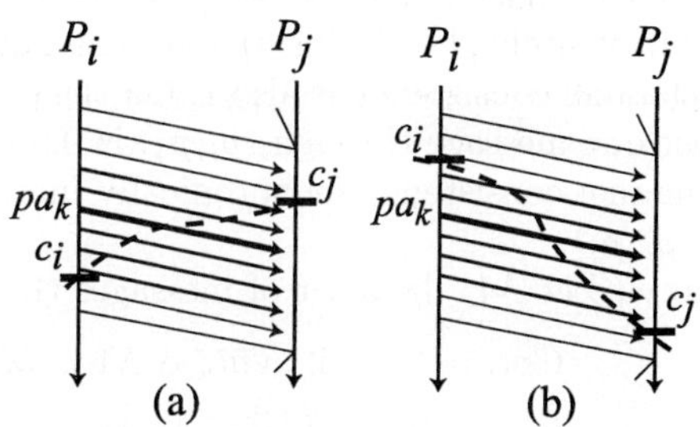

Fig. 1. Lost packet (a) and Orphan packet (b).

[**Lost and orphan packets.**] Suppose that c_i and c_j are between $s(pa_s)$ and $s(pa_{s+1})$ and between $r(pa_r)$ and $r(pa_{r+1})$ for $\overline{m_u} = \langle pa_1, \ldots, pa_l \rangle$, respectively. pa_k is a *lost packet* iff $s(pa_k)$ occurs before c_i in p_i and $r(pa_k)$ occurs after c_j in p_j. pa_k is an *orphan packet* iff $s(pa_k)$ occurs after c_i in p_i and $r(pa_k)$ occurs before c_j in p_j. $\square$

If $s = r$, there is no lost and orphan packet. Hence, $Mc_{ij}^u = 1$.

If $s > r$, $\{pa_{r+1}, \ldots, pa_s\}$ is a set of lost packets. These packets are not retransmitted after recovery. Lost packets in a conventional network are restored by logging them in failure-free execution. However, in a multimedia network, less overhead in failure-free execution is required since applications require time-constrained execution. In addition, even if part of a multimedia message is lost in recovery, an application accepts the message. The less packets are lost, the higher message consistency is achieved. A multimedia message is usually compressed for transmission. Thus, value of packets for a message is not unique. Therefore, message consistency depends on total value of lost packets as follows:

$$\frac{\partial Mc_{ij}^u}{\partial l\text{-}value} < 0 \quad \text{where } l\text{-}value = \sum_{lost\ packets\ pa_k} value(pa_k). \tag{2}$$

Here, a domain of Mc_{ij}^u is an open interval $(0, 1)$.

If $s < r$, $\{pa_{s+1}, \ldots, pa_r\}$ is a set of orphan packets. An orphan multimedia message might not be retransmitted after recovery. However, orphan packets are surely retransmitted since c_i and c_j are taken during transmission and receipt of $\overline{m_u}$ and the data of $\overline{m_u}$ being carried by a sequence $\langle pa_1, \ldots, pa_l \rangle$ of packets is not changed even after recovery. Hence, message consistency does not depend on orphan packets.

[**Message consistency**] Let $value(\overline{m_u})$ be total value of all packets in $\overline{m_u}$.

$$Mc_{ij}^u = 0 \qquad \text{if } l\text{-}value = value(\overline{m_u}).$$
$$Mc_{ij}^u = 1 \qquad \text{if } l\text{-}value = 0. \tag{3}$$
$$\frac{\partial Mc_{ij}^u}{\partial l\text{-}value} < 0 \qquad \text{otherwise. } \square$$

2.2 Channel Consistency

Channel Consistency. Cc_{ij} is calculated by Mc_{ij}^u for every message $\overline{m_u}$ through $\langle p_i, p_j \rangle$. For compatibility with (1), if message consistency for every message through $\langle p_i, p_j \rangle$ is 1, channel consistency is also 1. On the other hand, if message consistency for at least one message through $\langle p_i, p_j \rangle$ is 0, channel consistency is also 0. In addition, channel consistency monotonically increases for consistency of messages through $\langle p_i, p_j \rangle$.

[**Channel consistency.**] Let $\mathcal{M}_{ij}$ be a set of messages through $\langle p_i, p_j \rangle$.

$$Cc_{ij} = 1 \qquad \text{if } \forall \overline{m_u} \in \mathcal{M}_{ij} \quad Mc_{ij}^u = 1.$$
$$Cc_{ij} = 0 \qquad \text{if } \exists \overline{m_u} \in \mathcal{M}_{ij} \quad Mc_{ij}^u = 0. \tag{4}$$
$$\forall \overline{m_u} \in \mathcal{M}_{ij}, \ \frac{\partial Cc_{ij}}{\partial Mc_{ij}^u} > 0 \qquad \text{otherwise. } \square$$

2.3 Global Consistency

Gc is calculated by Cc_{ij} for every communication channel $\langle p_i, p_j \rangle$. For compatibility with (1), if channel consistency for every channel is 1, global consistency is also 1. On the other hand, if consistency for at least one communication channel is 0, global consistency is also 0. In addition, channel consistency monotonically increases for consistency of the communication channels.

[**Global consistency.**]

$$Gc = 1 \qquad \text{if } \forall \langle p_i, p_j \rangle \quad Cc_{ij} = 1.$$
$$Gc = 0 \qquad \text{if } \exists \langle p_i, p_j \rangle \quad Cc_{ij} = 0. \tag{5}$$
$$\forall \langle p_i, p_j \rangle, \ \frac{\partial Gc}{\partial Cc_{ij}} > 0 \qquad \text{otherwise. } \square$$

3 Conclusion

This paper has proposed a novel criteria for consistency of a global checkpoint in multimedia network systems. The authors will design QoS-based checkpoint protocols based on the criteria.

References

1. Chandy, K.M. and Lamport, L., "Distributed Snapshots: Determining Global States of Distributed Systems," ACM Trans. on Computer Systems, Vol. 3, No. 1, pp. 63–75 (1985).

Combining the Radon, Markov, and Stieltjes Transforms for Object Reconstruction

Annie Cuyt and Brigitte Verdonk

Dept of Mathematics and Computer Science, University of Antwerp
Middelheimlaan 1, B–2020 Antwerpen, Belgium
{annie.cuyt, brigitte.verdonk}@ua.ac.be

Abstract. In shape reconstruction, the celebrated Fourier slice theorem plays an essential role. By virtue of the relation between the Radon transform, the Fourier transform and the 2-dimensional inverse Fourier transform, the shape of an object can be reconstructed from the knowledge of the object's Radon transform. Unfortunately, a discrete implementation requires the use of interpolation techniques, such as in the filtered back projection.
We show how the need for interpolation can be overcome by using the relationship between the Radon transform, the Markov transform and the 2-dimensional Stieltjes transform. When combining the knowledge of an object's Radon transform for discrete angles θ, with the less well-known Padé slice theorem, the object under consideration can be reconstructed from the solution of a linear least squares problem.

1 The Radon, Markov, and Stieltjes Integral Transforms

The Radon transform $R_{\vec{\xi}}(u)$ of a square-integrable n-variate function $f(\vec{x})$ with $\vec{x} = (x_1, \dots, x_n)$ is defined as

$$R_{\vec{\xi}}(u) = \int_{\mathbb{R}^n} f(\vec{x})\, \delta(\vec{\xi}\vec{x} - u)\, d\vec{x} \qquad d\vec{x} = dx_1 \dots dx_n$$

with $||\vec{\xi}|| = 1$ and $\vec{\xi}\vec{x} = u$ an $(n-1)$-dimensional manifold orthogonal to $\vec{\xi}$. When $n = 2$, $\vec{\xi}$ is fully determined by an angle θ and

$$R_\theta(u) = \int_{-\infty}^{+\infty} \int_{-\infty}^{+\infty} f(t, s)\, \delta(t \cos \theta + s \sin \theta - u)\, dt\, ds$$

In the sequel of the text, to simplify notation, we mainly focus on the two-dimensional case, without loss of generality. Let the square-integrable function $f(t, s)$ be defined in a compact region A of the first quadrant $t \geq 0$, $s \geq 0$ of the plane. According to a fundamental property of the Radon transform $R_\theta(u)$ of $f(t, s)$ [5], the following relation holds for any square-integrable function $F(u)$:

$$\int_{-\infty}^{+\infty} R_\theta(u) F(u)\, du = \int_0^\infty \int_0^\infty f(t, s) F(t \cos \theta + s \sin \theta)\, dt\, ds \qquad (1)$$

M. Bubak et al. (Eds.): ICCS 2004, LNCS 3036, pp. 695–698, 2004.

If we take $F(u) = 1/(1 + zu)$, then

$$g_\theta(z) = \int_{-\infty}^{+\infty} \frac{R_\theta(u)}{1 + zu}\, du = \int_0^\infty \int_0^\infty \frac{f(t, s)}{1 + (t\cos\theta + s\sin\theta)z}\, dt\, ds \qquad (2)$$

A Markov function is defined to be a function with an integral representation

$$g(z) = \int_a^b \frac{f(u)}{1 + zu}\, du \quad -\infty < a \le 0 \le b < +\infty,\ z \notin\,]-\infty, -1/b] \cup [-1/a, +\infty[\tag{3}$$

where $f(u)$ is non-trivial and positive and the moments

$$c_i = \int_a^b u^i f(u)\, du \qquad i = 0, 1, \ldots \tag{4}$$

are finite. If f is nonzero in $[a, b]$ with $0 < a < b$ then (3) is considered on $[0, b]$. If f is nonzero in $[a, b]$ with $a < b < 0$, then (3) is considered on $[a, 0]$. A Markov series is defined to be a series

$$\sum_{i=0}^\infty (-1)^i c_i z^i \tag{5}$$

which is derived by a formal expansion of (3). The Markov function $g(z)$ is also called the Markov transform of the function $f(u)$. Furthermore, in case (5) is the formal series expansion of a Markov function with a nonzero radius of convergence, the Markov moment problem, in which one reconstructs $f(u)$ from the moments c_i, is determinate.

A bivariate Stieltjes function $g(z, w)$ is defined by the integral representation

$$g(z, w) = \int_0^\infty \int_0^\infty \frac{f(t, s)}{1 + (zt + ws)}\, dt\, ds \tag{6}$$

where $f(t, s)$ is non-trivial and positive. Its finite real-valued moments are given by

$$c_{ij} = \int_0^\infty \int_0^\infty t^i s^j f(t, s)\, dt\, ds$$

A formal expansion of (6) provides a bivariate Stieltjes series

$$\sum_{i,j=0}^\infty \binom{i + j}{i} (-1)^{i+j} c_{ij} z^i w^j \tag{7}$$

The function $g(z, w)$ is also called the bivariate Stieltjes transform of $f(t, s)$.

Now let us have another look at (2) and identify our object under reconstruction with its characteristic function. If $f(t, s)$ is the characteristic function of a compact set A lying in the first quadrant, then $g_\theta(z)$ is a Markov function, because $R_\theta(u)$ is zero outside a region of compact support. Furthermore, since $g_\theta(z) = g(z\cos\theta, z\sin\theta)$, there is a close link between the bivariate Stieltjes transform of the characteristic function of A and the Markov transform of its Radon transform. In order to translate these properties into an algorithm for the reconstruction of A from the knowledge of its Radon transform $R_\theta(u)$, we need to detail how its Markov transform can be computed.

2   Reconstruction Algorithm

Let the unknown object A which we identify with its characteristic function lie in the first quadrant and within the unit circle. This is a matter of shifting and scaling. The reconstruction of A then goes as follows.

- Input of the algorithm is some indirect information that is available on the object A, namely its Radon transform for a discrete number of angles θ_n (bivariate case). If the univariate moments $C_\ell^{(\theta)}$ of the Radon transform or the multivariate moments c_{ij} of $f(t,s)$ are given instead, one skips the first, respectively the first two steps of the algorithm.
- Compute the moments

$$C_\ell^{(\theta)} = \int_{a(\theta)}^{b(\theta)} u^\ell R_\theta(u)\, du$$

for a discrete number of angles $\theta = \theta_n$ with $0 \le n \le N$. From the parameterized moments

$$C_\ell^{(\theta_n)} = \sum_{i=0}^{\ell} \binom{\ell}{i} c_{i,\ell-i} \cos^i \theta_n \sin^{\ell-i}\theta_n \qquad \ell = 0,1,2,\dots \qquad (8)$$

the bivariate moments $c_{i,\ell-i}$ can be computed by solving (8), possibly in the least
- With the moments c_{ij} one computes, for successive m, the bivariate homogeneous Padé approximant [3,2] $r_{m-1,m}(z,w)$ of the Stieltjes transform $g(z,w)$. Increasing m to $m+1$, implies adding the moments $c_{i,2m-i}$ and $c_{i,2m+1-i}$ to the data. It is well-known [1, p. 228] that on each slice S_{θ_n} the sequence $\{r_{m-1,m}(z)\}_{m\in\mathbb{N}}$ converges rapidly to $g(z,w)$ restricted to that slice. The relationship between m and N is $N = 2m+1$ with N usually rather small.
- At the same time, for each $-\pi/2 < \theta_n \le \pi/2$ and each $0 \le z_j \le 1$, the value of the Stieltjes transform $g(z,w)$ evaluated at $(z_j \cos\theta_n, z_j \sin\theta_n)$ can be approximated to high accuracy by a cubature formula

$$\sum_{i=1}^{L} \frac{\omega_i}{1 + z_j(t_i \cos\theta_n + s_i \sin\theta_n)} f(t_i, s_i) \qquad n = 0,1,\dots, \qquad j = 0,1,\dots$$

with weights ω_i and nodes (t_i, s_i). Subsequently the values $f(t_i, s_i)$ are computed from the least squares problem

$$\sum_{i=1}^{L} \frac{\omega_i}{1 + z_j(t_i \cos\theta_n + s_i \sin\theta_n)} f(t_i, s_i) \approx g(z_j \cos\theta_n, z_j \sin\theta_n) \qquad (9a)$$

$$= \lim_{m\to\infty} r_{m-1,m}(z_j \cos\theta_n, z_j \sin\theta_n) \qquad (9b)$$

- The reconstruction of A is identified with $A \approx \{(t_i, s_i) \mid f(t_i, s_i) \ge 0.5\}$ where the threshold 0.5 is chosen because for the original shape $f(t,s) = 1$ inside A and $f(t,s) = 0$ outside A.

Since the homogeneous Padé approximant can be defined analogously in any number of variables, the procedure for three-dimensional shape reconstruction is entirely similar.

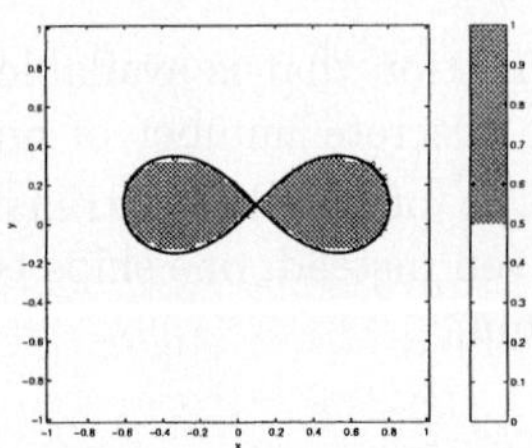
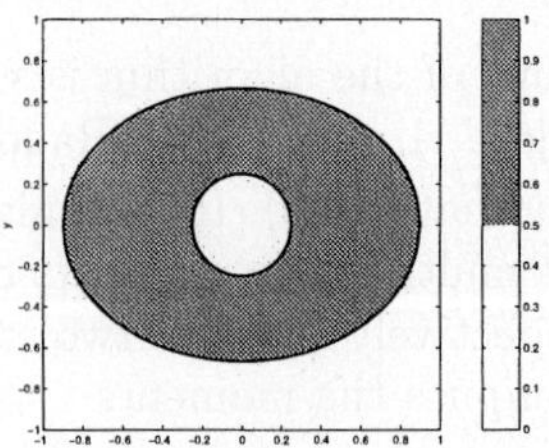

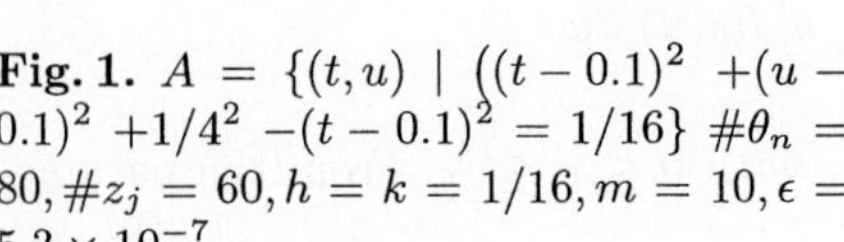

Fig. 1. $A = \{(t,u) \mid ((t-0.1)^2 +(u-0.1)^2 +1/4^2 -(t-0.1)^2 = 1/16\}$ $\#\theta_n = 80, \#z_j = 60, h = k = 1/16, m = 10, \epsilon = 5.3 \times 10^{-7}$

Fig. 2. $A = \{(t,u) \mid 81t^2/100 + 4u^2/9 \le 1\} \setminus \{(t,u) \mid t^2 + u^2 < 1/16\} \#\theta_n = 25, \#z_j = 15, h = k = 1/32, m = 10, \epsilon = 1.2 \times 10^{-4}$

Within the set of interesting objects A we present a non-convex example (Figure 1) and an example with non-connected boundary (Figure 2). We delimit the original shape in black, show the reconstructed area in grey and list the number of angles θ_n and the number of radial points z_j used in the least squares formulation (9), the degree m of the Padé denominator and the relative error $\epsilon = \max_{x^2+y^2\le 1} |r_{m-2,m-1} - r_{m-1,m}|/|r_{m-1,m}|$ in the computation of the Padé approximant. The value ϵ is an estimate of the relative error present in the right hand side of the linear least squares problem (9). The least squares problem (9), which is an inverse problem, is in general ill-conditioned and therefore a regularization technique must be applied. In all of the following examples we have found the technique known as truncated SVD [4] to do an excellent job.

For the approximation of $g(z_j \cos\theta_n, z_j \sin\theta_n)$ we use the simple compound 4-point Gauss-Legendre product rule [6, pp. 230–231] with $h = 1/16 = k$ for Figure 1 and $h = 1/32 = k$ for Figure 2.

References

1. G.A. Baker, Jr. and P. Graves-Morris. *Padé approximants (2nd Ed.).* Cambridge University Press, 1996.
2. A. Cuyt. A comparison of some multivariate Padé approximants. *SIAM J. Math. Anal.*, 14:195–202, 1983.
3. A. Cuyt. *Padé approximants for operators: theory and applications.* LNM 1065, Springer Verlag, Berlin, 1984.
4. P.C. Hansen. The truncated SVD as a method for regularization. *BIT*, 27:543–553, 1987.
5. A.C. Kak and M. Slaney. *Principles of Computerized Tomographic Imaging.* IEEE Press, 1988.
6. A.H. Stroud. *Approximate calculation of multiple integrals.* Prentice-Hall, 1971.

Author Index

Lecture Notes in Computer Science

For information about Vols. 1–2970

please contact your bookseller or Springer-Verlag

Vol. 3022: T. Pajdla, J. Matas (Eds.), Computer Vision - ECCV 2004. XXVIII, 621 pages. 2004.

Vol. 3021: T. Pajdla, J. Matas (Eds.), Computer Vision - ECCV 2004. XXVIII, 633 pages. 2004.

Vol. 3019: R. Wyrzykowski, J. Dongarra, M. Paprzycki, J. Wasniewski (Eds.), Parallel Processing and Applied Mathematics. XIX, 1174 pages. 2004.

Vol. 3016: C. Lengauer, D. Batory, C. Consel, M. Odersky (Eds.), Domain-Specific Program Generation. XII, 325 pages. 2004.

Vol. 3015: C. Barakat, I. Pratt (Eds.), Passive and Active Network Measurement. XI, 300 pages. 2004.

Vol. 3014: F. van der Linden (Ed.), Software Product-Family Engineering. IX, 486 pages. 2004.

Vol. 3012: K. Kurumatani, S.-H. Chen, A. Ohuchi (Eds.), Multi-Agnets for Mass User Support. X, 217 pages. 2004. (Subseries LNAI).

Vol. 3011: J.-C. Régin, M. Rueher (Eds.), Integration of AI and OR Techniques in Constraint Programming for Combinatorial Optimization Problems. XI, 415 pages. 2004.

Vol. 3010: K.R. Apt, F. Fages, F. Rossi, P. Szeredi, J. Váncza (Eds.), Recent Advances in Constraints. VIII, 285 pages. 2004. (Subseries LNAI).

Vol. 3009: F. Bomarius, H. Iida (Eds.), Product Focused Software Process Improvement. XIV, 584 pages. 2004.

Vol. 3008: S. Heuel, Uncertain Projective Geometry. XVII, 205 pages. 2004.

Vol. 3007: J.X. Yu, X. Lin, H. Lu, Y. Zhang (Eds.), Advanced Web Technologies and Applications. XXII, 936 pages. 2004.

Vol. 3006: M. Matsui, R. Zuccherato (Eds.), Selected Areas in Cryptography. XI, 361 pages. 2004.

Vol. 3005: G.R. Raidl, S. Cagnoni, J. Branke, D.W. Corne, R. Drechsler, Y. Jin, C.G. Johnson, P. Machado, E. Marchiori, F. Rothlauf, G.D. Smith, G. Squillero (Eds.), Applications of Evolutionary Computing. XVII, 562 pages. 2004.

Vol. 3004: J. Gottlieb, G.R. Raidl (Eds.), Evolutionary Computation in Combinatorial Optimization. X, 241 pages. 2004.

Vol. 3003: M. Keijzer, U.-M. O'Reilly, S.M. Lucas, E. Costa, T. Soule (Eds.), Genetic Programming. XI, 410 pages. 2004.

Vol. 3002: D.L. Hicks (Ed.), Metainformatics. X, 213 pages. 2004.

Vol. 3001: A. Ferscha, F. Mattern (Eds.), Pervasive Computing. XVII, 358 pages. 2004.

Vol. 2999: E.A. Boiten, J. Derrick, G. Smith (Eds.), Integrated Formal Methods. XI, 541 pages. 2004.

Vol. 2998: Y. Kameyama, P.J. Stuckey (Eds.), Functional and Logic Programming. X, 307 pages. 2004.

Vol. 2997: S. McDonald, J. Tait (Eds.), Advances in Information Retrieval. XIII, 427 pages. 2004.

Vol. 2996: V. Diekert, M. Habib (Eds.), STACS 2004. XVI, 658 pages. 2004.

Vol. 2995: C. Jensen, S. Poslad, T. Dimitrakos (Eds.), Trust Management. XIII, 377 pages. 2004.

Vol. 2994: E. Rahm (Ed.), Data Integration in the Life Sciences. X, 221 pages. 2004. (Subseries LNBI).

Vol. 2993: R. Alur, G.J. Pappas (Eds.), Hybrid Systems: Computation and Control. XII, 674 pages. 2004.

Vol. 2992: E. Bertino, S. Christodoulakis, D. Plexousakis, V. Christophides, M. Koubarakis, K. Böhm, E. Ferrari (Eds.), Advances in Database Technology - EDBT 2004. XVIII, 877 pages. 2004.

Vol. 2991: R. Alt, A. Frommer, R.B. Kearfott, W. Luther (Eds.), Numerical Software with Result Verification. X, 315 pages. 2004.

Vol. 2990: J. Leite, A. Omicini, L. Sterling, P. Torroni (Eds.), Declarative Agent Languages and Techniques. XII, 281 pages. 2004. (Subseries LNAI).

Vol. 2989: S. Graf, L. Mounier (Eds.), Model Checking Software. X, 309 pages. 2004.

Vol. 2988: K. Jensen, A. Podelski (Eds.), Tools and Algorithms for the Construction and Analysis of Systems. XIV, 608 pages. 2004.

Vol. 2987: I. Walukiewicz (Ed.), Foundations of Software Science and Computation Structures. XIII, 529 pages. 2004.

Vol. 2986: D. Schmidt (Ed.), Programming Languages and Systems. XII, 417 pages. 2004.

Vol. 2985: E. Duesterwald (Ed.), Compiler Construction. X, 313 pages. 2004.

Vol. 2984: M. Wermelinger, T. Margaria-Steffen (Eds.), Fundamental Approaches to Software Engineering. XII, 389 pages. 2004.

Vol. 2983: S. Istrail, M.S. Waterman, A. Clark (Eds.), Computational Methods for SNPs and Haplotype Inference. IX, 153 pages. 2004. (Subseries LNBI).

Vol. 2982: N. Wakamiya, M. Solarski, J. Sterbenz (Eds.), Active Networks. XI, 308 pages. 2004.

Vol. 2981: C. Müller-Schloer, T. Ungerer, B. Bauer (Eds.), Organic and Pervasive Computing – ARCS 2004. XI, 339 pages. 2004.

Vol. 2980: A. Blackwell, K. Marriott, A. Shimojima (Eds.), Diagrammatic Representation and Inference. XV, 448 pages. 2004. (Subseries LNAI).

Vol. 2979: I. Stoica, Stateless Core: A Scalable Approach for Quality of Service in the Internet. XVI, 219 pages. 2004.

Vol. 2978: R. Groz, R.M. Hierons (Eds.), Testing of Communicating Systems. XII, 225 pages. 2004.

Vol. 2977: G. Di Marzo Serugendo, A. Karageorgos, O.F. Rana, F. Zambonelli (Eds.), Engineering Self-Organising Systems. X, 299 pages. 2004. (Subseries LNAI).

Vol. 2976: M. Farach-Colton (Ed.), LATIN 2004: Theoretical Informatics. XV, 626 pages. 2004.

Vol. 2973: Y. Lee, J. Li, K.-Y. Whang, D. Lee (Eds.), Database Systems for Advanced Applications. XXIV, 925 pages. 2004.

Vol. 2972: R. Monroy, G. Arroyo-Figueroa, L.E. Sucar, H. Sossa (Eds.), MICAI 2004: Advances in Artificial Intelligence. XVII, 923 pages. 2004. (Subseries LNAI).

Vol. 2971: J.I. Lim, D.H. Lee (Eds.), Information Security and Cryptology -ICISC 2003. XI, 458 pages. 2004.